ANNUAL REVIEW OF
FLUID MECHANICS

ANNUAL REVIEW OF FLUID MECHANICS

VOLUME 26, 1994

JOHN L. LUMLEY, *Co-Editor*
Cornell University

MILTON VAN DYKE, *Co-Editor*
Stanford University

HELEN L. REED, *Associate Editor*
Arizona State University

ANNUAL REVIEWS INC   4139 EL CAMINO WAY   P.O. BOX 10139   PALO ALTO, CALIFORNIA 94303-0897

ANNUAL REVIEWS INC.
Palo Alto, California, USA

International Standard Serial Number: 0066-4189
International Standard Book Number: 0-8243-0726-7
Library of Congress Catalog Card Number: 74-80866

TYPESET BY BPCC-AUP GLASGOW LTD., SCOTLAND
PRINTED AND BOUND IN THE UNITED STATES OF AMERICA

PREFACE

How do you arrange the books on your shelves? One of our colleagues puts them in chronological order beginning, no doubt, with the Gutenberg Bible and ending, we trust, with this very volume. Others arrange them alphabetically by author; but we suspect that most of you—like ourselves—devise some scheme of arranging them by subject. But how is it possible, even within the range of *fluid mechanics*, to proceed in linear fashion? Does *viscous flow* come after *gasdynamics* or before—and where does that leave *compressible boundary layers*? Evidently it is impossible to force such a multifaceted subject logically into a one-dimensional array.

Yet that is the problem we face each November when we sit down to prune our individual lists of suggested topics and authors down to a final two dozen invitations for the next volume of the *Annual Review of Fluid Mechanics*. To be sure of covering the whole subject in one day's deliberations, we need a guide along the path through all of fluid mechanics. For this purpose we adopted many years ago a classification system freely adapted from *Mathematical Reviews* by sometime Editor John Wehausen. This slices the subject neatly into 32 categories, starting plausibly enough with "(1) History," and ending inevitably with "(32) Miscellaneous," while passing through "(11) Turbulence" and "(22) Numerical methods" at the one-third and two-third marks.

We have never been satisfied with this one-dimensional classification, so one of our number volunteered recently to construct a more logical and systematic scheme. He reported back a year later with a remarkable five-dimensional decimal system, which was at the same time so cumbersome yet incomplete that even he could not seriously propose its adoption. Evidently the fluid-mechanical world is even more than five-dimensional; perhaps it is a fractal. So we continue to use the old illogical classification; and you find in the pages that follow a one-dimensional bird's-eye view of the world of fluid mechanics, beginning with an article by Yaglom on both (1) History and (11) Turbulence—or, more precisely, on the contributions to turbulence research of A. N. Kolmogorov.

THE EDITORS

SOME RELATED ARTICLES IN OTHER *ANNUAL REVIEWS*

From the *Annual Review of Astronomy and Astrophysics*, Volume 31 (1993):

Theory of Interstellar Shocks, Bruce T. Draine and Christopher F. McKee

The Atmospheres of Uranus and Neptune, Jonathan I. Lunine

Jupiter's Great Red Spot and Other Vortices, Philip S. Marcus

From the *Annual Review of Earth and Planetary Science*, Volume 21 (1993):

Accretion and Erosion in Subduction Zones: The Role of Fluids, Xavier Le Pichon, Pierre Henry, and Siegfried Lallement

From the *Annual Review of Materials Science*, Volume 22 (1992):

Transient Liquid Phase Bonding, W. D. MacDonald and T. W. Eager

From the *Annual Review of Physical Chemistry*, Volume 43 (1992):

Transport Properties in Polymeric Fluids, R. Byron Bird and Hans Christian Öttinger

Atmospheric Ozone, Harold S. Johnston

Annual Review of Fluid Mechanics
Volume 26, 1994

CONTENTS

(*Left* to *right*) M. D. Millionshchikov, A. N. Kolmogorov, A. M. Yaglom, and R. Kraichnan at meeting at the Institut de Mécanique Statistique de la Turbulence, Marseille, 1961. (Photo courtesy J. L. Lumley.)

Annu. Rev. Fluid Mech. 1994. 26 : 1–22

A. N. KOLMOGOROV AS A FLUID MECHANICIAN AND FOUNDER OF A SCHOOL IN TURBULENCE RESEARCH

A. M. Yaglom

Institute of Atmospheric Physics, USSR Academy of Sciences, Moscow, Russia

Andrei Nikolaevich Kolmogorov, an extraordinary scientist and quite remarkable man, was born on April 25, 1903, and died on October 20, 1987. According to the Russian usage he was called Andrei Nikolaevich by his colleagues and students (except for a few close friends and relatives who called him simply Andrei); for short he will usually be called AN in what follows. Several expositions of AN's work and life have already been published; see e.g. the long comprehensive paper by Shiryayev (1989), the excellent collaborative obituary edited by Kendall (1990), three volumes of AN's selected works (in accordance with the author's recommendations) supplemented by short comments by AN himself and more detailed commentaries by some of his students and colleagues (Kolmogorov 1985, 1986, 1987; these three Russian books have now been published in English, see Tikomirov 1991, Shiryayev 1992, and Sossinsky 1992), and the Royal Society Volume compiled and edited by Hunt, Phillips, and Williams (1991) on the occasion of the 50th anniversary of Kolmogorov's papers on turbulence (1941a,c). This latter volume is devoted mostly to the modern development of the ideas presented in these papers. Nevertheless, many facts related to AN's scientific and social activities are not known to the majority of the fluid dynamics community; therefore it seems appropriate to publish a paper about him in this *Annual Reviews* volume.

AN is most famous as one of the greatest mathematicians of the 20th century: of the same caliber as such mathematical geniuses as Henry Poincaré, David Hilbert, Hermann Weyl, and John (János, Johann) von

1

0066–4189/94/0115–0001$05.00

Neumann. His mathematical works are amazing in their profundity and originality combined with unique diversity and breadth; they concern quite different topics which, together with a number of very abstract and sophisticated matters, include many quite practical and easily applicable ideas and results. His brilliant works in mechanics are also famous. AN's theory of locally isotropic turbulence fully transformed the most complicated and practically important part of modern fluid mechanics. His profound studies of ergodic and dynamical system theories gave birth to the splendid KAM-theory (after Kolmogorov, Arnold, and Moser) in the classical mechanics of systems with finite numbers of degrees of freedom— one of the greatest achievements in mathematical natural sciences of this century [see e.g. Arnold's commentary in Kolmogorov (1985) and Moffatt's section in Kendall (1990)]. It is therefore only natural that Abraham & Marsden (1978) include in their well-known book on mechanics AN's photo in a small portrait gallery of the greatest mechanicians of humanity (beginning with Archimedes' portrait). Moreover, many readers of this volume probably do not know that Kolmogorov's scientific interests were not limited to only mathematics and mechanics. His first scientific work was devoted to medieval (15th century) Russian history; this work has just recently been published (Kolmogorov 1992) but is considered very important by leading experts on ancient Russia (see Yanin 1988). Prosody was another part of humanities that attracted Kolmogorov's attention for many years. In Russian magazines and collections of papers on philology and linguistics he published (in part jointly with his students) about a dozen papers devoted to the profound study of style, metrical structure, and rhythm of poetry by various Russian poets (both classical and modern); these papers will be collected in the fourth volume of AN's selected works, which is being prepared now by Nauka Press.

Let us now go on to AN's works on turbulence which are of most interest to the readers of this volume. In his short comments on these works Kolmogorov (1985) wrote:

I took an interest in the study of turbulent flows of liquids and gases in the late thirties. It was clear to me from the very beginning that the main mathematical instrument in this study must be the theory of random functions of several variables (random fields) which had only then originated. Moreover, it soon also became clear to me that there was no chance to develop a closed purely mathematical theory. For lack of such a theory it was necessary to use some hypotheses based on the results of treatment of experimental data. Therefore it was important to find talented collaborators able to work in such a mixed regime who could combine theoretical studies with the analysis of experimental results. In this respect I was quite successful.

(AN goes on to mention his students A. M. Obukhov, M. D. Million-shchikov, A. S. Monin, and the present author.)

AN's comments require some additional explanation. It is characteristic of AN that, if possible, he always tried to supplement his purely mathematical works with some practical applications. In particular, his profound general theory of Markov processes was supplemented by the solution of some specific problems related to Brownian motion in physical systems (see e.g. Kolmogorov 1986). His study of fundamental theoretical problems of mathematical statistics led him (in the war years) to investigate applications of statistical methods to the theory of ballistics. Before and after the war he sought solutions for statistical problems of quality control, weather forecasting, and so on. In his famous book on foundations of probability theory (Kolmogorov 1933, 1956a) AN was the first to give a rigorous definition of a random function $X(t)$, where t passes through infinitely many values, as a probability measure in the infinite-dimensional space of real functions $x(t)$ specified by the infinite family of multi-dimensional probability distributions. Two years later Kolmogorov (1935) also provided another method to specify a random function by its characteristic functional. This was an infinite-dimensional generalization of the characteristic functions widely used in probability theory. These works formed the basis for the mathematical theory of random (or stochastic) functions which later became an extensive and highly developed part of modern probability theory.

It was typical of AN that his development of the mathematical theory of random functions aroused his interest in possible applications of such functions. For the case of a scalar (one-dimensional) variable t, some applications are obvious; they are provided by Brownian motion and irregular oscillations of physical systems (here t must be interpreted as time). However, for the case where t is multidimensional, the most natural applications are those related to the study of fields of fluid dynamic variables in turbulent flows. This was the primary reason for AN's interest in turbulence.

AN easily made contacts with people, and being interested in turbulence he participated in many discussions with Soviet experts in fluid mechanics such as L. G. Loitsyanskiy, I. A. Kibel [a former student and collaborator of A. A. Friedmann whose joint paper with Keller (1924) was highly valued by Kolmogorov], and many others. The study of Taylor's paper of 1935 on isotropic turbulence was especially inspiring for AN and led him to ideas about further study of such turbulence. AN felt that such a study had to involve a direct analysis of experimental data and he started searching for a student appropriate for such a task. He consulted a number of fluid mechanics professors and some of them recommended a young mechanical engineer named M. D. Millionshchikov. Millionshchikov had graduated from an oil industry engineering college in his native city of Grozny in the Caucasus (which has recently become famous for turbulent political activities) and in the mid-1930s taught at the Moscow college preparing aviation industry engineers. AN spoke with him and then agreed to be his scientific adviser during his graduate study at the Aviation Industry College. Thus, AN found his first graduate student specializing in the mechanics of turbulence.

AN's second graduate student of the same speciality was A. M. Obukhov. The latter was born in Saratov (a city on the Volga river) on May 5, 1918—exactly one hundred years after the birth of Karl Marx—and in 1935 he entered Saratov University to study mathematics. In 1937 a special competition dedicated to the 20th anniversary of the 1917 October Revolution was announced in Moscow for young scientists of the USSR. AN was one of the judges at the mathematics section. According to his suggestion the first prize in mathematics was given to the unknown young student Obukhov from Saratov University. The prize-winning paper by Obukhov was, in fact, extraordinary: It dealt with multivariate statistical analysis and proposed a new statistical technique which later became known as canonical correlation analysis. The technique is now presented in dozens of textbooks, manuals, and monographs and has many diverse applications; see e.g. the survey by Yaglom (1990) published in the journal issue devoted to the memory of Obukhov. AN naturally took a keen interest in the young student. He arranged Obukhov's transfer from Saratov to Moscow University, took care of him, and after his graduation recommended him for graduate study under his guidance. At that time AN's main interest focused on the mechanics of turbulence and he suggested to Obukhov a theme related to this field. This suggestion happened to be extremely successful. In fact, between high school and university Obukhov spent a year working at the Saratov Meteorological Observatory. (He was not admitted to the entrance exams at Saratov University in 1934 because he was too young: Russian provincial universities are often quite conservative.) Later Obukhov used the results of his meteorological observations as the basis for his first scientific work and henceforth meteorology became his permanent love. (The work on mathematical statistics mentioned above was inspired by Obukhov's reflections about possible ways of treating wind observations.) Since atmospheric air flows are always turbulent, the study of turbulence attracted Obukhov very much and he worked with great enthusiasm and energy. Simultaneously with the beginning of his graduate study Obukhov started working at the Institute for Theoretical Geophysics of the USSR Academy of Sciences which had just been organized by the famous Soviet mathematician (an expert in abstract algebra), polar explorer, theoretical astronomer (author of a new theory for the origin of planetary systems), and member of the USSR Academy of Sciences—Professor Otto Schmidt. This circumstance later proved to be of some importance for the development of turbulence studies in the USSR.

During the war years AN found two more graduate students interested in turbulence: A. S. Monin and the present author. Monin graduated from the Moscow University Mathematics Department in the spring of 1942 and was recommended for graduate study; AN agreed to be his scientific adviser. At that time Monin knew nothing about fluid mechanics and asked for a subject on pure mathematics. However, he was inducted into the army shortly after his graduation and had no chance to start his graduate work. In the army he was admitted to the military meteorological school and ended the war as an officer-meteorologist serving at military airfields. After the war he got in contact with AN who helped to arrange his transfer to the Central Weather Forecasting Institute of the Ministry of Defense in Moscow. There he continued his graduate study under the guidance of AN who naturally gave him a subject related to the study of atmospheric turbulence.

I studied at Moscow University simultaneously with A. S. Monin in both the Physics and Mathematics Departments. As the war came to the USSR in 1941, I tried to join the army but was rejected because of nearsightedness (probably great luck, since most of my friends admitted to the army were killed in the war). In the autumn of 1941 I went to the city of Sverdlovsk together with my parents and continued studying at Sverdlovsk University. AN spent the early war years partially in Moscow and partially in Kazan where the USSR Academy of Sciences was transferred. But he also made several visits to Sverdlovsk which was a big scientific center. During his first visit to Sverdlovsk in the winter of 1941–1942, AN delivered a lecture on the small-scale structure of turbulence at Sverdlovsk University. He already knew me a little since in the spring of 1938 he handed me a prize as one of the winners of the Moscow Mathematical Olympiad for high school students which he helped arrange. (AN was always very active in high school mathematical education and he never forgot the young students he met.) After the lecture he noticed me and invited me to visit him the next day at his hotel. When we met, AN told me that many talented young mathematicians had already perished in the war; therefore he wanted to be informed about any good students who had survived and would be happy to help them. (Later he helped me in arranging my admittance, after my graduation from the university, to the Main Geophysical Observatory—a famous meteorological research institution evacuated from Leningrad to Sverdlovsk during the war.) Then AN turned to a discussion of his lecture at the University and, in fact, gave me another impromptu lecture on turbulence. At that time I was very poorly educated in fluid mechanics and did not understand most of the lecture, though I felt that the topic was very interesting. This situation was not exceptional at all; AN often overestimated the level of his listeners and for them it was often difficult to understand everything he said. It was noted by somebody in Moscow that AN apparently assumed that all the world around him was inhabited by Kolmogorovs! However, for good students he was an excellent lecturer and supervisor since his lectures and speeches were always very interesting and challenging. At the end of our meeting he recommended that I read some papers on stochastic processes (also overestimating my abilities to some extent) and asked me to contact him during his subsequent visits to Sverdlovsk.

After his first visit I met AN again in Sverdlovsk two or three times. In the spring of 1943, when Moscovites who had left the city were invited back by some of the Moscow institutions, AN proposed that I go to Moscow and start graduate study under his guidance. I agreed joyfully and he sent me an invitation from the Steklov Mathematical Institute of the USSR Academy of Sciences. When all the necessary formalities had been overcome I returned to Moscow to become AN's graduate student. During our conversations in Sverdlovsk AN suggested that I study turbulence during my graduate studentship, but in Moscow he changed his mind and proposed a subject related to the theory of Brownian motion. The problem suggested to me was sketched by AN several years earlier (in Kolmogorov 1937) and apparently he selected it because he knew of my interest in statistical physics. (This, in fact, appeared to be important for finding the solution of the problem.) However, my discussions with AN in Sverdlovsk and the job at the Main Geophysical Observatory made me take an interest in turbulence. Therefore during my graduate studentship I participated in a seminar on turbulence headed by AN, and after my thesis defense I agreed to take a job related to research in this field (discussed below).

In 1951 AN acquired one more graduate student specializing in turbulence research: G. I. (Grisha) Barenblatt. He had graduated from the Fluid Dynamics Department of Moscow University and been recommended for graduate study in that department. Barenblatt asked AN to be his scientific adviser.

Let us now go on to the fluid mechanics work by AN and his students. The first short paper by such a student was written by Millionshchikov (1939). It was devoted to the study of asymptotic behavior, as $t \to \infty$, of the solution of an approximate equation for the longitudinal correlation function of isotropic turbulence $B_{LL}(r, t) = \langle u_L(\mathbf{x}+\mathbf{r}, t)u_L(\mathbf{x}, t) \rangle$ (where angular brackets symbolize averaging, u_L is the velocity component parallel to $\mathbf{r}$, and $r = |\mathbf{r}|$), obtained when the third moments of velocity fluctuations are neglected in comparison with the second moments. In the introduction to his paper Millionshchikov formulated for the first time a basic idea of AN: that the fields of fluid mechanical parameters in incompressible turbulent flows must be considered as random fields in a three-dimensional space of points $\mathbf{x}(t)$ and the velocity fields are specified by probability measures in the functional space of all solenoidal vector fields. Millionshchikov's paper of 1939 was mainly a continuation (to some extent a completion) of the results by von Kármán & Howarth (1938), related to the same equation. More original was the following paper by Millionshchikov (1941) where the exact form of the equations for the velocity correlations of isotropic turbulence was used. These equations were supplemented in this paper by equations for the third-order moments of velocity fluctuations, where the fourth-order moments of these fluctuations—entering the equations for third-order moments—were approximately expressed through correlation functions with the aid of the assumption that the random velocity field has a Gaussian (or normal) probability distribution. Such a procedure leads to a closed system of equations for second- and third-order correlation functions. It was the first two-point closure of dynamic equations for turbulent flows. The corresponding approximation is now usually called the *quasi-normal* or *zero-fourth-cumulant* approximation and the hypothesis leading to this approximation is called *Millionshchikov's* or the *zero-fourth-cumulant hypothesis*. When I first read Millionshchikov's paper, I was sure that the idea of applying such a hypothesis to the study of turbulence was due to Kolmogorov. In fact, Millionshchikov was not an expert in the theory of random functions. However, in his paper of 1935 AN gave the first rigorous specification of Gaussian random functions and emphasized that all the moments of such a function can be expressed through their first- and second-order moments. Later my late friend Professor S. V. Fomin—who was AN's graduate student simultaneously with Millionshchikov and was present at several conversations between AN and the latter—confirmed my guess about the originator of the zero-fourth-cumulant hypothesis.

AN himself occasionally used the term "Millionshchikov's hypothesis"; maybe it was even first proposed by him. This does not contradict the statement above since AN often attributed his results to some of his students or collaborators and was overly delicate in mentioning the contribution of others. As an example, AN's paper on information theory (Kolmogorov 1956b) gives an equation for ε-entropy of some signal which, according to the author, "was found by Yaglom." However, I first learned of this equation after reading this paper. I asked AN about this reference to me in his paper of 1956 and he answered with a vague remark that the given equation follows from some remark of mine made sometime in a discussion with him. However, I do not remember any such remark and think that he was mistaken. Moreover, in 1947 at a meeting of the special seminar on the theory of branching stochastic processes led by AN (the notion of "branching processes" first appeared at this seminar), AN showed me the manuscript of the fundamental paper on the subject (Kolmogorov & Dmitriev 1947). To my surprise he asked me to excuse him for not including my name in the list of authors saying: "Of course, you and E. B. Dynkin (another student of his) also contributed to this work by your remarks during the discussion, but the remark by Kolya Dmitriev was more important and he is younger than both of you and needs to be encouraged; therefore I included only his name in the list of authors." I do not remember the contribution of Dynkin, but I knew well that my remark was quite minor and of no real importance. As to the contribution by Dmitriev [a talented young student who shortly after 1947 changed his speciality to applied physics, played a rather important part in the development of the Soviet hydrogen bomb, and was in this respect mentioned in Sakharov's memoirs and wrote his own memoirs on Sakharov; see Lebedev Physics Institute (1991), pp. 167–70], though clearly more important than mine, it seemed to me that it was also insufficient for coauthorship. Some other examples of AN's overestimation of the works by his colleagues, which are very typical of him, can be found in Kendall (1990, p. 34).

AN's first papers on turbulence mechanics appeared in 1941. Two remarkable short notes (Kolmogorov 1941a,d) undoubtedly marked the greatest achievement in the theory of turbulence since the time of O. Reynolds. They cardinally changed this theory and are basic to all subsequent developments in the field during the second half of this century. The results of these notes are now well known since they are presented in detail and discussed in many dozens of monographs and surveys; see e.g. the first survey and the first monograph in this field by Batchelor (1947, 1953), the long handbook by Monin & Yaglom (1975), the survey prepared by Yaglom (1981) on the occasion of the 40th anniversary of Kolmogorov's notes, and a special volume compiled by Hunt, Phillips & Williams (1991) celebrating the 50th anniversary of these same works. It is not necessary here to consider the content of these papers in any detail. However, the relation between these works and the paper of Obukhov (1941b) requires some comments.

Before the completion of the investigation described in his papers of 1941 AN presented at a seminar some preliminary arguments about the self-similarity of the cascade process of breakdown of turbulent eddies. He stated that self-similarity must imply proportionality of the longitudinal structure function $D_{LL}(r) = \langle [u_L(\mathbf{x}+\mathbf{r}) - u_L(\mathbf{x})]^2 \rangle$, $r = |\mathbf{r}|$, to some power r^m of the distance r for a wide range of r values. However, at that time AN did not know how to determine the exponent m. These preliminary considerations were known to Obukhov but AN and his graduate student were at that time separated and continued working quite independently. This independent work led both men to the determination of the value of m, but each used a different method. AN formulated (in Kolmogorov 1941a) two similarity hypotheses describing the universal equilibrium regime of small-scale components in any turbulent flow at high enough Reynolds number. He justified these hypotheses by clearly formulated heuristic physical arguments, related to general ideas about the mechanism of developed turbulence stated during the 1920s by L. F. Richardson. The arguments seemed to be convincing but they contained no proof (there are no mathematical proofs at all in his paper) and required experimental verification. The Kolmogorov hypotheses severely restrict the form of all the statistical characteristics of small-scale turbulence and hence have many special corollaries. In particular, as applied to $D_{LL}(r)$ (this particular application was considered by AN in the 1941a paper) the hypotheses, supplemented by simple dimensional arguments, imply that $m = 2/3$ and also give a crude estimation of the range of r values where $D_{LL}(r) \sim r^{2/3}$. In another paper on the same subject (Kolmogorov 1941d) AN supplemented the results about $D_{LL}(r)$ given in his first paper by corollaries from the dynamic equations and also tried to compare the results obtained with the experimental data available in 1941 (which, as it became clear later, were not fully appropriate for this aim). On the other hand Obukhov (1941b) based his derivation on a model equation (which was established with the aid of a special semiempirical hypothesis) for the spectrum of turbulence $E(k)$. [The spectrum of disordered fluctuations was first introduced in the probabilistic context by Kolmogorov (1940) for the case of fluctuations represented by a random function of one variable; the spectrum of turbulence $E(k)$ related to the random velocity field $\mathbf{u}(\mathbf{x})$ first appeared in the paper by Obukhov (1941a).] Obukhov's semiempirical hypothesis was, of course, nonrigorous and Obukhov showed an amazing physical intuition by extracting from his model equation one consequence which is, in fact, universal and is independent of the particular form of the closure hypothesis used. Obukhov's result states that $E(k) \sim k^{-5/3}$ over a wide range of wave numbers k. It is easy to show that this is equivalent to the relation $D_{LL}(r) \sim r^{2/3}$. (This follows from the relation between $E(k)$ and $D_{LL}(r)$ which is a version of a Fourier transformation.) However, this is only one particular consequence of Kolmogorov's general theory, whose essence consists of two similarity hypotheses. True, Obukhov (1941b) also indicated at the end of his paper another important result of the Kolmogorov theory—namely the "4/3 power law" for the eddy diffusivity that characterized the relative turbulent diffusion of different length scales. This last law was empirically established much earlier by Richardson (1926), and in his theoretical derivation of it Obukhov referred not to general principles but only to his result as it related to the spectrum $E(k)$. Hence, strictly speaking, the content of the paper by Kolmogorov (1941a) was considerably more general than the (very interesting and important but nevertheless only particular) results given by Obukhov. However, AN here also demonstrated his high respect for the works of others, always stressing that the general theory of small-scale turbulence was a joint creation of his and Obukhov (see e.g. Kolmogorov 1949, 1962a,b).

In 1941 AN published one more paper (Kolmogorov 1941c) devoted to a special consequence of the general theory presented in Kolmogorov (1941a). This paper deals with an idealized model of isotropic turbulence in unbounded three-dimensional space and employs an assumption about the existence of the bounded and time-independent Loitsyanskii integral Λ. The last assumption now seems to be questionable (see e.g. Monin & Yaglom 1975, section 15); therefore the results implied by this assumption must now be considered as unreliable.

In early 1941 AN was absorbed in the study of turbulence, but on June 22 of that year Germany attacked the USSR. The war made working conditions in our country much worse for everybody. Nevertheless, AN continued his work and in January of 1942 he delivered a report on equations of turbulent motion at the first meeting of the Section of Physical and Mathematical Sciences of the USSR Academy of Sciences held in Kazan. (This Section was headed by Kolmogorov from 1939, when he was elected to the Academy, to the end of 1942. The USSR Academy of Sciences was transferred to the city of Kazan on the Volga shore in the Autumn of 1941 when the German armies were very close to Moscow.) The abstract of the report by AN was later published (Kolmogorov 1942) and it was quite remarkable. Here AN did not limit himself to investigation of only small-scale fluctuations, but he tried to construct an approximate (model) system of equations describing the large-scale properties of turbulent motions responsible for most of the turbulence effects that are interesting to engineers. However, in contrast with the early semiempirical theories of turbulence developed in the 1920s and early 1930s by L. Prandtl, T. von Kármán, and G. I. Taylor, AN did not limit himself to Reynolds equations for mean velocity components $\langle u_i \rangle$, $i = 1, 2, 3$, closed by hypothetical algebraic equations for Reynolds stresses. Instead he introduced additional differential equations for the energy of the fluctuating motion b and its typical frequency ω (related to the turbulence length scale L, eddy viscosity K, and energy dissipation rate ε by the equations $L \sim b^{1/2}/\omega$, $K \sim b/\omega$, and $\varepsilon \sim b\omega$). The resulting closed system of equations determined the variables $\langle u_i \rangle$, b, and ω (and hence also L, K, and ε) which are the main characteristics of turbulent motion.

In the pre-computer era it was natural to note, as AN did in his 1942 paper, that the solution of the proposed system of differential equations "presents great difficulties." However, AN stated that these equations were nevertheless solved by him and his collaborators for the case of plane Couette flow and that some preliminary results were also obtained for flow in a circular pipe. In his interesting commentary on AN's paper of 1942, D. B. Spalding (1991) noted that a source term was missing in the equation proposed by AN for ω. Spalding assumed that in the performed numerical solutions this lack was probably compensated by introducing a special boundary condition at the wall; if so, then AN had anticipated the practices of a number of later workers. However, it is also possible that the source term is missing because of a misprint in the text (there are a number of obvious misprints in this poorly printed war-publication of 1942). Spalding also noted the incompleteness and inappropriateness of the emphasis on the strong similarity between the works by Kolmogorov (1942) and Prandtl (1945) in many reviews of model equations of turbulent flows (e.g. in Harsha 1977 where the term "Prandtl–Kolmogorov models" is widely used). In fact, Prandtl developed a one-equation model of turbulence in which the equations for the mean velocity components were supplemented by an equation for the turbulent energy (and therefore the length scale L, or K, or ε, must be determined here by some crude guess), while AN was the first who proposed a two-equation model, which was closed and required no additional hypotheses. Of course, the models that use two additional equations are more complicated and require more difficult computational work than one-equation models but the former give a more accurate description of turbulent characteristics and therefore are the most popular at present. Apparently the first rather crude one-equation model was proposed by Zagustin (1938) [the reference to this early work was included in the book by Monin & Yaglom (1971) at the suggestion of AN]; after Prandtl (1945), one-equation models equivalent to his were independently proposed and used for turbulent computations by several authors during the 1950s and 1960s (see Spalding 1991). Note also that when after the war AN proposed to his graduate students Monin and Barenblatt the particular computational problems of mechanics of turbulence, they both used (at the recommendation of AN) the simplified one-equation version of equations given in Kolmogorov (1942) (see Monin 1950; Barenblatt 1953, 1955; or Monin & Yaglom 1971, Sections 6.6 and 6.7). However, in the late 1960s and 1970s a number of scientists (Harlow and Nakayama, Spalding, Saffman, and some others) working independently from each other and from AN (whose work of 1942 was then unknown in the West) proposed two-equation models quite close to that of AN (see again Spalding 1991). The high popularity of such models clearly shows that in his paper of 1942 wonderfully AN anticipated the future developments needed to satisfy practical workers. This penetrating insight, clearly seen in many of his papers, often places these papers much ahead of their time.

After 1942 much time passed before AN's next publication on turbulence appeared. The war years were, of course, not propitious at all for pure science but, nevertheless, AN continued to preserve his keen interest in the mechanics of turbulence. Around 1944, when it became clear that the war was going to end and Obukhov returned to Moscow from Kazan together with the Institute of Theoretical Geophysics where he worked, AN arranged a small seminar on turbulence which was held first at Moscow University (in the old building in downtown Moscow) and then at the Institute of Theoretical Geophysics. At the beginning the seminar was not regular and alternated with meetings devoted to other problems, but from 1946 onward meetings of the turbulence seminar were held regularly every fortnight. AN delivered a number of reports at the seminars—partially original and partially reviewing publications by other authors. I remember well his report on the spectral theory of isotropic turbulence where the spectral formulation of Millionshchikov's results of 1939 was presented and the spectral expression for the Loitsyanskii integral was given. This report was very helpful to me since it permitted me to simplify considerably the analysis of the final period of decay for isotropic turbulence in a compressible fluid. I started the investigation of this problem in Sverdlovsk during my work at the Main Geophysical Observatory, but without the spectral approach the analysis was very cumbersome and only after AN's report could I complete the work and publish it as my first paper on turbulence (Yaglom 1948). However, AN never published any results from this report of his; only his expression for the Loitsyanskii integral was presented in my paper of 1948 (of course, with a direct reference to AN) and still later it was independently found by Batchelor (1949). I remember also AN's review report on the derivation of skin friction laws by assuming the existence of an overlap of the wall and outer sublayers in turbulent flows. Such a derivation is, in fact, a simple combination of similarity and dimensional arguments; now it is quite popular and is presented in many textbooks (usually with a reference to Millikan 1939). However, AN delivered his report in the mid-1940s, when this approach was unknown to most researchers. Moreover, he based his talk on a rather unpopular paper by von Mises (1941) which contains a very clear and logical presentation of Millikan's results supplemented by some useful original remarks.

Together with AN's students interested in turbulence, many applied scientists and engineers participated in the turbulence seminar. Quite often these participants presented their experimental results and AN was always eager to discuss such reports, to explain the theoretical deduction related to the data, and to take part in comparison of experimental results with theory. Several publications of AN (Kolmogorov 1946a,b, 1949, 1952, 1954) were devoted to discussions of engineering papers, to criticism of errors committed by the authors, and to theoretical explanations of the results obtained. In particular, in Kolmogorov (1946b, 1954) AN criticized the views of the well-known Russian hydrologist M. A. Velikanov on the transport of suspended sediment by turbulent flows in rivers. In this respect he proposed a new theory of such transport which was then developed quantitatively by his student Barenblatt (1953, 1955). In Kolmogorov (1946a, 1952) he explains errors committed by some authors in the "theoretical derivation" of skin friction laws in circular tubes. Of great interest is his 1949 note (the only one of the papers cited above that AN felt should be included in the collection of his selected works published in 1985), which reflects AN's reaction to one purely experimental report at the seminar. It contains an elegant application of dimensional analysis and universal equilibrium theory from Kolmogorov (1941a) to the physical problem of dispersion of drops of a liquid in the turbulent flow of another liquid where the surface-tension stress plays an important part. Many other applications of the same theory were discussed at the seminar; AN participated actively in such discussions but let the others publish the results.

Note now that Obukhov tried to arrange experimental verifications of the laws, found in AN's and his own papers of 1941, immediately after the discovery of these laws; some preliminary results of the very first measurements were published in Obukhov (1942) but then the war intervened. When the war came to an end in 1945 Obukhov returned to his old idea which was also supported by AN and by Professor O. Yu. Schmidt, Director of the Institute of Theoretical Geophysics. It was then decided to arrange a special Laboratory of Atmospheric Turbulence at the Institute of Theoretical Geophysics and Schmidt proposed that AN head the laboratory. In short comments on his works on turbulence included in Kolmogorov (1985) AN gave the following description of this experience:

> In 1946 O. Yu. Schmidt proposed me as the head of the Laboratory of Atmospheric Turbulence at the Institute of Theoretical Geophysics, USSR Academy of Sciences. In 1949 I stopped working there and Obukhov replaced me as the head of the laboratory. I did not participate directly in the experimental work but spent much energy on computational and graphical treatment of the data obtained by other investigators.

Some details must be added to this short description. The Institute of Theoretical Geophysics was originally placed in the former private house of a rich Moscow merchant named Lepekhin, in a part of Moscow that was traditionally inhabited by merchants. The house was not large enough for a big scientific institute and the only room which could be found for the new laboratory was a former merchant's bathroom with all its walls covered with white tiles. Moreover, there was also a grid on the window since before the laboratory was placed there the room had been occupied by an office where classified information had been stored. Two desks and several chairs were the only furniture which could be placed in the small room. The first staff of the laboratory consisted of AN and only three other people: A. M. Obukhov, P. A. Kozulyaev (a former graduate student of AN whose candidate dissertation was devoted to the extrapolation of time series), and myself. This selection was quite accidental. Kozulyaev was jobless at that moment and AN invited him to help with the treatment of the data; one year later he left the laboratory and became a professor in one of the Moscow engineering colleges. I was then more interested in turbulence than Kozulyaev but my inclusion in the laboratory staff was also in some sense accidental since I had at that time quite another plan for my future. During my graduate study I solved the problem proposed by AN in one year. AN told me that this was enough for the candidate dissertation and therefore he could arrange my dissertation defense which would end my graduate study. I answered that I would prefer to preserve the position of a graduate student (having a small stipend and much free time) as long as possible; hence I asked him to delay my defense until the normal end of my studentship. (The normal period for a graduate study in the USSR is three years.) AN approved my decision and said: "I can understand you. I spent five years as an undergraduate student and four years as a graduate student and was happy to have that opportunity. By the end of my graduate study I had 18 published works." So I was permitted to do what I wanted for two years. During these years I participated in AN's seminars on turbulence and stochastic processes but also attended all the meetings of theoretical physics seminars led by L. D. Landau and by I. E. Tamm (both future Nobel laureates). Simultaneously I worked jointly with Professor I. M. Gelfand on the general theory of relativistic wave equations: Theoretical physics was then my main passion. After the end of my graduate studies AN proposed several possible jobs to me (there were few young candidates of sciences in 1946 and many free vacancies), but more attractive to me was the proposition by Tamm to join the Theory Department of the P. N. Lebedev Physics Institute. However, it upset me to learn that I would have to spend some of my time on "applied problems" (i.e. on the atomic bomb project) in this department. I have already written [in my memoirs about Sakharov; see Lebedev Physics Institute (1990), pp. 660–74] that I probably would have reacted negatively to any proposal to participate in such a project under any government, but in the case of the Stalin regime, which I considered as being very oppressive and dangerous to all mankind, such participation was impossible for me. Therefore, I agreed to take the job at the Institute of Theoretical Geophysics which had nothing to do with military problems. At the beginning I considered this job as a temporary one and planned to return to theoretical physics in a year or two when the atomic bomb problem ceased to be so urgent. However, it did not cease to be urgent for a long time and, moreover, political development in the USSR made changing jobs impossible for many years. Therefore, I remained working in the same laboratory (of the same Institute which, however, changed its name twice after 1946) till today. (Maybe it was luck. Who knows?)

Two months after organization of the Turbulence Laboratory three more people joined: two young women working as secretaries and laboratory assistants and a qualified technician from the P. N. Lebedev Physics Institute, USSR Academy of Sciences, who immediately began constructing a hot-wire anemometer under the direction of Obukhov and later was responsible for measurements of wind velocity and temperature fluctuations in the atmosphere. AN usually spent one day a week in the laboratory heading the seminar, discussing current work, and looking through the collected experimental data which always interested him very much. Between AN's visits to the laboratory, Obukhov and I often met him at Moscow University and sometimes visited him either at home or at his country house in the village of Komarovka near Moscow. We told AN about our work and discussed the most urgent problems. Such meetings with AN were during some periods more frequent and at other times less frequent but they continued until his last illness.

The replacement in 1949 of AN by Obukhov as the head of the Turbulence Laboratory at first changed nothing. It was a purely formal act aimed at increasing Obukhov's salary: The day by day laboratory work was directed by Obukhov from the very beginning and the general guidance continued to be AN's duty. However, in the early and mid-1950s topics unrelated to turbulence (superpositions of functions, ergodic theory, Hamiltonian mechanics) became AN's main scientific interests. He naturally began to give less attention to work at the Turbulence Laboratory and stopped attending the laboratory seminar.

It is, however, worth noting that AN's approach to the mechanics of turbulence and his general ideas directed the works of his students (and of the students of his students) even during periods when AN himself was engaged on other research problems. One of the most important achievements of the Turbulence Laboratory was the development by Obukhov (1949) of a universal equilibrium theory for scalar fields in turbulent flows. This theory is again based on intuitive physical arguments and the paper by Obukhov is very close in style to the classical paper by Kolmogorov

(1941a). Obukhov's paper of 1949 immediately stimulated an attempt to also apply the approach presented in Kolmogorov (1941d) (see Yaglom 1949) to scalar fields. The very useful similarity theory of turbulence in thermally-stratified boundary layers developed by Monin & Obukhov (1954) has many applications (see e.g. Monin & Yaglom 1971, chapter 4) and is a generalization of the classical wall laws for nonstratified turbulent boundary layers often used by AN in his reports and discussions. Knowledge of the theory by Monin & Obukhov prompted Kader & Yaglom (1978) to apply a similar approach to the analysis of pressure-gradient turbulent wall flows; AN's old report on the derivation of skin-friction laws inspired the derivation of universal heat- and mass-transfer laws for turbulent wall flows (Kader & Yaglom 1972). The paper by Novikov (1971), a former student of Monin, contains a direct development of the ideas given in Kolmogorov (1962a,b) (these ideas will be discussed below); the dimensional analysis of the dynamics of planetary atmospheres by Obukhov's student Golitsyn (1973) reflects the general approach to turbulence problems often popularized by AN.

Let us now return to works by AN himself. The next peak of AN's activity in turbulence research came in the early 1960s. In 1961 two big International Meetings on the Mechanics of Turbulence were arranged one after another in Marseille by the IUTAM (International Union of Theoretical and Applied Mechanics) and the IUGG (International Union of Geodesy and Geophysics). AN was invited to both meetings together with a group of his former students. AN was always fond of France, especially Southern France. (He was less enthusiastic about Germany; probably this was partially connected with memories of the Second World War but my impression was that apparently some experience from his first visit to Germany and France in the early 1930s also contributed to his attitude toward these two countries.) Therefore he was happy to have the opportunity to visit Marseille again. Scientifically the Marseille meetings were also of great interest to him. In preparation for these meetings he showed again great interest in the works of the Turbulence Laboratory he had founded fifteen years before.

Obukhov prepared for the Marseille meeting a report stimulated by the data of A. S. Gurvich who had been working in our laboratory and was engaged in measurements of the spectra of wind velocity fluctuations. Gorvich's data show that although high frequency spectra measured at the same point in the atmosphere during two closely-spaced time intervals both agree well with the relation $E(k) \sim k^{-5/3}$, the proportionality coefficients are often quite different. According to the theoretical results obtained by AN and Obukhov in 1941 the proportionality coefficient is equal to $C\varepsilon^{2/3}$, where C is a universal constant and ε is the rate of energy dissipation. Therefore the data by Gurvich showed that the dissipation rate ε has a strongly fluctuating time dependence in the Earth's atmosphere. It was clear to Obukhov that fluctuations in ε must affect the structure functions (i.e. the mean values of the squared velocity differences at two points) and the spectra (the Fourier transforms of the structure functions). Referring to Kolmogorov's paper (1941b) on the probability distribution of the sizes of particles under fragmentation, Obukhov assumed that the dissipation rate ε is a random variable having a logarithmically normal probability distribution. Moreover, he also assumed that the structure function $D_{LL}(r)$ depends on $\bar{\varepsilon}_r$—the dissipation rate averaged over a sphere of radius r. Under this assumption he evaluated the influence of the dissipation-rate fluctuations on the value of the velocity structure function $D_{LL}(r)$ over an inertial range of distances r [where $D_{LL}(r) \sim r^{2/3}$ according to the Kolmogorov theory of 1941]. Taking the fluctuations into account, Obukhov obtained a result that differed from the classical "2/3 power law" of the old theory by a correction factor of a special form. Obukhov told AN about this result and the latter was very interested in it and asked for a copy of the manuscript of the prepared report. Obukhov and Kolmogorov then parted with each other and were not to meet again until Marseille: AN could not fly because of an ear problem and therefore he went to France by train separately from the other Russian participants at the meetings. Arriving in Marseille AN showed Obukhov his own manuscript which contained an important development of Obukhov's ideas.

AN's paper (Kolmogorov 1962a,b) presented as a report at the Colloquium on Turbulence in Marseille is close in some respects to his earlier 1941 paper (Kolmogorov 1941a). The new paper is also very physical in its style; it contains no mathematical proofs or analytical derivations but instead suggests general similarity hypotheses justified by heuristic arguments. The first two new hypotheses stated more precisely the two similarity hypotheses from Kolmogorov (1941a); they refer to probability distributions not for velocity differences but for the ratios of such differences related to two pairs of points. (By the way, the mathematical theory necessary for the rigorous presentation of Kolmogorov's new theory sketched in AN's papers of 1962 has not yet been developed. This theory must deal with random fields that have probability distributions for the ratios of two differences of field values at different points that are invariant under all spatial similarity transformations.) These two hypotheses were supplemented by a third more special hypothesis postulating the logarithmic normality of the dissipation rate ε and indicating the form of the variance of log $\bar{\varepsilon}_r$. The three hypotheses imply corrected relations for spectra, structure functions of various orders, and many other statistical characteristics of turbulence. The corrected relations include, in particular, Obukhov's result for the second-order structure function (which was also presented in Marseille and later published; see Obukhov 1962a,b) and differ from the conclusions of AN's theory of 1941 by additional factors depending on the Reynolds number. Unlike AN's theory of 1941 the new theory proposed by AN and Obukhov twenty years later does not lead to any entirely new results but implies only some minor corrections to the known relations which have been, up until now, unimportant for real applications. However, this new theory was a very significant step toward understanding the physical mechanism of developed turbulence and therefore was crucial to most of the further developments in turbulence theory (see e.g. Hunt et al 1991 and Meneveau & Sreenivasan 1991).

Both Marseille meetings were a great success. They collected many brilliant scientists of all generations from many countries of the world including such giants as T. von Kármán, G. I. Taylor, and A. N. Kolmogorov. The American delegation included representatives from at least four generations (the great von Kármán, his student H. W. Liepmann, Liepmann's student S. Corrsin, and Corrsin's student J. L. Lumley). The USSR delegation included Kolmogorov (this was the first and only time when AN actively participated in an international meeting on turbulence), his two pre-war students M. D. Millionshchikov and A. M. Obukhov, and me—a war-years student. Such a composition had a flavor of Khrushchev's liberalization (for me it was the first time I was permitted to attend a meeting in a "capitalist country"), but nevertheless the authorities did not permit Monin to go with us—apparently to show that "high authorities" continue to exist. There were many very interesting reports at the meetings but my feeling was that a short report by AN and an accompanying paper by Obukhov (presented at the second geophysical meeting) were considered by many participants to be the most exciting and sensational. After these talks G. K. Batchelor said that they were worthy of publication not only in the Proceedings of the meetings but also in the Journal of Fluid Mechanics; due to his insistence both papers were published in 1962 twice (1962a are Proceedings publications, 1962b—publications in JFM).

His participation in many scientific discussions at the Marseille meetings and the great success of his report there again aroused AN's interest in the mechanics of turbulence. After returning home from Marseille AN organized the new USSR Seminar on Turbulence within the framework of the Laboratory of Statistical Methods, Moscow University, which he established in 1960 and then headed till his death. The seminars took place in a big hall of the new Moscow University Laboratory Building at Lenin (Vorobjev) Hills in Moscow. Often more than a hundred people participated in these meetings. Both original and review reports were presented at the Seminar and their influence on the development of the mechanics of turbulence in our country was quite substantial. AN led the discussions and always actively participated in them. Some of his deductions based on discussions at the seminar were formulated by AN in his survey lecture on "Experimental and theoretical methods in the study of turbulence," read in May 1965 at the meeting of the Moscow Mathematical Society (AN was then President of this Society). The last report presented by AN at the meetings of this Society was also devoted to turbulence; it was read in January 1978 and was entitled "Remarks on the statistical solutions of Navier-Stokes equations" (see Kolmogorov 1978).

At one of the specialized meetings at the Laboratory of Statistical Methods in the mid-1960s I presented a short review of an old paper by E. Hopf (1952) on the equation for the characteristic functional of the fluctuating velocity field. (This paper was for some reason poorly known in the USSR.) AN then told me that he understood long ago that the Navier-Stokes equations must imply a linear functional differential equation for the characteristic functional of the velocity field. (Let us recall that the concept of a characteristic functional was introduced by AN in his paper of 1935.) "However," continued AN, "I did not try to determine the exact form of this equation since I could not see what applications such an equation could have." AN also stressed that without any theorems about conditions guaranteeing the existence and uniqueness of the solution of an initial value problem for functional differential equations, such equations continue to be useless. Later AN tried to interest M. I. Vishik— a known expert on both functional analysis and partial differential equations—in mathematical problems related to Hopf's equation. Vishik was very interested in these problems, he enlisted some students of his, and began a profound study of the mathematical problems of turbulence. During this study he regularly consulted AN and their long conversations played a very important part in the development of the work. The results of these investigations are summarized in the monograph by Vishik & Fursikov (1988)—which never would have appeared without AN's inspiration.

Renewal of AN's interest in turbulence after 1961 also prompted him to accept Monin's proposal to take part in two voyages (in 1970 and 1971– 1972) of the big scientific research ship "Dmitrii Mendeleyev" which set out to study oceanic turbulence. AN had the position of a scientific supervisor on these voyages (together with A. S. Monin). He was mostly engaged during the voyages in the development and applications of improved methods of spectral analysis. These methods permitted one to minimize interference due to the noise at neighboring frequencies, and in principle could be applied even to the spectral analysis of nonstationary (slowly evolving) processes. Another contribution of AN to the joint experimental work consisted in estimating the necessary averaging times, duration of realizations, discretization intervals, etc. He also evaluated the reliability and statistical representativeness of experimental values obtained. AN very much enjoyed his participation in the oceanic voyages and always remembered them with great pleasure and some pride. After the first voyage he presented the scientific impressions which it produced in a survey report at the Moscow Mathematical Society entitled "Statistical fluid mechanics of the ocean."

In the late 1970s the scientific interests of AN again shifted to subjects unrelated to fluid mechanics. Mathematical logic and the foundations of probability and information theories, mathematical linguistics, and the problems of high school mathematical education took most of AN's time near the end of his life. Moreover, in old age he suffered from Parkinson's disease and from an eye illness that made him almost blind. Nevertheless, he tried to work practically until the end. The death of this great scientist on October 20, 1987 was undoubtedly a great loss to all mankind. For the fluid mechanics community his death marked the end of a whole epoch, for AN's essential contributions changed the very content and style of our science.

Literature Cited

Abraham, R., Marsden J. E. 1978. *Foundations of Mechanics*. Reading: Benjamin. 2nd ed.

Barenblatt G. I. 1953. Motion of suspended particles in a turbulent flow. *Prikl. Mat. Mekh.* 17: 261–74

Barenblatt, G. I. 1955. Motion of suspended particles in a turbulent flow occupying a half-space or plane channel of finite depth. *Prikl. Mat. Mekh.* 19: 61–68

Batchelor, G. K. 1947. Kolmogorov's theory of locally isotropic turbulence. *Proc. Cambridge Philos. Soc.* 43: 533–59

Batchelor, G. K. 1949. The role of big eddies in homogeneous turbulence. *Proc. R. Soc. London Ser. A* 195: 513–32

Batchelor, G. K. 1953. *The Theory of Homogeneous Turbulence*. Cambrige: Cambrige Univ. Press

Golitsyn, G. S. 1973. *An Introduction to Dynamics of Planetary Atmospheres*. Leningrad: Gidrometeoizdat. 104 pp.

Harsha P. T. 1977. Models of kinetic energy transfer. In *Handbook of Turbulence, Vol. 1: Fundamentals and Applications*, ed. W. Frost, T. H. Moulden, Chap. 8. New York: Plenum

Hopf E. 1952. Statistical hydromechanics and functional calculus. *J. Rat. Mech. Anal.* 1: 87–123

Hunt, J. C. R., Phillips, O. M., Williams, D., eds. 1991. *Turbulence and Stochastic Processes: Kolmogorov's Ideas 50 Years On*. London: Royal Society. 240 pp.

Kader, B. A., Yaglom, A. M. 1972. Heat and mass transfer laws for fully turbulent wall flows. *Int. J. Heat Mass Transfer* 15: 2329–51

Kader, B. A., Yaglom, A. M. 1978. Similarity treatment of moving-equilibrium turbulent boundary layers in adverse pressure gradients. *J. Fluid Mech.* 89: 305–42

Keller, L. V., Friedmann, A. A. 1924. Differentialgleichungen für die turbulente Bewegung einer kompressiblen Flüssigkeit. *Proc. 1st Int. Congr. Appl. Mech.*, *Delft*, pp. 395–405

Kendall, D. G., ed. 1990. Obituary: Andrei Nikolaevich Kolmogorov (1903–1987). *Bull. London Math. Soc.* 22: 31–100

Kolmogorov, A. N. 1933. *Grundbegriffe der Wahrscheinlichkeitsrechnung*. Berlin: Springer-Verlag

Kolmogorov, A. N. 1935. La transformation de Laplace dans les espaces linéaires. *C. R. Acad. Sci. Paris* 200: 1717–18

Kolmogorov, A. N. 1937. Zur Umkehrbarkeit der statistischen Naturgesetze. *Math. Ann.* 113: 766–72

Kolmogorov, A. N. 1940. Curves in a Hilbert space that are invariant under the one-parameter group of motions. *Dokl. Akad. Nauk SSSR* 26: 6–9

Kolmogorov, A. N. 1941a. Local structure of turbulence in an incompressible fluid at very high Reynolds numbers. *Dokl. Akad. Nauk SSSR* 30: 299–303

Kolmogorov, A. N. 1941b. Logarithmically normal distribution of the size of particles under fragmentation. *Dokl. Akad. Nauk SSSR* 31: 99–101

Kolmogorov, A. N. 1941c. Decay of isotropic turbulence in an incompressible viscous fluid. *Dokl. Akad. Nauk SSSR* 31: 538–41

Kolmogorov, A. N. 1941d. Energy dissipation in locally isotropic turbulence. *Dokl. Akad. Nauk SSSR* 32: 19–21

Kolmogorov, A. N. 1942. Equations of turbulent motion of an incompressible fluid. *Izv. Akad. Nauk SSSR Ser. Fiz.* 6: 56–58

Kolmogorov, A. N. 1946a. On the resistance law for turbulent flows in smooth tubes. *Dokl. Akad. Nauk SSSR* 52: 669–71

Kolmogorov, A. N. 1946b. Comments on the paper by A. M. Velikanov "Transfer of suspended alluvium by a turbulent flow." *Izv. Akad. Nauk SSSR, Otd. Tekh. Nauk* No. 5: 781–84

Kolmogorov, A. N. 1949. On the fragmentation of drops in a turbulent flow. *Dokl. Akad. Nauk SSSR* 66: 825–28

Kolmogorov, A. N. 1952. On the resistance law and velocity profile for turbulent flows in tubes. *Dokl. Akad. Nauk SSSR* 84: 29–30

Kolmogorov, A. N. 1954. New variant of the gravitational theory of motion of suspended alluviums. *Vestn. Mosk. Gosud. Univ.* No. 3: 41–45

Kolmogorov, A. N. 1956a. *Foundations of the Theory of Probability*. New York: Chelsea. 2nd English ed.

Kolmogorov, A. N. 1956b. On the Shannon theory of information transmission in the case of continuous signals. *IEEE Trans. Inform. Theory* IT-2: 102–8

Kolmogorov, A. N. 1962a. Précisions sur la structure locale de la turbulence dans un fluide visqueux aux nombres de Reynolds élevés (in French and Russian). In *Mécanique de la Turbulence* (Coll. Int. du CNRS à Marseille), pp. 447–58. Paris: CNRS

Kolmogorov, A. N. 1962b. A refinement of previous hypotheses concerning the local structure of turbulence in a viscous incompressible fluid at high Reynolds numbers. *J. Fluid Mech.* 13: 82–85

Kolmogorov, A. N. 1978. Remarks about statistical solutions of Navier—Stokes equations. *Usp. Mat. Nauk* 33(3): 124 (Abstr.)

Kolmogorov, A. N. 1985. *Mathematics and Mechanics*. Moscow: Nauka. 469 pp. (In Russian)

Kolmogorov, A. N. 1986. *Probability Theory and Mathematical Statistics*. Moscow: Nauka. 534 pp. (In Russian)

Kolmogorov, A. N. 1987. *Information Theory and the Theory of Alogorithms*. Moscow: Nauka. 304 pp. (In Russian)

Kolmogorov, A. N. 1992. *The Landholding in Novgorod in the 15th Century* (with comments by L. A. Bassalygo). Moscow: Nauka

Kolmogorov, A. N., Dmitriev, N. A. 1947. Branching stochastic processes. *Dokl. Akad. Nauk SSSR* 56: 7–10

Lebedev Physics Institute 1991. *Andrei Sakharov: Facets of a Life*. Collection of reminiscences prepared by P. N. Lebedev Physics Inst., USSR Acad. Sci. Gif-sur-Yvette: Editions Frontières. 730 pp.

Meneveau, C., Sreenivasan, K. R. 1991. The multifractal nature of turbulent energy dissipation. *J. Fluid Mech.* 224: 429–85

Millikan, C. B. 1939. A critical discussion on turbulent flows in channels and circular tubes. *Proc. 5th Intern. Congr. Appl. Mech.*, pp. 386–92. Cambridge, Mass.

Millionshchikov, M. D. 1939. Decay of homogeneous isotropic turbulence in a viscous incompressible fluids. *Dokl. Akad. Nauk SSSR* 22: 236–40

Millionshchikov, M. D. 1941. Theory of homogeneous isotropic turbulence. *Dokl. Akad. Nauk SSSR* 22: 241–42; *Izv. Akad. Nauk SSSR, Ser. Geogr. Geofiz.* 5: 433–46

Monin, A. S. 1950. Dynamic turbulence in the atmosphere. *Izv. Akad. Nauk SSSR Ser. Geogr. Geofiz.* 14: 232–54

Monin, A. S., Obukhov, A. M. 1954. Basic turbulent mixing laws in an atmospheric surface layer. *Tr. Geofiz. Inst. Akad. Nauk SSSR* 24 (151): 163–87

Monin, A. S., Yaglom, A. M. 1971. *Statistical Fluid Mechanics*, Vol. 1. Cambridge: MIT Press

Monin, A. S., Yaglom, A. M. 1975. *Statistical Fluid Mechanics*, Vol. 2. Cambridge: MIT Press

Novikov, E. A. 1971. Intermittency and scale similarity of the structure of turbulent flow. *Prikl. Mat. Mekh.* 35: 266–77

Obukhov, A. M. 1941a. Sound scattering in a turbulent flow. *Dokl. Akad. Nauk SSSR* 30: 611–14

Obukhov, A. M. 1941b. Spectral energy distribution in a turbulent flow. *Dokl. Akad.*

22   YAGLOM

Nauk SSSR 32: 22–24; *Izv. Akad. Nauk
SSSR Ser. Geogr. Geofiz.* 5: 453–66
Obukhov, A. M. 1942. On the theory of
atmospheric turbulence. *Izv. Akad. Nauk
SSSR Ser. Fiz.* 6: 59–63
Obukhov, A. M. 1949. Structure of the tem-
perature field in a turbulent flow. *Izv.
Akad. Nauk SSSR Ser. Geogr. Geofiz.* 13:
58–69
Obukhov, A. M. 1962a. Some specific fea-
tures of atmospheric turbulence. *J.
Geophys. Res.* 67: 311–14
Obukhov, A. M. 1962b. Some specific fea-
tures of atmospheric turbulence. *J. Fluid
Mech.* 13: 77–81
Prandtl, L. 1945. Über ein neues Formel-
system für die ausgebildete Turbulenz.
*Nachr. Ges. Wiss. Göttingen Math.-Phys.
Kl.*, pp. 6–19
Richardson, L. F. 1926. Atmospheric
diffusion shown on a distance-neighbour
graph. *Proc. R. Soc.London Ser. A* 110:
709–37
Shiryayev, A. N. 1989. Kolmogorov: Life
and creative activities. *Ann. Probab.* 17:
866–944
Shiryayev, A. N., ed. 1992. *Probability Theory
and Mathematical Statistics.* Dordrecht:
Kluwer. 598 pp.
Sossinsky, A., ed. 1992. *Information Theory
and the Theory of Alogorithms.* Dordrecht:
Kluwer
Spalding, D. B. 1991. Kolmogorov's two-
equation model of turbulence. In *Turbulence and Stochastic Processes: Kolmogorov's Ideas 50 Years On*, pp. 211–14.
London: Royal Society
Taylor, G. I. 1935. Statistical theory of turbulence, I-IV. *Proc. R. Soc. London Ser.
A* 151: 421–78
Tikomirov, V. M., ed. 1991. *Mathematics
and Mechanics.* Dordrecht: Kluwer. 552
pp.
Vishik, M. I., Fursikov, A. V. 1988. *Mathematical Problems of Statistical Hydrodynamics.* Dordrecht: Kluwer
von Kármán, T., Howarth L. 1938. On the
statistical theory of isotropic turbulence.
Proc. R. Soc. London Ser. A 164: 192–215
von Mises, R. 1941. Some remarks on the
laws of turbulent motion in tubes. In *T.
von Kármán Anniversary Volume*, pp. 317–
27. Pasadena: Calif. Inst. Technol. Press
Yaglom, A. M. 1948. Homogeneous and isotropic turbulence in a viscous compressible fluid. *Izv. Akad. Nauk SSSR Ser.
Geogr. Geofiz.* 12: 501–22
Yaglom, A. M. 1949. Local structure of temperature field in a turbulent flow. *Dokl.
Akad. Nauk SSSR* 69: 743–46
Yaglom, A. M. 1981. Laws of small-scale
turbulence in atmosphere and ocean. (In
commemoration of the 40th anniversary
of the theory of locally isotropic turbulence.) *Izv. Akad. Nauk SSSR Ser. Fiz.
Atmos. Oceana* 17: 1235–57 (pp. 919–35 in
English ed.)
Yaglom, A. M. 1990. Canonical correlation
analysis and its meteorological appli-
cations. *Izv. Akad. Nauk SSSR Ser. Fiz.
Atmos. Oceana* 26: 1248–66
Yanin, V. L. 1988. Kolmogorov as an his-
torian. *Usp. Mat. Nauk* 43(6): 189–95
Zagustin, A. I. 1938. Equations for the turbulent motions of fluids. *Tr. Voronezh.
Univ.* 10: 7–39

Annu. Rev. Fluid Mech. 1994. 26 : 23–63

LAGRANGIAN PDF METHODS FOR TURBULENT FLOWS

S. B. Pope

Sibley School of Mechanical and Aerospace Engineering,
Cornell University, Ithaca, New York 14853

KEY WORDS: turbulence modeling, stochastic methods, Langevin equation

1. INTRODUCTION

Lagrangian Probability Density Function (PDF) methods have arisen in the past 10 years as a union between PDF methods and stochastic Lagrangian models, similar to those that have long been used to study turbulent dispersion. The methods provide a computationally-tractable way of calculating the statistics of inhomogeneous turbulent flows of practical importance, and are particularly attractive if chemical reactions are involved. The information contained at this level of closure—equivalent to a multi-time Lagrangian joint pdf—is considerably more than that provided by moment closures.

The computational implementation is conceptually simple and natural. At a given time, the turbulent flow is represented by a large number of particles, each having its own set of properties—position, velocity, composition etc. These properties evolve in time according to stochastic model equations, so that the computational particles simulate fluid particles. The particle-property time series contain information equivalent to the multi-time Lagrangian joint pdf. But, at a fixed time, the ensemble of particle properties contains no multi-point information: Each particle can be considered to be sampled from a different realization of the flow. (Hence two particles can have the same position, but different velocities and compositions.)

It is generally acknowledged (e.g. Reynolds 1990) that many different approaches have important roles to play in tackling the problems posed by turbulent flows. Each approach has its own strengths and weaknesses.

23

0066–4189/94/0115–0023$05.00

At one end of the spectrum of approaches, Direct Numerical Simulations (DNS) offer unmatched accuracy, but their computational cost is high, and their range of applicability is extremely limited (Reynolds 1990). At the other end of the spectrum, simple turbulence models such as k-ε (Launder & Spalding 1972) offer essentially unrestricted applicability, moderate cost, but poor or uncertain accuracy (except in some simple flows). Compared to such turbulence models, Lagrangian PDF methods have the same wide applicability; their cost is greater (but less than DNS); and, for the reasons presented below, their potential for accuracy is greatly increased.

In going beyond simple turbulence models (i.e. moment closures), the challenge is to incorporate a fuller description of the turbulence, while retaining computational tractability. If appropriately chosen, the fuller description allows unsatisfactory modeling assumptions to be avoided, and at the same time provides more information to model the unavoidable.

It can be argued that the dominant process in turbulent flows is convection (by the instantaneous fluid velocity). At high Reynolds number, molecular diffusion makes a negligible contribution to spatial transport, and so convection dominates the transport of momentum, chemical species, and enthalpy. In moment closures, at some level, convection is modeled by a gradient diffusion assumption, which can lead to qualitatively incorrect behavior (see, for example, Deardorff 1978). In Lagrangian PDF methods, on the other hand, convection is treated simply and naturally in the Lagrangian frame with no modeling assumptions. Similarly, in reacting flows, finite-rate nonlinear reaction rates can present insurmountable difficulties to moment closures; whereas arbitrarily complex reactions can be handled naturally by Lagrangian PDF methods, without modeling assumptions (Pope 1985, 1990).

In any statistical approach to turbulence, modeling assumptions are required at some level. It is notoriously difficult to construct general and accurate models for turbulence, whereas there are many other stochastic physical phenomena for which simple statistical models are successful (see e.g. van Kampen 1983). There are two features of turbulent flows that go a long way to explaining the inherent difficulties. First, turbulence has a long memory: In free shear flows, it is readily deduced (from experimental data) that the characteristic time scale of the energy-containing turbulent motion is, typically, four times the characteristic mean-flow time scale. Second, through the fluctuating pressure field, the velocity field experiences long-range interactions. Among other effects, this can lead to large-scale organized motions, and to the boundary geometry influencing the turbulence structure in the interior of the flow.

Lagrangian PDF methods can take full account of the long memory of turbulence. For the fluid properties considered, the multi-time Lagrangian joint pdf completely describes the past history of all fluid particles that (on different realizations) pass through a given point at a given time. As discussed in Section 5, the long-memory incorporated in current stochastic Lagrangian models leads to fluid-particle motions that are consistent with large-scale turbulent structures.

In the next Section the relevant Eulerian and Lagrangian pdfs are introduced, and the different PDF methods are categorized. Section 3 is devoted to the Langevin equation. This equation provides a simple stochastic model for the velocity of a fluid particle: It is also a building block for other stochastic models. The particle representation of a turbulent flow, which is fundamental to the Lagrangian PDF approach, is described in Section 4. Then more recent and sophisticated stochastic models are reviewed in Section 5.

2. EULERIAN AND LAGRANGIAN PDF METHODS

The purpose of this section is to introduce the various pdfs considered, to categorize PDF methods, and to provide references to the relevant literature.

2.1 *Eulerian pdfs*

We start by considering a single *composition variable* (e.g. a species mass fraction) which, at position $\mathbf{x}$ and time t, is denoted by $\phi(\mathbf{x}, t)$. At fixed $(\mathbf{x}, t)$, ϕ is a random variable, corresponding to which we introduce the independent *sample-space variable* ψ. Then the *cumulative distribution function* (cdf) of ϕ is defined by

$$F_\phi(\psi, \mathbf{x}, t) \equiv \text{Prob}\{\phi(\mathbf{x}, t) < \psi\}, \tag{1}$$

and the *probability density function* (pdf) of ϕ is

$$f_\phi(\psi; \mathbf{x}, t) \equiv \frac{\partial}{\partial \psi} F_\phi(\psi, \mathbf{x}, t). \tag{2}$$

The fundamental significance of the pdf is that it measures the probability of the random variable being in any specified interval. For example, for $\psi_b > \psi_a$, from Equations (1) and (2) we obtain

$$\text{Prob}\{\psi_a \le \phi(\mathbf{x}, t) < \psi_b\} = \int_{\psi_a}^{\psi_b} f_\phi(\psi; \mathbf{x}, t) \, d\psi. \tag{3}$$

The pdf just defined, $f_\phi(\mathbf{x}, t)$, is the one-point, one-time Eulerian pdf of $\phi(\mathbf{x}, t)$. It completely describes the random variable ϕ at each $\mathbf{x}$ and t

separately; but it contains no joint information about ϕ at two or more space-time points. If $\phi(\mathbf{x}, t)$ is statistically homogeneous, then f_ϕ is independent of $\mathbf{x}$.

More generally, we may want to consider a set of σ composition variables $\phi(\mathbf{x}, t)$ where $\phi = \{\phi_1, \phi_2, \ldots, \phi_\sigma\}$. Then, with $\psi = \{\psi_1, \psi_2, \ldots, \psi_\sigma\}$ being corresponding sample-space variables, the joint pdf of ϕ is denoted by $f_\phi(\psi; \mathbf{x}, t)$. Further, with $\mathbf{U}(\mathbf{x}, t)$ being the Eulerian velocity of the fluid, we introduce sample-space velocity variables $\mathbf{V} = \{V_1, V_2, V_3\}$ and denote the (one-point, one-time Eulerian) joint pdf of velocity by $f_u(\mathbf{V}; \mathbf{x}, t)$. Finally, the velocity-composition joint pdf is denoted by $f(\mathbf{V}, \psi; \mathbf{x}, t)$.

By definition, *in a PDF method, a pdf (or joint pdf) in a turbulent flow is determined as the solution of a modeled evolution equation.*

2.2   Assumed PDF Methods

In spite of their name, *assumed PDF methods* are not PDF methods (according to the above definition). Instead of being determined from a modeled evolution equation, the pdf is assumed to have a particular shape that is parametrized (usually) by its first and second moments. The method has found application in combustion (e.g. Bilger 1980). For the pdf of a single composition, the suggested shapes include: a beta-function distribution (Rhodes 1975); a clipped Gaussian (Lockwood & Naguib 1975); and a maximum entropy distribution (Pope 1980). The extension to several composition variables, which is considerably more difficult, has been considered by, among others, Correa et al (1984), Bockhorn (1990), and Girimaji (1991).

Although assumed PDF methods are favored in some applications, compared to PDF methods they have two disadvantages. First, no account is taken of the influence of the dynamics (e.g. reaction) on the shape of the pdf. Second—and maybe surprising at first sight—assumed PDF methods are computationally more expensive (if not intractable) for the general case of many compositions.

2.3   Eulerian PDF Methods

One of the first cases studied using PDF methods, and one that continues to receive considerable attention, is reaction in constant-density homogeneous turbulence. In the simplest situation, the composition is characterized by a single passive scalar $\phi(\mathbf{x}, t)$, that evolves by

$$\frac{D\phi}{Dt} = \Gamma\nabla^2\phi + S(\mathbf{x}, t). \tag{4}$$

Here $D/Dt = \partial/\partial t + \mathbf{U}\cdot\nabla$ is the substantial derivative, Γ is the (constant) molecular diffusivity, and the reaction rate is a known function of the composition:

$$S(\mathbf{x}, t) = \hat{S}[\phi(\mathbf{x}, t)]. \tag{5}$$

For the statistically homogeneous case considered, the composition pdf $f_\phi(\psi; t)$ is independent of $\mathbf{x}$. Without further assumption, the evolution equation for f_ϕ can be deduced from Equation (4). This can be done using any one of several different techniques that have been developed over the years. These techniques are reviewed by Pope (1985), Dopazo (1993), and Kuznetsov & Sabel'nikov (1990), and are not described here. For the present case the result for $f_\phi(\psi; t)$ is

$$\frac{\partial}{\partial t} f_\phi = -\frac{\partial^2}{\partial \psi^2} [f_\phi \chi(\psi, t)] - \frac{\partial}{\partial \psi} [f_\phi \hat{S}(\psi)], \tag{6}$$

where

$$\chi(\psi, t) \equiv \Gamma \langle \nabla \phi \cdot \nabla \phi \,|\, \phi(\mathbf{x}, t) = \psi \rangle \tag{7}$$

is the conditional scalar dissipation.

An important observation is that the reaction term in the pdf equation (6) is in closed form, whereas the corresponding terms in moment closures [e.g. $\langle \hat{S}(\phi) \rangle$ and $\langle \phi \hat{S}(\phi) \rangle$] are not—hence the attraction of PDF methods for reactive flows.

The term involving χ in Equation (6) represents molecular mixing and requires modeling. In general—as exemplified by Equation (7)—*in Eulerian PDF methods, the quantities that have to be modeled are one-point one-time conditional expectations.*

Molecular mixing models (which model the term in χ in Equation 6) have a long history which is briefly reviewed in Section 5.5 and more thoroughly elsewhere (Pope 1982, 1985; Borghi 1988; Dopazo 1993); and there are several relevant recent works (Chen et al 1989, Sinai & Yakhot 1989, Valiño & Dopazo 1991, Pope 1991a, Gao 1991, Gao & O'Brien 1991, Fox 1992, Pope & Ching 1993).

The composition PDF method can be extended to several compositions, and to inhomogeneous flows. The latter necessitates modeling turbulent convection—generally as gradient diffusion. Recent applications to multi-dimensional flows are described by Chen et al (1990), Roekaerts (1991), and Hsu et al (1993).

For inhomogeneous flows, the method based on the velocity-composition joint pdf $f(\mathbf{V}, \psi; \mathbf{x}, t)$ has the advantage of avoiding gradient-diffusion modeling. For constant-density flow, the Navier-Stokes equations can be written

$$\frac{D\mathbf{U}}{Dt} = v\nabla^2\mathbf{U} - \nabla\langle p\rangle - \nabla p', \tag{8}$$

where v is the kinematic viscosity, and the pressure (divided by the density) p is subjected to the Reynolds decomposition. From this equation, and from Equation (4) written for each composition $\phi_\alpha(\mathbf{x}, t)$, the evolution equation for $f(\mathbf{V}, \boldsymbol{\psi}; \mathbf{x}, t)$ can be deduced to be

$$\frac{\partial f}{\partial t} + V_i\frac{\partial f}{\partial x_i} + \frac{\partial}{\partial \psi_\alpha}[f\hat{S}_\alpha(\psi)] - \frac{\partial \langle p\rangle}{\partial x_i}\frac{\partial f}{\partial V_i} = \frac{\partial}{\partial V_i}\left[f\left\langle\frac{\partial p'}{\partial x_i} - v\nabla^2 U_i|\mathbf{V}, \boldsymbol{\psi}\right\rangle\right]$$

$$- \frac{\partial}{\partial \psi_\alpha}[f\langle\Gamma\nabla^2\phi_\alpha|\mathbf{V}, \boldsymbol{\psi}\rangle], \tag{9}$$

where $\langle\nabla^2\phi_\alpha|\mathbf{V}, \boldsymbol{\psi}\rangle$ is written for the conditional expectation $\langle\nabla^2\phi_\alpha|\mathbf{U}(\mathbf{x}, t) = \mathbf{V}, \boldsymbol{\phi}(\mathbf{x}, t) = \boldsymbol{\psi}\rangle$, and the summation convention applies to α as well as to i.

The terms on the left-hand side of Equation (9) are in closed form and represent convection, reaction, and acceleration due to the mean pressure gradient. Those on the right-hand side contain one-point one-time conditional expectations. Models for the effects of the viscous stresses and the fluctuating pressure gradient are discussed below.

2.4   *Lagrangian pdfs*

Fundamental to the Lagrangian description is the notion of a *fluid particle*. Let t_0 be a reference time, and let $\mathbf{x}_0 = \{x_{01}, x_{02}, x_{03}\}$ be *Lagrangian coordinates*. Then $\mathbf{x}^+(t, \mathbf{x}_0)$ denotes the position at time t of the fluid particle that is at $\mathbf{x}_0$ at time t_0 [i.e. $\mathbf{x}^+(t_0, \mathbf{x}_0) = \mathbf{x}_0$].

For each Eulerian variable [e.g. $\mathbf{U}(\mathbf{x}, t)$] the corresponding Lagrangian variable (denoted by the superscript $+$) is defined by (for example)

$$\mathbf{U}^+(t, \mathbf{x}_0) = \mathbf{U}[\mathbf{x}^+(t, \mathbf{x}_0), t]. \tag{10}$$

By definition, a fluid particle moves with its own velocity. So, given the Eulerian velocity, $\mathbf{x}^+$ is determined as the solution of

$$\frac{\partial \mathbf{x}^+(t, \mathbf{x}_0)}{\partial t} = \mathbf{U}^+(t, \mathbf{x}_0), \tag{11}$$

with the initial condition

$$\mathbf{x}^+(t_0, \mathbf{x}_0) = \mathbf{x}_0. \tag{12}$$

To simplify the subsequent development we impose two restrictions. First, we consider flow domains that are (possibly time-dependent) material volumes. Thus fluid particles do not cross the boundary of the flow domain. Second, we consider constant-density flow so that the determinant of the Jacobian $\partial x_i^+ / \partial x_{0j}$ is unity. Both of these restrictions are readily removed (see Pope 1985).

The primary Lagrangian pdf considered is

$$f_L(\mathbf{V}, \mathbf{x}; t \,|\, \mathbf{V}_0, \mathbf{x}_0), \tag{13}$$

which is the joint pdf of the event

$$\{\mathbf{U}^+(t, \mathbf{x}_0) = \mathbf{V}, \mathbf{x}^+(t, \mathbf{x}_0) = \mathbf{x}\} \tag{14}$$

subject to the condition $\mathbf{U}^+(t_0, \mathbf{x}_0) = \mathbf{V}_0$. Thus f_L is the joint pdf of the fluid particle properties at time t, conditional upon their properties at time t_0. Note that, in the Lagrangian pdf f_L, $\mathbf{x}$ denotes the sample space corresponding to $\mathbf{x}^+(t, \mathbf{x}_0)$, while in the Eulerian pdf it is a parameter. Also, other fluid particle properties [e.g. $\phi^+(t, \mathbf{x}_0)$] can be included in the definition of f_L.

The Lagrangian pdf f_L is defined in terms of the reference initial time t_0 and a future time $t > t_0$. More generally, we may consider M times $t_0 < t_1 < t_2 \ldots < t_M$, and define the M-time Lagrangian pdf

$$f_{LM}(\mathbf{V}_M, \mathbf{x}_M; t_m \colon \mathbf{V}_{M-1}, \mathbf{x}_{M-1}; t_{M-1}; \ldots; \mathbf{V}_1, \mathbf{x}_1; t_1 \,|\, \mathbf{V}_0, \mathbf{x}_0) \tag{15}$$

as the joint pdf of the events

$$\{\mathbf{U}^+(t_k, \mathbf{x}_0) = \mathbf{V}_k, \mathbf{x}^+(t_k, \mathbf{x}_0) = \mathbf{x}_k; k = 1, 2, \ldots, M\} \tag{16}$$

subject to the same initial condition as before, $\mathbf{U}^+(t_0, \mathbf{x}_0) = \mathbf{V}_0$.

2.5 *Lagrangian PDF Methods*

In Eulerian PDF methods, the quantities to be modeled are one-point, one-time conditional expectations (see Equation 9). In Lagrangian PDF methods, the modeling approach is entirely different. Stochastic models are constructed to simulate the evolution of fluid particle properties. For example, Figure 1 shows the time series of one component of velocity (in stationary, homogeneous, isotropic turbulence) according to a simple stochastic model—the Langevin equation (which is the subject of the next Section).

It is useful to distinguish between the fluid-particle properties ($\mathbf{U}^+$ and $\mathbf{x}^+$) and the values obtained from the stochastic models. Thus $\mathbf{U}^*(t)$ and $\mathbf{x}^*(t)$ denote the modeled particle properties, with $\mathbf{x}^*$ evolving by

$$\frac{d\mathbf{x}^*(t)}{dt} = \mathbf{U}^*(t). \tag{17}$$

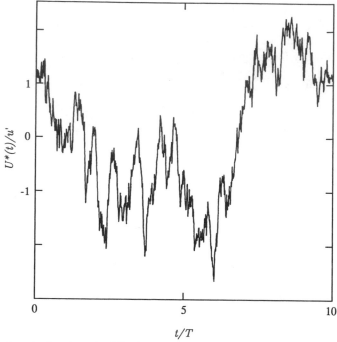

Figure 1 Sample of an Ornstein-Uhlenbeck process obtained as the solution of the Langevin equation (Equation 21).

Then $f_L^*(\mathbf{V}, \mathbf{x}; t \,|\, \mathbf{V}_0, \mathbf{x}_0)$ is the joint pdf to $\mathbf{U}^*(t)$ and $\mathbf{x}^*(t)$ subject to the initial condition

$$\mathbf{U}^*(t_0) = \mathbf{V}_0, \mathbf{x}^*(t_0) = \mathbf{x}_0. \tag{18}$$

If the stochastic model is accurate, then f_L^* is an accurate approximation to f_L.

Stochastic Lagrangian models—such as the Langevin equation—have long been used in studies of turbulent dispersion, where the quantities of interest are (or can be obtained from) Lagrangian pdfs. For example, if a pulse of a contaminant is released at time t_0 and location $\mathbf{x}_0$, then (if molecular diffusion can be neglected) the expected concentration of the contaminant at a later time t is proportional to the pdf of $\mathbf{x}^+(t, \mathbf{x}_0)$, which is modeled by $\mathbf{x}^*(t)$. In fact, as early as 1921, G. I. Taylor proposed a stochastic model for $\mathbf{x}^*(t)$ precisely for this application (Taylor 1921).

A direct numerical implementation of a stochastic model of turbulent dispersion is to release a large number N of particles at the source [i.e. $\mathbf{x}^*(t_0) = \mathbf{x}_0$] with initial velocities $\mathbf{U}^*(t_0)$ distributed according to the Eulerian pdf $f(\mathbf{V}; \mathbf{x}_0, t_0)$. Then the stochastic model equations are integrated forward in time to obtain $\mathbf{U}^*(t)$ and $\mathbf{x}^*(t)$. The expected particle number density (at any $\mathbf{x}$, t) is then proportional to the mean contaminant concentration.

In Lagrangian PDF methods, the stochastic models are used to determine both Lagrangian and Eulerian pdfs. This is achieved through the fundamental relation:

$$f(\mathbf{V}; \mathbf{x}, t) = \int\int f(\mathbf{V}_0; \mathbf{x}_0, t_0) f_{\mathrm{L}}(\mathbf{V}, \mathbf{x}; t | \mathbf{V}_0, \mathbf{x}_0) \, d\mathbf{V}_0 \, d\mathbf{x}_0, \tag{19}$$

where integration is over all velocities and over the entire flow domain at t_0. (A derivation of this equation is given by Pope 1985.) Thus the Lagrangian pdf f_{L} is the *transition density* for the turbulent flow: It determines the transition of the Eulerian pdf from time t_0 to time t.

Since f_{L} determines f, it also determines simple Eulerian means, such as the mean velocity $\langle \mathbf{U}(\mathbf{x}, t) \rangle$ and the Reynolds stresses $\langle u_i u_j \rangle$ (where $\mathbf{u} = \mathbf{U} - \langle \mathbf{U} \rangle$). Such means can therefore be used as coefficients in the stochastic Lagrangian models.

It is almost inevitable that computationally viable stochastic models are Markov processes. That is, with $t_{k-1} < t_k < t_{k+1}$, the joint pdf of $\mathbf{U}^*(t_{k+1})$ and $\mathbf{x}^*(t_{k+1})$ is completely determined by $\mathbf{U}^*(t_k)$, $\mathbf{x}^*(t_k)$ and the Eulerian pdf $f(\mathbf{V}; \mathbf{x}, t_k)$, independent of the particle properties at earlier times t_{k-1}. It then follows that (the model equivalent of) the M-time Lagrangian pdf (Equation 15) is given by the product of the M transition densities

$$f_{\mathrm{LM}}^*(\mathbf{V}_M, \mathbf{x}_M; t_M: \mathbf{V}_{M-1}, \mathbf{x}_{M-1}; t_{M-1}; \ldots; \mathbf{V}_1, \mathbf{x}_1; t_1 | \mathbf{V}_0, \mathbf{x}_0)$$

$$= \prod_{k=1}^{M} f_{\mathrm{L}}^*(\mathbf{V}_k, \mathbf{x}_k; t_k | \mathbf{V}_{k-1}, \mathbf{x}_{k-1}). \tag{20}$$

Thus for Markov models it is sufficient to consider the transition density f_{L}^* since this contains the same information as the M-time Lagrangian pdf.

Lagrangian PDF methods are implemented numerically as Monte Carlo/ particle methods. In contrast to implementations for dispersion studies, the large number of particles are at all times uniformly distributed in the flow domain. This particle representation is described in more detail in Section 4. Calculations based on this approach are described by Haworth & Pope (1987), Anand et al (1989, 1993), Haworth & El Tahry (1991), Taing et al (1993), and Norris (1993), for example.

3. LANGEVIN EQUATION

The Langevin equation is the prototypical stochastic model. The basic mathematical and physical concepts are introduced here in a simple setting. More general and advanced models are described in Section 5.

We begin by considering stationary homogeneous isotropic turbulence, with zero mean velocity, turbulence intensity u', and Lagrangian integral time scale T. The subject of the Langevin equation, $U^*(t)$, is a model for one component of the fluid-particle velocity $U^+(t)$.

Written as a stochastic differential equation (sde), the Langevin equation is

$$dU^*(t) = -U^*(t)\,dt/T + (2u'^2/T)^{1/2}\,dW(t), \tag{21}$$

where $W(t)$ is a Wiener process. The reader unfamiliar with sdes, can appreciate the meaning of the Langevin equation through the finite-difference approximation

$$U^*(t+\Delta t) = U^*(t) - U^*(t)\Delta t/T + (2u'^2\Delta t/T)^{1/2}\xi, \tag{22}$$

where ξ is a standardized Gaussian random variable ($\langle\xi\rangle = 0$, $\langle\xi^2\rangle = 1$) which is independent of the corresponding random variable on all other time steps. Thus the increment in the Wiener process $dW(t)$ can be thought of as a Gaussian random variable with mean zero, and variance dt.

The basic mathematical properties of the Langevin equation are now described, and then their relationship to the physics of turbulence is discussed.

The Langevin equation (Equation 21 or 22) describes a Markov process $U^*(t)$ that is continuous in time (see Gardiner 1990, for a more precise statement). Hence, in the terminology of stochastic processes, $U^*(t)$ is a *diffusion process*. Although it is continuous, it is readily seen that $U^*(t)$ is not differentiable: Equation (22) shows that $[U^*(t+\Delta t) - U^*(t)]/\Delta t$ varies as $\Delta t^{-1/2}$, and hence does not converge as Δt tends to zero.

For simplicity we consider the initial condition at time t_0 that $U^*(t_0)$ is a Gaussian random variable with zero mean and variance u'^2. Then, for $t > t_0$, $U^*(t)$ is the stationary random process known as the *Ornstein-Uhlenbeck (OU) process*, a sample of which is shown on Figure 1. The OU process is a stationary, Gaussian, Markov process, and hence is completely characterized by its mean ($\langle U^*(t)\rangle = 0$), its variance ($\langle U^*(t)^2\rangle = u'^2$), and its autocorrelation function, which is

$$\rho^*(s) \equiv \langle U^*(t+s)U^*(t)\rangle/u'^2 = e^{-|s|/T}, \tag{23}$$

(see e.g. Gardiner 1990). Notice that these results confirm the consistency of the specification of the coefficients in the Langevin equation: The rms fluid-particle velocity is u', and the Lagrangian integral time scale is

$$T = \int_0^\infty \rho^*(s)\,ds. \tag{24}$$

To what extent does $U^*(t)$ model the fluid particle velocity $U^+(t)$? The first and obvious limitation is that $U^+(t)$ is differentiable, whereas $U^*(t)$ is not. Hence the model is qualitatively incorrect if $U^*(t)$ is examined on an infinitesimal time scale.

But consider high Reynolds number turbulence in which there is a large separation between the integral time scale T and the Kolmogorov time scale τ_η; and let us examine $U^+(t)$ on inertial-range time scales s, $T \gg s \gg \tau_\eta$. This is best done through the Lagrangian structure function (see e.g. Monin & Yaglom 1975)

$$D_L(s) \equiv \langle [U^+(t+s) - U^+(t)]^2 \rangle. \tag{25}$$

The Kolmogorov hypotheses (both original 1941 and refined 1962) predict (in the inertial range)

$$D_L(s) = C_0 \langle \varepsilon \rangle s, \tag{26}$$

where C_0 is a universal constant, and $\langle \varepsilon \rangle$ is the mean dissipation rate. And the Langevin equation yields [for the structure function based on $U^*(t)$]:

$$D_L^*(s) = 2u'^2 s/T, \quad \text{for} \quad s/T \ll 1, \tag{27}$$

as is evident from Equation (22). Thus the Langevin equation is consistent with the Kolmogorov hypotheses in yielding a linear dependence of D_L on s in the inertial range. (Equation 27 corresponds to an ω^{-2} frequency spectrum (at high frequency), which in turn corresponds to white-noise acceleration.)

By comparing the coefficients in Equations (26) and (27) we obtain the relation

$$T^{-1} = C_0 \langle \varepsilon \rangle / (2u'^2) = \tfrac{3}{4} C_0 \langle \varepsilon \rangle / k, \tag{28}$$

where k is the turbulent kinetic energy; and the Langevin equation (Equation 21) can be rewritten in the alternative form:

$$dU^*(t) = -\frac{3}{4} C_0 \frac{\langle \varepsilon \rangle}{k} U^*(t)\,dt + (C_0 \langle \varepsilon \rangle)^{1/2}\,dW(t). \tag{29}$$

To date, Lagrangian statistics in high-Reynolds number flows have proven inaccessible both to experiment and to direct numerical simulation. Consequently, a direct test of Equation (26) has not been possible. However at low or moderate Reynolds number, both techniques have been used to measure the Lagrangian autocorrelation function $\rho(s)$. Figure 2 shows the results compared to the exponential (Equation 23) arising from the Langevin equation. At very small times s/T, the behavior is qualitatively different because $U^*(t)$ is not differentiable—correspondingly, $\rho^*(s)$ has negative slope at the origin. But for larger times, the exponential form provides a very reasonable approximation to the observed autocorrelations.

Measurements on turbulent dispersion provide an indirect test of the Langevin equation. For particles originating from the origin at time $t = 0$, their subsequent position is

$$x^*(t) = \int_0^t U^*(t')\,dt'. \tag{30}$$

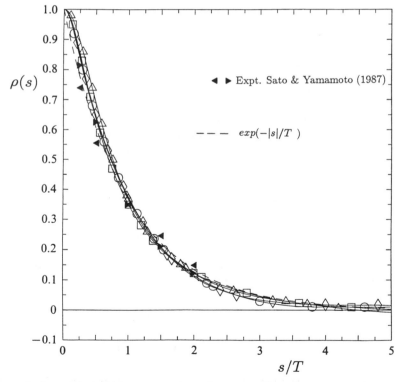

Figure 2 Lagrangian velocity autocorrelation function in isotropic turbulence: open symbols and lines from DNS of Yeung & Pope (1989); full symbols from experiments of Sato & Yamamoto (1987); dashed line exponential, Equation 23. (From Yeung & Pope 1989.)

This (according to the Langevin equation) is a Gaussian process, with zero mean, and variance

$$\langle x^*(t)^2 \rangle = 2u'^2 T[t - T(1 - e^{-t/T})], \tag{31}$$

which exhibits the correct short-time limit $[\langle x^*(t)^2 \rangle \approx (u't)^2]$ and long-time limit $[\langle x^*(t)^2 \rangle \approx 2u'^2 Tt]$ given by Taylor's (1921) theory.

The Langevin equation has been applied to dispersion behind a line source in grid turbulence by Anand & Pope (1985) with modifications to account for the decay of the turbulence and the first-order effects of molecular diffusion. The result shown on Figure 3 is in excellent agreement with the data.

Several refinements and extensions to the Langevin equation are described in Section 5, where further comparisons with DNS data are made.

We now describe the most rudimentary extension of the Langevin equation to inhomogeneous flows. The equation is a model for the evolution of all three components of velocity $\mathbf{U}^*(t)$ of a fluid particle with position $\mathbf{x}^*(t)$.

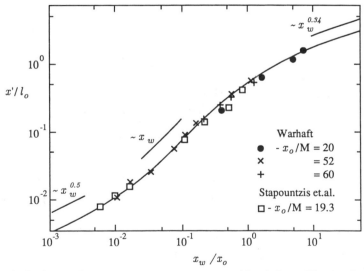

Figure 3 Turbulent dispersion behind a line source in grid turbulence. The rms dispersion $x' = \langle x^{*2} \rangle^{1/2}$ is normalized by the integral scale at the source l_o; the distance downstream of the source x_w is normalized by the distance from the grid to the source x_o. Experimental data of Warhaft (1984) and of Stapountzis et al (1986). Solid line is the calculation of Anand & Pope (1985) based on the Langevin equation.

There are three modifications to Equation (29)—all to the drift term. First, the velocity increment due to the mean pressure gradient $-dt\nabla\langle p\rangle$ is added. Second, the fluid particle velocity relaxes to the local Eulerian mean $\langle \mathbf{U}(\mathbf{x}^*[t], t)\rangle$ (rather than to zero). And, third, the coefficient of the drift term is altered. The result is

$$d\mathbf{U}^*(t) = -\nabla\langle p\rangle\, dt - \left(\frac{1}{2} + \frac{3}{4}C_0\right)\frac{\langle\varepsilon\rangle}{k}(\mathbf{U}^*(t) - \langle\mathbf{U}\rangle)\, dt$$

$$+ (C_0\langle\varepsilon\rangle)^{1/2}\, d\mathbf{W}(t). \quad (32)$$

In this (and subsequent) equations, it is understood that mean quantities (i.e. $\nabla\langle p\rangle$, $\langle\varepsilon\rangle$, k and $\langle\mathbf{U}\rangle$) are evaluated at the fluid-particle position $\mathbf{x}^*(t)$. The vector-valued Wiener process $\mathbf{W}(t)$ is simply composed of three independent components $W_1(t)$, $W_2(t)$, and $W_3(t)$. The increment $d\mathbf{W}$ has zero mean and covariance

$$\langle dW_i\, dW_j\rangle = dt\, \delta_{ij}. \quad (33)$$

[As discussed at greater length in Section 5.1, the coefficient $(\frac{1}{2} + \frac{3}{4}C_0)$ in Equation (32) (compared to $\frac{3}{4}C_0$ in Equation 29), correctly causes the turbulent kinetic energy to be dissipated at the rate $\langle\varepsilon\rangle$. The omission of the $\frac{1}{2}$ in Equation (29) is because that equation pertains to the hypothetical case of stationary (i.e. non-decaying) isotropic turbulence.]

A stochastic model for fluid-particle properties implies a modeled evolution equation for the corresponding Lagrangian joint pdf. In the present context, $f_L^*(\mathbf{V}, \mathbf{x}, t\,|\,\mathbf{V}_0, \mathbf{x}_0)$ is the joint pdf of $\mathbf{U}^*(t)$ and $\mathbf{x}^*(t)$, with the initial conditions $\mathbf{U}^*(t_0) = \mathbf{V}_0$, $\mathbf{x}^*(t_0) = \mathbf{x}_0$. Then with $\mathbf{U}^*(t)$ evolving by the extended Langevin equation (Equation 32), and with $\mathbf{x}^*(t)$ evolving by Equation (17), f_L^* evolves according to the *Fokker-Planck equation*

$$\frac{\partial}{\partial t}f_L^* = -V_i\frac{\partial f_L^*}{\partial x_i} + \frac{\partial\langle p\rangle}{\partial x_i}\frac{\partial f_L^*}{\partial V_i}$$

$$+ \left(\frac{1}{2} + \frac{3}{4}C_0\right)\frac{\langle\varepsilon\rangle}{k}\frac{\partial}{\partial V_i}[f_L^*(V_i - \langle U_i\rangle)] + \frac{1}{2}C_0\langle\varepsilon\rangle\frac{\partial^2 f_L^*}{\partial V_i\partial V_i}, \quad (34)$$

(see Gardiner 1990, Risken 1989), with the initial condition

$$f_L^*(\mathbf{V}, \mathbf{x}, t_0\,|\,\mathbf{V}_0, \mathbf{x}_0) = \delta(\mathbf{V} - \mathbf{V}_0)\delta(\mathbf{x} - \mathbf{x}_0). \quad (35)$$

The Eulerian joint pdf of velocity $f(\mathbf{V}; \mathbf{x}, t)$ is related to its Lagrangian counterpart by Equation (19). Hence the above evolution equation for f_L^* implies a corresponding evolution equation for the modeled Eulerian pdf $f^*(\mathbf{V}; \mathbf{x}, t)$. Indeed, since the differential operators in the Fokker-Planck equation are independent of t_0, $\mathbf{x}_0$, and $\mathbf{V}_0$, it follows immediately from Equation (19) that the Eulerian pdf $f^*(\mathbf{V}; \mathbf{x}, t)$ also evolves according to Equation (34).

The Eulerian pdf equation (i.e. Equation 34 written for f^*), together with a modeled equation for $\langle \varepsilon \rangle$, form a complete set of turbulence-model equations. They are complete in the sense that all the coefficients in Equation (34) are known in terms of f^* and $\langle \varepsilon \rangle$. The mean velocity $\langle \mathbf{U} \rangle$ and the kinetic energy k are determined as first and second moments of f^*, while the mean pressure field $\langle p \rangle$ is determined as the solution of a Poisson equation. The source in the Poisson equation involves $\langle \mathbf{U} \rangle$ and $\langle u_i u_j \rangle$ which are known in terms of f^*.

In Section 5.3 coupled stochastic models for $\mathbf{U}^*(t)$ and $\omega^*(t) \equiv \varepsilon^*(t)/k$ are described, which lead to an evolution equation for the joint pdf of $\mathbf{U}$ and ω. This single equation provides a complete model: All of the coefficients are known in terms of the pdf itself.

The above development illustrates the different use of the Langevin equation in turbulent dispersion and in PDF methods. In the former, the turbulent flow field is assumed known, and so the coefficients in the Langevin equation are specified; and the equation is used (at most) to deduce the Lagrangian pdf. In PDF methods, the Langevin equation is used to determine the Eulerian pdf, from which the coefficients are deduced.

4.   PARTICLE REPRESENTATION

Central to Lagrangian PDF methods is the idea that a turbulent flow can be represented by an ensemble of N fluid particles, with positions and velocities $\mathbf{x}^{(n)}(t)$, $\mathbf{U}^{(n)}(t)$, $n = 1, 2, \ldots, N$. The purpose of this section is to describe this particle representation, and to make precise the connection between particle properties and statistics of the flow.

4.1   *Basic Representation*

We begin by considering a single component of velocity U at a particular point and time. Thus U is a random variable, with pdf $f(V)$, which we consider to be known.

For a given ensemble size $N(N \geq 1)$, the particle velocities $\{U^{(n)}\}$ are specified to be independent random samples, each with pdf $f(V)$. It is conceptually useful (and legitimate) to think of $U^{(n)}$ as the value of U on the n-th (independent) realization of the flow. Note that the particle velocities are independent and identically distributed, and hence the numbering of the particles is irrelevant.

A fundamental question, to which we provide three answers, is: In what sense does the ensemble $\{U^{(n)}\}$ "represent" the underlying distribution $f(V)$?

The first answer is in terms of the discrete pdf $f_N(V)$, defined by

$$f_N(V) \equiv \frac{1}{N} \sum_{n=1}^{N} \delta(U^{(n)} - V). \tag{36}$$

It is readily shown (see, e.g. Pope 1985) that the *expected* discrete pdf equals $f(V)$ for *any* $N \geq 1$:

$$\langle f_N(V) \rangle = f(V). \tag{37}$$

In the analysis of PDF methods, this relation allows properties of the pdf [i.e. $f(V)$] to be deduced from the properties of a single particle ($U^{(1)}$, say).

The second answer involves ensemble averages. In numerical implementations of PDF methods it is necessary to *estimate* means such as $\langle U \rangle$ and $\langle U^2 \rangle$ from the ensemble $\{U^{(n)}\}$. Let $Q(U)$ be some function of the velocity U, then we have

$$\langle Q(U) \rangle = \int_{-\infty}^{\infty} f(V)Q(V)\, dV. \tag{38}$$

For example, the choices of V and V^2 for $Q(V)$ lead to $\langle U \rangle$ and $\langle U^2 \rangle$. The mean $\langle Q \rangle$ can be estimated from the ensemble simply as the ensemble average

$$\langle Q(U) \rangle_N \equiv \frac{1}{N} \sum_{n=1}^{N} Q(U^{(n)}) = \int_{-\infty}^{\infty} f_N(V)Q(V)\, dV. \tag{39}$$

Then (since $\{U^{(n)}\}$ are independent and identically distributed) a basic result from statistics is that $\langle Q \rangle_N$ is an unbiased estimator of $\langle Q \rangle$:

$$\langle \langle Q \rangle_N \rangle = \langle Q \rangle. \tag{40}$$

Further, if the variance of $Q(U)$ is finite, it follows from the central limit theorem that for large N the rms statistical error in $\langle Q \rangle_N$ tends to zero as $N^{-1/2}$.

Hence the second sense in which the ensemble $\{U^{(n)}\}$ "represents" the pdf $f(V)$ is that, for all functions Q [for which $Q(U)$ has finite mean and variance], the ensemble average $\langle Q \rangle_N$ converges in mean square to $\langle Q \rangle$. This is written

$$\lim_{N \to \infty} \langle Q \rangle_N = \langle Q \rangle. \tag{41}$$

[An additional convergence result is provided by the Glivenko-Cantelli theorem (e.g. Billingsley 1986): As N tends to infinity, the difference between the cdf $F(V)$ and the *empirical cdf* $F_n(V)$—i.e. the definite integral of $f_N(V)$—converges to zero with probability one.]

For almost all purposes, the two answers provided above are sufficient: Equation (37) is used in the analysis of PDF methods, while Equation (41) is used in numerical implementations. However, neither of these relations (nor the Glivenko-Cantelli theorem) provides an estimate of the pdf $f(V)$ in terms of $\{U^{(n)}\}$ that converges in mean square as N tends to infinity. The third answer, then, is that the techniques of *density estimation* can be used for this purpose (see e.g. Tapia & Thompson 1978, Silverman 1986). These are not reviewed here, since they have not played an important role in PDF methods. This is because in the implementation of PDF methods, an *explicit* representation of the pdf is not required.

The above considerations apply to any random variable U. Consider now $U^{(n)}(t)$ to be a model for the velocity of a fluid particle, obtained as the solution to a stochastic model equation—the Langevin equation, for example. At the initial time t_0, the values of $\{U^{(n)}(t_0)\}$ are sampled from the specified initial pdf $f(V; t_0)$.

The representations described above are readily extended. The one-time discrete pdf [representing $f(V; t)$] is

$$f_N(V; t) = \frac{1}{N} \sum_{n=1}^{N} \delta[U^{(n)}(t) - V], \tag{42}$$

while the discrete Lagrangian pdf is

$$f_{1,N}(V; t\,|\,V_0) = \frac{1}{N} \sum_{n=1}^{N} \{\delta[U^{(n)}(t) - V]\,|\,U^{(n)}(t_0) = V_0\}. \tag{43}$$

Multi-time Lagrangian statistics can be estimated as ensemble averages: for example,

$$\langle Q(U(t_1), U(t_2))\rangle_N \equiv \frac{1}{N} \sum_{n=1}^{N} Q[U^{(n)}(t_1), U^{(n)}(t_2)]. \tag{44}$$

4.2   *Inhomogeneous Flows*

The extension of this particle representation to inhomogeneous flows requires some new ingredients, and it leads to some subtle consistency conditions.

Throughout, for simplicity, we are restricting our attention to constant-density flows in a material volume. Hence the volume V of the flow domain D, and the mass of fluid within it, do not change with time.

At a given time t, an ensemble of N particles is constructed as follows to represent the joint pdf of velocity $f(V; x, t)$. The particle positions $x^{(n)}(t)$

are mutually independent, random, uniformly-distributed in D. [Hence the pdf of each $\mathbf{x}^{(n)}(t)$ is $1/V$.] Then the particle velocity $\mathbf{U}^{(n)}(t)$ is random, with pdf $f[\mathbf{V}; \mathbf{x}^{(n)}(t), t]$. In terms of these properties, the discrete pdf is defined by

$$f_N(\mathbf{V}; \mathbf{x}, t) \equiv \frac{V}{N} \sum_{i=1}^{N} \delta[\mathbf{x}^{(n)}(t) - \mathbf{x}]\delta[\mathbf{U}^{(n)}(t) - \mathbf{V}]. \tag{45}$$

The specification of $\mathbf{x}^{(n)}(t)$ (and also the constant in Equation 45) is determined by a consistency condition. We require the expectation of f_N to equal f; where f satisfies the normalization condition that its integral over all $\mathbf{V}$ is unity. Hence from Equation (45) we obtain

$$1 = \int \langle f_N \rangle \, d\mathbf{V} = V\langle \delta[\mathbf{x}^{(n)}(t) - \mathbf{x}] \rangle, \tag{46}$$

for any n (since $\{\mathbf{x}^{(n)}\}$ are independent and identically distributed). This condition is satisfied if, and only if, $\mathbf{x}^{(n)}(t)$ is uniformly distributed.

If this consistency condition is satisfied at an initial time t_0, will it remain satisfied as the particle properties evolve in time? The answer (established by Pope 1985, 1987) is yes, provided the mean continuity equation is satisfied. This in turn requires that the mean pressure gradient [affecting the evolution of $\mathbf{U}^{(n)}(t)$, Equation 32] satisfies the appropriate Poisson equation.

Both of these results are reflected in the Eulerian pdf equation [e.g. Equation 34 written for $f(\mathbf{V}; \mathbf{x}, t)$]. When this equation is integrated over all $\mathbf{V}$, all the terms on the right-hand side vanish, expect the first which is $-\nabla \cdot \langle \mathbf{U} \rangle$. If this is nonzero—in violation of the continuity equation—then the normalization condition on f is also violated. An evolution equation for $\nabla \cdot \langle \mathbf{U} \rangle$ is obtained from Equation (34) by multiplying by V_j, integrating over all $\mathbf{V}$, and then differentiating with respect to x_j. Equating the time rate of change of $\nabla \cdot \langle \mathbf{U} \rangle$ to zero, yields a Poisson equation for $\langle p \rangle$.

With $Q(\mathbf{V})$ being a function of the velocity, we now consider the estimation of the mean $\langle Q[\mathbf{U}(\mathbf{x}, t)] \rangle$ from the ensemble of particles. This is an important issue because Eulerian means such as $\langle \mathbf{U} \rangle$ and $\langle u_i u_j \rangle$ must be estimated from the particle properties in order to determine the coefficients in the modeled particle evolution equations, e.g. Equation (32).

Since (with probability one) there are no particles located at $\mathbf{x}$, it is unavoidable that an estimate of $\langle Q[\mathbf{U}(\mathbf{x}, t)] \rangle$ must involve particles in the vicinity of $\mathbf{x}$. We describe now the *kernel estimator* (see e.g. Eubank 1988, Härdle 1990), which is useful both conceptually and in practice (although a literal implementation is not efficient). It is assumed that $Q[\mathbf{U}(\mathbf{x}, t)]$

has finite mean and variance, and that the mean is twice continuously differentiable with respect to $\mathbf{x}$.

For simplicity we consider points $\mathbf{x}$ that are remote from the boundary of the domain; and for definiteness we take the kernel to be a Gaussian of specified width h. In D dimensions this is

$$K(\mathbf{r}, h) = (\sqrt{2\pi}h)^{-D} \exp(-\tfrac{1}{2}r^2/h^2), \tag{47}$$

(where $r = |\mathbf{r}|$). Then a kernel estimator of $\langle Q[\mathbf{U}(\mathbf{x}, t)]\rangle$ is

$$\langle Q[\mathbf{U}(\mathbf{x}, t)]\rangle_{N,h} \equiv \frac{V}{N} \sum_{n=1}^{N} K[\mathbf{x} - \mathbf{x}^{(n)}(t), h] Q[\mathbf{U}^{(n)}(t)]. \tag{48}$$

For small h, the bias in this estimate is

$$\langle\langle Q\rangle_{N,h}\rangle - \langle Q\rangle = \tfrac{1}{2}h^2\nabla^2\langle Q\rangle + O(h^4). \tag{49}$$

Hence, as h tends to zero, $K(\mathbf{r}, h)$ tends to $\delta(\mathbf{r})$ and $\langle\langle Q\rangle_{N,h}\rangle$ converges to $\langle Q\rangle$.

But as h becomes smaller, fewer particles have significant values of $K[\mathbf{x} - \mathbf{x}^{(n)}(t), h]$, and so the statistical error rises: The variance of $\langle Q\rangle_{N,h}$ varies as

$$V/(Nh^D) = (L/h)^D/N, \tag{50}$$

where $L \equiv V^{1/D}$ is a characteristic length of the domain. It is readily shown that, for large N, it is optimal for h/L to vary as $N^{-1/(4+D)}$. For then, the sum of the bias and the rms statistical error is minimized, each varying as $N^{-2/(4+D)}$. Thus, for such a choice of h we have

$$\lim_{N \to \infty} \langle Q\rangle_{N,h} = \langle Q\rangle. \tag{51}$$

These results have two major significances. First, Equation (51) shows the convergence of the particle representation. Second, it is likely that in a numerical implementation the error decreases with increasing N no faster than $N^{-2/(4+D)}$.

Although it is seldom done in practice, it is in principle possible to use the above ideas to extract multi-time Lagrangian statistics. It is important to realize, however, that (at a fixed time) the particle representation contains no two-point information. Recall that different particles can be viewed as being sampled from different, independent realizations of the flow.

5.   STOCHASTIC LAGRANGIAN MODELS

5.1   *Generalized Langevin Model*

The Langevin equation for inhomogeneous flows described in Section 3 (Equation 32) is referred to as the *Simplified Langevin Model* (SLM). It is the simplest possible extension of the basic Langevin model (Equation 29) that is consistent with momentum and energy conservation.

The *Generalized Langevin Model* (GLM)—proposed by Pope (1983a) and developed and demonstrated by Haworth & Pope (1986, 1987)—overcomes some of the qualitative and quantitative defects of the SLM. For the increment in the fluid-particle velocity $\mathbf{U}^*(t)$, the GLM is

$$dU_i^* = -\frac{\partial \langle p \rangle}{\partial x_i}\, dt + \mathscr{G}_{ij}(U_j^* - \langle U_j \rangle)\, dt + (C_0 \langle \varepsilon \rangle)^{1/2}\, dW_i, \tag{52}$$

where the drift coefficient tensor $\mathscr{G}_{ij}$ is a modeled function of the local mean velocity gradients $\partial \langle U_i \rangle / \partial x_j$, Reynolds stresses $\langle u_i u_j \rangle$, and dissipation $\langle \varepsilon \rangle$. It may be immediately observed that the SLM corresponds to the simple specification

$$\mathscr{G}_{ij} = \mathscr{G}_{ij}^0 \equiv -\left(\frac{1}{2} + \frac{3}{4} C_0\right)\frac{\langle \varepsilon \rangle}{k}\delta_{ij}. \tag{53}$$

Hence the GLM is distinguished by a more elaborate specification of $\mathscr{G}_{ij}$.

The first term in Equation (52) is uniquely determined by the mean momentum equations (at high Reynolds number, when the viscous term is negligible). The final term in Equation (52) (the diffusion term) has the same form as in homogeneous isotropic turbulence. This is justified (at high Reynolds number) by the Kolmogorov (1941) hypotheses: The term pertains to small time-scale (high frequency) processes that are hypothesized to be locally isotropic and characterized by $\langle \varepsilon \rangle$. (Implications of the Kolmogorov 1962 hypotheses are discussed in Section 5.3, and Reynolds-number effects in 5.4.)

In the construction of the drift term (involving $\mathscr{G}_{ij}$), the principal assumption made is that the term is linear in $\mathbf{U}^*$. For homogeneous turbulence, the assumption is fully justified, since this linearity is necessary and sufficient (Arnold 1974) for the joint pdf of velocity to be joint normal, in accord with experimental observations (e.g. Tavoularis & Corrsin 1981).

The observed joint normality of the one-point velocity pdf in homogeneous turbulence leads to several important results. For this case, the joint pdf is fully determined by the (known) mean velocity, and by the covariance matrix, namely the Reynolds stresses; and, according to the GLM (Equation 52), the Reynolds stresses evolve by

$$\frac{d}{dt}\langle u_k u_l \rangle = \mathscr{P}_{kl} + \mathscr{G}_{ki}\langle u_i u_l \rangle + \mathscr{G}_{li}\langle u_i u_k \rangle + C_0\langle \varepsilon \rangle \delta_{kl}, \tag{54}$$

where $\mathscr{P}_{kl}$ is the production tensor:

$$\mathscr{P}_{kl} \equiv -\langle u_i u_k \rangle \frac{\partial \langle U_l \rangle}{\partial x_i} - \langle u_i u_l \rangle \frac{\partial \langle U_k \rangle}{\partial x_i}. \tag{55}$$

Thus for any choice of $\mathscr{G}_{ij}$ there is a corresponding modeled Reynolds-stress equation, which (as shown by Pope 1985) is realizable. The known behavior of the Reynolds stress equation in certain limits (e.g. rapid distortion, or two-component turbulence) can then be invoked to impose constraints on $\mathscr{G}_{ij}$.

Using these constraints and experimental data on homogeneous turbulence, Haworth & Pope (1986) determined a specific form of $\mathscr{G}_{ij}$ that accurately describes the evolution of the Reynolds stresses (and hence the velocity joint pdf) in these flows. As an example, Figure 4 shows the evolution of the anisotropy tensor

$$b_{ij} \equiv \langle u_i u_j \rangle / \langle u_l u_l \rangle - \tfrac{1}{3}\delta_{ij}, \tag{56}$$

for the plane strain experiment of Gence & Mathieu (1979).

As is customary in turbulence modeling, with simplicity as the main justification, the same model is used in inhomogeneous flows. Haworth &

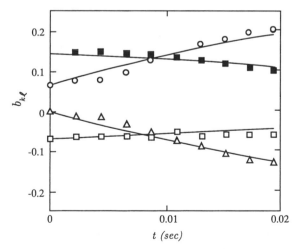

Figure 4 Reynolds stress anisotropies $b_{kl} \equiv \langle u_k u_l \rangle / \langle u_i u_i \rangle - \tfrac{1}{3}\delta_{kl}$ against time for transverse plane strain of homogeneous turbulence. Symbols: experimental data of Gence & Mathieu (1979) □ b_{11}, △ b_{22}, ○ b_{33}, ■ b_{23}. Lines: GLM calculations. (From Haworth & Pope 1986.)

Pope (1987) describe the successful application of the GLM to a range of free shear flows.

Finally, we observe that there is a Reynolds-stress equation corresponding to the SLM. Specifically, for homogeneous turbulence, Equations (52) and (53) lead to:

$$\frac{d}{dt}\langle u_k u_l \rangle = \mathscr{P}_{kl} - (2 + 3C_0)\langle \varepsilon \rangle b_{kl} - \tfrac{2}{3}\langle \varepsilon \rangle \delta_{kl}, \tag{57}$$

which is Rotta's (1951) model. Thus, in Reynolds-stress-closure terminology, the GLM is superior to the SLM in allowing for a nonlinear return-to-isotropy, and for incorporating "rapid pressure" effects. Recently Pope (1993) considered in detail the relationship between the GLM and Reynolds-stress models, and thereby deduced specifications corresponding to the isotropization of the production model (IPM, Naot et al 1970) and to the SSG model (Speziale et al 1991).

5.2   *Stochastic Model for Frequency*

The Generalized Langevin Model, just described, leads to a modeled transport equation for the velocity joint pdf $f(\mathbf{V}; \mathbf{x}, t)$. This equation does not provide a complete model, because the mean dissipation rate $\langle \varepsilon \rangle$ (or equivalent information) must be supplied separately—from a modeled transport equation for $\langle \varepsilon \rangle$, for example. This shortcoming motivated the development of a complete closure based on the joint pdf of velocity and dissipation (Pope & Chen 1990), which required the development of a stochastic model for dissipation.

In fact, rather than the instantaneous dissipation rate $\varepsilon(\mathbf{x}, t)$ the model developed by Pope & Chen (1990) is based on the *turbulence frequency* defined by

$$\omega(\mathbf{x}, t) \equiv \varepsilon(\mathbf{x}, t)/k(\mathbf{x}, t). \tag{58}$$

It should be noted that this is a mixed quantity in that $\varepsilon(\mathbf{x}, t)$ is random whereas $k(\mathbf{x}, t)$ is not. Thus the probability distribution of ω is the same as that of ε, to within a scaling. The mean frequency $\langle \omega \rangle$ has been used previously as a turbulence-model variable by, for example, Kolmogorov (1942) and Wilcox (1988).

For homogeneous turbulence, Pope & Chen (1990) developed a stochastic model $\omega^*(t)$ for the turbulent frequency following a fluid particle, $\omega^+(t)$. Their model is constructed by reference to the Lagrangian statistics of dissipation extracted from direct numerical simulations by Yeung & Pope (1989).

The simulations show that (to a very good approximation) the one-point one-time distribution of ε is log-normal. That is, for fixed t, the random variable

$$\chi^+(t) \equiv \ln[\varepsilon^+(t)/\langle\varepsilon\rangle] = \ln[\omega^+(t)/\langle\omega\rangle], \tag{59}$$

is Gaussian, with variance denoted by σ^2. Further, except near the origin, the autocorrelation function of $\chi^+(t)$, $\rho_\chi(s)$, is well approximated by the exponential

$$\rho_x(s) = e^{-|s|/T_\chi}. \tag{60}$$

where T_χ is the corresponding integral time scale. The simulations support the approximation

$$T_\chi^{-1} = C_\chi\langle\omega\rangle, \tag{61}$$

with C_χ being a constant.

Given that $\chi^+(t)$ has a Gaussian pdf and an exponential autocorrelation, it is obvious to model it as an OU process. The appropriate stochastic differential equation is

$$d\chi^*(t) = -[\chi^*(t)-\langle\chi^*(t)\rangle]\,dt/T_\chi + (2\sigma^2/T_\chi)^{1/2}\,dW, \tag{62}$$

(cf Equation 21).

The modeled frequency is related to χ^* by

$$\omega^*(t) = \langle\omega(t)\rangle e^{\chi^*(t)}, \tag{63}$$

(cf Equation 59). Consequently, in order to obtain a model equation for ω^*, it is necessary also to model the evolution of $\langle\omega\rangle$. With the nondimensional rate of change S_ω defined by

$$\frac{d\langle\omega\rangle}{dt} = -\langle\omega\rangle^2 S_\omega, \tag{64}$$

the standard model equation for $\langle\varepsilon\rangle$ (Launder & Spalding 1972) implies

$$S_\omega = (C_{\varepsilon 2}-1)-(C_{\varepsilon 1}-1)P/\langle\varepsilon\rangle, \tag{65}$$

where $C_{\varepsilon 1}$ and $C_{\varepsilon 2}$ are standard model constants, and P is the rate of production of turbulence kinetic energy.

The stochastic model for ω^* proposed by Pope & Chen (1990) is then obtained from Equations (62)–(64):

$$d\omega^* = -\omega^*\langle\omega\rangle\,dt\{S_\omega + C_\chi[\ln(\omega^*/\langle\omega\rangle)-\tfrac{1}{2}\sigma^2]\}$$
$$+\omega^*(2C_\chi\langle\omega\rangle\sigma^2)^{1/2}\,dW. \tag{66}$$

The above development pertains to homogeneous turbulence. The extension of the model to inhomogeneous flows is considered by Pope (1991b).

The only significant modification required to Equation (66) is the addition of a term that (under appropriate circumstances) causes nonturbulent fluid (characterized by $\omega^* = 0$) to become turbulent ($\omega^* > 0$). As described in the next subsection, with this modification, the stochastic model for frequency is successful in describing the intermittent turbulent/nonturbulent regions of free shear flows.

5.3   Refined Langevin Model

The stochastic model for frequency $\omega^*(t)$ (Equation 66) can be combined with the Generalized Langevin Model (Equation 52) to provide a closed modeled joint pdf equation. However, if the frequency $\omega^*(t)$ following a fluid particle is known, it is possible to incorporate this information in a stochastic model for velocity so as to increase its physical realism. Such a refined Langevin model has been developed by Pope & Chen (1990) and Pope (1991b).

According to all of the Langevin equations described above, for a small time interval s ($s/T \ll 1$), the modeled Lagrangian velocity increment

$$\Delta_s \mathbf{U}^*(t) \equiv \mathbf{U}^*(t+s) - \mathbf{U}^*(t), \tag{67}$$

is an isotropic Gaussian random vector with covariance

$$\langle \Delta_s U_i^*(t) \Delta_s U_j^*(t) \rangle = C_0 \langle \varepsilon \rangle s \delta_{ij} + \mathrm{O}(s^2). \tag{68}$$

This covariance is consistent with the refined Kolmogorov (1962) hypotheses; but the Gaussianity of $\Delta_s \mathbf{U}^*(t)$ is clearly at odds with notions of internal intermittency. In the spirit of Kolmogorov's refined hypotheses, it is natural to model $\Delta_s \mathbf{U}^*(t)$ in terms of the particle dissipation $\varepsilon^*(t) = k\omega^*(t)$. This is simply achieved by replacing the diffusion coefficient $C_0 \langle \varepsilon \rangle$ in the Langevin equation by $C_0 \varepsilon^*$. Then, the *conditional* covariance of $\Delta_s \mathbf{U}(t)$ is

$$\langle \Delta_s U_i^*(t) \Delta_s U_j^*(t) | \varepsilon^*(t) = \hat{\varepsilon} \rangle = C_0 \hat{\varepsilon} s \delta_{ij} + \mathrm{O}(s^2), \tag{69}$$

while, correctly, the unconditional variance is again given by Equation (68).

Since the performance of the GLM is completely satisfactory for homogeneous turbulence, Pope & Chen (1990) developed the *Refined Langevin model* (RLM) to retain this behavior (while replacing $C_0 \langle \varepsilon \rangle$ by $C_0 \varepsilon^* = C_0 k\omega^*$ in the diffusion term). For homogeneous turbulence the model is

$$dU_i^* = -\frac{\partial \langle p \rangle}{\partial x_i} dt + \mathcal{L}_{ij}(U_j^* - \langle U_j \rangle) dt + (C_0 k\omega^*)^{1/2} dW_i, \tag{70}$$

where

$$\mathscr{L}_{ij} = \mathscr{G}_{ij} - \tfrac{3}{4}C_0(\omega^* - \langle\omega\rangle)\mathscr{B}_{ij}, \tag{71}$$

and the tensor $\mathscr{B}_{ij}$ is the inverse of $\langle u_i u_j \rangle/(\tfrac{2}{3}k)$. Notice that, compared to the GLM (Equation 52), the additional term in $\mathscr{B}_{ij}$ is needed to produce the correct Gaussian joint pdf of velocity (in homogeneous turbulence). Additional modifications for inhomogeneous flows are described by Pope (1991b).

The combination of the stochastic model for $\omega^*(t)$ (Equation 66) and the RLM for $U^*(t)$ provides a closed modeled evolution equation for their joint pdf. This is the most advanced pdf model currently available. It has been applied to several different flows by Pope (1991b), Anand et al (1993), and Norris (1993). Calculations for a plane mixing layer are now briefly reported in order to illustrate several features of the model.

The calculations pertain to the statistically plane, two-dimensional, self-similar mixing layer formed between two uniform streams of different velocities. The dominant flow direction is x_1; the lateral direction is x_2; and the flow is statistically homogeneous in the spanwise direction x_3. The free-stream velocities are U_∞ (at $x_2 = \infty$) and $2U_\infty$ (at $x_2 = -\infty$), so that the velocity ratio is 2, and the velocity difference is $\Delta U = U_\infty$. At large axial distances the flow spreads linearly and is self-similar. Consequently, statistics of $U(\mathbf{x}, t)/\Delta U$ depend only on x_2/x_1 [where $(x_1, x_2) = (0, 0)$ is the virtual origin of the mixing layer]. Lang (1985) provides experimental data on this flow. The calculations are performed by integrating the stochastic differential equations for the properties $(\mathbf{x}^{(n)}, \mathbf{U}^{(n)}, \omega^{(n)}; n = 1, 2, \ldots, N)$ of $N \approx 50,000$ particles. A comparison of the mean and rms velocities with experimental data shows good agreement (see Pope 1991b).

Figure 5 is a scatter plot of the axial velocity and lateral position. For this flow, with extremely high probability, there is no reverse flow (i.e. $U_1^{(n)}(t) > 0$ for all n and t). Hence the axial location $x_1^{(n)}(t)$ of each particle increases monotonically with time. Figure 5 is constructed by plotting the points $(U_1^{(n)}/\Delta U, x_2^{(n)}/x_1)$ for about one fifth of the particles (selected at random) as they pass a particular axial location x_1. (In view of self-similarity, the value of x_1 is immaterial.)

At large and small values of x_2^*/x_1, the points are dense at $U_1^*/\Delta U = 1$ and 2, respectively, and so appear as horizontal straight lines. These points correspond to fluid with the free-stream velocity. At the center of the layer (e.g. $x_2^*/x_1 = 0$), the points are broadly scattered in $U_1^*/\Delta U$, indicative of turbulent fluctuations with rms of order 0.2. Toward the edges of the layer, bimodal behavior is evident: with increasing distance from the layer, a band of points tends to the free-stream velocity, while other points exhibit fluctuations of order 0.1, but with decreasing probability. This reflects the turbulent/nonturbulent nature of these regions.

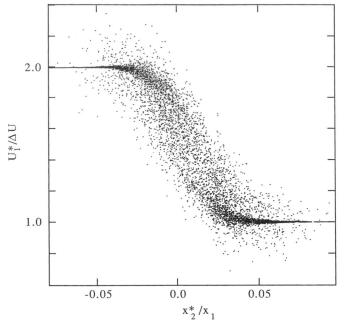

Figure 5 Scatter plot of axial velocity and lateral position from joint pdf calculations of the self-similar plane mixing layer (from Pope 1991b).

The intermittent nature of the edges of the mixing layer is yet more evident in Figure 6, which is a scatter plot of frequency and lateral position. The frequency is normalized by its maximum mean value $\langle\omega\rangle_{max}$ (at the axial location considered) and is shown on a logarithmic scale. At the edges of the layer the bimodal nature of ω^* is clear: There is a diffuse band of points centered around $\omega^* \sim 0.3\langle\omega\rangle_{max}$, with a second denser band with ω^* values two or three orders of magnitude less. These bands correspond to turbulent and nonturbulent fluid respectively.

For inhomogeneous flows, experimental data on Lagrangian quantities are essentially nonexistent. For this reason, there has been little impetus to extract Lagrangian statistics from pdf calculations. However, as an illustration of the type of information that is available in Lagrangian PDF methods, shown on Figure 7 are the fluid particle paths of five particles whose initial positions were selected at random near the center of the self-similar mixing layer. It may be observed that several of these trajectories traverse the layer monotonically, and that the trajectories are devoid of high wave number fluctuations. From this we conclude that the motion

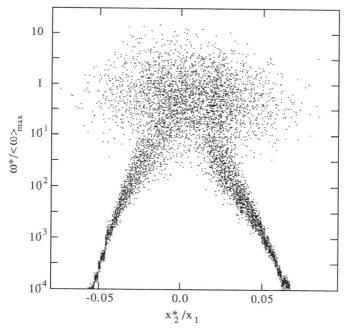

Figure 6 Scatter plot of turbulence frequency and lateral position from joint pdf calculations of the self-similar plane mixing layer (from Pope 1991b).

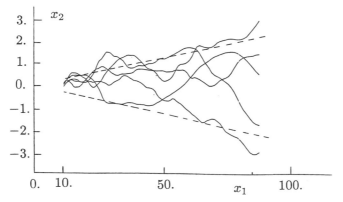

Figure 7 Fluid particle paths in the self-similar plane mixing layer according to stochastic models: x_1 and x_2 have arbitrary units. The dashed lines show the nominal edge of the layer, where the mean velocity differs from the free-stream velocity by 10% of the velocity difference. (From Pope 1991b.)

implied by the model is consistent with the large-scale coherent motions observed experimentally in mixing layers; and, conversely, it does not resemble the small-scale random motion analogous to molecular diffusion or Brownian motion.

5.4   *Stochastic Model for Acceleration*

When examined in detail, the basic Langevin model (described in Section 3) can be justified on physical grounds only in the limit of infinite Reynolds number *Re*. Sawford (1991) presents a stochastic model for the fluid particle acceleration which is extremely valuable and successful in incorporating Reynolds number effects. One virtue of the model is that it can be directly related to Lagrangian statistics obtained from direct numerical simulations—which are found to depend strongly on Reynolds number. In the limit of infinite Reynolds number, the model reverts to the Langevin equation. [As Sawford shows, his model is equivalent to a different formulation given earlier by Krasnoff & Peskin (1971).]

As in Section 3 we consider stationary homogeneous isotropic turbulence with zero mean velocity. The turbulence is characterized by its intensity u' (or kinetic energy $k = \frac{3}{2}u'^2$), the mean dissipation rate $\langle \varepsilon \rangle$, and by the kinematic viscosity v. In terms of these quantities, the Reynolds number is defined by:

$$Re = \frac{k^2}{\langle \varepsilon \rangle v}. \tag{72}$$

It is instructive to relate the Reynolds number to time scales. As usual, the eddy-turnover time T_E and the Kolmogorov time scale τ_η are defined by

$$T_E \equiv k/\langle \varepsilon \rangle = \frac{3}{2}u'^2/\langle \varepsilon \rangle, \tag{73}$$

and

$$\tau_\eta \equiv (v/\langle \varepsilon \rangle)^{1/2}. \tag{74}$$

Hence we obtain

$$Re = (T_E/\tau_\eta)^2. \tag{75}$$

The Langevin equation contains the single time scale T_E; whereas Sawford's stochastic model for acceleration contains two time scales, T_∞ and τ. These time scales (precisely defined below) scale as T_E and τ_η, respectively, at high Reynolds number.

Let $U^*(t)$ and $A^*(t)$ denote the model for one component of velocity and acceleration following a fluid particle. Then the velocity evolves by

$$\frac{d}{dt}U^*(t) = A^*(t).$$ (76)

With a'^2 defined by

$$a'^2 = u'^2/(T_\infty \tau),$$ (77)

Sawford's stochastic model for acceleration can be written

$$dA^*(t) = -\left\{1 + \frac{\tau}{T_\infty}A^*(t)\right\}\frac{dt}{\tau} - \frac{U^*(t)}{T_\infty}\frac{dt}{\tau}$$

$$+ \left\{2a'^2\left(1 + \frac{\tau}{T_\infty}\right)\right\}^{1/2} dW(t)/\tau^{1/2}, \quad (78)$$

where $W(t)$ is a Wiener process.

An analysis of this model (see Sawford 1991 or Priestly 1981) reveals that $U^*(t)$ and $A^*(t)$ are stationary processes with zero means and variances u'^2 and a'^2, respectively. The autocorrelation function of $U^*(t)$ is

$$\rho^*(s) = \left[e^{-|s|/T_\infty} - \left(\frac{\tau}{T_\infty}\right)e^{-|s|/\tau}\right]\bigg/\left(1 - \frac{\tau}{T_\infty}\right),$$ (79)

from which it follows that the (modeled) Lagrangian integral time scale is

$$T = T_\infty + \tau.$$ (80)

The principal features of this model are most clearly seen at high (but finite) Reynolds number, at which there is a complete separation of scales, i.e. $\tau \ll T_\infty$. For all times s much larger than τ, the velocity autocorrelation function is $\rho^*(s) \approx \exp(-|s|/T_\infty)$—the same as for the Langevin model. Consequently, in the inertial range ($\tau \ll s \ll T_\infty$), the Lagrangian velocity structure function varies linearly with s, in accord with the Kolmogorov hypotheses (Equations 25–27). Correspondingly, the Lagrangian velocity frequency spectrum varies as ω^{-2}. But for times s comparable to τ, this model is quite different from the Langevin model. Because $U^*(t)$ is a differentiable function of time, the autocorrelation function has zero slope at the origin. For not-too-large s/τ, the autocorrelation function of acceleration is $\rho_A^*(s) \approx \exp(-|s|/\tau)$. Correspondingly, the Lagrangian velocity frequency spectrum varies as ω^{-4} at high frequency ($\omega\tau \gg 1$).

In order to complete the model, two specifications are required to fix T_∞ and τ in terms of T_E and Re. Sawford (1991) used the Lagrangian DNS data of Yeung & Pope (1989) to achieve this. Here we do the same, but in a slightly different way. First, the DNS data on the Kolmogorov-scaled acceleration variance

$$a_o \equiv a'^2 \tau_\eta / \langle \varepsilon \rangle, \tag{81}$$

can be well approximated (for not too small R_λ) by

$$a_o \approx 3(1 - 22/R_\lambda), \tag{82}$$

where $R_\lambda = (20/3Re)^{1/2}$ is the Taylor-scale Reynolds number. [This form of correlation can be justified in terms of the inertial-range pressure fluctuation spectrum (M. S. Nelkin 1991, private communication; George et al 1984).] Second, for each value of R_λ studied in the DNS, the quantity

$$C_T(R_\lambda) \equiv \frac{4}{3} \frac{T_E}{T_\infty}, \tag{83}$$

can be determined by matching T/τ_η between DNS and the model. The values obtained are between 6 and 7, with a least-squares fit yielding

$$C_T \approx C_T(\infty)(1 + 4/R_\lambda), \tag{84}$$

with $C_T(\infty) = 6.2$. In this case there is no justification for the form of the correlation, and the data exhibit significant scatter around it. Given the empirical correlations for a_o and C_T the two time scales are determined as

$$T_\infty = T_E \left(\frac{4}{3C_T} \right), \tag{85}$$

and

$$\tau = \tau_\eta \left(\frac{C_T}{2a_o} \right). \tag{86}$$

The ability of this model to describe Lagrangian statistics is impressive. Figure 8 shows a comparison of the acceleration autocorrelation functions $\rho_A(s)$ obtained from the model and from DNS. The agreement indicates that the model provides a good approximation to the short-time behavior [although, because $A^*(t)$ is not differentiable, $\rho_A^*(s)$ has finite slope at the origin].

A revealing plot is of the Lagrangian velocity structure function $D_L(s)$ (Equation 25) normalized by $\langle \varepsilon \rangle s$. As may be seen from Figure 9, the model is in good agreement with the DNS data, and correctly shows that the peak value—denoted by C_0^*—increases with R_λ. According to the Kolmogorov hypotheses, at high Reynolds number, and for inertial-range times s ($\tau_\eta \ll s \ll T$), the quantity $D_L(s)/(\langle \varepsilon \rangle s)$, adopts a constant value C_0. It is readily shown that the model has this property, with $C_0 = C_T(\infty)$. But, as may be seen on Figure 10, the peak value C_0^* of $D_L(s)/(\langle \varepsilon \rangle s)$

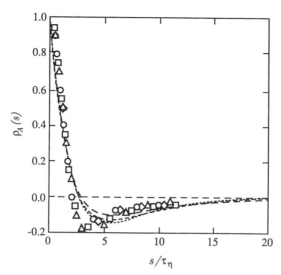

Figure 8   Acceleration autocorrelation function against Kolmogorov-scaled time lag. Symbols: DNS data (Yeung & Pope 1989); lines: Sawford's model. $R_\lambda = 38$: △ ---; $R_\lambda = 63$: □ ----; $R_\lambda = 90$: ○ ——; $R_\lambda = 93$: ◇. (From Sawford 1991, with permission.)

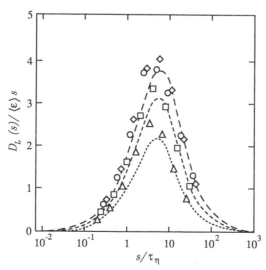

Figure 9   Lagrangian velocity structure function $D_L(s)$ divided by $\langle \varepsilon \rangle s$ against Kolmogorov-scaled time, s/τ_η. Symbols and lines, same as Figure 8. (From Sawford 1991, with permission.)

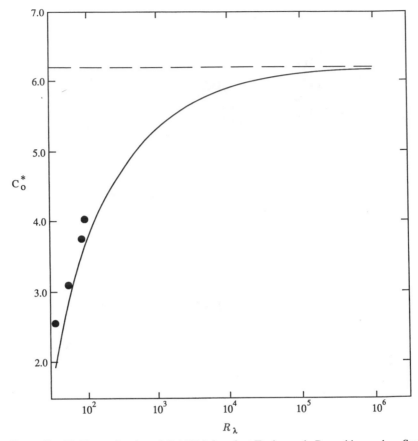

Figure 10 C_0^* [the peak value of $D_L(s)/\langle\varepsilon\rangle s$] against Taylor-scale Reynolds number. Symbols: DNS data (Yeung & Pope 1989); full line: from stochastic model for acceleration, dashed line: model asymptote $C_T(\infty)$.

approaches C_0 slowly as R_λ increases: At the relatively high value $R_\lambda = 1000$, C_0^* is only 85% of C_0.

    Figure 11 shows the ratio of the Lagrangian to Eulerian time scales. It may be seen that this ratio varies appreciably over the range of R_λ accessible to DNS and wind tunnel experiments.

    A question of some interest and importance is the value of the Kolmogorov constant C_0. The estimate from the above model $[C_0 = C_T(\infty) = 6.2]$ is, in essence, obtained by extrapolating from DNS data in the R_λ range 40–90. Other values given in the literature are:

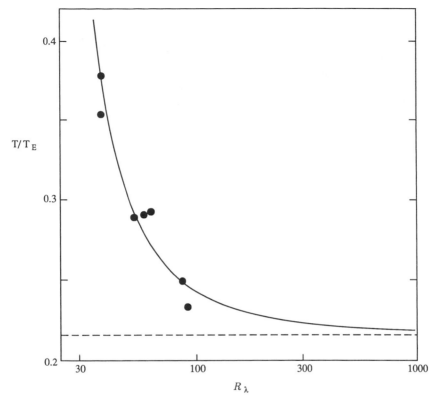

Figure 11 Ratio of Lagrangian to Eulerian time scales against Taylor-scale Reynolds number. Symbols: DNS (Yeung & Pope 1989); line: stochastic model for acceleration, Equations (83–86).

$C_0 \approx 3.8 \pm 1.9$ from measurements in the atmospheric boundary layer (Hanna 1981); $C_0 \approx 5.0$ from kinematic simulations (Fung et al 1992); $C_0 \approx 5.9$ from the Lagrangian renormalized approximation theory (Kaneda 1992); and $C_0 = 5.7$ based on the Langevin equation and further assumptions applied to the constant-stress region of the neutral atmospheric boundary layer (Rodean 1991). It is not unreasonable to suppose, therefore, that C_0 is in the range 5.0–6.5.

With the Langevin and refined Langevin models (which contain no Reynolds-number dependence) it is found that values of $C_0 = 2.1$ (Anand & Pope 1985) and $C_0 = 3.5$ (Pope & Chen 1990), respectively, are required to calculate accurately the dispersion behind a line source in grid turbulence at $R_\lambda \approx 70$. It is now apparent that these values—while being appropriate values of the model constants at moderate Reynolds number—do not correspond to the value of the Kolmogorov constant C_0.

A stochastic model for the fluid particle acceleration $\mathbf{A}^*(t)$ (such as Equation 78), combined with the equations $\dot{\mathbf{x}}^* = \mathbf{U}^*$ and $\dot{\mathbf{U}}^* = \mathbf{A}^*$, leads to a modeled equation for the Eulerian joint pdf of velocity and acceleration, f_{UA}. Such a model equation has not, to date, been applied to inhomogeneous flows. Compared to the velocity joint pdf equation (stemming from a Langevin equation), the equation for f_{UA} has the advantages of incorporating Reynolds-number effects and of representing Kolmogorov-scale processes. This may be of particular value in the study of near-wall flows.

5.5 Other Stochastic Models

Table 1 summarizes the stochastic Lagrangian models that have been proposed for various fluid properties.

Table 1 Stochastic Lagrangian models of turbulence

Subject of model	Authors
Fluid particle position	Taylor (1921)
Fluid particle velocity (single-particle dispersion)	Novikov (1963), Chung (1969), Frost (1975), Reid (1979), Wilson et al (1981a–c), Legg & Raupach (1982), Durbin (1983), Ley & Thomson (1983), Wilson et al (1983), Thomson (1984), Anand & Pope (1985), Van Dop et al (1985), De Baas et al (1986), Haworth & Pope (1986), Sawford (1986), Thomson (1986a), Pope (1987), Thomson (1987), MacInnes & Bracco (1992)
Fluid particle acceleration	Sawford (1991)
Relative velocity between fluid particle pairs (two-particle dispersion)	Novikov (1963), Durbin (1980), Lamb (1981), Gifford (1982), Sawford (1982), Durbin (1982), Lee & Stone (1983), Sawford & Hunt (1986), Thomson (1986b), Thomson (1990)
Dissipation	Pope & Chen (1990), Pope (1991b)
Velocity-gradient tensor	Pope & Cheng (1988), Girimaji & Pope (1990)
Scalar (e.g. species concentration)	Valiño & Dopazo (1991)
Scalar and scalar gradient	Fox (1992)

Taylor (1921) proposed a stochastic model for a component of fluid particle position $x^*(t)$ in which successive increments $\Delta x^* \equiv x^*(t+\Delta t) - x^*(t)$ are correlated. It is interesting to observe that the statistics implied by this model are identical to those from the Langevin equation (Durbin 1980).

For the fluid particle velocity $U^*(t)$, early proposals for the use of the Langevin equation were made by Novikov (1963), Chung (1969), and Frost (1975). The works cited in Table 1 from the period 1979–1984 reflect active use of stochastic modeling of atmospheric dispersion. These models are essentially of the Langevin type, with the primary issue being the specification of the coefficients. In inhomogeneous flows, if the coefficients are specified incorrectly, stochastic models can predict (incorrectly) that an initially uniform distribution of particles becomes nonuniform. Most of the works since 1985 address this issue; a complete explanation is provided by Pope (1987).

The concentration variance of a contaminant in a turbulent flow can be studied in terms of the relative dispersion of fluid particle pairs (Batchelor 1952). Hence stochastic models have been developed (see Table 1) for the relative velocity between particles. These models have had some notable successes in predicting and explaining experimental observations. For example, Durbin (1982) shows that a two-particle dispersion model accounts for the observed sensitivity of the scalar variance in decaying grid turbulence to the initial scalar-to-velocity length scale ratio; and Thomson (1990) shows that his model accounts for the nontrivial evolution of the correlation coefficient between scalars emanating from a pair of line sources in grid turbulence, which has been studied experimentally by Warhaft (1984).

A key quantity in the specification of two-particle model coefficients is the separation distance between the particles. Only recently (Yeung 1993) have DNS results that can be used to develop and test such models become available. Some insight is also provided by kinematic simulations (Fung et al 1992).

The local deformation of material lines, surfaces, and volumes in a turbulent flow is determined by the velocity gradient tensor following the fluid (see e.g. Monin & Yaglom 1985). This motivated the development of stochastic Lagrangian models for the velocity gradient tensor by Pope & Cheng (1988) and Girimaji & Pope (1990). One use of such models is in the calculation of the area density of premixed turbulent flame sheets (Pope & Cheng 1988).

An important yet difficult topic is stochastic Lagrangian models $\phi^*(t)$ for a set of scalars $\phi^+(t)$—such as temperature and species concentrations. In conjunction with a Langevin model, a stochastic model for $\phi^*(t)$ leads to a modeled equation for the Eulerian joint pdf of velocity and composition which can be used to study turbulent reactive flows. Examples of applications of this approach can be found in Anand & Pope (1987), Masri & Pope (1990), Haworth & El Tahry (1991), Correa & Pope (1992), Noms (1993), Taing et al (1993), and elsewhere.

A set of compositions ϕ has certain properties that are very different from these of velocity $\mathbf{U}$. Among these are: boundedness; localness of interactions in composition space; and (in important limiting cases) linearity and independence (Pope 1983b, 1985). These properties make the modeling of $\phi^*(t)$ different and more difficult than the modeling of $\mathbf{U}^*(t)$. Currently there is no model that is even qualitatively satisfactory in all respects.

The simplest model—proposed in several different contexts and with different justifications—is the linear deterministic model:

$$\frac{d\phi^*}{dt} = -C_\phi \langle \omega \rangle (\phi^* - \langle \phi \rangle) \tag{87}$$

(Chung 1969, Yamazaki & Ichigawa 1970, Dopazo & O'Brien 1974, Frost 1975, Borghi 1988). Although (in application to inhomogeneous turbulent reactive flows) the model is not without merit, because it is deterministic, it clearly provides a poor representation of time series of the fluid particle composition $\phi^+(t)$.

Also widely employed are *stochastic mixing models* (e.g. Curl 1963, Dopazo 1979, Janicka et al 1979, Pope 1982). In the terminology of stochastic processes, these models are point processes: The value of $\phi^*(t)$ is piecewise constant, changing discontinuously at discrete time points. Again, these models have their uses, but clearly the time series they generate, $\phi^*(t)$, are qualitatively different to those of turbulent fluid, $\phi^+(t)$.

Shown in Table 1 are the only proposed models that are stochastic, that generate continuous time series, and that preserve the boundedness of scalars. It is possible that a completely satisfactory model at this level will not be achieved. Instead, it may be necessary to incorporate more information, particularly that pertaining to scalar gradients (see e.g. Meyers & O'Brien 1981, Pope 1990, Fox 1992).

6. CONCLUSION

Lagrangian PDF methods are based on stochastic Lagrangian models— that is, stochastic models for the evolution of properties following fluid particles. For example, stochastic models for the fluid particle velocity $\mathbf{U}^*(t)$ (Equation 70) and for the turbulence frequency $\omega^*(t)$ (Equation 66) lead to closed model equations for both the Lagrangian and Eulerian joint pdfs of these quantities. The Eulerian pdf equation can be used as a turbulence model to calculate the properties of inhomogeneous turbulence flows. This equation is solved numerically by a Monte Carlo method which is based, naturally, on the tracking of a large number of particles.

The primary stochastic models reviewed here are for velocity (based on the Langevin equation), for the turbulent frequency (or dissipation), and for the fluid particle acceleration. Lagrangian statistics extracted from direct numerical simulations of homogeneous turbulence have played a central role in the development of these models. Similar statistics at higher Reynolds numbers and in inhomogeneous flows are needed to develop and test the models further.

Other fluid properties—most importantly the composition ϕ—can be adjoined to the PDF method. This requires stochastic models for the quantities involved. In spite of considerable efforts, deficiencies remain in stochastic models for composition.

There is a close connection between Lagrangian PDF methods and Reynolds-stress closures. This connection can be used to benefit both approaches. In particular, new ideas in Reynolds-stress modeling (e.g. Durbin 1991, Lumley 1992, Reynolds 1992) can be readily incorporated in PDF methods.

ACKNOWLEDGMENTS

I am grateful to Dr. B. L. Sawford for permission to reproduce Figures 8 and 9.

For comments and suggestions on the draft of this paper I thank M. S. Anand, R. O. Fox, D. C. Haworth, J. C. R. Hunt, B. L. Sawford, D. J. Thomson, C. C. Volte, and P. K. Yeung.

This work was supported in part by the US Air Force Office of Scientific Research (grant number AFOSR-91-0184), and by the National Science Foundation (grant number CTS-9113236).

Literature Cited

Anand, M. S., Pope, S. B. 1985. Diffusion behind a line source in grid turbulence. In *Turbulent Shear Flows* 4, ed. L. J. S. Bradbury, F. Durst, B. E. Launder, F. W. Schmidt, J. H. Whitelaw, pp. 46–61. Berlin: Springer-Verlag

Anand, M. S., Pope, S. B. 1987. Calculations of premixed turbulent flames by pdf methods. *Combust. Flame* 67: 127–42

Anand, M. S., Pope, S. B., Mongia, H. C. 1989. A pdf method for turbulent recir-culating flows. In *Turbulent Reactive Flows, Lect. Notes in Engrg.* 40: 672–93. Berlin: Springer-Verlag

Anand, M. S., Pope, S. B., Mongia, H. C. 1993. PDF calculations of swirling flows. *AIAA Pap. 93-0106*

Arnold, L. 1974. *Stochastic Differential Equations: Theory and Applications.* New York: Wiley. 228 pp.

Batchelor, G. K. 1952. Diffusion in a field of homogeneous turbulence. II The relative motion of particles. *Proc. Cambridge Phil. Soc.* 48: 345–62

Bilger, R. W. 1980. Turbulent flows with nonpremixed reactants. In *Turbulent Reacting Flows*, ed. P. A. Libby, F. A. Williams, pp. 65–113. Berlin: Springer-Verlag

Billingsley, P. 1986. *Probability and Measure*. New York: Wiley. 622 pp.

Bockhorn, H. 1990. Sensitivity analysis based reduction of complex reaction mechanisms in turbulent non-premixed combustion. *Symp. (Int.) Combust. 23rd*, pp. 767–74. Pittsburgh: Combust. Inst.

Borghi, R. 1988. Turbulent combustion modelling. *Prog. Energy Combust. Sci.* 14: 245–92

Chen, H., Chen, S., Kraichnan, R. H. 1989. Probability distribution of a stochastically advected scalar field. *Phys. Rev. Lett.* 63: 2657–60

Chen, J.-Y., Dibble, R. W., Bilger, R. W. 1990. PDF modeling of turbulent non-premixed CO/H$_2$/N$_2$ jet flames with reduced mechanisms. *Symp. (Int.) Combust. 23rd*, pp. 775–80. Pittsburgh: Combust. Inst.

Chung, P. M. 1969. A simplified statistical model of turbulent chemically reacting shear flows. *AIAA J.* 7: 1982–91

Correa, S. M., Drake, M, C., Pitz, R. W., Shyy, W. 1984. Prediction and measurement of a non-equilibrium turbulent diffusion flame. *Symp. (Int.) Combust. 20th*, pp. 337–43. Pittsburgh: Combust. Inst.

Correa, S. M., Pope, S. B. 1992. Comparison of a Monte Carlo PDF finite-volume mean flow model with bluff-body Raman data. *Symp. (Int.) Combust. 24th*, pp. 279–85. Pittsburgh: Combust. Inst.

Curl, R. L. 1963. Dispersed phase mixing: I. Theory and effects of simple reactors. *AIChE J.* 9: 175–81

Deardorff, J. W. 1978. Closure of second- and third-moment rate equations for diffusion in homogeneous turbulence. *Phys. Fluids* 21: 525–30

De Boas, A. F., Van Dop, H., Nieuwstadt, F. T. M. 1986. An application of the Langevin equation for inhomogeneous conditions to dispersion in a convective boundary layer. *Q. J. R. Meteorol. Soc.* 112: 165–80

Dopazo, C. 1979. Relaxation of initial probability density functions in the turbulent convection of scalar fields. *Phys. Fluids* 22: 20–30

Dopazo, C. 1993. Recent developments in PDF methods. In *Turbulent Reacting Flows*, ed. P. A. Libby, F. A. Williams. New York: Academic. In press

Dopazo, C., O'Brien, E. E. 1974. An approach to the autoignition of a turbulent mixture. *Acta Astronaut.* 1: 1239–66

Durbin, P. A. 1980. A stochastic model of two-particle dispersion and concentration fluctuations in homogeneous turbulence. *J. Fluid Mech.* 100: 279–302

Durbin, P. A. 1982. Analysis of the decay of temperature fluctuations in isotropic turbulence. *Phys. Fluids* 25: 1328–32

Durbin, P. A. 1983. Stochastic Differential Equations and Turbulent Dispersion. *NASA Ref Publ.* 1103

Durbin, P. A. 1991. *Theoret. Comput. Fluid Dyn.* 3: 1–13

Eubank, R. L. 1988. *Spline Smoothing and Nonparametric Regression*. New York: Marcel Dekker. 438 pp.

Fox, R. O. 1992. The Fokker-Planck closure for turbulent molecular mixing: passive scalars. *Phys. Fluids A* 4: 1230–44

Frost, V. A. 1975. Model of a turbulent, diffusion-controlled flame jet. *Fluid Mech. Sov. Res.* 4: 124–33

Fung, J. C. H., Hunt, J. C. R., Malik, N. A., Perkins, R. J. 1992. Kinematic simulation of homogeneous turbulence by unsteady random Fourier modes. *J. Fluid Mech.* 236: 28 1–83

Gao, F. 1991. Mapping closure and non-Gaussianity of the scalar probability density function in isotropic turbulence. *Phys. Fluids A* 3: 2438–44

Gao, F., O'Brien, E. E. 1991. A mapping closure for multispecies Fickian diffusion. *Phys. Fluids A* 3: 956–59

Gardiner, C. W. 1990. *Handbook of Stochastic Methods for Physics Chemistry and Natural Sciences*. Berlin: Springer-Verlag. 442 pp. 2nd ed.

Gence, J. N., Mathieu, J. 1979. On the application of successive plane strains to grid-generated turbulence. *J. Fluid Mech.* 93: 501–13

George, W. K., Beuther, P. D., Arndt, R. E. A. 1984. Pressure spectra in turbulent free shear flows. *J. Fluid Mech.* 148: 155–91

Gifford, F. A. 1982. Horizontal diffusion in the atmosphere: a Lagrangian dynamical theory. *Atmos. Environ.* 16: 505–12

Girimaji, S. S. 1991. Assumed β-pdf model for turbulent mixing: validation and extension to multiple scalar mixing. *Combust. Sci. Technol.* 78: 177–96

Girimaji, S. S., Pope, S. B. 1990. A stochastic model for velocity gradients in turbulence. *Phys. Fluids A* 2: 242–56

Hanna, S. R. 1981. Lagrangian and Eulerian time-scale relation in the daytime boundary layer. *J. Appl. Meteorol.* 20: 242–49

Härdle, W. 1990. *Applied Nonparameteric Regression*. Cambridge: Cambridge Univ. Press. 333 pp.

Haworth, D. C., El Tahry, S. H. 1991. Probability density function approach for multidimensional turbulent flow calculations with application to in-cylinder flows in reciprocating engines. *AIAA J.* 29: 208–18

Haworth, D. C., Pope, S. B. 1986. A generalized Langevin model for turbulent flows. *Phys. Fluids* 29: 387–405

Haworth, D. C., Pope, S. B. 1987. A pdf modelling study of self-similar turbulent free shear flow. *Phys. Fluids* 30: 1026–44

Janicka, J., Kolbe, W., Kollmann, W. 1977. Closure of the transport equation for the probability density function of turbulent scalar fields. *J. Non-equilib. Thermodyn.* 4: 47–66

Kaneda, Y. 1993. Lagangian and Eulerian time correlations in turbulence. Submitted

Kolmogorov, A. N. 1941. Local structure of turbulence in an incompressible fluid at very high Reynolds numbers. *Dokl. Akad. Nauk SSSR* 30: 299–303

Kolmogorov, A. N. 1942. Equations of turbulent motion of an incompressible fluid. *Izv. Acad. Sci. USSR Phys.* 6: 56–58

Kolmogorov, A. N. 1962. A refinement of previous hypotheses concerning the local structure of turbulence in a viscous incompressible fluid at high Reynolds number. *J. Fluid Mech.* 13: 82–85

Krasnoff, E., Peskin, R. L. 1971. The Langevin model for turbulent diffusion. *Geophys. Fluid Dyn.* 2: 123–46

Kuznetsov, V. R., Sabel'nikov, V. A. 1990. *Turbulence and Combustion.* New York: Hemisphere. 362 pp.

Lamb, R. G. 198 1. A scheme for simulating particle pair motions in turbulent fluid. *J. Comput. Phys.* 39: 329–46

Lang, D. B. 1985. *Laser Doppler velocity and vorticity measurements in turbulent shear layers.* PhD thesis. Calif. Inst. Technol.

Launder, B. E., Spalding, D. B. 1972. *Mathematical Models of Turbulence.* New York: Academic

Lee, J. T., Stone, G. L. 1983. The use of Eulerian initial conditions in a Lagrangian model of turbulent diffusion. *Atmos. Environ.* 17: 2477–81

Legg, B. J., Raupach, M. R. 1982. Markov-chain simulation of particle dispersion in inhomogeneous flows: the mean drift velocity induced by a gradient in the Eulerian velocity variance. *Boundary-Layer Meteorol.* 24: 3–13

Ley, A. J., Thomson, D. J. 1983. A random walk model of dispersion in the diabatic surface layer. *Q. J. R. Meteorol. Soc.* 109: 867–80

Lockwood, F. C., Naguib, A. S. 1975. The prediction of the fluctuations in the properties of free, round jet, turbulent, diffusion flames. *Combust. Flame* 24: 109–24

Lumley, J. L. 1992. Some comments on turbulence. *Phys. Fluids A* 4: 203–11

MacInnes, J. M., Bracco, F. V. 1992. Stochastic particle dispersion modeling and the tracer-particle limit. *Phys. Fluids A* 4: 2809–24

Masri, A. R., Pope, S. B. 1990. PDF calculations of piloted turbulent non-premixed flames of methane. *Combust. Flame* 81: 13–29

Meyers, R. E., O'Brien, E. E. 1981. The joint pdf of a scalar and its gradient at a point in a turbulent fluid. *Combust. Sci. Technol.* 26: 123–34

Monin, A. S., Yaglom, A. M. 1975. *Statistical Fluid Mechanics*, Vol. 2. Cambridge, Mass: MIT Press. 874 pp.

Naot, D., Shavit, A., Wolfshtein, M. 1970. Interactions between components of the turbulent velocity correlation tensor due to pressure fluctuations. *Israel J. Technol.* 8: 259–69

Norris, A. T. 1993. *The application of PDF methods to piloted diffusion flames.* PhD thesis, Cornell Univ.

Novikov, E. A. 1963. Random force method in turbulence theory. *Sov. Phys. JETP* 17: 1449–54

Pope, S. B., 1980. Probability distributions of scalars in turbulent shear flows. In *Turbulent Shear Flows* 2, ed. L. J. S. Bradbury, F. Durst, B. E. Launder, F. W. Schmidt, J. H. Whitelaw, pp. 7–16. Berlin: Springer-Verlag

Pope, S. B., 1982. An improved turbulent mixing model. *Combust. Sci. Technol.* 28: 13 1–45

Pope, S. B. 1983a. A Lagrangian two-time probability density function equation for inhomogeneous turbulent flows. *Phys. Fluids* 26: 3448–50

Pope, S. B. 1983b. Consistent modeling of scalars in turbulent flows. *Phys. Fluids* 26: 404–8

Pope, S. B. 1985. PDF methods for turbulent reactive flows. *Prog. Energy Combust. Sci.* 11: 119–92

Pope, S. B. 1987. Consistency conditions for random-walk models of turbulent dispersion. *Phys. Fluids* 30: 2374–79

Pope, S. B. 1990. Computations of turbulent combustion: progress and challenges. *Symp. (Int.) Combust.* 23rd, pp. 591–612. Pittsburgh: Combust. Inst.

Pope, S. B. 1991a. Mapping closures for turbulent mixing and reaction. *Theoret. Comput. Fluid Dyn.* 2: 255–70

Pope, S. B. 1991b. Application of the velocity-dissipation probability density function model to inhomogeneous turbulent flows. *Phys. Fluids A* 3: 1947–57 (See also Erratum: *Phys. Fluids A* 1992. 4: 1088)

Pope, S. B. 1993. On the relationship between stochastic Lagrangian models of turbulence and second-moment closures. *Phys. Fluids A*. 2: to be published

Pope, S. B., Chen, Y. L. 1990. The velocity-dissipation probability density function model for turbulent flows. *Phys. Fluids A* 2: 1437–49

Pope, S. B., Cheng, W. K. 1988. Statistical calculations of spherical turbulent flames. *Symp. (Int.) Combust. 21st*, pp. 1473–82. Pittsburgh: Combust. Inst.

Pope, S. B., Ching, E. S. C. 1993. Stationary probability density functions in turbulence. *Phys. Fluids A*. In press

Priestley, M. B. 1981. *Spectral Analysis and Time Series*. New York: Academic

Reid, J. D. 1979. Markov chain simulations of vertical dispersion in the neutral surface layer for surface and elevated releases. *Boundary-Layer Meteorol.* 16: 3–22

Reynolds, W. C. 1990. The potential and limitations of direct and large eddy simulations. In *Whither Turbulence? Turbulence at the Crossroads*, ed. J. L. Lumley, pp. 313–43. Berlin: Springer-Verlag

Reynolds, W. C. 1992. Towards a structure-based turbulence model. *Bull. Am. Phys. Soc.* 37: 1727

Rhodes, R. P. 1975. A probability distribution function for turbulent flows. In *Turbulent Mixing in Nonreactive and Reactive Flows*, ed. S. N. B. Murthy, pp. 235–41. New York/London: Plenum. 464 pp.

Risken, H. 1989. *The Fokker-Planck Equation: Methods of Solution and Applications*. Berlin: Springer-Verlag. 472 pp. 2nd ed.

Rodean, H. C. 1991. The universal constant for the Lagrangian structure function. *Phys. Fluids A* 3: 1479–80

Roekaerts, D. 1991. Use of a Monte Carlo PDF method in a study of the influence of turbulent fluctuations on selectivity in a jet-stirred reactor. *Appl. Sci. Res.* 48: 27 1–300

Rotta, J. C. 195 1. Statistische Theorie nichthomogener Turbulenz. *Z. Phys.* 129: 547–72

Sawford, B. L. 1982. Lagrangian Monte Carlo simulation of the turbulent motion of a pair of particles. *Q. J. R. Meteorol. Soc.* 108: 207–13

Sawford, B. L. 1986. Generalized random forcing in random-walk turbulent dispersion models. *Phys. Fluids* 29: 3582–85

Sawford, B. L., Hunt, J. C. R. 1986. Effects of turbulence structure, molecular diffusion and source size on scalar fluctuations in homogeneous turbulence. *J. Fluid Mech.* 165: 373–400

Sawford, B. L. 1991. Reynolds number

effects in Lagrangian stochastic models of turbulent dispersion. *Phys. Fluids A* 3: 1577–86

Silverman, B. W. 1986. *Density Estimation for Statistics and Data Analysis*. New York: Chapman and Hall

Sinai, Ya. G., Yakhot, V. 1989. Limiting probability distribution of a passive scalar in a random velocity field. *Phys. Rev. Lett.* 63: 1962–64

Speziale, C. G., Sarkar, S., Gatski, T. B. 1991. Modeling the pressure-strain correlation of turbulence: an invariant dynamical systems approach. *J. Fluid Mech.* 227: 245–72

Taing. S., Masri, A. R., Pope, S. B. 1993. PDF calculations of turbulent non-premixed flames of H_2/CO_2 using reduced chemical mechanisms. *Combust. Flame* In press

Tapia, R. A., Thompson, J. R. 1978. *Nonparametric Density Estimation*. Baltimore: Johns Hopkins Press

Tavoularis, S., Corrsin, S. 1981. Experiments in nearly homogeneous turbulent shear flow with a uniform mean temperature gradient. Part 1. *J. Fluid Mech.* 104: 311–47

Taylor, G. I. 1921. Diffusion by continuous movements. *Proc. London Math. Soc.* 20: 196–212

Thomson, D. J. 1984. Random walk modelling of diffusion in inhomogeneous turbulence. *Q. J. R. Meteorol. Soc.* 110: 1107–20

Thomson, D. J. 1986a. A random walk model of dispersion in turbulent flows and its application to dispersion in a valley. *Q. J. R. Meteorol. Soc.* 112: 511–30

Thomson, D. J. 1986b. On the relative dispersion of two particles in homogeneous stationary turbulence and the implications for the size of the concentration fluctuations at large times. *Q. J. R. Meteorol. Soc.* 112: 890–94

Thomson, D. J. 1987. Criteria for the selection of stochastic models of particle trajectories in turbulent flows. *J. Fluid Mech.* 180: 529–56

Thomson, D. J. 1990. A stochastic model for the motion of particle pairs in isotropic high-Reynolds-number turbulence, and its application to the problem of concentration variance. *J. Fluid Mech.* 210: 113–53

Valiño, L., Dopazo, C. 1991. A binomial Langevin model for turbulent mixing. *Phys. Fluids A* 3: 3034–37

Van Dop, H., Nieuwstadt, F. T. M., Hunt, J. C. R. 1985. Random walk models for particle displacements in inhomogeneous unsteady turbulent flows. *Phys. Fluids* 28: 1639–53 van Kampen, N. G. 1981. *Stochastic Processes in Physics and Chemistry*. Amsterdam: North-Holland

Wilcox, D. C. 1988. Multiscale model for turbulent flows. *AIAA J.* 26: 1311–20

Wilson, J. D., Thurtell, G. W., Kidd, G. E. 1981a. Numerical simulation of particle trajectories in inhomogeneous turbulence, I: Systems with constant turbulent velocity scale. *Boundary-Layer Meteorol.* 21: 295–313

Wilson, J. D., Thurtell, G. W., Kidd, G. E. 1981b. Numerical simulation of particle trajectories in inhomogeneous turbulence, II: Systems with variable turbulent velocity scale. *Boundary-Layer Meteorol.* 21: 423–41

Wilson, J. D., Thurtell, G. W., Kidd, G. E. 1981c. Numerical simulation of particle trajectories in inhomogeneous turbulence, III: Comparion of predictions with experimental data for the atmospheric surface layer. *Boundary-Layer Meteorol.* 21: 443–63

Wilson, J. D., Legg, B. J., Thomson, D. J. 1983. Calculation of particle trajectories in the presence of a gradient in turbulent-velocity variance. *Boundary-Layer Meteorol.* 27: 163–69

Yamazaki, H., Ichigawa, A. 1970. *Int. Chem. Eng.* 10: 471–78

Yeung, P. K. 1993. Direct numerical simulation of relative diffusion in stationary isotropic turbulence. *Ninth Symp. in Turbulent Shear Flows*, Kyoto, Japan

Yeung, P. K., Pope, S. B. 1989. Lagrangian statistics from direct numerical simulations of isotropic turbulence. *J. Fluid Mech.* 207: 531–86

Annu. Rev. Fluid Mech. 1994. 26:65–102

DYNAMICS OF DROP DEFORMATION AND BREAKUP IN VISCOUS FLUIDS

Howard A. Stone

Division of Applied Sciences, Harvard University, Cambridge, Massachusetts 02138

KEY WORDS: capillary instability, satellite drops, chaotic flows, surfactants, tip streaming

1.   INTRODUCTION

This article describes the dynamics of drop deformation and breakup in viscous flows at low Reynolds numbers. An attempt has been made to bring together a wide range of studies in the drop deformation literature, as well as to provide a large number of references to potential applications. In particular, a summary is provided of experimental, numerical, and theoretical investigations that examine drop breakup in externally-imposed flows, e.g. uniaxial extensional fluid motion or more complicated time-periodic flows. For well-characterized flow conditions that lead to breakup, the effects of flow and material parameters on the drop size distribution are summarized. Also, a short discussion is given of the stability of the shapes of translating drops.

The subject of deformation of neutrally buoyant drops in viscous shear flows at low particle Reynolds numbers was summarized by Acrivos (1983) and was reviewed in this series by Rallison (1984). The Acrivos and Rallison papers present (*a*) theoretical descriptions of steady, nearly spherical shapes and steady, long slender shapes, (*b*) a description of efficient boundary integral numerical methods, and (*c*) a summary of the experimental work performed prior to 1984. As documented in these review articles, many of the important ideas necessary for understanding drop

65

0066–4189/94/0115–0065$05.00

deformation can be traced to three articles by G. I. Taylor (1932, 1934, 1964).

During the past ten years there have been several notable advances in the understanding of drop dynamics in well-characterized flow fields. Not only have several gaps in the literature been filled, while the limits of existing theoretical predictions based upon small (nearly spherical) and large (slender body) deformation analyses have been clarified, but the recent investigations have focused upon several new problems (e.g. breakup) and physical influences (e.g. surfactants and complex flows). Our goal here is to provide the reader with an understanding of these recent studies that focus upon drop breakup.

We will first outline the problem area (Section 2) and describe experimental studies that contribute to an improved understanding of drop deformation in shear flows (Section 3). Inevitably there is overlap with the Acrivos and Rallison review articles, which is necessary for this review to be self-contained. Section 4 provides a discussion of different aspects of the drop breakup problem, including an overview of effects introduced by flows more complicated than steady shear flow. Section 5 describes a number of studies examining the influences of surfactants on the degree of drop deformation and breakup and implications for tip streaming.

Section 6 gives a discussion of the low Reynolds number buoyancy-driven translation of nonspherical drops. As is well known (Batchelor 1967), an initially spherical drop is a steady solution to the Stokes equations, independent of the magnitude of the interfacial tension. However, only recently has the *stability* of this solution been investigated. Indeed, translating drops may assume highly distorted shapes with long tails or large cavities. All research, however, has focused on either of the driving forces of buoyancy or imposed shear acting alone, and a quantitative study of the dynamics when the buoyancy and shearing motions are comparable has not been reported.

There are a number of topics with connection to drop deformation in viscous flows that cannot be discussed because of lack of space. For example, drop breakup in electric and magnetic fields, the effect of multiple particles or nearby boundaries, non-Newtonian effects, and high Reynolds number motions are all outside the scope of this review.

In all the results presented in this paper, the degree of drop deformation and the character of the breakup are largely determined by the magnitude of interfacial tension stresses relative to the magnitude of the flow-generated viscous stresses. Some frequently mentioned applications which motivate, or serve as extensions of, many of the studies of drop deformation and breakup are summarized in Table 1. The typical problems span a wide variety of length scales, from the submicron scale to hundreds

Table 1   Typical applications for which a drop deformed or fragmented in an extensional flow is a useful prototype

Application	Reference
Dispersion of color concentrates	Grace (1971)
Processing of microscale dispersions such as margarine and ice cream	de Bruijn (1989)
Mixing in multiphase viscous systems	Ottino (1990), Meijer & Janssen (1993)
Blending of molten polymers	Elemans (1989)
Emulsion formation and rheology	Han (1981)
Ink-jet printers	Döring (1982)
Viscous sintering (cohesion) of adjacent fluid regions	Kruiken (1990)
Deformation of biological cells	Greenspan (1978), Zinemanas & Nir (1988)
Deformation of liquids encapsulated inside elastic membranes	Li et al (1988), Skalak et al (1989)
Manufacture of two-phase glass	Seward (1974)
Drawing of glass sheets	Wilmott (1989a)
Measurement of interfacial tension from time-dependent shape changes	Elemans et al (1990), Carriere et al (1989), Tjahjadi et al (1993)
Stretching of viscous inclusions in geophysical studies of mantle dynamics	Spence et al (1988)

of kilometers, hence provide a common theme for discussion among fluid dynamicists interested in a variety of applications.

2.   PROBLEM DESCRIPTION AND GOVERNING EQUATIONS

The prototypical two-phase flow problem we wish to consider consists of a drop of Newtonian fluid of density $\rho - \Delta\rho$ and viscosity $\lambda\mu$ suspended in an unbounded Newtonian fluid, density ρ and viscosity μ (Figure 1). The undeformed radius of the drop is denoted by a. The interfacial tension acting between the two fluid phases is denoted by γ and may vary along the interface S owing to a nonuniform temperature or the presence of a nonuniform distribution of surfactants with surface concentration Γ. In the presence of surfactant (or temperature) gradients, an equation of state $\gamma(\Gamma)$ must also be introduced, though other possible rheological influences of surfactant are neglected (e.g. surface elasticity; see Edwards et al 1991).

At low Reynolds numbers, fluid motion is governed by the Stokes and continuity equations, which have the dimensionless form (ˆ denotes drop fluid variables)

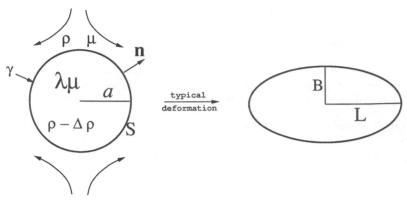

Figure 1   Schematic of drop deformation in a straining flow.

$$\nabla^2 \mathbf{u} = \nabla p, \quad \nabla^2 \hat{\mathbf{u}} = \nabla \hat{p},$$

$$\nabla \cdot \mathbf{u} = 0, \quad \nabla \cdot \hat{\mathbf{u}} = 0, \tag{1}$$

where the pressure p is the dynamic pressure (i.e. the actual fluid pressure minus the background hydrostatic pressure). All distances are scaled by a, velocities are scaled by u_c, which depends on the nature of the forcing present in the problem, and pressures are scaled by $\mu u_c/a$ and $\lambda \mu u_c/a$, respectively, outside and inside the drop.

Continuity of velocity requires $\mathbf{u} = \hat{\mathbf{u}}$ along the fluid-fluid interface S. Viscous and pressure stresses generated by the fluid motion tend to deform the drop while interfacial tension stresses tend to resist deformation. The dimensionless form of the stress boundary condition depends on the specific problem of interest and the choice of the characteristic velocity scale. In addition to the viscosity ratio λ, two dimensionless parameters typically appear in the stress boundary conditions:

$$\mathsf{C} = \frac{\mu u_c}{\gamma} \quad \text{and} \quad \mathsf{B} = \frac{\Delta \rho g a^2}{\gamma}, \tag{2}$$

where g is the gravitational acceleration. The capillary number C represents a measure of viscous stresses relative to interfacial tension stresses and the Bond number B represents typical hydrostatic pressure variations relative to interfacial tension stresses.

Fluid motions internal and external to the drop arise from one of four sources:

1. An externally-imposed velocity field $\mathbf{u}^\infty(\mathbf{x})$ at large distances from the drop, with typical shear rate G in the neighborhood of the drop. With

$u_c = Ga$, the stress boundary condition, for constant γ is

$$\mathbf{n} \cdot \mathbf{T} - \lambda \mathbf{n} \cdot \mathbf{T} = \frac{1}{C}(\nabla_s \cdot \mathbf{n})\mathbf{n} - \frac{B}{C}\hat{\mathbf{g}} \cdot \mathbf{xn} \quad \text{on } S, \tag{3}$$

where $\mathbf{T}$ is the stress tensor defined in terms of the dynamic pressure, $\mathbf{n}$ is the unit normal directed outward from the drop, $\nabla_s \cdot \mathbf{n}$ is the local interface curvature, $\hat{\mathbf{g}}$ denotes a unit vector in the direction of gravity and $\mathbf{x}$ denotes the position vector measured relative to the drop center. Theoretical and numerical studies in this limit have been restricted to neutrally buoyant drops $(B = 0)$.

2. A deformed drop for which the finite interfacial tension generates a flow. Here the fluid is assumed quiescent at large distances, so that with $u_c = \gamma/\mu(1 + \lambda)$ the normal stress balance takes the form

$$\mathbf{n} \cdot \mathbf{T} - \lambda \mathbf{n} \cdot \hat{\mathbf{T}} = (1 + \lambda)(\nabla_s \cdot \mathbf{n})\mathbf{n} - (1 + \lambda)B\hat{\mathbf{g}} \cdot \mathbf{xn} \quad \text{on } S. \tag{4}$$

If we introduce the viscosity ratio into the definition of the characteristic velocity, the larger fluid viscosity controls the typical drop deformation rate (Rallison 1984). The classical capillary breakup of an infinite fluid thread provides a common example where interfacial tension is the driving force for motion and causes amplification of disturbances with wavelength greater than the thread circumference, eventually leading to fragmentation of the thread into a series of drops. Hence, comparing cases (1) and (2), it is apparent that interfacial tension has different dynamical influences at different stages of the drop breakup process.

3. Buoyancy-driven motions in an otherwise quiescent fluid. Here the fluid is assumed quiescent at large distances and with $u_c = \Delta\rho ga^2/\mu(1 + \lambda)$ (alternatively one can use the Hadamard-Rybcyński rise speed of a drop in an infinite fluid which simply changes u_c by a constant), the normal stress balance is

$$\mathbf{n} \cdot \mathbf{T} - \lambda \mathbf{n} \cdot \hat{\mathbf{T}} = \frac{1 + \lambda}{B}(\nabla_s \cdot \mathbf{n})\mathbf{n} - (1 + \lambda)\hat{\mathbf{g}} \cdot \mathbf{xn} \quad \text{on } S. \tag{5}$$

4. Flows either produced or affected by interfacial tension variations (Marangoni motions) caused by temperature variations and/or the presence of surfactants (e.g. see Levich & Krylov 1969). A typical velocity induced by variations in interfacial tension has magnitude $u_c = \Delta\gamma/\mu(1 + \lambda)$, where $\Delta\gamma$ characterizes the initial variation in interfacial tension. A tangential stress imbalance occurs along the surface and, appropriately nondimensionalized, $-\nabla_s\gamma$ must be added to the right-hand side of Equations (3–5). The effects of surfactants on deformation and breakup in extensional flows are summarized in Section 5.

The problem statement is completed by the kinematic condition, which we may write as $dS/dt = \mathbf{u} \cdot \mathbf{n}$ for points on the interface.

In case (1) most research has focused on neutrally buoyant drops immersed in locally linear flows $\mathbf{u}^{\infty}(\mathbf{x}) = \boldsymbol{\Gamma} \cdot \mathbf{x}$, where $\boldsymbol{\Gamma}$ is a second order tensor characterizing the local velocity gradient. This flow approximation is expected to be representative of many applications so long as the largest dimension of the drop is smaller than the typical scale over which variations of the velocity gradient occur. The components of $\boldsymbol{\Gamma}$ in the neighborhood of the drop must be specified, which requires specifying the ratio of the local rate-of-strain to the local vorticity of the applied fluid motion. Any time-dependence of the external flow must also be specified. This *flow-type* classification is important for characterizing the degree of drop deformation (Section 3).

Recent studies have considered the case of time-periodic flow fields, defined over dimensions much larger than a typical drop size (see Section 4). Such a flow is globally more complicated than a linear flow, but even in such circumstances the local flow field in the neighborhood of the drop is often well-approximated by a linear flow with a time-dependent velocity gradient tensor $\boldsymbol{\Gamma}(t)$.

The *initial drop shape* is important for completely characterizing drop dynamics in either shear flows or buoyancy-driven motions and, in particular, plays an important role for determining whether or not transient effects eventually lead to breakup. From the standpoint of applications this is not as restrictive as it may first appear since the details of the shape are typically not important, but rather it is the initial degree of deformation (whether the drop begins nearly spherical or as a highly extended thread)—information which is expected to be available—that is relevant.

In order to characterize the degree of drop distortion it is convenient for modest shape changes to use the deformation parameter $D = (L - B)/(L + B)$, where $2L$ and $2B$ are the drop length and breadth, respectively ($0 \leq D < 1$). For more highly extended drop shapes, a dimensionless length L/a is an appropriate measure of deformation (Figure 1). A complete description of configurations in shear flows also necessitates knowledge of the drop's orientation relative to say the principal axes of strain of the imposed flow (for example, see Rallison 1984).

The Reynolds numbers for the fluid motions considered here are assumed small so that we require $\rho a u / \mu \ll 1$ in both fluid phases, where u is chosen as the largest velocity characteristic of the fluid motion. For the majority of our discussion, drops are to be considered neutrally buoyant, $\Delta \rho = 0$, though in Section 6, drop deformation is a consequence of fluid motion associated with buoyancy. We will summarize those aspects of the flow problems (1)–(4) concerned with steady isolated drop shapes, transient deformation, breakup in simple and complex flows, and surfactant effects.

3. STEADY DROP DEFORMATION IN LINEAR FLOWS

An excellent discussion of drop deformation, including experimental observations and theoretical predictions of steady drop shapes and orientations, is provided by Rallison (1984) and is not reviewed here. In this section we describe qualitatively the dependence of the degree of drop deformation on the viscosity ratio. We also summarize recent experimental observations spanning a wide range of flow conditions, which add substantially to the understanding of the steady state deformation problem and the prediction of the critical capillary number for breakup as a function of the viscosity ratio and flow type. Effects of surfactants are discussed in Section 5.

Qualitative Features of Deformation

When a neutrally buoyant drop is placed in a shear flow it will deform. For all viscosity ratios, the drop shape will be nearly spherical provided the capillary number is sufficiently small. In this limit a small deformation analysis is valid and predicts the drop shape $D(C)$ and orientation in the flow (e.g. see Figures 2 and 3).

When $C \ll 1$, the deformation D is linear in C, as first demonstrated by Taylor (1932). In particular, $D = (19\lambda + 16)C/(16\lambda + 16)$ for an axisymmetric extensional flow, which shows that the effect of the viscosity ratio λ is small. Even if $C > O(1)$, a drop will maintain a nearly spherical shape provided $\lambda \gg 1$ *and* the imposed flow has sufficient vorticity so that the drop fluid spins almost as a rigid body. On the other hand, for low viscosity ratios, $\lambda \ll 1$, the application of a sufficiently large shear rate ($C \gg 1$) forms steady, highly elongated slender drop shapes with nearly pointed ends.

As the capillary number is increased the drop becomes increasingly elongated. The critical capillary number for breakup, C_c, corresponds to the smallest steady shear rate G_c for which the drop, beginning with a nearly spherical shape, is unable to maintain a steady shape and consequently undergoes a transient, continuous stretching. The drop attains a thread-like shape and eventually breaks into smaller drops. The rate of transient stretching is controlled by the larger of the two fluid viscosities. The actual fragmentation of the thread into small drops is described in Section 4.

Prior to the transient stretching and breakup, drops are characterized by their maximum deformation for capillary numbers slightly below C_c. For $\lambda > 0.1$, the maximum steady ellipsoidal deformation is rather modest and for viscosity ratios greater than about 5 the final steady shapes prior to breakup are not very different from a sphere. Proceeding beyond the linear small deformation analysis in order to describe these more distorted shapes is straightforward in principle, but, in practice, is rather difficult algebraically. The higher order analysis for nearly spherical distortions is developed by Barthès-Biesel & Acrivos (1973) and summarized by Rallison (1980). Comparisons of the predictions of small deformation theory with experiments and numerical simulations, including the variation of C_c with viscosity ratio and flow-type Γ, are typically quite good for $\lambda > 0.1$. On the other hand, for $\lambda < 0.01$, slender body theory (Hinch & Acrivos 1979) predicts the drop shapes and critical capillary number with typical errors less than 20% (Bentley & Leal 1986b).

The effect of vorticity of the external flow plays a critical role in determining whether or not drop breakup is possible in a shear-type flow. This fact was recognized by Taylor (1934), who showed that if $\lambda > 4$, and the drop begins with a nearly spherical shape, drop breakup is not possible in a simple shear flow, independent of the magnitude of the shear rate. In this case, it is not possible to deform a drop beyond a modest distortion, so that upon application of higher shear rates the deformation remains unchanged and the drop fluid simply circulates faster. Breakup, however, is always possible in a planar extensional flow for any viscosity ratio λ. A pure extensional flow is, of course, one with zero vorticity, while a simple shear flow is special in that it has equal parts vorticity and strain rate.

This *flow-type* effect provides a dramatic example of the role of vorticity for inhibiting drop breakup. For a simple shear flow, drops become more susceptible to breakup for $\lambda < 4$ due to a coupling of deformation and orientation in the flow: At lower viscosity ratios (but comparable capillary numbers) drops remain oriented closer to the extensional axis of the shear flow (Rumscheidt & Mason 1961, Bentley & Leal 1986b) and, hence, experience a stronger flow; as the shear rate increases, increased deformation thus occurs to a larger degree for small viscosity ratios eventually leading to breakup.

Because of experimental difficulties, experimental investigations of flows intermediate to simple shear and two-dimensional extensional flow—flows that have variable ratios of vorticity to strain rate—were investigated only recently and are now summarized.

Deformation and Critical Capillary Number in Linear Flows With Vorticity

Bentley & Leal (1986a,b) used a computer-controlled four-roll mill to study drop deformation experimentally in the undisturbed two-dimensional linear flow

$$\mathbf{u}^{\infty}(\mathbf{x}) = \mathbf{\Gamma} \cdot \mathbf{x} \quad \text{where} \quad \mathbf{\Gamma} = \frac{G}{2} \begin{pmatrix} 1+\alpha & 1-\alpha & 0 \\ -1+\alpha & -1-\alpha & 0 \\ 0 & 0 & 0 \end{pmatrix}. \tag{6}$$

The velocity gradient tensor $\mathbf{\Gamma}$ is known by specifying the shear rate G and a flow-type parameter α. The ratio of vorticity to strain rate in these flows is given by $(1-\alpha)/(1+\alpha)$ and is controlled independent of G. Simple shear flow corresponds to the limit $\alpha = 0$, while a two-dimensional irrotational flow has $\alpha = 1$. The four-roll mill was first used by Taylor (1934) for drop deformation studies, and was also utilized in a large number of studies by Mason and coworkers (e.g. Rumscheidt & Mason 1961). The particular flow described by Equation (6) has been studied in detail by Giesekus (1962). With the exception of an experimental study by Hakimi & Schowalter (1980) that was confined to vorticity dominated flows, and hence limited to the study of small drop deformations only, all experimental studies prior to 1986 were concerned with either simple shear or two-dimensional extensional flows.

Owing to the implementation of a computer-control strategy, hence a degree of patience not possible in earlier experiments, Bentley & Leal's study provides a wealth of experimental data. Improved data were obtained for extensional flows, especially for viscous drops which require considerable times to deform. It should be noted that previous extensional flow experiments, and all published plots summarizing the dependence of the critical shear rate on viscosity ratio (e.g. Grace 1971) report that the critical capillary number for breakup C_c increases for larger viscosity ratios, $\lambda > O(1)$. However, the Bentley & Leal study clarified that, in fact, C_c remains nearly constant for $\lambda > 5$ (see Figure 4). It appears that earlier studies probably did not wait long enough to establish a steady state, as such waiting times can be $O(\mu a/\gamma)$ which is quite long for high viscosity ratio drops.

Specifically, Bentley & Leal's (1986b) investigation includes:

1. Comparison of small deformation and slender body theories with steady state experiments covering a wide-range of viscosity ratios and flow-types, $10^{-3} < \lambda < 10^2$ and $0.2 \leq \alpha \leq 1$.
2. Measurements of the critical capillary number and critical degree of drop deformation for breakup as a function of λ and α, i.e. $C_c(\lambda, \alpha)$ and $D_c(\lambda, \alpha)$.

We now briefly examine typical results for low and high viscosity ratio drops.

LOW VISCOSITY RATIOS   For small viscosity ratios, $\lambda \ll 1$, the largest steady drop shapes are long and slender, provided that the applied shear rate is sufficiently large. In this limit some simple scaling arguments are possible (Acrivos 1983). In particular, if the slender drop has volume V, steady length $2L$, and a narrow breadth εL ($\varepsilon \ll 1$), then conservation of mass requires $2\pi\varepsilon^2 L^3 = O(V = 4\pi a^3/3)$. An applied shear rate G exerts a viscous stress μG along the interface, whose normal component must be balanced by the interfacial stress $\gamma/(\varepsilon L)$. Combining these two arguments yields an estimate for the slenderness $\varepsilon = C^{-3}$ and a steady dimensionless half-length $L/a = O(C^2)$, where we continue to use the definition $C = \mu Ga/\gamma$. Hence, slender shapes ($\varepsilon \ll 1$) require large capillary numbers, $C \gg 1$. These arguments suggest that for an air or vapor bubble ($\lambda = 0$) of uniform internal pressure, a stable steady shape is always possible and indeed detailed analysis and experiments show this to be true.

However, for a steady shape of a low viscosity ratio drop ($\lambda \ll 1$) exposed to a high shear rate G, an internal pressure gradient $\Delta p/L \approx \lambda\mu(GL)/(\varepsilon L)^2$ is required to push the fluid from the end of the drop back towards the center with typical speed GL. As the magnitude of the pressure gradient is $\gamma/(\varepsilon L)$, we see that $\lambda = O(\varepsilon^2)$ for the slenderness approximation to hold, i.e. $\lambda^{1/2} \ll 1$. Also, it follows that $C\lambda^{1/6} = O(1)$, which with a more detailed analysis may be shown to be the form of a criterion for the critical capillary number necessary for breakup as a function of viscosity ratio.

In Figure 2, drop deformation as a function of capillary number is shown for $\lambda = 10^{-3}$ and two-dimensional flows corresponding to $\alpha = 0.6$ and 0.8. This figure is typical of results from the very low viscosity ratio experiments. In Figure 2a theoretical predictions for the drop shape are shown below the corresponding experimental photographs, while Figure 2b displays the deformation parameters D and L/a as functions of capillary number. For small distortions from a spherical shape, the shapes predicted by a small deformation analysis are calculated, while larger drop distortions are compared with shapes calculated using slender body analysis (Acrivos & Lo 1978).

The small deformation analysis is in reasonable agreement with the overall shape for $C < 0.25$, though clearly it has a limited range of utility as it predicts unrealistic lobes (Figure 2a). The drop shapes become increasingly elongated as the capillary number increases, but the effect of flow type in the low viscosity ratio limit is rather small. Since the elongated drop is oriented by the flow along a direction where the effective strain rate is $G\alpha^{1/2}$, the slender body theory estimate (Hinch & Acrivos 1979) of the critical capillary number is modified to read $C_c\lambda^{1/6} = 0.145/\alpha^{1/2}$, which is in good agreement with the data. Overall, slender body theory generally lies within 20% of the experiment results for the most highly deformed drops.

The most distorted drop shapes provide a vivid example of nearly pointed ends which may arise even in systems with finite interfacial tension. Two-dimensional free-surface experiments and theory illustrate that cusp-like shapes are indeed possible even if $C = O(1)$; the reader is referred to Jeong & Moffatt (1992) and Joseph et al (1991). The analysis has not been extended to three dimensions but it appears that for a low viscosity ratio drop the end is not a cusp, and appears on the macroscale to be sharp (i.e. there is a finite but very large curvature; see also Buckmaster 1972 and Sherwood 1981).

For the large distortions typical of low viscosity ratio drops it is reasonable to question the validity of the linear flow approximation. Sherwood (1984), motivated by Taylor's (1934) experimental observation of tip streaming of small droplets from the end of stretched drops, has addressed the effect of higher order terms in the undisturbed flow field within the context of a slender body analysis. Cubic terms in the external flow, if sufficiently strong, give rise to rapid growth at the end of the drop. Though this result may have some bearing on breakup when velocity gradients occur on the same scale as the highly distorted drop shape, there is now strong evidence that tip streaming is a consequence of surfactant gradients (Section 5; de Bruijn 1989).

HIGH VISCOSITY RATIOS   In Figure 3, drop deformation is examined as a function of flow type for $\lambda = 57$. It is observed experimentally, in agreement with a prediction using the $O(C^2)$ theory of Barthès-Biesel & Acrivos (1973), that for both the $\alpha = 0.2$ and $\alpha = 0.4$ flows, it is impossible to fragment the drop when beginning with a spherical shape and thus a limiting steady deformation is reached. The effect of vorticity is clearly to inhibit breakup for $\alpha \leq 0.4$. Overall, there is excellent agreement between theory and experiment, including prediction of the drop shape and the critical capillary number for onset of transient stretching, which is a precursor to eventual fragmentation.

Figure 3 is typical of the high viscosity ratio data. For drops with lower viscosity ratios, $\lambda = O(1)$, the deformation is similar to Figure 3, except that no limiting deformation is obtained when the vorticity of the undisturbed flow is increased ($0 < \alpha \leq 1$).

CRITICAL CAPILLARY NUMBER VS VISCOSITY RATIO AND FLOW TYPE   Figure 4 shows the critical capillary number for breakup (i.e. the onset of continuous deformation) as a function of viscosity ratio. All data for two-dimensional flows spanning simple shear to planar extensional flow are presented. The solid and dashed curves indicate theoretical predictions based upon small deformation theory and slender body analysis, respectively. A similar plot illustrating the maximum stable deformation as a

	$C = 0.0$	
	D	ϑ
Expt	0.0	0.0
Theory	0.0	0.0

	$C = 0.175$	
	D	ϑ
Expt	0.315	−1.0
Theory	0.312	−0.9

	$C = 0.263$	
	D	ϑ
Expt	0.531	−1.5
Theory	0.516	−1.4

	$C = 0.321$	
	L/a	ϑ
Expt	2.70	−2.0
Theory	1.67	—

	$C = 0.409$	
	L/a	ϑ
Expt	3.77	−3.0
Theory	2.77	—

	$C = 0.453$	
	L/a	ϑ
Expt	4.51	−3.0
Theory	3.51	

	$C = 0.511$	
	L/a	ϑ
Expt	6.45	−3.0
Theory	5.14	—

(a)

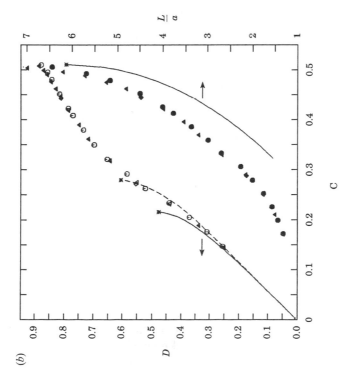

Figure 2 Drop deformation as a function of capillary number for a low viscosity ratio, $\lambda = 10^{-3}$. (*a*) $\alpha = 0.8$; theoretical predictions for the drop shape are given below the corresponding experimental photographs. Small deformation theory is used for $C < 0.27$ and slender body theory is used for $C > 0.32$. The orientation angle relative to the extensional axis is denoted by θ. (*b*) $\alpha = 0.6$; open symbols correspond to the deformation parameter D and are compared with the $O(C)$ (*solid curve*) and $O(C^2)$ (*dashed curve*) predictions of small deformation theory. The filled-in symbols correspond to the deformation parameter L/a and are compared with the predictions of slender body theory (*solid curve*). An asterisk terminating a theoretical line indicates a prediction of breakup, i.e. unsteady stretching (Bentley & Leal 1986b).

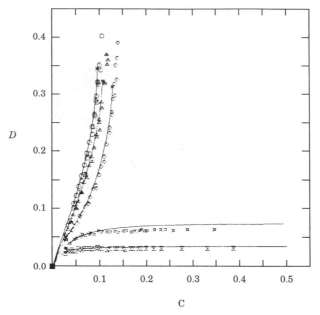

Figure 3   Drop deformation as a function of capillary number for $\lambda = 57$ and flow-types $\alpha = 1.0$ (○), 0.8 (△), 0.6 (◇), 0.4 (⊠), 0.2 (⊠). An asterisk terminating a theoretical curve indicates a prediction of breakup, i.e. unsteady stretching (Bentley & Leal 1986b).

function of viscosity ratio is given by Bentley & Leal (1986b). Overall, the very high capillary numbers necessary to produce large deformation of low viscosity ratio drops indicate that such surfactant-free drops will be difficult to fragment in a flow.

4.   DROP BREAKUP

The focus of most drop dynamics research has been the prediction of the critical conditions beyond which no steady drop shape exists. A practical question from the standpoint of applications is to ask about the final state of a multiphase system, in particular what the final drop size distribution might be. Clearly, this is a formidable task. In order to develop an improved understanding, it is necessary to determine which variables control the time-dependent dynamics of drop breakup, how changes in the flow affect the drop behavior, etc. Several investigations have combined experimental and numerical studies in an attempt to answer some of these questions, though the current state of knowledge is still largely qualitative.

In this section we provide a summary of the basic modes of breakup of

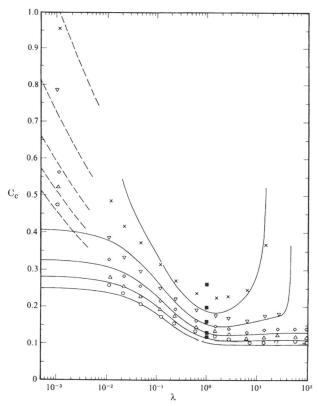

Figure 4 Effects of flow-type on the critical capillary number for drop breakup. Experimental results are presented for flow types $\alpha = 1.0$ (○), 0.8 (△), 0.6 (◇), 0.4 (▽), 0.2 (×). The solid curves are the prediction of the $O(C^2)$ small deformation theory and the dashed curves are the prediction of slender body theory (Bentley & Leal 1986b). The ■ symbols are the boundary integral numerical results of Rallison (1981).

isolated drops, the effect on drop breakup of simple time-dependent flows as well as more complex (in this case chaotic) flows, the role of capillary waves during breakup, and the formation of satellite drops. A description is also given of the dynamics associated with the response of a stretchable (dumbbell-like) microstructure in time-periodic flows. Finally, time-dependent drop stretching is discussed for the case of vanishing interfacial tension, which is a limit with geophysical and material science applications.

As described in Section 3, if an initially spherical drop is placed in a weak *steady* flow, the drop deforms eventually reaching a state where the viscous and pressure forces exerted by the flow on the drop are balanced by the resistance due to interfacial tension forces. When the critical capillary number is exceeded, the drop begins a transient elongation. The term "drop breakup" often refers to this condition, even though actual fragmentation of the droplet is rarely discussed. The drop subsequently stretches and thins, eventually reaching a thread-like shape where it stretches passively, identical to a fluid element of the undisturbed flow. There are cases in which, if the drop radius becomes small enough, capillary waves can cause the thread to fragment during flow (Mikami et al 1975, Tjahjadi & Ottino 1991), though this response requires very large elongations, typically greater than 20 times the initial drop radius. Also the drop may fragment near the middle of the thread, forming a number of smaller droplets; the details of this mode of breakup depend on the nature (and time dependence) of the externally applied flow.

The most comprehensive experimental work focusing upon drop dynamics is the study by Grace (1971), which describes model transient experiments including the use of a programmed gradual reduction of shear rate to produce drop fragmentation without significant stretching and a systematic study of the drop size distributions measured following breakup at supercritical values of the capillary number. Grace's work suggested several fundamental problems involving time-dependent effects which could provide insight into details of the breakup process. We begin our discussion of breakup with the response of a modestly deformed drop to some simple flow changes.

Time-Dependent Effects: Step Changes in Flow Conditions

Stone et al (1986) and Stone & Leal (1989a,b) present an experimental and numerical investigation of the effect of simple time-dependent flows on drops that have been stretched initially in flows at slightly supercritical values of the capillary number C_c. A computer-controlled four-roll (Bentley & Leal 1986a,b) is used to perform the experiments, which allows control over a wide range of parameters. Only step changes in the flow to subcritical conditions are examined. For flows with constant flow type (α), the experiments correspond to step reductions in shear rate, or $C(\alpha) < C_c(\alpha)$, where the notation indicates that the final value of the capillary number is less than the critical value at the specified flow type. In another series of experiments, abrupt changes of the flow type are studied, as are flows where the flow type and shear rate are changed simultaneously so long as the final state corresponds to subcritical flow conditions. Such simple flow modifications are expected to be representative of situations where rather abrupt flow changes occur (Grace 1971); for example, abrupt flow changes occur wherever the apparatus geometry changes substantially such as at entrances and exits, baffles, etc, and at the entrance to the repetitive mixing sections of a static mixer.

Breakup and the details of the drop size distribution depend on the time history of the applied flow. For example, consider the effect of an abrupt halt of the flow for a drop deformed beyond its maximum steady shape D_c with a flow at the critical capillary number. Provided the drop has been stretched sufficiently beyond D_c, then after flow stoppage, the drop first relaxes back towards a spherical shape, but subsequently fragments, forming a number of smaller droplets. Capillary wave instabilities play only a minor role for such modest extensions. This relaxation and breakup process following cessation of the flow is documented for several different viscosity ratios in Figures 5 and 6, illustrating the results of typical experiments and boundary integral calculations, respectively. The results demonstrate that (a) the ends of the drop become rounded and eventually pinch off from the central filament if the initial deformation is large enough, (b) capillary waves do not play a significant role in the dynamics for these modest extensions, and (c) high viscosity ratios clearly tend to inhibit the breakup process.

The breakup mechanism illustrated in Figures 5 and 6 is termed "end-pinching" and was observed previously in the experiments of Taylor (1934) and Grace (1971). Grace mentioned capillary wave instabilities as an explanation for this mode of fragmentation, though it is proper to describe the breakup of modestly deformed drops in terms of a deterministic interfacial-tension-driven flow dominated by end effects, as discussed below.

There are a number of different situations where, following an abrupt change in flow conditions, drop relaxation perhaps leading to this "end-pinching" mode of breakup is relevant. For example, the retraction of extended drops occurs following the rapid ejection of drops from a nozzle during ink-jet printing (Döring 1982); similar relaxation of finitely extended fluid drops is observed during processing of polymer composites (Carriere et al 1989).

End-pinching is a consequence of an interfacial-tension-driven flow associated with curvature variations along the surface of the finite drop. The drop attempts to return to a spherical shape, though fluid motions produced by internal pressure gradients lead to breakup. Because higher drop viscosities damp internal flows, while lower external viscosities increase the speed with which the ends of the drop retract through the surrounding fluid, it follows that higher-λ drops are able to sustain much higher initial elongations prior to breakup during the relaxation process. This behavior is clearly illustrated in Figures 5 and 6. The available experiments suggest that the maximum length needed to ensure breakup for $\lambda > 1$ increases roughly linearly with λ as the qualitative arguments above would suggest (see Stone & Leal 1989a, Figure 10).

In all the experiments reported by Stone & Leal it is possible to explain

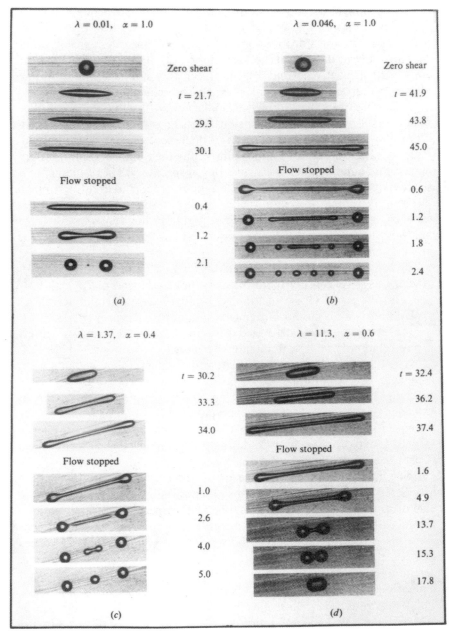

Figure 5 Experiments illustrating the effect of viscosity ratio on the relaxation and fragmentation of initially stretched drops after the flow has been stopped (Stone et al 1986).

$t = 0.0, 15.0, 21.0, 23.4$ $\lambda = 0.05$

$t = 0.0, 15.0, 20.36, 26.34, 29.67$ $\lambda = 0.1$

$t = 0.0, 24.0, 38.0, 50.0$ $\lambda = 1.0$

$t = 0.0, 27.28, 39.28, 46.72, 49.2$ $\lambda = 7.5$

$t = 0.0, 15.8, 36.0, 60.0$ $\lambda = 10.0$

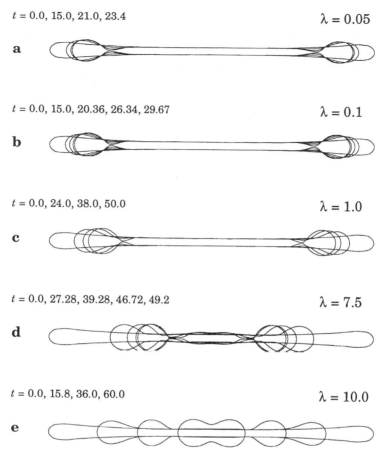

Figure 6 Numerical simulations illustrating the effect of viscosity ratio on the relaxation of initially extended drops in an otherwise quiescent fluid. The same initial shape is used in each simulation. The initial drop shape is taken from an experiment for $\lambda = 11.3$ and is similar to all experimental results for $\lambda > 0.05$ (Stone & Leal 1989a).

the qualitative and quantitative dynamics assuming the interfacial tension is constant. We expect that both large and small viscosity ratio drops will be difficult to fragment. In the case of $\lambda < 0.01$, high capillary numbers are necessary to cause a transient elongation as very elongated steady shapes are possible (Figures 2 and 4). For $\lambda > O(10)$, highly stretched drops are capable of relaxing back to a spherical shape following cessation of the flow because of the significant internal flow resistance. Even for $\lambda = O(1)$, a deformation (L/a) approximately three times the maximum steady deformation is necessary to ensure breakup; the fact that rather large deformations relative to the steady state shape are necessary to promote breakup was first indicated by Grace (1971).

One further observation that follows from the above experiments is that the initial fragmentation of the drop is controlled by global features of the shape, rather than being dependent on small scale, or local, features. However, local details of the shape, which vary with viscosity ratio, and the experimental details (e.g. asymmetries present in the initial configuration), affect the formation of satellite and subsatellite drops.

If the drop is stretched sufficiently, $L/a > O(20)$, then capillary waves, which grow to finite amplitude slowly owing to the small initial disturbances present in carefully controlled experiments, have time to evolve to a sufficiently large amplitude to affect the drop dynamics. In such circumstances, capillary waves play a significant, if not dominant, role in controlling the final drop size distribution (Tjahjadi & Ottino 1991); this behavior is commonly observed in polymer blending processes (e.g. Elemans 1989).

Finally, for step changes to subcritical conditions with $C \neq 0$, it is possible to break drops without producing large-scale stretching of the drop. Such a model time-dependent flow is illustrated in the numerical simulations shown in Figure 7. The detailed velocity fields and transient drop shapes following a step change to $0.5\,C_c$ are displayed in Figure 7b. Comparable experiments have also been performed (Stone & Leal 1989b). This manner of breakup adds just a few—though rather large and uniformly sized—drops to the drop size distribution. Also, it is possible to fragment very viscous drops in flows with significant vorticity (for example, a drop with $\lambda > 4$ in a simple shear flow), if the *initial deformation* is sufficiently large, as may be produced by first applying a low vorticity, high shear rate flow. An example is provided by Stone & Leal (1989b, Figure 13).

Complex Flows

It is worthwhile to consider more complicated flows, especially flows more representative of industrial applications. However, it is difficult to characterize three-dimensional flows in a manner that could be combined with fundamental studies of drop breakup to produce a useful picture of real mixing processes. Also, it remains a challenging numerical problem to describe even modest, low capillary number, three-dimensional distortions of drops in a simple shear flow. The reader interested in numerical investigations of three-dimensional drop deformation at low Reynolds numbers is referred to Rallison (1981), de Bruijn (1989), and Pozrikidis (1992). The calculation of the fragmentation of highly deformed drop shapes has not

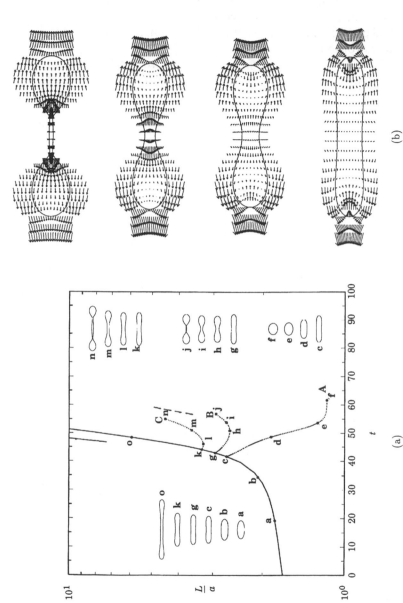

Figure 7 $\lambda = 1.0$; axisymmetric extensional flow. (*a*) Numerical simulations of step changes in capillary number from C_c to the subcritical value $0.5C_c$. The solid curve is the time-dependent stretching of the drop at the critical capillary number, while the three dashed curves illustrate the response to the step change. The solid (dashed) almost vertical line is the asymptotic limit of stretching like a fluid element in the original (subcritical) extensional flow. (*b*) The flow field and drop shape as a function of time following a step reduction in capillary number. The simulations correspond to the intermediate shapes shown along curve B in (*a*) (Stone & Leal 1989b).

yet been implemented satisfactorily. The two elements of (a) a complicated bulk flow and (b) dynamics of drop deformation, are in general coupled— the flow field from the Lagrangian view point of the drop will be time-dependent and perhaps changing on a time-scale comparable to the typical time for drop distortion, $a\mu(1 + \lambda)/\gamma$.

Tjahjadi & Ottino (1991) present a first step towards bridging the gap between model studies of drop breakup and applications involving the mixing of immiscible viscous fluids. These authors describe an experimental investigation of the stretching and breakup of filaments in two-dimensional low Reynolds number chaotic flows generated via periodic modulation of the boundaries of an eccentric journal bearing apparatus. The time-dependent boundary motion is responsible for the appearance, in at least some part of the flow domain, of chaotic particle paths, which indicate that two nearby fluid particles separate exponentially in time for some (often large) range of initial conditions in the flow. In the experiments a fluid drop ($0.01 < \lambda < 2.8$) is placed in the chaotic region of the flow where the critical capillary number for breakup is exceeded. The drop is stretched rapidly into a slender filament, which then behaves as a passive tracer. The length of the thread-like shape generally increases exponentially with time, often reaching end-to-end stretches 10^3 times the initial drop radius. The final drop size distributions are also determined.

A typical example of the evolution of the drop, along with its subsequent fragmentation, is shown in Figure 8. Drop breakup primarily occurs as a result of capillary wave instabilities. Near the end of the filament and at intermediate locations where the filament is broken, end-pinching occurs producing rather large drops. The chaotic character of the flow produces *folds* in the filament, where at later times interfacial tension causes breakup again leading to large satellite drops.

A simplified model which assumes that the filament behaves passively in the flow accounts for the overall large-scale stretching. When combined with the linearized analysis for the growth of perturbations on cylindrical threads, the model can predict regions of potential breakup and approximate the size of the largest satellite drops. In the experiments, the lower viscosity ratio systems typically lead to more uniform drop size distributions since the breakup occurs rather quickly and the flow can generally break up the larger drops a second or third time (see also Figure 5a). For the higher viscosity ratio systems $\lambda > O(1)$, internal fluid motions required for breakup are damped, capillary instabilities thus occur less readily, and therefore the filament stretches and thins more prior to breakup. The small drops thus formed upon fragmentation of the thread are characterized by capillary numbers less than the critical value and hence do not typically break a second time. Breakup near folds and end-

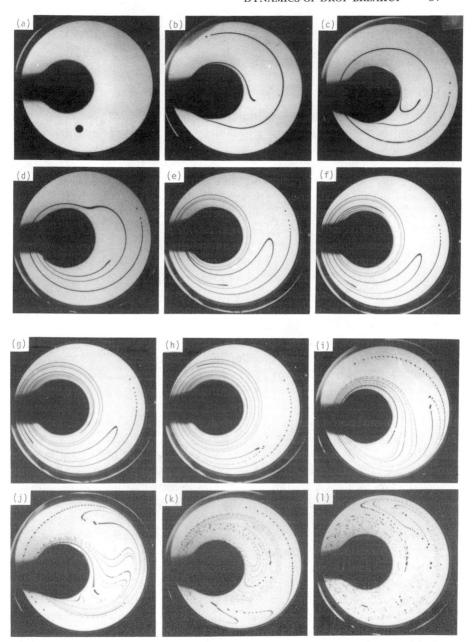

Figure 8   Dynamics of elongation, folding, and breakup of a drop placed in the globally chaotic flow produced in a time-periodically modulated eccentric journal bearing. The times corresponding to the different snapshots of the flow are given in the original publication. $\lambda = 0.067$ (Tjahjadi & Ottino 1991).

pinching lead to rather large drops that will likely fragment again. Overall, the experiments indicate that the drop size distribution in high viscosity ratio systems is more nonuniform, with a smaller mean drop size, than the lower viscosity ratio systems.

Satellite Drops

Drop or fluid filament fragmentation, whether by growth of capillary instabilities or via end-pinching because of the finite size of the drop, results in a distribution of drop sizes. In between the largest visible drops are smaller satellite drops and on still smaller scales there are subsatellite drops. There are, of course, similarities between the fragmentation of a viscous fluid filament and the breakup of weakly viscous fluid jets (Bogy 1979). Linear stability theory is unable to predict the existence of satellite drops and consequently, at least in its simplest form, provides no insight into the detailed evolution of the satellite drop size distribution.

Tjahjadi et al (1992) report a detailed study documenting satellite and subsatellite drop formation that result from the growth of capillary instabilities on the surface of a long, otherwise stationary, viscous fluid thread. In a set of experiments with viscosity ratios $0.01 < \lambda < 2.8$, fluid threads with radius 0.1 cm fragment; this leads to an array of smaller drops, where the smallest visible drops have radii of about 10 μm. The basic features of the breakup are illustrated in Figure 9. Complementary boundary integral calculations also shown in Figure 9 illustrate the small-scale flow and breakup processes (see Tjahjadi et al for a description of the numerical approximations introduced for treating the multiple breakup events). Linear stability theory for this viscous flow problem dates back to Tomotika (1935).

The viscosity ratio λ between the two fluids plays the most significant role in determining the eventual drop size distribution. For $\lambda > O(1)$, internal flow processes are damped leading to a smaller number of more uniformly sized drops, while for the smallest viscosity ratio studied, $\lambda = 10^{-2}$, many small satellites and subsatellites are formed between the largest mother drops. In general, the disturbance wavelength observed is within 20% of the optimum value predicted by linear stability theory on the basis of the largest disturbance growth rate.

For cases where drop breakup occurs in model chaotic flows, it appears that the drop size distribution generated at long times (i.e. after several splittings) can be described with a scaling analysis that yields the fraction of the population with a given radius (Muzzio et al 1991). The available experiments fall into two classes depending on whether $0.01 < \lambda < O(1)$ (experiments with smaller viscosity ratios were not performed) or whether $\lambda > O(1)$. The observation that the drop size distribution appears to be

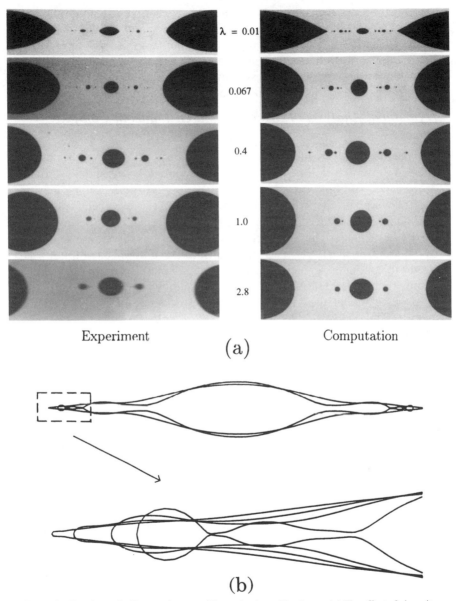

$\lambda = 0.01$

0.067

0.4

1.0

2.8

Experiment   Computation

(a)

(b)

Figure 9 Breakup of a filament into satellite and subsatellite drops. (*a*) The effect of viscosity ratio. Shapes of the filament after the last fragmentation are shown. The dimensionless times at which the experiments and simulations are compared are within 4%. (*b*) Numerical simulation of the breakup of the narrow thread-like filament that appears between the large parent drops. An enlargement is shown illustrating the relaxation and breakup of the narrow end (Tjahjadi et al 1992).

controlled by the viscosity ratio has a relatively simple explanation. For the higher viscosity ratios, very large stretchings and thin threads are produced by the flow. Upon capillary fragmentation of the thread the small drops formed do not typically fragment a second time. On the other hand, the lower viscosity ratio systems fragment more quickly forming larger drops which thus undergo subsequent stretching and breakup.

Dynamical Analysis of Microstructured Fluids in Complex Flows

The detailed analysis of drop deformation and breakup in other complicated flows is formidable owing to the coupling of three-dimensional drop shapes with time-dependent velocity fields, as experienced by a drop as it traverses a given mixing device. Some progress is nevertheless possible if a model is introduced for the drop dynamics while maintaining complexity of the underlying flow field. A first step in this direction was made by Olbricht et al (1982) who introduced an evolution equation for a stretchable, orientable microstructure. In its simplest form the state of the microstructure is described by an axial vector which may be used for example to model the average length and orientation of a deforming droplet. This work establishes a strong flow criterion for steady flows which gives an estimate for the shear rate and flow conditions required to initiate stretching [see also Ottino (1990), Sections 9.3 and 9.4; and Khakhar & Ottino (1986), who discuss how the model vector equation may be interpreted to obtain the approximate dynamics of low viscosity ratio, highly stretched drops].

Szeri et al (1992) establish a strong flow criterion for the case of unsteady, spatially inhomogeneous flows which illustrates the effects of time dependence on the evolution of the microstructural element. Specifically, the vector model equation of Olbricht et al is used to describe the dynamics of a stretchable, orientable microstructure and is coupled to a time-dependent two-dimensional flow. The time evolution of the configuration of the microstructure is determined for time-periodic flows and the analysis takes advantage of several advances in understanding dynamical systems. A time-periodic flow history is experienced by a drop in a number of quite plausible circumstances, including recirculating particle paths in flows with closed streamlines, a steady flow in a spatially periodic domain, or a flow near a stagnation point that varies periodically in time.

Via several examples Szeri et al show that the neglect of time dependence may lead to incorrect characterization of the flow if instead the flow field is modeled by a sequence of steady states. An important assumption in the analysis is the neglect of any particle-particle interactions. Implications for stretched droplets are difficult to infer directly since at low capillary numbers, interfacial tension plays an important role in the dynamics and introduces another time scale for the stretching dynamics.

Stretching of Thin Viscous Inclusions

The limit of zero interfacial tension is of practical significance to geophysical flow problems as well as to such industrial problems as glass manufacturing. The slender body analysis of the stretching of very viscous, neutrally buoyant inclusions has been presented by Spence et al (1988) and Wilmott (1989a,b). No steady drop shape is possible in these extensional fluid motions since there is nothing to resist the viscous stresses exerted on the inclusion by the external flow, and thus the inclusion stretches and thins on a time scale given approximately by the inverse shear rate G^{-1}. On this time scale the internal and external flows are coupled provided the slender inclusion, with slenderness ratio ε, is sufficiently viscous, i.e. $\lambda = O(\varepsilon^{-2})$ for axisymmetric inclusions and $\lambda = O(\varepsilon^{-1})$ for two-dimensional sheets. In general, the inclusion stretches exponentially, with small corrections from behavior identical to a fluid element owing to the viscosity contrast and finite inclusion length.

The analyses demonstrate analytically that an inclusion with smooth initial data cannot fragment in finite time, which is a result that might be expected owing to the absence of interfacial tension or any other destabilizing mechanism. For arbitrary initial shapes, it is shown how the slender inclusion will progressively thin; which may be useful for understanding the role of other effects, e.g. intermolecular forces that will become important at very short length scales (approximately 1000 Å).

5.   EFFECTS OF SURFACTANTS

The presence of impurities has noticeable effects on the deformation of fluid-fluid interfaces simply because the impurities lower the interfacial tension. Also, surfactant gradients along the interface cause tangential stresses that produce fluid motions (Marangoni flows). Consequently, the presence of surfactants alters drop shapes and changes the critical capillary number for drop breakup. An experimental study by de Bruijn (1989, 1993) documents the importance of surfactants for explaining the phenomenon of tip streaming.

In this section, we summarize the effects of surfactants on the deformation and breakup of drops in externally imposed straining flows. It is assumed that the only effect of the surfactant is to modify the magnitude of the interfacial tension, which is quite reasonable provided the surfactant is present at low concentrations. Other interfacial rheological effects such as surface shear and dilatational viscosities are neglected (Edwards et al

1991). The surface rheology characteristic of thin elastic membranes is discussed, for example, by Li et al (1988).

The effects of surfactants on the motion of translating drops in unbounded fluids or in capillaries are well studied and will not be discussed here. A complete understanding of the influences of surfactants on the motion of fluid-fluid interfaces may require knowledge of interfacial transport processes since Stebe et al (1991) demonstrate the remobilization of a surfactant retarded interface in the limit where desorption kinetics and diffusion away from the interface are fast. Surfactants are also useful in the hydrodynamic modeling of cell cleavage (e.g. Greenspan 1978, Zinemanas & Nir 1988).

Deformation in Shear Flow

In order to quantify the effect of surfactants on free-boundary problems it is necessary to introduce a constitutive equation relating surfactant concentration Γ to the local value of interfacial tension γ. Typically, a linear equation of state is chosen with $\gamma(\Gamma)$ given by the two-dimensional gas law, $\gamma_s - \gamma = RT\Gamma$, where γ_s is the interfacial tension of a clean interface ($\Gamma = 0$), R is the gas constant, and T is the absolute temperature. Milliken et al (1993) have examined numerically some influences of a nonlinear equation of state on drop deformation and breakup. We let Γ_0 denote the (equilibrium) uniform surfactant concentration along the interface in the absence of flow; $\gamma_0 = \gamma_s - RT\Gamma_0$.

Surfactants have two independent effects on free-boundary motions. First, surfactants modify the mean interfacial tension, which alters the normal stress jump across the interface. A capillary number based upon a mean value of the interfacial tension $\mathbf{C}_0 = Ga\mu/\gamma_0$ is thus indicative of the degree of deformation caused by an external flow. Second, Marangoni stresses affect the fluid motion and produce a tangential stress jump. The dimensionless stress balance accounting for these two influences of surfactants is (we have chosen the characteristic velocity $u_c = Ga$)

$$\mathbf{n} \cdot \mathbf{T} - \lambda \mathbf{n} \cdot \hat{\mathbf{T}} = \frac{\gamma(\Gamma)/\gamma_0}{\mathbf{C}_0}(\nabla_s \cdot \mathbf{n})\mathbf{n} - \frac{\mathbf{B}_0}{\mathbf{C}_0}\hat{\mathbf{g}} \cdot \mathbf{x}\mathbf{n} + \frac{\beta}{\mathbf{C}_0 \gamma_0/\gamma_s}\nabla_s(\Gamma/\Gamma_0) \quad \text{on } S,$$

(7)

where $\beta = -(d\gamma/d\Gamma)(\Gamma_0/\gamma_s)$ provides a dimensionless measure of interfacial tension variations produced by the surfactant. For the familiar two-dimensional gas law, $\beta = RT\Gamma_0/\gamma_s$. Here, $\mathbf{B}_0$ is the Bond number based upon γ_0 (see Equation 2).

Predicting the degree of drop deformation in shear flows requires coupling the Stokes equations with equations describing surfactant transport and the surface constitutive relation. Surfactant transport along the fluid-fluid interface is governed by a convective-diffusion equation (for a straightforward derivation see Stone 1990)

$$\frac{\partial(\Gamma/\Gamma_0)}{\partial t} + \nabla_s \cdot \left[\frac{\Gamma}{\Gamma_0} \mathbf{u}_s - \frac{1}{C_0 \delta} \nabla_s \frac{\Gamma}{\Gamma_0} \right] + \frac{\Gamma}{\Gamma_0} (\nabla_s \cdot \mathbf{n})(\mathbf{u} \cdot \mathbf{n}) = j_n, \tag{8}$$

where $\mathbf{u}_s$ is the velocity vector directed tangent to the surface, j_n is the dimensionless net flux of surface-active material to and from the surface, $\delta = \gamma_0 a/(\mu D_s)$, and D_s is the surface diffusivity. The surface Péclet number is $C_0 \delta$.

Drop deformation now depends on C_0, λ, β (i.e. the constitutive equation), and δ. From a theoretical viewpoint, not only must a free-boundary problem be considered, but the presence of surfactants complicates the calculation further, since their distribution is coupled to the drop shape, which, of course, depends upon the detailed surfactant distribution. In the simplest case the surfactant is treated as insoluble ($j_n = 0$), which is a useful approximation when most of the surfactant resides along the interface (as expected at low concentrations for cases such that the partition coefficient favors surface adsorption). The study of insoluble surfactant and the analogous clean interface problem thus provide bounds for the possible interface distortions. If surfactant resides in the bulk fluids, then additional transport equations are required and boundary conditions are needed relating the surface and bulk concentrations (e.g. Stebe et al 1991).

The qualitative deformation behavior of drops in extensional flows can be understood by accounting for (a) *convection* of surfactant toward stagnation points at the end of the drop—this tends to lower interfacial tension, hence tends to increase the observed deformation; (b) *dilution* of surfactant owing to increased interfacial area which accompanies drop deformation—this tends to increase the interfacial tension, hence acts to decrease the observed deformation; and (c) additional *adsorption* and *desorption* of surfactant to and from the surface owing to fluxes from the bulk phases as deformation increases interfacial area.

Items (a) and (b) are discussed by Stone & Leal (1990). Because surface Péclet numbers are typically large (small surface diffusivities), convective transport effects dominate the surfactant distribution so deformation is enhanced over the uniform surfactant, or clean interface, case. Also, items (b) and (c) tend to offset one another. Nevertheless, if the time-scale characteristic of adsorption k_a is long compared with both G^{-1} and $(1 + \lambda)\mu a/\gamma_0$, it may be expected that dilution effects play a more prominent role, tending to inhibit drop deformation.

SMALL DEFORMATION ANALYSIS   An analysis of nearly spherical drop shapes ($C_0 \ll 1$) was performed by Flumerfelt (1980), who incorporated interfacial shear and dilatational viscosities, and an accompanying experimental study was presented by Phillips et al (1980). Stone & Leal (1990) also considered the near-sphere limit analytically as part of a numerical investigation of finite drop deformation. Assuming that the deformation is small and the concentration of surfactant is nearly uniform (low surface Péclet numbers), the leading order deformation of the drop D may be calculated as (Stone & Leal 1990)

$$D \approx \frac{3C_0 b_r}{4 + C_0 b_r} \quad \text{where} \quad b_r = \frac{5}{4} \frac{(16+19\lambda)+4\beta\delta/(1-\beta)}{10(1+\lambda)+2\beta\delta/(1-\beta)}. \tag{9}$$

This equation reduces to Taylor's (1932) well-known result in the limits $\beta \to 0$ or $\delta \to 0$, which correspond, respectively, to no effect of surfactant on the interfacial tension or a uniform distribution of surfactant as convective effects are weak.

Drop deformation increases as the magnitude of b_r increases, which occurs at a fixed dimensionless shear rate C_0 if δ increases or β increases. In the small deformation analysis, the effect of surfactant convection appears via the combined parameter $\beta\delta/(1-\beta)$, which is proportional to the magnitude of the interfacial tension variations. The flow convects surfactant toward the end of the drop, resulting in a higher surfactant concentration and lower interfacial tension, thus requiring larger deformations to balance viscous forces.

NUMERICAL STUDIES   Stone & Leal (1990) present a numerical study of surfactant effects on finite drop deformation in extensional flows. In particular, for the case of the insoluble surfactant it is demonstrated that dilution of the local surface concentration, owing to increases in interfacial area, may produce situations where the interfacial tension is actually higher than its original value and consequently further deformation, leading to breakup, tends to be inhibited. Also, the critical capillary number for breakup $(C_0)_c$ is calculated as a function of the two dimensionless parameters, β and δ. Milliken et al (1993) have extended this study to examine the manner in which surfactants alter the transient motion of elongating drops and the relaxation of initially stretched drops. Generally, Marangoni stresses reduce the interfacial velocity and cause the drop to behave as if it has a higher viscosity. Surfactant effects are more pronounced at smaller viscosity ratios, $\lambda < 1$, generally producing drop shapes that have higher curvatures than in the absence of surfactants. This behavior is suggestive of the tip streaming phenomenon (tip streaming, however, was not observed in the numerical simulations), which is now discussed in detail.

Tip Streaming

Several experiments (Taylor 1934, Grace 1971) show that a low viscosity ratio drop, typically $\lambda < O(0.1)$, may establish a steady deformation with a shape that has nearly pointed ends, yet very small drops are ejected from the ends. The ejected drops typically have radii less than 50 μm. This behavior is called "tip streaming." Because of the generation of large numbers of very small droplets, it is of interest to understand what factors control this phenomenon. The tip streaming phenomenon was unresolved at the time of the earlier reviews of Acrivos (1983) and Rallison (1984).

Recently, de Bruijn (1989, 1993), via a comprehensive and careful set of experiments, has shed light on this subject. A convincing qualitative explanation is presented describing tip streaming as the result of surfactants being swept toward the narrow ends of the drop, so that at high enough surface concentrations the ends become rigid and viscous stresses tear the narrow end regions away. A corresponding theoretical description or numerical simulation is still lacking however.

De Bruijn (1989) documents most of the experimental references to tip streaming. An important observation is that this behavior only occurs for $\lambda < 0.1$. Also, tip streaming occurs for capillary numbers lower than C_c and when surfactant is purposefully added to the system, tip streaming is generally (though not always) observed.

The main conclusions to be drawn from de Bruijn's experimental study are:

1. Contrary to studies reporting that tip streaming stops due to a decrease in drop volume, the actual volume changes are too small for this to be true. It appears that tip streaming stops when most of the surfactant on the interface is swept away.
2. Controlled addition of surfactant leads to tip streaming for $\lambda < O(0.1)$ provided the surfactant concentration is large enough. However, surface impurity concentrations that are too high also inhibit tip streaming and the drop breaks by the usual mode, presumably because the interface is uniformly covered.
3. Inferred interfacial tensions of the small tip droplets (determined using a small deformation experiment) are lower, often much lower, than the interfacial tension of the original mother drop. The corresponding surfactant concentrations are close to the saturation value.

As a final note, Milliken & Leal (1991) observed tip streaming in an experimental study of the deformation of viscoelastic drops in extensional flows. Viscoelastic materials are often made by dissolving a small amount of a polymer in water or some other solvent. The polymers modify the bulk rheology, but may also exhibit a tendency to accumulate along the interface due to the different chemical interactions of the polymer with the two bulk fluids. In Milliken & Leal's experiments with viscoelastic drops, tip streaming was observed when the Deborah number was larger than one and the viscosity ratio was small (the Deborah number is the ratio of the fluid relaxation time to the flow time scale). Hence, it appears likely that the tip streaming may be a consequence of the polymer acting as a surfactant rather than being caused by some modification of the bulk rheology (Milliken et al 1993).

6.   INSTABILITY OF TRANSLATING DROPS

We have discussed in detail deformation in flows where the drops have been assumed to be neutrally buoyant ($B = 0$). Buoyancy-driven translation of spherical drops at low Reynolds numbers has been studied extensively. In this section we summarize studies, which have been reported only recently, of the stability and flow-induced deformation of nonspherical drop shapes.

When a spherical drop of one fluid undergoes low Reynolds number buoyancy-driven translation through a second fluid, the translational velocity is higher than that of the equivalent rigid sphere by the factor $3(1+\lambda)/(2+3\lambda)$. The Hadamard-Rybczyński flow field is derived using the Stokes equations to describe the velocity field in both fluid phases, along with the boundary conditions of continuity of velocity and tangential stress. The solution is remarkable in that the normal stress boundary condition is satisfied identically, as the pressure and viscous forces are in exact balance everywhere along the interface, and this result is independent of the magnitude of the interfacial tension (including zero interfacial tension). Hence, a spherical shape is an exact steady solution to the low Reynolds number drop translation problem (Batchelor 1967).

Translation in Unbounded Fluids

The shape evolution of nonspherical drops in simple translation was first examined by Kojima et al (1984). Using a linear stability analysis, these authors show that (*a*) in the absence of interfacial tension the spherical shape is *unstable* to infinitesimal disturbances and (*b*) at finite Bond numbers the drop will recover a steady spherical shape provided that the initial nonspherical distortion is not too large (the Bond number and capillary number are equivalent for buoyancy-driven flows). When an instability is predicted to occur, the analysis demonstrates that perturbations near the front stagnation point are diminished in amplitude due to the local elongational flow, thus returning the front of the drop to a spherical shape. Perturbations near the rear stagnation point are amplified in a manner which suggests (*i*) the formation of a tail for initially prolate distortions and (*ii*) the formation of a dimple, or cavity, for initially oblate distortions. The question of whether these conclusions carry over to the case of finite initial deformation has been examined in detail via numerical boundary integral calculations by Koh & Leal (1989) and Pozrikidis (1990).

Koh & Leal (1989) present an investigation of the critical initial deformation beyond which no steady spherical shape is possible and thus the drop undergoes continuous distortion. The axisymmetric initial shapes are chosen to be either oblate or prolate ellipsoids of revolution. A companion experimental study (Koh & Leal 1990) shows that the numerical results are in excellent agreement with laboratory observations which allows the further conclusion that provided global shape information is given, then the evolution of the interface shape can be predicted. Pozrikidis (1990) presents a careful study of the limit of vanishing interfacial tension; very large distortions typified by either long tails or large cavities are common in this limit. Numerical simulations showing typical shape evolutions characteristic of different Bond numbers and viscosity ratios are shown in Figure 10 for sedimenting drops.

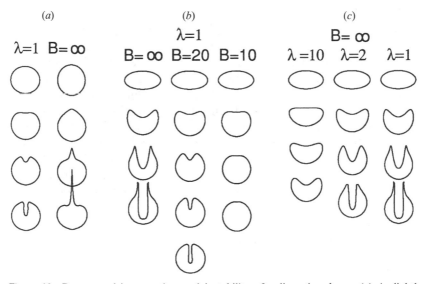

Figure 10   Buoyancy-driven motion and instability of sedimenting drops. (*a*) A slightly nonspherical oblate (prolate) initial shape translates and evolves a cavity (tail). The interfacial tension is zero. (*b*) The effect of interfacial tension on the stability of an initially oblate drop shape. For sufficiently strong interfacial tension (B $\lesssim$ 10) the drop recovers a spherical shape, while drops with lower interfacial tensions develop cavities. (*c*) The effect of viscosity ratio on the evolution of oblate initial shapes. (Numerical computations performed by M. Manga.)

Overall, we see a connection with drop deformation in extensional flows: For each nonzero Bond number, there is a critical initial distortion such that the drop is unable to recover the spherical shape, and so deforms continuously (this defines a critical Bond number). In particular, sufficiently distorted prolate shapes evolve into drops with nearly spherical fronts and long narrow tails. Initially oblate shapes develop large cavities at the back with the internal circulatory flow causing the cavity to continually increase in depth until it approaches the front interface; a nearly steady ring shape, or torus, is thus formed. In the experiments of Koh & Leal (1990), oblate drops that evolve cavities eventually close upon themselves, encapsulating the continuous phase fluid, and form double emulsion drop. Qualitatively the viscosity ratio has only a small influence on the evolution (and stability) of prolate shapes. However, for oblate shapes that evolve cavities, the higher the viscosity ratio (thus higher internal viscous stresses), the lower the critical Bond number for instability.

In the original experiments of Kojima et al (1984) the drops develop cavities and rapidly evolve into a toroidal shape with the radius of the torus increasing with time. An analysis incorporating inertial effects was proposed to explain this observation.

Other Driving Forces

Thermocapillary migration of drops in a temperature gradient occurs owing to the temperature dependence of the interfacial tension. For this flow situation, Ascoli & Lagnado (1992) use a linear stability analysis to show that a spherical drop shape may also be unstable to infinitesimal perturbations even if the interfacial tension is finite. The instability is a consequence of the interfacial tension gradient that accompanies the nonuniform temperature field. For properties characteristic of air bubbles in silicon fluids, the temperature gradients required to produce instability of centimeter-sized bubbles is estimated to be $200°C$ cm^{-1}, a rather unrealistic value. The effect of finite distortions, which requires a numerical solution, has not been examined, so may yet show that moderate temperature gradients can produce instability of finitely distorted drops. The corresponding problem of translation of dielectric fluid drops in nonuniform electric fields has not been examined.

7. CONCLUSIONS

The subject of drop deformation in viscous flows has found recent application in a large number of diverse disciplines (Table 1) spanning a wide range of length scales. A common theme of research during the past ten years has been the understanding of (a) transients leading to drop fragmentation and (b) drop deformation for systems containing sur-factants or characterized by more complex surface rheologies.

Future research directions are difficult to anticipate. One important and challenging problem which awaits an adequate solution is to incorporate the basic elements of the research described here into a model of a viscous multiphase flow containing a large number of dispersed droplets, with the goal to predict accurately the drop size distribution. A first step in this direction was taken by Tjahjadi & Ottino (1991)—see Section 4. The incorporation of microscale dynamics into a global, or large scale, model of a dispersed multiphase flow is now appropriate, not the least because many of the research problems described in the drop breakup literature are motivated at least in part by industrial processes. A useful model should include the possibility of coalescence of two drops in the presence of shear, a problem which is not well understood at this time. A second area where drop deformation and breakup studies may prove useful is the quantitative analysis of the motion and deformation of drops in complex media, such as fibrous beds or other porous materials. In this case the drop experiences a time-dependent shear and extensional motion as it passes through the pore space (this problem area was described to the author by E. S. G. Shaqfeh). A third area that deserves more attention, notably because many practical problems involve polymeric dispersed materials, includes both the response of viscoelastic drops to shear and an improved understanding of how the dynamics of breakup are affected by a viscoelastic continuous phase.

ACKNOWLEDGMENTS

I would like to thank E. J. Hinch, L. G. Leal, M. Manga, J. M. Ottino, J. M. Rallison, J. D. Sherwood, and M. Tjahjadi for providing a critical reading of the original version of this paper. Their penetrating remarks and suggestions led to a large number of improvements in the final version of the manuscript. Also, I am thankful to B. J. Bentley and W. J. Milliken for their collaborations on some of the work reported here. E. J. Hinch is thanked for initially describing de Bruijn's research to me and M. Manga is thanked for performing the numerical simulations shown in Figure 10. Finally, I would like to thank the National Science Foundation for their support of this project and the editors of Annual Reviews for their patience.

Literature Cited

Acrivos, A. 1983. The breakup of small drops and bubbles in shear flows. 4th Int Conf. on Physicochemical Hydrodynamics, Ann. N. Y. Acad. Sci. 404: 1–11

Acrivos, A., Lo, T. S. 1978. Deformation and breakup of a single slender drop in an extensional flow. J. Fluid Mech. 86: 641–72

Ascoli, E. P., Lagnado, R. R. 1992. The linear stability of a spherical drop migrating in a vertical temperature gradient. *Phys. Fluids A* 4: 225–33

Barthès-Biesel, D., Acrivos, A. 1973. Deformation and burst of a liquid droplet freely suspended in a linear shear field. *J. Fluid Mech.* 61: 1–21

Batchelor, G. K. 1967. *An Introduction to Fluid Dynamics.* Cambridge: Cambridge Univ. Press. 615 pp.

Bentley, B. J., Leal, L. G. 1986a. A computer-controlled four-roll mill for investigations of particle and drop dynamics in two-dimensional linear shear flows. *J. Fluid Mech.* 167: 219–40

Bentley, B. J., Leal, L. G. 1986b. An experimental investigation of drop deformation and breakup in steady two-dimensional linear flows. *J. Fluid Mech.* 167: 241–83

Bogy, D. B. 1979. Drop formation in a circular liquid jet. *Annu. Rev. Fluid Mech.* 11: 207–28

Buckmaster, J. D. 1972. Pointed bubbles in slow viscous flow. *J. Fluid Mech.* 55: 385–400

Carriere, C. J., Cohen, A., Arends, C. B. 1989. Estimation of interfacial tension using shape evolution of short fibers. *J. Rheol.* 33: 681–89

de Bruijn, R. A. 1989. *Deformation and breakup of drops in simple shear flows.* PhD thesis. Tech. Univ. Eindhoven. 274 pp.

de Bruijn, R. A. 1993. Tipstreaming of drops in simple shear flows. *Chem. Eng. Sci.* 48: 277–84

Döring, M. 1982. Ink-jet printing. *Philips Tech. Rev.* 40: 192–98

Edwards, D. A., Brenner, H., Wasan, D. T. 1991. *Interfacial Transport Processes and Rheology.* Boston: Butterworth-Heinemann. 558 pp.

Elemans, P. H. M. 1989. *Modelling of the processing of incompatible polymer blends.* PhD thesis. Tech. Univ. Eindhoven. 207 pp.

Elemans, P. H. M., Janssen, J. M. H., Meijer, H. E. H. 1990. The measurement of interfacial tension in polymer/polymer systems: the breaking thread method. *J. Rheol.* 34: 1311–25

Flumerfelt, R. W. 1980. Effects of dynamic interfacial properties on drop deformation and orientation in shear and extensional flow fields. *J. Colloid Interface Sci.* 76: 330–49

Giesekus, H. 1962. Strömungen mit konstantem Geschwindigkeitsgradienten und die Bewegung von darin suspendierten Teilchen. Teil II: Ebene Strömungen und eine experiementelle Anordnung zu ihrer Realisierung. *Rheol. Acta* 2: 113–21

Grace, H. P. 1971. Dispersion phenomena in high viscosity immiscible fluid systems and application of static mixers as dispersion devices in such systems. *Eng. Found., Res. Conf. Mixing, 3rd, Andover, N. H.* Republished in 1982 in *Chem. Eng. Commun.* 14: 225–77

Greenspan, H. P. 1978. On fluid-mechanical simulations of cell division and movement. *J. Theoret. Biol.* 70: 125–34

Hakimi, F. S., Schowalter, W. R. 1980. The effects of shear and vorticity on deformation of a drop. *J. Fluid Mech.* 98: 635–45

Han, C. D. 1981. *Multiphase Flow in Polymer Processing.* New York: Academic

Hinch, E. J., Acrivos, A. 1979. Steady long slender droplets in two-dimensional straining motion. *J. Fluid Mech.* 91: 401–14

Jeong, J.-T., Moffatt, H. K. 1992. Free-surface cusps associated with flow at low Reynolds number. *J. Fluid Mech.* 241: 1–22

Joseph, D. D., Nelson, J., Renardy, M., Renardy, Y. 1991. Two-dimensional cusped interfaces. *J. Fluid Mech.* 223: 383–409

Khakhar, D. V., Ottino, J. M. 1986. A note on the linear vector model of Olbricht, Rallison, and Leal as applied to the breakup of slender axisymmetric drops. *J. Non-Newtonian Fluid Mech.* 21: 121–27

Koh, C. J., Leal, L. G. 1989. The stability of drop shapes for translation at zero Reynolds number through a quiescent fluid. *Phys. Fluids A* 1: 1309–13

Koh, C. J., Leal, L. G. 1990. An experimental investigation on the stability of viscous drops translating through a quiescent fluid. *Phys. Fluids A* 2: 2103–9

Kojima, M., Hinch, E. J., Acrivos, A. 1984. The formation and expansion of a toroidal drop moving in a viscous fluid. *Phys. Fluids* 27: 19–32

Kuiken, H. K. 1990. Viscous sintering: the surface-tension-driven flow of a liquid form under the influence of curvature gradients at its surface. *J. Fluid Mech.* 214: 503–15

Levich, V. G., Krylov, V. S. 1969. Surface-tension-driven phenomena. *Annu. Rev. Fluid Mech.* 1: 293–316

Li, X. Z., Barthès-Biesel, D., Helmy, A. 1988. Large deformations and burst of a capsule freely suspended in an elongational flow. *J. Fluid Mech.* 187: 179–96

Meijer, H. E. H., Janssen, J. M. H. 1993. Mixing of immiscible liquids. In *Mixing and Compounding—Theory and Practice,* ed. I. Mana-Zloczower, Z. Tadmor, Progress in Polymer Processing Ser. Munich: Carl Hanser

Mikami, T., Cox, R. G., Mason, S. G. 1975. Breakup of extending liquid threads. *Int. J. Multiphase Flow* 2: 113–38

Milliken, W. J., Leal, L. G. 1991. Deformation and breakup of viscoelastic drops in planar extensional flows. *J. Non-Newtonian Fluid Mech.* 40: 355–79

Milliken, W. J., Stone, H. A., Leal, L. G. 1993. The effect of surfactant on the transient motion of Newtonian drops. *Phys. Fluids* 5: 69–79

Muzzio, F. J., Tjahjadi, M., Ottino, J. M. 1991. Self-similar drop-size distributions produced by breakup in chaotic flows. *Phys. Rev. Lett.* 67: 54–7

Olbricht, W. L., Rallison, J. M., Leal, L. G. 1982. Strong flow criteria based on microstructure deformation. *J. Non-Newtonian Fluid Mech.* 10: 291–318

Ottino, J. M. 1990. *The Kinematics of Mixing: Stretching, Chaos and Transport.* Cambridge: Cambridge Univ. Press. 364 pp.

Phillips, W. J., Graves, R. W., Flumerfelt, R. W. 1980. Experimental studies of drop dynamics in shear fields: role of dynamic interfacial effects. *J. Colloid Interface Sci.* 76: 350–70

Pozrikidis, C. 1990. The instability of a moving viscous drop. *J. Fluid Mech.* 210: 1 21

Pozrikidis, C. 1992. *Boundary Integral and Singularity Methods for Linearized Viscous Flow.* Cambridge: Cambridge Univ. Press. 259 pp.

Rallison, J. M. 1980. Note on the time-dependent deformation of a viscous drop which is almost spherical. *J. Fluid Mech.* 98: 625–33

Rallison, J. M. 1981. A numerical study of the deformation and burst of a viscous drop in general shear flows. *J. Fluid Mech.* 109: 465–82

Rallison, J. M. 1984. The deformation of small viscous drops and bubbles in shear flows. *Annu. Rev. Fluid Mech.* 16: 45–66

Rumscheidt, F. D., Mason, S. G. 1961. Particle motions in sheared suspensions. XII. Deformation and burst of fluid drops in shear and hyperbolic flow. *J. Colloid Sci.* 16: 238–61

Seward III, T. P. 1974. Elongation and spheroidization of phase-separated particles in glass. *J. Non-Crystalline Solids* 15: 487–504

Sherwood, J. D. 1981. Spindle-shaped drops in a viscous extensional flow. *Math. Proc. Cambridge Philos. Soc.* 90: 529–36

Sherwood, J. D. 1984. Tip streaming from slender drops in a nonlinear extensional flow. *J. Fluid Mech.* 144: 281–95

Skalak, R., Özkaya, N., Skalak, T. C. 1989. Biofluid mechanics. *Annu. Rev. Fluid Mech.* 21: 167–204

Spence, D. A., Ockenden, J. R., Wilmott, P., Turcotte, D. L., Kellogg, L. H., 1988. Convective mixing in the mantle: the role of viscosity differences. *Geophys. J.* 95: 79–86

Stebe, K. J., Lin, S.-Y., Maldarelli, C. 1991. Remobilizing surfactant retarded fluid particle interfaces. I. Stress-free conditions at the interfaces of micellar solutions of surfactants with fast sorption kinetics. *Phys. Fluids* 3: 3–20

Stone, H. A. 1990. A simple derivation of the time-dependent convective-diffusion equation for surfactant transport along a deforming interface. *Phys. Fluids A* 2: 111–12

Stone, H. A., Bentley, B. J., Leal, L. G. 1986. An experimental study of transient effects in the breakup of viscous drops. *J. Fluid Mech.* 173: 131–58

Stone, H. A., Leal, L. G. 1989a. Relaxation and breakup of an initially extended drop in an otherwise quiescent fluid. *J. Fluid Mech.* 198: 399–427

Stone, H. A., Leal, L. G. 1989b. The influence of initial deformation on drop breakup in subcritical time-dependent flows at low Reynolds numbers. *J. Fluid Mech.* 206: 223–63

Stone, H. A., Leal, L. G. 1990. The effects of surfactants on drop deformation and breakup. *J. Fluid Mech.* 220: 161–86

Szeri, A. J., Wiggins, S., Leal, L. G. 1991. On the dynamics of suspended microstructure in unsteady, spatially inhomogeneous, two-dimensional fluid flows. *J. Fluid Mech.* 228: 207–41

Taylor, G. I. 1932. The viscosity of a fluid containing small drops of another fluid. *Proc. R. Soc. London Ser. A* 138: 41–48

Taylor, G. I. 1934. The formation of emulsions in definable fields of flow. *Proc. R. Soc. London Ser. A* 146: 501–23

Taylor, G. I. 1964. Conical free surfaces and fluid interfaces. *Proc. Int. Congr. Appl. Mech., 11th, Munich*, pp. 790–96

Tjahjadi, M., Ottino, J. M. 1991. Stretching and breakup of droplets in chaotic flows. *J. Fluid Mech.* 232: 191–219

Tjahjadi, M., Stone, H. A., Ottino, J. M. 1992. Satellite and subsatellite formation in capillary breakup. *J. Fluid Mech.* 243: 297–317

Tjahjadi, M., Stone, H. A., Ottino, J. M. 1993. Estimating interfacial tension via relaxation of moderately extended droplets and breakup of highly extended fluid filaments. *AIChE J.* In press

Tomotika, S. 1935. On the stability of a cylindrical thread of a viscous liquid surrounded by another viscous fluid. *Proc. R. Soc. London Ser. A* 150: 322–37

Wilmott, P. 1989a. The stretching of a thin viscous inclusion and the drawing of glass sheets. *Phys. Fluids A* 1: 1098–103

Wilmott, P. 1989b. The stretching of a slender, axisymmetric, viscous inclusion—Part I: Asymptotic analysis. *SIAM J. Appl. Math.* 49: 1608–16

Zinemanas, D., Nir, A. 1988. On the viscous deformation of biological cells under anisotropic surface tension. *J. Fluid Mech.* 193: 217–41

Annu. Rev. Fluid Mech. 1994. 26 : 103–36

WAVE EVOLUTION ON A FALLING FILM

Hsueh-Chia Chang

Department of Chemical Engineering, University of Notre Dame, Notre Dame, Indiana 46556

KEY WORDS: hydrodynamic instability, bifurcation, solitary waves, interfacial turbulence, coherent structures

INTRODUCTION

Since the pioneering experiment by the father-son team of the Kapitza family during their house arrest in the late forties (Kapitza & Kapitza 1949), wave evolution on a falling film has intrigued many researchers. One of its main attractions is its simplicity—it is an open-flow hydrodynamic instability that occurs at very low flow rates. It can hence be studied with the simplest experimental apparatus, an obviously important factor for the Kapitzas. Yet, it yields a rich spectrum of fascinating wave dynamics, including a very unique and experimentally well-characterized sequence of nonlinear secondary transitions that begins with a selected monochromatic disturbance and leads eventually to nonstationary and broad-banded (in both frequency and wave number) "turbulent" wave dynamics. (Turbulence here is used interchangeably with irregular spatio-temporal fluctuations.) While this transition to "interfacial turbulence" or "spatio-temporal chaos" seems to be quite analogous to other classical instabilities at first glance, there are subtle but important differences that have recently come to light. The pertinent nonlinear mechanisms behind these secondary transitions are the focus of the present review.

We shall be mostly concerned with transitions on a free-falling vertical film. Wave dynamics on an inclined plane is quite analogous to the vertical limit and most experiments and theories have focused on the latter. For the vertical film, the problem is defined by two independent dimensionless parameters and we prefer the Russian convention of using the Reynolds

103

number $R = \langle u \rangle h_N / v$ and the Kapitza number $\gamma = \sigma / \rho v^{4/3} g^{1/3}$ where $\langle u \rangle$ is the average velocity, h_N is the Nusselt flat-film film thickness such that $h_N \langle u \rangle$ is the flow rate per unit width and $\langle u \rangle = g h_N^2 / 3v$, v the kinematic viscosity, ρ the density, σ the interfacial tension and g the gravitational acceleration. If one uses the "natural" scalings on the full Navier-Stokes equation, the resulting two parameters would be R and the Weber number $W = \sigma / \rho \langle u \rangle^2 h_N$ which are the ones used in some literature. However, the Kapitza number offers the advantage that it is only a function of the physical properties of the liquid and not the flow rate. (For water, $\gamma = 2850$ at 15°C.) The two conventions are interchangeable by $\gamma = W R^{5/3} / 3^{1/3}$.

At very large Reynolds numbers ($R > 1000$), the waves observed on the falling film are of the shear-wave variety with wavelengths comparable to or shorter than h_N. [See the linear stability analysis of Floryan et al (1987) and the experiment of Bertschy et al (1983) for inclined films.] Such high-flow-rate conditions typically yield turbulent films (turbulent in the classical sense) dominated by internal Tollmien-Schlichting disturbances. The interfacial dynamics is simply enslaved by the internal turbulence. At moderately high R ($300 < R < 1000$), long interfacial waves characteristic of gravity-capillary instabilities begin to appear (Chu & Dukler 1974). However, the wave dynamics is extremely nonstationary, especially for the persisting short waves which seem to be generated by a vortex shedding mechanism from the long waves. At extremely low flow rates ($R \ll 1$), the film becomes so thin that intermolecular forces and contact line dynamics become important as the film ruptures. We are concerned with the intermediate region ($1 < R < 300$) such that the instability consists of long interfacial waves dominated by gravity-capillary effects.

Wave evolution by natural excitation in this region of Reynolds numbers is shown in the schematic of Figure 1. Four distinct wave regions have been observed. In the inception region (region I), infinitesimal disturbances at the inlet are amplified downstream to form a monochromatic wave at the end of the region, indicating that the instability is a convective one and not of the absolute variety. If the initial disturbance is sufficiently monochromatic in frequency, the emerging wave inherits the forcing frequency. If the disturbance has a wide band of frequency, as is true with natural noise, a highly selective linear filtering process in region I yields a unique monochromatic wave field for all wide-band disturbances. In particular, transverse disturbances are selectively damped in this inception region. Within this region, the amplitude of the monochromatic wave grows exponentially downstream as in all linear excitation processes of convectively unstable systems. Beginning in region II, however, the exponential growth is arrested by weakly nonlinear effects as the amplitude of the monochromatic wave saturates to a finite value dependent on the wave

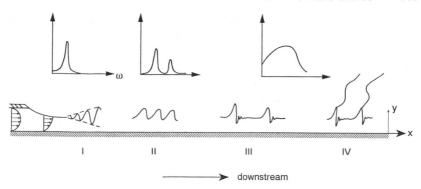

Figure 1 Schematic of the four wave regimes in a naturally excited evolution on a falling film from a slit. The wave spectra from localized probes at the four regions are also shown. The parameter ω is the wave frequency. For higher R values ($R > 50$), region III may not be present.

number, R, and γ. Due to this weakly nonlinear interaction between an unstable fundamental Fourier mode and a stable second harmonic, the monochromatic wave of region I begins to develop a finite overtone in region II as its sinusoidal shape steepens downstream. There is also a negative nonlinear correction to the wave speed of the inception region such that the waves actually slow down as they grow (Lin 1983). Some experimental evidence indicates that even the wave number of the saturated periodic wave exiting from region II, which is still a very uniform wave field, is different from that of the monochromatic wave emerging from region I due to a nonlinear selection mechanism. In a periodically forced experiment with a sufficiently large forcing amplitude, both region I and region II may be bypassed and the first uniform wave field that emerges contains large-amplitude waves whose wave frequencies are close to the forcing frequency except at very low values (Kapitza & Kapitza 1949, Alekseenko et al 1985). The periodic finite-amplitude disturbances in a forced experiment can hence entrain waves that would not have been selected by the linear and nonlinear selection mechanisms of regions I and II. For small-amplitude and broad-banded natural or artificially introduced disturbances, however, the uniform wave fields emerging from regions I and II are independent of the disturbances present. They are uniquely selected by the linear and nonlinear mechanisms in regions I and II.

Both finite-amplitude wave fields generated by periodic forcing and naturally excited wave fields emerging from region II travel a long distance (~ 10 wavelengths) in a stationary manner, e.g. without visible changes in their shape or speed, before they undergo another slow evolution in region III. Here, two dominant instabilities of the finite-amplitude waves have been observed. The best data are recorded by Liu & Gollub (1993) although they are taken for inclined films. Neighboring waves coalesce at intermittent locations due to a subharmonic instability (Prokopiou et al 1991) or a long-wave modulation appears characteristic of sideband instabilities (Cheng & Chang 1992a). These two instabilities create intermittent patches of defects to the otherwise uniform field of waves. Within these patches, the distorted waves grow in wavelength, amplitude, and speed and evolve into characteristic spatially localized teardrop humps. These humps have steep fronts which are relaxed by a series of front-running bow waves whose wavelength is close to the monochromatic waves at inception. These larger and faster humps begin to expand the patches in the downstream direction as they overtake the original slower waves, so much so that all of the latter may vanish eventually. The wave frequency or wave number spectrum now becomes very broad, in contrast to the monochromatic spectrum in region I and the fundamental-overtone pair in region II. This, however, does not imply that a continuous band of dispersive waves dominate the interface as in turbulent channel flow. Instead, a large portion of the band is locked into the characteristic hump of the evolved waves and travels in synchrony. The broad-banded spectrum is due mostly to the localized shape of the humps which is strikingly identical. These robust humps have a characteristic length that is much shorter than the typical separation between humps where the film is essentially flat. They are hence referred to as *solitary waves*, or more appropriately, *solitary humps*. The separation between the individual humps are typically nonuniform and time-varying, indicating nonstationary interaction among them and reflecting the spatio-temporal irregularity of their births. However, the humps themselves remain nearly identical and do not alter their shape significantly during the interaction within region III.

Finally, in region IV, transverse variation begins to develop on the wave crests of the solitary humps. The dynamics of the transverse variation is nonstationary and these transverse variations grow to such amplitude (not in height but in the direction parallel to the wall) that adjacent crests merge at various points and pinch off. However, the wave shape in the flow direction (x in Figure 1) retains the solitary shape except near the pinch points. For vertical water films at low R, regions I to III occupy 30 to 40 inception wavelengths with each region spanning about 10 wavelengths. (The wavelength selected at inception is about 1 cm for water at the Reynolds number of interest.) Region IV seems to persist indefinitely downstream. If the introduced disturbance at the feed, or elsewhere in the channel, contains significant transverse variation such that it is not entirely filtered in the inception region I, then region III may be negligible in length or may be skipped entirely.

MODEL EQUATIONS AND WAVE INCEPTION

Although the wave transition sequence shown in Figure 1 seems to be quite similar to transitions in other classical hydrodynamic instabilities like free-convection boundary layer and shear instabilities in channel flow, there are several important factors that make this instability unique and also render it more amenable to analysis. It could well be the first transition sequence to be fully understood. The first factor is the difference in the streamwise and normal length scales. In all regions in Figure 1 and for $R < 300$, the film height at the wave crest never exceeds $1/10$ of the characteristic wavelength or the characteristic length of the solitary humps. At low R, this long-wave characteristic of the interfacial disturbances motivates a lubrication-type expansion with respect to the "film parameter" ε defined as the ratio between the characteristic wave height and the characteristic wavelength. This long-wave expansion yields the following evolution equation

$$\frac{\partial h}{\partial t} + 3h^2 h_x + \left(\frac{4}{5} Rh^6 h_x + \frac{RWh^3}{3} h_{xzz} + \frac{RWh^3}{3} h_{xxx} \right)_x$$

$$+ \frac{RW}{3} (h^3 h_{zzz} + h^3 h_{zxx})_z = 0, \quad (1)$$

where the characteristic length and time used to nondimensionalize the variables are h_N and $h_N/\langle u \rangle$, respectively. In a more popular version of this equation, the interfacial velocity is used in the time scale. This corresponds to multiplying all coefficients in (1) except the first one by a factor of $2/3$. In this other version $RW/3$ is also replaced by a different Weber number $W' = \sigma/\rho h_N^2 g$. Roskes (1970) is responsible for the $O(\varepsilon^2)$ version of (1) for strong surface-tension fluids $[W \sim O(\varepsilon^{-2})]$. Dispersion, which will be shown to be important in both the linear and nonlinear instability mechanisms, has been omitted in the equation. It was, however, included in a higher-order equation that Nakaya (1975) derived for both $W \sim O(\varepsilon^0)$ and $W \sim O(\varepsilon^{-2})$. Lower-order equations, for various relative orders of W with respect to ε, have been derived by different investigators beginning with the work of Benney (1966). All these evolution equations, however, require the Reynolds number to be $O(\varepsilon^0)$, corresponding to lubrication-flow conditions.

Further simplification of (1) is possible by carrying out an expansion in the amplitude η, where $h = 1 + \eta$. This weakly nonlinear expansion for strong surface-tension fluids was first carried out by Lin (1974) and Nepomnyaschy (1974) independently for falling-film waves and it yields the following $O(\eta^2)$ equation for two-dimensional waves without z variation, known as the Kuramoto-Sivashinsky (KS) equation:

$$H_\tau + 4HH_x + H_{xx} + H_{xxxx} = 0, \tag{2}$$

where both R and W have been conveniently absorbed by the moving-coordinate, slow-time, long-wave, and small-amplitude expansion:

$$t = \tau \bigg/ \left(\frac{48R}{25W} \right), \quad \eta = 4 \left(\frac{8R}{25W} \right) \left(\frac{5W}{12} \right)^{1/2} H,$$

and

$$x \to \left(\frac{5W}{12} \right)^{1/2} (x - 3t).$$

The parameter-independent property of the KS equation and its simplicity in retaining only the most dominant nonlinear term HH_x, which arises from interfacial kinematics in the present problem, have made it a popular generic model for numerical and mathematical scrutiny.

One should, however, be cognizant of the assumptions that have been made in deriving both (1) and (2). The former assumes a small wave height to wavelength ratio ε and $R \sim O(\varepsilon^0)$ while the latter imposes the additional stipulation that the deviation amplitude $\eta = h - 1$ must be small. Note that the small η approximation is much stronger than the small-ε long-wave approximation since the wave height can be small relative to the wavelength and still be the same order or larger than the Nusselt film thickness h_N or the average film thickness. The observed waves typically obey the long-wave approximation but not necessarily the small-amplitude assumption. Both equations are also strictly valid only for strong surface-tension fluids unless higher order terms are added to (1). Nevertheless, with careful consideration concerning their validity, they can be welcomed substitutes for the full Navier-Stokes equation since the latter is still beyond exhaustive numerical analysis. (See, however, the recent work of Kheshgi & Scriven 1987 and Ho & Patera 1990. There is also an ongoing effort at MIT.)

Extensive numerical analysis of the "strongly" nonlinear equation in (1) has been carried out by Pumir et al (1983) and by the Northwestern group (Joo et al 1991; Joo & Davis 1992a,b). Pumir et al first demonstrated and Rosenau et al (1991) and Joo et al (1991) recently confirmed that finite-time blow-up and wave breaking, which violate the small ε long-wave expansion, are often encountered during the integration of (1). Since the actual film does not exhibit such behavior, the strongly nonlinear evolution equation in (1) must have omitted certain important nonlinear growth-arresting mechanisms in its long-wave expansion. On the other hand, one can easily show by the energy method that the KS equation (2), with reasonable boundary conditions, always yields bounded solutions for all time. Since the small-amplitude evolution of (1) is also described by (2), while the large-amplitude evolution of (1) quite often yields blow-up solutions that are not consistent with the long-wave expansion, the advantage of the more complex evolution equation over the KS equation is rather limited.

For $R \gg 1$, both (1) and (2) are definitely not valid and the only available simplification of the Navier-Stokes equation is offered by the boundary-layer equation (BL) first studied by Shkadov et al (1970). Unlike the derivation that leads to (1) and (2), the BL equation is derived with only the long-wave expansion without overly restrictive stipulations on the order for R, W, and the deviation wave amplitude η (Chang et al 1993a). Inertia-induced instability and dispersion are fully captured. The BL equation is

$$\frac{\partial u}{\partial t} + u\frac{\partial u}{\partial x} + v\frac{\partial u}{\partial y} = \frac{1}{5\delta}\left(h_{xxx} + h_{xzz} + \frac{1}{3}\frac{\partial^2 u}{\partial y^2} + 1\right)$$

$$\frac{\partial w}{\partial t} + u\frac{\partial w}{\partial x} + v\frac{\partial w}{\partial y} + w\frac{\partial w}{\partial z} = \frac{1}{5\delta}\left(h_{xxz} + h_{zzz} + \frac{1}{3}\frac{\partial^2 w}{\partial y^2}\right)$$

$$\frac{\partial u}{\partial x} + \frac{\partial v}{\partial y} + \frac{\partial w}{\partial z} = 0 \tag{3}$$

$$y = h(x,z) \quad h_t = v - uh_x - wh_z \quad \frac{\partial u}{\partial y} = \frac{\partial w}{\partial y} = 0$$

$$y = 0 \quad u = v = w = 0.$$

We have compared the linear stability result of (3) to the exact results of the full Orr-Sommerfeld equation and found it to be accurate for $R < 300$ for most realistic fluids (Chang et al 1993a). Unlike the KS equation which contains no explicit parameters, (3) yields a scaled Reynolds number

$$\delta = R^{11/9}/5\gamma^{1/3}3^{7/9}$$

which introduces the destabilizing and dispersive effects of inertia at higher flow rates. Nevertheless, it is still more convenient to study than the full Navier-Stokes equation which contains two parameters. At vanishing δ, it can be readily shown that (3) reduces to (1) and then to (2) if a proper scaling for δ in terms of the film parameter is assigned. An ad hoc but convenient simplification of (3) can be made by arbitrarily assuming a self-similar parabolic flow profile beneath the film (Shkadov 1967, 1968). This reduces (3) to the greatly simplified form of the "integral boundary-layer" equation (IBL) or averaged equation for two-dimensional waves,

$$\frac{\partial q}{\partial t} + \frac{6}{5}\frac{\partial}{\partial x}(q^2/h) - \frac{1}{5\delta}(hh_{xxx}+h-q/h^2) = 0$$

$$\frac{\partial h}{\partial t} + \frac{\partial q}{\partial x} = 0 \qquad (4)$$

where $q = \int_0^h u\,dy$ is the volumetric flow rate per unit span width.

In numerical simulations of (3) and (4) (Demekhin & Shkadov 1985, Trifonov & Tsvelodub 1991, Chang et al 1993a), it is found that the boundary-layer or long-wave approximation is always obeyed, e.g. the blow-up behavior of (1) is never observed and ε remains small. This is true even at low δ conditions where (1) supposedly applies. We believe higher order ε terms omitted in (1) are responsible for arresting the blow-up phenomenon. This is supported by our effort to reduce Equation (3) to Equation (1) (Chang et al 1993a). Since there are now two small parameters, the dispersion δ and the "film" parameter ε, relative order between the two must be established. It is found that $\delta \sim O(\varepsilon^{2/3})$ for the reduction to (1). This is equivalent to assigning R and W orders in ε in the derivation of (1) and (2). Since ε is not really a free parameter but one determined by the solution, the reduction essentially permits large-amplitude solutions like blow-up solutions by artificially reducing the effect of δ-related mechanisms like dispersion. Equivalently, a blow-up solution of (1) would trigger higher order ε effects such as dispersion to suppress further growth. Since these terms are not included in (1), blow-up occurs. The weakly nonlinear version of the KS equation suppresses unbounded growth by limiting itself to small-amplitude evolutions of (1) that do not trigger blow up. Since dispersion and other higher order terms in ε are included in the BL equation—the order of δ is not stipulated to be artificially small at a specific order of ε, it yields the proper description of wave evolution even at low δ. (Quantitative agreement with measured wave tracings will be presented in the next section.) It should hence be considered the model equation of choice, short of the complete Navier-Stokes equation, for long waves at $R < 300$. Equations (1) and (2) should be reserved for small-amplitude waves at low δ. In this limit, the KS equation is a far simpler equation to study than Equation (1). Although the IBL equation is derived in an ad hoc manner, it yields the correct leading-order linear (Prokopiou et al 1991) and nonlinear (Trifonov &

Tsvelodub 1991, Tsvelodub & Trifonov 1992) behavior in the low-δ limit ($\delta < 0.05$). It is hence a good substitute for the KS equation to include dispersive waves and an excellent simplification of the BL equation at low δ.

The danger in a priori assigning relative orders of R, W, and η with respect to ε in a long-wave expansion was recognized by Benney (1966). He found that his weakly nonlinear theories yield either the Burger equation, the KdV equation, or others depending on the specific assignments made. Since R and W are independent parameters that specify the wave height and wavelength in ε, these a priori assignments often yield equations that cannot describe the full range of waves. As a result, shock formation for the Burger equation and blow-up solutions for others occur when these equations are integrated as the waves attempt to evolve into ones beyond their description. One should only assume ε is small without specifying the orders of R, W, and η as in the derivation of the BL equation.

Another welcomed feature of wave evolution on a falling film at $R < 300$ is the difference in time scale between the characteristic time of wave evolution and the wave period. This is seen in the locally stationary and uniform two-dimensional periodic waves in region II where the waves are nearly identical over 10 wavelengths and a particular crest can travel the same distance without changing its shape or speed appreciably. The evolution in region III is more localized in space, especially at the beginning of the region. Nevertheless, one can still see small patches of stationary waves that span one or two wavelengths. The solitary humps of regions III and IV are also stationary although the separation between them can often be time-varying. Even here, the fluctuations can probably be modeled as local dynamics near a periodic train of stationary solitary humps. It is hence quite reasonable to construct stationary periodic waves by a Lagrangian transformation $x \rightarrow x - ct$ and study the steady-state version of any of the model equations in the Lagrangian frame. This introduces an additional parameter, the wave speed c, but converts the initial value problem into a far simpler boundary value problem. Linear stability analysis and weakly nonlinear theories for the dynamics near these stationary waves can then determine which wave will be selected in the various transitions. In a sense, these stationary waves, which include the Nusselt flat film as a stationary wave at all speeds c, are "fixed points" of the governing equations which are strictly unstable but the dynamics seem to approach them on a stable manifold and leave them rather reluctantly on an unstable manifold after a finite lifetime. The complete transition sequence then consists of evolution from one stationary-wave fixed point to another, possibly along a heteroclinic orbit. Such a scenario has also been envisioned for other instabilities. For example, stationary finite-amplitude traveling waves are speculated to be the intermediate between the primary instability and the 3D tertiary phase in the transition sequence to turbulence in plane Poiseuille flow (Pugh & Saffmann 1988). The role of stationary waves is fully confirmed experimentally, however, on a falling film. This approach has spurred some recent activities to construct stationary wave families and analyze their stability. The latest results have already offered a good understanding of the evolution in regions II and III.

The last unique feature of wave evolution on a falling film is the dominance of the robust solitary-hump structure in the dynamics of regions III and IV. This feature promises to yield a very rational and quantitative description of the chaotic dynamics in these regions unavailable for turbulence in other instabilities. The existence of identical coherent units during spatio-temporal chaos in regions III and IV motivates a "coherent structure" theory describing the weak interaction among a finite number of such indestructible units. The dimension of the strange attractor in region IV, if the dynamics there is indeed governed by an attractor, can be relatively small. This is contrasted to the noise-like turbulence of high Reynolds number shear instabilities with continuous bands of length and time scales from unsynchronized waves and astronomical attractor dimensions. In this respect, "interfacial turbulence" in region IV of a falling film may indeed be related to low-dimensional chaos.

The next three sections summarize our current understanding of the transition sequence in Figure 1 and possible future development, especially along the line of "coherent structure" theory for the dynamics of regions III and IV. The dynamics of the inception region in I is well understood and we refer the readers to the earlier review of Lin (1983) and the latest computation of Floryan et al (1987). It should be mentioned, however, that although the linear dynamics at inception is clearly a convective instability, a complete confirmation of this has not been carried out. Pierson & Whitaker (1977) have computed the spatial amplification rate by presupposing the instability is convective. Most of the other literature on linear instability, beginning with the classical long-wave results of Yih (1963) and Benjamin (1957), actually pertain to the inappropriate temporal stability problem. The only attempts to ascertain convective instability are by Joo & Davis (1992b) and Liu ct al (1992) who used the simplified equations of (1) and (2). However, their results yield a prediction of absolute instability at a sufficiently large R which has never been observed. The inconsistency was attributed to the failure of (1) and (2) in the absolutely unstable region. More precisely, it may be due to the missing high order (in ε) dispersion effect in these equations which becomes more pronounced at higher R. Since convective instability can be viewed as a competition between local growth and disturbance propagation, dispersion of wave speed should have a profound influence on it. The erroneously predicted absolute instability is perhaps the linear precursor to the aphysical nonlinear blow-up behavior of (1). Both can then be attributed to the absence of dispersion in the equation. The inconsistencies of (1) and (2) with experimental data in this respect again limit their application. The confirmation of convective instability can be easily addressed by tackling the full Orr-Sommerfeld equation for the linearized Navier-Stokes equation or the linearized BL equation and this remains a worthy open problem.

We shall only cite a few important results from the linear study pertinent to subsequent discussion on nonlinear theories. These results remain valid in general despite their temporal stability formulation. Instability for the vertical film occurs beyond $R_c = 0$ and in an analog of Squire's theorem, the two-dimensional disturbances of the form $f(y) \exp(i\alpha x + \lambda t)$ are found to be more unstable than three-dimensional disturbances. The growth rate λ_r, shown in Figure 2, is a parabolic one that encompasses the range $\alpha \in (0, \alpha_0)$ where in the limit of low flow rate $(\delta \to 0)$, the neutral wave number α_0 approaches $\sqrt{18\delta}$. Destabilization of long waves is due to gravity-driven inertia and stabilization of short waves is due to capillary effects. In the same limit of low δ, the fastest growing mode is $\alpha_m = \alpha_0/\sqrt{2} = \sqrt{9\delta}$, which seems to agree with the wave number of the naturally excited monochromatic wave emerging from region I. Again, a complete analysis of this linear filtering mechanism due to convective instability is still lacking. Also at low δ, the normalized phase velocity $c = -\lambda_i/\alpha$ is exactly 3 for all wave numbers α, e.g. all Fourier modes travel at three times the

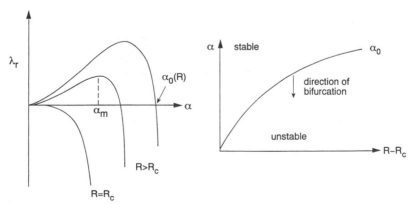

Figure 2   Schematic of the growth rate λ_r as a function of streamwise wave number α and the neutral curve for the neutral wave number α_0 as a function of Reynolds number R. The critical Reynolds number is R_c. The direction of bifurcation of the stationary periodic wave family is also indicated.

average velocity $\langle u \rangle$. With increasing δ, dispersion sets in and a mode near α_m travels slower than any other mode and is significantly lower than 3.

STATIONARY WAVES

Like other nonlinear theories for hydrodynamic instabilities, stationary waves and their stability can be studied with either a constant-flux or a constant-average thickness formulation. (These are analogous, but not identical, to constant-flux and constant-pressure-gradient formulations of other open-flow instabilities.) If a weakly nonlinear expansion about the flat-film is used, the latter formulation is usually implemented. For the full problem, however, either formulation can be used. The two results are qualitatively the same and quantitatively transformable (Chang et al 1993a).

The earliest attempts to construct stationary waves were by Benney (1966) for weak surface-tension fluids and Lin (1969) for strong ones. The constructed stationary waves are not necessarily the ones selected at the end of region II. Instead, they correspond to all possible stationary waves that are observed in a periodically forced experiment. Benney and Lin used the classical multi-scale formalism of Stuart-Watson to derive a Stuart-Landau (SL) equation

$$\frac{\partial a}{\partial t} = \lambda a - \sigma |a|^2 a \qquad (5)$$

for the complex amplitude of a monochromatic wave with wave number α, $a \exp(i\alpha x)$. The complex coefficients λ and σ are simply the linear growth rate and the Landau interaction coefficient due to cubic self-interaction and quadratic interaction between the fundamental α and the overtone 2α. In the classical formalism, λ is expanded to leading order with respect to a bifurcation parameter like R at the maximum ("nose") of the growth rate curve near criticality $\lambda \sim (R - R_c)$ and σ is evaluated at the "nose." However, the growth rate of the falling-film problem in Figure 2 does not have a nose near criticality. More accurately, its nose at $\alpha = 0$ near criticality is too complex to allow an expansion that yields (5). As a result, (5) actually corresponds to an expansion carried out at the neutral curve and not at criticality, i.e. $\lambda \sim (\alpha - \alpha_0)$ and σ is evaluated at α_0 for all values of R and W. Hence, the solution of (5) is not limited to near-critical conditions as in other SL formulations. However, the most unstable waves near α_m of Figure 2 are not well-resolved at large R since $\alpha_m \ll \alpha_0$ at higher flow rates as the band of unstable waves grow with increasing inertia. This is best clarified by a new derivation of (5) using modern Center Manifold theories (Cheng & Chang 1990). In the limit of small R (or δ), σ is a real parameter since the system is nondispersive. We also note that, by definition, λ_r vanishes at $\alpha = \alpha_0$ and is positive for $\alpha < \alpha_0$. The real part of the Landau constant σ_r is found to be positive, corresponding to a supercritical bifurcation, by all studies to date. This then implies that, at a given R and δ, there is a one-parameter family of nearly monochromatic periodic stationary waves parameterized by α with wave number decreasing continuously from α_0 (see Figure 2). The amplitude of each member of this wave family is given by $|a|^2(\alpha) = |\lambda_r(\alpha)/\sigma_r|$ which increases with increasing wavelength. The speed of each member is $-[\lambda_i(\alpha_0)/\alpha_0 - \sigma_i|a|^2/\alpha_0]$ where the second term represents the nonlinear correction to the linear phase speed $-\lambda_i/\alpha$ which is close to 3 at low R. Because the classical multi-scale formalism is extremely complex to apply, the computed speeds often deviated from investigator to investigator (Lin 1983). However, it is generally agreed that the nonlinear correction to speed is negative and longer waves tend to travel slower even though they have larger amplitudes. There were also attempts to construct stationary waves with wave numbers near zero, which are presumably near the infinitely long wavelength limit of the same wave family that begins at $\alpha = \alpha_0$. However, it is now known that the SL equation is only valid when the fundamental α is weakly unstable (i.e. just below α_0) while the overtone 2α is stable (Cheng & Chang 1990) and all prior results for α near zero such that $2\alpha < \alpha_0$ are incorrect. Even the maximum growing linear mode at $\alpha = \alpha_m$ is barely within this range and is hence poorly described by (5). Some investigators concluded correctly that whenever 2α lies within the unstable band of wave numbers in Figure 2, more than one amplitude equation must be considered because of the multiplicity of dominant Fourier modes. Lin (1974) suggested instead to use the long-wave evolution equations of (1) and (2) which can accommodate a large band of Fourier modes.

The first numerical studies of the stationary waves of the KS equation by Tsvelodub (1980), Demekhin (1983), Chen & Chang (1986), and Demekhin & Shkadov (1986) reveal the fate of the wave family that bifurcates off the neutral wave number at α_0 as predicted by the local analysis of the SL equation (5). As shown in the schematic of Figure 3, in the limits described by the KS equation, the wave family that bifurcates from α_0 is a standing wave family (5) with no nonlinear correction to the linear phase speed of 3. This is consistent with the local analysis of (5) since at low δ, σ_i vanishes and $-\lambda_i/\alpha$ approaches 3. The constructed amplitude is also in approximate agreement with the local theories of (5). However, at $(\alpha/\alpha_0) = 0.5547$, this standing-wave family undergoes a pitchfork bifurcation and yields two traveling-wave (relative to the linear phase speed)

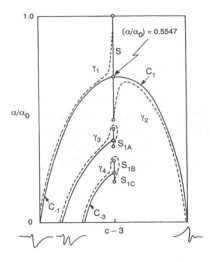

Figure 3 Schematic of the wave families of the KS equation in the wave-number/ speed $\alpha - c$ parameter space (*solid curves*). There is symmetry across $c = 3$ which is broken by inertia at finite δ (*broken curves*). Each wave family ends with a solitary wave at $\alpha = 0$ with different numbers of humps. The S wave families contain standing waves and the C families contain traveling waves of the KS equation. The γ families are the traveling waves at non-zero δ.

families $C_{\pm 1}$. This bifurcation is due to a "1-2 resonance" between the fundamental and its overtone which can be studied analytically if the amplitude equation of the now dominant overtone is added to the lone amplitude equation of the fundamental in (5) (Armbruster et al 1988). This bifurcation is hence beyond the description of the local Stuart-Landau equation (5). The two traveling-wave families are identical upon shape inversion ($\eta \rightarrow -\eta$) and reflection in space across the maximum ($x \rightarrow -x$) and in speed with respect to the linear phase speed [$c - 3 \rightarrow -(c - 3)$], transformations with respect to which the stationary KS equation with constant mean thickness

$$H_{xxx} + H_x - \mu H + 2H^2 = Q \qquad \langle H \rangle = 0 \qquad (6)$$

remains invariant. The constant $Q = \langle 2H^2 \rangle$ is the deviation flux and $\langle \; \rangle$ denotes a spatial average over one wavelength. The parameter μ is related to the deviation speed, $(c - 3) \sim (8Q + \mu^2)^{1/2}$. [A convenient representation of the stationary wave solutions of (6) can be obtained by transforming (6) into a dynamical system $(x_1, x_2, x_3) = (H, H_x, H_{xx})$, such that $\dot{\mathbf{x}} = \mathbf{f}(\mathbf{x})$. In this formulation, the stationary waves become closed trajectories (limit cycles) in the three-dimensional phase space. Numerical construction is then achieved by using now-standard continuation routines to construct the limit-cycle branches and their bifurcations.] The two traveling-wave families, which possess stationary-wave members with a wide-band Fourier spectrum and speeds different from 3, then extend indefinitely to vanishing α without further bifurcations. The limiting wave member with an infinitely long wavelength on the faster traveling-wave branch C_1 has a shape that has a striking resemblance to the solitary humps seen in regions III and IV of Figure 1. The limiting "negative" solitary wave on C_{-1} is inverted and reflected. Some members of these two families are shown in Figure 4.

All stationary wave families of the KS equation have now been constructed (Kevrekidis et al 1990, Demekhin et al 1991, Chang et al 1993a). There are actually an infinite number of such families. In the parameter space of μ and α, additional standing-wave families bifurcate from $\alpha = \alpha_0/n$, $n = 2, 3, \ldots$, as shown in Figure 5. Near the bifurcation points, they are identical to S and are, in fact, indistinguishable in the $c - \alpha$ bifurcation diagram of Figure 3. This is because a periodic standing wave with wavelength $2\pi/\alpha$ is also one with wavelength $2\pi n/\alpha$. However, further from the bifurcation points, they begin to deviate from S as perturbations much longer than $2\pi/\alpha$ are seen in the shapes (see the last wave family of Figure 4). Such distortions correspond to finite-amplitude manifestation of the classical subharmonic instabilities (Cheng & Chang 1992b) of the wave members on S and period-doubling in the dynamical-systems formulation. In the phase space, they correspond to a period-n limit cycle close to the period-1 limit cycle of S. All these subharmonic branches will undergo bifurcations to yield two symmetric traveling-wave families each as seen in Figures 3 and 5. These traveling-wave families terminate in solitary-hump-like waves in the vanishing-α limit but these limiting waves, in contrast to the one on C_1, have multiple humps. The faster multi-hump solitary waves are shown in Figure 6. (In Figure 3, the symmetry across $c = 3$ implies that whenever a family with speed slower than 3 exists, an inverted and reflected twin also exists with speed faster than 3. For clarity, only the slower twin is shown for some cases in Figure 3.) The n-hump solitary wave originates from the standing wave branch that bifurcates from $\alpha = \alpha_0/n$. It is convenient to use these solitary waves to represent the infinite number of traveling-wave families they head. They are not only physically important, as they resemble the humps seen in regions II and III, but also mathematically convenient since they allow the application of many new theories on homoclinic orbits.

It was realized recently (Pumir et al 1983) that the solitary waves correspond to homoclinic orbits in the phase-space formulation of any evolution and that the approach towards the solitary wave limit by the traveling-wave families in Figures 3 to 5 corresponds to a homoclinic bifurcation of a limit-cycle solution. Such homoclinic bifurcations and the resulting homoclinic orbits can be resolved *analytically* if the linearized Jacobian at the origin possesses eigenvalues close to two zeros with a geometric index of one (a double-zero singularity) or a zero and a purely imaginary conjugate pair (a $\{0, \pm i\}$ singularity). We (Chang 1986, 1987)

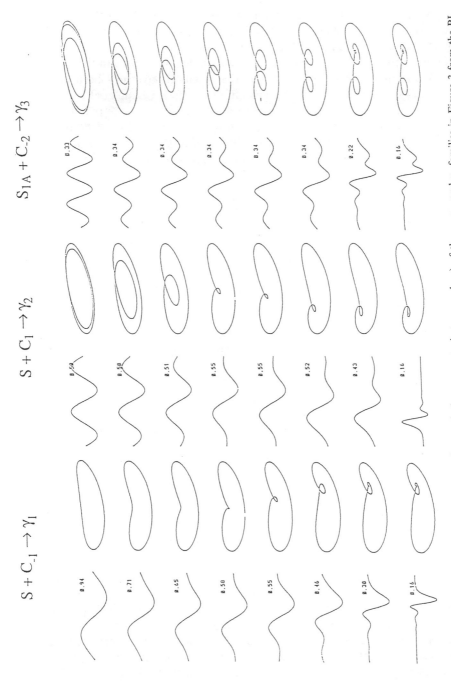

Figure 4   Individual wave members (wave tracings and phase-space trajectory analogs) of the γ_1, γ_2, and γ_3 families in Figure 3 from the BL equation at $\delta = 0.01$. The scales are not identical. The values of (α/α_0) are marked. The phase-space trajectories are also shown. The stationary waves range from a closed trajectory to a homoclinic orbit.

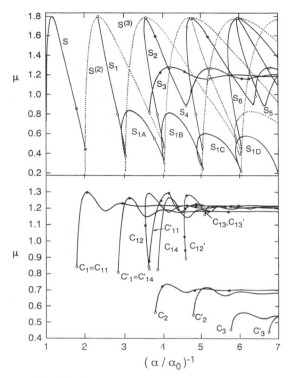

Figure 5   Fast wave families ($c > 3$) of the KS equation in the parameter space of $\mu - \alpha$ where μ is related to the speed and amplitude.

showed that for strong surface-tension fluids [$W \sim O(\varepsilon^{-2})$], both (1) and (2) have eigenvalues close to the latter singularity if the speed c is close to 3. In this limit, the wave amplitude is small and both equations yield the same results. We carried out the necessary nonlinear coordinate transformation to convert the dynamical system into the normal form of that singularity. The lowest order nonlinear coordinate transformation smears the small differences in speed among the various small-amplitude solitary waves in Figure 5 and describes them as a continuous family of solitary waves parameterized by c. The resolution is also insufficient to distinguish the number of humps. Nevertheless, it yields the generic solitary wave shape shown in Figure 7. It corresponds to the homoclinic orbit shown in the same figure with a large loop followed by damped oscillations toward the fixed point connected to the loop. This corresponds to the gentle back and the steep front relaxed by the bow waves observed in the solitary humps of Figure 1. Since the bow waves correspond to local dynamics

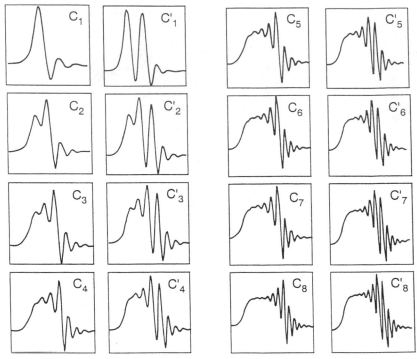

Figure 6 Multi-hump solitary waves which terminate the fast wave families ($c > 3$) of the KS equation at $\alpha = 0$. The traveling-wave family that each solitary wave belongs to is also indicated.

near the fixed point, a simple linear analysis shows that the bow waves have a wave number close to the neutral wave number α_0 which is in excellent agreement with experimental observation. Another interesting result is that the amplitude-speed correlation of all solitary waves or nearly solitary waves should be close to

$$c - 3 = 3(h - 1). \tag{7}$$

This simple prediction is consistent with the numerical result (Chang et al 1993a) and to the experimental speed and amplitude data of the solitary humps both in forced experiments and in regions III and IV of naturally excited waves as shown in Figure 7.

Recently, we have carried out a high-order resolution of the solitary-wave solutions (Chang et al 1993b) by adding a fictitious dispersion term to the KS equation,

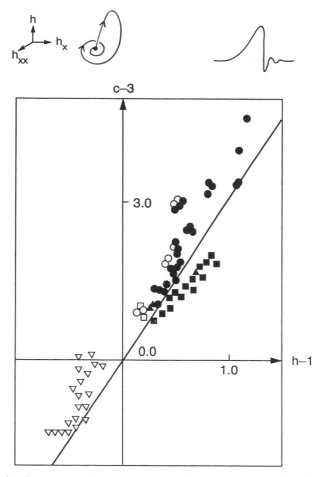

Figure 7   A solitary wave and its phase-space analog of a homoclinic orbit. Comparison of Equation (7) to the measured data of solitary humps (Chang 1986).

$$H_\tau + 4HH_x + H_{xx} + \delta' H_{xxx} + H_{xxxx} = 0. \tag{8}$$

(This equation can also be derived for inclined films with moderate surface tensions of a specific order.) With the unfolding provided by the additional dispersive parameter δ', the solitary-wave solutions indeed form a continuous family as shown in Figure 8. The limiting winding behavior near $\delta' = 0$ with a negligible distance between the speed c of individual solitary waves is the region we resolved earlier with our normal-form analysis which yielded Equation (7). For the KS equation with $\delta' = 0$, the solitary-

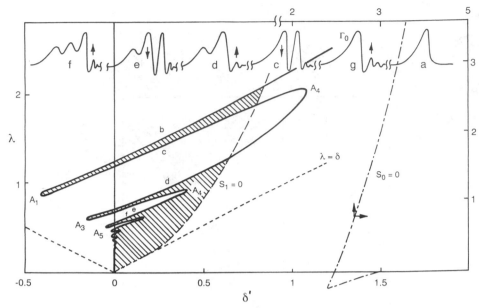

Figure 8 Unfolding of the solitary-wave solution branch of the extended KS equation of (8) with a fictitious dispersion parameter δ'. The solitary wave structure at each branch is also shown. The parameter λ is the deviation speed $\lambda \sim c-3$. Intersections with the λ axis correspond to solitary waves of the KS equation shown in Figure 6. The dotted lines and shaded areas are related to the global Silnikov bifurcation that generates these solitary waves (Chang et al 1993b).

wave members are discrete with the wave speeds forming a geometric series with decreasing separation. Each member closer to $c = 3$ has an extra hump corresponding to an additional traverse of the major loop in the phase space of Figure 7. This sequence of solitary waves is also shown in the vanishing α limit of Figure 3 and it corresponds to a peculiar bifurcation of the homoclinic orbits related to the Silnikov mechanism (Glendinning & Sparrow 1984). The primary solitary wave with one hump, corresponding to the limiting wave member of the traveling-wave family C_1, has the largest speed of all solitary waves. It has a negative twin with an inverted and reflected profile which is the slowest solitary wave. All other solitary waves in Figure 8 can be considered as derivatives which bifurcate off these two primary one-hump solitary waves. In fact, one can construct the genealogy of the infinite number of wave families in Figures 3 and 5 by the following sequence of bifurcations. One begins with the one-hump solitary wave of Figures 7 and 8. The speed and its shape are known in detail (Toh 1987). Through a Silnikov bifurcation, unfolded by the fictional dispersion term in Figure 8, it generates an infinite number of n-hump solitary waves. Each solitary wave then undergoes a reverse homoclinic bifurcation to generate a family of limit cycles (traveling waves). This sequence is followed in mirror image by the inverted twin of the primary one-hump solitary wave with the slowest speed. The one-hump twins are then the parents of all stationary traveling waves of the KS equation.

That the addition of dispersion can allow a more detailed resolution of the solitary waves is consistent with the analysis of the $O(\varepsilon^3)$ evolution equation of Nakaya for weak surface tension fluids (Chang 1989) which includes the higher-order dispersion effect in a rigorous manner. His equation was shown to be close to the double-zero singularity and the solitary waves form a continuous branch resolved by our analysis. It yields another amplitude-speed correlation for the near-solitary stationary waves which is also in favorable agreement with waves on glycerin or glycerin-water solutions. In general, the improved resolution of the stationary waves with the addition of either real dispersion or the artificial one in Equation (8) is consistent with our earlier discussions on the importance of higher-order terms like dispersion on linear instability and large-amplitude growth. To further support this, we point out that as δ approaches zero, the small-amplitude stationary-wave solutions of the BL equation collapse into the solutions of the KS equation and the small-amplitude waves of the evolution equation (1). However, the large-amplitude solitary-wave solutions of the strongly nonlinear evolution equation constructed by Pumir et al (1983) are not approached by the solitary waves of the BL equation in the same limit. This underscores the argument that the large-amplitude solutions of (1) are not correct because of the omission of dispersion and other high-order terms.

Most wave experiments for strong-surface-tension fluids like water are carried out in the intermediate-flow-rate region of $10 < R < 300$ where the KS equation is invalid. In this region, the BL equation (3) must be used and the elegant analysis available to the KS equation must now be replaced by brute-force numerical construction of the stationary waves. Nevertheless, many of the analytical and numerical results for the KS equation, which can be very clearly classified (for example, slow-fast symmetry and Silnikov bifurcations of solitary waves), are still useful in deciphering and organizing the bifurcations of the wave families of the BL equation from the KS solutions at vanishing δ.

The first attempts to construct finite-δ stationary waves focused on the ad hoc but simple IBL equation (Shkadov 1967, 1968; Demekhin & Shkadov 1985; Trifonov & Tsvelodub 1991; Tsvelodub & Trifonov 1992). The solutions are found to reduce to KS solutions at vanishing δ. However, some of the constructed wave shapes are not in good agreement with the measured ones even at small δ values ($\delta > 0.05$). For the most common experimental conditions, only the IBL equation and the full Navier-Stokes equation are sufficiently accurate. While construction of stationary solution branches for the latter remain formidable, the former can now be analyzed completely to show good quantitative agreement with experimental data. We (Chang et al 1993c) have developed a spectral-element domain-decomposition numerical method for resolving the stationary waves of the BL equation. It is a routine specifically designed for the difficult free-surface problem of the falling film and we summarize below its resolution of the stationary waves.

At finite δ, the symmetries of the KS equation are broken and the pitchfork bifurcation of the standing-wave branch S which gives rise to the twin traveling-wave families becomes the imperfect pitchfork bifurcation shown in Figure 3. The standing-wave family S is linked with the slow primary traveling-wave family C_{-1} to form a new family γ_1. The primary fast traveling-wave family is isolated from this branch and forms γ_2. This perturbation of the pitchfork is also shown in Figure 4. As a result, the γ_1 family becomes a traveling-wave family with speed less than 3 and with wave number ranging from α_0 to zero. The segment of this slow wave family γ_1 near α_0 is the wave family resolved by the SL equation of (5). As shown in Figure 9, it is indeed slower than the linear phase speed due to nonlinear effects. The local resolution of the near-monochromatic waves by the Stuart-Landau equation is hence extended to the broad-banded stationary waves of Figure 11 at vanishing α. The γ_2 family, as shown in Figure 10, is a fast traveling-wave family with speed in excess of 3. This family could never have been resolved by local theories like the SL equation since it does not bifurcate off the neutral curve. Its wave number begins at $\alpha/\alpha_0 \sim 0.5$ and ends at $\alpha = 0$ in the positive one-hump solitary wave shown in Figure 7. The near-solitary waves ($\alpha \to 0$) of both the γ_1 and γ_2 families are still described by the analytical amplitude-speed correlation of Equation (8) up to $\delta = 0.3$. This is also confirmed in Figure 7 where the data correspond mostly to finite-δ values. Other pitchfork bifurcations of the subharmonic standing-wave branches are likewise broken to form an infinite number of traveling-wave families γ_i. However, as δ increases slightly from zero, these higher wave families begin to shift down in the $\alpha - c$ plane (as shown in Figure 9) and many of them actually coalesce and eliminate each other. By the moderate value of $\delta = 0.1$, corresponding to $R \sim 10$ for water, only a handful remain for $\alpha > 0.2$, including the primary γ_1 and γ_2 families. In fact, stability analysis described in the next section shows that the most stable (least unstable) waves of all branches lie only on the primary slow and fast families, γ_1 and γ_2, and the other wave families are essentially invisible. In retrospect, the two primary traveling

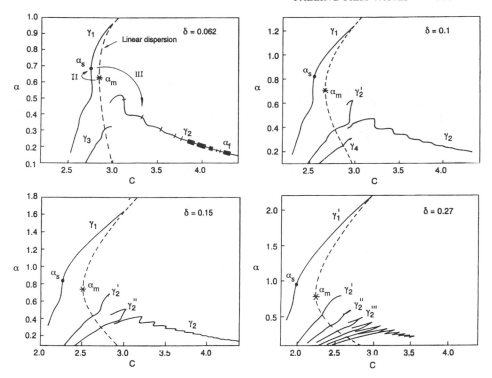

Figure 9 Actual wave families of the BL equation of various δ values. The top value of α in each figure corresponds to $\alpha_0 \sim \sqrt{18\delta}$. At moderate δ values, only the γ_1 and γ_2 families are important (See Figure 3 for the creation of these two families.) The narrow segments on these 2 families that are stable to two-dimensional disturbances are marked in the first figure. There is only one segment on γ_1, but multiple segments which correspond to a geometric series in α appear on γ_2. The wave transitions in a natural evolution are also marked. For $\delta > 0.09$, the high-α end of the fast γ_2 family degenerates into various families, none of which yield any stable segment.

waves $C_{\pm 1}$ of the KS equation, which arise due to a 1-2 resonance of the standing-wave family, and the two primary 1-hump solitary waves that terminate $C_{\pm 1}$ are actually the key stationary-wave families and solitary humps of the falling-film problem. The other infinite number of wave families are never reached in reality. This is hence a considerable simplification of the confusion of wave families shown in Figures 3 to 5. For $\delta > 0.09$, the short-wave (high α) end of the γ_2 family also undergoes a sequence of bifurcations which give rise to new γ_2'' families as shown in Figure 9. These new families are found to be extremely unstable, however. The long-wave end of the γ_2 family persists even at large δ and some

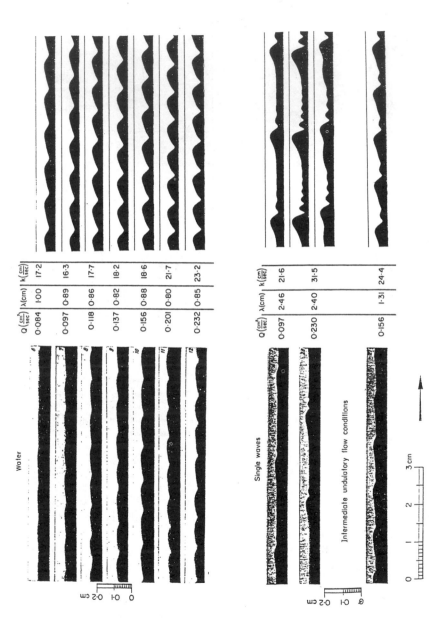

Figure 10 Comparison of constructed γ_1 (*top*) and γ_2 (*bottom*) waves of the BL equation (*right*) to Kapitza's photographs of periodically excited waves of the same wavelength (*left*). The second tracing at the bottom shares the same wavelength with two constructed waves, one on γ_2 and one γ_2'.

selected waves, including ones exhibiting the characteristic solitary humps, are always found on the surviving segment.

The two primary families γ_1 and γ_2 are not only physically distinct in their speeds: In a constant-flux formulation, the slower γ_1 family has a smaller average thickness than h_N while the faster γ_2 has a larger average film thickness (Chang et al 1993a). (This is evident from the second γ_1 tracing and the first γ_2 tracing of Figure 10 which have the same flow rate.) This conclusion would seem counterintuitive unless one remembers that the waves are not mass-carrying—they travel faster than the fluid elements. Since region II in Figure 1 will be shown to contain γ_1 waves, this would imply that the averaged film thickness in region II is smaller than h_N if its flow rate is the same as in region I, i.e. the feed flow rate.

In Figure 10, some nearly monochromatic waves on γ_1 and some nearly solitary waves on γ_2 from our construction are compared to the classical photographs of Kapitza & Kapitza (1949) of the first waves generated by their periodically forced experiment. The waves are constructed to be the same wavelength as the observed ones. A slightly smaller amplitude is seen in the observed waves which can be attributed to the curvature effect of their cylindrical wall. Much better agreement is seen in Figure 11 where the forced-wave tracings from a more accurate measurement of Nal-

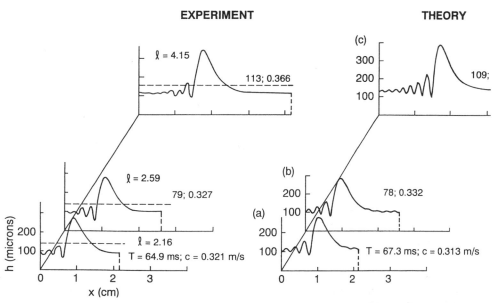

Figure 11   Comparison of constructed γ_2 waves to the ones recorded in the forced experiment of Nakoryakov et al (1985).

koryakov et al (1985) are used. Their waves are γ_2 waves generated by periodic forcing. Excellent agreement has also been found with the naturally excited waves measured by Stainthrop & Allen (1965). The inverted one-hump solitary waves on the vanishing α limit of γ_1 have never been recorded for a vertical film, but our constructed ones in Figure 4 resemble the waves measured by Liu et al (1993) for an inclined film. All measured stationary waves reported in the literature are in good agreement with the constructed ones from the BL equation (Chang et al 1993a).

WAVE SELECTION

P. Kapitza (1948) suggested at the very beginning that only some of the stationary waves generated by periodic forcing will be selected in a natural setting. All waves are expected to be unstable but the least unstable ones are presumed to be the chosen ones. He suggested a wave with the maximum absolute energy dissipation rate as the observed one. If the competing periodic stationary waves all have the same average thickness, which can be experimentally imposed, the selected one by Kapitza's criterion can be shown to carry the highest flow rate at low R (Chang et al 1993a). The earliest attempts to rigorize Kapitza's physical argument involved resolving the sideband stability of the near-neutral γ_1 waves estimated by the SL equation (5). The approach is to use the Ginzbury-Landau (GL) equation

$$\frac{\partial a}{\partial t} = \lambda a + \beta \frac{\partial^2 a}{\partial x^2} - \sigma |a|^2 a \tag{9}$$

by including the sideband effects in the complex coefficient β. However, as mentioned earlier, unlike other instabilities whose growth rate curve has a simple nose, the coefficients of the present SL equation are expanded about the neutral wave number α_0 and not at the nose near criticality. This implies that the linear growth rate has a nonzero slope $(\partial \lambda / \partial \alpha)(\alpha_0)$ and there should be a convective $\partial / \partial x$ term absent in the classical GL equation. Perhaps due to this, the GL equation (9) yields the erroneous prediction that slow γ_1 stationary waves near α_0 are stable to sideband disturbances which contradicts the recent experimental results of Liu et al (1993). A reexamination of the problem (Cheng & Chang 1992a), which uses center manifold techniques on the KS equation (2), reveals that, analogous to the Eckhaus bound for nondispersive GL equations, the near-neutral waves on the γ_1 (actually S) family are indeed unstable to sideband disturbances. The same waves have also been subjected to subharmonic instabilities by using the IBL equation for inclined films (Prokopiou et al

1991). Here, one needs to analyze the coupled fundamental and subharmonic amplitude equations (Cheng & Chang 1992b). The near-neutral waves on γ_1 are also found to be unstable to subharmonic disturbances. A more detailed analysis (Cheng & Chang 1993) has recently allowed us to compare these two dominant instabilities of the γ_1 stationary waves. It is found that the sideband instability is dominant near α_0 while the subharmonic instability takes over at a critical wave number below α_0. Our prediction agrees quantitatively with Liu & Gollub's experimental demarcation (1993) of these two instabilities. The boundary shifts down towards longer waves with increasing δ since the subharmonic instability is weakened by the detuning effect of dispersion (Cheng & Chang 1992b) while the sideband instability can be enhanced by dispersion (Cheng & Chang 1990).

We have extended the above stability analysis of γ_1 waves near α_0 by numerically imposing general two-dimensional and three-dimensional disturbances of arbitrary wavelengths in both the x and z direction to our constructed γ_1 waves (Chang et al 1993a). The resulting Floquet calculation confirms all the predictions of the local theories. Moreover, it shows that there is a finite-amplitude wave with wavenumber α_s on γ_1 (marked in Figure 9) that is the least unstable wave of the family. For small δ, this wave is actually stable to two-dimensional disturbances and only slightly unstable to three-dimensional ones. Beyond $\delta \sim 0.1$, however, it also becomes unstable to two-dimensional disturbances. Its growth rate never exceeds 30% of the dominant primary disturbance of the flat film at $\alpha = \alpha_m$ for $\delta < 3.0$. This confirms the long lifetime of the stationary waves. The dominant instabilities of this selected γ_1 wave are either a sideband mode ($\alpha_s \pm \Delta$) or a subharmonic ($\alpha_s/2$) in the x direction and a long transverse sideband instability in the z direction. Amazingly, Kapitza's criterion yields an accurate estimate of α_s at small δ after the viscous dissipation rate of our constructed waves is determined. This selected wave at α_s has a very physical characteristic—it is the wave with the highest flow rate (maximum dissipation) among all γ_1 waves in the constant-thickness formulation. It is a continuation of the stable segment at $(\alpha/\alpha_0) \in (0.77, 0.84)$ on the S family of the KS equation (Nepomnyaschy 1974, Demekhin & Kaplan 1989) which is actually stable to all disturbances. For all δ values, this selected wave has a wave number that is higher than the maximum growing linear mode α_m. This immediately suggests that a naturally excited wave which emerges with wave number α_m and the corresponding linear phase speed at the end of the inception region I will evolve in region II into a slower wave on γ_1 with a higher wave number α_s. This transition (shown in Figure 9) corresponds to a nonlinear deceleration and compression of the infinitesimally small wave field selected by the linear mechanism. The deceleration is qualitatively consistent with many observations (Lin 1983) and we (Chang et al 1993a) have obtained some quantitative confirmation with Stainthrop & Allen's (1965) wave speed data. Wave number data are, however, scarce and we verify the above transition scenario from region I to region II with a numerical experiment shown in Figure 12. We solve the initial-value problem of the BL equation in a frame moving at the linear phase speed to minimize wave translation. To bypass the long (in both space and time) inception region and to avoid boundary effects in the small computation domain, a small-amplitude

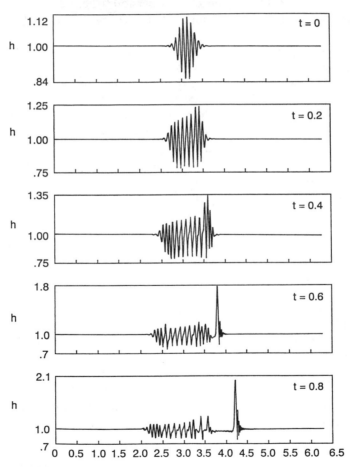

Figure 12 Numerical experiment on the transient evolution of a small-amplitude wave packet by using the BL equation at $\delta = 0.05$. The time scale t is actually stretched, αt, and the length scale αx.

wave packet with wave number close to α_m is inserted at the middle of the domain at $t = 0$. Our initial condition is hence spatially localized. The development of the overtone due to weakly nonlinear excitation is clearly visible at $t = 0.2$ and 0.4 as these waves form a lagging wave patch with more compressed waves. The wave number in this patch is close to α_s. If a uniform wave field could be used as the initial condition, one would see uniform compression and deceleration over a large domain.

We have also extended the stability analysis to the fast γ_2 waves and all other wave families. Here the Kapitza criterion is again found to be accurate at low δ. Interestingly, there are now multiple segments of local stability to two-dimensional disturbances on the γ_2 family as shown in Figure 9. These stable segments appear only at finite δ since the C_1 branch of the KS equation in Figure 3 is always unstable to two-dimensional disturbances. Members within these segments are again weakly unstable to three-dimensional disturbances that are sidebands or subharmonic in the streamwise direction and long sideband disturbances in the transverse direction. All other wave families are unstable to two-dimensional disturbances and hence are unlikely to be observed. The stable segments on γ_2 possess unique structures. The segment with the shortest waves is located near the turning point of the isolated branch γ_2 after the pitchfork is broken at $\delta = 0$. It hence possesses a wave number of about $(\alpha/\alpha_0) = 0.5$, i.e. the waves are twice as long as the neutral ones. They are much larger, longer, and faster than the waves near α_s on the γ_1 wave. Subsequent segments correspond to even longer and faster waves in the γ_2 family. As sketched in Figure 9, these narrow segments center at the wave number α_n and $|\alpha_{n+1} - \alpha_n|$ decreases geometrically as α_n decreases. For example, for $\delta = 0.062$, $\alpha_n = (0.48, 0.38, 0.31, 0.26, \ldots)$. An accumulation point α_f exists at about $(\alpha/\alpha_0) \sim 0.1$ for $\delta < 1.0$ below which the flat film between the humps becomes too long for the weak interaction between them to stabilize the flat-film primary instabilities of Figure 2. However, as shown in Figure 4, by $(\alpha/\alpha_0) \sim 0.2$, the periodic stationary waves in the γ_2 family have already taken on the shape of a solitary wave with the characteristic hump of Figure 7.

The important effect of dispersion is again seen here in the stability of γ_2 waves. The C_1 traveling-wave family of the nondispersive KS equation is unstable to even two-dimensional disturbances and yet its derivative γ_2 exhibits multiple stable segments with respect to the same disturbances. This is due to two effects, both related to the increased inertia in the BL equation. The destabilizing mechanism of inertia yields larger solitary humps which, in turn, intensify the interaction between humps that stabilizes the primary instability of the flat-film region between them. This interaction can be described with a coherent structure theory (Chang et al

1993a). However, the interaction between the solitary humps is stabilizing only if inertia-induced dispersion is present. This stabilizing effect of dispersion has been qualitatively confirmed by a recent study of the fictitious equation (8) (Chang et al 1993b). It was shown that not only does dispersion stabilize the long stationary waves, it also enhances their domains of attraction such that as δ' increases, more initial conditions approach the stationary waves corresponding to periodic trains of solitary humps. In the limit of infinite δ', (8) approaches the integrable KdV equation and periodic trains of its cnoidal solitary wave are approached for all initial conditions. The wavelength of the solitary wave train or correspondingly, the particular stable stationary waves on γ_2, is determined or selected by the initial condition. However, there is no question that, in the absence of transverse disturbances, stationary periodic trains of solitary humps will be selected if there is significant inertia-induced dispersion.

The stability analysis of the γ_1 and γ_2 wave families also suggests some possible scenarios for the transition from region II to region III in Figure 1. In essence, it corresponds to the system departing from the α_s wave on γ_1 to a stable one on γ_2. If significant transverse disturbances are present, this transition would lead to a stable wave on γ_2 with a wave number close to $\alpha_s/2$ since the most unstable three-dimensional disturbance has a subharmonic streamwise component. However, departure from α_s can also be triggered by finite-amplitude two-dimensional disturbances even when it is stable to infinitesimal two-dimensional disturbances at low δ. The most unstable (least stable) purely two-dimensional disturbance is often a sideband one and we would expect transition to a long wave in the γ_2 family. By the same token, if the domains of attraction of the stable waves on γ_2 are small, almost continuous transitions from one stable wave to another can be expected in the presence of finite-amplitude two-dimensional disturbances or three-dimensional disturbances. This is apparently the case but the observed transition in region III is not nearly as simple as the approach towards α_s in region II. The transition is first triggered by an intermittent and localized instability of the α_s wave which renders it difficult to discern a large patch of distinct periodic γ_2 waves. Consequently, the evolution along the γ_2 family towards the near-solitary wave at α_f tends to occur locally over one or two wavelengths rather than as a uniform wave field. Even in our idealized numerical experiment of Figure 12, each individual fast wave that leads the entire packet for $t > 0.6$ lies approximately within a different stable segment on γ_2. The large solitary hump that is dramatically emitted at $t = 0.6$ is close to the α_f wave of the γ_2 periodic stationary wave in Figure 4, suggesting an eventual evolution towards the solitary limit of this branch. The distinct difference in speed, shape, amplitude, and wavelength between this hump and the lagging wave field underscores the distinction between the short γ_1 waves near α_s and the long γ_2 waves near α_f. Although this hump travels in a stationary manner and subsequent humps that leave the field are all almost identical to the first one, the separation between the humps is not uniform due to the intermittent nature of their creation process. We are unable to follow the humps further downstream to examine if they interact weakly with each other to form a periodic train or if the interaction remains nonstationary.

The more gradual and spatially uniform approach towards the α_s wave and the localized and intermittent departure from it are probably due to the different nonlinear mechanisms for the creation and destruction of the α_s wave. It is born by a short-range fundamental-overtone interaction involving only disturbances within one wavelength which are always present. Its death, on the other hand, requires long-range subharmonic and sideband streamwise disturbances which do not exist uniformly in space. Long-wave disturbances are definitely localized in our numerical experiment of Figure 12. Hence, coalescence of adjacent waves and long modulations tend to occur intermittently as localized defects. As a result, uniform γ_1 waves are observed during natural excitation while uniform γ_2 waves can only be observed with entrainment by periodic excitation. Even in the latter experiment, the primary instability of the flat film separating adjacent solitary humps prevents very long γ_2 waves to be sustained by periodic forcing (Alekseenko et al 1985, Liu & Gollub 1993). We (Chang et al 1993a) have shown that α_f is a good estimate of the lower bound on the wave number of the excitable γ_2 waves.

TURBULENT WAVE DYNAMICS AND FUTURE DEVELOPMENT

The only remaining wave dynamics that still escape understanding in this difficult but intriguing instability are the two-dimensional spatio-temporal chaos of region III and the three-dimensional interfacial turbulence of region IV. It is quite clear that the solitary-hump structures play an important role in these regions. It would be consistent with the underlying simple elegance of this difficult problem that the same solitary wave in Figure 7 that generates the infinite families of stationary waves via Silnikov and homoclinic bifurcations is also responsible for the turbulent dynamics. There are some preliminary results supporting this view. The long periodic stationary waves on the γ_2 family that are stable to two-dimensional disturbances must arrest the primary instability on the flat film region through a weak interaction of the two bounding solitary humps. This interaction is dominated by the small-amplitude ends of the solitary humps, the front bow waves of the back hump, and the smooth slope of the front one (Kawahara & Toh 1988). In the phase space analogy, these two regions are described by the linear dynamics near the fixed point that the homoclinic orbit is attached to, and are hence easily deciphered by linear analysis. A weakly interacting coherent structure theory has been able to use these resolved ends to faithfully reproduce the sequence of stable wave segments on γ_2, including the geometric series in α (Chang et al 1993a).

These stable segments are unstable to three-dimensional disturbances. Consequently, transverse variation should be included in the coherent structure theory. The wavy crests seen in regions III and IV correspond to the classical "phase instability" of fronts described by a nonlinear diffusion equation. (In the more extreme cases, the phase evolution of the front is described by the KS equation!) One can hence envision a set of coupled nonlinear diffusion equations for the position of the wave crests as a function of t, x, and z. The coupling between the destabilizing transverse variation and the stabilizing streamwise interaction between the humps is responsible for the nonstationary dynamics. It is quite possible that a statistical theory with only nearest-neighbor interaction is sufficient to describe the dynamics of the entire interface in region IV. Dispersion will again play an important role here as it does in determining the convective instability of the inception region, in arresting blow-up behavior to form saturated two-dimensional stationary waves and in promoting the stability of stationary waves, especially periodic trains of solitary humps. The nonstationary dynamics is likely due to a competition between dispersion for the stabilizing streamwise interaction and Rayleigh capillary effects near the crest which cause the transverse instability.

The construction of weakly interacting theory for the solitary humps is equivalent to a perturbation analysis of a homoclinic orbit. The homoclinic orbit corresponds to a single solitary hump and the perturbations come from neighboring humps and transverse variation. Such perturbations can be studied with the new technique from Dynamical Systems theory for bifurcations of a homoclinic orbit, such as the Melnikov and Silnikov theories. Bifurcations of a homoclinic orbit are known to yield horseshoe maps and chaotic dynamics. It is hence very likely that the irregular spatio-temporal dynamics of regions III and IV can be described by chaos theory. It would then offer a direct contact between hydrodynamic turbulence, albeit a low Reynolds number one, and low-dimensional chaos. Research in this direction should be very fruitful and exciting.

ACKNOWLEDGMENTS

Our work on falling-film waves has been supported by NSF-PYI, ACS-PRF, the Notre Dame Center for Applied Mathematics, and DOE. I have been privileged to work with a group of uniquely talented colleagues and students on this subject. The contributions of E. A. Demekhin, M. Cheng, S.-H. Hwang, M. Sangalli, and S. Kalliadasis have been especially important and I would like to dedicate this review to them.

Literature Cited

Armbruster, D., Guckenheimer, J., Holmes, P. 1988. Heteroclinic cycles and modulated travelling waves in systems with O(2) symmetry *Physica D* 29: 257–82
Alekseenko, S. V., Nakoryakov, V. E., Pokusaev, B. G. 1985. Wave formation on a vertical falling liquid film. *AIChE J.* 31: 1446–60
Benjamin, T. B. 1957. Wave formation in laminar flow down an inclined plane. *J. Fluid Mech.* 2: 554–74
Benney, B. J. 1966. Long waves in liquid films. *J. Math. Phys.* 45: 150–55
Bertschy, J. R., Chin, R. W., Abernathy, F. W. 1983. High-strain-rate free-surface boundary-layer flows. *J. Fluid Mech.* 126: 443–61
Chang, H.-C. 1986. Traveling waves in fluid interfaces: normal form analysis of the Kuramoto-Sivashinsky equation. *Phys. Fluids* 29: (10) 3142–47
Chang, H.-C. 1987. Evolution of nonlinear waves on vertically falling films—a normal form analysis. *Chem. Engrg. Sci.* 42: 515–33
Chang, H.-C. 1989. Onset of nonlinear waves on falling films. *Phys. Fluids A* 1: (9) 1314–27
Chang, H.-C., Demekhin, E. A, Kopelevich, D. I. 1993a. Nonlinear evolution of waves on a falling film. *J. Fluid Mech.* 250: 433–80
Chang, H.-C., Demekhin, E. A., Kopelevich, D. I. 1993b. Laminarizing effects of dispersion in an active-dissipative nonlinear medium. *Physica D* 63: 299–320
Chang, H.-C., Demekhin, E. A., Kopelevich, D. I. 1993c. Construction of stationary waves on a falling film. *Comput. Mech.* 11: 313–32
Chen, L.-H., Chang, H.-C. 1986. Nonlinear waves on liquid film surfaces—II. Bifurcation of the long-wave equation. *Chem. Engrg. Sci.* 41: 2477–86
Cheng, M., Chang, H.-C. 1990. A generalized sideband stability theory via center manifold projection. *Phys. Fluids* 2: (8) 1364–79
Cheng, M., Chang, H.-C. 1992a. Stability of axisymmetric waves on liquid films flowing down a vertical column to azimuthal and streamwise disturbances. *Chem. Engrg. Comm.* 118: 327–40

Cheng, M., Chang, H.-C. 1992b. Subharmonic instabilities of finite-amplitude monochromatic waves. *Phys. Fluids A* 4: (3) 505–23
Cheng, M., Chang, H.-C. 1993. Competition between subharmonic and sideband secondary instabilities of film flows. Preprint
Chu, K. J., Dukler, A. E. 1974. Statistical characteristics of thin wavy films. *AIChE J.* 20: 695–706
Demekhin, E. A. 1983. Bifurcation of the solution to the problem of steady traveling waves in a layer of viscous liquid on an inclined plane. *Izv. Akad. Nauk SSSR, Mekh. Zhidk. Gaza* 5: 36–44
Demekhin, E. A., Kaplan, M. A. 1989. Stability of stationary traveling waves on the surface of a vertical film of viscous fluid *Izv. Akad. Nauk SSSR, Mekh. Zhidk. Gaza* 3: 34–41
Demekhin, E. A., Shkadov, V. Ya. 1985. Two-dimensional wave regimes of a thin liquid films. *Izv. Akad. Nauk SSSR, Mekh. Zhidk. Gaza* 3: 63–67
Demekhin E. A., Shkadov V. Ya. 1986. Theory of solitons in systems with dissipation *Izv. Akad. Nauk SSSR, Mekh. Zhidk Gaza* 3: 91–97
Demekhin, E. A., Tokarev, G. Yu, Shkadov, V. Ya. 1991. Hierarchy of bifurcations of space-periodic structures in a nonlinear model of active dissipative media. *Physica D* 52: 338–61
Floryan, J. M., Davis, S. H., Kelly, R. E. 1987. Instabilities of a liquid film flowing down a slightly inclined plane. *Phys. Fluids* 30: (4) 983–89
Glendinning, P., Sparrow, C. 1984. Local and global behavior near homoclinic orbits. *J. Stat. Phys.* 35: 645–96
Ho, L.-W., Patera, A. T. 1990. A Legendre spectral element method for simulation of unsteady incompressible viscous free-surface flow. *Comput. Methods Appl. Mech. Eng.* 80: 355–66
Joo, S. W., Davis, S. H. 1992a. Irregular waves on viscous falling films. *Chem. Engrg. Comm.* 118: 111
Joo, S. W., Davis, S. H. 1992b. Instabilities of three-dimensional viscous falling films. *J. Fluid Mech.* 242: 529–47

136 CHANG

Joo, S. W., Davis, S. H., Bankoff, S. G., 1991. On falling film instabilities and wave breaking. *Phys. Fluids* A3: 231–32

Kapitza, P. L. 1948. Wave flow of thin viscous fluid layers. *Zh. Eksp. Teor. Fiz.* 18: 1, 3–28; also in *Collected Works of P. L. Kapitza*, ed. D. Ter Haar. Oxford: Pergamon (1965)

Kapitza, P. L., Kapitza, S. P. 1949. Wave flow of thin fluid layers of liquid. *Zh. Eksp. Teor. Fiz.* 19: 105–20; also in *Collected Works of P. L. Kapitza*, ed. D. Ter Haar. Oxford: Pergamon (1965)

Kawahara, T., Toh, S. 1988. Pulse interaction in an unstable dissipative-dispersive nonlinear system. *Phys. Fluids* 31: 2103–11

Kevrekides, I. Nicolaenko, B., Scovel, J. C. 1990. Back in the saddle again: a computer-assisted study of the Kuramoto-Sivashinsky equation. *SIAM J. Appl. Math.* 50: 760–90

Kheshgi, H. A., Scriven, L. E. 1987. Disturbed film flow on a vertical plate. *Phys. Fluids* 30: (4) 990–997

Lin, S. P. 1969. Finite amplitude stability of a parallel flow with a free surface. *J. Fluid Mech.* 36: 113–26

Lin, S. P. 1974. Finite amplitude side-band stability of a viscous film. *J. Fluid Mech.* 63: (3) 417–429

Lin, S. P., 1983. Film waves. In *Waves on Fluid Interfaces*, ed. R. E. Meyer, pp. 261–90. New York: Academic

Liu, J., Paul, J. D., Banilower, E., Gollub, J. P., 1992. Film flow instabilities and spatiotemporal dynamics. *Proc. First Expt. Chaos Conf.*, ed. S. Vohra, M. Spano, M. Shlesigner, L. M. Pecora, W. Ditto, pp. 225–39. River Edge, NJ: World Sci.

Liu, J., Paul, J. D., Gollub, J. P., 1993. Measurement of the primary instabilities of film flows. *J. Fluid Mech.* 220: 69–101

Liu, J., Gollub, J. P. 1993. Onset of spatially chaotic waves on flowing films. *Phys. Rev. Lett.* 70: 2289–92

Nakaya, C. 1975. Long waves on thin fluid layer flowing down an inclined plane. *Phys. Fluids* 18: 1407–12

Nakoryakov, V. E., Pokusaev, B. G., Radev, K. B. 1985. Influence of waves on convective gas diffusion in falling down liquid film. In: *Hydrodynamics and Heat and Mass Transfer of Free-Surface Flows*, pp. 5–32. Novosibirsk: Inst. Heat Phys., Siberian Branch USSR Acad. Sci. (In Russian)

Nepomnyaschy, A. A. 1974. Stability of wave regimes in a film flowing down on inclined plane. *Izv. Akad. Nauk SSSR, Mekh. Zhidk. Gaza* 3: 28–34

Pierson F. W., Whitaker, S. 1977. Some theoretical and experimental observation of wave structure of falling liquid films. *Ind. Engng. Chem. Fundam.* 16: (4) 401–8

Prokopiou, Th., Cheng, M., Chang, H.-C. 1991. Long waves on inclined films at high Reynolds number. *J. Fluid Mech.* 222: 665–91

Pugh, J. D., Saffman, P. G. 1988. Two-dimensional superharmonic stability of finite-amplitude waves in plane Poiseuille flow. *J. Fluid Mech.* 194: 295–307

Pumir, A., Manneville P., Pomeau Y. 1983. On solitary waves running down an inclined plane. *J. Fluid Mech.* 135: 27–50

Rosenau, P., Oron, A., Hyman, J. M. 1992. Bounded and unbounded patterns of the Benney equation. *Phys. Fluid A* 4: (6) 1102–4

Roskes, G. J. 1970. Three-dimensional long waves on a liquid film. *Phys. Fluids* 13: 1440–45

Shkadov, V. Ya. 1967. Wave conditions in the flow of thin layer of a viscous liquid under the action of gravity. *Izv. Akad. Nauk SSSR, Mekh. Zhidk. Gaza* 1: 43–50

Shkadov, V. Ya. 1968. Theory of wave flows of a thin layer of a viscous liquid. *Izv. Akad. Nauk SSSR, Mekh Zhidk. Gaza* 2: 20

Shkadov, V. Ya., Kholpanov, L. P., Malyusov, V. A., Zhavoronkov, No, M. 1970. Nonlinear theory of wave flows of liquid films. *Teor. Osn. Khim. Tekhnol.* 4: 859–67

Stainthorp, F. P., Allen, J. M. 1965. The development of ripples on the surface of liquid film flowing inside a vertical tube. *Trans. Inst. Chem. Eng.* 43: 85–91

Toh, S. 1987. Statistical model with localized structures describing the spatio-temporal chaos of Kuramoto-Sivashinsky equation. *J. Phys. Soc. Jpn.* 56: 949–62

Trifonov, Yu. Ya, Tsvelodub, O. Yu. 1991. Nonlinear waves on the surface of a falling liquid film part I. *J. Fluid Mech.* 229: 531–54

Tsvelodub, O. Yu. 1980. Steady traveling waves on a vertical film of fluid. *Izv. Akad. Nauk SSSR, Mekh. Zhidk. Gaza* 4: 142–46

Tsvelodub, O. Yu., Trifonov, Yu. Ya. 1992. Nonlinear waves on the surface of a falling film part II. *J. Fluid Mech.* 244: 149–69

Yih, C.-S. 1963. Stability of liquid flow down an inclined plane. *Phys. Fluids* 6: (3) 321–34

Annu. Rev. Fluid Mech. 1994. 26 : 137–68

HIGH RAYLEIGH NUMBER CONVECTION

Eric D. Siggia

Laboratory of Atomic and Solid State Physics, Cornell University, Ithaca, New York 14853-2501

KEY WORDS: turbulence, boundary layer, heat flux, plumes

1. INTRODUCTION

Turbulent convection exemplifies many of the startling aspects of turbulent flows that have been uncovered in the past two decades, but frequently exhibits a novel twist. Thus, as in the case of free shear flows, convection can organize into large-scale vortical structures, but these then react back in subtle ways on the boundary layers which ultimately sustain them. Thermal plumes are a coherent mode of heat transport, analogous to boundary layer bursts, yet their overall effect can be surprisingly close to the structureless predictions of mixing length theory. Convection cells are closed, which facilitates their experimental control, but fluctuations never exit and there is a dynamically determined bulk forcing. While the single-pass mode characteristic of wind tunnel experiments seems simpler, the convection cell is, in ways to be discussed, more constrained.

This review aims to familiarize the turbulence researcher with convergent lines of investigation in convection and also to remind those working in convection that turbulence is not a new subject. To situate convection within the gamut of other turbulent flows, let us by way of introduction contrast the directions in which convection has developed with research on the turbulent boundary layer.

From the onset of convection up to Rayleigh numbers $Ra \sim 10$ times critical, there is a great wealth of information about flow structures (which can be visualized from above), and their relative stabilities (Busse 1981). Turbulence, in the sense of many coupled modes, and not just sensitive dependence on initial conditions, can arise for low Ra in large aspect ratio

137

0066–4189/94/0115–0137$05.00

systems where it reduces to modulations of the basic cellular patterns (Newell et al 1993, Cross & Hohenberg 1993). For Ra 10^2–10^3 times critical and beyond, the vertical structure of the flow becomes important and organized motions have received less attention. In conformity with turbulence usage, "high" Ra will be loosely defined by where power-law scaling applies. In this regime mixing length ideas have heavily influenced what is studied. For instance, the behavior of the heat flux with Ra has monopolized attention to a far greater extent than the drag as a function of Reynolds number has dominated studies of the boundary layer. Both measurements are of direct technical importance but neither is local enough to reveal the details of the flow that are essential to advance phenomenological theory.

The recognition of thermal plumes as a coherent component of turbulent convection (Turner 1973), predates most studies of boundary layer bursts; yet subsequent quantitative studies of plumes have languished by comparison. For instance, Lu & Willmarth (1973) have measured the Reynolds stress (the analogue of the heat flux), conditioned upon the values of the component fields to display the contribution from the bursts. There are no comparable measurements of the local heat flux conditioned on temperature and vertical velocity. As for the nucleation of plumes at the walls, and the relative roles of buoyancy and shear, our knowledge is mostly qualitative, in contrast to the many studies in boundary layers where bursts are triggered and followed downstream.

The salient new feature that has enlivened convection studies in the past 10 years, and prompted this review, is the coherent large-scale buoyancy-driven circulation that in certain cases can persist to the highest Ra attained in the laboratory ($\sim 10^{14}$). This flow modifies, via its shear, the thermal boundary layer, which ultimately is responsible for the heat flux and the buoyancy, thereby posing a subtle self-consistency problem. An incidental consequence of this coupling is that the primitive theoretical tools extant are unable to select between two turbulent states (basically with and without shear), each of which appears internally consistent.

With the decision to review convection as a branch of turbulence comes the restriction to high Rayleigh numbers (and generally large Reynolds number also), and the focus on laboratory experiments in the Boussinesq limit. This is not to minimize the importance of geophysical problems and the recent successes in understanding solar convection (Spruit et al 1990). However, most progress in ordinary turbulence has been achieved through precisely controlled laboratory experiments and visualizations. Since, for the reasons alluded to, convection involves subtle competitions, laboratory investigations are essential, and the need for this review would not have arisen without them. Non-Boussinesq convection is not one problem but many—the Earth's mantle, the Sun, and a fluid near its critical point are not easily treated together. Our viewpoint, however, does not have wholly exclusionary implications since convection experiments, extended to small scales where buoyancy is not of direct relevance, blend naturally into the subject of passive scalars in turbulence.

To organize our concepts and expectations, we begin with a section on theory and its limitations with an eye towards experiment, which remains the preeminent mode of inquiry in this field. Experimental results and technique are reviewed in Section 3, generally in order of decreasing Prandtl number and increasing shear effects. Attention is focused on the best measurements of a given quantity and no attempt is made at historical completeness. Section 4 summarizes numerical simulations which have mushroomed in response to recent experimental findings.

2. THEORETICAL BACKGROUND

2.1   *Exact Results and Background*

In this section we formally define the problem under consideration, extract several integral relations from the basic equations, and recall some general results from incompressible turbulence necessary for the sequel. Rigorous bounds on flow quantities are also collected here.

We will deal exclusively with the Boussinesq approximation (Tritton 1988, p. 188) for a fluid between rigid horizontal conducting plates across which a constant temperature difference is maintained and the heat flux is measured or visa versa. The lateral walls are insulating and rigid in the case of experiment, or possibly free slip or periodic in the case of simulations. Nondimensionalization is conventionally done by setting the depth d, thermal diffusivity κ, and total temperature differences Δ, to one. The largest lateral dimension then becomes the aspect ratio A; the heat flux is given by the Nusselt number Nu (e.g. $Nu = 1$ for conduction); and the viscosity v, is replaced by the Prandtl number Pr. The Boussinesq equations then read (gravity along $\hat{z}$)

$$(\partial_t v + v \cdot \nabla v + \nabla p)/Pr = \nabla^2 v + Ra\theta\hat{z} \tag{2.1a}$$

$$\partial_t \theta + v \cdot \nabla \theta = \nabla^2 \theta, \tag{2.1b}$$

where θ is the temperature equal to $\pm\frac{1}{2}$ on the boundaries. The Rayleigh number Ra is $g\alpha\Delta d^3/\kappa v$, where g is the gravitational acceleration and α the thermal expansion coefficient. The relevant length and velocity scales needed to define the Reynolds number Re will depend on context.

If an overbar denotes an average over a plane $z = $ constant, and angular brackets a volume average, then (2.1b) implies

$$Nu \equiv \overline{v_z\theta} - \partial_z\bar{\theta} = \langle(\nabla\theta)^2\rangle. \tag{2.2}$$

The first equality, obtained by integrating over all $z' < z$, states that the heat flux is independent of z; the second, obtained by integrating the equation for $\frac{1}{2}\partial_t\theta^2$ over the entire volume, gives the dissipation rate of temperature variance, i.e. ε_θ in the turbulence literature. The analogous equation for the balance between buoyant forcing and kinetic energy dissipation follows by averaging an equation for the time derivative of total energy, $\partial_t(v^2/2 - z\theta\,Ra\,Pr)$, under stationary conditions,

$$(Nu-1)Ra = \langle(\nabla v)^2\rangle. \tag{2.3}$$

It should be emphasized that Equations (2.2) and (2.3) make no assumptions about any convectively driven flow.

Rigorous upper bounds on $Nu(Ra)$ [or lower bounds on $Ra(Nu)$] have been derived by maximizing the defining expression for Nu over all $\mathbf{v}$ and θ subject to incompressibility, (2.2), and (2.3). Since the techniques employed have been the subject of their own review (Howard 1972, Busse 1978), we merely state the results. For general Pr

$$Nu \leq (Ra/1035)^{1/2}. \tag{2.4a}$$

If $Pr \to \infty$ in (2.1a), then retaining only the piece of the pressure necessary to ensure incompressibility,

$$Nu \leq 0.152\,Ra^{1/3}. \tag{2.4b}$$

An interesting extension of this analysis was made by Howard (1990), who includes mean shear which puts a lower bound on Ra given Nu and the momentum flux $\tau = \partial_z\bar{v}_x(z=0)$ (supposing $\bar{v}_y=0$). Then for $\tau > 8(Nu-1)^2/Pr^{1/3}$ (among other cases), Howard finds

$$(Nu-1)Ra \geq \frac{8\sqrt{2}}{3}\tau^{3/2}Pr^{1/2}, \tag{2.5}$$

which will prove relevant to what follows.

In the same context the empirically verified scaling relations for a turbulent boundary layer are needed and we collect the pertinent formulae here (Tennekes & Lumley 1972, Hinze 1975). If we denote the friction velocity $(\tau\,Pr)^{1/2}$ by u_* and define $z_* = Pr/u_*$, then

$$v_x(z) = \begin{cases} u_*z/z_* & 0 \leq z \lesssim z_v \tag{2.6a}\\ u_*(2.5\ln(z/z_*)+5) & z_v \lesssim z \ll 1, \tag{2.6b} \end{cases}$$

where the viscous-buffer layer thickness, $z_v \sim (7\text{–}12)z_*$, and u_* is given implicity in terms of the large-scale Reynolds number as

$$u_* = Pr\, Re/[2.5\ln(u_*/Pr) + 6.0].\tag{2.7}$$

To the extent fluctuations away from the boundaries are homogeneous and isotropic, it will be useful in what follows to recall the von Karman-Howarth analysis (Landau & Lifshitz 1987, p. 139), which generalizes (2.3):

$$\langle \mathbf{v}\cdot\nabla v_i(r)v_i(0)\rangle/Pr + Ra\langle v_z(r)\theta(0)\rangle = \langle\nabla v_i(r)\nabla v_i(0)\rangle.\tag{2.8}$$

2.2 *Mixing Length Theories*

Mixing length theory refers to a class of arguments in which the assumption of a single temperature and velocity scale as a function of distance to the nearest boundary is used to balance (2.1a,b). While the applications to convection are well known and seemingly compelling, we will stress their limitations so as to lessen the surprise when other scaling laws arise.

Perhaps the most direct argument for the asymptotics of $Nu(Ra, Pr)$ is that of Priestly (cf Spiegel 1971), who supposes that the heat flux is regulated by processes confined to the region near the horizontal plates, and in addition that the two boundary layers do not communicate. It follows that the spacing between the plates should not enter into the expression for the physical heat flux, $(\kappa\Delta/d)Nu$, so assuming $Nu \sim f(Pr)Ra^x$ implies $x = \frac{1}{3}$. A rather more dangerous argument further asserts $f \sim Pr^{1/3}$ based on the supposition that for small Pr the zero viscosity limit of Nu exists (Spiegel 1971). (One should not suppose, however, that for $\kappa \to 0$ and Pr large, $f \sim Pr^{-1/3}$.)

These arguments are circumvented in one instance by the rather surprising experimental fact that a persistent mean flow can exist at high Ra, through which the boundary layers communicate. While this could not have been foreseen theoretically, the more careful mixing length treatment which follows strongly suggests that $Nu \sim Ra^{1/3}$ cannot be asymptotic, particularly from small Pr (Kraichnan 1962). The plate spacing is re-introduced into the problem via the Reynolds number and the invasion of the thermal boundary layers by the turbulent shear flow driven by the largest eddies.

A more thorough argument for $x = \frac{1}{3}$ proceeds slightly differently depending on Prandtl number (Kraichnan 1962). For $Pr \gtrsim 1$, there will be a thermal boundary layer within which we can ignore the velocity. Its thickness δ is related to the heat flux via $\delta = 0.5/Nu$ (half the temperature drop is across each boundary layer). If one imagines that the boundary layer thickens and then remains at the threshold for convection, the δ-based Rayleigh number, $Ra_\delta \sim 10^3$, or $Ra/(16\,Nu^3) \sim 10^3$ and $Nu \sim 0.05\,Ra^{1/3}$. [To infer whether this mechanism is operative, it is more revealing to compare $Ra/(16\,Nu^3)$ with 10^3 than worry about small discrepancies in the exponent.]

To estimate a characteristic velocity and temperature scale $v(z)$, $\theta(z)$ for $z > \delta$ (for smaller z one uses continuity and boundary conditions), two regimes are distinguished. Depending on whether the local Reynolds number $zv(z)/Pr$ is small or large, either the viscous term $\sim v/z^2$, or the advective term $\sim v^2/(z\,Pr)$, balances the buoyant force in (2.1a). Estimating $\theta(z)$ generally as $\sim Nu/v(z)$ yields for the two cases respectively,

$$v \sim (Ra\,Nu)^{1/2}z, \tag{2.9a}$$

$$v \sim (Ra\,Nu\,Pr\,z)^{1/3}. \tag{2.9b}$$

For small Pr the boundary layer thickness is defined by where the Peclet number $Pe_\delta = \delta v(\delta) \sim 1$ (actually ~ 10 in experiments). The approximate equality of diffusive and convective fluxes at $z \sim \delta$ then implies $Nu = 0.5/\delta \sim v(\delta)$. Since $Re_\delta \sim Pr^{-1}$ is now large, we use (2.9b) for $v(\delta)$ together with $\delta\,Nu \sim 1$ to yield $Nu \sim v(\delta) \sim (Ra\,Pr)^{1/3}$.

In summary, the basic mixing length theory yields

$$Nu \sim 0.18(Ra\,Pr)^{1/3} \quad Pr \lesssim 0.1$$

$$Nu \sim 0.066\,Ra^{1/3} \quad Pr \gtrsim 0.1. \tag{2.10}$$

The first coefficient comes from Globe & Dropkin (1959) though there is some question whether a $\frac{1}{3}$ regime exists at low Pr (cf Rossby 1969, p. 321); the second is from Goldstein et al (1990) and is very close to the naive estimate $Ra/(16\,Nu^3) \sim 10^3$.

Provided that the advection dominates $\nabla^2 v$ in the center of the cell in the $Pr \gtrsim 0.1$ case, (2.9b) and (2.10) imply

$$Re \sim 0.4\,Ra^{4/9}\,Pr^{-2/3}, \tag{2.11}$$

where Re is defined in terms of the cell depth and the rms vertical velocity in the center, and the numerical coefficient follows from either Garon & Goldstein (1973) or Tanaka & Miyata (1980), mutatis mutandis.

The Re scaling in (2.11) permits us to check an important consistency condition for the $Ra^{1/3}$ regime—namely that the thickness of the viscous boundary layer set up by the large scales, z_v from (2.6–2.7), should exceed $\delta \sim 0.1\,Ra^{-1/3}$. Since z_v scales as $Ra^{-4/9}$ times logarithmic terms, ultimately it becomes smaller than δ; which for $Pr \sim 1$ (6) occurs for $Ra \gtrsim 10^{18}$ (4×10^{23}). [Kraichnan (1962), extrapolating from less extensive data, found yet larger Ra for the crossover.]

For these truly asymptotic Ra, and $Pr \lesssim O(1)$, mixing length theory

(Kraichnan 1962) again adopts a Peclet criterion for the thermal boundary layer thickness $\delta = 0.5/Nu$ using u_* as the velocity, i.e. $u_*/Nu \sim 1$. Then (2.9b) fixes the large-scale velocity and thus Re, which (2.7) ties to u_*. The result (Kraichnan 1962) is

$$Nu \sim (Ra\,Pr)^{1/2}/(\ln Ra)^{3/2}, \qquad\qquad (2.12)$$

for $Pr \lesssim 1$, and (2.12) times $Pr^{-3/4}$ otherwise. Therefore to within logarithmic factors, this result attains the rigorous bound in (2.4a). When (2.12) applies, buoyancy only acts indirectly, via the large-scale shear on the thermal boundary layer which itself remains convectively stable, i.e. $Ra/(16\,Nu^3) \ll 10^3$.

Since the crossover to (2.12) is well out of reach of laboratory experiments except perhaps in mercury (below), it is worthwhile to note the related problem for the torque G as a function of Reynolds number for Taylor-Couette flow where the logic leading to (2.12) has recently been confirmed (Lathrup et al 1992). Namely, the conventional skin friction result, $G \sim u_*^2$, fits experiment much better than the marginally stable boundary layer prediction, $G \sim Re^{5/3}$.

A novel variant on mixing length theory, motivated by experiment (Castaing et al 1989), supposes three layers: 1. a mixed core with velocity and temperature fluctuations v, θ independent of z; 2. a conventional thermal boundary layer with $\delta \sim Nu^{-1}$ and, 3. an intermediate plume-dominated regime with thin ($\sim \delta$), hot ($\theta \sim 1$), sheets of fluid ejected from the wall. As usual, in the core, $v \sim (Nu\,Ra\,Pr)^{1/3}$, which is matched to the velocity in the plume regime, $v_{\text{plume}} \sim Ra\,\delta^2$, derived under the assumption of small Reynolds number (viscous balance). One then finds

$$Nu \sim Ra^{2/7}\,Pr^{-1/7}, \quad Re \sim Ra^{3/7}\,Pr^{-5/7}, \quad \theta_{\text{core}} \sim Ra^{-1/7}\,Pr^{-3/7}, \quad (2.13)$$

where θ in the core was estimated as Nu/v. [The Pr dependence was not reported in Castaing et al (1989) but is no less plausible than the Ra dependence.]

2.3   Effects of Large-Scale Flow

This section establishes that when a buoyancy-driven large-scale shear flow develops for large enough Ra in a convection cell, the familiar relation (2.10) is superseded by (2.13) times logarithmic corrections. The internal bulk flow is turbulent, but the Reynolds number is not so large as to prevent the thermal boundary layer from lying within the viscous one (n.b. 2.12 will only apply at higher Ra). Also the "wind" makes the thermal boundary layer thinner than the stability limit, $Ra/(16\,Nu^3) \sim 10^3$, thereby evading the argument leading to (2.10).

The global dynamics of the flow are accounted for by estimating the energy dissipation in terms of the Reynolds number and substituting into (2.3). The $Nu(Re)$ relation can be convincingly established when the boundary layers are nested as assumed (a condition to be verified subsequently), by solving

$$(z/\gamma)\partial_x\theta = \partial_z^2\theta, \tag{2.14}$$

where γ^{-1} is the shear rate near the wall. The solution is $\theta = \frac{1}{2}-0.27I[z/(\gamma x)^{1/3}]$ with $I(y) = \int_0^y e^{-y^3/9}$ or $Nu \sim 0.5/(\gamma A)^{1/3}$. The dependence on the aspect ratio A for $A \gg 1$ is contingent on the shear originating from a single extended circulation, while for $A \ll 1$ details of the corner flow enter, and in all cases the origin of x should be offset to avoid problems at $x = 0$. With these assumptions, we arrive at two relations among Nu, Ra, and Re; thus fixing $Nu(Ra)$, and $Re(Ra)$.

To illustrate how this argument works in a simple context (Shraiman & Siggia 1990), consider the limit $Ra \gg 1$, $Re \ll 1$. In this case the velocity has a single scale, so $Nu\,Ra \sim \langle(\nabla v)^2\rangle \sim \langle v^2\rangle \sim (Re\,Pr)^2$, and $\gamma^{-1} = Re\,Pr$. Therefore, $Nu(\gamma)$ can be reexpressed as $Nu \sim Ra^{1/5}A^{-2/5}$ which agrees with Roberts (1979) (cf his Equation 5.23) for $A \lesssim 2$; otherwise multiple cells form. For completeness we note that for a Blasius boundary layer profile, $Nu \sim Re^{1/2}Pr^{1/3}A^{-1/2}$ and $\langle(\nabla v)^2\rangle \sim Re^3Pr^2A^{-1/2}$, implying

$$Nu \sim Ra^{1/4}Pr^{-1/6}A^{-1/2}. \tag{2.15}$$

[The similarity in the Ra exponent between (2.15) and the analogous relationship of Busse & Clever (1981) seems coincidental in view of the differing assumptions.]

The most interesting application of this reasoning (Shraiman & Siggia 1990) arises when the velocity boundary layer is turbulent with a characteristic velocity u_* (cf 2.7). We can then estimate the kinetic energy dissipation as

$$Pr\langle(\nabla v)^2\rangle \sim 100\,u_*^3. \tag{2.16}$$

The coefficient in (2.16) is large because the viscous layer, where the dissipation peaks, is rather thick in comparison with $z_* = Pr/u_*$, e.g. $z_v/z_* \sim 10$, and boundary layers are present along all the walls. Note also that the dissipation scales with u_* not Re (Hinze 1975). The shear rate is just $\gamma^{-1} = u_*/z_*$ so

$$Nu = 0.22\,Ra^{2/7}Pr^{-1/7}A^{-3/7}, \tag{2.17a}$$

$$Re = 0.052\,Ra^{3/7}[2.5\ln(u_*/Pr)+6.0]\,Pr^{-5/7}A^{-1/7}, \tag{2.17b}$$

where u_*/Pr is determined from Re via (2.7), and in analogy with (2.6b),

$$\bar{\theta}(z) \sim \text{constant}+(Nu/u_*)\ln(z/z_*). \tag{2.17c}$$

The numerical prefactors in (2.17a,b) are from the $A = 1$ cell of Wu & Libchaber (1992) [the theoretical numbers based on (2.16) would have been 0.27 and 0.18 respectively], and the cell height is used in defining Re. The experiment used the same fixed probe at all Ra, so the maximum velocity may be larger than that given in (2.17b); however using the coefficient in (2.17a) and (2.6–2.7), all Ra dependence disappears from (2.5) to yield a rigorous upper bound of 0.39 on the coefficient in (2.17b).

The lower Ra limit on the validity of (2.17) is set by $Re \gtrsim 3 \times 10^3$ which assumes that the maximum velocity occurs at a height equal to a substantial fraction of the depth. This lower limit ensures that the boundary layer is turbulent while the upper limit is the consistency requirement that the thermal boundary layer nests within the viscous one:

$$10^8 \, Pr^{5/3} \lesssim Ra \lesssim (3 \times 10^{12} – 2 \times 10^{14}) Pr^4. \tag{2.18}$$

The range of upper values corresponds to the choices $z_{\mathrm{v}} = (7–12) z_*$ when used along with the numerical coefficients in (2.17a,b). There is a large uncertainty in the upper limit because, although the boundary layers must ultimately cross, their thickness ratio has a small effective exponent, i.e. $\sim Ra^{0.2}$. (The A dependence of either limit requires additional assumptions and is not worth quoting.)

It should be emphasized that Equation (2.18) is merely a necessary condition for (2.17) to hold; it is not sufficient since we have assumed that the mean flow is dominant. When the upper limit in (2.18) is exceeded, the physics suggests a crossover to (2.12) which can be rederived with mean shear by combining the passive scalar transport estimate, $Nu \sim u_* / |\ln z_*|$, with (2.16) and (2.3). Clearly this level of theory cannot preclude a return to (2.10) and evidently the dependence on Pr is the best way to test (2.18).

Another potential limitation on (2.17)—namely that the thermal boundary layer be convectively stable, i.e. $Ra/(16 \, Nu^3) \lesssim 10^3$, or from (2.17a) $Ra \lesssim 7 \times 10^{14} Pr^{-3}$—is not a limitation in practice [n.b. the Pr dependence of this expression comes entirely from the $Pr^{-1/7}$ in (2.17a) and has not been checked experimentally]. In other words, the heat flux in (2.17a) is greater than in (2.10) for the accessible Ra, in spite of the unfavorable exponent, as one would expect intuitively from the shear.

Convection with a uniform imposed shear presents an interesting stability problem; for recent references see Clever & Busse (1991, 1992). The relevance of this literature to the thermal boundary layer associated with (2.17) subject to (2.18) is minimal because the Reynolds number based on $\bar{v}(z)$ at the edge of the viscous layer and hence throughout the thermal boundary layer is $\sim 10 z_{\mathrm{v}}/z_* \sim 100$ (cf 2.6 and below) and as just noted, the boundary layer Ra is less than 10^3. In a strictly two-dimensional flow a small amount of shear will suppress convection (Castaing et al 1989), but this restriction is unphysical since the rolls align with the mean flow. Convection and shear may interact in subtle ways on larger scales (cf Section 3.3) but this should not upset (2.16) or (2.17a,b) (for an alternative view see Zaleski 1991).

2.4 Buoyancy vs Shear

To understand the small-scale fluctuations and spectra in turbulent convection, one is forced to address the relative importance of the energy input directly into a given velocity mode by the buoyancy compared with the energy cascaded from larger scales. The analogous question for the temperature is: When does it reduce to a passive scalar?

Let us first consider the center of the cell and suppose that the Reynolds number is large enough, say $\gtrsim 10^5$ for the bulk to be reasonably homogeneous and isotropic and exhibit some scaling as a function of the spacing $r \ll 1$ over which differences, δv or $\delta\theta$, are defined. Take the aspect ratio large enough so that there is negligible convective heat transfer along the side walls. Under the hypothesis of no energy cascade and direct buoyant forcing, mixing length ideas (cf 2.9b) imply $\delta v^2/r \sim \delta\theta \, Ra \, Pr$. The scalar should cascade if the velocity does not, because turbulence should dissipate, and the dissipation rate for θ^2, ε_θ, is of order $\delta v \delta\theta^2/r$ which furnishes a second equation for δv, $\delta\theta$. For consistency one must suppose $\delta v \, \delta\theta \ll Nu$, since otherwise Kolmogorov scaling for the velocity would reemerge in the guise $\delta v^3/r \sim Nu \, Ra \, Pr$ (cf 2.3). This inequality is physically reasonable because the heat flux is carried by the largest scales accessible (cf Fitzjarrald 1976). This chain of reasoning results in the Bolgiano-Obukhov scaling (Procaccia & Zeitak 1989):

$$\delta v \sim (\varepsilon_\theta \, Ra^2 \, Pr^2)^{1/5} \, r^{3/5}$$

$$\delta\theta \sim (\varepsilon_\theta^2/(Ra, Pr))^{1/5} \, r^{1/5}, \tag{2.19}$$

for r larger than a diffusive cutoff $l_\theta \sim (Ra^2 \, Pr^2 \varepsilon_\theta)^{-1/8}$ set by $\delta\theta^2(l_\theta)/l_\theta^2 \sim \varepsilon_\theta$. In addition, if (2.19) applies up to $r \sim 1$ then the bound $\delta v \delta\theta < Nu$ together with (2.17a) requires $\varepsilon_\theta < Ra^{1/7} \, Pr^{-4/7}$.

The partitioning of heat flux over scales is defined by the correlation function $\langle v_z(r)\theta(0) \rangle$ appearing in (2.8). Experiment (Fitzjarrald 1976) suggests that this correlation function sticks at its $r = 0$ value, Nu, until $r \gtrsim 1$. Then we can invoke (2.8) for $r \ll 1$ (which is the formal statement that the velocity must cascade and is exact given our symmetries) to derive the Kolmogorov law $\delta v^3 = \varepsilon_v r$, with the energy dissipation rate $\varepsilon_v = Ra \, Nu \, Pr$. Therefore we regard (2.19) as untenable as an asymptotic statement. Some caution is in order as regards to numerical factors since as remarked in conjunction with (2.16), most of the kinetic energy dissipation takes place in the boundary layers, not the bulk. Therefore ε_v deduced from (2.8) is too large, and this can be rationalized most readily in an $A \sim 1$ cell by supposing that heat is predominately transported along the side walls and that $v_z(r)\theta(0)$ averaged in the bulk is substantially less than Nu.

To quantify the importance of direct buoyant forcing within the Kolmogorov picture, we need the analogous Corrsin-Obukhov prediction for the scalar, $\delta\theta \sim \varepsilon_\theta^{1/2}\varepsilon_v^{-1/6}r^{1/3}$, and an estimate for ε_θ in the bulk. If we use the upper bound for ε_θ of Nu given by (2.2), then applying (2.17a) gives $\delta\theta(r \sim 1) \sim Ra^{-1/14}$. This exceeds the scaling limit $\sim Nu/u_*$ (cf 2.17c) so we adjust ε_θ downwards to $\sim Ra^{1/7}$ reasoning that most of the dissipation occurs in the boundary layers. Finally the ratio of shear to buoyancy in (2.1a) becomes $r^{-1}\delta v^2/(Ra\,Pr\,\delta\theta) \sim r^{-2/3} \gtrsim 1$.

While scaling arguments rule out (2.19), small numerical factors do occur which favor direct buoyant forcing. For instance within the log-layer, the rate of vorticity production by the shear is no greater than $u_*^2/(z_* z_v) \sim 10^{-6}Ra^{12/7}Pr^{-6/7}$ via (2.6–2.7) and (2.17b); while the buoyancy contribution is $Ra\,Pr\,\partial_x\theta$. Although the exponents favor the shear, the buoyancy dominates, particularly when the boundary layer separates and a plume is ejected, during which time $\partial_x\theta$ could scale as Nu.

More generally, averaged estimates such as (2.10) or (2.17), are a priori suspect, and higher moments more so, when plumes play an important role in the dynamics. A very analogous situation exists for the boundary layer where mixing length arguments have proven successful although the bursts contribute appreciable Reynolds stress (Lu & Willmarth 1973). With buoyancy and shear both present, even their relative contributions to the instability leading to plume emission is out of reach theoretically.

For comparison with experiment, two standard scaling results for plumes at high Re should be noted (Turner 1973). The first presupposes a constant source of heat (analogous to fixed Nu) and gives a velocity $\sim z^{1/3}$ as in (2.9b) and a diameter $\sim z$. The second assumes a detached bubble with size l and temperature contrast $\Delta\theta$ obeying $l^3\Delta\theta = Q = $ const. Then $v \sim (Ra\,Pr\,Q)^{1/2}/l$. However experiments by Moses et al (1993), including one run under reasonably turbulent conditions, find a constant plume velocity.

3. EXPERIMENTS

3.1 *Large-Scale Structure of the Flow*

In 1981 Krishnamurti & Howard discovered, in a variety of large aspect ratio cells (cf Table 1), an internally generated coherent flow, moving with a steady mean along the top plate and returning along the bottom.

Table 1 Synopsis of the principle experiments discussed in the text[a]

Reference	A	Pr	Mean flow	(Range for flow)	Nu	(Range for Nu)
Goldstein et al 1990	1.3–9	2750	?		$0.0659\,Ra^{1/3}$	$(3\times10^9\text{–}5\times10^{12})$
Krishnamurti & Howard 1981	3.5, 10, 24	6, 10^3	y	$(>3\times10^6,\ >10^7)$		—
Garon & Goldstein 1973	2.5, 4.5	5.5	?	—	$0.130\,Ra^{0.293}$	$(10^7\text{–}3\times10^9)$
Goldstein & Tokuda 1980	0.57–2.5	6.5	?	—	$0.0556\,Ra^{1/3}$	$(10^9\text{–}2\times10^{11})$
Tanaka & Miyata 1980	3.5–1.4	6.8	?	—	$0.145\,Ra^{0.290}$	$(3\times10^7\text{–}4\times10^9)$
Zocchi et al 1990	1	5.6	y	$(\sim10^9)$	—	—
Threlfall 1975	2.5	0.8	?	—	$0.173\,Ra^{0.280}$	$(4\times10^5\text{–}2\times10^9)$
Castaing et al 1989	0.5	0.64–1.5	y	$(>10^8)$	$0.17\,Ra^{0.290}$	$(10^8\text{–}10^{14})$
Wu & Libchaber 1992	1.0	0.64–1.14	y	$(>4\times10^7)$	$0.22\,Ra^{0.285}$	$(4\times10^7\text{–}10^{12})$
Wu 1991	6.7	0.64–1.0	y	?	$0.146\,Ra^{0.286}$	$(10^4\text{–}10^{10})$
Belmont & Libchaber (private communication)	1	0.7	y	$(10^7\text{–}10^{11})$	—	—
Rossby 1969	7.4, 12.3	0.025	?	—	$0.147\,Ra^{0.247}$	$(2\times10^4\text{–}5\times10^5)$

[a] The experiments are ordered by Pr. The highest Prandtl numbers correspond to oil or electroconvection, those ~6 are water experiments, helium and other gases comprise the $Pr\sim1$ group, and mercury the last. A mean flow was marked "y" even if there were occasional reversals (see text for details).

Although the *Re* based on mean flow and depth did not exceed 50, the effect in retrospect presaged more recent higher *Ra* experiments, already noted in connection with (2.17a,b) where the shear generated by the coherent flow alters the heat flux. The sign of the coherent flow is random, as in a symmetry breaking bifurcation, yet the flow is turbulent when this occurs. Particular attention was paid to obvious biases such as tilt or side wall heating so the effect is intrinsic and not understood (but see Howard & Krishnamurti 1986 and Busse 1983).

The same nonzero averaged circulation was also observed at elevated *Ra* in a water experiment by Zocchi et al (1990) and in $A = 1$ helium cells (Wu 1991, Castaing et al 1989) as shown in Table 1. [Their $A = 0.5$ and 6.7 cells exhibited mean flow but with occasional reversals (Wu & Libchaber 1992).] The helium measurements are both more surprising than the earlier ones, in the sense that the *Ra* is higher and *Pr* lower, but less surprising since the aspect ratio is smaller and the large-scale mode is reminiscent of the "fly wheel" effect proposed at low *Pr* (Proctor 1977). However the precise conditions required for the coherent flow are a bit obscure since Tanaka & Miyata (1980), in an experiment similar to Zocchi et al (1990), do not note the effect.

The balance of this section is organized in order of decreasing Prandtl number and roughly parallel with the theoretical development in Section 2. Traditional mixing length ideas are generally confirmed in water with $Pr \sim 5$–6 and for all higher *Pr*, while shear effects will be seen to predominate for $Pr \lesssim 1$.

3.2   *Successes of Mixing Length Theory for* $Pr \gtrsim 5$

By far the highest Rayleigh numbers in this regime were attained by Goldstein et al (1990) who employed an electro-chemical analogue to convection to reach $Ra \sim 5 \times 10^{12}$ with $Pr = 2750$. A flux of Cu^{++} ions in a $CuSO_4$ solution between two Cu electrodes is established by applying a current. Sulfuric acid is present in much greater concentrations to screen out the electric field and eliminate any ion current that would result. If the voltage is set at the "plateau" of the current-voltage plot, the Cu^{++} concentration at the cathode is zero and that at the anode is assumed to be twice the solution average by symmetry (though the justification for this assumption under all flow conditions is unclear). The local density is proportional to Cu^{++} concentration. Assuming isothermal conditions, this determines the "Rayleigh" number, and the "Nusselt" number follows from the current after corrections are made. [It would be reassuring to see this technique accurately reproduce *Nu(Ra)* obtained by conventional means for a cell of identical shape and *Pr*.] The authors' obtain an accurate fit to Equation (2.10) (Table 1).

Although no velocity measurements were made, it is reasonable to estimate their Reynolds number by extrapolating from a fit of mixing length theory to data in water (2.11) which implies $Re \sim 900$ at $Ra \sim 5 \times 10^{12}$. A Re this low precludes any contradiction with (2.17a,b) (and by implication the helium experiments in Section 3.4), which by (2.18) only apply for $Ra \gtrsim 5 \times 10^{13}$ at $Pr = 2750$.

Some of the most detailed investigations of mixing length ideas come from water at $Pr = 5–6$. For $Ra \lesssim 10^9$, all experiments in water and air give exponents for $Nu(Ra)$ around 0.29 (Table 1 and Goldstein et al 1990), as shown in Figure 1. At high Ra Figure 1 shows a crossover to an exponent of $\frac{1}{3}$, but there is not much overlap between data at different aspect ratios.

The theory leading to (2.17a) only applies in water for $Ra > 2 \times 10^9$,

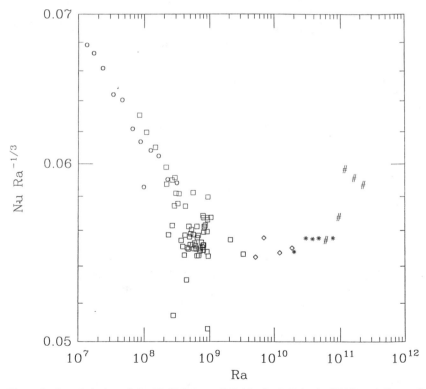

Figure 1 A scaled plot of the $Nu(Ra)$ data of Goldstein & Tokuda (1980) and Garon & Goldstein (1973). A fit to the data below 3.3×10^9 gave $Nu = 0.130\ Ra^{0.293}$. The symbols— *circle, square, diamond, star,* and *pound*—denote the aspect ratios 4.5, 2.5, 0.98, 0.67, and 0.57, respectively. The points at $Ra = 1.18$ and 1.64×10^{11} are "suspect."

provided that there is a mean flow—although (2.13) could apply at lower *Ra* and nicely accounts for the exponent of 0.29. Why the $\frac{1}{3}$ exponent occurs only above 10^9, and whether the variable aspect ratio in Figure 1 masks some other scaling are open questions.

In my view, direct measurements of the velocity should provide a more compelling test of mixing length ideas than small differences in the $Nu(Ra)$ exponent. The scaling of Re, defined around the midpoint of the cell, with Ra was already noted in (2.11) from Garon & Goldstein (1973) and Tanaka & Miyata (1980). The Prandtl number of water is generally felt to be large enough for the velocity fluctuations to evolve from (2.9a) to (2.9b) as z moves beyond the boundary layer. Garon & Goldstein (1973) provide evidence for (2.9b), while Chung et al (1992) find evidence for a crossover in the temperature data $[\theta(z) \sim Nu/v(z)]$. Far more extensive data should soon be available from particle image velocimetry (Adrian et al 1990). Finally, by using 0.05 mm diameter thermocouple wire, Tanaka & Miyata (1980) measured the mean temperature well within the boundary layer and found it to be accurately linear out to $z = 0.5/Nu$.

3.3 Plumes and Visualizations in Water

Plumes—hot (cold) parcels of fluid fed from the lower (upper) boundary layers—and thermals—detached parcels of marked fluid—have long been associated with convection particularly at high *Pr*. Experimental capabilities have advanced in recent years due to the application of digital imaging (Hesselink 1988). Gluckman et al (1993) provide a nice illustration of a technique for simultaneous measurement of isothermal surfaces and the velocity field (in a plane). They seed the flow with small (50–100 μm) encapsulated spheres of liquid crystals. This material undergoes a transition around 25°C to a phase which acts like a diffraction grating with temperature-dependent properties. At fixed scattering angle, there is a sigmoidal map of temperature to hue (peak scattered wavelength); thus the resolution of a given isotherm in scaled temperature can be optimized by adjusting the plate temperatures to place the isotherm on the inflection point of the response curve. The image created by plane illumination is recorded on a 480×512 grid in three color bands with 256 levels each. The projected velocity field on a $\sim 30^2$ mesh could also be simultaneously reconstructed by particle tracking. [Higher resolution measurements of velocity alone are given in Adrian et al (1990).]

For $10^7 < Ra < 3 \times 10^8$ in water, the statistics of the mean temperature isotherm from Gluckman et al (1993) were substantially the same in vertical and horizontal sections at the center of the cell. One could also hope to measure the heat flux carried by the more prominent plumes and the $v_z - \theta$ spectra by extensions of this technology, as was earlier done in air with a moving hot wire (Fitzjarrald 1976). An intriguing picture of ridge-like undulations of the thermal boundary layer was obtained with the super-imposed velocity field. These images should stimulate a fruitful dialog with those engaged in numerical simulations.

Zocchi et al (1990) use the same liquid crystal technique to study the dynamics of the thermal boundary layer in water. At $Ra = 1.2 \times 10^9$, they observed a mean flow with a definite sense of circulation but slow, bounded fluctuations in direction, and a maximum horizontal velocity of ~ 6 mm/sec. They report semi-quantitative statistics on "wave like" disturbances which have the appearance of unstable modes growing downstream in the large-scale flow. The measured dispersion relation, $\omega^2 \propto k$, is appropriate for Rayleigh-Taylor instability of the boundary layer and the numerical coefficient is reasonable. In a vertical slice through the fluid, they observe in the temperature field spirals (suggestive of vorticity ejected from the wall perpendicular to the plane of view) and tilted mushrooms (naturally interpreted as plumes). The mushrooms also resemble smoke visualizations of hairpin vortices viewed with the mean flow normal to the page (Head & Bandyopadhyay 1981). Since the "mushrooms" are seen ~ 1 cm from the wall (the thermal boundary layer is confined to 1.1 mm and the cell is 19 cm on a side), the applicable Reynolds number is under 100—a bit too low for shear effects to predominate. However, Shelley & Vinson (1992), based on an idealized two-dimensional model, conclude that the "spirals" and "mushrooms" originate from the same instability and differ in the degree of imposed shear.

Unfortunately there is insufficient theory available to guide interpretation. Numerical studies of plane Couette flow with heating (e.g. Domaradzki & Metcalfe 1988, Clever & Busse 1992) could be redesigned to model just the viscous boundary layer of experiments at high Ra and large enough Pr so that the velocity is linear over an interesting range of scales. Although the simulations do reach $Re \sim 10^3$, their Ra is $\lesssim 2 \times 10^5$; thus the top and bottom boundary layers interact directly, contrary to what happens at the higher Ra of the experiments where plumes do not span the cell. Also, the experiments seem to call for an initial value problem with a local disturbance allowed to grow downstream.

Moses et al (1993) study plume formation experimentally in the context of thermally-driven turbulence, and also review the earlier literature.

3.4 *Convection in Helium and the Effects of Shear*

The technical virtues of gaseous helium at low temperatures (4–5 K) as a medium for high Ra convection studies have been appreciated for some time (Threlfall 1975). They include: the ability to minimize the thermal noise in bolometers, the ease of thermal isolation by means of a vacuum enclosure and radiation shields, and the low heat capacity of solids at helium temperatures.

Another advantage, most thoroughly exploited by the Libchaber group (Heslot et al 1987, Wu 1991), is the ability to achieve most of the Rayleigh number variation by adjusting the fluid properties, i.e. pressure and mean temperature. In this way Ra can be varied by 10^6 or more for the same ΔT in a single cell (with minimal variation in Pr; e.g. 0.64–1.07 for $Ra = 10^7$–10^{13}). The heat source and temperature measurements can then be optimized within a more limited range (ΔT from 50 mk to 700 mk only). Unfortunately, visualizations are not possible. Non-Boussinesq effects are most pronounced in the thermal diffusivity which varied by only 12% between the two plates at $Ra = 10^{13}$ in Libchaber's aspect ratio 0.5 cell.

The high Ra and small viscosity imply very thin boundary layers, e.g. $\lesssim 0.1$ mm, with consequent demands on technique when point measurements are made. The Libchaber group has employed cubic bolometers 0.2 mm on a side suspended on thin wires, from which they extract the probability distribution function (PDF), temporal power spectrum, and also the large-scale velocity (by cross-correlating two nearby bolometers to determine a transit time).

Three cylindrical cells of aspect ratio 0.5, 1.0, and 6.7 were used. For the two smaller cells there is a change around $Ra \gtrsim 10^7$ in $Nu(Ra)$ from an ill-defined exponent near $\frac{1}{3}$ to one indistinguishable from 2/7 (Figures 2, 3). This break correlates with the appearance of a quasi-steady large-scale flow and a change in the temperature PDFs in the center of the cell from Gaussian to exponential which we discuss below.[1]

This crossover coincides with the point where $Re \sim 2000$, in accord with the assumptions made in deriving (2.17–18). In the aspect ratio 6.7 cell a reasonable fit to 2/7 extends to such low Ra in Figure 4 so that the Reynolds number must be $< 10^3$—though it was not measured. Data in the wider cell agree well with those of Threlfall (1975), who obtained an exponent of 0.2800 ± 0.0005 (Table 1).

The mean vertical velocity, measured 1 cm from the side wall in the $A = 1$ cell (diameter $= 8.7$ cm) (Sano et al 1989, Wu 1991, Wu & Libchaber 1992) defined the large-scale Re, and the velocity at other points was checked to verify the coherence of the flow. Although the experiments were first fit as $Re = 0.31\ Ra^{0.485 \pm 0.005}$, Figure 5 shows the fit to (2.17b)

---

[1] We do not use the terms "soft" and "hard" convection (Castaing et al 1989) to refer to the states on opposite sides of this transition because the definitions are irreparably muddled. For the original $A = 1$ cell, a mean flow, exponential probability distributions, and the 2/7 scaling of Nu all occurred together and constituted hard turbulence. In other cells and experiments each of these properties can occur singly. The "hard" regime is also unlikely to be asymptotic.

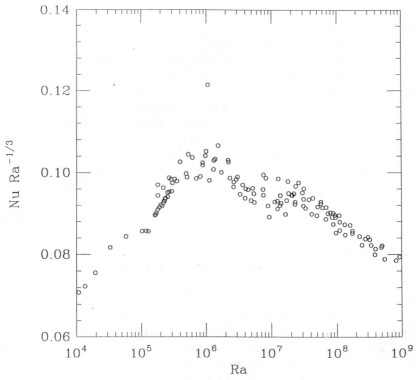

Figure 2 The low *Ra* portion of the *Nu* vs *Ra* data for the $A = 1$ helium cell from Castaing et al (1989) and Wu (1991). The *Pr* is 0.64–0.66. There is no well-defined 1/3 exponent but rather a cross over to (2.17a) for $Ra \gtrsim 10^7$.

with only one free parameter, the overall coefficient. The data are sufficiently precise to show that the *Pr* dependence in (2.17b) is definitely necessary as is the logarithmic factor; the mixing length expression (2.11) would not fit.

The rms temperature fluctuations in the center measured with a single bolometer, scaled consistently as $\sim Ra^{-1/7}$ for $A = 0.5$ and 1; and as $Ra^{-0.20 \pm 0.01}$ for $A = 6.7$. These exponents are reasonably close to the scaling estimates of Nu/Re, or Nu/u_*, e.g. (2.13) or (2.17c).

In addition, temporal spectra at the center and side probes, if taken literally, suggest a transition within the 2/7 regime (Sano et al 1989, Wu et al 1990, Procaccia et al 1991). The low-frequency cutoff in physical units is set by the circulation time of the mean flow around the box. [A more elaborate interpretation was given to this "ω_p" in Castaing et al (1989)

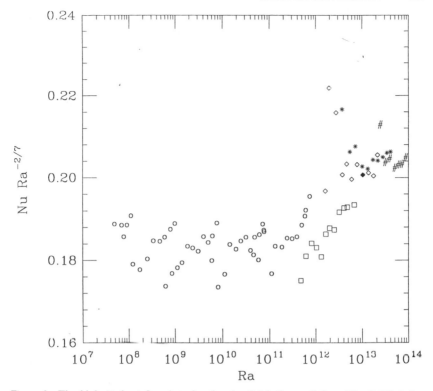

Figure 3 The high *Ra* heat flux data for the $A = 0.5$ helium cell from Wu & Libchaber (1992) and Wu (1991) scaled according to (2.17a) (inclusion of $Pr^{-1/7}$ gives a slightly worse fit). The *circle*, *square*, *diamond*, *star*, and *pound* correspond to $Pr \simeq 0.67, 0.80, 0.97, 1.07$, and 1.52. For the solid diamond near $Ra \sim 10^{13}$, there is a 12% difference in thermal diffusivity between the temperature extremes, and somewhat greater non-Boussinesq effects at higher *Ra*. The fit to (2.17a) for $A = 1$ is comparable.

which has been superseded.] For $Ra < 10^{11}$ the spectrum is well fit by $(f_0/f)^s \exp(-f/f_h)$ (Figure 6) and s is found to increase from ~ 1 to ~ 1.4 with *Ra*. At higher *Ra*, Grossmann & Lohse (1993) have made a good case that the bolometer response for physically relevant frequencies is limited by the thermal response time of the fluid viscously locked to it, thereby calling into question the transitions observed beyond $Ra \sim 10^{11}$.

The similarity of the low-frequency exponent in Figure 6 with the Bolgiano-Obukhov exponent in (2.19) has been noted (Procaccia & Zeitak 1989, Procaccia et al 1991), but is probably a coincidence. The mean flow is not sufficient, particularly in the cell center, to permit a linear relation between wave number and frequency, and since the effective velocity of

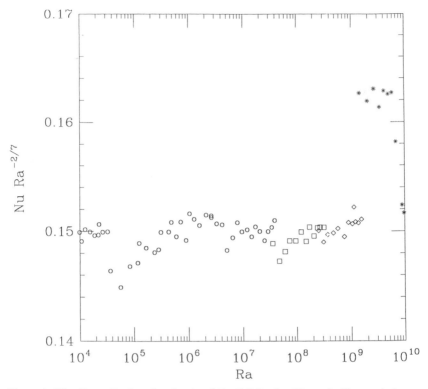

Figure 4 The *Nu* vs *Ra* data for the $A = 6.7$ cell following Figure 3. The symbols now correspond to $Pr = 0.65$, 0.7, 0.8, and 1.1 with comparable non-Boussinesq effects at the highest *Ra*.

advection increases with frequency, the wave number spectrum will be steeper than the frequency spectrum. The lower frequency limit to the $f^{-1.4}$ scaling corresponds to the circulation time around the cell and is implausibly low for inertial range scaling. Even if the wave number exponent were 1.4, over less than 2 decades, it would be dangerous to infer that the physics leading to (2.19) is responsible. For a passive scalar, equally convincing exponents of 1.35 behind a cylinder (figure 5 of Sreenivasan 1991) and ~1.45 in a turbulent boundary layer (figure 18 of Mestayer 1982) have been found.

To further quantify and visualize the new mode of convection discovered in helium, A. Belmonte & A. Libchaber (private communication) have employed a cubic cell with a variety of room temperature gases as the convecting fluid. By varying the pressure P, the Rayleigh number which

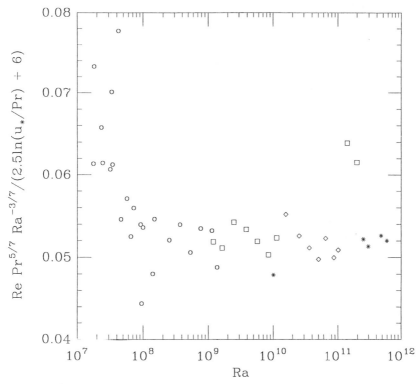

Figure 5   The Reynolds number scaled according to (2.17b) for the *A* = 1 data in Figure 2. The symbols—*circle, square, diamond, star,* and *pound*—correspond to *Pr* ≃ 0.65, 0.7, 0.9, and 1.4.

scales as P^2, could be varied from $\sim 10^6$ to 10^{11} with ΔT maintained in the range 10–30°C, and $Pr = 0.7$. A mean flow was clearly evident at all *Ra*.

In addition, a bolometer of the same style and size as used in the helium experiments could be positioned continuously along a 0.9 mm vertical rod under the center of the top plate. The peak in $\overline{\theta^2}$ falls at a value of z that correlates well with the thermal boundary layer thickness as inferred from the helium data, or the break in $\bar{\theta}(z)$ measured directly. Some measure of the vertical penetration of the plumes is provided by the shape of the θ distribution which together with inferences about the velocity from the time dependence of θ and two-point correlations should help quantify how convection drives the mean flow.

To more directly address the role played by the mean flow in the helium

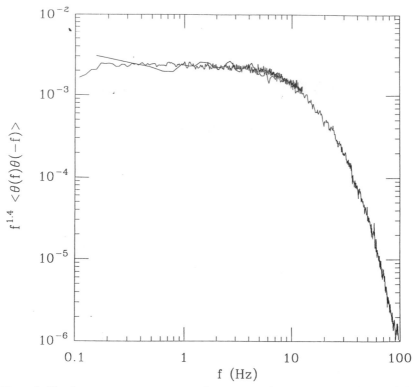

Figure 6 The frequency power spectrum of temperature fluctuations in the center of the $A = 0.5$ helium cell at $Ra = 7.3 \times 10^{10}$ from Wu et al (1990). Two traces at different sampling rates are superimposed. The measured large-scale velocity is 11 cm/sec at this Ra and the cell diameter is 20 cm.

experiments, Solomon & Gollub (1990, 1991) devised a way to apply an external shear to a convection system (water with $Ra \lesssim 3 \times 10^8$) while measuring $Nu(Ra)$ and visualizing with temperature sensitive liquid crystals. The shear was created by using, as a lower "plate," mercury that was set in motion by running a current through it in the presence of fixed magnetic field. In one study, oscillatory solid body rotation was induced in the mercury so that the shear zone in the water had a thickness comparable to the thermal boundary layer and a maximum velocity at that height of ~ 4 times the natural one. The shear drastically changed the number and size of plumes, but left $Nu(Ra)$ unchanged. In a second experiment, a fixed 4×4 grid of alternating vertical vortex columns in the mercury created, via Ekman pumping, a roughly fixed array of secondary flows in the overlying water layer. The heat flux then scaled with the imposed forcing. The magnitude of the effect was sufficient to account for the change from the mixing length expression for *Nu(Ra)* to (2.17a), a result that supports the importance of the mean flow.

3.5 *Probability Distribution Function of Temperature and its Derivatives*

In the first systematic measurements of the temperature probability distribution function, PDF, in convection by Castaing et al (1989), universal (*Ra* independent when scaled to unit variance), nearly exponential, distributions appeared at the center of the cell within the 2/7 regime (Figure 7). More extensive measurements by the Libchaber group and others, plus numerical simulations, have made the systematics for the occurrence of these unexpected distributions more obscure rather than less. The subject has taken on a life of its own which we now recount.

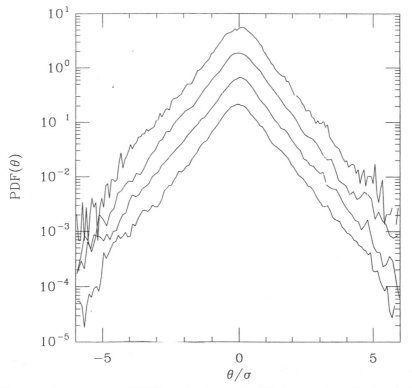

Figure 7   The temperature PDF from Castaing et al (1989) for $Ra = 2.7 \times 10^8$, 2.3×10^9, 4.2×10^{10}, and 5.4×10^{11}, increasing downwards. The data at each *Ra* are offset by a factor of 3 and the temperature is normalized to the variance at each *Ra*.

Measurements at probes located off center, and particularly near the wall (Sano et al 1989, Solomon & Gollub 1990, Tanaka & Miyata 1980) show distributions that are skewed toward or have tails in the direction of the nearest wall (Figure 8). This is indicative of occasional large-temperature excursions due to plumes. Under some conditions, when the data taken off center are high-pass filtered, exponentials reemerge (Libchaber et al 1990). In their $A = 6.7$ cell, Wu & Libchaber (1992) find universal but nonexponential distributions in the center which only become exponential when the low frequencies are removed. If an exponential PDF is characteristic of temperature fluctuations in the center of a roll, then the unsteadiness of the large-scale flow for $A = 6.7$ would account for the change in PDF with filtering.

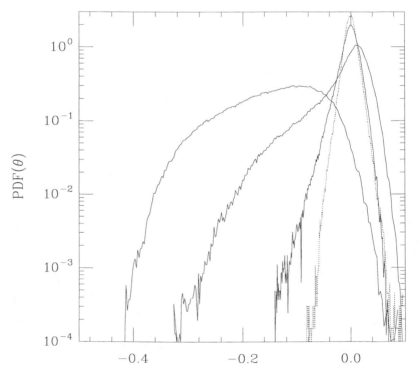

Figure 8   The temperature PDF from the pressurized gas cell of A. Belmonte & A. Libchaber (personal communication) at $Ra = 4.8 \times 10^7$. The traces from the left correspond to the bolometer at $z = 0.014$ [maximum of $\overline{\theta^2}(z)$ and at the edge of the thermal boundary layer as defined by $\bar{\theta}(z)$], $z = 0.047$, $z = 0.24$, $z = 0.49$ (*dashed*, cell center). The position is measured from the top plate and normalized to unit depth and the temperature is scaled to $[-0.5, 0.5]$. At higher Ra, the variance of the central exponentials would shrink as $Ra^{-1/7}$ while those at $z \sim Nu^{-1}$ would be roughly fixed.

Solomon & Gollub (1990) at $Ra \sim 10^8$ in water find exponentials in the center irrespective of the shear they impose on the boundary layer which qualitatively alters the emission of plumes. The global flow state at these Ra is probably not the same in water as in helium, because of the higher Pr and lower Re. Numerical simulations of the temperature PDF are discussed in Section 4.

There is a general expectation that the PDFs of large-scale quantities in turbulence are Gaussian (Tennekes & Lumley 1972). So it is of interest to note two recent experiments, which following a suggestion of Pumir et al (1991), demonstrate one mechanism for the production of exponentials. In both cases the temperature was passive, the turbulence was homogeneous, and its integral scale was much less than the lateral dimension along which a constant temperature gradient was imposed. Gollub et al (1991) obtained a stationary distribution in water by forcing with an oscillatory grid and found a remarkably exponential temperature distribution (flatness $\gtrsim 5.5$) for large enough Re. Their velocity field was nonisotropic but reasonably Gaussian in the direction along the scalar gradient.

Jayesh & Warhaft (1991, 1992) studied conventional grid turbulence where the temperature PDF had exponential tails but a smaller flatness factor of 3.8–5 throughout the tunnel. They also introduced temperature variance with no mean gradient and found a strictly Gaussian distribution. The relevance of any of this to convection will be decided by whether measurements of $\bar{\theta}(z)$ reveal an appreciable linear gradient [on the scale of $(\theta - \bar{\theta})^2$] in the cell center.

The Jayesh-Warhaft experiment does make implausible a theory of Sinai & Yakhot (1989) which yielded an algebraic PDF (with a Gaussian limit) for a passive decaying scalar in the absence of a mean gradient (the experiment gave Gaussians). Their theoretical reexpression of the PDF in terms of a conditional probability is rigorous but subsequent arguments that close this expression are probably superfluous if they only function to rederive a Gaussian. Also flawed is the theory Yakhot (1989), who factored out the z-dependent quantity, $\partial_z \bar{\theta}$, from a volume average, and made other ad hoc decouplings.

Ching (1991) fits the PDFs of temporal temperature differences from the center probe in the $A = \frac{1}{2}$ cell of Wu (1991), to $P(x) = a \exp(-b|x|^c)$. For short time differences, which approximate a derivative, and all $Ra \gtrsim 10^8$, she found $c = 0.5 \pm 0.05$ followed by a crossover to $c \sim 1.7 \pm 0.1$ for large differences. The later data are also consistent with a Gaussian core plus exponential tails and would be more noticeably exponential if it were high-pass filtered. There are no theoretical predictions for the derivative PDF, though it has cusp-like behavior near the origin down to scales much less than the variance.

3.6   *Convection for* Pr ≪ *1*

The effects of large-scale flow and shear should all be magnified at low Pr. In mercury, with $Pr \sim 0.02$, it is feasible under Boussinesq conditions to exceed the upper limit in (2.18), and perhaps also the threshold for the crossover from (2.10) to (2.12) (S. Ciliberto, private communication). The range of 2/7 scaling predicted by (2.18) is rather small but may extend to lower Ra in large aspect ratio cells if the helium data are reliable. Whether $Nu(Ra)$ goes directly to (2.12) or proceeds via (2.10) when Ra exceeds the upper limit in (2.18) is unclear.

To date, the best measurements of Nu are those of Rossby (1969) (Table 1), which agree very well with the analytic calculations of Busse & Clever (1981) for $Nu-1$, i.e. $3\pi/64(2\,Ra)^{1/4}$, as regards prefactor and exponent. However, the difference between Nu (the quantity fit in the experiments) and $(Nu-1)$ is quite material when the highest Nu attained in the lab was 4! If the theory of Busse & Clever (1981) is used to infer the Rayleigh number at which $Re \sim 3000$, and where a transition to (2.17a,b) may occur; one finds $Ra \sim 10^5$, near the highest value attained by Rossby, and in good agreement with the lower limit in (2.18) obtained by fitting to the helium data. Suffice it to say, more extensive measurements in mercury should be very rewarding.

4.   NUMERICAL SIMULATIONS

Most of the experiments we have discussed are well beyond the capabilities of current computers, so serious compromises are required if simulations are to contribute at all to the discussion. The optimal trade-off between computational efficiency and fidelity to experiment are spectral methods that assume periodic or free-slip boundaries laterally and use Chebyshev polynomials with no-slip boundary conditions in the vertical. Since derivatives are not reduced to multiplications by the Chebyshev transform, imposing rigid boundaries laterally would significantly complicate the Poisson inversion for the pressure.

The number of grid points should scale as the cube of the Kolmogorov wave number, $(Nu\,Ra/Pr^2)^{1/4}$, for $Pr \lesssim 1$ (Grötzbach 1983) or the Batchelor cutoff for the temperature, $(Nu\,Ra)^{1/4}$ (Tennekes & Lumley 1972) otherwise. The boundary layers should be well resolved, for large Ra, since if the bulk grid spacing is ε, near the walls it becomes ε^2. The inverse time step is limited to Re times the cutoff wave number, though large eruptions from the boundary layers may strain this limit. Therefore the operation count required for one circulation time scales as $\sim Ra^{16/9}$ assuming (2.9, 2.10). To set the coefficient, Kerr (1993) for instance needs $48^2 \times 96$ modes per unit aspect ratio at $Ra \sim 10^7$ and ~ 8 Cray YMP hours to simulate one circulation time at this resolution.

The most realistic simulations currently are those of Kerr (1993) who treats an $A = 6$ cell at Ra up to 2×10^7; but the integration times are $\sim$ half a circulation time which is cause for concern. He finds the velocity spectra and derivative skewness to be comparable to conventional turbulence simulations at similar large-scale Re ($\lesssim 10^3$). This finding argues against the unconventional scaling (2.19), which suppresses the energy cascade. The spectrum of $\langle \theta(k) v_z(-k) \rangle$ is of interest in the same context and easily accessible numerically. Contour plots at fixed z of θ and v_z show circular structures within the thermal boundary layer and more sheet-like entities further away. Plumes can penetrate substantially through the cell, as shown by the skewness of the temperature fluctuations, and the θ PDF is only reasonably exponential in the cell center at the highest Ra. Kerr's Re is not high enough to honestly apply (2.17a,b) but perhaps the arguments leading to (2.13) could be checked.

Extensive statistics are also reported by Balachandar et al (1989) and Balachandar & Sirovich (1991) in a three-dimensional $A = 2\sqrt{2}$ cell with Ra up to 10^7 but with free slip top and bottom boundaries. Christie & Domaradzki (1993), with a code similar to Kerr's, find substantial dependence of the θ PDF on aspect ratio at fixed Ra, but this may be due to the low values of Ra ($\lesssim 6.3 \times 10^5$). The tendency, however, for small A to favor exponentials, accords with the experiments in helium discussed in Section 3.5.

Low Pr three-dimensional simulations are reviewed and extended by Thual (1992), who is primarily interested in the onset of turbulence. He also reports simulations done for $Pr = 0$, the limit being taken as if $Pe \to 0$.

Werne et al (1991), DeLuca et al (1990), and J. Werne (private communication) work in two dimensions with rigid horizontal and free slip vertical walls and $A = 1$, and thus are able to reach $Ra = 1.6 \times 10^8$ and integrate for $\sim 10^2$ convective times. They see a transition to exponential PDFs which are symmetrical within the well-mixed core, and find exponents for $Nu(Ra)$ and $Re(Ra)$ of 0.290 ± 0.005 and 0.55 ± 0.02, respectively, for $5 \times 10^6 < Ra < 1.6 \times 10^8$. They have a coherent flow (which was not seen in any of the three-dimensional simulations) and most of the heat transport is along the vertical walls. Some caution is in order because the simulations are at $Pr = 6$, and at $Pr = 1$ they are very much less turbulent, contrary to experiment. Also, real turbulent boundary layers are inherently three dimensional, so the assumptions made in deriving (2.17a,b) do not apply in detail.

The interesting problem of infinite Pr convection was addressed in Hansen et al (1990, 1992) in two dimensions with free slip top and bottom boundaries, lateral periodicity, with $A \lesssim 10$ for $Ra \lesssim 10^8$ and $A = 1.8$ at $Ra = 10^9$. The turbulence consists of random thermals plus a coherent flow. The $Nu(Ra)$ exponent is 0.315, consistent with (2.4b), and that for the energy spectrum, treated as isotropic, is -2.

5. CONCLUSION AND FUTURE PROSPECTS

Turbulent convection in the past five years has become an exciting subject because experimental advances have firmly established an alternative to the critical boundary theory and the $Nu \sim Ra^{1/3}$ scaling. The new physics brings with it all the complexities of turbulent shear flow, and argues for a convergence between conventional engineering turbulence approaches and convection research. The introduction of shear makes Pr as important a parameter as Ra; conventional mixing length ideas work well only for large Pr where the $\frac{1}{3}$ exponent applies, whereas the new work utilized helium with $Pr \sim 1$. The appearance of a large-scale flow—the source of the shear—also accentuates the relevance of the aspect ratio parameter.

The flow-dominated 2/7 regime (i.e. $Nu \sim Ra^{2/7}$) has a greater heat flux than the 1/3 scaling for all accessible Ra as is intuitively reasonable;[2] and only paradoxical when one ignores the coefficient [i.e. (2.10) vs (2.17a)]. In fact, numerical factors are indispensable to those indulging in pheno-menological considerations since numbers of order one are frequently ~ 0.1 or ~ 10, and are then raised to the 7th or 9th power when terms are balanced (cf Sections 2.3–2.4). That the 2/7 regime is not asymptotic—most probably only (2.12) is—does not detract from its importance, since by a numerical conspiracy it occurs over a wide range of Rayleigh numbers.

Our present understanding of anything turbulent is at best phenomeno-logical, so a multiplicity of measurements on the same system is essential to advance our understanding. Thus $Re(Ra)$ is no less interesting than $Nu(Ra)$ and the various temperature PDFs have a wealth of unexploited information. In the near future we can expect a detailed picture of the two-dimensional projected velocity via particle tracking (Adrian et al 1990) which should resolve questions about Equation (2.19) (see also Tong & Shen 1992). Better measurements of the $v_z - \theta$ correlation function will also help determine heat transport scales and quantify the amount of heat that moves along the side walls. It may also be of interest to intentionally tilt the cell and thus introduce a new parameter which couples directly to

---

[2] For lower Ra ($\lesssim 10^6$) the effects of an imposed shear on convection are more complex, see Domaradzki & Metcalfe (1988).

the mean flow. Expected too are measurements of $\overline{\theta''}(z)$ for Ra up to 10^{11} (A. Belmonte & A. Libchaber) which should provide a challenge to those pursuing one-point closures. Finally, one can hope that the very qualitative picture we have for plume formation in the presence of shear can be quantified via the methodology developed for turbulent bursts.

In contrast to this experimental cornucopia, theory can offer only a few crumbs. Two mechanisms [i.e. the arguments preceding (2.13) and (2.17a,b)] for the 2/7 regime have been offered by way of "postdiction." The $Re(Ra)$ exponent in (2.13) is close to, but definitely distinct from, experiment, whereas (2.17b) works much better. However for large aspect ratios, the 2/7 exponent is seen at much lower Ra than (2.18) would suggest. Perhaps the turbulent boundary layer is not a consequence of a bulk-driven mean flow as was assumed in deriving (2.18), but rather the buoyancy and plumes force the boundary layer directly and render it turbulent. The bulk flow would then derive from the boundary layers, which is an inversion of the usual situation encountered in engineering problems. Leaving aside numerical coefficients, the Pr dependence of the upper and lower limits in (2.18) is a legitimate prediction which should be tested by experiments in mercury. The more fundamental problem, to explain from first principles why there is a mean flow and a 2/7 exponent, is beyond reach.

While the 22 year gap between this review and its predecessor in these series (Spiegel 1971), seems to have inconvenienced no one, the current rate of experimental advances will soon antiquate this summary and humiliate its author where he was rash enough to make predictions.

ACKNOWLEDGMENTS

Whatever new perspective this review brings to convection stems from my continuing collaboration with Boris Shraiman whose contributions extend far beyond the few brief articles we coauthored. My understanding of the experiments owes much to discussions with Albert Libchaber, Jerry Gollub and their students. R. Kerr and J. Werne provided similar assistance with the simulations. All figures were replotted from the original data which was kindly supplied by A. Belmonte, E. Ching, and X. Wu. Belmonte, Gollub, Libchaber, and Shraiman commented on a draft of this review. A substantial fraction of the authors cited provided copies of their papers and responded to my questions. Space permits only a collective thanks. Financial support was supplied by the Air Force Office of Scientific Research grant no. 91-0011, and the National Science Foundation grant no. DMR-9012974.

166   SIGGIA

Literature Cited

Adrian, R. J., Offitt, P. W., Liu, Z.-C., Hanratty, T. J., Landreth, C. C. 1990. Studies of liquid turbulence using double-pulsed particle correlation. In *Application of Laser Techniques to Fluid Mechanics*, ed. R. J. Adrian, D. F. G. Durão, F. Durst, M. Maeda, J. H. Whitelaw, pp. 435–51. New York: Springer-Verlag. 479 pp.

Balanchandar, S., Maxey, M. R., Sirovich, L. 1989. Numerical simulation of high Rayleigh number convection. *J. Sci. Comput.* 4: 119–236

Balanchandar, S., Sirovich, L. 1991. Probability distribution functions in turbulent convection. *Phys. Fluids A* 3: 919–27

Busse, F. H. 1978. The optimum theory of turbulence. *Adv. Appl. Mech.* 18: 77–121

Busse, F. H. 1981. Transition to turbulence in Rayleigh-Bénard convection. In *Topics in Applied Physics*, ed. H. L. Swinney, J. P. Gollub, 45: 97–137. Berlin/Heidelberg/New York: Springer-Verlag

Busse, F. H. 1983. Generation of mean flows by thermal convection. *Physica D* 9: 287–99

Busse, F. H., Clever, R. M. 1981. An asymptotic model of two-dimensional convection in the limit of low Prandtl number. *J. Fluid Mech.* 102: 75–83

Castaing, B., Gunaratne, G., Heslot, F., Kadanoff, L. Libchaber, A., et al. 1989. Scaling of hard thermal turbulence in Rayleigh-Bénard convection. *J. Fluid Mech.* 204: 1–30

Ching, E. S. C. 1991. Probabilities for temperature differences in Rayleigh-Bénard convection. *Phys. Rev. A* 44: 3622–29

Christie, S. L., Domaradzki, J. A. 1993. Numerical evidence for nonuniversality of the soft/hard turbulence classification for thermal convection. *Phys. Fluids A* 5: 412–21

Chung, M. K., Yun, H. C., Adrian, R. J. 1992. Scale analysis and wall-layer models for the temperature profile in turbulent thermal convection. *Int. J. Heat Mass Transfer* 35: 43–51

Clever, R. M., Busse, F. H. 1991. Instabilities of longitudinal rolls in the presence of Poiseuille flow. *J. Fluid Mech.* 229: 517–29

Clever, R. M., Busse, F. H. 1992. Three-dimensional convection in a horizontal fluid layer subjected to a constant shear. *J. Fluid Mech.* 234: 511–27

Cross, M. C., Hohenberg, P. C. 1993. Pattern formation outside of equilibrium. *Rev. Mod. Phys.* To appear

DeLuca, E. E., Werne, J., Rosner, R. 1990. Numerical simulations of soft and hard turbulence: preliminary results for two-dimensional convection. *Phys. Rev. Lett.* 64: 2370–73

Domaradzki, J. A., Metcalfe R. W. 1988. Direct numerical simulations of the effects of shear on turbulent Rayleigh-Benard convection. *J. Fluid Mech.* 193: 499–531

Fitzjarrald, D. E. 1976. An experimental study of turbulent convection in air. *J. Fluid Mech.* 73: 693–719

Garon, A. M., Goldstein, R. J. 1973. Velocity and heat transfer measurements in thermal convection. *Phys. Fluids* 16: 1818–25

Globe, S., Dropkin, D. 1959. Natural-convection heat transfer in liquids confined by two horizontal plates and heated from below. *J. Heat Transfer Trans. ASME* 81: 24–28

Gluckman, B. J., Willaime, H., Gollub, J. P. 1993. Geometry of Isothermal and Isoconcentration surfaces in thermal turbulence. *Phys. Fluids A* 5: 647–61

Goldstein, R. J., Chiang, H. D., See, D. L. 1990. High-Rayleigh-number convection in a horizontal enclosure. *J. Fluid Mech.* 213: 111–26

Goldstein, R. J., Tokuda, S. 1980. Heat transfer by thermal convection at high Rayleigh numbers. *Int. J. Heat Mass Transfer* 23: 738–40

Gollub, J. P., Clarke, J., Gharib, M., Lane, B., Mesquita, O. N. 1991. Fluctuations and transport in a stirred fluid with a mean gradient. *Phys. Rev. Lett.* 67: 3507–10

Grossmann, S., Lohse, D. 1993. Characteristic scales in Rayleigh-Benard turbulence. *Phys. Lett. A* 173: 58

Grötzbach, G. 1983. Spacial resolution requirements for direct numerical simulations of the Rayleigh-Benard Convection. *J. Comput. Phys.* 49: 241–64

Hansen, U., Yuen, D. A., Kroening, S. E. 1990. Transition to hard turbulence in thermal convection at infinite Prandtl number. *Phys. Fluids A* 2: 2157–63

Hansen, U., Yuen, D. A., Malevsky, A. V. 1992. Comparison of steady-state and strong chaotic thermal convection at high Rayleigh number. *Phys. Rev. A* 46: 4742–54

Head, M. R., Bandyopadhyay, P. 1981. New aspects of turbulent boundary-layer structure. *J. Fluid Mech.* 107: 297–338

Heslot, F., Castaing, B., Libchaber, A. 1987. Transitions to turbulence in Helium gas. *Phys. Rev. A* 36: 5870–73

Hesselink, L. 1988. Digital image processing in flow visualization. *Annu. Rev. Fluid Mech.* 20: 421–85

Hinze, J. O. 1975. *Turbulence.* New York: McGraw Hill. 790 pp. 2nd ed.

Howard, L. 1972. Bounds on flow quantities. *Annu. Rev. Fluid Mech.* 4: 473–94

Howard, L. N. 1990. Limits on the transport of heat and momentum by turbulent convection with large-scale flow. *Stud. Appl. Math.* 83: 273–85

Howard, L. N., Krishnamurti, R. 1986. Large scale flow in turbulent convection a mathematical model. *J. Fluid Mech.* 170: 385–410

Jayesh, Warhaft, Z. 1991. Probability distribution of a passive scalar in grid-generated turbulence. *Phys. Rev. Lett.* 67: 3503–6

Jayesh, Warhaft, Z. 1992. Probability distribution, conditional dissipation, and transport of passive fluctuations in grid-generated turbulence. *Phys. Fluids A* 4: 2292–307

Kerr, R. M. 1993. Simulations of high Rayleigh number convection. *J. Fluid Mech.* Submitted

Kraichnan, R. H. 1962. Turbulent thermal convection at arbitrary Prandtl number. *Phys. Fluids* 5: 1374–89

Krishnamurti, R., Howard, L. N. 1981. Large-scale flow generation in turbulent convection. *Proc. Natl. Acad. Sci.* 78: 1981–85

Landau, L. D., Lifshitz, E. M. 1987. *Fluid Mechanics.* Elmsford, NY: Pergamon. 539 pp. 2nd ed.

Lathrop, D. P., Fineberg, J., Swinney, H. L. 1992. Transition to shear-driven turbulence in Couette-Taylor flow. *Phys. Rev. A* 46: 6390–405

Libchaber, A., Sano, M., Wu, X. 1990. About thermal turbulence. *Physica A* 163: 258–64

Lu, S. S., Willmarth, W. W. 1973. Measurements of the structure of the Reynolds stress in a turbulent boundary layer. *J. Fluid Mech.* 60: 481–511

Mestayer, P. 1982. Local isotropy and anisotropy in a high Reynolds number turbulent boundary layer. *J. Fluid Mech.* 125: 475–503

Moses, E., Zocchi, G., Libchaber, A. 1993. A localized heat source: an experimental study of laminar plumes. *J. Fluid Mech.* To appear

Newell, A. C., Passot, T., Lega, J. 1993. Order parameter equations for patterns *Annu. Rev. Fluid Mech.* 25: 399–453

Procaccia, I., Ching, E. S. C., Constantin, P., Kadanoff, L. Libchaber, A., et al. 1991. Transitions in convective turbulence: the role of thermal plumes. *Phys. Rev. A* 44: 8091–102

Procaccia, I. Zeitak, R. 1989. Scaling exponents in nonisotropic turbulence. *Phys. Rev. Lett.* 62: 2128–31

Proctor, M. R. E. 1977. Inertial convection at low Prandtl number. *J. Fluid Mech.* 82: 97–114

Pumir, A., Shraiman, B. I., Siggia, E. D. 1991. Exponential tails and random advection. *Phys. Rev. Lett.* 66: 2984–87

Roberts, G. O. 1979. Fast viscous Benard convection. *Geophys. Astrophys. Fluid Dyn.* 12: 235–72

Rossby, H. T. 1969. A study of Benard convection with and without rotation. *J. Fluid Mech.* 36: 309–35

Sano, M., Wu, X., Libchaber, A. 1989. Turbulence in helium-gas free convection. *Phys. Rev. A* 40: 6421–30

Sreenivasan, K. 1991. On local isotropy of passive scalars in turbulent shear flows. *Proc. R. Soc. London Ser. A* 434: 165–82

Shelley, M. J., Vinson, M. 1992. Coherent structures on a boundary layer in Rayleigh-Benard turbulence. *Nonlinearity* 5: 323–51

Shraiman, B. I., Siggia, E. D. 1990. Heat transport in high-Rayleigh-number convection. *Phys. Rev. A* 42: 3650–53

Sinai, Y. G., Yakhot, V. 1989. Limiting probability distributions of a passive scalar in a random velocity field. *Phys. Rev. Lett.* 63: 1962–64

Solomon, T. H., Gollub, J. P. 1990. Sheared boundary layers in turbulent Rayleigh-Benard Convection. *Phys. Rev. Lett.* 64: 2382–85

Solomon, T. H., Gollub, J. P. 1991. Thermal boundary layers and heat flux in turbulent convection: the role of recirculating flows. *Phys. Rev. A* 45: 1283

Spiegel, E. A. 1971. Convection in stars. *Annu. Rev. Astron. Astrophys.* 9: 323–52

Spruit, H. C., Nordlund, A., Title, A. M. 1990. Solar convection. *Annu. Rev. Astron. Astrophys.* 28: 263–301

Tanaka, H., Miyata, H. 1980. Turbulent natural convection in a horizontal water layer heated from below. *Int. J. Heat Mass Transfer* 23: 1273–81

Tennekes, H., Lumley, J. L. 1972. *A First Course in Turbulence.* Cambridge/London: MIT press. 300 pp.

Threlfall, D. C. 1975. Free convection in low-temperature gaseous helium. *J. Fluid Mech.* 67: 17–28

Thual, O. 1992. Zero Prandtl number convection. *J. Fluid Mech.* 240: 229–58

Tong, P., Shen, Y. 1992. Relative velocity fluctuations in turbulent Rayleigh-Benard convection. *Phys. Rev. Lett.* 67: 2066–69

Tritton, D. J. 1988. *Physical Fluid Dynamics.* Oxford: Oxford Univ. Press. 519 pp. 2nd ed.

Turner, J. S. 1973. *Buoyancy Effects in Fluids.* London: Cambridge Univ. Press. 367 pp.

Werne, J. 1993. The structure of hard-turbulent convection in two dimensions:

numerical evidence. *Phys. Rev. E* To appear

Werne, J., DeLuca, E. E., Rosner, R., Cattaneo, F. 1991. Development of hard-turbulent convection in two dimensions: numerical evidence. *Phy. Rev. Lett.* 67: 3519–22

Wu, X. 1991. *Along a road to developed turbulence: free thermal convection in low temperature helium gas.* Thesis. Univ. Chicago, Dept Phys.

Wu, X., Kadanoff, L., Libchaber, A., Sano, M. 1990. Frequency power spectrum of fluctuations in free convection. *Phys. Rev. Lett.* 64: 2140–43

Wu, X., Libchaber, A. 1992. Scaling relations in thermal turbulence: the aspect-ratio dependence. *Phys. Rev. A* 45: 842–45

Yakhot, V. 1989. Probability distributions in high-Rayleigh-number Benard convection. *Phys. Rev. Lett.* 63: 1965–67

Zaleski, S. 1991. Boundary layer stability and heat flux in Rayleigh-Benard experiments. *C. R. Acad. Sci. Paris* 313 Ser. B: 1099–103

Zocchi, G., Moses, E., Libchaber, A. 1990. Coherent structures in turbulent convection, an experimental study. *Physica A* 166: 387–407

Annu. Rev. Fluid Mech. 1994. 26: 169–89

VORTEX RECONNECTION

S. Kida

Research Institute for Mathematical Sciences, Kyoto University,
Kyoto 606-01, Japan

M. Takaoka

Department of Mechanical Engineering, Faculty of Engineering Science,
Osaka University, Osaka 560, Japan

KEY WORDS: vorticity, scalar, helicity, turbulence, coherent structure

1. INTRODUCTION

Fluid flow has great variability as a result of vortical motions. Either ordered or disordered, vortical motions in nature, such as a tornado, a whirlpool, or a raging flow in the basin of a waterfall, etc have long attracted attention. Understanding the dynamics and the mutual interaction among various types of vortical motions, including *vortex reconnection*, is a key ingredient in clarifying and controlling fluid motions. The first analytical work on vortex reconnection goes back to the study of the sinusoidal instability in a pair of counter-rotating vortex tubes shed from the wing tips of an airplane (Crow 1970).

The systematic study of vortex reconnection was initiated by observations and laboratory experiments of highly ordered vortical motions. Special and constant attention has been paid to the topological structure of high-vorticity concentrated regions. One of the simplest and most fundamental experiments on vortex reconnection is the interaction of two colliding circular vortex rings (Fohl & Turner 1975, Oshima & Asaka 1977, Schatzle 1987, Oshima & Izutsu 1988). Two successive reconnections were observed in visualizations with dye and smoke. Two vortex rings that are ejected side by side from nozzles approach each other to merge into a single elongated vortex ring. After a complicated three-dimensional motion this ring splits into two rings that align perpendicular to the original two

169

0066–4189/94/0115–0169$05.00

rings. The motion of an elliptic vortex ring was also studied by Hussain & Husain (1989). They found a bending motion, a self-collision, and a reconnection in the case of larger ellipticity.

It is not easy to measure the vorticity directly nor to analyze the three-dimensional data of a flow field. In most experiments a passive scalar, such as dye or smoke, has been used as a tracer of vortex tubes. But special care should be taken in interpreting these results because the evolution of a passive scalar is quite different from that of the vorticity field (Section 2.3).

Thanks to the rapid development of supercomputers, the direct numerical study of vortex reconnection has become possible in the past few years. Complete data of field quantities, such as velocity, vorticity, and concentration of a passive scalar, at any time and at any position can be stored in the computer memory. There have been many numerical simulations published so far with various motivations ranging from fundamental physics to engineering application. But for understanding the fundamental mechanism of vortex interaction, which is our main concern here, it is better to investigate the simplest situation possible that still retains some essential mechanism. Among others, the interaction of two vortex tubes, either straight or curved (Pumir & Kerr 1987; Melander & Zabusky 1988; Melander & Hussain 1988, 1989, 1990; Zabusky & Melander 1989; Meiron et al 1989; Zabusky et al 1991; Boratav et al 1992; Shelly et al 1993), and two vortex rings (Ashurst & Meiron 1987; Kida et al 1989, 1990, 1991a; Aref & Zawadzki 1990) have been extensively studied. With relation to helicity dynamics (Section 2.2) the time-evolution of a trefoiled vortex tube (Kida & Takaoka 1987, 1988) and the interaction of two elliptical vortex rings (Aref & Zawadzki 1991) have also been simulated. Computer simulation offers the great advantage that the visualization of a complicated three-dimensional flow field can be relatively easily obtained. The iso-surfaces of vorticity magnitude have often been used for visualization of vortical structure and for discussion of reconnection. Low-pressure regions (Douady et al 1991) and high-strain rate regions (Tanaka & Kida 1993) also characterize vortical structure.

Despite much attention to this phenomenon, however, the concept of "vortex reconnection" does not seem to be well understood—as pointed out by the present authors (Kida & Takaoka 1991a). Moreover, the word "vortex" or "eddy" as well as "vortex reconnection" itself has often been used without clarifying its exact definition and its relation to *vorticity*. Indeed vorticity is defined unambiguously by the curl of the velocity, $\omega = \nabla \times \mathbf{u}$, but usually the word "vortex" seems to refer vaguely to a large magnitude of vorticity and the word "eddy" to a rotational motion of fluid.

As stated above, vortex reconnection in laboratory experiments is usually understood as a change of topology of a passive scalar like dye or smoke, which behaves quite differently from the vorticity field (Section 2.3). In numerical simulations, on the other hand, reconnection is usually discussed for the iso-surface of vorticity magnitude, which is also different from the vorticity surface which is composed of vorticity lines.[1] It is, therefore, necessary to distinguish three types of reconnection: *scalar*, *vortex*, and *vorticity* reconnections. Scalar and vortex reconnections will respectively be referred to when the topology of iso-surfaces of a passive scalar and of the vorticity magnitude has changed, whereas vorticity reconnection applies when the topology of vorticity lines has changed. Recall that in the case of an inviscid flow the last reconnection is prohibited, as stated by the Kelvin-Helmholtz theorems, though the first two are possible. Viscous effects are therefore indispensable for *vorticity* reconnection to occur. As will be discussed in Section 2.4, however, we encounter an inherent difficulty in defining vorticity reconnection owing to the lack of a method of identification of vorticity lines at different times in a viscous flow.

In this review we describe the current understanding of vortex (vorticity or scalar) reconnection and the related phenomena from a fundamental point of view. In Section 2, we consider the dynamical behavior of the vorticity field in an incompressible Navier-Stokes flow. We discuss the global invariants of motion in an inviscid flow, the problem of selective dissipation in a slightly viscous flow (Section 2.1), the dynamics of helicity as an indicator of topological structure of the velocity field (Section 2.2), the difference in the effects of vorticity stretching on a passive scalar and on the vorticity field, the generation of singularities in the vorticity field (Section 2.3), and the breakdown of the frozen motion of vorticity lines due to viscous effects, which leads to difficulty in defining vorticity reconnection (Section 2.4). In Section 3, we deal with the mechanism of vortex (or vorticity) reconnection. Topics chosen are: the motion of thin vortex tubes and the importance of a core deformation during vortex interaction (Section 3.1), a detailed mechanism of vorticity reconnection (Section 3.2), and physical models and analytical solutions that may explain the physical process of vorticity reconnection (Section 3.3). A quantification of the degree of reconnection will be discussed in Section 4.

2. VORTICITY DYNAMICS

2.1 *Inviscid Conservation*

The time-development of the vorticity field is described by the vorticity equation which is derived by taking the curl of the Navier-Stokes equation as

---

[1] In the usual textbook the term "vortex lines" is used instead.

$$\frac{D\omega}{Dt} \equiv \frac{\partial\omega}{\partial t} + (\mathbf{u}\cdot\nabla)\omega = (\omega\cdot\nabla)\mathbf{u} + v\nabla^2\omega = \omega\cdot\mathscr{S} + v\nabla^2\omega, \tag{2.1}$$

where $\mathscr{S}$ is the strain rate tensor, v is the kinematic viscosity of the fluid, and D/Dt is the Lagrangian derivative. The fluid is assumed to be incompressible so that the velocity field is solenoidal,

$$\nabla\cdot\mathbf{u} = 0. \tag{2.2}$$

The dynamics of a fluid of *zero viscosity* is constrained by several *global* invariants of motion. In an unbounded flow with vorticity localized in a finite domain or decreasing exponentially with distance, the kinetic energy $\frac{1}{2}\int \rho|\mathbf{u}|^2 \, d\mathbf{x}$, the virtual momentum $\mathbf{P} = \frac{1}{2}\int \rho(\mathbf{x}\times\omega) \, d\mathbf{x}$, the virtual angular-momentum $\mathbf{M} = \frac{1}{3}\int \rho[\mathbf{x}\times(\mathbf{x}\times\omega)] \, d\mathbf{x}$, and the helicity (Moffatt 1969)

$$H = \int \mathbf{u}\cdot\omega \, d\mathbf{x} \tag{2.3}$$

are inviscid invariants of motion, where the integration is carried out over the whole domain occupied by the fluid.[2] These quantities, if defined appropriately, are also conserved for a flow contained in a periodic box. Notice that the kinetic energy and the helicity, which are quadratic in velocity, play an important role in the dynamical property of fluid motion (see below). Incidentally, in a two-dimensional flow the squared vorticity (called enstrophy) $\int |\omega|^2 \, d\mathbf{x}$ serves as another inviscid conserved quantity, while the helicity is identically zero.

As opposed to the above global conserved quantities, there are an infinite number of *local* conserved quantities for an inviscid barotropic fluid flow. The circulation $\Gamma(C)$ along any arbitrary closed curve C that is advected with the fluid is invariant in time (Kelvin 1871).[3] It is also known that a vorticity tube, the boundary of which is composed of vorticity lines that pass through a closed curve, remains a vorticity tube for all time, preserving its identity (Helmholtz 1858). This implies that vorticity reconnection is prohibited in inviscid flow (Section 2.4).

The plurality of inviscid conserved physical quantities has special importance in the generation of an ordered structure in a slightly viscous

---

[2] If $|\mathbf{u}|$ decreases as $O(1/r^3)$ with distance r from the region of distributed vorticity, the virtual momentum is rewritten as $\mathbf{P} = \int \rho\mathbf{u} \, d\mathbf{x}$; and if $|\mathbf{u}|$ decreases as $O(1/r^4)$, the virtual angular-momentum is rewritten as $\mathbf{M} = \int \rho\mathbf{x}\times\mathbf{u} \, d\mathbf{x}$.

[3] Conversely, the requirement of conservation of circulation along any closed line and of mass of any fluid element under an arbitrary infinitesimal displacement leads to the vorticity and continuity equations, (2.1) and (2.2).

flow (see Hasegawa 1985). Suppose that two or more global quantities are conserved in the inviscid flow; and suppose that in a slightly viscous flow only one of the inviscid conserved quantities dissipates substantially and the others remain nearly conserved. Then, the flow may relax to such a state that this particular quantity is minimized with the others being kept constant. A well-known example of this *relaxation process* is found in a two-dimensional incompressible flow in which two quadratic global quantities, i.e. the kinetic energy and the enstrophy, are conserved in the inviscid case. If a small viscosity is introduced, the enstrophy dissipates drastically while the kinetic energy is still nearly conserved. Then, the flow field may relax to the minimum enstrophy state with a given kinetic energy.

In the above argument it is essential to find which inviscid conserved quantity is dissipated substantially in a slightly viscous flow. This problem of *selective dissipation* is not always trivial and special consideration is needed in each case. In the above example of a two-dimensional flow the inverse cascade of energy and the normal cascade of enstrophy may support the selective dissipation of enstrophy with energy being a quasi-conserved quantity. The same reasoning was successfully applied to three-dimensional MHD turbulence in which the magnetic field dominates the flow motion (the low-β limit). The system relaxes to a state of minimum magnetic energy for a given magnetic helicity (Taylor 1974, Horiuch & Satoh 1986).

In a three-dimensional Navier-Stokes flow it is not known which of the two inviscid conserved quantities—the kinetic energy or the helicity—is dissipated in a slightly viscous case. It may be natural to expect that the helicity dissipates more quickly than the kinetic energy because the former involves higher-order spatial derivatives, namely, the smaller scales that are easier to dissipate. On the contrary, however, there is numerical evidence that the modal helicity is not transferred to the smaller scales, in contrast with the kinetic energy (K. Ohkitani 1992, private communication). It is therefore possible that only the kinetic energy dissipates while the helicity is nearly conserved in the inviscid limit.

The invariance of total kinetic energy and helicity in inviscid flow may constrain the cascade transfer of modal excitations in turbulence (Brissaud et al 1973). It was shown by using the Eddy-Damped Quasi-Normal Markovian (EDQNM) closure theory that a helicity cascade suppresses the energy transfer towards larger wavenumbers (André & Lesieur 1977). The inverse cascade of energy and the generation of large-scale structures in helical turbulence are controversial (Levich 1987). Waleffe (1992) recently investigated the nonlinear interactions in homogeneous turbulence by using a helical decomposition. He found both inverse and forward

· cascades depending on whether the small-scale helical modes have helicity of the same sign or opposite sign in the fundamental triad interaction. Since the helicity is related to the topological structure of vorticity lines (Section 2.2), the above problem may be related to the vorticity reconnection in the inviscid limit.

2.2   Helicity Dynamics

The helicity (2.3) is related to the topological structure, e.g. the knottedness and linkage, of vorticity lines and represents the skewed structure of a velocity field (Moreau 1961, Moffatt 1969). The invariance in time of the helicity follows from the invariance of knottedness and linkage of vorticity lines and that of the circulation of vorticity tubes, which holds for an inviscid barotropic fluid flow under a conservative body force. This then gives a strong constraint to the topological structure of the velocity field in the relaxation process of a slightly viscous flow (Moffatt 1990, 1992; Section 2.1).

It should be stressed here that only the topology of vorticity lines is not sufficient to express the helicity, but the intensity of vorticity lines also does matter. In order to see this let us take a viscous ABC flow,

$$\mathbf{u} = \boldsymbol{\omega} = (A \sin x_3 + C \cos x_2, B \sin x_1 + A \cos x_3, C \sin x_2 + B \cos x_1),$$

$$(2.4)$$

where

$$A = A_0 e^{-vt}, \quad B = B_0 e^{-vt}, \quad C = C_0 e^{-vt},$$

$$(2.5)$$

and A_0, B_0, C_0 are constants. This is an exact solution to the Navier-Stokes equation. The topology of vorticity lines in this flow is invariant in time but the helicity changes because the magnitude of vorticity does (Kida & Takaoka 1991a). Recall that vorticity reconnection concerns only the change of topology of vorticity lines.

In addition to the total helicity (2.3) the *local helicity*

$$H_V = \int_V \mathbf{u} \cdot \boldsymbol{\omega} \, d\mathbf{x},$$

$$(2.6)$$

where $V(t)$ is the volume of any closed vorticity surface, is also invariant in time for the inviscid case. Thus, there are as many invariants as the number of closed vorticity surfaces. It is interesting to see how these invariants affect turbulence dynamics (Levich et al 1984). It should be noted, however, that closed vorticity surfaces are unlikely to exist in a general three-dimensional flow (Moffatt & Tsinober 1992).

If a flow field is made up of only closed thin vortex tubes that do not overlap each other, the helicity (2.3) is written as a sum of the contributions from the writhe (W_i), the torsion (T_i), and the intrinsic twist (N_i) of each tube, and the linkage (L_{ij}) among different tubes as

$$H = \sum_i (W_i + T_i + N_i)\kappa_i^2 + \sum_{i,j} L_{ij}\kappa_i\kappa_j, \tag{2.7}$$

where κ_i is the circulation of the i-th vortex tube (Moffatt & Ricca 1992). This quadratic integral, however, does not always detect linkage involving three or more tubes. For example, the helicity vanishes for a Borromean ring because all the Gaussian integrals do, and therefore it cannot be distinguished from three unlinked rings (Berger 1990). The third- or higher-order integral invariants are then needed to characterize the topological structure. Note that the above argument is based on the assumption of the closure of vorticity tubes, which cannot be expected in a general three-dimensional distributed vorticity field.

Recently, Moffatt & Tsinober (1992) proposed an interesting decomposition of the helicity. A vorticity field is expressed as the sum of three overlapping two-dimensional ones, each of which is made up of only a single velocity component, e.g. $\omega_1 = (0, \partial u_1/\partial x_3, -\partial u_1/\partial x_2)$. The other components ω_2 and ω_3 are similarly defined and $\omega = \omega_1 + \omega_2 + \omega_3$. If the vorticity field has finite support, the ω_i-lines are always closed and the helicity is written as

$$H = \sum_{i,j} H_{ij}, \quad H_{ij} = \int \mathbf{u}_i \cdot \omega_j \, d\mathbf{x}, \quad (i, j = 1, 2, 3). \tag{2.8}$$

Thus, the total helicity is again written in terms of expression (2.7) for these ω_i-lines even if the original vorticity ω-lines are unclosed or chaotic. Note that the relation between the topology of these ω_i-lines and that of real three-dimensional vorticity ω-lines is not clear.

In a viscous flow the helicity is no longer invariant in time. The change in time of the helicity is written as

$$\frac{d}{dt}H = -2v \int \omega \cdot \nabla \times \omega \, d\mathbf{x}. \tag{2.9}$$

The integrand on the rhs of (2.9) represents the skewed structure of a *vorticity* field (Kida et al 1991a). The helicity may increase or decrease depending on the sense of this skewed structure, which is in contrast to the monotonic decrease of kinetic energy. A velocity field with nonzero helicity lacks reflectional symmetry. In a numerical simulation of the collision of two unlinked elliptical vortex tubes Aref & Zawadzki (1991) demonstrated the generation of the linkage of vortex tubes and therefore of helicity. The helicity, or the linkage and the knottedness of vorticity lines, is generated through vorticity reconnections, but the reverse is not always true. It would be interesting to see whether there is any difference between reconnections with and without helicity change. It is known that some non-Galilean invariant forcing to small-scale motions can generate large-scale structures with nonzero helicity in turbulent flow (Krause & Rudiger 1974, Frisch et al 1987). This is called the AKA-effect, and is formally similar to the α-effect for magnetic fields. This breaking of the Galilean invariance is closely related to the lack of reflectional symmetry.

2.3 Vorticity Stretching

The vorticity field evolves under three dynamical actions, that is, advection $(\mathbf{u} \cdot \nabla)\omega$, stretching $\omega \cdot \mathscr{S}$, and viscous dissipation $\nu\nabla^2\omega$. The action of stretching magnifies the differences in the scalar, vortex, and vorticity reconnections. The evolution equations of a passive scalar θ and of the squared vorticity $|\omega|^2$ are respectively written as

$$\frac{D\theta}{Dt} = \kappa\nabla^2\theta, \tag{2.10}$$

$$\frac{D|\omega|^2}{Dt} = 2\omega \cdot \mathscr{S} \cdot \omega + \nu\nabla^2|\omega|^2 - 2\nu\nabla\omega : \nabla\omega, \tag{2.11}$$

where κ is the diffusion coefficient for the passive scalar. The first term on the rhs of (2.11) represents the effect of vorticity stretching on $|\omega|^2$. The vorticity is strengthened (or weakened) in the direction of the eigenvector of $\mathscr{S}$ with a positive (or negative) eigenvalue. The passive scalar field, on the other hand, is only advected by the velocity field. The stretching term in (2.1) changes both the direction and the magnitude of vorticity, whereas that in (2.11) changes only the projected magnitude of $\mathscr{S} \cdot \omega$ in the direction of vorticity. The three kinds of surfaces, i.e. the scalar-, vortex- and vorticity-surfaces, therefore, develop differently from each other due to these different effects of stretching on the respective quantities.

Figure 1 shows how differently the passive scalar and vorticity fields develop. Here, the iso-surfaces of the scalar and of the vorticity magnitude are drawn with green and red, respectively. The overlapped surfaces, inside of which both values are above the respective thresholds, appear as yellow. This is a snapshot at the time of the first reconnection in the interaction of two identical circular vortex rings that start side by side on a common plane (Kida et al 1991a). The initial distributions of magnitude of the scalar and of the vorticity are exactly the same and the Schmidt number $\nu/\kappa = \frac{1}{2}$. Nevertheless, these two iso-surfaces exhibit quite different shapes.

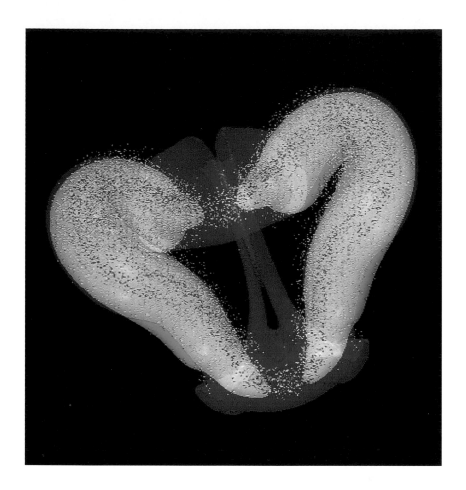

Figure 1 The shape of vortex tubes during the first reconnection of two identical circular vortex tubes which start side by side on a common plane [cf Fig. 4$a(v)$ in Kida *et al* 1991a]. The Schmidt number $\nu/\kappa = \frac{1}{2}$. The iso-surfaces of a passive scalar and of the vorticity magnitude are drawn with green and red, respectively. The overlapped surfaces appear as yellow. The particles plotted have been advected passively with fluid starting with exactly the same distribution as the passive scalar field simulated. The cloud of these particles should represent the spatial distribution of the passive scalar at infinite Schmidt number. The color of the particles indicates the density of the passive scalar. The highest level (*red*) is ten times as large as the lowest (*green*).

A couple of thin tubes (called *threads*) are seen at the central part in the vortex surface, but not in the scalar surface. This difference in appearance mainly arises from the intensification of the vorticity by stretching not seen in the scalar field.

A number of particles, which were distributed with the same density (indicated by color) as the scalar field at the initial instant and which have been advected with the fluid, are also plotted to illustrate the Schmidt number dependence. The spatial distribution of the particles should represent that of a passive scalar in the limit of infinite Schmidt number ($\kappa = 0$). It is more uniform along the vortex tube than the distribution of the iso-surface of the passive scalar and is compared with the dye distribution in Schatzle's (1987) experiments.[4] Because of the different behavior of the vorticity and the passive scalar fields, we must take special care in interpreting the results by visualization with dye or smoke. The role of stretching in the reconnection mechanism will be discussed in Section 3.2.

One of the most basic and difficult open problems is whether or not the solution to the Navier-Stokes equation becomes singular in the inviscid limit. As stated in Section 2.1, vorticity reconnection is prohibited in the inviscid case if the flow field remains regular, so that the possibility of vortex reconnection in the inviscid *limit* depends on the occurrence of a finite-time generation of a singularity. The competition between stretching and viscous effects plays a crucial role in the singularity problem. It is known that if a solution to the Euler equation loses any regularity at some time instant, then the maxima of both the vorticity magnitude and the strain rate tensor diverge simultaneously at a rate greater than $1/t$, where t is time (Beale et al 1984, Ponce 1985).

As for solutions to the Navier-Stokes equation with finite viscosity—regularity, analyticity, and uniqueness have been established for a smooth initial condition when the Laplacian operator $\nu \nabla^2$ in the viscous term is replaced by $-\nu'(-\nabla^2)^p$ with arbitrary $\nu' > 0$ and $p \geq \frac{5}{4}$ (Ladyzenskaya 1963, Lions 1969). Therefore, although a double Laplacian type of viscous term is sometimes used to gain numerical accuracy, care should be taken in using this to discuss vorticity reconnection.

No singularities are expected in either two-dimensional or cylindrically symmetric flows because in these cases the stretching term disappears. One of the simplest systems with possible stretching is a rotationally symmetric

---

[4] A comparable result with respect to the *global* shape of vortex rings has been obtained by the vortex-in-cell method (Aref & Zawadzki 1990). Note that their plots of the "vortex particles" should represent the spatial distribution of a passive scalar rather than that of vorticity.

flow, that is, an axisymmetric flow with swirl. Using this symmetric flow, the generation of finite-time singularities in an Euler flow was investigated numerically by Gauer & Sideris (1991) and Pumir & Siggia (1992). Kerr (1993), on the other hand, directly simulated the interaction of two antiparallel vortex tubes to find a sign of divergence of the maximum of the vorticity magnitude. At the moment, however, this issue is still controversial.

2.4 Breakdown of Frozen Motion

A difficulty in defining vorticity reconnection was recently pointed out by the present authors (Kida & Takaoka 1991a). In order to see whether a particular vorticity line experiences reconnection or not, it is necessary to track it in *time*. In inviscid flow ($v = 0$) the identification in time of vorticity lines can be made by following fluid elements because each vorticity line is frozen to the fluid. Vorticity reconnection, however, is impossible in an inviscid flow as stated by the Helmholtz vortex theorem (Helmholtz 1858). In other words, viscous effects are indispensable for vorticity reconnection to take place. But, for a viscous flow a difficulty arises from the breakdown of the frozen motion of vorticity lines. As will be shown below, any pair of fluid elements which are on a common vorticity line at some time instant will not, in general, be on a common vorticity line at later times. Since the vorticity lines cannot be tracked in time, we cannot identify which vorticity line has experienced a topological change. This kind of difficulty was also noticed before in magnetic reconnection (Axford 1983).

Let us consider how the frozen motion of vorticity lines is destroyed in a viscous flow. Take two fluid elements separated by an infinitesimal distance $\delta\mathbf{x}$ on a single vorticity line at time t, i.e. $\delta\mathbf{x} \parallel \boldsymbol{\omega}(\mathbf{x}, t)$ (see Figure 2). These two fluid elements are advected by the velocity field so that the relative position $\delta\mathbf{x}'$ after infinitesimal time Δt is estimated to be

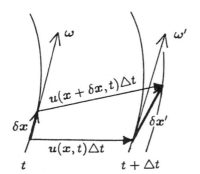

Figure 2 Breakdown of frozen motion of vorticity lines. A fluid element at position $\mathbf{x}$ at time t moves to $\mathbf{x}' = \mathbf{x} + \mathbf{u}(\mathbf{x}, t)\Delta t$ at time $t' = t + \Delta t$. Here, ω and $\delta\mathbf{x}$ are the vorticity and a line element at $\mathbf{x}$ and t, and ω' and $\delta\mathbf{x}'$ are at $\mathbf{x}'$ and t'.

$$\delta \mathbf{x}' = \delta \mathbf{x} + [\mathbf{u}(\mathbf{x} + \delta \mathbf{x}, t) - \mathbf{u}(\mathbf{x}, t)]\Delta t$$

$$= \delta \mathbf{x} + (\delta \mathbf{x} \cdot \nabla)\mathbf{u}(\mathbf{x}, t)\Delta t. \tag{2.12}$$

The vorticity at time $t + \Delta t$ is expressed as

$$\omega' \equiv \omega[\mathbf{x} + \mathbf{u}(\mathbf{x}, t)\Delta t, t + \Delta t]$$

$$= \omega(\mathbf{x}, t) + [\mathbf{u}(\mathbf{x}, t) \cdot \nabla]\omega(\mathbf{x}, t)\Delta t + \frac{\partial}{\partial t}\omega(\mathbf{x}, t)\Delta t$$

$$= \omega(\mathbf{x}, t) + [\omega(\mathbf{x}, t) \cdot \nabla]\mathbf{u}(\mathbf{x}, t)\Delta t + \nu\nabla^2\omega(\mathbf{x}, t)\Delta t, \tag{2.13}$$

where use has been made of (2.1). Equations (2.12) and (2.13) tell us that for an inviscid flow the time-evolutions of $\delta \mathbf{x}$ and ω are exactly the same if $\delta \mathbf{x} \parallel \omega$ at some time instant. In other words, vorticity lines move as if they were frozen to the fluid. This frozen motion of vorticity lines is destroyed in a viscous ($\nu > 0$) flow through the viscous term $\nu\nabla^2\omega$. The parallel component to vorticity of the viscous term, $\nu(\nabla^2\omega)_\parallel$, represents the difference in the stretching rate between the vorticity and the fluid lines, while the perpendicular component, $\nu(\nabla^2\omega)_\perp$, represents the rate of deviation in the direction of the two kinds of lines. Therefore, the vorticity lines can no longer be tracked by fluid elements in a viscous flow when the perpendicular component exists. Rather, this perpendicular component is used for quantifying the degree of vorticity reconnection (Section 4). As was confirmed in Kida & Takaoka (1991a) by applying it to several analytical and numerical flows, this quantification may lead to results consistent with our intuitive concept of vorticity reconnection that the vorticity lines undergo rapid changes in direction during a reconnection process. The same idea was applied to express the deviation due to viscosity of iso-scalar surfaces and iso-vorticity surfaces from frozen motion (Kida & Takaoka 1991b).

3.   RECONNECTION MECHANISM

3.1   *Motion of Vortex Filaments*

The velocity field $\mathbf{u}$ is in general represented as the sum of rotational and irrotational parts: $\mathbf{u} = \mathbf{u}_r + \mathbf{u}_i$. The rotational part $\mathbf{u}_r$ is completely determined by the vorticity field ω as

$$\mathbf{u}_r(\mathbf{x}) = \frac{1}{4\pi} \int \frac{\omega(\mathbf{x}') \times (\mathbf{x} - \mathbf{x}')}{|\mathbf{x} - \mathbf{x}'|^3} d\mathbf{x}', \tag{3.1}$$

which is known as the Biot-Savart law. The irrotational part $\mathbf{u}_i$, on the other hand, is determined by the boundary condition as well as the vorticity field. It can be taken as zero in an appropriate moving frame in the case of either an unbounded flow with a constant velocity at infinity or a flow in a periodic box.

Suppose that a velocity field is composed of many thin vortex tubes. Then, each vortex tube is advected by the velocity induced by all the vortex tubes including the one concerned. In the limit of vanishing core the motion of a vortex tube (now called a *vortex filament*) is described, in the leading order, by the so-called localized induction equation (LIE),

$$\frac{\partial \mathbf{X}}{\partial t} = \frac{\partial \mathbf{X}}{\partial s} \times \frac{\partial^2 \mathbf{X}}{\partial s^2}, \tag{3.2}$$

where $\mathbf{X}(s, t)$ is the position vector of a vortex filament, s is the arc length measured from an arbitrary point on the filament, and t is a rescaled time (Da Rios 1906, Hama 1962). This equation tells us that any element of vortex filament moves in the local bi-normal direction with velocity proportional to local curvature. Hasimoto (1972) obtained a solitary wave solution of the LIE, and Kida (1981, 1982) found a family of steady solutions and investigated their stability.

A strong vortex filament induces a swirling motion around it and any part of the filament moves in the local bi-normal direction. From these facts it can be deduced that two strong vortex filaments tend to approach each other in *antiparallel* fashion. Siggia (1985) demonstrated this antiparallel approach numerically and gave it a simple explanation.

The problem of singularity generation in a solution to the Euler equation, which was mentioned in Section 2.3, was attacked by Siggia (1985) with a modified Biot-Savart formula for thin vortex tubes that was regularized by introducing an effective core size to find a self-similar collapse of two antiparallel pieces of vortex tubes. However, this Biot-Savart model, in which the shape of the cores of vortex tubes is assumed to be essentially circular, becomes inaccurate near the time when a singularity develops. In reality, the vortex core is flattened more and more as two vortex tubes approach each other (Kida et al 1991b, Shelly et al 1993), which drastically retards the growth of vorticity magnitude. The possibility of a finite-time singularity generation is still inconclusive because the effects of core deformation are not clear.[5] The singularity problem in the Euler equation is closely related to the possibility of vorticity reconnection in the inviscid limit (Section 2.3).

---

[5] A finite-time singularity generation was observed in a system of three point vortices in a two-dimensional Euler flow (Kimura 1987).

3.2   Reconnection Process—Bridging

As stated in the introduction, there have been many numerical simulations of the reconnection of various shapes of vortex tubes with various initial configurations. Even if not introduced artificially, strong vortex tubes are generated from a random vorticity field perhaps through Kelvin-Helmholtz-type instabilities (Ruetsch & Maxey 1992, Kida & Tanaka 1993).

There are some common mechanisms in the reconnection process of two vortex tubes with nearly the same intensity.[6] A typical sequence of physical events is as follows. First, the tubes tend to approach each other in antiparallel fashion advected by the mutual- and self-induction velocity (Section 3.1).[7] As the two vortex tubes get closer, the shape of the vortex core is deformed typically into the so-called *head-tail* structure (see Kida et al 1991b). Then, viscous cancellation of opposite signed vorticity in the interaction zone initiates vorticity reconnection. Advected by a complicated three-dimensional velocity field, the vorticity lines now experience a cross-linking, or *bridging*, as explained shortly below. In the process of bridging, vorticity is created in the direction perpendicular to the original vortex tubes. The intensity of the new vorticity is strengthened by vorticity stretching to surpass that of the original vortex tubes, resulting in a change of global topology of the vortex tubes.

A simple and naive explanation of the mechanism of vortex reconnection of a pair of antiparallel vortex tubes may be a simple viscous cancellation of vorticity of opposite sign in a converging flow induced by either the rest of the vortex tubes or an external flow (Figure 3a). Actually, however, a slightly more complicated cross-linking of vorticity lines is taking place which makes the vorticity lines twisted and tangled (Figure 3b).

When two vortex tubes come into contact, the closest vorticity lines (denoted by p) which are in opposite directions, are cancelled by viscous diffusion in the interaction zone (the hatched area in the top panel of Figure 3b). At the same time, they are connected with their counterparts in the other tube at the ends of the interaction zone. The vorticity lines are rotating around each other in the vortex core and the angular velocity of the rotation is smaller near the interaction zone because of flattening of the core. Therefore, the vorticity lines p and q must be tangled at the ends

---

[6] More complicated physical processes manifest themselves in the interaction of vortex tubes with much different strength. A weaker vortex tube will wind around a stronger one accompanied by strong deformation of the shape of the core (Zabusky & Melander 1989).

[7] This antiparallel approach of vortex tubes has been clearly observed in the motion of a trefoiled vortex tube (Kida & Takaoka 1987; see Figure 4a) and two originally orthogonal straight vortex tubes (Melander & Zabusky 1988).

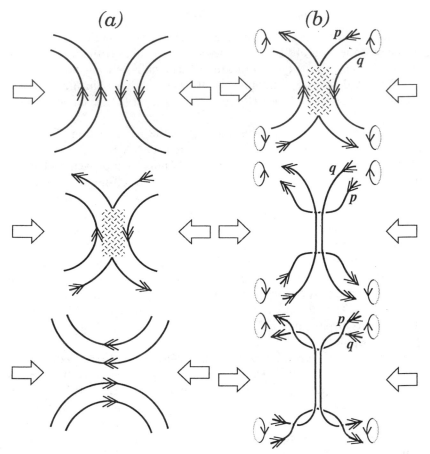

Figure 3 Mechanism of vorticity reconnection. (*a*) A simple viscous cancellation. (*b*) Bridging. Two vortex tubes are pushed from left and right by a converging flow designated by blank arrows. Double and single arrows indicate the directions of vorticity and the rotation of vorticity lines, respectively. Viscous cancellation is taking place in hatched areas.

of the interaction zone (lower two panels in Figure 3*b*). As a result, the vorticity lines are strongly twisted especially in the vicinity of the ends of the interaction zone either like a right-handed or a left-handed screw. During the reconnection the interaction zone is bent in the direction perpendicular to the plane of Figure 3*b*, which makes the viscous cancellation weaker. The portions of vorticity lines *p* that link the two interaction tubes are called *bridges* and this process of cross-linking of vorticity lines is called *bridging* (Melander & Hussain 1988, Kida et al 1991a). The bridges are born in the vicinity of either a forward stagnation line in the frame moving with the interacting vortex tubes (Melander & Hussain 1988) or in the vicinity of where the largest stretching rate occurs (Kida et al 1991a). The vorticity in the bridges is intensified by stretching in the direction of the eigenvector of the strain rate tensor with a positive eigenvalue (Kida et al 1991a, Boratav et al 1992).

This process of bridging has been observed in the interaction of two anti-parallel vortices with a sinusoidal perturbation (Melander & Hussain 1988) and in the interaction of two vortex rings (Kida et al 1991a). In these rather "symmetric" configurations there is no drastic change in the shape of the vortex tubes before reconnection starts. In contrast with such a quiet initiation of reconnection, a violent ejection of vorticity lines is observed in cases with less symmetric initial conditions such as for two straight orthogonally offset vortex tubes (Melander & Zabusky 1988) or a trefoiled vortex tube (Kida & Takaoka 1987, 1988).

Figure 4a is a close-up of the ejected portion (also called the *bridge* in Kida & Takaoka 1987) of a trefoiled vortex tube, where the iso-vorticity surfaces are drawn on which the magnitude of vorticity takes 10% of its

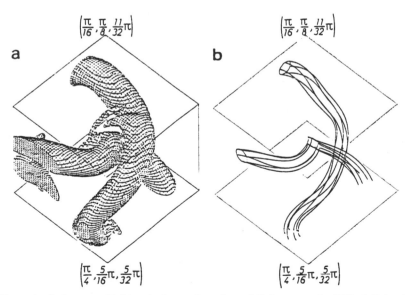

Figure 4   A close-up of bridges in the motion of a trefoiled vortex tube (Kida & Takaoka 1988). (*a*) The iso-surface of vorticity magnitude at level of 10% of the maximum. (*b*) The vorticity lines which go through circles on the faces of the box. The vorticity line attached with letter *a* goes from one part of the vortex tube to another through a bridge.

maximum. The vorticity points from bottom to top through the vertical tube and from right to left through the horizontal one, so that it is nearly antiparallel around the closest approach. The structure of the vorticity lines is seen in Figure 4b in which groups of such vorticity lines go through circles on the faces of the box. The vorticity lines are pulled out of the main vortex tube at the bridges as a hairpin vortex. The directions of vorticity are opposite in the two bridges seen in Figure 4a. The vorticity line marked with letter a goes from one part of the vortex tube to another through the bridge.

This abrupt ejection of vorticity lines may be due to a strong and localized straining velocity induced by the vortex tube itself. This localized jet velocity stretches and intensifies vorticity lines in the direction of the eigenvector of the strain rate tensor with a positive eigenvalue and pulls them out of the main tube. The "fingering" observed in the case of ortho-gonally offset straight vortex tubes (Boratav et al 1992) seems to occur by a similar mechanism.

3.3 Model and Analytical Solution

The mechanism of vortex reconnection is now mainly understood through experimental and numerical observations. There have been several attempts to explain each of the sequence of physical events in the recon-nection process (Section 3.2). The antiparallel approach of two vortex tubes is an almost inviscid process. Thus, it is well explained with the Biot-Savart law as mentioned in Section 3.1 (Siggia 1985). The formation of the head-tail structure was found in the numerical simulation of the Batchelor dipole by adding curvature as disturbance (Pumir & Kerr 1987). Kida et al (1991b) showed that the head-tail structure was formed in a two-dimensional stagnation flow even if the Batchelor dipole had no curvature. The formation of this structure was also reported in the simulation of the interaction of two opposite-signed vortex tubes in a three-dimensional uniform straining flow (Buntine & Pullin 1989). Although the effects of curvature of the vortex tubes were modeled as a stagnation flow, there was no feedback mechanism from deformation of the core of the vortex dipole. The assumption of constancy of the external strain causes the difference between the models and the numerical simulations (see the next paragraph). Greene (1990) focused on the role of the vorticity nulls and the separatrices in vortex reconnection. Indeed there exists a vorticity null at a reconnection point in a purely two-dimensional flow field, but both the two-dimensional structure and the vorticity null are structurally unstable in a general three-dimensional flow. The importance of the third component was stressed by Yamada et al (1990) in the magnetic reconnection problem (see Figure 1 in their paper).

A simple model to explain the commonly observed phenomenon in vortex reconnection is desirable for understanding the physics of the reconnection process. Takaki & Hussain (1985, 1988) first gave a simple model by locally expanding the flow field in a series of polynomials around a symmetric center. A special functional form was assumed for the unknown coefficients to close the hierarchy of equations. The numerical solution of this model, however, did not realize any head-tail structure nor any bridge. Saffman (1990) proposed a set of simultaneous equations as a reconnection model based on the idea that vorticity cancellation weakens the centrifugal force, leading to a local increase in pressure which accelerates fluid elements along the axial direction and enhances the reconnection. This model was critically examined by direct numerical simulation of the interaction of two orthogonally offset straight vortex tubes of equal strength (Boratav et al 1992) and two sinusoidally perturbed counter-rotating vortex tubes (Shelly et al 1993). The differences between the numerical results and the model are pointed out to be due to the assumption about the strain and pressure terms.

Analytical solutions to the Navier-Stokes equation, if not so general but constrained with some special symmetries, would be helpful for understanding the fundamental and detailed mechanism of reconnection. One of the present authors (Takaoka 1991) discovered a class of exact solutions in a three-dimensional uniform straining flow. This flow field depends on *two* spatial variables. The vorticity has *two* components in the plane of these two variables, whereas the velocity has a single component perpendicular to this plane. In this particular flow the iso-surface of velocity is also the vorticity surface. The vorticity surfaces as well as the iso-vorticity surfaces are easily constructed to see different effects of vorticity stretching on these two quantities. Furthermore, we can show explicitly that the iso-scalar surface and the vorticity surface develop similarly if the scalar and the single component of velocity are identical at some time instant. However, it is not easy to generalize this correspondence to the three-dimensional vortex structure.

4. CONCLUDING REMARKS

The complexity in fluid motions, e.g. in turbulence, originates from various types of vortical motions. In recent years, the study of vortex reconnection has been extensively performed either experimentally, numerically, or theoretically as a part of the central subject of vorticity dynamics. Nevertheless, it seems to us that the concept of vortex reconnection has been understood rather vaguely in the literature. Dye or smoke (a passive scalar) used for visualization of vortex tubes does not represent the vorticity distribution faithfully. Moreover, the iso-surface of vorticity magnitude often used in numerical studies to visualize the vortex tubes is in general different from the vorticity surface. Therefore, we should consider separately the reconnections of these three quantities, that is, the scalar, the vortex, and the vorticity reconnections.

The definition of reconnection of any of these quantities has an inherent difficulty. Let us define reconnection by a topological change in time of some quantity, e.g. the iso-surface of a passive scalar or of the vorticity magnitude, or the vorticity lines. Since the topology of the iso-surface of some quantity can be different at different levels, there is an ambiguity in discussing the topological change unless the level of the iso-surface is specified beforehand. But, a choice of constant level is not necessarily meaningful except that the quantity concerned is simply advected without change of its intensity (like a passive scalar without diffusion). As for the vorticity reconnection, on the other hand, as discussed in Section 2.4, another difficulty arises because of the breakdown of frozen motion of vorticity lines due to viscosity. The perpendicular component to vorticity of the viscous term in the vorticity equation, $v(\nabla^2\omega)_\perp$, then represents the rate of deviation in the direction of the vorticity relative to the material lines.

For discussion of vorticity reconnection it is highly desirable to quantify the *degree of reconnection*, if possible (Kida & Takaoka 1991a). Is there such a criterion for vorticity reconnection that (a) represents no reconnection in the inviscid case, (b) is Galilean invariant, (c) is defined *locally* in space, and (d) is consistent with our intuitive concept of vorticity reconnection, e.g. a rapid rotation of vorticity lines? One of the promising candidates that satisfy these conditions may be $v(\nabla^2\omega)_\perp$ mentioned above.

Since vorticity reconnection is a local phenomenon, it may be hard to characterize in terms of integral quantities such as kinetic energy, enstrophy, and helicity, etc. Helicity, which is often referred to in discussion of vorticity reconnection, is not a good indicator of vorticity reconnection. Helicity is determined not only by the topology of vorticity lines but also by the distribution of vorticity. There is no clear one-to-one correspondence between vorticity reconnection and helicity change. For example, in a collision of two identical circular vortex rings, the helicity is preserved throughout the reconnection process. In a viscous ABC flow (Equations 2.4 and 2.5), on the other hand, the helicity changes in time while the vorticity field decays without change of its structure. Since the helicity is a global quantity, it cannot describe a local phenomenon such as vorticity reconnection.

With relation to vortex reconnection we here have discussed the selective dissipation in a slightly viscous flow and the generation of singularities in Euler flow. Both of these problems, which are of fundamental importance, are still open and are waiting for further study.

ACKNOWLEDGMENTS

The authors would like to express their cordial gratitude to Professor H. K. Moffatt who kindly read the manuscript and made invaluable suggestions for improvement.

Literature Cited

André, J. C., Lesieur, M. 1977. Influence of helicity on the evolution of isotropic turbulence at high Reynolds number. *J. Fluid Mech.* 81: 187–207

Aref, H., Zawadzki, I. 1990. Comment on Vortex Ring Reconnections. In *Topological Fluid Mechanics, Proc. IUTAM Symp., Cambridge*, ed. H. K. Moffatt, A. Tsinober, pp. 535–39. Cambridge: Cambridge Univ. Press

Aref, H., Zawadzki, I. 1991. Linking of vortex rings. *Nature* 354: 50–53

Ashurst, W. T., Meiron, D. 1987. Numerical study of vortex reconnection. *Phys. Rev. Lett.* 58: 1632–35

Axford, W. I. 1983. Magnetic field reconnection. In *Magnetic Reconnection in Space and Laboratory Plasmas*, ed. E. W. Hones Jr., *Geophys. Monogr. Ser. Vol. 30*, pp. 1–8. Washington, DC: Am. Geophys. Union

Beale, J. T., Kato, T., Majda, A. 1984. Remarks on the breakdown of smooth solutions for the 3-D Euler equations. *Comm. Math. Phys.* 94: 61–66

Berger, M. A. 1990. Third-order link integrals. *J. Phys. A: Math. Gen.* 23: 2787–93

Boratav, O. N., Peltz, R. B., Zabusky, N. J. 1992. Reconnection in orthogonally interacting vortex tubes: direct numerical simulations and quantifications. *Phys. Fluids A* 4: 581–605

Brissaud, A., Frisch, U., Leorat, J., Lesieur, M., Mazure, A. 1973. Helicity cascade in fully developed isotropic turbulence. *Phys. Fluids* 16: 1366–67

Buntine, J. D., Pullin, D. I. 1989. Merger and cancellation of strained vortices. *J. Fluid Mech.* 205: 263–95

Crow, S. C. 1970. Stability theory for a pair of trailing vortices. *AIAA J.* 8: 2172–79

Da Rios, L. S. 1906. Sul moto di un liquido indefinito con un fileto vorticoso di forma qualunque. *Rend. Circ. Mat. Palermo* 22: 117–35

Douady, S., Couder, Y., Brachet, M. E. 1991. Direct observation of the intermittency of intense vorticity filaments in turbulence. *Phys. Rev. Lett.* 67: 983–86

Fohl, T., Turner, J. S. 1975. Colliding vortex rings. *Phys. Fluids* 18: 433–36

Frisch, U., She, Z. S., Sulem, P. L. 1987. Large-scale flow driven by the anisotropic kinetic alpha effect. *Physica D* 28: 382–92

Grauer, R., Sideris, T. C. 1991. Numerical computation of 3-D incompressible ideal fluids with swirl. *Phys. Rev. Lett.* 67: 3511–14

Greene, J. M. 1990. Vortex nulls and magnetic nulls. *Proc. IUTAM Symp. Topological Fluid Mech.*, pp. 478–84. Cambridge: Cambridge Univ. Press

Hama, F. R. 1962. Progressive deformation of a curved vortex filament by its own induction. *Phys. Fluids* 5: 1156

Hasegawa, A. 1985. Self-organization processes in continuous media. *Adv. Phys.* 34: 1–42

Hasimoto, H. 1972. A soliton on a vortex filament. *J. Fluid Mech.* 51: 477–85

Helmholtz, H. 1858. Über diskontinuierliche Flüssigkeitsbewegungen. *Monatsber. K. Akad. Wiss. Berlin*, 23: 215–28. See *Philos. Mag.* 4(36): 337–46 (1968)

Horiuchi, R., Satoh, T. 1986. Self-organization and energy relaxation in a three-dimensional magnetohydrodynamics plasma. *Phys. Fluids* 29: 1161–68

Hussain, F., Husain, H. S. 1989. Elliptic jets. Part 1. Characteristics of unexcited and excited jets. *J. Fluid Mech.* 208: 257–320

Kelvin, Lord 1871. Hydrokinematic solutions and observations. *Philos. Mag.* 4(42): 362–77

Kerr, R. M. 1993. The role of singularities in Euler. In *Unstable and Turbulent Motion of Fluid*, ed. S. Kida. River Edge, N.J.: World Sci. In press

Kida, S. 1981. A vortex filament moving without change of form. *J. Fluid Mech.* 112: 397–409

Kida, S. 1982. Stability of a steady vortex filament. *J. Phys. Soc. Jpn.* 51: 1655–62

Kida, S., Takaoka, M. 1987. Bridging in vortex reconnection. *Phys. Fluids* 30: 2911–24

Kida, S., Takaoka, M. 1988. Reconnection of vortex tubes. *Fluid Dyn. Res.* 3: 257–61

Kida, S., Takaoka, M. 1991a. Breakdown of frozen motion and vorticity reconnection. *J. Phys. Soc. Jpn.* 60: 2184–96

Kida, S., Takaoka, M. 1991b. Vortex reconnection. *Kokyuroku, RIMS Kyoto U.* 769: 200–8 (In Japanese)

Kida, S., Takaoka, M., Hussain, F. 1989. Reconnection of two vortex rings. *Phys. Fluids A* 1: 630–32

Kida, S., Takaoka, M., Hussain, F. 1990. Reconnection of two vortex rings. In *Topological Fluid Mechanics, Proc. IUTAM Symp., Cambridge*, ed. H. K. Moffatt, A. Tsinober, pp. 525–34. Cambridge: Cambridge Univ. Press

Kida, S., Takaoka, M., Hussain, F. 1991a. Collision of two vortex rings. *J. Fluid Mech.* 230: 583–646

Kida, S., Takaoka, M., Hussain, F. 1991b. Formation of head-tail structure in a two-dimensional uniform straining flow. *Phys. Fluids A* 3: 2688–97

Kida, S., Tanaka, M. 1993. Generation of vortical structure in a homogeneously sheared turbulence. In *Unstable and Turbulent Motion of Fluid*, ed. S. Kida. River Edge, N.J.: World Sci. In press

Kimura, Y. 1987. Similarity solution of two-dimensional point vortices. *J. Phys. Soc. Jpn.* 56: 2024–30

Krause, F., Rudiger, G. 1974. On the Reynolds stresses in mean-field hydrodynamics. *Astron. Nachr.* 295: 93–99

Ladyzenskaya, O. A. 1963. *The Mathematical Theory of Viscous Incompressible Flow.* New York: Gordon and Breach. 2nd ed.

Levich, E. 1987. Certain problems in the theory of developed hydrodynamical turbulence. *Phys. Rep.* 151: 129–238

Levich, E., Levich, B., Tsinober, A. 1984. Helical structures, fractal dimensions and renormalization group approach in homogeneous turbulence. In *Turbulence and Chaotic Phenomena in Fluids*, ed. T. Tatsumi, pp. 309–17. Amsterdam: Elsevier

Lions, J. L. 1969. *Quelques Méthodes de Résolution des Problémes aux Limites Nonlinéaires.* Paris: Dunod-Gauthier-Villars

Meiron, D. I., Schelly, M. J., Ashurst, W. T., Orszag, S. A. 1989. *Mathematical Aspects of Vortex Dynamics.* Philadelphia: SIAM

Melander, M. V., Hussain, F. 1988. Cut-and-connect of two antiparallel vortex tubes. *Center for Turbulence Res., Proc.* of the Summer Program 1988, ed. P. Moin, W. C. Reynolds, J. Kim, pp. 257–86

Melander, M. V., Hussain, F. 1989. Cross-linking of two antiparallel vortex tubes. *Phys. Fluids A* 1: 633–36

Melander, M. V., Hussain, F. 1990. Topological aspects of vortex reconnection. *Proc. IUTAM Symp. Topological Fluid Mech.*, pp. 485–99. Cambridge: Cambridge Univ. Press

Melander, M. V., Zabusky, N. J. 1988. Interaction and "apparent" reconnection of 3D vortex tubes via direct numerical simulations. *Fluid Dyn. Res.* 3: 247–50

Moffatt, H. K. 1969. The degree of knottedness of tangled vortex lines. *J. Fluid Mech.* 159: 117–29

Moffatt, H. K. 1990. The energy spectrum of knots and links. *Nature* 347: 367–69

Moffatt, H. K. 1992. Relaxation under topological constraints. In *Topological Aspects of the Dynamics of Fluids and Plasmas*, ed. H. K. Moffatt et al, pp. 3–28. Dordrecht: Kluwer

Moffatt, H. K., Ricca, R. L. 1992. Helicity and the Călugăreanu Invariant. *Proc. R. Soc. London Ser. A* 439: 411–29

Moffatt, H. K., Tsinober, A. 1992. Helicity in laminar and turbulent flow. *Annu. Rev. Fluid Mech.* 24: 281–312

Moreau, J. J. 1961. Constants d'un ilot tourbillonnaire en fluide parfait barotrope. *C. R. Acad. Sci. Paris Ser. A* 252: 2810–12

Oshima, Y., Asaka, S. 1977. Interaction of two vortex rings along parallel axes in air. *J. Phys. Soc. Jpn.* 42: 708–13

Oshima, Y., Izutsu, N. 1988. Cross-linking of two vortex rings. *Phys. Fluids* 31: 2401

Ponce, G. 1985. Remarks on a paper by J. T. Beale, T. Kato, and A. Majda. *Comm. Math. Phys.* 98: 349–53

Pumir, A., Kerr, R. M. 1987. Numerical simulation of interacting vortex tubes. *Phys. Rev. Lett.* 58: 1636–39

Pumir, A., Siggia, E. D. 1992. Finite-time singularities in the axisymmetric three-dimensional Euler equations. *Phys. Rev. Lett.* 68: 1511–14

Ruetsch, G. R., Maxey, M. R. 1992. The evolution of small-scale structures in homogeneous isotropic turbulence. *Phys. Fluids A* 4: 2747–60

Saffman, P. G. 1990. A model of vortex reconnection. *J. Fluid Mech.* 212: 395–402

Schatzle, P. R. 1987. *An experimental study of fusion of vortex rings.* PhD thesis. Calif. Inst. Technol.

Shelly, M. J., Meiron, D. I., Orszag, S. A. 1993. Dynamical aspects of vortex reconnection of perturbed anti-parallel vortex tubes. *J. Fluid Mech.* 246: 613–52

Siggia, E. D. 1985. Collapse and amplification of a vortex filament. *Phys. Fluids* 28: 794—805

Takaki, R., Hussain, F. 1985. Recombination of vortex filaments and its role in aerodynamic noise. *Turb. Shear Flow V, Cornell Univ.*, ed. F. Durst, B. E. Launder, J. L. Lumley, F. W. Schmidt, J. H. Whitelaw, pp. 3.19–3.25. New York: Springer-Verlag

Takaki, R., Hussain, A. K. M. F. 1988. Singular interaction of vortex filaments. *Fluid Dyn. Res.* 3: 251–56

Takaoka, M. 1991. Straining effects and vortex reconnection of solutions to the 3-D Navier-Stokes equation. *J. Phys. Soc. Jpn.* 60: 2602–12

Tanaka, M., Kida, S. 1993. Characterization of vortex tubes and sheets. *Phys. Fluids.* In press

Taylor, J. B. 1974. Relaxation of toroidal plasma and generation of reversed magnetic fields. *Phys. Rev. Lett.* 33: 1139

Waleffe, F. 1992. The nature of triad interactions in homogeneous turbulence. *Phys. Fluids A* 4: 350–63

Yamada, M., Ono, Y., Hayakawa, A., Katsurai, M. 1990. Magnetic reconnection of plasma toroids with cohelicity and counterhelicity. *Phys. Rev. Lett.* 65: 721–24

Zabusky, N. J., Melander, M. V. 1989. Three-dimensional vortex tube reconnection: morphology for orthogonally-offset tubes. *Physica D* 37: 555–62

Zabusky, N. J., Boratav, O. N., Pelz, P. B., Gao, M., Silver, D., Cooper, S. P. 1991. Emergence of coherent patterns of vortex stretching during reconnection: a scattering paradigm. *Phys. Rev. Lett.* 67: 2469–72

Annu. Rev. Fluid Mech. 1994. 26 : 191–210

THREE-DIMENSIONAL LONG WATER-WAVE PHENOMENA

T. R. Akylas

Department of Mechanical Engineering, Massachusetts Institute of Technology, Cambridge, Massachusetts 02139-4307

KEY WORDS: Kadomtsev-Petviashvili equation, solitons, nonlinear waves, propagation

INTRODUCTION

Water-wave motion is a fascinating subject of fluid mechanics. Apart from being important in various branches of engineering and applied science, many water-wave phenomena are also familiar from everyday experience; obtaining a thorough understanding of the relevant physical mechanisms, however, presents fluid dynamicists with great challenges.

Since the early work of Airy and Stokes in the middle of the previous century, it has been recognized that, even under the assumption of potential flow, the water-wave equations are analytically intractable in general. The main difficulty stems from the nonlinear boundary conditions that apply on the free surface, which itself is unknown and is to be determined as part of the solution.

In recent years, considerable progress has been made in illuminating some aspects of water-wave propagation—most notably the interplay of weak finite-amplitude and dispersive effects—using approximate (model) equations, valid asymptotically in certain limits. Combined with experimental observations, these asymptotic theories reveal useful insights into the relevant physics. Moreover, in many instances, they have provided the impetus for related computational efforts.

Perhaps the most well known model equation for water waves, first proposed about a century ago [see Miles (1981) for a review including historical details], is the Korteweg–de Vries (KdV) equation (Korteweg & de Vries 1895):

191

$$\eta_{t^*}^* + c_0 \eta_{x^*}^* + c_0 \left(\frac{3}{2h_0} \eta^* \eta_{x^*}^* + \tfrac{1}{6} h_0^2 \eta_{x^*x^*x^*}^* \right) = 0, \tag{1}$$

where $\eta^*(x^*, t^*)$ denotes the free-surface elevation from the undisturbed equilibrium level, $c_0 = (gh_0)^{1/2}$ is the speed of infinitesimal-amplitude (linear), long-crested gravity waves, h_0 is the undisturbed water depth (assumed constant), and g the gravitational acceleration.

In deriving (1), a key assumption is that waves are unidirectional—they propagate to the right along the x^*-direction (see Whitham 1974, Section 13.11 for details). Furthermore, it is assumed that disturbances are long (the horizontal lengthscale l is long compared with h_0), and weakly nonlinear (the typical amplitude a is small compared with h_0).

In terms of the dimensionless variables

$$x^* = lx, \quad t^* = (l/c_0)t, \quad \eta^* = a\eta,$$

the KdV equation (1) becomes

$$\eta_t + \eta_x + \tfrac{3}{2}\varepsilon\eta\eta_x + \tfrac{1}{6}\mu^2\eta_{xxx} = 0, \tag{2}$$

where

$$\varepsilon = a/h_0, \quad \mu = h_0/l.$$

Physically, the parameters ε and μ control nonlinear and dispersive effects, respectively. The assumption of weakly nonlinear, long waves implies that both ε and μ are small, and it is customary to take

$$\varepsilon = \mu^2, \quad \varepsilon \ll 1; \tag{3}$$

this choice specifies the lengthscale l.

In view of (3), it is clear that the KdV equation (2) expresses a balance of leading-order nonlinear and dispersive effects. In a reference frame moving with the linear-long-wave speed, $X = x - t$, these weak effects come into play after a long time, $t = O(\varepsilon^{-1})$, so that in terms of the scaled time $T = \varepsilon t$, (2) obtains the standard form

$$\eta_T + \tfrac{3}{2}\eta\eta_X + \tfrac{1}{6}\eta_{XXX} = 0. \tag{4}$$

In addition to water waves, the KdV equation also applies in various other physical contexts where weak nonlinearity balances weak dispersion (Benney 1966). Apart from its wide applicability, however, the KdV equation owes much of its fame to the fact that it can be solved exactly by the inverse-scattering transform (see, for example, Ablowitz & Segur 1981; Newell 1985). In general, a localized initial disturbance gives rise to a finite number of solitary waves:

$$\eta = a_0 \operatorname{sech}^2 [(3a_0/4)^{1/2}(X - a_0 T/2)], \tag{5}$$

where a_0 is the peak amplitude. These locally confined waves of permanent form may also be considered as long-wavelength limits of nonlinear periodic-wave solutions (cnoidal waves) of the KdV equation. Furthermore, KdV solitary waves have the remarkable property that they retain their shape when colliding with each other and hence are also solitons. Water-wave solitons were first recorded experimentally by Russell (1845).

The KdV equation is limited to straight-crested (two-dimensional) wave disturbances. These are more likely to be observed under controlled conditions in the laboratory than in the field where depth variations, non-uniform currents, variable wind, and other effects in general give rise to three-dimensional wave patterns.[1]

The propagation of three-dimensional, finite-amplitude, long waves has been studied extensively during the past few years. Central to this research activity is the Kadomtsev-Petviashvili (KP) equation, an extension of the KdV equation that allows for weak three-dimensional effects; while the KP equation was first proposed over twenty years ago (Kadomtsev & Petviashvili 1970), it is only recently that its physical significance has been fully appreciated.

The present article, using the KP model as the common link, reviews recent advances in understanding certain three-dimensional, nonlinear long-wave phenomena—namely oblique interactions of solitary and periodic waves, the propagation of long waves in channels of variable depth, and the effect of rotation on KdV solitary waves. Although closely related work has been done in other physical settings—for example internal waves in stratified fluids—here the discussion concentrates on surface waves. Also, in this review, emphasis is placed on relating theoretical predictions with experimental observations, and no attempt is made to include work of more mathematical nature, such as the application of inverse-scattering methods to the KP equation (see Krichever 1989, Ablowitz & Clarkson 1991, and Fokas & Sung 1992 for recent surveys).

THE KP MODEL

The KP equation describes weakly three-dimensional, small-amplitude, long waves that propagate predominantly in one direction (the positive x-direction, say); nonlinear, dispersive and three-dimensional effects are assumed to be weak and of equal importance.

---

[1] According to the terminology adopted here, a wave disturbance is three-dimensional if the associated free-surface displacement varies in both horizontal directions; of course, the corresponding velocity field varies in all three directions.

Derivation

Formal derivations of the KP equation using perturbation expansions of the full water-wave equations were presented by, among others, Johnson (1980) and Katsis & Akylas (1987a). It is more straightforward, however, to follow an intuitive procedure akin to the one suggested in the original paper (Kadomtsev & Petviashvili 1970).

The linearized version ($\varepsilon = 0$) of the KdV equation (2) is consistent, to leading order in the dispersion parameter μ, with the gravity-wave dispersion relation of plane waves propagating in the x-direction, $\eta = \exp[i(kx - \omega t)]$:

$$\omega = \mu^{-1/2}(k \tanh \mu k)^{1/2} = k - \tfrac{1}{6}\mu^2 k^3 + O(\mu^4).$$

Similarly, keeping only the dominant dispersive term, the dispersion relation of oblique waves, $\eta = \exp[i(kx + qz - \omega t)]$, is

$$\omega = \kappa - \tfrac{1}{6}\mu^2 \kappa^3, \tag{6}$$

where $\kappa^2 = k^2 + q^2$.

Focusing now on weakly three-dimensional disturbances, $q/k \ll 1$, for which

$$\kappa = k\left(1 + \frac{q^2}{2k^2} + \cdots\right),$$

it is clear that weak three-dimensional and dispersive effects are of equal importance if $q = \mu m$. Hence, (6) becomes to leading order

$$\omega = k - \mu^2\left(\frac{k^3}{6} - \frac{m^2}{2k}\right), \tag{7}$$

and, working in the reverse way, it is easy to see that the three-dimensional counterpart of the KdV equation (4), consistent with (3) and (7), reads

$$(\eta_T + \tfrac{3}{2}\eta\eta_X + \tfrac{1}{6}\eta_{XXX})_X + \tfrac{1}{2}\eta_{ZZ} = 0, \tag{8}$$

where $Z = \mu z$ is the (scaled) spanwise variable. This is, in fact, the KP equation for gravity waves in a reference frame moving in the x-direction with the linear-long-wave speed.

The effect of surface tension can be readily incorporated by starting from the gravity-capillary dispersion relation,

$$\omega^2 = k \tanh \mu k(1 + \gamma \mu^2 k^2)/\mu,$$

where γ is the surface-tension parameter, $\gamma = \tau/(\rho g h_0^2)$, ρ being the fluid density and τ the surface-tension coefficient. Considering then weakly three-dimensional, long waves

$$\omega = k - \tfrac{1}{2}\mu^2[(\tfrac{1}{3}-\gamma)k^3 - m^2/k] + O(\mu^4),$$

and working as before, the KP equation for gravity–capillary waves is found to be

$$[\eta_T + \tfrac{3}{2}\eta\eta_X + \tfrac{1}{2}(\tfrac{1}{3}-\gamma)\eta_{XXX}]_X + \tfrac{1}{2}\eta_{ZZ} = 0. \tag{9}$$

It is important to note that as γ increases past the critical value $\gamma = 1/3$, the linear-long-wave speed changes from a local maximum (for $\gamma < 1/3$) into a local minimum (for $\gamma > 1/3$), and the coefficient of the dispersive term in (9) reverses sign. For $\gamma < 1/3$ ($\gamma > 1/3$), the KP equation sometimes is said to have negative (positive) dispersion. It turns out that the behavior of three-dimensional disturbances is affected dramatically by the sign of dispersion, as indicated by the fact that the KdV solitary wave (5) is stable (unstable) to three-dimensional perturbations when dispersion is negative (positive) (Kadomtsev & Petviashvili 1970). However, for water waves, positive dispersion ($\gamma > 1/3$) requires the water depth to be less than 2 mm in which case dissipation cannot be neglected and the KP equation in not a realistic model. On the other hand, it has been proposed (Zufiria 1987) that the positive dispersion regime could be realized in a low-gravity environment.

Initial Conditions

The choice of appropriate initial conditions for the KP equation involves certain subtleties, as pointed out by Grimshaw (1985), and requires careful discussion. Integrating (8) with respect to X from $-\infty$ to ∞, assuming that the wave disturbance is locally confined so that η and its X-derivatives vanish as $X \to \pm\infty$, it is readily concluded that

$$\frac{\partial^2}{\partial Z^2} \int_{-\infty}^{\infty} \eta(X, Z, T)\, dX = 0. \tag{10}$$

In particular, in the case of a laterally unbounded domain, $-\infty < Z < \infty$, Equation (10) together with mass conservation yield

$$\int_{-\infty}^{\infty} \eta(X, Z, T)\, dX = 0, \tag{11}$$

assuming that the disturbance is locally confined along Z as well. This constraint appears to imply a restriction on allowed initial conditions, which raises some doubts about the validity of the KP equation in case $\eta(X, Z, T = 0)$ is not consistent with (11).

To clarify this point, following Katsis & Akylas (1987b), note that, according to the dispersion relation (7), the group velocity tends to $-\infty$ as $k \to 0$, unless $m = 0$ as well. Hence, the constraint (11) is entirely equivalent to the hypothesis that the disturbance remains locally confined; in case the initial condition does not observe (11), there are wave components with $k = 0$, $m \neq 0$ that have infinite group velocity and, therefore, the wave disturbance extends to $X = -\infty$ for any $T > 0$. This intuitive argument was carried further by Grimshaw & Melville (1989); their more complete analysis confirms that, in general, the constraint (11) does not apply because disturbances do not remain locally confined owing to radiation of three-dimensional waves behind.

This property of three-dimensional disturbance also bears on numerical procedures for solving the KP equation. As computational domains have finite extent, difficulties are expected to arise when the constraint (11) is not met, and details will be given later in terms of specific examples.

In integrating the KP equation numerically, Katsis & Akylas (1987a) and, independently, Pierini (1986) pointed out that it is advantageous first to convert (8) to an integral-differential equation

$$\eta_T + \tfrac{3}{2}\eta\eta_X + \tfrac{1}{6}\eta_{XXX} - \tfrac{1}{2}\int_X^\infty \eta_{ZZ}\, dX' = 0,$$

invoking the group-velocity argument noted above. In this form, then, one may discretize using numerical schemes applicable to the KdV equation (see, for example, Vliegenthart 1971, Fornberg & Whitham 1978).

SOLITARY-WAVE DYNAMICS

As already remarked, solitary waves are essential to the long-time evolution of two-dimensional disturbances. In the three-dimensional problem, however, this is not the case in general, although in some instances—when the wave motion is confined by rigid walls, for example—solitary waves still can arise from a three-dimensional initial disturbance.

If no boundaries are present, three-dimensional (locally confined) solitary waves have to obey the constraint (11) which imposes a rather severe restriction on their mass distribution; this suggests that such waves, if possible at all, are not likely to play a major role in general. In fact, in the negative-dispersion case ($\gamma < 1/3$) which is relevant to water waves, there are no three-dimensional solitary waves and a three-dimensional initial disturbance eventually disperses out. For positive dispersion ($\gamma > 1/3$), the KP equation (9) admits locally confined "lump" solutions, consistent with

(11), but so far they have not been observed experimentally [see Ablowitz & Segur (1981) for a comprehensive account].

As expected, the KdV solitary wave (5) and its obliquely propagating counterpart

$$\eta = a_0 \operatorname{sech}^2 [(3a_0/4)^{1/2}(X+mZ-(a_0+m^2)T/2)], \qquad (12)$$

of course, are solutions of the KP equation. But it is clear that, since they are not truly localized, these oblique KdV solitary waves cannot emerge from a three-dimensional disturbance of finite mass in an unbounded domain. On the other hand, KdV solitary waves and their oblique interactions turn out to be important in understanding the dynamics of three-dimensional disturbances in channels bounded by rigid sidewalls.

Oblique interactions of two-dimensional solitary waves have been studied by, among others, Benney & Luke (1964), Byatt-Smith (1971), Satsuma (1976), and Miles (1977a,b); a detailed review of this work can be found in Freeman (1980). Briefly, when the propagation directions of two solitary waves are not too close, the interaction is weak (the interaction time is relatively short) and, to leading order, linear superposition holds. As the propagation directions of the two waves approach each other, however, the interaction becomes intrinsically nonlinear and is described by a two-soliton solution of the KP equation. Moreover, Miles (1977a,b) pointed out that there is a regime in which this two-wave interaction is singular; as a result, regular reflection of a KdV solitary wave at a rigid wall becomes impossible when the angle of incidence is less than $(3a_0/h_0)^{1/2}$. Within this range of incidence angles, in analogy with the similar shock-wave phenomenon in gas dynamics, regular reflection is replaced by Mach reflection: The incident and reflected waves now meet away from the wall along with a third wave (the Mach stem) that is normal to the wall. Miles (1977b) also proposed a theoretical model for Mach reflection in terms of a resonant interaction of three solitary waves, consistent with the boundary condition at the wall, and calculated the associated run-up as a function of the angle of incidence. The predictions of Miles' model are not in quantitative agreement with the laboratory experiments of Melville (1980). However, computations that take into account transient effects give more satisfactory results (Kirby et al 1987).

In more recent laboratory experiments, Ertekin (1984) studied the transformation of a KdV solitary wave, propagating along a channel with vertical sidewalls, as it encountered an abrupt increase in the channel width. He observed that, during a brief transition period, the three-dimensional disturbance resulting from the boundary discontinuity adjusted to form a new KdV soliton (of lower amplitude and speed) spanning the entire channel width.

In an attempt to understand how this transformation from the initial to the final soliton state takes place, Katsis & Akylas (1987a) studied a model problem which, although it does not mimic the experimental set-up precisely, exhibits similar behavior and is simpler to handle theoretically. In a uniform channel ($0 \leq Z \leq W$) of width $W = 6.0$, they used as initial condition a KdV solitary wave (5) (with peak amplitude $a_0 = 1$) spanning only half of the channel width ($0 \leq Z \leq W/2$), and studied the subsequent development of the disturbance by numerically solving the KP equation (8), subject to the boundary conditions

$$\eta_Z = 0 \qquad (Z = 0, W) \tag{13}$$

at the sidewalls.

It is worth pointing out that, as this initial condition violates the constraint (10), the disturbance will tend to propagate far behind instantly, and some difficulties are to be expected at the edge of the computational domain owing to artificial reflections. Nevertheless, according to Katsis (1986), the corresponding error is relatively small in most of the computational domain and mass is conserved to a reasonable approximation.

Results from the computations of Katsis & Akylas (1987a) at three different times are shown in Figure 1(a–c). After the initial disturbance is released, it bends backwards and slows down but, as soon as it feels the presence of the sidewall at $Z = W$, it quickly adjusts to form a KdV soliton of lower amplitude spanning the entire channel width. The role of Mach reflection in this adjustment process is rather crucial, as pointed out by Ertekin (1984) and further elaborated by Pedersen (1988). Owing to the presence of a Mach stem normal to the sidewall at $Z = W$ (see Figure 1b), the disturbance near the sidewall speeds up and quickly catches up with the rest of the wave to form a straight-crested soliton.

A similar mechanism is responsible for the upstream generation of straight-crested solitons in channels by moving three-dimensional sources (see, for example, Wu 1987, Akylas 1988).

THREE-DIMENSIONAL PERIODIC WAVES

As expected from the above discussion of KdV solitary-wave interactions, the KP equation also describes oblique interactions of cnoidal waves that give rise to three-dimensional periodic wave patterns.

Mathematically, three-dimensional, periodic solutions of the KP equation can be expressed in terms of Riemann theta functions of genus N (Krichever 1977). For $N = 1$ one recovers the two-dimensional cnoidal waves, while $N = 2$ yields doubly periodic waves that may be interpreted as two obliquely interacting cnoidal waves (Segur & Finkel 1985). These

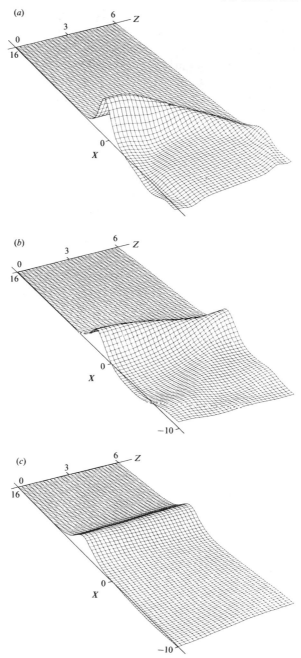

Figure 1   Transformation of a KdV solitary wave in a channel with vertical sidewalls. The initial condition consists of a KdV solitary wave spanning only half of the channel. (*a*) $T = 1$, (*b*) $T = 5$, (*c*) $T = 20$ (adapted from Katsis & Akylas 1987a).

genus-2 solutions are periodic in two independent horizontal directions and in time.

In related experimental work, Hammack et al (1989) studied doubly periodic wave disturbances in the laboratory and made a comparison with the corresponding genus-2 solutions of the KP equation. In particular, they focused on waves symmetric about the X-axis (the propagation direction) and periodic in both X and Z. Experimentally, these waves were generated by the interaction of two identical cnoidal waves, one directed at an angle β and the other at an angle $-\beta$ to the X-axis. The corresponding KP solution family, apart from translations in space that introduce two phase constants, depends on three dynamical parameters that specify the two wavelengths and the maximum wave amplitude; in comparing the experimental results with the predictions of the KP equation, these five parameters were chosen so as to obtain the best possible fit.

Figure 2 shows a symmetric KP solution of genus 2 corresponding to one of the experiments of Hammack et al (1989). The hexagonal surface patterns are characteristic of all the observed waves and result from the interaction of the two underlying cnoidal waves. It is worth noting that this is a genuinely nonlinear interaction rather than mere superposition of cnoidal waves, as evidenced by the long crests of relatively uniform amplitude—the analogs of Mach stems for periodic waves—that are normal to the propagation direction, and by the phase shifts of the underlying waves before and after interaction. Furthermore, these features are also apparent in shallow-water wave patterns observed in the field (Figure

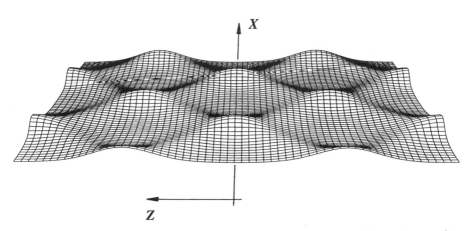

Figure 2   Genus-2 solution of the KP equation corresponding to one of the experiments of Hammack et al (1989) (adapted from Hammack et al 1989).

3), supporting the view of Segur & Finkel (1985) that genus-2 solutions of the KP equation represent typical three-dimensional periodic wave phenomena on shallow water.

Quantitative agreement of the wave profiles observed by Hammack et al (1989) with the corresponding KP solutions is quite favorable, even well outside the formal range of validity of the KP theory—most of the experimentally generated waves were near breaking. Also, the experiments indicate that, unlike small-amplitude (sinusoidal), two-dimensional waves

Figure 3 Aerial photograph of three-dimensional, shallow-water wave pattern near Jones Beach, Long Island, New York (kindly provided by Dr. N. Scheffner).

that are unstable to oblique perturbations on shallow water (Benney & Roskes 1969), finite-amplitude, periodic waves are stable.

VARIABLE-DEPTH EFFECTS

Quite often—for example, in marine straits (Alpers & Salusti 1983, Fu & Holt 1984)—three-dimensional effects on shallow-water waves are induced by depth variations. For this reason, several studies have aimed at extending the KP model to account for the effect of nonuniform depth.

Intuitively, it is clear that the horizontal lengthscale and typical amplitude of bottom variations are essential to the wave evolution and, depending on the relative magnitudes of these scales, various theoretical models have been proposed.

Channels of Nonrectangular Cross Section

The propagation of long waves in a uniform channel of arbitrary cross section is governed by the KdV theory, as long as the channel width is comparable to the water depth (Peregrine 1968, Fenton 1973, Teng & Wu 1992). Under this condition, the wave disturbance is quasi-two-dimensional—a relatively small spanwise variation in the wave elevation is present, but the wave crests remain straight—and the KdV equation (with constant coefficients that depend on the cross-section shape) remains valid to leading order. These theoretical predictions were confirmed experimentally by Peregrine (1969) for solitary waves propagating in channels of trapezoidal cross section.

Peregrine (1969) also noted, however, that the KdV theory breaks down when the channel width is much larger than the water depth because then spanwise variations are relatively large. This was also brought out by the experimental work of Sandover & Taylor (1962), who drew attention to the three-dimensional structure of undular bores in trapezoidal channels of moderately large width (about six to seven times the water depth). Particularly, the presence of wave-crest curvature across the channel and the tendency for the observed waves to have irregular forms (except at very low discharge) cannot be explained by the KdV theory (Fenton 1973).

In the further development of the theory, David et al (1987) studied the propagation of long waves in channels of slowly varying cross section, modeling marine straits with width much larger than depth. They derived a generalized KP equation with variable coefficients (depending on the local cross section) along with boundary conditions that include the effect of the sides of the channel. Under certain conditions on depth and width variations, this model admits solitary-wave solutions with curved wave crests (David et al 1989). Independently, assuming again gradual depth and width variations in wide channels, Kirby et al (1987) suggested a system of two KP equations with variable coefficients that also accounts for the waves generated by reflections owing to depth and width changes along the channel.

In more recent theoretical work, Mathew & Akylas (1990a), motivated by the experiments of Sandover & Taylor (1962), developed a theoretical model for long, three-dimensional water waves propagating along a channel bounded by sloping sidewalls. Assuming that the channel has uniform cross section and is wide ($h_0/w \ll 1$, where w is the channel width at the undisturbed free surface and h_0 the uniform water depth away from the sidewalls), they derived the approximate governing equation and boundary conditions using matched asymptotic expansions. Specifically, the main body of the fluid, away from the sloping boundaries, forms an "outer" region where waves satisfy the KP equation (8) and three-dimensional effects become appreciable if $h_0/w = O(\varepsilon)$. Close to each sidewall, where the depth varies, there is an "inner" region. Matching between the corresponding inner and outer expansions gives the appropriate asymptotic boundary condition at each sidewall, depending on the relative size of the inner region.

In particular, when the wall slope is $O(1)$, as in the experiments of Sandover & Taylor (1962), the extent of the inner region is $O(h_0)$ and the sidewall boundary condition obtains a particularly simple form:

$$\eta_Z = A\eta_{XX}. \tag{14}$$

Here the parameter $A = A/h_0^2$, A being the area under the depth profile; for a channel of trapezoidal cross section, $A = (\tan \theta)/2$, where θ is the angle that the wall makes with the vertical. Note that according to (14), the outer flow is insensitive to the exact geometry of the sidewalls; of course, if the sidewalls are vertical, $A = 0$ and the familiar no-flux boundary condition (13) is recovered.

For wall slopes $O(\varepsilon^{1/2})$, on the other hand, the inner region is relatively thicker—its size is now comparable with the horizontal lengthscale l, but still small compared with w. Hence, the outer flow is affected more seriously by the details of the depth variation close to the wall, and it turns out that the corresponding boundary condition takes an integral-differential form (Mathew & Akylas 1990a).

Undular Bore

In terms of the model suggested by Mathew & Akylas (1990a), the development of an undular bore in a symmetric channel of width W is studied by solving the KP equation in half the channel, $0 \le Z \le W/2$, subject to the wall boundary condition (14) at $Z = 0$ and the symmetry condition (13) at $Z = W/2$. The appropriate initial conditions are

$$\eta(X, Z, T = 0) = 0 \quad (X > 0), \qquad \eta(X, Z, T = 0) = \eta_\infty \quad (X \leq 0),$$

where η_∞ is the initial difference in water levels. Note that these initial conditions are consistent with the constraint (10), so that no difficulties are expected in the numerical treatment of this initial-boundary-value problem.

In the experiments of Sandover & Taylor (1962), the channel sidewalls were supported at various angles θ to the vertical, while the width of the bed and the still-water depth were kept fixed at 12 inches and 3 inches, respectively. Wave-height measurements were taken for a range of different discharges using gauges located at several points across the channel. In relating the theory to the experiments, it is convenient to take the dimensionless channel width at the free surface, W, equal to unity; this normalization and the cross-section geometry, then, specify ε. For instance, the values of ε corresponding to the sidewall inclinations $\theta = \pi/4$ and $\pi/3$ ($A = 0.5, 0.866$) are found to be 0.167 and 0.134, respectively.

Figure 4 is a snapshot of the computed free surface for $\theta = \pi/4$ and $\eta_\infty = 1.08$ (this value of η_∞ is equivalent to the corresponding discharge in the experiment) at $T = 12.5$, when the leading crest of the bore crosses the streamwise position at which the gauges were placed in the experiment. The wave disturbance is clearly three-dimensional: Wave crests are curved across the channel and steeper close to the sidewalls [consistent with the boundary condition (14)]; moreover, there is a tendency for individual waves to form a double-hump structure, with two crests separated by a shallow trough in the middle. These features are also apparent, at least qualitatively, in the photographs from the experiments of Sandover & Taylor (1962).

Mathew & Akylas (1990a) also report good quantitative agreement between computed and experimentally observed wave profiles, as long as the discharge, or equivalently η_∞, is kept below the critical value at which wave breaking occurred in the experiments. Given that the values of ε corresponding to the experiments of Sandover & Taylor (1962) are only moderately small, this seems to reinforce the conclusion of Hammack et al (1989) that the KP model may often be useful even outside its formal range of validity.

Edge Waves

The KP model for waves in channels with sloping sidewalls also applies to the propagation of shallow-water waves along a beach, and Mathew & Akylas (1990b) have pursued some of the details.

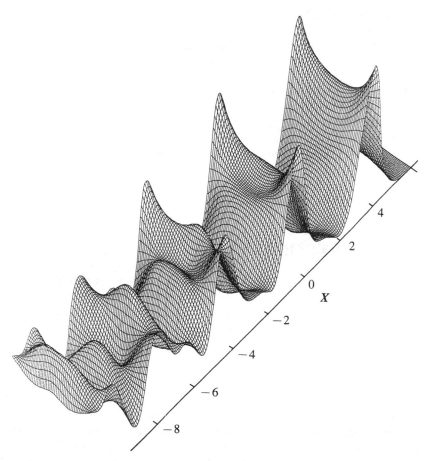

Figure 4 Computed free surface at the head of undular bore in a trapezoidal channel with sidewall slope $\theta = \pi/4$; water-level difference $\eta_\infty = 1.08$, time $T = 12.5$ (adapted from Mathew & Akylas 1990a).

Briefly, the model formulation consists of the KP equation (8) along with the wall condition (14) at the shoreline $Z = 0$ [for a beach of $O(1)$ slope] and the condition that η is bounded out to sea $(Z \to \infty)$. In particular, the linearized problem admits two kinds of infinitesimal traveling-wave disturbances

$$\eta = H(Z) \exp[i(KX - \Omega T)].$$

There is a continuous spectrum of obliquely incident and reflected waves at the beach with $\Omega > -K^3/6$, and in addition, an edge-wave mode,

$$H = \exp(-K^2 AZ),$$

that is trapped close to the shoreline $Z = 0$ and obeys the dispersion relation

$$\Omega = -\tfrac{1}{2}K^3(\tfrac{1}{3}+A^2),$$

in agreement with earlier work by Grimshaw (1974).

Finite-amplitude traveling edge waves, however, do not remain trapped; rather, they are slowly attenuated by radiating oblique waves out to sea. This interaction mechanism can also transfer energy from incoming oblique waves to subharmonic edge waves and is not limited to shallow-water edge waves (see Mathew & Akylas 1990b for details).

ROTATION EFFECTS

The effects of rotation on the propagation of long waves are of interest in various geophysical applications and have received considerable attention, mainly in the context of internal waves in stratified fluids. However, entirely analogous results hold for surface waves.

Rotation-Modified KP

In the course of laboratory experiments, Maxworthy (1983) first reported that KdV solitary internal waves propagating along a rotating channel acquire marked three-dimensional features: The wave amplitude varies exponentially across the channel (like a Kelvin wave) and, most surprisingly, the wave crest is curved backwards. Later, Renouard et al (1987) conducted experiments in two-layer flow confirming the findings of Maxworthy (1983). They further noticed, however, a train of small-amplitude waves trailing the main wave disturbance, and recognized that the solitary-wave amplitude was slowly attenuated presumably due to viscous dissipation.

From the theoretical point of view, it seems clear that the effect of rotation, as it amounts to wave-crest curvature, has to balance with weak nonlinear and dispersive effects. Along this line of thought, Grimshaw (1985) arrived at the rotation-modified KP (RMKP) equation for internal waves in a rotating channel.

For surface waves, the corresponding RMKP equation can be readily inferred from the linear dispersion relation of weakly three-dimensional, long waves in the presence of rotation. Specifically, including the effect of rotation, (7) becomes

$$\omega = k - \mu^2 \left(\frac{k^3}{6} - \frac{m^2}{2k} - \frac{v^2}{2k} \right),$$ (15)

on the assumption

$$\frac{fl}{c_0} = v\mu, \qquad v = O(1),$$

where f is the rotation rate, so that rotation is as important as weak three-dimensional variation and dispersion. In view of (8) and (15), then, the RMKP equation takes the form

$$(\eta_T + \tfrac{3}{2}\eta\eta_X + \tfrac{1}{6}\eta_{XXX})_X + \tfrac{1}{2}(\eta_{ZZ} - v^2\eta) = 0.$$ (16)

Furthermore, if the fluid is contained in a channel with vertical sidewalls (located at $Z = 0, W$) the transverse velocity should vanish at the walls, and the boundary conditions (13) are replaced by

$$\eta_Z + v\eta = 0 \quad (Z = 0, W).$$ (17)

For disturbances that are locally confined as $X \to \pm \infty$, it is straightforward to derive, from (16), (17), and mass conservation, a constraint analogous to (11):

$$\int_{-\infty}^{\infty} \eta(X, Z, T)\, dX = Ce^{-vZ},$$ (18)

where C is a constant. As expected, this constraint ensures that no wave-number components with infinite group velocity [according to (15)] are present, and the disturbance remains locally confined (Katsis & Akylas 1987b, Grimshaw & Melville 1989).

Radiation Damping

Motivated by the experiments of Maxworthy (1983), Grimshaw (1985) looked for, but was not able to find, locally confined solitary-wave solutions of the RMKP equation (16), subject to the wall conditions (17). This is not surprising in view of the rather stringent condition that (18) places on such solutions of the RMKP equation, and Grimshaw (1985) raised some doubt as to whether this model equation could describe the experimental observations, without taking into consideration viscous or strongly nonlinear effects.

In a related numerical study, Katsis & Akylas (1987b) explored the development of an initially straight-crested Kelvin solitary wave—the KdV solitary wave (5) multiplied with $\exp(-vZ)$ so that (17), (18) are met—along a rotating channel, in terms of the RMKP model. They confirmed that rotation gives rise to wave-front curvature, in qualitative agreement with the experiments; however, the wave amplitude decays slowly as the disturbance propagates downstream, indicating that this is not a wave of permanent form and the observed attenuation is only partly caused by viscous damping.

The mechanism responsible for wave attenuation and wave-crest curvature was further elucidated by Melville et al (1989), who drew attention to the role of Poincaré waves—the small-amplitude (linear) cross-modes consistent with (16) and (17):

$$\eta_n = \left(\cos \frac{n\pi Z}{W} - \frac{vW}{n\pi} \sin \frac{n\pi Z}{W} \right) \exp\left[i(kX - \omega T) \right] \qquad (n = 1, 2, \ldots),$$

with

$$\omega = -\tfrac{1}{6}k^3 + b_n^2/k, \qquad b_n^2 = [v^2 + (n\pi)^2/W^2]/2.$$

On the basis of an approximate analysis, valid for small times, and numerical results, Melville et al (1989) argued that an initially straight-crested Kelvin solitary wave generates Poincaré waves downstream (having the same phase speed as the main disturbance) which, when superposed on the Kelvin wave, give rise to an apparent wave-front curvature. The same conclusion was reached, using a small-time asymptotic analysis of the RMKP equation, by Grimshaw & Tang (1990) who also demonstrated that the generation of Poincaré waves results from decay of the Kelvin-wave amplitude through radiation damping, consistent with the observations of Renouard et al (1987). Furthermore, Akylas (1991) investigated the long-time evolution of a solitary wave in a slowly rotating channel ($v \ll 1$); this analysis indicates that radiation damping is amplitude-dependent and that the trailing disturbance is dominated by the first Poincaré mode ($n = 1$), in agreement with the numerical solutions of Melville et al (1989).

Finally, it is worth pointing out that, although there is no experimental confirmation as yet, radiation damping of solitary waves is also expected to occur in a (nonrotating) channel with sloping sidewalls. For slightly sloping sidewalls (A ≪ 1), Mathew & Akylas (1990a) found that, owing to cross-modes, small-amplitude oscillations develop at the tails of KdV solitary waves. This suggests that, during the transient evolution of a KdV solitary wave, such tails are bound to appear; their position is determined by the sign of the cross-mode group velocity relative to the speed of the main disturbance.

ACKNOWLEDGMENTS

The author would like to thank Professor H. Segur for valuable discussions on this topic and Dr. N. Scheffner for providing the figures of doubly periodic waves. Part of the results reported here were obtained in the course of research supported by the Office of Naval Research (Project No. NR062-742), the National Science Foundation (grants MSM-8451154, DMS-9202064), and the Air Force Office of Scientific Research (grant F49620-92-J0086).

Literature Cited

Ablowitz, M. J., Clarkson, P. A. 1991. *Solitons, Nonlinear Evolution Equations and Inverse Scattering.* Cambridge: Cambridge Univ. Press

Ablowitz, M. J., Segur, H. 1981. *Solitons and the Inverse Scattering Transform.* Philadelphia: SIAM

Akylas, T. R. 1988. Nonlinear forced wave phenomena. In *Nonlinear Wave Interactions in Fluids*, ed. R. W. Miksad, T. R. Akylas, T. Herbert, pp. 157–63. New York: ASME

Akylas, T. R. 1991. On the radiation damping of a solitary wave in a rotating channel. In *Mathematical Approaches in Hydrodynamics*, ed. T. Miloh, pp. 175–81. Philadelphia: SIAM

Alpers, W., Salusti, E. 1983. Scylla and Charybdis observed from space. *J. Geophys. Res.* 88: 1800–8

Benney, D. J. 1966. Long nonlinear waves in fluid flows. *J. Math. Phys.* 45: 52–63

Benney, D. J., Luke, J. C. 1964. On the interactions of permanent waves of finite amplitude. *J. Math. Phys.* 43: 309–13

Benney, D. J., Roskes, G. J. 1969. Wave instabilities. *Stud. Appl. Math.* 48: 377–85

Byatt-Smith, J. G. B. 1971. An integral equation for unsteady surface waves and a comment on the Boussinesq equation. *J. Fluid Mech.* 49: 625–33

David, D., Levi, D., Winternitz, P. 1987. Integrable nonlinear equations for water waves in straits of varying depth and width. *Stud. Appl. Math.* 76: 133–68

David, D., Levi, D., Winternitz, P. 1989. Solitons in shallow seas of variable depth and in marine straits. *Stud. Appl. Math.* 80: 1–23

Ertekin, R. C. 1984. *Soliton generation by moving disturbances in shallow water: theory, computation and experiment.* Doctoral dissertation. Univ. Calif., Berkeley

Fenton, J. D., 1973. Cnoidal waves and bores in uniform channels of arbitrary cross-section. *J. Fluid Mech.* 58: 417–34

Fokas, A. S., Sung, L.-Y. 1992. On the solvability of the *N*-wave, Davey-Stewartson and Kadomtsev-Petviashvili equations. *Inverse Probl.* 8: 673–708

Fornberg, B., Whitham, G. B. 1978. A numerical and theoretical study of certain nonlinear wave phenomena. *Phil. Trans. R. Soc. London Ser. A* 289: 373–404

Freeman, N. C. 1980. Soliton interactions in two dimensions. *Adv. Appl. Mech.* 20: 1–37

Fu, L.-L., Holt, B. 1984. Internal waves in the Gulf of California: observations from a space borne radar. *J. Geophys. Res.* 89: 2053–60

Grimshaw, R. 1974. Edge waves: a long-wave theory for oceans of finite depth. *J. Fluid Mech.* 62: 775–91

Grimshaw, R. 1985. Evolution equations for weakly nonlinear, long internal waves in a rotating fluid. *Stud. Appl. Math.* 73: 1–33

Grimshaw, R., Melville, W. K. 1989. On the derivation of the modified Kadomtsev-Petviashvili equation. *Stud. Appl. Math.* 80: 183–202

Grimshaw, R., Tang, S. 1990. The rotation-modified Kadomtsev-Petviashvili equation: an analytical and numerical study. *Stud. Appl. Math.* 83: 223–48

Hammack, J., Scheffner, N., Segur, H. 1989. Two-dimensional periodic waves in shallow water. *J. Fluid Mech.* 209: 567–89

Johnson, R. S. 1980. Water waves and Korteweg–de Vries equations. *J. Fluid Mech.* 97: 701–19

Kadomtsev, B. B., Petviashvili, V. I. 1970. On the stability of solitary waves in weakly dispersing media. *Sov. Phys. Dokl.* 15: 539–41

Katsis, C. 1986. *An analytical and numerical study of certain three-dimensional nonlinear wave phenomena.* Doctoral dissertation. MIT

Katsis, C., Akylas, T. R. 1987a. On the excitation of long nonlinear water waves by a moving pressure distribution. Part 2. Three-dimensional effects. *J. Fluid Mech.* 177: 49–65

Katsis, C., Akylas, T. R. 1987b. Solitary internal waves in a rotating channel: a numerical study. *Phys. Fluids* 30: 297–301

Kirby, J. T., Philip, R., Vengayil, P. 1987. One dimensional and weakly two-dimensional waves in varying channels: numerical examples. In *Nonlinear Water Waves*, ed. K. Horikawa, H. Maruo, pp. 357–64. Berlin, Heidelberg: Springer-Verlag

Korteweg, D. J., de Vries, G. 1895. On the change of form of long waves advancing in a rectangular canal, and on a new type of long stationary waves. *Phil. Mag.* 39: 422–43

Krichever, I. M. 1977. Methods of algebraic geometry in the theory of non-linear equations. *Russ. Math. Surv.* 32: 185–313

Krichever, I. M. 1989. Spectral theory of two-dimensional periodic operators and its applications. *Russ. Math. Surv.* 44(2): 145–225

Mathew, J., Akylas, T. R. 1990a. On three-dimensional long water waves in a channel with sloping sidewalls. *J. Fluid Mech.* 215: 289–307

Mathew, J., Akylas, T. R. 1990b. On the radiation damping of finite-amplitude progressive edge waves. *Proc. R. Soc. London Ser. A* 431: 419–31

Maxworthy, T. 1983. Experiments on solitary internal Kelvin waves. *J. Fluid Mech.* 129: 365–83

Melville, W. K. 1980. On the Mach reflection of a solitary wave. *J. Fluid Mech.* 98: 285–97

Melville, W. K., Tomasson, G. G., Renouard, D. P. 1989. On the stability of Kelvin waves. *J. Fluid Mech.* 206: 1–23

Miles, J. W. 1977a. Obliquely interacting solitary waves. *J. Fluid Mech.* 79: 157–69

Miles, J. W. 1977b. Resonantly interacting solitary waves. *J. Fluid Mech.* 79: 171–79

Miles, J. W. 1981. The Korteweg–de Vries equation: a historical essay. *J. Fluid Mech.* 106: 131–47

Newell, A. C. 1985. *Solitons in Mathematics and Physics*. Philadelphia: SIAM

Pedersen, G. 1988. Three-dimensional wave patterns generated by moving disturbances at transcritical speeds. *J. Fluid Mech.* 196: 39–63

Peregrine, D. H. 1968. Long waves in a uniform channel of arbitrary cross-section. *J. Fluid Mech.* 32: 353–65

Peregrine, D. H. 1969. Solitary waves in trapezoidal channels. *J. Fluid Mech.* 35: 1–6

Pierini, S. 1986. Solitons in a channel emerging from a three-dimensional initial wave. *Nuovo Cimento* 9C: 1045–61

Renouard, D. P., Chabert d'Hières, G., Zhang, X. 1987. An experimental study of strongly nonlinear waves in a rotating system. *J. Fluid Mech.* 177: 381–94

Russell, J. S. 1845. Report on waves. *Rep. Meet. Brit. Assoc. Adv. Sci.*, 14th, York, 1844, pp. 311–90. London: John Murray

Sandover, J. A., Taylor, C. 1962. Cnoidal waves and bores. *Houille Blanche* 7: 443–55

Satsuma, J. 1976. *N*-soliton solution of the two-dimensional Korteweg–de Vries equation. *J. Phys. Soc. Jpn.* 40: 286–90

Segur, H., Finkel, A. 1985. An analytical model of periodic waves in shallow water. *Stud. Appl. Math.* 73: 183–220

Teng, M. H., Wu, T. Y. 1992. Nonlinear water waves in channels of arbitrary shape. *J. Fluid Mech.* 242: 211–33

Vliegenthart, A. C. 1971. On finite-difference methods for the Korteweg–de Vries equation. *J. Engng. Math.* 5: 137–55

Whitham, G. B. 1974. *Linear and Nonlinear Waves*. New York: Wiley-Intersci.

Wu, T. Y. 1987. Generation of upstream advancing solitons by moving disturbances. *J. Fluid Mech.* 184: 75–99

Zufiria, J. A. 1987. Symmetry breaking in periodic and solitary capillary-gravity waves on water of finite depth. *J. Fluid Mech.* 184: 183–206

Annu. Rev. Fluid Mech. 1994. 26 : 211–54

COMPRESSIBILITY EFFECTS ON TURBULENCE

Sanjiva K. Lele

Department of Mechanical Engineering and Department of Aeronautics and Astronautics, Stanford University, Stanford, California 94305-4035

KEY WORDS: shock-turbulence interaction, supersonic mixing, turbulence modeling

1. INTRODUCTION

Turbulence is a macroscopic state of flow in which the instantaneous flow variables exhibit a seemingly random variation in time and in the three spatial coordinates. The temporal and spatial disorder has finite decorrelation time and length scales. This article reviews those features of the turbulent state that depend on compressibility. Effects associated with the volume changes of fluid elements in response to changes in pressure are regarded as compressibility effects. These are contrasted with variable inertia effects associated with either variable composition or volume changes due to heat transfer. The behavior of incompressible turbulent flows has been frequently reviewed in this series and compressibility effects on turbulent shear layers have been reviewed by Bradshaw (1977). Morkovin (1961, 1992) has reviewed compressibility effects on wall-bounded and free-shear flows; Smits (1991) and Spina et al (1994) have recently reviewed the turbulent boundary layer structure in supersonic flow.

In recent years the renewed interest in high-speed civil transport aircraft and supersonic combustion ramjet engines for high altitude hypersonic propulsion has invigorated fundamental research on compressible turbulent flows. New experimental information on shear flows under compressible conditions has been gathered. Direct numerical simulations (DNS) of turbulence in compressible regimes have also been performed. Despite these research efforts the phenomenology of compressible turbulence is far from mature. This article is as much a review of what is not known about compressible turbulence vis-à-vis its incompressible counter-

211

0066–4189/94/0115–0211$05.00

part, as it is a review of more commonly held views. Only macroscopic phenomena are reviewed and in most of the discussion the flow medium is an ideal gas. Chemical and thermodynamic nonequilibrium effects arising in hypersonic and rarefied flow regimes are not discussed; these flow regimes have been reviewed by Muntz (1989) and Stalker (1989). Issues related to supersonic combustion have been reviewed by Ferri (1973), Walthrup (1987), and Billig (1988). Compressible flows near the thermodynamic critical point have been reviewed by Kutateladze et al (1987); shock wave phenomena have been reviewed by Zel'dovich & Razier (1969) and Hornung (1986); and shock wave boundary layer interactions have been reviewed by Green (1970) and Adamson & Messiter (1980).

The organization of this article is as follows. Different facets of compressibility effects on turbulence and the dimensionless parameters characterizing them are summarized in Section 2. Homogeneous compressible flows are discussed in Section 3, followed by a discussion of simple inhomogeneous flows in Section 4. The review is thus limited to the simplest possible flows.

2.   THE NATURE OF TURBULENCE IN A COMPRESSIBLE MEDIUM

2.1   *Modes of Fluctuations*

When a compressible medium is in turbulent motion the thermodynamic fluid properties such as density (mass per unit volume), specific entropy, as well as the thermophysical fluid properties such as viscosity coefficients and specific heats, undergo fluctuations. Measures of these fluctuations need to be included in a specification of the turbulent state. Yaglom (1948) and Moyal (1952) appear to be the first to formulate the equations governing the two-point correlations of these fluctuations. Kovasznay (1953) decomposed the turbulent fluctuations into vorticity, acoustic, and entropy modes of fluctuations. This decomposition is defined for (weak) turbulent fluctuations about a flow state of uniform mean velocity and spatially uniform mean thermodynamic properties. The three modes of fluctuations are dynamically decoupled from each other to first order in the fluctuation amplitude. At second order, mode couplings arise and a combination of any two modes (including self-interaction) generates other modes (Chu & Kovasznay 1958, Monin & Yaglom 1971).

If linearized perturbations about a nonuniform mean state are considered, mode couplings arise in the linearized equations. However, for linear unsteady disturbances about an arbitrary potential flow (over a body) the disturbance velocity can be decomposed into a "vortical" and "acoustic" part (Goldstein 1978a). The vortical disturbance is generally not solenoidal and can be calculated without a specific knowledge of the acoustic part. Analogous decomposition of unsteady disturbances about transversely sheared mean flows can also be obtained (Goldstein 1978b, 1979a). Such decompositions have been extended to flows with mean entropy gradient orthogonal to mean vorticity (Debieve 1978a, Dussauge et al 1988). Linear rapid-distortion theories have exploited such decompositions in studies as turbulence passing through a Prandtl-Meyer fan (Goldstein 1978a), turbulence generated by entropy fluctuations in a non-uniform flow (Goldstein 1979b), homogeneous turbulence subject to bulk compression (Durbin & Zeman 1992), and shock turbulence interaction (Ribner 1953, 1954, 1969, 1987; Moore 1953; Chang 1957; Kerrebrock 1956; McKenzie & Westphal 1968; Anyiwo & Bushnell 1982). Hot-wire measurements taken at a point in high-speed flow are commonly discussed in terms of fluctuation modes (Kovaznay 1950, Morkovin 1956, Smits & Dussauge 1988). In that context the "modes" are locally defined and associated with small deviations from a uniform state. Local fluctuations of mass flow and total temperature are obtained by exploiting the varying sensitivity of the hot-wire signal, as its operating conditions (such as overheat ratio) are varied and then decomposed into "modes."

2.2   Evolution of Turbulent Kinetic Energy

Nondimensional parameters characterizing the influence of compressibility on turbulence are contained in equations governing the turbulent flow. In this article the density-weighted form (Favre 1965a) is adopted, primarily for its compactness of notation. The density-weighted averaged velocity, internal energy per unit mass, and enthalpy per unit mass are denoted by $\tilde{u}_i$, $\tilde{e}$, and $\tilde{h}$, while the fluctuations from these are denoted by u_i'', e'', and h''. Ensemble- or Reynolds-averaged pressure and density are $\bar{p}$ and $\bar{\rho}$, and the fluctuation relative to the ensemble averages are p' or ρ' (denoted with a single prime). As there is no mass flux across the Favre-averaged streamlines (a property not shared by the conventional Reynolds-averaged streamlines) the equations for mean density, mean velocity, and mean enthalpy are more compact. Further discussion of the merits/limitations of Favre-averaging and its transformation properties can be found in Lele (1993).

As indicated in Figure 1 the evolution of turbulent kinetic energy (TKE) is coupled to the evolution of mean kinetic energy and the mean internal energy, and several pathways exist for these exchanges. Parameters measuring the influence of compressibility are defined by considering the importance of those pathways that exist only in a compressible flow relative to those occurring under incompressible conditions.

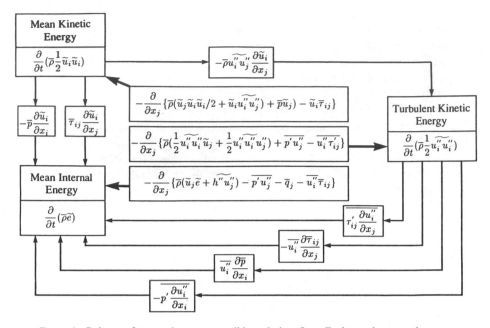

Figure 1   Balance of energy in a compressible turbulent flow. Exchange between the mean kinetic energy, mean internal energy, and turbulent kinetic energy occurs via the terms indicated by small arrows. All arrows pointing into the energy balance for some quantity Q are taken with a positive sign on the right-hand side; outward pointing arrows require multiplication by a negative sign.

Four terms couple TKE changes to the changes in $\bar{\rho}\tilde{e}$ (mean internal energy) and are shown in the bottom four paths in Figure 1. The viscous dissipation rate (per unit volume) of TKE, $\varepsilon = \overline{\tau'_{ij}u''_{i,j}}$ ($u''_{i,j}$ is shorthand for $\partial u''_i/\partial x_j$), is a significant contributor to the TKE budget in all turbulent flows (compressible or incompressible), with the possible exception of flows undergoing a rapid distortion (e.g. flow through a shock wave). Using the Stokes (Cauchy-Poisson) constitutive relation the viscous dissipation rate is

$$\Phi = \tau_{ij}S_{ij} = \mu_v\Theta^2 + 2\mu S^d_{ij}S^d_{ij},$$

where $S_{ij} = (u_{i,j}+u_{j,i})/2$ is the rate of strain, $\Theta = u_{i,i}$ is the dilatation, $S^d_{ij} = S_{ij}-\delta_{ij}\Theta/3$ is the deviatoric part of S_{ij}, μ_v is the bulk viscosity, and μ the (shear) viscosity. Thus if μ_v is not zero the volume changes of fluid elements lead to an additional dissipation. It is possible to rearrange the above as

$$\Phi = (\mu_v + \tfrac{4}{3}\mu)\Theta^2 + \mu\omega_k\omega_k + \mu(u_iu_j)_{,ij} + \mu(u_ju_{i,j})_{,i} - 3\mu(u_j\Theta)_{,j},$$

where ω_k is the vorticity. Under conditions of homogeneity (discussed in Section 3) the last three terms vanish for averaged quantities. The first of the surviving terms is called compressible or dilatational dissipation (Favre 1969, Moyal 1952, Zeman 1990, Sarkar et al 1991) and the second is regarded as incompressible or solenoidal dissipation. The compressible dissipation term is identically zero for a strictly incompressible flow but otherwise finite (even for $\mu_v = 0$).

The relative magnitude of ε compared to the rate of production of TKE (P) may depend on compressibility, but a priori estimates cannot be made with sufficient accuracy. For homogeneous flows ε may be approximated as

$$\varepsilon = \varepsilon_s + \varepsilon_c = \varepsilon_s(1 + \chi_\varepsilon), \quad \varepsilon_s = \bar{\mu}\overline{\omega'_k\omega'_k}, \quad \varepsilon_c = (\bar{\mu_v} + \tfrac{4}{3}\bar{\mu})\overline{\theta'^2},$$

$$\chi_\varepsilon = \left(\frac{4}{3} + \frac{\bar{\mu_v}}{\bar{\mu}}\right)\overline{\theta'^2}/(\overline{\omega'_k\omega'_k}).$$

Terms arising from variations of μ and μ_v have been neglected; numerical simulations (Lee et al 1991, Blaisdell et al 1991, Erlebacher et al 1990) consistently show that the neglected terms are small. If ε_s is largely independent of the influence of compressibility, the ratio χ_ε may be regarded as a measure of the extra dissipation resulting from compressibility. Numerical simulations of decaying compressible isotropic turbulence (Sarkar et al 1991, Kida & Orszag 1992, Lee et al 1991, Blaisdell et al 1991) support this view. The simulations show that χ_ε depends on both the turbulence Mach number $M_t = q/C$ (q is a velocity scale characteristic of the turbulence and C is a representative speed of sound) and initial conditions such as initial density variance $\overline{\rho'^2}/\bar{\rho}^2$, and the initial partition ratio of the kinetic energy in the dilatational mode, χ_c (defined in Section 3). Simulations of compressible homogeneous shear flow (Blaisdell et al 1991, Sarkar et al 1993) also suggest that ε_s is largely independent of compressibility. Although the dissipation of TKE occurs at small scales, its rate ε is set by the energy-containing range of scales. It is unclear what attributes of the energy-containing scales, in a high Reynolds number flow, determine ε_s and ε_c. Whether ε_s/P is insensitive to compressibility in high Reynolds number flows, as found in the DNS at low Reynolds number, is an open question.

The second term $\overline{u''_i\tau_{ij,j}}$ coupling TKE evolution to $\bar{\rho}\tilde{e}$, in Figure 1, is an artifact of Favre-averaging and, in principle, can have either sign. The third coupling term involving a Reynolds-averaged value of Favre velocity fluctuation is $\overline{u''_i\bar{p}_{,i}}$, which has been called enthalpic production of turbulence by exchange with enthalpic energy (Favre 1969) and may cause a gain or loss of TKE. It is natural to associate this exchange term with variable inertia in the flow rather than compressibility. When a body force or a mean pressure gradient acts on the fluid the regions with smaller inertia (per unit volume) can respond more quickly to the imposed change. Work done by the force is a gain in the kinetic energy of the flow, part of which contributes to TKE. Due to the acceleration reaction (Pope 1987, Batchelor 1967) the differential acceleration is $O(-\nabla \bar{p}/\bar{\rho})$, even when $\rho_{\mathrm{rms}}/\bar{\rho}$ is not small, which is consistent with associating $\overline{u_i'' \bar{p}_{,i}}$ with the rate of energy exchange to TKE. It appears that a formulation similar to Goldstein (1979b) may be useful in estimating this energy exchange (Hunt, cited in Fulachier et al 1989). The exact transport equation for $\overline{u_j''}$ (Taulbee & Van Osdol 1991) can also be used to estimate $\overline{u_j''}$. The leading order terms of this equation show that $\overline{u_j''}$ depends not only on $\bar{\rho}_{,k}$ as commonly assumed, but also on $\bar{p}_{,k}$ and $\tilde{u}_{k,k}$. An identity relating the energy fluxes

$$\bar{\rho}\widetilde{h''u_j''} = \bar{\rho}\widetilde{e''u_j''} + \overline{p'u_j''} + \bar{p}\overline{u_j''}$$

is also relevant to the models of $\overline{u_j''}$. For an ideal gas this identity implies that once models for $\widetilde{T''u_j''}$ and $\overline{p'u_j''}$ have been chosen the turbulent mass flux $\overline{u_j''}$ is already determined as $\overline{u_j''} = \widetilde{T''u_j''}/\tilde{T} - \overline{p'u_j''}/\bar{p}$. This consistent choice of $\overline{u_j''}$ is not always recognized in turbulence modeling. In recent work (Speziale & Sarkar 1991) models for $\widetilde{T''u_j''}$ and $\overline{u_j''}$ are chosen and $\overline{p'u_j''}$ is then consistently evaluated. However, the model forms are such that for inhomogeneous, uniform density, isothermal flows $\overline{p'u_j''}$ is always zero.

In external flows involving significant pressure gradients—such as shock-wave boundary layer interactions or streamwise acceleration/ deceleration due to imposed pressure gradients or curved walls (Jayaram et al 1987, Fernando & Smits 1990, Dussauge et al 1988)—the turbulent mass flux $\overline{u_j''}$ is usually not associated with volume fluctuations of a fluid element as long as M_t is small. Hence effects associated with the enthalpic production of TKE, which commonly occur in compressible flows, are not compressibility effects but better regarded as effects of variable inertia. The importance of separating compressibility effects from effects of variable inertia cannot be overemphasized (Morkovin 1961, 1992). Enthalpic production of TKE is also important in Rayleigh-Taylor and Richtmeyer-Meshkov flows (Youngs 1989, Besnard et al 1989, Benjamin 1992, Meshkov 1992, Gauthier & Bonnet 1990). In premixed flames the differential acceleration is responsible for countergradient diffusion of species (Pope 1987). As M_t increases the distinction between variable inertia effects and compressibility effects gets blurred and the enthalpic production of TKE may introduce compressibility effects.

The fourth term coupling the evolutions of TKE and $\bar{\rho}\tilde{e}$, in Figure 1, is the net rate of work done by the pressure fluctuations due to the simultaneous fluctuations in dilatation, or the pressure dilatation (correlation), $\Pi_d = \overline{p'u''_{i,i}} = \overline{p'u'_{i,i}}$. In a compressible flow Π_d may take either sign and when negative represents a loss from TKE and gain of mean internal energy. This energy exchange does not change the mean entropy and thus can be regarded as reversible. In regions with negligible mean acceleration and negligible mean volume change, Π_d can also be interpreted as an exchange with acoustic potential energy (Section 2.5). The relative importance of Π_d compared to the Reynolds stress production of TKE is a natural measure of compressibility.

Estimates of Π_d can be obtained by considering the density ρ as a thermodynamic function of pressure, p, and specific entropy, s, (e.g. Thompson 1988) and by taking a material derivative to give (for an ideal gas)

$$\frac{\partial u_i}{\partial x_i} = -\frac{1}{\gamma p}\frac{Dp}{Dt} + \frac{1}{C_p}\frac{Ds}{Dt}. \tag{1}$$

Decomposing Equation 1 into its Reynolds-averaged mean and fluctuation gives (for high Reynolds number)

$$\overline{p'u'_{i,i}} \approx -\frac{\bar{p}}{2\gamma}\frac{\bar{D}}{\bar{D}t}\frac{\overline{p'^2}}{\bar{p}^2} - \frac{\bar{p}_{,i}}{\gamma\bar{p}}\overline{p'u'_i} - \frac{\gamma-1}{\gamma}\frac{\overline{p'^2}}{\bar{p}^2}(\varepsilon-\bar{q}_{j,j}+Q) + \ldots, \tag{2}$$

where only the leading terms are explicitly written. In Equation 2, $\bar{D}/\bar{D}t$ stands for the material derivative following the mean flow and Q is the external rate of energy addition per unit volume (e.g. due to chemical reactions). The first term of this equation can also be derived by considering linearized disturbance equations for homogeneous turbulence (Durbin & Zeman 1992). The approach adopted here (which is similar to that of Zeman & Blaisdell 1991) is more general. The expressions given here for an ideal gas are easily extended to an arbitrary equation of state.

The part of the third term in Equation 2 associated with the TKE dissipation rate ε always converts TKE into internal energy and is thus akin to *extra* dissipation due to compressibility, but does not cause an entropy increase. In a flow with mean deformation time scale $1/S$ and turbulence length scale l, the pressure fluctuation generated by the mean deformation p_r can be estimated as ρSlq, which leads to $\overline{p'^2}/\bar{p}^2 \sim \gamma^2 M^2 M_t^2$, where $M = Sl/C$. Thus the third term of Equation 2 can behave as extra dissipation which is proportional to M_t^2 when $M \sim O(1)$. Terms representing compressible dissipation which depend on M_t have recently been proposed for use in turbulence models (Zeman 1990, Sarkar et al 1991). The first term of Equation 2 is $O(M^2)$ relative to the TKE production (P) if $\bar{D}/\bar{D}t$ is estimated as $O(S)$, and can be appreciable even when M_t is small. These rough estimates for Π_d are consistent with the rapid distortion theory (RDT) predictions (Durbin & Zeman 1992, Sabel'nikov 1975) applicable in the limit of $S\bar{\rho}q^2/\varepsilon \gg 1$. The estimate of Π_d identifies M as an important compressibility parameter (in addition to M_t). This parameter can be regarded as the Mach number difference across an eddy (Durbin & Zeman 1992, who use the symbol Δm for it) or as a ratio of the acoustic propagation time across an eddy to the mean deformation time scale. With the latter interpretation if may be expected that when M is not small the loss of acoustic communication across an eddy may have a significant influence on the flow (Morkovin 1992, Papamoschou & Lele 1993). The importance of M has been underscored by Cambon et al (1992) who advocate a *pressure released* limit for $M \gg 1$: Pressure fluctuations are ignored and the velocity fluctuations undergo a pure kinematic distortion due to the mean flow deformation.

The pressure fluctuation associated with the large eddies (in the absence of mean deformation), p_s, is expected to be ρq^2. The first term in (2) is then $\Pi_d/P \sim M_t^2$ compared to $\Pi_d/P \sim M^2$ obtained earlier. Furthermore the ratio of p_r to p_s is $O(Sl/q)$ which is also the ratio of TKE production to its dissipation. In an equilibrium shear flow $Sl/q \sim O(1)$ or $M \sim M_t$. Hence M and M_t can be regarded as independent parameters only in nonequilibrium flows, e.g. flows subject to rapid distortions.

The pressure fluctuation estimate ($p' \sim \rho Slq$) leading to $\Pi_d/P \sim M^2$ is valid only for the solenoidal fluctuations ($M_t \approx 0$) and should be revised when acoustic pressure fluctuations are dominant, e.g when M is large or when M_t is not small. If the characteristic acoustic particle velocity is $q_c = \chi_c^{1/2}q$, and the acoustic pressure $p_c \sim \rho C q_c$, it follows that $\Pi_d/P \sim \chi_c$ (χ_c defined in Section 3). This suggests a decreasing importnce of Π_d in a nearly solenoidal flow subjected to a deformation with large M (Cambon et al 1992). A different approach using the Poisson equation for pressure (Sarkar 1992a) also yields similar estimates for Π_d.

The second term in Equation 2 is non-zero only for inhomogeneous flows. It may be significant in rapidly distorted flows, such as a boundary layer interacting with a shock wave. In hypersonic boundary layers, even over a flat surface, or in supersonic boundary layers on curved surfaces, the static pressure variation across the boundary layer is not negligible (Fernholz & Finley 1977). The second term may be important in these circumstances as well.

It should be stressed that while the parameter M_t naturally arises in nondimensional estimates of compressibility effects, it is generally not possible to set its value by design. The situation is analogous to not being able to set a value for the flux Richardson number in a turbulent flow involving buoyancy effects. Adjusting the speed of sound would provide the desired control if the turbulence was not affected by compressibility, but then M_t would also be a nonessential parameter. Just as the flux Richardson number can be indirectly varied by changing the imposed velocity difference (or temperature difference) the turbulence Mach number can be changed indirectly. Prescribing a turbulence Mach number M_t at a given turbulence Reynolds number is also equivalent to prescribing a Knudsen number: the ratio of microscopic to macroscopic length scales (see Lele 1993 for a discussion). Increasing M_t naturally increases the density fluctuations $\rho_{rms}/\bar{\rho}$, but this compressibility-associated increase is separate from the density fluctuations associated with temperature (entropy) fluctuations. The latter is dominant in turbulent flames, in turbulent boundary layers with $M_\infty \leq 5$ (Bradshaw 1977), and in low-speed buoyancy driven flows. The parameter M_t allows the compressibility effects to be separated from variable inertia effects and thus is preferred here. Furthermore it should be kept in mind that M_t is not the only parameter characterizing the compressibility effects. For nonequilibrium flows the deformation rate Mach number M is an independent parameter.

2.3 Coupling between Momentum and Energy Exchanges

The coupling between the momentum and energy exchanges in a compressible turbulent flow can be illustrated by considering the compressible mixing layer flow between two streams with free-stream speeds U_1 and U_2; where $\Delta U \equiv U_1 - U_2$, sound speeds are characterized by C, the width of the mixing layer at a station x is $\delta(x)$, the averaged flow speed $U_c = \frac{1}{2}(U_1 + U_2)$, the magnitude of turbulent velocity fluctuations is q, and magnitude of temperature fluctuations is T'. As the fluid particles that originate in the high speed stream exchange their excess momentum with the slower stream they are also compressed, which requires that work be done on them. Likewise the slower stream fluid gains momentum and loses work as it is entrained and mixed. The turbulent eddies affect this momentum and energy exchange between the two streams. The averaged rate of dilatation, $\bar{u}_{i,i}$, can be estimated using Equation (1) combined with the momentum balance and the entropy balance (Thompson 1988). This yields

$$\frac{\partial u_i}{\partial x_i} = \frac{1}{c^2} u_j \frac{\partial(\frac{1}{2}u_k u_k)}{\partial x_j} + \frac{1}{c^2}\left[\frac{\partial(\frac{1}{2}u_k u_k)}{\partial t} - \frac{1}{\rho}\frac{\partial p}{\partial t}\right]$$

$$-\frac{u_k f_k}{c^2} + \frac{\gamma-1}{\gamma p}\left[\Phi + \frac{\partial q_k}{\partial x_k}\right], \quad (3a)$$

where u_i, p, c, f_k, q_k, and Φ, represent the instantaneous velocity, pressure, sound speed, external body force (per unit mass), heat flux vector, and viscous dissipation rate, respectively. Decomposing the variables into their Reynolds averages and fluctuations (for a stationary flow without external flow forces) gives

$$\bar{c}^2 \frac{\partial \bar{u}_i}{\partial x_i} = \frac{1}{2} \bar{u}_j \frac{\partial \bar{u}_k \bar{u}_k}{\partial x_j} + \frac{\partial \bar{u}_j}{\partial x_k} \overline{u'_j u'_k} + \frac{1}{2} \bar{u}_j \frac{\partial \overline{u'_k u'_k}}{\partial x_j} + \overline{\bar{u}_k u_j \frac{\partial u'_k}{\partial x_j}} + \overline{u'_j u'_k \frac{\partial u'_k}{\partial x_j}}$$

$$- \frac{1}{\bar{T}} \left(\bar{u}_k \frac{\partial \bar{u}_k}{\partial x_j} \overline{u'_j T'} + \bar{u}_j \frac{\partial \bar{u}_k}{\partial x_j} \overline{u'_k T'} + \bar{u}_j \bar{u}_k \overline{u'_k \frac{\partial T'}{\partial x_j}} \right) + \frac{\gamma - 1}{\bar{\rho}} \left(\bar{\Phi} - \frac{\partial \bar{q}_j}{\partial x_j} \right) + \dots,$$

(3b)

where T denotes the absolute temperature, $\bar{c}^2 = \gamma R \bar{T}$, and a Taylor series expansion of c^{-2} about $\bar{c}$ is used. Individual terms of Equation (3b) have been discussed in Lele (1993).

Terms 4, 5, 8, and 10 on the right-hand side of (3b) are expected to be small in high Reynolds number flows (Tennekes & Lumley 1972). The magnitude of the remaining terms (1–3, 6, 7, and 9) of (3b) may be estimated for the mixing layer example as

$$\frac{\partial \bar{u}_i}{\partial x_i} \Big/ \frac{\partial \bar{u}}{\partial y} \sim O\left(\frac{d\delta}{dx} \frac{U_c^2}{C^2} \right) + O\left(\frac{q^2}{C^2} \right) + O\left(\frac{d\delta}{dx} \frac{q^2}{C^2} \right)$$

$$+ O\left(\frac{U_c}{C} \frac{q}{C} \frac{T'}{\bar{T}} \right) + O\left(\frac{d\delta}{dx} \frac{\Delta U}{C} \frac{q}{C} \frac{T'}{\bar{T}} \right) + \left(\frac{q}{\Delta U} \frac{q^2}{C^2} \right).$$

Evidently, the mean dilatation relative to mean shear $\partial \bar{u}/\partial y$ is the largest of $O(d\delta/dx \, U_c^2/C^2)$, $O(q^2/C^2)$, and $O[(U_c/C)(q/C)(T'/\bar{T})]$. From Equation (2) it follows that $d\delta/dx \sim O[q^2/(U_c \Delta U)]$ (Brown & Roshko 1974). With this, $d\delta/dx \, U_c^2/C^2 \sim O[q^2/C^2(U_1 + U_2)/(U_1 - U_2)]$ which makes the first term in Equation (3b) at least as large as the second term and the seventh term smaller than the sixth. Under the assumption of the Strong Reynolds Analogy (SRA) which is valid if $T_{01} \approx T_{02}$ (discussed in Section 4.3) the temperature fluctuations are estimated as $T'/\bar{T} \sim -(\gamma - 1)(\bar{u}^2/C^2)(u'/\bar{u})$ so that $(U_c/C)(q/C)(T'/\bar{T}) \sim -(\gamma - 1)(U_c^2/C^2)(q^2/C^2)$. Common to these estimates of the mean flow divergence is the parameter $M_t = q/C$. The mean divergence relative to mean shear is proportional to M_t^2 in a canonical mixing layer. Since *extra strain rates* are known to produce effects on turbulence that may be substantially larger than the measures of the extra strain rates (Bradshaw 1974), it may be anticipated that compressibility effects may become significant when M_t exceeds 0.2–0.3.

2.4   *Entropy Changes*

From a thermodynamic point of view the entropy changes associated with the different exchange processes are fundamental. The equation for the rate of change of specific entropy, s, can be obtained from the primitive (nonaveraged) mass, momentum, and energy balances. Its Favre-averaged form is (Favre 1969)

$$\frac{\partial}{\partial t}\bar{\rho}\tilde{s} + \frac{\partial}{\partial x_j}(\bar{\rho}\tilde{s}\tilde{u}_j + \overline{\rho s''u_j''}) - \frac{\partial}{\partial x_j}\overline{(kT_{,j}/T)} = \Upsilon = \overline{\Phi/T} + \overline{(k|\nabla T/T|^2)},$$

where the Fourier law of heat conduction $q_j = -kT_{,j}$ with thermal conductivity k has been used. The irreversible increase of mean entropy Υ may be approximated by $\bar{\Phi}/\bar{T} + k(\bar{T})|\nabla\bar{T}/\bar{T}|^2$, if $T_{\mathrm{rms}}/\bar{T}$ is small. The mean rate of viscous dissipation, $\bar{\Phi}$, is the sum of dissipation due to the mean flow $\bar{u}_i$ and the mean dissipation of TKE, ε. The dissipation rate due to the mean flow (deformation) is typically R^{-1} relative to ε, where R is a Reynolds number based on the energy-containing eddies, and thus relatively small except in the immediate vicinity of a no-slip surface or in the interior of a shock wave. It may seem that the entropy flux due to turbulent motions, $\overline{\rho s''u_j''}$, is difficult to estimate and may limit the usefulness of an entropy analysis. Under certain assumptions, given in Section 2.7, useful estimates can be made. It is shown there that $\overline{s''u_j''}/C_{\mathrm{v}} \approx \overline{p'u_j''}/\bar{p} - \gamma\overline{\rho'u_j'}/\bar{\rho}$. In a steady flow an integration of the mean entropy equation along the Favre-averaged streamlines gives

$$\Delta\tilde{s}/C_{\mathrm{v}} = \int_{\xi_{\mathrm{a}}}^{\xi_{\mathrm{b}}}(\bar{\rho}\tilde{u}_{\mathrm{s}}C_{\mathrm{v}})^{-1}\left\{\Upsilon - \frac{\partial}{\partial x_j}[\overline{\rho s''u_j''} - \overline{(kT_{,j}/T)}]\right\}d\xi,$$

where ξ and $\tilde{u}_{\mathrm{s}}$ denote the position and mean velocity along the mean streamline, respectively. The contribution of the energy dissipation due to turbulence, $\varepsilon \approx \bar{\rho}q^3/\delta$, towards an entropy increase (over a streamwise distance L) may be estimated as $\gamma(\gamma-1)(q/U_{\mathrm{s}})(L/\delta)M_{\mathrm{t}}^2$. In a mixing layer flow $(q/U_{\mathrm{s}})(L/\delta) \sim (U_1-U_2)/q$ making the entropy rise $\Delta\tilde{s}/C_{\mathrm{v}} \sim M_{\mathrm{t}}^2(U_1-U_2)/q$. At moderate and high $M_{\mathrm{c}} = (U_1-U_2)/(C_1+C_2)$, this entropy rise and the associated total pressure loss, $\Delta p_0/p_0 = \exp[-(\Delta\tilde{s}/C_{\mathrm{v}})/(\gamma-1)]$, may be significant. A similar conclusion about total pressure losses in a compressible mixing layer has been obtained using a different method of analysis (Papamoschou 1993).

2.5   *Role of Pressure Fluctuations*

The energy equation expressed in terms of mean pressure $\bar{p}$ serves to emphasize the important role of the pressure transport term in compressible flows. From the ideal gas relation, $p = \rho h(\gamma - 1)/\gamma$, the equation governing enthalpy can be rearranged to give

$$\frac{\bar{D}}{\bar{D}t}\bar{p} = -\gamma\bar{p}(\bar{u}_{j,j}) - (\gamma - 1)\overline{p'u'_{j,j}} - \frac{\partial}{\partial x_j}(\overline{p'u'_j}) + (\gamma - 1)\overline{u_{i,j}\tau_{ij}} + (\gamma - 1)\bar{q}_{j,j}, \quad (4)$$

$$\frac{\bar{D}}{\bar{D}t}\overline{p'^2}/2 = -\frac{\partial\bar{p}}{\partial x_j}\overline{p'u'_j} - \gamma\overline{p'^2}\bar{u}_{j,j} - \gamma\bar{p}\overline{p'u'_{j,j}} - (2\gamma - 1)/2\overline{p'^2u'_{j,j}}$$

$$- \frac{\partial}{\partial x_j}(\overline{p'^2u'_j})/2 + (\gamma - 1)\overline{p'u_{i,j}\tau_{ij}} + (\gamma - 1)\overline{p'q_{j,j}}, \quad (5)$$

where $\bar{D}/\bar{D}t = \partial/\partial t + \bar{u}_j\partial/\partial x_j$. An approximate boundary layer form of Equation (4) was used by Brown & Roshko (1974) and the terms causing deviations from incompressible behavior (i.e. $\bar{u}_{j,j} \approx 0$) were estimated. From Equation (4) it follows that $\bar{u}_{j,j}/S \sim O(M_t^2 q/\Delta U) + O(M_t^2 M^2)$, where S is mean deformation rate. The first estimate is based upon the pressure transport and/or dissipation term and the second is based on the pressure dilatation term. In a thin shear flow the static pressure variation across the flow is $O(\bar{\rho}\overline{v'^2})$ so that the contribution of $\bar{D}\bar{p}/\bar{D}t$ to $\bar{u}_{j,j}/S$ becomes $O(M_t^2 d\delta/dx)$ and is small relative to the right-hand side of Equation (4). To correctly predict the compressible mean flow it is necessary to correctly predict the $O(M_t^2)$ relative variation of static pressure in the direction normal to the shear layer.

Equation (5) governing the pressure variance is exact (Sarkar 1992a) and is related to the approximate Equation (2) discussed earlier. The appearance of the pressure dilatation term in this equation, also anticipated by classical acoustic theory, motivates associating $\overline{p'^2}/(2\gamma\bar{p})$ with the *potential* energy (PE) of the compressible fluctuations (Zeman 1991, Sarkar et al 1993). This identification is rigorous for small-amplitude acoustic waves when $\nabla\bar{p} = 0$ and $\bar{u}_{j,j} = 0$. In this situation (for example, the turbulence evolution downstream of a shock wave discussed in Section 4.1) the sum of TKE and PE change along the mean streamlines primarily due to a net imbalance between the production of TKE and its dissipation, with pressure dilatation Π_d representing an exchange between TKE and PE. Terms in Equation 5 involving $\nabla\bar{p}$ or $\bar{u}_{j,j}$ do not represent an exchange with kinetic energy. Generalized definitions of acoustic wave energy density and acoustic energy flux for arbitrary time-independent mean flows are available (Myers 1991, Pierce 1981, Morfey 1971, Cantrell & Hart 1964). These definitions have not, so far, been systematically exploited in studies of turbulent flows.

2.6 *Acoustic Energy Loss*

In a compressible flow an additional mechanism of energy loss due to acoustic radiation to the far field becomes active. Estimates of such losses can be made following the pioneering work of Lighthill (1952). In a turbulent jet the total acoustic power lost to the far field increases as M^8, where M is the Mach number of the jet. The ratio of the radiated acoustic power to the rate at which the turbulence extracts energy from the mean flow is a measure of acoustic efficiency, which increases as M^5 for small M, but because the coefficient in front is $O(10^{-4})$ only a small fraction of TKE production is lost to acoustic waves. The acoustic radiation remains an insignificant contributor to the TKE budget in compressible mixing layers at least up to convective Mach number $M_c = (U_1 - U_2)/(C_1 + C_2) \approx 0.6$ (Lele & Ho 1993). For M near unity the acoustic efficiency rises with M more gradually than M^5. Experimental data on supersonic jets summarized by Goldstein (1976) show that for $M \geq 1$ the acoustic efficiency becomes independent of M. It has been suggested (Liepmann 1979) that the energy loss due to acoustic radiation may be significant for hypersonic turbulent boundary layers but conclusive evidence for it is not available. Laufer (1962) has estimated that in a boundary layer at $M_\infty = 5$ the acoustic energy loss is of the order of 1% of the net work $(\tau_w U_\infty)$ done by the wall shear stress τ_w.

Measurements of the acoustic radiation from supersonic boundary layers were made by Laufer (1962, 1964) who emphasized the Mach wave radiation as its primary mechanism. Mach wave radiation requires that the acoustic sources move at a supersonic convection velocity relative to the free stream. In the far field the distance r between the observation point and the acoustic source associated with a radiating eddy is large compared to both the eddy size and the distance traveled by the eddy over its radiation life. As a result the Mach waves in the far field take the form of spherical wave fronts and the acoustic intensity due to Mach wave radiation from a volume of size l^3 varies as $\overline{\rho'^2}/\bar{\rho}^2 \sim M^3 l^2/r^2$ (Ffowcs Williams 1966). In the near field, however, the Mach waves appear as straight wave fronts lying on Mach cones generated by the relative supersonic motion of the radiating eddies. Mach wave radiation has also been identified as a primary radiation mechanism in supersonic jets (Tam & Morris 1980, Tam & Burton 1984) and is consistent with acoustic efficiency becoming independent of the Mach number. Theoretical estimates of the energy loss have also been made (Ffowcs Williams & Maidanik 1965). For Mach wave radiation an acoustic efficiency of 1% is also reported for supersonic hot jets (Seiner 1992).

2.7 Thermodynamic Relations

It is useful to know the extent to which the thermodynamic relations applying to instantaneous state variables may be applied to the mean or averaged values of the state variables and if the fluctuations in state variables may be related in some fashion. In the following discussion the equation of state for an ideal gas will be commonly used. Relations holding for fluids with an arbitrary equation of state may similarly be constructed. Taking the instantaneous density ρ and entropy s as two independent state variables, the pressure p and temperature T may be expressed by thermodynamic relations $p = \vartheta(\rho, s)$ and $T = \mathsf{T}(\rho, s)$, which for an ideal gas become $\vartheta(\rho, s) = p_r e^{s/C_v}(\rho/\rho_r)^\gamma$ and $\mathsf{T}(\rho, s) = \vartheta(\rho, s)/(R\rho)$, where p_r and ρ_r are reference state values. Since the functions ϑ and T are generally nonlinear, $\bar{p} = \overline{\vartheta(\rho, s)} \neq \vartheta(\bar{\rho}, \bar{s})$, etc. However, if the fluctuations in ρ and s are small, a Taylor series expansion of $\vartheta(\rho, s)$ about the state $(\bar{\rho}, \bar{s})$ yields

$$\bar{p} = \vartheta(\bar{\rho}, \bar{s})[1 + \tfrac{1}{2}\gamma(\gamma - 1)\overline{\rho'^2}/\bar{\rho}^2 + \tfrac{1}{2}\overline{s'^2}/C_v^2 + \gamma\overline{\rho's'}/(\bar{\rho}C_v) + \ldots].$$

Treating entropy as a dependent variable gives

$$\bar{s}/C_v = S_{\rho p}(\bar{\rho}, \bar{p}) - \frac{1}{2}\overline{p'^2}/\bar{p}^2 + \frac{\gamma}{2}\overline{\rho'^2}/\bar{\rho}^2 + \ldots,$$

where $S_{\rho p}$ is a thermodynamic entropy function with the indicated arguments for an ideal gas. Evidently, the leading-order error in applying the thermodynamic relations to the mean state variables are quadratic in the fluctuation. Relations applying to thermodynamic fluctuations can be similarly obtained. Such relations are useful in eliminating a chosen variable in terms of other variables. The entropy flux associated with turbulence may be evaluated using these as $\overline{s'u_j'}/C_v \approx \overline{p'u_j'}/\bar{p} - \gamma\overline{\rho'u_j'}/\bar{\rho}$ and using $\widetilde{s''u_j''} \approx \overline{s'u_j'}$. Similarly it follows that $\overline{p'u_j'}/\bar{p} \approx \overline{T'u_j'}/\bar{T} + \overline{\rho'u_j'}/\bar{\rho}$.

Relations analogous to the Gibbs relations, Maxwell relations, etc. which hold for the averaged thermodynamic state variables can also be obtained by a Taylor series expansion.

3.   COMPRESSIBLE HOMOGENEOUS TURBULENCE

A turbulent flow is considered homogeneous if the statistics of turbulent fluctuations are independent of position (Batchelor 1953, Monin & Yaglom 1971). For a compressible flow the restrictions on the mean fields necessary to maintain spatial homogeneity (Blaisdell et al 1991, Cambon et al 1992, Feiereisen et al 1981, Delorme 1985, Dang & Morchoisne 1987) are more stringent than the restrictions for an incompressible flow. Volume-preserving mean deformations, except for a pure shearing motion, are excluded. Pure rotation is also excluded. Suitable combinations of time-dependent rotation and dilatational straining flows are permitted. Pure shearing is the only time-independent deformation allowed. A flow not strictly satisfying the homogeneity conditions, but with small deviations from them (over an integral scale), is termed quasi-homogeneous (Goldstein & Durbin 1980, Hunt & Carruthers 1990, Durbin & Zeman 1992).

The set of equations referred to as the *low Mach number equations* or *anelastic approximation* (Spiegel 1971, Rehm & Baum 1978, Majda & Sethian 1985) is intermediate to the incompressible and compressible equations. The low Mach number equations describe the effects of variable fluid inertia but do not contain acoustic wave propagation. The restrictions for maintaining homogeneity for velocity and density fluctuations (but inhomogeneous temperature fluctuations) within this approximation are similar to the strict incompressible case. Studies of variable density (near zero Mach number) homogeneous turbulence are relevant to low-speed combustion flows. Many compressible flows also have accompanying variable inertia effects which are distinct from compressibility effects. Studies of variable density homogeneous turbulence are presently not available. Discussion is, hereafter, limited to the fully compressible case.

For homogeneous turbulence there is no distinction between the Reynolds-averaged and Favre-averaged velocity (Blaisdell et al 1991) making the velocity fluctuations relative to these averages also identical. The velocity fluctuations, u_i, can be uniquely decomposed (Moyal 1952) via Helmholtz decomposition into a solenoidal part, u_i^s (whose Fourier-Stieltjes transform, $\hat{u}_i^s(\mathbf{k})$, is orthogonal to the wavenumber vector $\mathbf{k}$), and a dilatational part, u_i^c [with $\hat{u}_i^c(\mathbf{k})$, parallel to $\mathbf{k}$]. Since the solenoidal and dilatational projections of u_i are orthogonal with respect to the L_2 norm, the variance $u_i u_i$ (not the kinetic energy $\overline{\rho u_i u_i}/2$) can be split into a sum of solenoidal and dilatational parts. The dilatational fraction of the variance $\chi_c = \overline{u_i^c u_i^c}/\overline{u_i u_i}$ is a diagnostic of compressibility in the energy-containing range (Blaisdell et al 1991, Kida & Orszag 1990a, Erlebacher et al 1990). Splitting the Reynolds stress $\overline{\rho u_i u_j}$ on the basis of u_i^s and u_i^c leads to solenoidal, dilatational, and cross Reynolds stresses (Blaisdell et al 1991). A related split, which accounts for the density variations, is developed (Kida & Orszag 1990a,b, 1992) by decomposing $w_i = \sqrt{\rho}u_i$ into its solenoidal, dilatational, and mean parts. If the flow is nearly incompressible the fluctuations in thermodynamic quantities can also be split by a systematic appeal to a low Mach number expansion. Given the nonuniqueness of the low Mach number expansions (Zank & Matthaeus 1991, Bayly et al 1992) the choice must be based on physical information about the flow regime. Separating the pressure fluctuations into vortical and compressible parts— a decomposition distinct from the solenoidal/dilatational split discussed above—has been helpful in developing turbulence models for terms containing pressure fluctuations (Sarkar et al 1993, Durbin & Zeman 1992, Blaisdell et al 1991).

Theoretical prediction of compressibility corrections to Kolmogorov-Obukhov inertial range spectra has been surveyed elsewhere (Passot & Pouquet 1987, Passot et al 1988). The estimated correction to the inertial range slope is significant only for $M_t \sim O(1)$, conditions which may be impossible to realize in laboratory experiments. To our knowledge these predictions have remained unconfirmed. The equation governing the two-point correlations of density fluctuations resembles a wave equation (Chandrasekhar 1951). For high Reynolds number nearly-incompressible flows, density fluctuations are predicted to possess a $-\frac{5}{3}$ inertial subrange (Bayly et al 1992, Montgomery et al 1987) and are offered as a possible explanation of the Kolmogorov spectra observed in the interstellar media (Armstrong et al 1981, Goldstein & Siscoe 1972).

3.1 Isotropic Turbulence

In isotropic turbulence the statistical correlations are invariant to an arbitrary rotation of the coordinates (Hinze 1975, Batchelor 1953). Additional reflectional symmetry which makes the turbulence free of helicity (Lesieur 1990) is also commonly imposed. Isotropic turbulence (in absence of any mean flow deformation) is devoid of any mechanism to produce TKE and must decay in the absence of external forcing. Viscous dissipation of TKE is accompanied by a decrease of the turbulence Mach number M_t while the total energy $\bar{\rho}(C_v\tilde{T} + \widetilde{u_k''u_k''}/2)$ is conserved. The evolution of density (and pressure) fluctuations depends strongly on the initial conditions, which in turn influence the kinetic energy evolution. Initial conditions include specifying the partition of the velocity field into its solenoidal and dilatational parts; specifying the density, pressure fluctuations; and specifying the values of M_t, Pr (the Prandtl number). In the limit of small M_t the qualitative behavior of the system depends sensitively on χ_c: When χ_c is $O(M_t)$ the vortical and compressible parts (defined as the deviation from the incompressible limit) of the flow evolve independently at leading order with the compressible part obeying linearized acoustic equations; for larger χ_c nonlinearity (wave steepening) cannot be ignored in the evolution of u_i^c and shocks form when $\chi_c \sim O(1)$ (Erlebacher et al 1990, Ghosh & Matthaues 1992). Numerical simulations of forced isotropic turbulence (Kida & Orszag 1990a) demonstrate that the coupling between the vortical and compressible parts of the flow is weak.

The sensitivity of u_i^c to the initial conditions is maintained even when M_t is not small (Blaisdell et al 1991, Sarkar et al 1991). In the numerical simulations the decay rate of TKE slightly increases with M_t. The contribution of dilatation to TKE decay, defined by χ_ε, increases with M_t and depends sensitively on the initial values of χ_c and $\rho_{rms}/\bar{\rho}$ (Blaisdell et al 1991, Sarkar et al 1991) which makes its parameterization difficult. The pressure dilatation, $\Pi_d = \overline{p'u_{i,i}}$ (which is always an exchange term between TKE and $\bar{\rho}\tilde{e}$), is also an exchange term between the kinetic energy, $\overline{\rho u_i^c u_i^c}/2$, and the potential energy, $\overline{p_c^2}/(2\gamma\bar{p})$, in a small-amplitude acoustic field (Morse & Ingard 1968, Lighthill 1978). The numerical simulations suffer from a poor statistical ensemble of the largest scales in the flow. This is particularly severe for the acoustic fluctuations generated by the vortical motions at small M_t (Blaisdell et al 1991). As a result, when the contribution of these scales to the statistics being studied becomes important, temporal oscillations with a time scale commensurate with the acoustic propagation time over these spatial scales become evident. The oscillations provide a mechanism of exchange between the kinetic and potential energy of the acoustic field with the total acoustic energy staying constant, and are also observed in TKE when χ_c is appreciable. When entropy fluctuations are small, i.e. $s_{rms}/C_v \ll \gamma\rho_{rms}/\bar{\rho}$, then $\Pi_d \approx \overline{\rho'u_{i,i}}R\bar{T}$ so that for temporally decaying turbulence Π_d is expected to be positive (Sarkar et al 1991) due to the appearance of $-\overline{\rho'u_{i,i}}$ in the exact equation for density variance. Numerical simulations consistently show that for isotropic-decaying turbulence, pressure dilatation is positive on average (Blaisdell et al 1991, Sarkar et al 1991) and its contribution to TKE decay is small. However, in shear flows and straining flows it can be a significant contributor. These flows are discussed later in Sections 3.2, 3.3, and 4.

Models proposed for the compressible dissipation ε_c (Zeman 1990, Sarkar et al 1991) have been shown to give better predictions of the spreading rate of a compressible mixing layer (Zeman 1990, 1992b; Sarkar & Lakshmannan 1991; Viegas & Rubesin 1991). Calibration of ε_c from numerical simulations of decaying isotropic turbulence is not reliable (Blaisdell et al 1991, Zeman 1992a) due to the sensitivity of χ_ε to initial conditions and the low values of the microscale Reynolds number R_λ. However, the suggested model form of Sarkar et al (1991) is consistent with numerical simulations having initial $\chi_c = 0$ (Sarkar et al 1991, Blaisdell et al 1991). Figure 2 reproduces the data from Lee et al (1991). Even though χ_ε varies over almost two decades (highest in case A and lowest in case E) this variation collapses to within a factor of two when plotted as χ_ε/M_t^2. Similar trends are reported by Blaisdell et al (1991). The DNS data, as in Figure 2, is limited to low values of R_λ and does not allow a test of specific Reynolds number dependence of χ_ε/M_t^2 based on spectral-similarity arguments (Zeman 1992a). The collapse in Figure 2 suggests that χ_ε might be modeled as αM_t^2 with α in the range 0.1–0.2; however, in homogeneous shear flow α is in the range 0.5–1. How much of this variation reflects a dependence on the flow type is presently unclear.

The M_t^2 dependence of χ_ε (Sarkar et al 1991) rests upon the expectation that χ_ε is proportional to χ_c, which is in turn estimated by invoking the

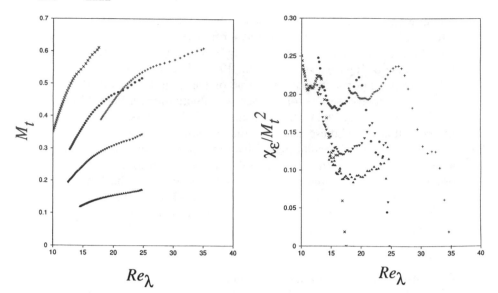

Figure 2 Evolution of isotropic compressible turbulence. The data of Lee et al (1991) are plotted showing the turbulence Mach number M_t against the Reynolds number Re_λ for cases A ($+$), B ($\times$), C ($\bullet$), D ($\blacktriangledown$), and E ($\blacktriangle$). The corresponding values of χ_ε/M_t^2 are also plotted against Re_λ.

equipartition of kinetic and potential energy for the acoustic mode (Morse & Ingard 1968, Lighthill 1978) and assuming that pressure fluctuations due to vortical and acoustic modes are comparable. The proportionality between χ_ε and χ_c is observed in numerical simulations only when the acoustic mode is not prescribed independently from the vortical mode (Blaisdell et al 1991). In the simulations of Lee et al (1991) the condition of no initial pressure fluctuations ensures that p^s and p^c are comparable and the observed M_t^2 dependence is thus not unexpected. Lee et al also report the occurrence of eddy shocklets for cases with high M_t. Parameterization of the dissipation due to the eddy shocklets is the centerpiece of the Zeman (1990) model. The detailed shocklet modeling of Zeman has not been confirmed by numerical simulations since the shocklets observed in the simulations contribute only a small fraction of the observed compressible dissipation; their contribution appears to increase with R_λ and M_t. Most of the compressible dissipation in numerical simulations occurs at large scales (Sarkar et al 1991, Blaisdell & Zeman 1992) and the proposed models are in reasonable agreement with it. Whether the same would hold in high Reynolds number flows remains an open question. In this context it may be noted that the spectral equations governing the dilatational velocity field $u_i^c(k)$ contain terms analogous to those describing passive scalar advection by the solenoidal velocity. This may cascade the acoustic energy to smaller scales. The associated spectral behavior does not seem to have been studied.

Measurements of decaying compressible isotropic turbulence are non-existent due to the practical difficulties in generating a quasi-homogenous turbulent flow with significant levels of M_t. Space-time correlations (Favre et al 1965b; Wills 1964, 1971) are commonly used to provide statistical measures of the spatial scales in the flow and their temporal persistence. They also provide a statistical definition of the convection velocity of various scales (Wills 1964, 1971), a notion well suited for the vortical mode. The acoustic mode is, by definition, associated with wave propagation rather than convection which makes its space-time correlation rather distinct from the vortical mode (Lee et al 1992). In a regime of significant coupling between the acoustic and vortical modes (e.g. with significant shock-generated vorticity) the space-time correlations of the vortical mode may also show some change. The phenomenology of such regimes are presently unknown. In flows with high M_t the influence of compressibility on spectral energy transfer and influence on local isotropy (e.g. due to the additional large scale–small scale coupling due to eddy shocklets) have not been examined. A complete theory of the final period of decay of compressible isotropic turbulence is not available. The degeneracy of the nearly incompressible flows will need to be resolved in such a theory. Unlike its incompressible counterpart the constraints governing the slopes of the low wavenumber energy spectra are also not available.

3.2   Homogeneous Sheared Turbulence

Spatially-uniform, time-independent shearing is the only nontrivial volume-preserving mean deformation that strictly satisfies the homogeneity constraints for a compressible flow. Mean flow shear is common to most turbulent flows and the homogeneous idealization allows a study of some effects of compressibility that are shared by many turbulent flows. The idealization is unable to retain other aspects critical to the inhomogeneous shear flow and boundary layer turbulence: Energy loss by acoustic radiation to the far field, which may be important under hypersonic free-stream conditions, and the strong mean density stratification near a no-slip surface in supersonic and hypersonic turbulent boundary layers are two such effects. Measurements realizing quasi-homogeneous compressible uniformly sheared turbulence are not available, unlike its incompressible counterpart (Tavoularis & Corrsin 1981, Tavoularis & Karnik 1989), and the following discussion is largely based on numerical simulations.

Feiereisen et al (1981) appear to have conducted the first numerical simulations of homogeneous sheared compressible turbulence. Due to their particular choice of initial conditions no significant effects of compressibility were reported. Delorme (1985) and Dang & Morchiosne (1987) also performed DNS and LES (Large Eddy Simulations) of this flow. Recently Blaisdell et al (1991) and Sarkar et al (1993) have conducted extensive studies of this flow for values of M_t exceeding the range covered in previous work. The simulations impose a spatially uniform shear, $S = dU/dy$, on an isotropic field of fluctuations with prescribed energy spectra. Under the influence of mean shear the fluctuations develop toward a "realistic" state of turbulence. Typically this adjustment takes a time $St \approx 4$–6 depending upon the initial conditions (developed isotropic turbulence requires a smaller adjustment time). After this the fluctuation quantities display an approximately exponential growth.

The incompressible homogeneous shear flow contains two time scales, i.e. $\bar{\rho}q^2/\varepsilon$ and $1/S$. The ratio of these, $S_* = S\bar{\rho}q^2/\varepsilon$, is also proportional to production $\mathsf{P} = -\bar{\rho}S\overline{uv}$ divided by the dissipation ε, if the structure parameter $a_1 = -\overline{uv}/q^2$ is a constant. In the exponential growth phase a_1 and S_* are approximately constant and an almost constant fraction, σ, of P goes towards increasing TKE. With increasing compressibility a larger fraction of P is "dissipated," lowering the normalized growth rate, σ/S. Analysis of the TKE budget indicates (Blaisdell et al 1991, Sarkar et al 1993) that the increased "dissipation" is a result of the loss due to Π_d and dissipation due to ε_c. The energy supply rate P also depends on turbulence (Rogers et al 1986 have shown the influence of initial conditions) and appears to decrease with increasing compressibility for large St (Blaisdell et al 1991, Sarkar et al 1993).

Numerical simulations (Blaisdell et al 1991, Sarkar et al 1993) show that the compressible dissipation fraction χ_ε becomes independent of the initial values of χ_c and $\rho_{\mathrm{rms}}/\bar{\rho}$ in an evolution time $St \approx 8$–10; χ_c, $\rho_{\mathrm{rms}}/\bar{\rho}$ also show a similar trend. In this evolved flow χ_ε, χ_c, $\rho_{\mathrm{rms}}/\bar{\rho}$, and $p^c_{\mathrm{rms}}/\bar{p}$ all exhibit an M_t^2 dependence (Sarkar et al 1993). There is some uncertainty in the value of χ_ε/M_t^2; Sarkar et al suggest a value of 0.5, revising their previous estimate of 1.0 which was chosen to model the decay of isotropic compressible turbulence. Blaisdell et al (1991) find the latter to be consistent with their simulations, although no best fit is attempted. DNS of other homogeneous flows (e.g. compressed turbulence, discussed in Section 3.3) suggest that an algebraic closure of χ_ε in terms of M_t alone is too restrictive.

Structural parameters such as the Reynolds stress anisotropy, $b_{ij} = \overline{\rho u_i'' u_j''}/\overline{\rho u_k'' u_k''} - \delta_{ij}/3$, structure tensor y_{ij} (Reynolds 1989), and ratios of integral length scales, which characterize the sheared turbulence, are similar in value to their incompressible analogs (Blaisdell et al 1991). Small-scale features, such as the preferred alignment between the intense vorticity and the intermediate principal strain rate direction, and most probable principal strain rate ratios in intense dissipating regions are also largely similar to the incompressible case (Blaisdell et al 1991, Sarkar et al 1993, Erlebacher & Sarkar 1992). Splitting the velocity field, however, reveals that the dilatational part of the velocity field has a structure distinct from the solenoidal part. Unlike u_i^s for which $\overline{u_s^2} > \overline{w_s^2} > \overline{v_s^2}$ is generally observed in shear flows, u_i^c shows that $\overline{v_c^2} > \overline{u_c^2} \approx \overline{w_c^2}$ (Blaisdell et al 1991). The contribution of $\overline{v_c^2}$ to $\overline{v^2}$ becomes significant with increasing M_t even when the contribution of u_i^c to other covariances is still small. Since u_i^c contributes little to scalar mixing (Blaisdell et al 1991) the mixing across the imposed shear is preferentially decreased as M_t increases (Blaisdell et al 1991, Sarkar et al 1993). This distinct anisotropy of u_i^c is also reflected in the anisotropy of the compressible dissipation which scales better with b_{ij}^c than with b_{ij} (Blaisdell et al 1991). Both ε_c and Π_d, whose importance is an indicator of compressibility, preferentially oppose the growth of transverse fluctuations. This in turn reduces the growth of Reynolds shear stress, and finally the energy supply rate P.

The turbulence Mach number M_t grows with time in the simulations and it is natural to ask if the late time evolution of homogeneous shear turbulence is always influenced by compressibility even when the initial M_t is quite small. The energy balance equations can be rewritten to yield an equation for M_t which for homogeneous shear flow at high Reynolds number simplifies to

$$\frac{\partial M_t^2}{\partial t} = M_t^2 S(\sigma_* - \hat{\gamma} M_t^2/\hat{S}),$$

where $\hat{\gamma} = \gamma(\gamma - 1)$, $\sigma_* = \sigma/S = (\partial q^2/\partial t)/(Sq^2)$, and $\hat{S} = S\bar{\rho}q^2/(\varepsilon - \Pi_d)$. Simulations have shown that σ_* is a decreasing function of M_t, while the dependence of $\hat{S}$ on M_t is not certain but it seems probable that $M_t^2/\hat{S}$ increases with M_t: Thus a fixed point of M_t (such that $\partial M_t/\partial t = 0$) may be approached in the long term evolution of the flow, although with initially small M_t the time for this approach may be very long. An equilibrium value of M_t has been noted by Zeman & Blaisdell (1991) and Zeman (1992a) but the DNS data, at present, do not show a limiting behavior (the long time behavior of DNS may suffer from inadequate sampling of the large scales). Asymptotic values of M_t have also been reported in turbulence model based calculations of compressible mixing layers (Zeman 1990, 1992b; Sarkar & Lakshmanan 1991).

To date the simulations have been limited to parameter ranges $S_* = S\bar{\rho}q^2/\varepsilon \approx 5\text{–}15$ and $M_t \approx 0\text{–}0.65$. Estimates of S_* in typical shear flows (e.g. in the logarithmic and outer regions of a boundary layer and channel flow) range between 5 and 10, a range spanned by the simulations. Self-preserving incompressible mixing layers have $S_* \approx 8$ (Rogers & Moser 1993). In the immediate vicinity of a no-slip wall, however, S_* can be as large as 30 (Lee et al 1990). Reliable estimates for the near-wall region of compressible boundary layers are currently not available. Incompressible homogeneous shear turbulence with large S_* has been successful in explaining several structural features of near-wall turbulence (Lee et al 1990). In the compressible case large S_* may also imply large M, since M $\sim S_* M_t$, and as noted in Section 2 additional effects of compressibility associated with loss of acoustic communication may arise for large M.[1] Such strongly sheared compressible turbulence has as yet not been explored. The importance of the shearing rate Mach number M has also been recently stressed by Sarkar (1992b).

3.3   *Homogeneous Compressed Turbulence*

Irrotational, time-dependent mean-deformation can still satisfy the compressible homogeneity constraints when the mean dilatation is not zero. The principal strain rates can be used to classify the dilatational straining mode as either one-dimensional, planar, axisymmetric, spherically-symmetric, or three-dimensional. One-dimensional compression arises in shock-turbulence interaction and in the compression stroke of a piston engine; planar compressions and expansions (generally inhomogeneous) are common in planar mean flows; axisymmetric strains arise in axisymmetric nozzle and diffusor flows or in attached axisymmetric external flows; spherically-symmetric strain is an idealization of implosions; and three-dimensional strains arise in complex flow fields. Bulk expansions generally dampen the turbulence and if sufficiently strong may even lead to relaminarization (Morkovin 1955, Narasimha & Sreenivasan 1979, Dussauge & Gaviglio 1987). Bulk expansions are not discussed in detail in this review; reference can be made to Bradshaw (1974, 1977), Dussauge & Gaviglio (1987), Jayaram et al (1989), and Smith & Smits (1991) for further details.

The influence of compressibility on compressed turbulence depends on the turbulence Mach number M_t, the deformation (rate) Mach number M, and on the mode of compression. The rapidity of compression characterized by $S_* = S\bar{\rho}q^2/\varepsilon$ relates the two characteristic Mach numbers: M $\sim S_* M_t$. Thus nearly solenoidal turbulence ($M_t \approx 0$) when compressed sufficiently rapidly [so that M $\sim O(1)$] can exhibit significant com-

---

[1] In boundary layers the effects associated with strong mean density stratification are large and compressibility effects are difficult to isolate.

pressibility effects (Zeman & Coleman 1993). When the fluctuations are themselves compressible ($M_t \neq 0$), even moderate values of S_* are associated with $M \geq O(1)$. When M is large an approximation reflecting the lack of acoustic communication should become applicable. In the context of homogeneous turbulence a *pressure released* RDT (Cambon et al 1992) has been proposed. Response of incompressible turbulence to bulk compression has been studied theoretically in the limit of rapid compression (Batchelor & Proundman 1952, Ribner & Tucker 1953, Batchelor 1955, Hunt 1978, Lee 1989, Coleman & Mansour 1993). Numerical simulations of solenoidal compressed turbulence (Wu et al 1985, Coleman & Mansour 1991) and compressible compressed turbulence (Coleman & Mansour 1991, Zeman & Coleman 1993, Cambon et al 1992) have considered both rapid and slow compressions for spherically-symmetric and one-dimensional compression modes. Theoretical analysis of compressible rapid compressions (Durbin & Zeman 1992, Cambon et al 1992, Sabelnikov 1975) has been used to aid the modeling of compressed turbulence.

The evolution of TKE for isotropic turbulence subjected to a rapid one-dimensional compression with a fixed initial compression rate S_* presented by Zeman & Coleman (1993) shows the interesting result that the TKE increase for a given value of mean density change is smaller when the initial M_t is small. The DNS results (Cambon et al 1992) show that in compressions with $S_* \gg 1$ the rapidity of compression (S_*) continues to influence the TKE change unless $M \gg 1$ in which case the pressure released limit is approached. Cambon et al (1992) use $\Delta m = M_t S/\omega$, where ω is the rms vorticity, as the important compressibility parameter. The Mach number Δm defined using the acoustic time based on a Taylor microscale is related to M by $\Delta m \sim M R^{-1/2}$, where $R = \rho q^4/(v \varepsilon)$. Compressibility parameters such as M or Δm are better suited than M_t in characterizing the compressibility influence. This is illustrated by Figure 3 (courtesy of G. Coleman) which shows the TKE change as a function of the net compression for different S_* (with initial $S_* > 47$ and $S/\omega > 2$) and M_t. Evidently the TKE change for a given compression depends monotonically on Δm but the dependence on M_t is not monotonic.

For small M but $S_* \gg 1$ the incompressible RDT limit is obtained, and for large M with large S_* the pressure released limit is achieved. As shown in Figure 3 the TKE growth for a given volume compression increases with compressibility (measured by Δm). This is opposite to the trend observed in homogeneous shear flow (Section 3.2). Cambon et al (1992) have emphasized the two different roles played by the pressure fluctuations, i.e. through Π_d in overall TKE balance and through the pressure strain rate correlation. For small but nonzero M the pressure dilatation Π_d is the dominating influence and as a result the TKE growth approaches the

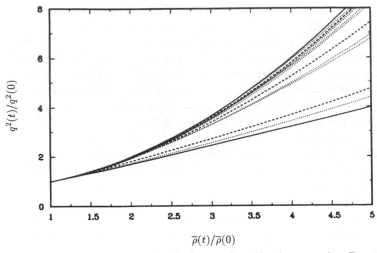

Figure 3 Turbulent kinetic energy histories for one-dimensional compression. Data taken from Cambon et al (1992). The lower solid line is the solenoidal RDT limit and the upper solid line is the *pressure released* RDT limit. DNS data lie between these limits varying monotonically with initial Δm; dotted curves are DNS with initial Δm ranging from 0.3 (lower) to 8.0 (upper), dashed curves are DNS cases: lower $(M_t, \Delta m) = (0.03, 0.3)$, upper (0.1, 7.0), middle (0.3, 1.0).

corresponding solenoidal case with the same S_* (toward the bottom curve in Figure 3). The RDT analysis (Durbin & Zeman 1992) also shows the dependence of Π_d on the mode of compression: The relative importance of Π_d decreases as the compression mode is changed from one dimensional to planar to spherically symmetric. At larger M the pressure strain rate term dominates; the DNS results show that the component energy redistribution is less effective as M increases resulting in a larger TKE increase (Cambon et al 1992). A limitation of the pressure-released RDT arises from the fact that flows that acquire large values of deformation rate Mach number may also be expected to show some effects of chemical/thermodynamic nonequilibrium. Such influences are not discussed in this article.

One-dimensional rapid compressions lead to negative Π_d (loss of TKE) regardless of the initial conditions, but for slow compressions Π_d displays significant positive and negative oscillations (G. Coleman 1993, private communication), reminiscent of the isotropic decay case (Section 3.1). For rapid spherically-symmetric compressions the sign of Π_d depends on the initial conditions (G. A. Blaisdell 1993, private communication). The inability of irrotational mean straining flows to erase the memory of the initial conditions rules out simple parametrizations of the compressible dissipation parameter χ_ε (Coleman & Mansour 1993). Rapid deformations also influence the overall TKE dissipation rate ε. For compressed flows, the need to account for the variations of kinematic viscosity in the transport equation for ε has been stressed by Coleman & Mansour (1991, 1993). Removing an imposed one-dimensional compression (G. Coleman 1993, private communication) shows a relaxation on an acoustic time scale but the return towards isotropy remains incomplete while the turbulence continues to decay.

3.4   Homogeneous Complex Shear Flow Turbulence

Homogeneous turbulence subjected to simultaneous mean dilatation and shear may be viewed as an idealized flow pertinent to distorted boundary layer turbulence. The passage of sheared turbulence through a shock suggests the idealization of homogeneous turbulence subjected to simultaneous shear and one-dimensional compression. Analysis using incompressible RDT (Mahesh et al 1993) has shown that the resulting amplification of TKE is influenced by the initial Reynolds stress anisotropy and the ratio of shearing rate to compression rate. The dependence arises through the pressure strain rate terms. For sufficiently strong normal compressions the Reynolds shear stress changes its sign. It is presently not known how M_t may influence these results. Other combinations of bulk compression and shear—e.g. shear with plane compressive strain, shear with lateral compression/divergence—are also relevant to distorted boundary layer applications. In light of the reported nonlinear behavior when multiple extra strain rates such as bulk compression, lateral divergence, and streamline curvature are combined with canonical boundary layers (Fernando & Smits 1990, Smits & Wood 1985, Smits et al 1979), the homogeneous idealizations (incompressible or compressible) of complex shear flows may prove useful for conceptual understanding, even though they do not contain the inhomogeneous transport effects of the large eddies.

4.   SIMPLE INHOMOGENEOUS FLOWS

4.1   Idealized Shock-Turbulence Interactions

The interaction of turbulence with shocks is a common element of external aerodynamic flows. In experiments it is common to study the interaction of an oblique shock (created by a wedge in a supersonic flow) with a turbulent boundary layer or the supersonic flow in a compression ramp (Settles et al 1979, Dolling & Or 1985, Andreopoulos & Muck 1987, Smits & Muck 1987, Kuntz et al 1987, Selig et al 1989). Although these flows are an idealization of shock-turbulence interactions they still constitute a complex turbulent flow (Green 1970). To help unravel the basic processes occurring in such flows a further idealization may be undertaken: The simplest is to consider the interaction of a normal shock with turbulence that is homogeneous in directions transverse to the mean shock front. Theoretical analyses linearize the upstream turbulence into vorticity, and acoustic and entropic modes and impose linearized Rankine-Hugoniot jump conditions across the shock. Predictions are then made of the downstream statistics in terms of the upstream statistics. Experiments attempting to study the idealized interaction are difficult (Jacquin et al 1991, Honkan & Andreopoulos 1992, Keller & Merzkirch 1990, Debieve & Lacharme 1986) and the currently available data are limited. Recent numerical simulations (Lee et al 1993, Rotman 1991) have confirmed some of the predictions of the linear analyses and identified nonlinear aspects of the problem. Linear analysis has also been used to estimate the turbulent corrections to the classical shock jump relations and shock speed formulae (Lele 1992).

When the turbulent pressure fluctuations are small compared to the pressure rise across the shock $[M_t^2 < 0.1(M_t^2 - 1)]$ the shock front is weakly distorted and the linearization of the Rankine-Hugoniot conditions used in theoretical analyses is justified. For stronger turbulence the shock structure is strongly modified and the instantaneous pressure rise along the mean streamlines does not remain monotonic (Lee et al 1993). A similar strong interaction is reported in experiments (Hesselink & Sturtevant 1988) on propagation of weak shocks in a medium with strong density fluctuations. In passing across the shock the vorticity fluctuations are strongly amplified and consequently the dissipation rate is enhanced (Lee et al 1993, Jacquin et al 1991). Linear analysis predicts the vorticity amplification accurately. Components of potential vorticity (vorticity divided by density) lying in the plane of the shock remain unaltered and the potential vorticity normal to the shock is diminished by the bulk compression. The net result is that the vorticity components in the plane of the shock increase in proportion to the density ratio across the shock and the normal component of vorticity remains unchanged. The linear analysis also predicts an amplification of the TKE across the shock (see Jacquin & Cambon 1992 for a comparison of different linear analyses).

The DNS results to data have been limited to weak shocks. Simulations using two-dimensional Euler equations (Rotman 1991) cover a wider range of shock strengths. The reported TKE amplification for small M_t is in reasonable agreement with a two-dimensional version of Ribner's analysis (S. Lee 1992, private communication). The TKE amplification across a shock of a prescribed strength is reported to decrease as M_t (i.e. turbulence intensity) is increased. This trend is shared by DNS (Lee et al 1993) and experiments (Hokan & Andreopoulos 1992). However, the DNS shows that the TKE increases not simply across the intermittent zone occupied by the shock but in a region downstream of it. Numerical simulations for stronger shocks using a shock-capturing ENO scheme in the region surrounding the shock (Lee 1992) also show a clear TKE rise but, as in DNS of weak-shock cases (Lee et al 1993), the increase occurs in a region downstream of the oscillating shock.

This downstream increase of TKE is due to an energy exchange between the potential energy in the pressure fluctuations just downstream of the shock and the kinetic energy of the acoustic mode. Pressure fluctuations are amplified across the shock and decay downstream. The acoustic energy density (KE plus PE) remains approximately constant downstream of the shock, causing a downstream rise of TKE. In the TKE balance of Figure 1 the potential energy of acoustic modes is not explicitly accounted for, and the PE to KE exchange mechanism is manifested via the pressure dilatation and pressure transport terms.

Experiments of Jacquin et al (1991) report no significant TKE rise (for $M_1 = 1.4$) for isotropic upstream turbulence but a clear rise when (anisotropic) jet turbulence interacts with the shock cells of an over-expanded jet. The latter flow is significantly inhomogeneous and comparisons with homogenous rapid distortion theory can at best be qualitative. Other experiments (Honkan & Andreopoulos 1992, Debieve & Lacharme 1986) report clear TKE amplification. To help clarify these trends, RDT calculations have been performed (Mahesh et al 1993) for the interaction of anisotropic turbulence with normal and oblique shocks. The anisotropy is observed to make a substantial difference in the TKE amplification as anticipated by Jacquin et al (1991). The dependence of TKE amplification on the initial anisotropy decreases with the obliquity of the shock.

Conflicting claims about the changes in the length scales of turbulence when passing across a shock have appeared in the literature. Honkan & Andreopoulos (1992) in their experiment ($M_t = 1.24$) suggest that $L_{\varepsilon 1} = \bar{\rho}(\overline{u^2})^{3/2}/\varepsilon_{11}$ increases across the shock; this is in conflict with DNS. Taylor microscales for velocity fluctuations decrease across the shock in DNS which is in agreement with Ribner's (1953, 1987) theory. An apparent conflict with earlier experiments (Debieve & Lacharme 1986) has been attributed to a false microscale increase due to shock front intermittency (Lee et al 1993). Figure 4 (courtesy of S. Lee) shows the evolution of different physical length scales across the shock from a shock-capturing simulation. For the case shown $M_1 \doteq 2.0$, $M_t = 0.08$, and $R_\lambda = 18$ just upstream of the shock. The region blanked out from the plot corresponds

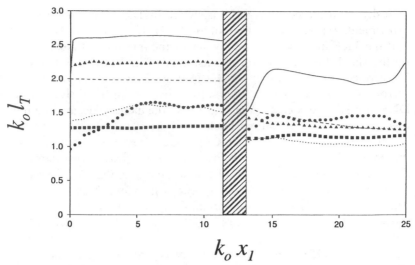

$k_o l_T$

$k_o x_1$

Figure 4 Evolution of several physical length scales in a shock-turbulence interaction. Figure taken from Lee et al (1994). Isotropic turbulence with $M_t = 0.08$, $R_\lambda = 18$ just upstream of the shock interacts with a $M_1 = 2.0$ normal shock; dissipation length $\bar{\rho}q^3/\varepsilon$ (*solid line*), longitudinal velocity microscale $\lambda_2(u_2)$ (*dashed line*), density microscale λ_2 (*dotted line*), longitudinal velocity integral scale $\Lambda_2(u_2)$ (▲), lateral velocity integral scale $\Lambda_2(u_1)$ (■), and integral scale for density fluctuations Λ_2 (●).

to the intermittent region occupied by the shock whose extent is defined by $\bar{u}_{i,i} = 0$ (Lee et al 1993). The simulation shows, in agreement with the DNS of weak-shock cases, that characteristic length scales—including longitudinal and lateral velocity integral scales, longitudinal velocity microscale, dissipation length scale $\bar{\rho}q^3/\varepsilon$, as well as the integral scale and microscale of density fluctuations—decrease across the shock. The result for density microscale is in conflict with the experimental observation of Keller & Merzkirch (1990) who claim that the Taylor microscale of density fluctuations increase across the shock. At present this conflict remains unresolved. It is unclear how closely the experiments approximate the homogeneous idealization of numerical simulations, and there may very well be practical limitations in improving them. In the future, as better diagnostics improve the experimental data base, numerical simulations that retain more of the physical effects occurring in laboratory flows are likely to further our understanding of shock-turbulence interaction phenomena (i.e. DNS of idealized shock-turbulence interactions which include the effects of mean shear and mean density gradient, LES of shock wave/turbulent boundary layer interaction, LES of compression corner flows, etc).

4.2 Compressible Mixing Layers

The turbulent mixing between two streams of different velocity has served as the most vivid experimental demonstration of the effect of compressibility on turbulence (Birch & Eggers 1972, Brown & Roshko 1974). The notion of convective Mach number, M_c (Papamoschou & Roshko 1988, Bogdanoff 1983), has been moderately successful in collapsing the normalized growth rate data onto a single curve. The collapse represented in Figure 5 (assembled by D. Papamoschou) still has significant scatter, perhaps serving as a reminder of the sensitivity of the mixing layer flow to background disturbance environment. Some of the scatter may also be due to the variable degree to which a self-preserving flow is realized in different experiments (Clemens & Mungal 1992) and to uncertainty in the normalizing incompressible mixing layer growth rate used in Figure 5.

It has also been suggested that a single compressibility parameter M_c may not be sufficient (Dimotakis 1991, Hall 1991) particularly when data from mixing layers with one subsonic stream are included (Viegas & Rubesin 1991). It is possible that the anomalous behavior of the small ρ_2/ρ_1 data (e.g. Hall 1991) in Figure 5 may be due to an inaccurate

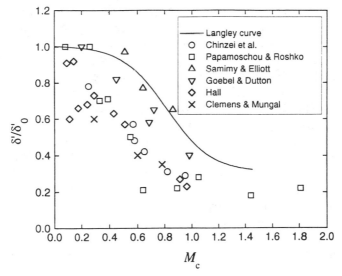

Figure 5 Normalized compressible mixing layer growth rates plotted against the convective Mach number M_c. Data assembled by D. Papamoschou. Different experiments employ different definitions for the mixing layer thickness and to compare them estimates of the factors relating one definition to another are needed. The factors recommended by the original authors were used when available.

extrapolation of the incompressible mixing layer growth rate data to extreme values of ρ_2/ρ_1. If it is assumed that the incompressible mixing layer growth rate is proportional to the maximum spatial amplification rate of a linearized disturbance, an alternate extrapolation of mixing layer growth rates to small ρ_2/ρ_1 can be plotted. With the new normalization the anomalous behavior is no longer evident (Lu & Lele 1993). Experimental data on low speed mixing layers with small ρ_2/ρ_1 are needed to satisfactorily resolve this issue. Data for $M_c \gg 1$ are also very limited; it is unclear if the limiting behavior suggested by Figure 5 is influenced by walls or possible pressure wave systems in the confining channel (Dimotakis 1991).

Turbulence statistics in a compressible mixing layer have been measured with hot wires (Barre et al 1992) and laser Doppler anenometers (LDA) (Elliott & Samimy 1990, Samimy & Elliott 1990, Goebel & Dutton 1991, Debisschop & Bonnet 1993). They exhibit a decrease of turbulent fluctuations (normalized with ΔU) with increasing M_c. Limited measurements of the third moments $\overline{v'(u'^2+v'^2)}/(\Delta U)^3$ and $\overline{v'^3}/(\Delta U)^3$ also show a decrease with M_c. The correlation coefficient $r_{uv} = \overline{u'v'}/u'_{rms}v'_{rms}$ appears to be unchanged from its incompressible value. The data on anisotropy v'_{rms}/u'_{rms} have conflicting trends in different experiments. Elliott & Samimy (1990) and Debisschop & Bonnet (1993) claim that both $u'_{rms}/\Delta U$ and $v'_{rms}/\Delta U$ decrease with M_c, while Goebel & Dutton (1991) claim that only the v fluctuations diminish with M_c leading to increasing anisotropy at higher M_c. The peak values of $k [= \frac{1}{2}(\overline{u'^2}+\overline{v'^2}+\overline{w'^2})]$ in the three cases ($M_c = 0.51$, 0.64, 0.86) studied by M. Samimy (private communication, 1992) correspond to M_t values of 0.27, 0.32 to 0.43 (defined as $\sqrt{2k}/C_1$); the values relative to the low speed stream sound speed C_2 would be 0.22, 0.24, and 0.26. The deformation rate Mach number $M = Sl/C = S_* M_t$ may be estimated to be in the range 2–3 for these flows where S_* is taken as 8. Measurements of density fluctuations in the range $0.3 \leq M_c \leq 0.9$ where a significant growth rate reduction occurs are not available. Measurements of scalar mixing are also limited. Clemens et al (1991) report a decrease in normalized scalar variance (unmixedness parameter) with increasing M_c but express caution in interpretation due to the limited signal-to-noise ratio and spatial resolution available in the experiment. Hall (1991) finds the opposite trend of a decrease in the fraction of mixed fluid with M_c in the compressible reacting mixing layers. Further experiments using "cold-chemistry" as well as a passive scalar to measure the molecular mixing (Clemens & Paul 1993) find that compressibility has only a small effect on mixing efficiency even though the turbulence structure visualized in the images change significantly with M_c.

The decrease of mixing layer growth rate with M_c is widely agreed upon;

but its cause is widely debated. Linear stability growth rates exhibit a similar reduction with M_c (Gropengeisser 1970, Ragab & Wu 1989, Jackson & Grosch 1989, Sandham & Reynolds 1990) which persists into the nonlinear phase (Sandham & Reynolds 1990, 1991; Lele 1989). Although the connection of the linear instability properties to the turbulent flow evolution (Morkovin 1992) is not mathematically rigorous, models based upon it (Planche & Reynolds 1992, Morris et al 1990, Tam & Hu 1990) do indeed explain some of the observations. Figure 6 (from Lu & Lele 1993) shows the correlation between the maximum spatial amplification rate and the mixing layer growth rates measured in different experiments. As in Figure 5, suitable scaling factors K are necessary to compare different experiments that use different definitions of mixing layer thickness. Acoustic radiation properties of supersonic jets have also been predicted using instability wave based models (Tam & Morris 1980, Seiner 1992). Recent observations of supersonic reacting mixing layers (Hall 1991, Miller et al 1993) have also been interpreted using linear stability analysis and nonlinear simulations (Planche & Reynolds 1992).

These approaches study the compressibility effects by examining inviscid energy exchanges. One alternative point of view centers on the dissipative

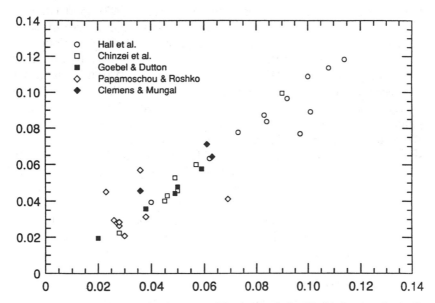

Figure 6 Comparison of the maximum amplification rate $|\alpha_i|_{max}/K$ with the growth rate δ'_{exp} measured by Hall et al (1991) ($K = 0.31$); Chinzei et al and Goebel & Dutton (1991) ($K = 0.51$); Papamoschou & Roshko (1988) and Clemens & Mungal (1992) ($K = 0.38$).

process (Zeman 1990, Sarkar et al 1991) and invokes additional compressible dissipation of TKE and pressure-dilatation. Another alternative is an interpretation in terms of the reduced communication between different (streamwise) zones of the flow (Papamoschou 1991b), a process necessary to establish the instability eigenfunctions or vortex interactions (Morkovin 1992). Model problems studied by Papamoschou & Lele (1993) have further supported this view. In the context of a turbulent flow, communication effects enter through terms involving pressure perturbations. A further observation is that numerical simulation of homogeneous shear flow (Blaisdell et al 1991) shows a preferential suppression of v'_{rms}, which in turn decreases $-\overline{u'v'}$, TKE and, for inhomogenous flow, turbulent momentum transport.

Laser sheet images (Clemens et al 1991, Clemens & Mungal 1992, Elliott et al 1992) show the instantaneous structures to be highly three dimensional in high M_c flows, with some tendency towards quasi two-dimensional structures when M_c is small. The lack of any noticeable two-dimensional structures for $M_c > 0.5$ is consistent with oblique wave linear instabilities being the most amplified waves at these M_c (Sandham & Reynolds 1990, 1991). No dominant oblique structures have been observed. Some difference between the high- and low-speed edges of the mixing layers have also been noted as M_c increases (Clemens & Paul 1993, Bonnet & Debisschop 1993).

The increased three-dimensionality is also evident in the two-point correlation measurements of static pressure: Samimy et al (1992) report a decrease of the spanwise coherence with M_c. The convection speed of large structures measured by double exposure photography (Papamoschou 1991a, Fourguette et al 1990, McIntyre & Settles 1991) by pressure records from transducers in tunnel walls (Hall et al 1991) have consistently differed from an isentropic theoretical estimate (Papamoschou & Roshko 1989) when $M_c > 0.5$. Two alternative views have been offered to reconcile these differences: an interpretation using linear stability theory (Jackson & Grosch 1989, Sandham & Reynolds 1990, Tam & Hu 1991) relates the discrepancy to fast and slow supersonic instability modes and possible obliquity of these waves; Papamoschou (1991a) and Dimotakis (1991) offer a different interpretation, proposing that total pressure losses due to eddy shocklets alter the convection velocity. Mach waves generated by supersonically moving disturbances have been widely reported but shocklets associated with large structures have not been observed. Numerical simulations of two-dimensional mixing layers (Lele 1989, Sandham & Reynolds 1990, Soetrisno et al 1989) display eddy shocklets for $M_c \geq 0.8$, but in three-dimensional simulations (Sandham & Reynolds 1991, Chen 1993) shocklets are not observed up to $M_c = 1$. This difference between two- and three-dimensional cases is not unexpected, as the critical M_c for shocklet formation in three-dimensional flows is expected to be higher than the critical M_c for two-dimensional flows. Whether shocklets occur in the highly three-dimensional flows at higher M_c is presently unknown.

A representative convection velocity is needed to determine the entrainment ratio in the mixing layer (Dimotakis 1991). Statistical estimates of the convection velocity U_c can be obtained from space-time correlations. Such techniques have been widely used in compressible boundary layer studies (Spina et al 1991) but space-time correlation data on compressible mixing layers are very limited. The convection velocity of the strong local extrema of pressure have been presented in the form of histograms (Samimy et al 1992); they show a significant spread in the U_c values and a dependence on the lateral coordinate. Near the free-stream edges the average U_c is closer to the free-stream speed, but in the middle of the shearing zone it is well approximated by the isentropic formula (for $\gamma_1 = \gamma_2$) $U_{c(\text{isentropic})} = (C_2 U_1 + C_1 U_2)/(C_1 + C_2)$. Measurements of the convection velocity of the large structures (Papamoschou 1991a and others, summarized by Dimotakis 1991) have differed consistently from the isentropic formula and have motivated suggestions of eddy structures with embedded shocks. Sufficient measurements are presently unavailable to judge if the overall entrainment ratio and chemical product formation are significantly influenced by the deviation of the convection speed from $U_{c(\text{isentropic})}$.

The influence of chemical heat release on compressible mixing layers and jets has not been well explored but is a subject of current studies. In low-speed studies (or incompressible flows) the primary effect of heat release is the displacement effect resulting from thermal expansion. The spreading rate of the mixing layer shows only a mild change and decreases with increasing heat release (Hermanson & Dimotakis 1989). The maximum shear stress is also observed to decrease with heat release with most of the decrease attributed to the lowered mean density. This is analogous to the density stratification effect common to most compressible boundary layers (Morkovin 1961). The entrainment ratio also changes with heat release. Dimotakis (1991) summarizes the effects of chemical kinetics (finite reaction rates), Reynolds number, and streamwise pressure gradient. Under compressible conditions the entrainment, stirring, and the subsequent heat release are intimately linked. The instability characteristics are qualitatively different from the nonreacting case: The heat release establishes local maxima in $\bar{\rho}(d\bar{U}/dy)$ which occur close to the free-stream edges (Planche & Reynolds 1992, Shin & Ferziger 1992). The most unstable disturbances are less oblique, dispersive (i.e. not subject to pairing/subharmonic instability), and appear to limit their growth by baroclinic vorticity generation and acoustic radiation (Planche & Reynolds 1992). The mixing in such a flow has been described in terms of two co-layers (Planche & Reynolds 1992). Current experimental evidence (Miller et al

1993, Hall 1991) is consistent with such an interpretation, but is too limited to draw any firm conclusions.

4.3 *Compressible Boundary Layers*

Bradshaw (1977), Morkovin (1961, 1992), and most recently Spina et al (1994) have reviewed the information on compressible turbulent boundary layers. A density gradient caused by the dissipative heating near the no-slip wall is the primary effect of increasing the mean flow Mach number. The importance of the dissipative heating is appreciated by noting that for boundary layers on adiabatic walls the ratio of wall temperature to the free-stream temperature rises from $T_w/T_\infty = 1.9$ at $M_\infty = 2.2$ to $T_w/T_\infty = 4.7$ at $M_1 = 4.5$ and is nearly 20 at $M_\infty = 10$. The variable mean density is largely responsible for a decreased skin-friction coefficient, smaller turbulence intensity, viscous effects, and for modifications to the incompressible law of the wall. [Due to the large variation in fluid properties a single Reynolds number is not sufficient to characterize the flow (Smits 1991).]

Accounting for the passive effects of the mean density variation via transformations of the low-speed analyses (Van Driest 1951, Coles 1964, Rotta 1960, Morkovin 1961) [summarized in Bradshaw (1977) and Fernholz & Finley (1977, 1980)] allows engineering prediction of the canonical adiabatic flat-plate boundary layers. For hypersonic, cooled boundary layers an additional modification is introduced. The variable density transformations are similar to the Howarth-Dordonitsyn transformation for compressible laminar boundary layers. Due to the closure problem their application to turbulent flow requires assumptions that are not obtained by first principles. The experimental data summarized by Fernholz & Finley (1980) show that such transformations are effective. The measurements of fluctuations are limited and accurate estimates of peak M_t or M for different M_∞ are currently not available. Spina et al (1994) provide a recent perspective on the available experimental data and Zeman (1993) addresses the modeling of compressibility effects inherent in them. One way to identify the variable inertia effects imbedded in compressible boundary layers is to contrast the high Mach number boundary layers with strongly heated, low-speed boundary layers (Cheng & Ng 1982, 1985; Thunker 1991). With a heated wall at 1100 K the primary effect, in the low-speed case, is the reduction of the Reynolds shear stress $-\overline{\rho uv}$ due to the change in $\bar{\rho}$ across the boundary layer (Cheng & Ng 1985) while the kinematic Reynolds stress $-\overline{uv}$ is close to the isothermal case. It is possible that the mean velocity profiles in these flows can be predicted using mean density-weighted transformations. Such comparisons should help in isolating the high-speed effects in compressible boundary layers.

An approximation called the Strong Reynolds Analogy (SRA) is introduced in the context of adiabatic flat-plate boundary layers (Gaviglio 1987, Bradshaw 1977, Morkovin 1961). SRA has been applied to other configurations (Barre et al 1992) and is used for turbulence modeling purposes (Rubesin 1990). The assumptions needed to derive SRA can only be validated by the experimental measurements of the flow under consideration. Gaviglio (1987) has examined these assumptions and proposed an extension. In Morkovin's proposal for SRA two assumptions are required: 1. negligible total temperature fluctuation and 2. negligible pressure fluctuation.

The following analysis is helpful in assessing the conditions under which these assumptions may apply. Consider small-disturbance inviscid equations for perturbations to a sheared mean flow, $[\bar{U}(y), 0, 0]$, with mean entropy variation due to TKE dissipation or heat transfer across the mean flow streamlines. If the disturbance velocity is nearly solenoidal the density and entropy fluctuations are expressible as $\rho' \sim -\eta(\partial\bar{\rho}/\partial y)$ and $s' \sim -\eta(\partial\bar{s}/\partial y)$, where η is a Lagrangian displacement. The thermodynamic relations (Section 2.7) connect the mean density and entropy gradients as

$$\frac{\partial\bar{s}}{\partial y}\bigg/ C_v \approx \frac{-\gamma}{\bar{\rho}}\frac{\partial\bar{\rho}}{\partial y}[1+O(\overline{\rho'^2}/\bar{\rho}^2)+O(M_t^2)]$$

which leads to

$$s'/C_v = -\gamma\rho'/\bar{\rho}+O(\overline{p'^2}/\bar{\rho}^2)+O(M_t^2).$$

Since the equation of state provides

$$p'/\bar{p} = \gamma\rho'/\bar{\rho}+s'/C_v+\dots$$

it follows that $p'/\bar{p} \ll \rho'/\bar{\rho}$. Thus in flows with small M_t, small mean streamline curvature, and small $\rho'_{rms}/\bar{\rho}$, but with a significant mean entropy variation across the flow, a large negative correlation coefficient between s' and ρ', or between T' and ρ', is expected. If such a flow is subjected to some further distortion (extra strain rate) the pressure fluctuations in the distorted flow may no longer be negligible (as noted in Sections 2.2 and 3.3, bulk compression can generate significant pressure fluctuations).

The justification for neglecting the total temperature fluctuation in SRA is less clear. From the definition of total temperature as $C_pT_0 = C_pT + u_ku_k/2$ it follows that

$$T_0'/\bar{T} = T'/\bar{T}+\bar{U}u_1'/(C_p\bar{T})+[u_k'u_k'-\overline{u_k'u_k'}]/(2C_p\bar{T}).$$

Estimating the streamwise velocity and temperature fluctuation (Gaviglio

1987) as $u_1' \sim -\eta(\partial \bar{U}/\partial y)$ and $T' \sim -\eta(\partial \bar{T}/\partial y)$ (which ignore the fluctuations of the x-direction pressure gradient), it follows that $T_0'/\bar{T} \ll T'/\bar{T}$ if the "mean" total temperature, $\bar{T} + \bar{U}^2/(2C_p)$, is constant across the flow. Thus under the strict SRA assumptions $T'/\bar{T} \approx -\bar{U}u_1'/(C_p\bar{T})$; or, T' and u_1' have a correlation coefficient R_{uT} close to -1. For a known variation of $\bar{T}_0$ across the flow this formulation predicts a value of R_{uT} (Gaviglio 1987) close to but not equal to -1. Compressible boundary layer measurements over adiabatic walls commonly show R_{uT} between -0.8 and -1, and the measured $\bar{T}_0$ variation is consistent with the values of R_{uT} or the extended SRA relation between T' and u_1' (Gaviglio 1987). Conditions under which the variation of $\tilde{T}_0$ across the flow is small can be obtained from the equation governing $\tilde{T}_0$ and associated boundary conditions (see Lele 1993 for a discussion).

The similarity between the x-momentum transport and the transport of heat (passive scalar) added at the wall is almost perfect in an incompressible flow near a no-slip wall (Guezennec et al 1990, Kim & Moin 1989) and progressively becomes imperfect in regions away from the wall. Extending such studies to compressible flows may shed further light on the SRA relations. In flows subjected to sudden distortions or that have curved mean streamlines the acceleration of fluid elements cannot be ignored and neglecting total temperature fluctuation is inaccurate. Likewise in flows with high M_t or high density fluctuations SRA may break down (Bradshaw 1977).

5.   CLOSURE

Some miscellaneous topics of fundamental importance that are not discussed in this article include: (*a*) combustion instability phenomena arising from the coupling between the unsteady compressible flow and chemical reactions, (*b*) the behavior of acoustic turbulence, e.g. the turbulent field dominated by a random distribution of shock waves (Porter et al 1992), (*c*) the spectral or two-point closure theories applied to compressible turbulence, and (*d*) the related topic of subgrid-scale models for compressible turbulent flows. Recent formulations of dynamic subgrid-scale models (Moin et al 1991) have suggested an approach for applying the customary parametrizations (Yoshizawa 1991, Erlebacher et al 1992, Zang et al 1992) to a wide variety of physical problems. If the subgrid models prove to be robust they will lead the way for future large eddy simulations of compressible turbulent flows.

At present measurements that allow the compressibility effects on turbulence to be isolated are very limited. The development of new measurement techniques combined with numerical simulations of compressible turbulent flows will continue to test our perspective. Faced with this data explosion, the efforts aimed at physical modeling of the compressible flow phenomena will be even more critical to advancing our knowledge.

ACKNOWLEDGMENTS

Discussions with Profs. Bradshaw, Morkovin, Smits, Papamoschou, Samimy, Moin, Reynolds, and Drs. Zeman, Durbin, Sarkar, Lee, Coleman, Blaisdell, Mansour, Dussauge, Buckingham, and Leith during various stages of writing this review were most helpful. Drs. Papamoschou, Lee, and Coleman generously provided some of the figures. Help from Mr. S. Collis in composing Figure 1 is gratefully appreciated. Detailed comments from Profs. Bradshaw, Cantwell, Moin, and Sarkar, Drs. Durbin, Zeman, and Coleman, and Mr. K. Mahesh on a preliminary draft are highly appreciated. Support from AFOSR, ONR, and the NSF-PYI program in conducting some of the research described and in writing this review is gratefully acknowledged. A longer version of this article is available as a Center for Turbulence Research Manuscript No. 145, 1993.

Literature Cited

Adamson, T. C. Jr., Messiter, A. F. 1980. Analysis of two-dimensional interactions between shock waves and boundary layers. *Annu. Rev. Fluid Mech.* 12: 103–38

Andreopoulos, J., Muck, K. C. 1987. Some new aspects of the shock-wave boundary layer interaction in compression ramp corner. *J. Fluid Mech.* 180: 405–28

Anyiwo, J. C., Bushnell, D. M. 1982. Turbulence amplification in shock-wave boundary-layer interaction. *AIAA J.* 20: 893–99

Armstrong, J. W., Cordes, J. M., Rickett, B. J. 1981. Density power spectrum in the local interstellar medium. *Nature* 291: 561–64

Barre, S., Quine, C., Dussauge, J. P. 1992. Compressibility effects on the structure of supersonic mixing layers. *Note IMST* 1/92, Inst. Mec. Statistique de la Turbulence, Marseille, France

Batchelor, G. K. 1953. *The Theory of Homogeneous Turbulence.*, p. 17. Cambridge: Cambridge Univ. Press

Batchelor, G. K. 1955. The effective pressure exerted by a gas in turbulent motion. In *Vistas in Astronomy*, ed. A. Beer. London: Pergamon

Batchelor, G. K. 1967. *An Introduction to Fluid Dynamics*, p. 615. Cambridge: Cambridge Univ. Press

Batchelor, G. K., Proudman, I. 1950. The effect of rapid distortion of a fluid in turbulent motion. *Q. J. Mech. Appl. Math.* 7: 83–103

Bayley, B. J., Levermore, C. D., Passot, T. 1992. Density variations in weakly compressible flows. *Phys. Fluids A* 4: 945–54

Benjamin, R. F. 1992. Experimental observations of shock stability and shock-induced turbulence. In *Advances in Compressible Turbulent Mixing*, ed. W. P. Dannevik, A. C. Buckingham, C. E. Leith, pp. 341–48. *Lawrence Livermore Lab. Rep., Conf*-8810234

Besnard, D. C., Haas, J. F., Rauenzahn, R. M. 1989. Statistical modeling of shock-interface interaction. *Physica D* 37: 227–47

Billig, F. S. 1988. Combustion processes in supersonic flow. *J. Propul. Power* 4: 209–16

Birch, S. L., Eggers, J. M. 1972. A critical review of the experimental data on turbulent shear layers. *NASA SP* 321: 943–49

Blaisdell, G. A., Mansour, N. N., Reynolds, W. C. 1991. Numerical simulation of compressible homogeneous turbulence. *Rep. No. TF*-50, Thermosci. Div., Mech. Engrg., Stanford Univ.

Blaisdell, G. A., Zeman, O. 1992. Investigation of the dilatational dissipation in compressible homogeneous shear flow.

Proc. 1992 *Summer Program*, pp. 231–45, Center for Turbulence Res., Stanford Univ.

Bogdanoff, D. W. 1983. Compressibility effects in turbulent shear layers. *AIAA J.* 21: 926–27

Bonnet, J. P., Debisschop, J. R. 1993. Experimental studies of the turbulent structure of supersonic mixing layers. *AIAA Pap.* 93-0217

Bradshaw, P. 1974. The effect of mean compression or dilatation on the turbulence structure of supersonic boundary layers. *J. Fluid Mech.* 63: 449–64

Bradshaw, P. 1977. Compressible turbulent shear layers. *Annu. Rev. Fluid Mech.* 9: 33–54

Brown, G., Roshko, A. 1974. On density effects and large structure in turbulent mixing layers. *J. Fluid Mech.* 64: 775–816

Cambon, C., Coleman, G., Mansour, N. N. 1992. Rapid distortion analysis and direct simulation of compressible homogeneous turbulence at finite Mach number. *Proc.* 1992 *Summer Program*, pp. 199–230, Center for Turbulence Res., Stanford Univ.

Cantrell, R. H., Hart, R. W. 1964. Interaction between sound and flow in acoustic cavities: mass, momentum, energy considerations. *J. Acoust. Soc. Am.* 36: 697–706

Chandrasekhar, S. 1951. The fluctuations of density in isotropic turbulence. *Proc. R. Soc. London Ser. A* 210: 18

Chang, C. T. 1957. Interaction of a plane shock and oblique plane disturbances with special reference to entropy waves. *J. Aeronaut. Sci.* 24: 675–82

Chen, J. H. 1993. The effect of compressibility on conserved scalar entrainment in a plane free shear layer. In *Turbulent Shear Flows* 8, ed. F. Durst, R. Friedrich, B. E. Launder, F. W. Schmidt, U. Schumann, J. H. Whitelaw, pp. 297–311. Berlin: Springer-Verlag

Cheng, R. K., Ng, T. T. 1982. Some aspects of strongly heated turbulent boundary layer flow. *Phys. Fluids* 25: 1333–41

Cheng, R. K., Ng, T. T. 1985. Conditional Reynolds stress in a strongly heated turbulent boundary layer with premixed combustion. *Phys. Fluids* 28: 473–88

Chinzei, N., Masuya, G., Komuro, T., Murakami, A., Kudou, K. 1986. Spreading of two-stream supersonic turbulent mixing layers. *Phys. Fluids* 29: 1345–47

Chu, B. T., Kovasznay, L. S. G. 1958. Nonlinear interactions in a viscous heat-conducting compressible gas. *J. Fluid Mech.* 3: 494–514

Clemens, N. T., Mungal, M. G. 1992. Two-and three-dimensional effects in the supersonic mixing layer. *AIAA J.* 30: 973–81

Clemens, N. T., Paul, P. H. 1993. Scalar measurements in compressible axisymmetric mixing layers. *AIAA Pap.* 93-0220

Clemens, N. T., Paul, P. H., Mungal, M. G., Hanson, R. K. 1991. Scalar mixing in the supersonic shear layer. *AIAA Pap.* 91-1720

Coleman, G. N., Mansour, N. N. 1991. Modeling the rapid spherical compression of isotropic turbulence. *Phys. Fluids A* 3: 2255–59

Coleman, G., Mansour, N. N. 1993. Simulation and modeling of homogeneous compressible turbulence under isotropic mean compression. In *Turbulent Shear Flows* 8, ed. F. Durst, R. Friedrich, B. E. Launder, F. W. Schmidt, U. Schumann, J. H. Whitelaw, pp. 269–82. Berlin: Springer-Verlag

Coles, D. 1964. The turbulent boundary layer in a compressible fluid. *Phys. Fluids* 7: 1403–23

Dang, K., Morchoisne, Y. F. 1987. Numerical simulation of homogeneous compressible turbulence. *2nd Int. Symp. on Transport Phenomena in Turbulent Flows*, Tokyo, *Oct.* 25–29, 1987

Debieve, J. F. 1986. Problemes de distorsion rapide en ecoulement compressible. *Proc. Colloq. ONERA/DRET Ecoulements turbulents compressibles*, Poitiers-Marseille

Debieve, J. F., Lacharme, J. P. 1986. A shock wave/free turbulence interaction. In *Turbulent Shear Layer/Shock Wave Interactions*, ed. J. Delery, pp. 393–403. Berlin: Springer-Verlag

Debisschop, J. R., Bonnet, J. P. 1993. Mean and fluctuative velocity measurements in supersonic mixing layers. In *Engineering Turbulence Modeling and Experiments* 2, ed. W. Rodi, F. Martelli. Amsterdam: Elsevier. Also see Debisschop, J. R. 1992. *Comportement de la turbulence en couches de mélange supersoniques*. PhD thesis. Univ. Poitiers, France

Delorme, P. 1985. PhD thesis. Univ. Poitiers. Available in English as: Numerical simulation of compressible homogeneous turbulence. *Eur. Space Agency Tech. Transl. ESA-TT-1030*, 1988, and *NASA Rep.* N89-15365

Dimotakis, P. E. 1991. Turbulent free shear layer mixing and combustion. In *High-speed Flight Propulsion Systems*, Progr. Astronaut. Aeronaut., ed. S. N. B. Murthy, E. T. Curran, pp. 265–340. Washington, DC: AIAA

Dolling, D. S., Or, C. T. 1985. Unsteadiness of the shock wave structure in attached and separated compression ramp flowfields. *Exp. Fluids* 3: 24–32

Durbin, P. A., Zeman, O. 1992. Rapid distortion theory for homogeneous compressed turbulence with application to modeling. *J. Fluid Mech.* 242: 349–70

Dussauge, J. P., Debieve, J. F., Smits, A. J. 1988. Rapidly distorted compressible boundary layers. *AGARDograph* AG 315: 2-1, 2-11

Dussauge, J. P., Gaviglio, J. 1987. The rapid expansion of a supersonic turbulent flow: role of bulk dilatation. *J. Fluid Mech.* 174: 81–112

Elliot, G. S., Samimy, M. 1990. Compressibility effects in free shear layers. *Phys. Fluids A* 2: 1231–40

Elliott, G. S., Samimy, M., Arnette, S. A. 1992. Study of compressible mixing layers using filtered Reyleigh scattering based visualizations. *AIAA J.* 30: 2567–68

Erlebacher, G., Hussaini, M. Y., Kreiss, H. O., Sarkar, S. 1990. The analysis and simulation of compressible turbulence. *Theoret. Comput. Fluid Dyn.* 2: 73–95

Erlebacher, G., Hussaini, M. Y., Speziale, C. G., Zang, T. A. 1992. Towards the large eddy simulation of compressible turbulent flows. *J. Fluid Mech.* 238: 155–85

Erlebacher, G., Sarkar, S. 1992. Statistical analysis of the rate of strain tensor in compressible homogeneous turbulence. *ICASE Rep. No.* 92-18, also *NASA Contractor Rep.* 189640

Favre, A. 1965a. Equations des gaz turbulents compressibles. *J. Mec.* 4: 361–421

Favre, A. J. 1965b. Review on space-time correlations in turbulent fluids. *Trans. ASME: J. Appl. Mech.* 32: 241–57

Favre, A. 1969. Statistical equations of turbulent gases. In *Problems of Hydrodynamics and Continuum Mechanics*, pp. 231–66. Philadelphia: SIAM

Ferri, A. 1973. Mixing controlled supersonic combustion. *Annu. Rev. Fluid Mech.* 5: 301–38

Feireisen, W. J., Reynolds, W. C., Ferziger, J. H. 1981. Numerical simulation of a compressible turbulent shear flow, *Mech. Engrg. Rep. No. TF*-13, Stanford Univ.; also see Feireisen, W. J., Shirani, E., Ferziger, J. H., Reynolds, W. C. 1982. Direct simulation of homogeneous turbulent shear flows on the Illiac IV computer. *Turbulent Shear Flows* 3, ed. L. J. S. Bradbury, F. Durst, B. E. Launder, F. W. Schmidt, J. H. Whitelaw, pp. 309–19. Berlin: Springer-Verlag

Fernando, E. M., Smits, A. J. 1990. A supersonic turbulent boundary layer in an adverse pressure gradient. *J. Fluid. Mech.* 211: 285–307

Fernholz, H. H., Finley, P. J. 1977. A critical compilation of compressible turbulent boundary layer data. *AGARDograph* AG 223, NATO

Fernholz, H. H., Finley, P. J. 1980. A critical commentary on mean flow data for two-dimensional compressible turbulent boundary layers. *AGARDograph* AG 253, NATO

Ffowcs Williams, J. E. 1966. On the development of Mach waves radiated by small disturbances. *J. Fluid Mech.* 22: 49–55

Ffowcs Williams, J. E., Maidanik, G. 1965. The Mach wave field radiated by supersonic turbulent shear flows. *J. Fluid Mech.* 21: 641–57

Fourguette, D. C., Mungal, M. G., Dibble, R. W. 1990. Time evolution of the shear layer of a supersonic axisymmetric jet at matched conditions. *AIAA J.* 29: 1123–30

Fulachier, L., Borghi, R., Anselmet, F., Paranthoen, P. 1989. Influence of density variations on the structure of low-speed turbulent flows—a report on Euromech 237. *J. Fluid Mech.* 203: 577–93

Gauthier, S., Bonnet, M. 1990. A k-ε model for turbulent mixing in shock-tube flows induced by Rayleigh-Taylor instability. *Phys. Fluids A* 2: 1685–94

Gaviglio, J. 1987. Reynolds analogies and experimental study of heat transfer in the supersonic boundary layer. *Int. J. Heat Mass Transfer* 30: 911–26

Ghosh, S., Matthaues, W. H. 1992. Low Mach number two-dimensional hydrodynamic turbulence: energy budgets and density fluctuations in a polytropic fluid. *Phys. Fluids A* 4: 148–64

Goebel, S. G., Dutton, J. C. 1991. Experimental study of compressible turbulent mixing layers. *AIAA J.* 29: 538–46

Goldstein, B., Siscoe, G. L. 1972. Spectra and cross spectra of solar wind parameters from Mariner 5. In *Solar Wind*, ed. C. P. Sonett, P. J. Colemaan, J. M. Wilcox. *NASA SP*-308

Goldstein, M. E. 1976. *Aeroacoustics*, p. 100. New York: McGraw-Hill

Goldstein, M. E. 1978a. Unsteady vortical and entropic distortions of potential flows round arbitrary obstacles. *J. Fluid Mech.* 89: 433–68

Goldstein, M. E. 1978b. Characteristics of unsteady motion on transversely sheared mean flows. *J. Fluid Mech.* 84: 305–29

Goldstein, M. E. 1979a. Scattering and distortion of the unsteady motion on transversely sheared mean flows. *J. Fluid Mech.* 91: 601–32

Goldstein, M. E. 1979b. Turbulence generated by the interaction of entropy fluctuations with non-uniform mean flows. *J. Fluid Mech.* 93: 209–24

Goldstein, M. E., Durbin, P. A. 1980. The

effect of finite turbulence spatial scale on the amplification of turbulence by a contracting stream. *J. Fluid Mech.* 98: 473–508

Green, J. E. 1970. Interaction between shock waves and turbulent boundary layers. *Prog. Aerosp. Sci.* 11: 235–340

Gropengeisser, H. 1970. Study of the stability of boundary layers in compressible fluids. *NASA TT-F-12*

Guezennec, Y., Stretch, D., Kim, J. 1990. The structure of turbulent channel flow with passive scalar transport. *Proc.* 1992 *Summer Program*, pp. 127–38, Center for Turbulence Res., Stanford Univ.

Hall, J. L. 1991. *Experimental investigation of structure, mixing and combustion in compressible turbulent shear layers.* PhD dissertation. Calif. Inst. Technol.

Hall, J. L., Dimotakis, P. E., Rosemann, H. 1991. Experiments in non-reacting compressible shear layers. *AIAA Pap.* 91-0629

Hermanson, J. C., Dimotakis, P. 1989. Effects of heat release in a turbulent reacting shear layer. *J. Fluid Mech.* 199: 333–75

Hesselink, L., Sturtevant, B. 1988. Propagation of weak shocks through a random medium. *J. Fluid Mech.* 196: 513–53

Hinze, J. O. 1975. *Turbulence.* New York: McGraw-Hill. 2nd ed.

Honkan, A., Andreopoulos, J. 1992. Rapid compression of grid generated turbulence by a moving shock wave. *Phys. Fluids A* 4: 2562–72

Hornung, H. 1986. Regular and Mach reflection of shock waves. *Annu. Rev. Fluid Mech.* 18: 33–58

Hunt, J. C. R. 1978. A review of the theory of rapidly distorted turbulent flow and its applications. *Fluid Dyn. Trans.* 9: 121–52

Hunt, J. C. R., Carruthers, D. J. 1990. Rapid distortion theory and the "problems" of turbulence. *J. Fluid Mech.* 212: 497–532

Jackson, T. L., Grosch, C. E. 1989. Inviscid spatial stability of a compressible mixing layer. *J. Fluid Mech.* 208: 609–37

Jacquin, L., Blin, E., Geffroy, P. 1991. Experiments of free turbulence/shock wave interaction. In *Proc. Turbulent Shear Flows* 8 Munich, Germany, 1-2-1 to 1-2-6.

Jacquin, L., Cambon, C. 1992. Turbulence amplification by a shock wave and rapid distortion theory. *Phys. Fluids A.* Submitted

Jayaram, M., Donovan, J. F., Dussauge, J. P., Smits, A. J. 1989. Analysis of a rapidly distorted supersonic turbulent boundary layer. *Phys. Fluids* 11: 1855–64

Jayaram, M., Taylor, M. W., Smits, A. J. 1987. The response of a compressible turbulent boundary layer to short regions of concave surface curvature. *J. Fluid Mech.* 175: 343–62

Keller, J., Merzkirch, W. 1990. Interaction of a normal shock wave with a compressible turbulent flow. *Exp. Fluids* 8: 241–48

Kerrebrock, J. L. 1956. *The interaction of flow discontinuities with small disturbances in a compressible fluid.* PhD thesis. Calif. Inst. Technol.

Kida, S., Orszag, S. A. 1990a. Energy and spectral dynamics in forced compressible turbulence. *J. Sci. Comput.* 5: 85–125

Kida, S., Orszag, S. A. 1990b. Enstrophy budget in decaying compressible turbulence. *J. Sci. Comput.* 5: 1–34

Kida, S., Orszag, S. A. 1992. Energy and spectral dynamics in decaying compressible turbulence. *J. Sci. Comput.* 7: 1–34

Kim, J., Moin, P. 1989. Transport of passive scalars in a turbulent channel flow. In *Turbulent Shear Flows* 6: 85–96. Berlin: Springer-Verlag

Kovasznay, L. S. G. 1950. The hot-wire anemometer in supersonic flow. *J. Aeronaut. Sci.* 17: 565–84

Kovasznay, L. S. G. 1953. Turbulence in supersonic flow. *J. Aeronaut. Sci.,* 20: 657–74

Kuntz, D. W., Amatucci, V. A., Addy, A. L. 1987. Turbulent boundary layer properties downstream of the shock wave/boundary layer interaction. *AIAA J.* 25: 668–75

Kutateladze, S. S., Nakoryakov, V. E., Borisov, A. A. 1987. Rarefaction waves in liquid and gas-liquid media. *Annu. Rev. Fluid Mech.* 19: 571–600

Laufer, J. 1962. Sound radiation from a turbulent boundary layer. In *Mecanique de la Turbulence,* ed. A. Favre, pp. 367–80. Paris: CNRS, also Gordon & Breach 1964

Laufer, J. 1964. Some statistical properties of the pressure field radiated a turbulent boundary layer. *Phys. Fluids* 7: 1191–97

Lee, M. J. 1989. Distortion of homogeneous turbulence by axisymmetric strain and dilatation. *Phys. Fluids A* 1: 1541–57

Lee, M. J., Kim, J., Moin, P. 1990. Structure of turbulence at high shear rate. *J. Fluid Mech.* 216: 561–83

Lee, S. 1992. Large eddy simulation of shock-turbulence interaction. Center for Turbulence Res., *Annu. Res. Briefs* 1992, pp. 73–84

Lee, S., Lele, S. K., Moin, P. 1991. Eddy shocklets in decaying compressible turbulence. *Phys. Fluids A* 3: 657–64

Lee, S., Lele, S. K., Moin, P. 1992. Simulation of spatially evolving compressible turbulence and the applicability of Taylor's hypothesis. *Phys. Fluids A* 4: 1521–30

Lee, S., Lele, S. K., Moin, P. 1993. Direct numerical simulation of isotropic turbulence interacting with a weak shock wave. *J. Fluid Mech.* 251: 533–62; also see *Rep. TF*-52, Mech. Engrg., Stanford Univ. 1992

Lee, S., Lele, S. K., Moin, P. 1994. Interaction of isotropic turbulence with a strong shock wave. *AIAA Conf., Reno,* 1994. Submitted

Lele, S. K. 1989. Direct numerical simulation of compressible free shear flows. *AIAA Pap.* 89-0374

Lele, S. K. 1992. Shock jump conditions in a turbulent flow. *Phys. Fluids A* 4: 2900–5

Lele, S. K. 1993. Notes on the effects of compressibility on turbulence. *Center for Turbulence Res. Manuscr. No. 145,* Stanford Univ.

Lele, S. K., Ho, C. M. 1993. Acoustic radiation from temporally evolving compressible mixing layers. *J. Fluid Mech.* Submitted

Lesieur, M. 1990. *Turbulence in Fluids.* Dordrecht: Kluwer. 2nd ed.

Liepmann, H. W. 1979. The rise and fall of ideas in turbulence. *Am. Sci.* 67: 221–28

Lighthill, M. J. 1952. On sound generated aerodynamically. *Proc. R. Soc. London Ser. A* 211: 564–87

Lighthill, M. J. 1978. *Waves in Fluids,*, pp. 76–85, and pp. 11–16. Cambridge: Cambridge Univ. Press

Lu, G., Lele, S. K. 1993. A note on the density ratio effect on the growth rate of a compressible mixing layer. *Phys. Fluids A.* In press

Mahesh, K., Lele, S. K., Moin, P. 1993. Shock turbulence interaction in presence of mean shear: an application of rapid distortion theory. *AIAA Pap.* 93-0663

Majda, A., Sethian, J. 1985. The derivation and numerical solution of the equations for zero Mach number combustion. *Combust. Sci. Technol.* 42: 185–205

McIntyre, S.S., Settles, G. S. 1991. Optical experiments on axisymmetric compressible turbulent mixing layers. *AIAA Pap.* 91-0623

McKenzie, J. F., Westphal, K. O. 1968. Interaction of linear waves with oblique shock waves. *Phys. Fluids* 11: 2350–62

Meshkov, E. E. 1992. Instability of shock-accelerated interface between two media. In *Advances in Compressible Turbulent Mixing,* ed. W. P. Dannevik, A. C. Buckingham, C. E. Leith, pp. 473–503. Lawrence Livermore Lab. Rep., Conf-8810234

Miller, M. F., Island, T. C., Yip, B., Bowman, C. T., Mungal, M. G., Hanson, R. K. 1993. An experimental study of the structure of a compressible reacting mixing layer. *AIAA Pap.* 93-0354

Moin, P., Squires, K., Cabot, W., Lee, S. 1991. A dynamic subgrid scale model for compressible turbulence and scalar transport. *Phys. Fluids A* 3: 2746–57

Monin, A. S., Yaglom, A. M. 1971. *Statistical Fluid Mechanics,* Vol. 1, Chap. 2. Cambridge: MIT Press

Montgomery, D., Brown, M. R., Matthaeus, W. H. 1987. Density fluctuation spectra in magnetohydrodynamic turbulence. *J. Geophys. Res.* 92: 282–84

Moore, F. K. 1953. Unsteady oblique interaction of a shock wave with a plane disturbances. *NACA TN*-2879; Also as *NACA Rep.* 1165

Morfey, C. L. 1971. Acoustic energy in nonuniform flows. *J. Sound Vib.* 14: 159–69

Morkovin, M. V. 1955. Effects of high acceleration on a turbulent supersonic shear layer. *Proc.* 1955 *Heat Transfer and Fluid Mech. Inst.,* Los Angeles

Morkovin, M. V. 1956. Fluctuations and hot-wire anemometry in compressible flows. *AGARDograph* AG 24

Morkovin, M. V. 1961. Effects of compressibility on turbulent flows. In *Mecanique de la Turbulence,* ed. A. Favre, pp. 367–80. Paris: CNRS; also Gordon & Breach 1964

Morkovin, M. V. 1992. Mach number effects on free and wall turbulent structures in light of instability flow interactions. In *Studies in Turbulence,* ed. T. B. Gatski, S. Sarkar, C.G. Speziale, pp. 269–84. New York: Springer-Verlag

Morris, P. J., Giridharan, M. G., Lilley, G. M. 1990. On the turbulent mixing of compressible free shear layers. *Proc. R. Soc. London Ser. A* 431: 219–43

Morse, P. M., Ingard, K. U. 1968. *Theoretical Acoustics.* Princeton: Princeton Univ. Press

Moyal, J. E. 1952. The spectra of turbulence in a compressible fluid; eddy turbulence and random noise. *Proc. Cambridge Phil. Soc.* 48: 329–44

Muntz, E. P. 1989. Rarefied gas dynamics. *Annu. Rev. Fluid Mech.* 21: 387–417

Myers, M. K. 1991. Transport of energy by disturbances in arbitrary steady flows. *J. Fluid Mech.* 226: 383–400

Narasimha, R., Sreenivasan, K. R. 1979. Relaminarization in fluid flows. *Adv. Appl. Mech.* 19: 449–64

Papamoschou, D. 1991a. Structure of the compressible turbulent shear layer. *AIAA J.* 29: 680–81

Papamoschou, D. 1991b. Effect of Mach number on communication between regions of a shear layer. *Proc. Eighth*

Symp. on Turbulent Shear Flows 21-5-1 to 21-5-6

Papamoschou, D. 1993. Total pressure loss in supersonic parallel mixing. *AIAA Pap.* 93-0216

Papamoschou, D., Lele, S. K. 1993. Vortex-induced disturbance field in a compressible shear layer. *Phys. Fluids A* 5: 1412–19

Papamoschou, D., Roshko, A. 1988. The compressible turbulent shear layer: an experimental study. *J. Fluid Mech.* 197: 453–77

Passot, T., Pouquet, A. 1987. Numerical simulation of compressible homogeneous flows in the turbulent regime. *J. Fluid Mech.* 181: 441–66

Passot, T. , Pouquet, A., Woddward, P. 1988. The plausibility of Kolmogorov-type spectra in molecular clouds. *Astron. Astrophys.* 197: 228–34

Pierce, A. D. 1981. *Acoustics: An Introduction to its Physical Principles and Applications.* New York: McGraw-Hill

Planche, O., Reynolds, W. C. 1992. A numerical investigation of the compressible reacting mixing layer, Mechanical Engineering Report No. TF-56

Pope, S. B. 1987. Turbulent premixed flames. *Annu. Rev. Fluid Mech.* 19: 237–70

Porter, D. H., Pouquet, A., Woodward, P. R. 1992. A numerical study of supersonic turbulence. *Theoret. Comput. Fluid Dynamics* 4: 13–49

Ragab, S. A., Wu, J. L. 1989. Linear instability in two-dimensional compressible mixing layers. *Phys. Fluids A* 1: 957–66

Rehm, R. G., Baum, H. R. 1978. The equations of motion for thermally driven buoyant flows. *J. Research Nat. Bureau Stand.* 83: 297–308

Reynolds, W. C. 1989. Effects of rotation on homogeneous turbulence. *Proc. 10th Australian Fluid Mech. Conf.*, Univ. Melbourne, Australia, Dec. 1989

Ribner, H. S. 1953. Convection of a pattern of vorticity through a shock wave. *NACA TN*-2864. Also as NACA Report 1164

Ribner, H. S. 1954. Shock-turbulence interaction and the generation of noise. *NACA TN*-3255. Also as NACA Report 1233

Ribner, H. S. 1969. Acoustic energy flux from shock-turbulence interaction. *J. Fluid Mech.* 35: 299–310

Ribner, H. S. 1987. Spectra of noise and amplified turbulence emanating from shock-turbulence interaction. *AIAA J.* 25: 436–42

Ribner, H. S., Tucker, M. 1953. Spectrum of turbulence in a contracting stream. *NACA Rep.* 1113

Rogers, M. M., Moin, P., Reynolds, W. C. 1986. The structure and modeling of the hydrodynamic and passive scalar fields in homogeneous turbulent shear flow. *Rep. No. TF*-25, Mech. Eng., Stanford Univ.

Rogers, M. M., Moser, R. D. 1993. Direct simulation of a self-similar turbulent mixing layer.*Phys. Fluids A.* Submitted

Rotman, D. 1991. Shock wave effects on a turbulent flow. *Phys. Fluids A* 3: 1792–1806

Rotta, J. C. 1960. Turbulent boundary layers with heat transfer in compressible flow. *AGARD Rep.* 281, NATO

Rubesin, M. W. 1990. Extra compressibility terms for Favre-averaged two-equation models of inhomogeneous turbulent flows. *NASA Contractor Rep.* 177556

Sabel'nikov, V. A. 1975. Pressure fluctuations generated by uniform distortion of homogeneous turbulence. *Fluid Mechanics—Soviet Research* 4: 46–56

Samimy, M., Elliot, G. S. 1990. Effects of compressibility on the characteristics of free shear layers. *AIAA J.* 28: 439–45

Samimy, M., Reeder, M. F., Elliott, G. S. 1992. Compressibility effects on large structures in free shear flows. *Phys. Fluids A* 4: 1251–1258

Sandham, N. D., Reynolds, W. C. 1990. Compressible mixing layer: linear theory and direct simulation. *AIAA J.* 28: 618–24

Sandham, N. D., Reynolds, W. C. 1991. Three dimensional simulations of large eddies in the compressible mixing layer. *J. Fluid Mech.* 224: 133–58

Sarkar, S. 1992a. The pressure-dilatation correlation in compressible flows. *Phys. Fluids A* 4: 2674–82

Sarkar, S. 1992b. The stabilizing effect of compressibility in homogeneous shear flow. *Bull. Am. Phy. Soc.* 37(8): 1769

Sarkar, S., Erlebacher, G., Hussaini, M. Y. 1993. Compressible homogeneous shear: simulation and modeling. *Tubulent Shear Flows* 8, Springer-Verlag, pp. 249–67

Sarkar, S., Erlebacher, G., Hussaini, M. Y., Kreiss, H. O. 1991. The analysis and modeling of dilatational terms in compressible turbulence. *J. Fluid Mech.* 227: 473–93

Sarkar, S., Lakshmanan, B. 1991. Application of a Reynolds stress turbulence model to the compressible shear layer. *AIAA J.* 29: 743–49

Seiner, J. M. 1992. Fluid dynamics and noise emission associated with supersonic jets, In *Studies in Turbulence*, ed. Gatski, T. B., Sarkar, S., Speziale, C. G., pp. 297–323. New York: Springer-Verlag

Selig, M. S., Andreapoulos, J., Muck, K. C., Dussauge, J. P., Smits, A. J. 1989. Turbulence structure in shock wave/turbulent boundary layer interaction. *AIAA J* 27: 862–69

Settles, G. S., Fitzpatrick, T. J., Bogdonoff,

S. M. 1979. Detailed study of attached and separated compression corner flowfields in high Reynolds number supersonic flow. *AIAA J* 17: 579–85

Shin, D. S., Ferziger, J. H. 1992. Stability of the compressible reacting mixing layer. *Mech. Engrg. Rep. No. TF*-53, Stanford Univ.

Smith, D. R., Smits, A. J. 1991. The rapid expansion of a turbulent boundary layer in a supersonic flow. *Theor. Comput. Fluid Dyn.* 2: 319–28

Smits, A. J. 1991. Turbulent boundary-layer structure in supersonic flow. *Phil. Trans. R. Soc. London Ser. A* 336: 81–93

Smits, A. J., Dussauge, J. P. 1988. Hot-wire anemometry in supersonic flow. *AGARDograph AG* 315: 5-1, 5-14

Smits, A. J., Eaton, J. A., Bradshaw, P. 1979. The response of a turbulent boundary layer to lateral divergence. *J. Fluid Mech.* 94: 243–68

Smits, A. J., Muck, K. C. 1987. Experimental study of three shock wave/turbulent boundary layer interactions. *J. Fluid Mech.* 182: 294–314

Smits, A. J., Wood, D. 1985. The response of turbulent boundary layers to sudden perturbations. *Annu. Rev. Fluid Mech.* 17: 321–58

Soetrisno, M., Eberhardt, S., Riley, J. R. 1989. A study of inviscid supersonic mixing layers using a second-order total variation diminishing scheme. *AIAA J.* 27: 1770–78

Speigel, E. A. 1971. Convection in stars, I. Basic Boussinesq convection. *Annu. Rev. Astron. Astrophys.* 9: 323–52

Speziale, C. G., Sarkar, S. 1991. Second order closure models for supersonic turbulent flows. *ICASE Rep. No.* 91-9; also *NASA Contractor Rep.* 187508

Spina, E. F., Donovan, J. F., Smits, A. J. 1991. Convection velocity in supersonic turbulent boundary layers. *Phys. Fluids A* 3: 3124–27

Spina, E., Smits, A. J., Robinson, S. 1994. Supersonic turbulent boundary layers. *Annu. Rev. Fluid Mech.* 26: 287–319

Stalker, R. J. 1989. Hypervelocity aerodynamics with chemical non-equilibrium. *Annu. Rev. Fluid Mech.* 21: 37–60

Tam, C. K. W., Burton, D. E. 1984. Sound generated by instability waves of supersonic flows, Part I. *J. Fluid Mech.* 138: 249–71; Part II. *J. Fluid Mech.* 138: 273–95

Tam, C. K. W., Hu, F. Q. 1990. A Wave model of the observed large structures of confined supersonic mixing layers. *Bull. Am. Phys. Soc.* 35: 2302

Tam, C. K. W., Morris, P. J. 1980. Radiation of sound by instability waves of a com-

pressible plane turbulent shear layer. *J. Fluid Mech.* 98: 349–81

Taulbee, D., Van Osdol, J. 1991. Modeling turbulent compressible flows: the mass fluctuating velocity and squared density. *AIAA Pap.* 91-0524

Tavoularis, S., Corrsin, S. 1981. Experiments in nearly homogeneous turbulent shear flow with uniform mean temperature gradient. Part I, *J. Fluid Mech.* 104: 311–47; Part II, *J. Fluid Mech.* 104: 349–67

Tavoularis, S., Karnik, U. 1989. Further experiments on the evolution of turbulent stresses and scales in uniformly sheared turbulence. *J. Fluid Mech.* 204: 457–78

Tennekes, H., Lumley, J. L. 1972. *A First Course in Turbulence*, pp. 21–24. Cambridge: MIT Press

Thompson, P. A. 1988. *Compressible Fluid Dynamics*, pp. 137–46; also published by McGraw-Hill, NY (1972)

Thunker, R. 1991. *Investigation of turbulent momentum and heat transfer in strongly non-adiabatic two-dimensional boundary layers.* PhD thesis. Tech. Univ. Berlin (In German)

Van Driest, E. R. 1951. Turbulent boundary layer in compressible fluids. *J. Aeronaut. Sci.* 18: 145–60

Viegas, J. R., Rubesin, M. W. 1991. A comparative study of several compressibility corrections to turbulence models applied to high-speed shear layers. *AIAA Pap.* 91-1783, AIAA 22nd Fluid Dyn. Plasma Dyn. & Lasers Conf., Honolulu; also see *AIAA J.* 30: 2369–70

Walthrup, P. J. 1987. Liquid-fueled supersonic combustion ramjets: a research prespective. *J. Propul. Power* 3: 515–24

Wills, J. A. B. 1964. On convection velocities in turbulent shear flow. *J. Fluid Mech.* 20: 417–32

Wills, J. A. B. 1971. Measurements of the wavenumber/phase velocity spectrum of wall pressure beneath a turbulent boundary layer. *J. Fluid Mech.* 45: 65–90

Wu, C. T., Ferziger, J. H., Chapman, D. R. 1985. Simulation and modeling of homogeneous compressed turbulence. *Dept. Mech. Engrg. Rep. No. TF*-21, Stanford Univ.

Yaglom, A. M. 1948. Homogeneous and isotropic turbulence in viscous compressible fluids. *Izv. Akad. Nauk SSSR. Ser. Geograf. Geofiz.* 12(6): 501–22

Yoshizawa, A. 1991. Subgrid-scale model of compressible turbulent flows. *Phys. Fluids A*, 3: 714–16

Youngs, D. L. 1989. Modelling turbulent mixing by Rayleigh-Taylor instability. *Physica D* 37: 270–87

Zang, T. A., Dahlburg, R. B., Dahlburg, J. P. 1992. Direct and large eddy simulations of three-dimensional compressible Navier-Stokes turbulence. *Phys. Fluids A* 4: 127–40

Zank, G. P., Matthaeus, W. H. 1991. The equations of nearly incompressible fluids. I. Hydrodynamics, turbulence, and waves. *Phys. Fluids A* 3: 69–82

Zel'dovich, Ya. B., Raizer, Yu. P. 1969. Shock waves and radiation. *Annu. Rev. Fluid Mech.* 1: 385–412

Zeman, O. 1990. Dilatation dissipation: the concept and application in modeling compressible mixing layers. *Phys. Fluids A* 2: 178–88

Zeman, O. 1991. On the decay of compressible isotropic turbulence. *Phys. Fluids A* 3: 951–55

Zeman, O. 1992a. Towards a constitutive relation in compressible turbulence. In *Studies in Turbulence*, ed. T. B. Gatski, S. Sarkar, C. G. Speziale, pp. 285–96. New York: Springer-Verlag

Zeman, O. 1992b. Similarity in supersonic mixing layers. *AIAA J.* 30: 1277–83

Zeman, O. 1993. A new model for super/hypersonic turbulent boundary layers. *AIAA Pap.* 93-0897

Zeman, O., Blaisdell, G. A. 1991. New physics and models for compressible turbulent flows. In *Advances in Turbulence* 3, ed. A. V. Johansson, P. H. Alfredsson, pp. 445–54. Berlin: Springer-Verlag

Zeman, O., Coleman, G. 1993. Compressible turbulence subjected to shear and rapid compression. *Turbulent Shear Flows* 8: 283–96. Berlin: Springer-Verlag

Annu. Rev. Fluid Mech. 1994. 26 : 255–85

DOUBLE DIFFUSION IN OCEANOGRAPHY

Raymond W. Schmitt

Woods Hole Oceanographic Institution, Woods Hole, Massachusetts 02543

KEY WORDS: salt fingers, convection, mixing, ocean microstructure, thermo-
              haline circulation

INTRODUCTION

The modern study of double-diffusive convection began with Melvin Stern's article on "The Salt Fountain and Thermohaline Convection" in 1960. In that paper, he showed how opposing stratifications of two component species could drive convection if their diffusivities differed. Stommel et al (1956) had earlier noted that there was significant potential energy available in the decrease of salinity with depth found in much of the tropical and subtropical ocean. While they suggested that a flow (the salt fountain) would be driven in a thermally-conducting pipe, it was Stern who realized that the two orders of magnitude difference in heat and salt diffusivities allowed the ocean to form its own pipes. These later came to be known as "salt fingers." Stern also identified the potential for the oscillatory instability when cold, fresh water overlies warm, salty water in the 1960 paper, though only in a footnote. Turner & Stommel (1964) demonstrated the "diffusive-convection" process a few years later.

From these beginnings in oceanography over three decades ago, double diffusion has come to be recognized as an important convection process in a wide variety of fluid media, including magmas, metals, and stellar interiors (Schmitt 1983, Turner 1985). However, it is interesting to note that about one hundred years before Stern's paper, W. S. Jevons (1857) reported on the observation of long, narrow convection cells formed when warm, salty water was introduced over cold, fresh water. He correctly attributed the phenomenon to a difference in the diffusivities for heat and

255

0066–4189/94/0115–0255$05.00

salt in a qualitative way, though his attempt to apply the results to cloud convection was misguided. It is a curious twist of history that Jevons' work motivated Rayleigh (1883) to first derive the expression for the frequency of internal waves in a stratified fluid. However, he ignored the effects of diffusion and thus missed the opportunity to initiate the theoretical study of double-diffusive phenomena. Thus, we could have had over a century of progress to report on, rather than only three decades!

Nonetheless, it is fitting that the ocean should be the impetus for the discovery of this fundamental convection process. As will be discussed below, the ocean is strongly unstable to double-diffusive processes and seems to be profoundly affected by their presence. Only recent advances in ocean instrumentation have enabled us to begin to understand the complicated nature of double-diffusive mixing in the large, under-explored fluid which covers most of the Earth. Given the vital role of mixing in control of ocean heat storage, the thermohaline circulation, climate, carbon dioxide absorption, and pollutant dispersal, it is increasingly important that we achieve a more complete understanding of oceanic double diffusion.

In this review, I attempt to summarize our current knowledge of oceanographic double diffusion for the salt finger, diffusive convection, and intrusive instabilities. Due to my own interests, there is a more complete discussion of salt fingers. In addition, I note some of the potential implications for large-scale modeling of the ocean thermohaline circulation.

SALT FINGERS

Early efforts to understand double-diffusive convection drew on extensive experience with classic Rayleigh-Bénard convection, in which a fluid confined between two boundaries is heated from below. However, at least for the salt finger case, boundaries are largely irrelevant, as the instability defines its own internal length scale. Down-flowing warm, salty water loses heat but not salt to surrounding cold, fresh water, thereby becoming denser and accelerating its flow. Cold, fresh plumes gain thermal buoyancy and are accelerated upward. The scale of the fingers is set by a balance between thermal diffusion (which releases energy in the salt stratification and is greatest at small scales) and viscous drag (which limits the smallest scales of motion). In the situation where opposing linear profiles (constant gradients) of temperature (or fast diffusing substance) and salinity (slow diffusing substance) are found, the simplest and most useful salt finger model is the unbounded, depth-independent similarity solution. Discussed for the asymptotic case of large Prandtl number and diffusivity ratio by Stern (1975), it has proven very valuable in understanding the length scale of ocean and laboratory salt fingers and the ratio of heat to salt transport. Here we examine the solutions given by Schmitt (1979a, 1983) which are valid for all Prandtl numbers and diffusivity ratios. Because of the purely vertical flow, the similarity solutions are valid at finite amplitude (it is a linear system, but not linearized). The evolution of the temperature (T), salinity (S), and vertical velocity (w) fields in salt fingers is described by the equations:

$$\frac{\partial \alpha T'}{\partial t} + w' \alpha \bar{T}_z = \kappa_T \nabla_2^2 \alpha T'$$

$$\frac{\partial \beta S'}{\partial t} + w' \beta \bar{S}_z = \kappa_S \nabla_2^2 \beta S'$$

$$\frac{\partial w'}{\partial t} + g(\beta S' - \alpha T') = \nu \nabla_2^2 w' \tag{1}$$

where the thermal expansion and haline contraction coefficients $\alpha [= -(1/\rho)(\partial \rho/\partial T)]$ and $\beta [= (1/\rho)(\partial \rho/\partial S)]$ are assumed constant, κ_T and κ_S are the molecular heat and salt diffusivities, ν is the kinematic viscosity, $(')$ represents the finger perturbation away from the horizontal average $(^-)$, and ∇_2^2 is the horizontal Laplacian. The requirement for instability is $\alpha \bar{T}_z / \beta \bar{S}_z \equiv R_\rho < \kappa_T/\kappa_S \simeq 100$ in the ocean. This requirement for the density ratio (R_ρ) to be less than the diffusivity ratio $(\kappa_T/\kappa_S = \tau)$ means that the vertical salt gradient necessary for fingers is only $1/100$ that of the temperature, when both are compared in density units. This criterion is satisfied over vast regions of the ocean. In order to maintain the larger scale static stability of the water column, we also require $R_\rho > 1$.

In an unbounded fluid, these equations have depth-independent solutions of the form:

$$(w', T', S') = (\hat{w}, \hat{T}, \hat{S}) \exp(\lambda t) \phi(x, y), \tag{2}$$

where λ is the growth rate and ϕ is a horizontal planform function. Schmitt (1979a, 1983) provides a comprehensive look at the parametric dependence of these solutions.

In order to yield a separable problem, the function ϕ must satisfy the Helmholtz equation, $\nabla_2^2 \phi + m^2 \phi = 0$ (Proctor & Holyer 1986). A solution for rectangular fingers is $\phi = \sin(m_1 x) \sin(m_2 y)$; a solution for sheets is $\phi = \sin(my)$. Equal amounts of horizontal diffusion (and the same growth rates) are realized in squares $(m_1 = m_2)$, rectangles $(m_1 \neq m_2)$, and sheets, provided $m_1^2 + m_2^2 = m^2$. Laboratory observations have confirmed the existence of the rectangular (Shirtcliffe & Turner 1970) and sheet (Linden 1974a) modes. Linden showed that parallel sheets are preferred when a vertical shear damps the downstream modes, leaving the cross-stream modes unaffected. The fact that horizontal diffusion in one dimension can compensate for the lack of it in the other dimension allows us to construct a variety of interesting planforms. The total horizontal wavenumber (m) is given by

$$m = \left(\frac{g\alpha\bar{T}_z}{\nu\kappa_T}\right)^{1/4} M \tag{3}$$

where M is a nondimensional function of R_ρ, the heat/salt buoyancy flux ratio (γ), Prandtl number ($\nu/\kappa_T = \sigma$), and diffusivity ratio ($\kappa_T/\kappa_S = \tau$).

The growth rate of the fingers is scaled with the local vertical temperature gradient,

$$\lambda = (g\alpha\bar{T}_z)^{1/2}G, \tag{4}$$

where G is the nondimensional growth rate. The dependence of G and M on σ, τ, γ, and R_ρ is given in Schmitt (1979a, 1983). The nondimensional growth rate can be expressed as:

$$G = \frac{M^2}{\sigma^{1/2}} \frac{(\gamma - R_\rho/\tau)}{(R_\rho - \gamma)}. \tag{5}$$

For heat–salt fingers ($\sigma = 10$, $\tau = 100$), an accurate expression for $\gamma(M)$ for $R_\rho > 1.15$ is: $\gamma = \frac{1}{2}[b - (b^2 - 4R_\rho)^{1/2}]$, where $b = R_\rho(1 + M^4) + 1$. This relation for $G(M)$ can be used to construct an evolving spectrum of salt fingers, given an initial "seed" spectrum, as in Schmitt (1979a) (Figure 1). Gargett & Schmitt (1982) compared such spectra with towed micro-structure data from the Pacific. Certain portions of the data with narrow bandwidth, limited amplitude structure (distinct from the wide band, spiky nature of turbulence patches), displayed a remarkably good fit to the theoretical spectrum with 2–4 buoyancy periods of growth. Later obser-vational work by Lueck (1987), Marmorino (1987), and Fleury & Lueck (1992) has also found a spectral peak in horizontal temperature gradient records at the predicted wavenumber of the fastest growing finger. Lab-oratory experiments by Taylor (1991) in which salt fingers grow after disruption by grid turbulence similarly reveal that the fastest growing finger dominates the structure. The fingers reform rapidly after the passage of the grid, within a few e-folding periods. He notes that exponential growth may be limited to a few e-folding times, before reaching a self limiting amplitude (which is expected to be a function of available salt contrast). In the ocean, where the time between turbulence events is long compared to the finger e-folding time (about one buoyancy period for $R_\rho = 2$), fingers should have sufficient time to establish modest fluxes. Shen

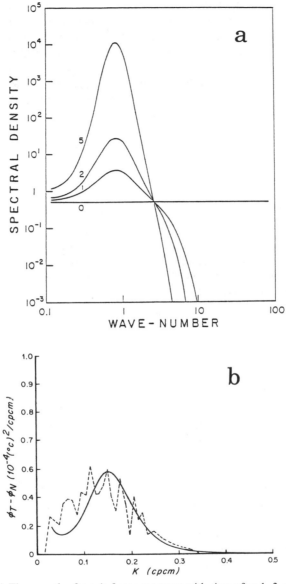

Figure 1 (*a*) The growth of a salt finger spectrum with time after 1, 2, and 5 buoyancy periods at $R_\rho = 2$. The "seed" spectrum is white. Both axes are nondimensional; when scaled to typical ocean conditions the spectral peak is usually at a wavelength of 5–7 cm. (*b*) Comparison of observed and theoretical salt finger spectra from Gargett & Schmitt (1982). Here a red seed spectrum ($-5/3$ slope) has been allowed to grow for 3.5 buoyancy periods at $R_\rho = 2.4$. The temperature microstructure data come from a towed body operating at a depth of 290 m in the North Pacific.

(1989) has performed numerical experiments on two-dimensional salt fingers which also show the dominance of the fastest growing finger.

The relationship between wavenumber and flux ratio in salt fingers is particularly strong. Wide fingers (low m) will lose less of their temperature anomaly than thin fingers, and thus advect more heat. The close dependence of flux ratio on wavenumber helps provide additional evidence that the fastest growing finger dominates. That is, the different flux ratios observed in laboratory experiments in both the heat/salt ($\gamma = 0.5$–0.7) and salt/sugar ($\gamma = 0.9$) systems agree well with those of the fastest growing finger (Schmitt 1979a). Similarly, ocean observations of the horizontal variations in layer properties within a thermohaline staircase require a substantial contribution from the fastest growing finger (Schmitt et al 1987).

In theory, any function satisfying the Helmholtz equation is a possible salt finger planform. Most work has focused on the square or sheet solutions. While squares appear to be the preferred mode in long-term laboratory experiments (Williams 1975), in the presence of vertical shear, sheets are observed (Linden 1974a). According to the theory of Proctor & Holyer (1986), sheets should be preferred over squares at finite amplitude, even in the absence of mean shear, contrary to the laboratory results. Schmitt (1993) has pointed out that a rich variety of finger planforms is possible when sheet and rectangular modes are combined. By choosing one component of the rectangular mode to be a subharmonic of the sheet wavenumber, interesting tesselations can be generated, including asymmetric [$\phi = 0.5\cos(my) + \sin(\sqrt{3}/2\,mx)\sin(\frac{1}{2}my)$] and triangular [$\phi = 0.5\sin(my) + \sin(\sqrt{3}/2\,mx)\sin(\frac{1}{2}my)$] modes. The asymmetric mode, in which a narrow jet of up- or down-going fluid is surrounded by a broader, weaker counterflow, appears to have been observed by Osborn (1991) in down-going, warm (and presumably salty) plumes observed from a submarine near the surface under calm conditions. Scales of the phenomena he discovered seem to be about right for the fastest growing asymmetric mode, though the density ratio was 10–20, much higher than typical salt finger observations. Perhaps exceptionally calm conditions permitted growth in this case. However, though the observations fit a possible mode of the system, the cause of the asymmetry remains obscure. It could result from vertical variations in the background gradients, asymmetries in the perturbations seeding the fingers or in the nonlinearity of the equation of state for seawater. No laboratory work has yet been done on this topic, though it seems ripe for exploration.

Another very puzzling aspect of salt finger planforms stems from microstructure observations during the Caribbean-Sheets and Layers Transects (C-SALT) experiment in 1985. Very strong temperature and salinity steps were observed in the main thermocline east of Barbados over an area of 1 million square kilometers (Schmitt et al 1987). Well mixed layers 5–30 m thick were separated by high gradient interfaces with temperature contrasts up to 1°C (Figure 2). Such "thermohaline staircases" are the expected finescale signature of active double diffusion, as demonstrated in the laboratory by Stern & Turner (1969). In a staircase, fingers within the high gradient interfaces provide an unstable buoyancy flux which drives large-scale overturning in the adjacent mixed layers. Marmorino et al (1987) documented both finger-scale microstructure within the interfaces and convective plumes in the mixed layers for the C-SALT staircase.

However, shadowgraph images from C-SALT (Kunze et al 1987) revealed small-scale laminae with a nearly horizontal orientation in the finger-favorable interfaces (Figure 3). This was surprising, as earlier shadowgraphs (Williams 1975, 1981; Schmitt & Georgi 1982) revealed more vertically oriented structure similar to that seen in the laboratory. A shadowgraph is an indicator of the Laplacian of the index of refraction. The images are sensitive to the focal length of the system; the C-SALT system was designed to best resolve relatively short focal-length structure. The earlier systems of Williams had focal lengths 2 to 10 times greater. It seems likely that the images represent fingers tilted by shear, or sheets initially aligned with the shear but tilted by the rotating shear vector (Kunze 1990). As discussed by Kunze, the centimeter scale laminae must represent primarily salt structure. The model he developed indicates that the temperature structure should have a more vertical orientation. It would also have a larger scale and thus a longer focal length. This perhaps explains the more vertical structure seen in previous shadowgraph imagery. Shear tilted fingers or sheets might also explain the low vertical coherence found in data from closely spaced temperature probes on the towed microstructure instrument described by Lueck (1987) and Fluery & Lueck (1992) and the low dissipations reported by Gregg & Sanford (1987). If most of the small-scale shear is due to inertial waves (as is usually found in the ocean), alignment of sheets with the shear should not be expected unless the time scale for finger growth is sufficiently short compared to the inertial period. This can only be achieved if the interfacial gradients are sharp. Most of the C-SALT interfaces were much thicker (2–5 m) than predicted from laboratory experiments (20–50 cm), so could not support very rapid finger growth. Marmorino (1989) and Fluery & Lueck (1991) find that the flux may be independent of the interface thickness, contrary to theory. However, the Kunze (1987, 1990) models do appear to explain most other features of the C-SALT microstructure: the low dissipation rate, the horizontal laminae, and the low vertical coherence of towed microstructure. In the Kunze (1990) model, even the slow turning of the inertial shear at this latitude is sufficient to limit sheet growth, so that fluxes are surprisingly sensitive to the Coriolis frequency. This predicted effect may help to rationalize some of the observed properties of ocean thermohaline staircases, discussed below.

Persistent staircases have been well documented in the Tyrrhenian Sea within the Mediterranean, below the warm, salty Mediterranean outflow water in the eastern North Atlantic, and in the western tropical North

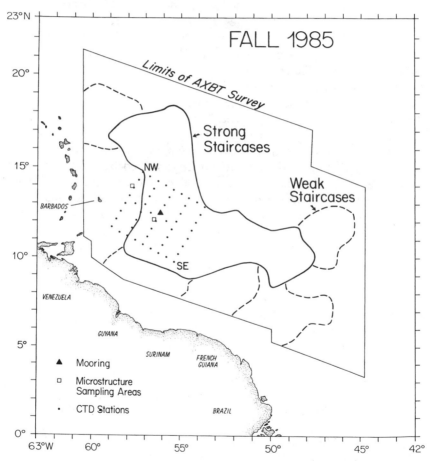

Figure 2 (*a*) The region of the western tropical North Atlantic surveyed during the C-SALT (Caribbean-Sheets and Layers Transects) field program. The AXBT is an Air deployable eXpendible BathyThermograph, a device used to obtain ocean temperature profiles from an airplane. An area of over 10^6 km^2 (roughly equivalent to the combined areas of Texas and California) was found to have significant thermohaline layering in the main thermocline.

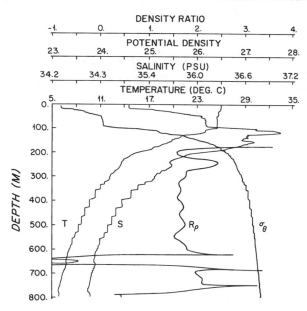

Figure 2 (*b*) The vertical profiles of temperature, salinity, potential density anomaly (σ_θ, kg/m³), and R_ρ from a station in the C-SALT area, from the surface to 800 m depth. Temperature contrasts across the steps are typically 0.5–1.0°C. Mixed layers are 5–30 m thick. The layered structure in the 300–600 m depth range has $1.5 < R_\rho < 1.8$.

Atlantic. In the Tyrrhenian layers, the temperature and salinity of the layers are observed to remain constant for many years (Johannessen & Lee 1974, Molcard & Williams 1975, Molcard & Tait 1977). The Mediterranean Outflow steps appear to have a lateral coherence of 50–100 km, and have been identified several times (Tait & Howe 1968, Elliot et al 1974). In the western tropical North Atlantic, the layers appear to be a permanent feature of the thermocline, with observations available over the past 3 decades (Mazeika 1974, Lambert & Sturges 1977, Boyd & Perkins 1987, Boyd 1989, Schmitt et al 1987). Lambert & Sturges (1977) suggested that the vertical flux convergence due to salt fingers was sufficient to balance the lateral advection in a thermohaline staircase in a Caribbean passage. However, when Lee & Veronis (1991) assume an advective-diffusive balance for the C-SALT layers in an inverse model, they find tracer velocities that differ from geostrophic, and mixing rates well above those derived from the microstructure measurements. Schmitt et al (1987) found that the layer temperature and salinity properties changed horizontally, with a remarkably constant lateral density ratio of $\alpha T_x / \beta S_x = 0.85 (\pm 0.03)$

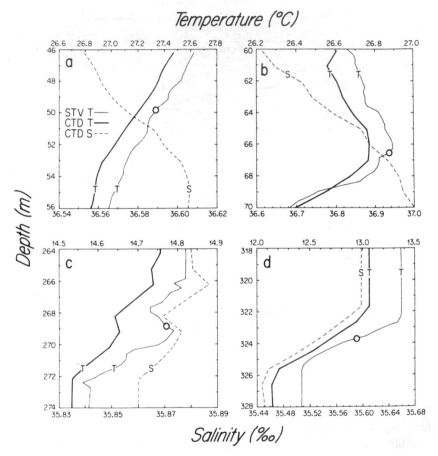

Figure 3 (*A*) Vertical thermohaline structure in the vicinity of shadowgraph images (*open circles*) shown in Figure 3*B*. Plotted are the shadowgraph temperature (offset) and temperature and salinity profiles from a simultaneous Conductivity-Temperature-Depth cast. Note that temperature and salinity scalings differ in the four panels. Both temperature and salinity profiles are gravitationally stable for the images in (*a*) and (*b*). In (*c*) the unstable temperature gradient is compensated by salinity, conditions which favor the "diffusive" form of double diffusion. In (*d*) an unstable salinity gradient across a 2-m thick interface between two homogeneous layers is compensated by temperature, conditions which are favorable to salt fingers.

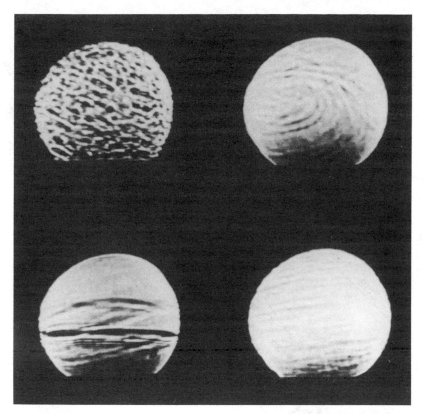

Figure 3   (*B*) Four images of optical microstructure from a free-fall shadowgraph profiler deployed in the C-SALT regions east of Barbados. Each image is 10 cm in diameter and represents a view through 60 cm of water. The four examples are thought to correspond to stratified turbulence (upper left, panel (*a*) of Figure 3*A*), a shear billow (upper right, (*b*) of Figure 3*A*), a diffusive interface (lower left, (*c*) of Figure 3*A*), and shear tilted salt fingers (lower right, (*d*) of Figure 3*A*).

(Figure 4). Equally remarkable, historical data from the region show layer T/S properties falling along the same lines. The C-SALT layers seem immune to disruption by internal waves and the strong eddy field of the region, persisting for decades like geologic strata over about one million square kilometers. That such extensive features could be maintained by small-scale salt fingers is testament to the large amount of energy available in the unstable salt gradient; indeed, it is comparable to the available potential energy of the general circulation (Schmitt & Evans 1978).

The value of the lateral density ratio of 0.85 is worth some discussion. It represents a clear signature of salt fingers, as data on the time evolution

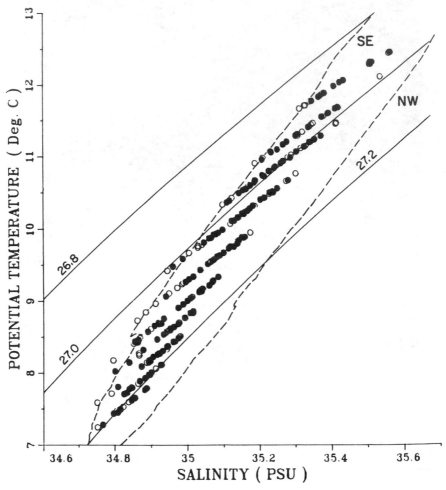

Figure 4 Potential temperature-salinity values of the mixed layers observed in the CTD stations occupied during the spring of 1985 for C-SALT. The solid circles are from mixed layers more than 10 m thick; the open circles are from layers 5–10 m thick. Also shown are the potential temperature-salinity relationships from stations at the northwest and southeast corners of the survey area (*dashed curves*). The evolution of properties across the region is such that individual layers become warmer, saltier, and denser from southeast to northwest as would be expected from the vertical convergence of salt finger fluxes. The layer properties cross isopycnals (the 26.8, 27.0, and 27.2 potential density surfaces are shown) with an apparent heat/salt density flux convergence ratio of 0.85 (± 0.03).

of layer properties in laboratory experiments show variations strikingly similar to those of Figure 4 (Lambert & Demenkow 1972). The simplest interpretation is that the layer temperature and salinity values are given by a balance between along-layer advection and the divergence of the vertical heat and salt fluxes. The fact that the horizontal density ratio is 0.85 rules out a dominance of vertical turbulence, which would require heat and salt to change in the same ratio as the vertical density ratio, $R_\rho = 1.6$. It also rules out a purely isopycnal process, which would yield a horizontal density ratio of 1.0. It is consistent with a dominance of salt fingers, though slightly higher than might be expected from the salt finger flux ratio in theory (Schmitt 1979a) and experiments (Turner 1965, Schmitt 1979b). The small discrepancy may be due to the occasional patches of turbulence noted by Marmorino (1990) and Fleury & Lueck (1991). More likely though, the elevation above that expected for simple salt finger flux convergence is due to the upward migration of interfaces caused by the nonlinear equation of state (McDougall 1991). The "densification on mixing" causes asymmetric entrainment of the bottom of fingering interfaces, leading to a weak but significant contribution to the flux divergence ratio that is proportional to $R_\rho = 1.6$. The interface migration effect has been observed in the laboratory by Schmitt (1979b) and McDougall (1981).

One interesting aspect of the lateral variation in layer properties is that the changes are sometimes concentrated at internal fronts. These fronts take the form of tilted temperature inversions within the mixed layers. As documented by Marmorino (1991), the temperature inversions will support diffusive convection, with a vertical density ratio of 0.85. These intrusive structures may be an intrinsic part of the staircase system—perhaps a self-propagating mechanism for lateral exchange. The flux convergence between the finger and diffusive interfaces should set up lateral pressure gradients capable of driving the intrusions. Their origin and propagation mechanism remain obscure though, and the fronts are more simply explained by eddy stirring of the mean horizontal gradients in the region.

A full understanding of the detailed structure of thermohaline staircases must await further observational and theoretical developments. However, the long-term persistence of ocean staircases, and the relatively modest mixing rates observed within the C-SALT layers, lead me to propose the following rationalization of salt fingering staircases.

1. Persistent staircases must be maintained against the smoothing effects of shear-induced mixing by an "up-gradient" density flux due to double diffusion. Since turbulence due to breaking internal waves in the thermocline is generally weak (Gregg 1989) (and even weaker in the C-SALT staircase), the double-diffusive flux needed may be small, corresponding to a salt vertical eddy diffusivity of only 0.1 cm^2/s or so.

2. Once a staircase is formed, it is a simple matter for it to greatly increase its salt flux in response to increased advection of salt (Schmitt 1990), by thinning the interfaces and increasing the temperature and salinity contrasts. Such effects are suggested by some of the C-SALT steps (Schmitt 1988), where thin interfaces were found in association with lower R_ρ (though without accompanying microstructure measurements). The enhanced transfer of salt over heat would tend to increase R_ρ (Schmitt 1981), driving the staircase back to a state where the double-diffusive fluxes are just sufficient to maintain the layers. The actual flux in a staircase may be a stronger function of the salt forcing caused by differential advection than the instantaneous density ratio.

3. If we accept the Kunze (1990) model of salt sheet limitation by inertial waves, which seems to explain many of the C-SALT observations, then we might expect that the critical value of the density ratio necessary for maintenance of a staircase is a function of inertial-wave shear and latitude. That is, the Kunze model depends on a balance between sheet growth and disruption by inertial shear. If the inertial period decreases, then fingers or sheets have less growth time before being limited by the turning shear vector. Since the growth rate is a function of R_ρ, this suggests that a lower R_ρ would be required at higher latitudes, to maintain sufficient separation between the finger growth time and the local inertial period. Interestingly, density ratios in known staircases do seem to vary in this manner. The Tyrhennian Sea steps at 40°N have a density ratio of 1.2; the Mediterranean Outflow steps at 35°N have a density ratio of 1.3; and the C-SALT steps at 12°N have a density ratio of 1.6. The ratio of finger growth rate to inertial frequency is found to be roughly constant. While the data are sketchy at best, the correspondence with what could be expected from the Kunze model is certainly encouraging. Staircase observations from other latitudes, with careful shear and turbulence measurements, would be valuable for extending the comparison.

Two further issues concern the interpretation of ocean microstructure measurements. The first approach involves the analysis of towed temperature microstructure data; the second utilizes coincident measurements of the dissipations of both thermal and velocity variance.

The detection of salt fingers in data from a towed instrument was first achieved by Magnell (1976) in the Mediterranean Outflow steps at the time of Williams' optical detection of salt fingers. As mentioned previously, Gargett & Schmitt (1982) found good agreement between towed temperature spectra and theory. Further progress was reported by Mack (1985), who found that some thermal microstructure was well correlated with values of R_ρ near one. Holloway & Gargett (1987) noted that the narrow band, limited amplitude character of salt fingers compared to the broad band, high amplitude nature of turbulence leads to very different kurtosis signatures in towed temperature or conductivity gradient data. The kurtosis (4th moment) of the signal can be computed efficiently and thus is a useful discriminator between salt fingers (with low kurtosis) and turbulence (with high kurtosis). Marmorino (1987) noted the similarity of towed gradient spectra with the Schmitt (1979a) spectral model and suggested that a slope of $+2$ described the conductivity gradient spectra at wavenumbers below that of the fastest growing. Mack (1989) reported further evidence for an association of microstructure with R_ρ near 1 and also found gradient spectral slopes of $+2$, as contrasted with a slope of $+1$ in the turbulent patches. Marmorino & Greenewalt (1988) used both kurtosis and spectral slope to discriminate salt fingers from turbulence. They found it was easy to discriminate the two processes in data from the seasonal thermocline, but curiously, a less clear picture emerges from the C-SALT data. There the interfaces seem to contain both salt fingering and turbulent signatures, perhaps consistent with the elevation of the horizontal density ratio above the simple flux ratio for salt fingers. This turbulence could arise from convection in the layers, interface erosion due to the nonlinear equation of state (McDougall 1991), or shear instability across the interfaces (Marmorino 1990). Mack & Schoeberlein (1993) develop a discrimination technique, the log-likelihood ratio, based on the two-dimensional probability distributions of the slope and kurtosis. It appears that high slope, low kurtosis salt finger regions can be efficiently distinguished from low slope, high kurtosis turbulent patches with this approach. It would be useful to see such techniques applied to increasingly common towed CTD (Conductivity-Temperature-Depth) vehicles, if they can be fitted with fast response microstructure probes.

The second major approach to salt finger identification has been applied to data from vertically profiling vehicles. From these, measurements of the dissipation rates of turbulent kinetic energy (ε) and thermal variance (χ) can be made. Such measurements can be combined with information on background gradients to provide estimates of the vertical eddy diffusivity. In the case of ordinary turbulence, such as that caused by breaking internal waves, the effective vertical diffusivity is given by the relations

$$K_T = \frac{\chi}{2(\bar{T}_z)^2},$$ (6)

due to Osborn & Cox (1972) and

$$K_\rho = \left[\frac{R_f}{(1 - R_f)}\right]\frac{\varepsilon}{N^2} = \Gamma^t\frac{\varepsilon}{N^2},$$ (7)

due to Osborn (1980), in which R_f is the efficiency of conversion of kinetic to potential energy, and N is the local buoyancy frequency, $N = \sqrt{-g(1/\rho)(\partial\rho/\partial z)}$. Laboratory data suggest that R_f is about 0.15–0.2, giving an expected Γ^t of about 0.18–0.25.

However, for salt fingers, the relation between diffusivity and ε/N^2 is much different. McDougall (1988), Schmitt (1988), and Hamilton et al (1989, 1993) point out that the equivalent relation for salt fingers is

$$K_S = \frac{[R_\rho - 1]\varepsilon}{[1-\gamma]N^2} = \Gamma^f \frac{\varepsilon}{N^2}. \tag{8}$$

This expression can give a diffusivity over ten times the Osborn formula for the same dissipation rate due to the high efficiency of salt fingers in converting haline to thermal potential energy. Turbulence, however, dissipates most of its kinetic energy, converting only a small fraction to potential energy. As Schmitt (1988) points out, this relation is very sensitive to the value of the flux ratio, leading to some uncertainty in the effective diffusivity expected in salt finger regions.

Oakey (1985) has used ocean turbulence measurements to estimate Γ as the scaled ratio of χ to ε:

$$\Gamma = \frac{\chi N^2}{2\varepsilon \bar{T}_z^2}. \tag{9}$$

He found that Γ is around 0.17–0.26 in non-double diffusive regimes in the ocean. Hamilton et al (1989) and McDougall & Ruddick (1992) show how the measured Γ should differ for turbulence and salt fingers. Hamilton et al (1993) report that Γ is indeed above the value expected for turbulence, and consistent with salt finger theory, in a finger-favorable ocean staircase. McDougall & Ruddick (1992) address the problem of interpreting such measurements in environments where both turbulence and salt fingers may be active. The sensitivity to flux ratio is somewhat problematic; the asymptotic theory of Stern (1975) differs from the complete solution of Schmitt (1979a) for both high and low values of R_ρ. In addition to differences based on theoretical models, Γ is a particularly difficult variable to estimate, because of (a) questions about the isotropy of microstructure, (b) the fact that both χ and ε have non-Gaussian distributions, and (c) the variability of ocean hydrodynamic regimes over vertical scales of a few meters. Even so, the consistency of the Hamilton et al (1993) data with salt finger theory is encouraging. We can expect to see further advances with this approach, as new instrumentation capable of defining the background Richardson numbers as well as R_ρ, χ, and ε (Schmitt et al 1988), is deployed in a variety of ocean environments.

DIFFUSIVE CONVECTION

As mentioned above, Stern (1960) noted the possibility of an oscillatory instability when a stable solute distribution compensates for an unstable temperature profile. That is, when warm, salty water (sitting beneath cold, fresh) is displaced upward, the loss of heat causes an enhanced restoring force (an overstability) and a growing oscillation. The experiments of Turner & Stommel (1964) showed that a steady convection is realized in such a system when thin interfaces separate well-mixed layers. Thermal diffusion across the interfaces drives the convection; weaker salt diffusion acts as a brake on the system. When a stable salt gradient is heated from below, a series of mixed layers and interfaces forms a staircase in temperature and salinity profiles analogous to the salt finger case. Walin (1964) studied the stability properties of different thermohaline stratifications. The diffusive convection system is more profitably analyzed in terms of classic Rayleigh-Bénard convection than salt fingers, since the limiting effect of diffusive fluxes through the boundaries is an accurate analog, and the convection cell size is of the order of the layer depth. Veronis (1965, 1968) examined the finite amplitude behavior of diffusive convection using numerical integration of a truncated spectral model. Relative to the non-double-diffusive problem, the presence of the solute delays the onset of convection. However, at sufficiently high Rayleigh number, fluxes are nearly as large. Also of interest is the salt/heat buoyancy flux ratio, which Veronis showed tends toward $\tau^{-1/2}$. Nield (1967) and Baines & Gill (1969) have also examined the stability of such systems for various boundary conditions.

The low value of the flux ratio compared to salt fingers is of interest and can be readily understood, as it reflects the differing roles of the two components in the two systems. That is, energy in the salt field is released in salt fingers by diffusion of heat, and advection is the primary vertical flux agent. In the diffusive convection case it is the thermal stratification that contains the energy, and the vertical flux is diffusive at the interface. As in Linden & Shirtcliffe (1978), we suppose that the diffusion of heat and salt across a diffusive interface yields penetration distances into the adjacent mixed layer proportional to the one half power of the diffusivity. Since heat penetrates further, an unstable region forms which convects away from the interface when sufficiently thick. The relative portions of heat and salt carried away will reflect their different penetration scales, giving a salt/heat buoyancy flux ratio of $\tau^{-1/2}$.

Shirtcliffe (1967) found that an oscillatory instability does indeed first appear when a stable salt gradient is heated from below. Continued heating drives overturning cells at finite amplitude, which generate well-mixed layers in the previously stratified fluid (Turner 1968). The initial bottom layer grows to a height given by scaling arguments, then additional layers form in succession. This transition to a series of high gradient interfaces separating well-mixed layers enables the fluid to support much higher fluxes. In two-layer experiments Turner (1965), Marmorino & Caldwell (1976), and Linden (1974b) have shown that the flux ratio is indeed low for high density ratios far from one, though modestly above the $\tau^{-1/2}$ expected from theory. As the density ratio approaches one, however, the flux ratio approaches one due to disruption of the interface by mixed layer turbulence. That is, turbulent fluid parcels will penetrate and disrupt the interface so that direct fluid transfer is achieved. The laboratory experiments provide a "4/3" flux law, based on the temperature difference across the interface. Recently, Kelley (1990) has questioned the validity of the 4/3 power law, suggesting that an exponent between 4/3 and 5/4 (depending on Rayleigh number) may be more accurate.

Whereas salt fingers tend to be found in the low to mid-latitude evaporative regions of the ocean, conditions favorable to diffusive convection are more frequently found in high-latitude precipitation zones. The penetration of warmer, saltier water from low latitudes beneath cold, fresh surface waters, sets up diffusive convection sites over vast regions of the Arctic Ocean and various sites around Antarctica. Salt stratified lakes with geothermal or solar heating also provide diffusively favorable sites (Hoare 1966), as do hot, salty brines in ocean deeps at spreading centers (Swallow & Crease 1965). Steps under the Arctic ice were first reported by Neal et al (1969) and Neshyba et al (1971). More recent observations have been provided by Padman & Dillon (1987, 1988, 1989) in other parts of the Arctic. A diffusive thermohaline staircase between about 200 and 400 m depth appears to be a ubiquitous feature under most of the Arctic ice field away from boundaries. There is also a weak internal wave field there (Levine et al 1987) perhaps due to the rigid lid, but also possibly due to enhanced wave decay in the convectively mixed staircase (Padman 1991). Areas near topography associated with stronger wave energy and turbulence show less propensity to form steps, probably due to the down-gradient buoyancy flux (an up-gradient buoyancy flux is required to maintain a staircase).

The scales of layers in diffusive convective staircases have been addressed by Huppert & Linden (1979), Kelley (1984, 1988), and Federov (1988). All suggest that the layer thickness H must scale as $H \sim (\kappa_T/N)^{1/2}$. Kelley (1984) finds that the proportionality constant is a function of R_ρ. His analysis indicates that both laboratory and oceanic staircase heights can be incorporated into the same scaling—a rather remarkable result. In the salt finger case there was great discrepancy between lab and ocean data in the application of similar scaling laws by Stern & Turner (1969). Kelley (1988) maintains that the R_ρ dependence of the scaling can be determined from a model of the competing effects of layer splitting and merging. We also note that Fernando (1987) has proposed a different scaling for the layer thickness. Kelley (1988) provides an estimate of the dependence of the effective vertical diffusivity on R_ρ, a parameterization which should prove useful for understanding the larger scale effects of diffusive convection. The eddy diffusivities are in the range of 10^{-2}–10^{-1} cm^2/s. In general, vertical diffusive fluxes in the Arctic may be small compared to lateral advection, and seem unable to account for the evolution of the warm, salty Atlantic water within the Arctic basin (Padman & Dillon 1987), which may be dominated by lateral processes.

Staircase layering in the diffusive sense is also a prominent feature of the ocean stratification around Antarctica (Foster & Carmack 1976, Middleton & Foster 1980, Muench et al 1990). Muench et al find staircases to be a common feature over much of the Weddell Sea. The layers were much thicker (10–100 m) than commonly found in the Arctic, and may support a larger flux. They estimate upward heat flux using laboratory flux laws to be about 15 W/m^2 in open waters. This is sufficient to be important in the upper ocean heat budget and may help to maintain ice-free conditions in the summer.

INTRUSIONS

Whereas salt fingers and diffusive convection have distinct geographical regimes in terms of the large-scale mean stratification of the ocean (low to mid-latitude versus high latitude), there are widespread opportunities for the transient occurence of both types of instability on smaller scales. This arises from the horizontal contrast in temperature and salinity along density surfaces, and the propensity for interleaving of water masses in such conditions. Finescale horizontal intrusions, in which temperature inversions are stabilized by a density compensating salinity profile, are found within water mass fronts at all latitudes (Figure 5). The vertical scales range from 5–100 m; a typical horizontal scale is of order 10 km. The distinct role of double diffusion in mixing, and probably driving such intrusions, is now well recognized.

The potential for double-diffusive generation of horizontal intrusions was first recognized by Stern (1967). In the presence of isopycnal gradients of temperature and salinity, lateral displacements can set up sites for enhanced double-diffusive mixing which generate pressure gradients that serve to reinforce the motion. The greater transfer of salt than heat by fingers means that warm, salty intrusions should rise across isopycnals as

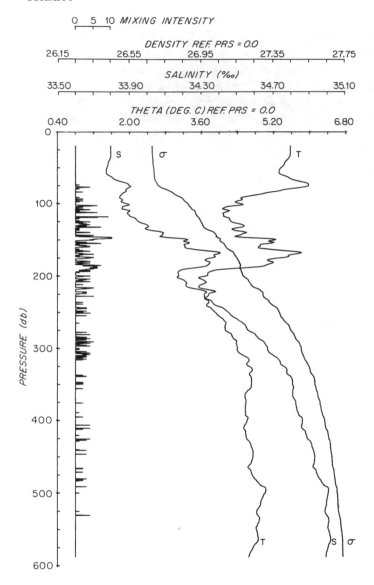

Figure 5 Temperature, salinity, and potential density profiles from the North Atlantic Current east of Newfoundland, from the surface to about 600 m depth (1 db in pressure is approximately 1 m in depth). The prominent intrusions found in this region arise because of the strong horizontal gradients in temperature and salinity along density surfaces. The profile of mixing intensity *(left)*, derived from an optical shadowgraph, indicates that the boundaries of the intrusions have the most microstructure. These are regions where R_ρ is near 1, and double-diffusive mixing should be strong. (From Schmitt & Georgi 1983.)

they cool and freshen; cold, fresh intrusions should sink. Stern suggested that this mechanism could help maintain the tightness of the temperature-salinity relationship. That is, intrusive anomalies would be the sites of enhanced double diffusion which would cause the intrusions to migrate across density surfaces until their T-S properties matched those of the surrounding fluid. In Stern's model, small vertical scales were preferred for the instability, because the enhanced flux convergence could more effectively accelerate the motions. Toole & Georgi (1981) extended the model to include viscosity, and found that the fastest growing features had an intermediate scale. Turner (1978) found that finger-dominated intrusions in the laboratory did indeed cross density surfaces as Stern predicted, and those dominated by diffusive interfaces were inclined in the opposite sense. Ruddick & Turner (1979) further demonstrated the intrusion phenomena in a beautiful set of experiments. They suggested that the vertical scale of the intrusions was set by the potential energy available across the initial sharp front of the experiments, giving a linear dependence on lateral salt contrast. The model of Toole & Georgi predicts an inverse square root dependence of vertical scale on horizontal salt gradient, giving a vertical intrusion scale of order 10s of meters.

Further advances in the theory of double-diffusive intrusions have been made by: Posmentier & Hibbard (1982), who examined the role of tilt on intrusion growth; Ruddick (1984), who studied the evolution of an intrusive layer subject to both finger and diffusive fluxes; McDougall (1985a,b), who developed linear-stability and finite-amplitude intrusion models; and Niino (1986), who studied the stability of a finite width front. In addition, Richards (1991) extended the linear stability problem to the equatorial beta plane, Posmentier & Kirwan (1985) suggested that intrusions could have dynamical effects on mesoscale eddies, McDougall (1986) pointed out the limitations of the Ruddick & Turner (1979) intrusion model, and Yoshida et al (1989) presented a generalized model that quantified the applicability of the Ruddick & Turner or Toole & Georgi scalings for the intrusion thickness. Attempts have been made to estimate the effects of intrusions on larger scales by Joyce (1977), who suggested a formalism to infer the effective lateral diffusivity due to intrusive interleaving, and Garrett (1982), who parameterized the diapycnal fluxes due to the migration of intrusions across density surfaces.

Oceanographers have reported on the existence of finescale thermohaline interleaving at a variety of ocean fronts. Joyce et al (1978) and Toole (1981a) reported on intrusions in the Antarctic polar front. They found that intrusions did slope as predicted by Stern. Gregg & McKenzie (1979) and Gregg (1980) found similar behavior in mid-latitude intrusions. Toole (1981a) reported reasonable agreement with the length scale of the

Toole and Georgi model, though eddy diffusivities and horizontal gradients are difficult to specify. Toole (1981b) described different intrusions in the equatorial Pacific, dominated by temperature variations, for which the vertical scale was more accurately given by the Ruddick & Turner expression.

Williams (1981) used his shadowgraph device to find evidence for active double diffusion in a Gulf Stream intrusion. Schmitt & Georgi (1982) also used the Williams shadowgraph in strong intrusions in the North Atlantic Current east of Newfoundland, and reported evidence for salt finger activity (Figure 5). They found plausible agreement of the intrusion scale with the Toole & Georgi model and also applied the Joyce scheme for estimating the effective lateral diffusivity of intrusions on larger scales. They suggest that intrusions may give lateral diffusivities two orders of magnitude larger than predicted from the shear dispersion of internal waves by Young et al (1982). Schmitt et al (1986) found intrusions around a warm core ring of the Gulf Stream which appeared to be dominated by diffusive-convection based on intrusion slopes and the relative intensity of fingering and diffusive microstructure. They also point out that the advective velocities and relative shears found around strong eddies significantly complicate our capacity to map intrusions and may contribute sufficient turbulence to mask the finescale effects of double diffusion.

Finally, Ruddick (1992) has analyzed data from a lens of warm, salty Mediterranean water in the eastern Atlantic (a "Meddy"). He finds diffusively-dominated intrusions at the upper edges of the eddy with a typical vertical scale of 13 m, and finger-dominated intrusions at the lower edges with vertical scales near 25 m. The slopes in the respective cases are as expected for the double-diffusive driving mechanisms and clearly inconsistent with the McIntyre (1970) process of differential diffusion of momentum and mass. This data set comprises the best evidence to date that double-diffusive intrusions play a significant role in ocean mixing. Hebert (1989) and Hebert et al (1990) have estimated that the intrusions were the most important dissipation mechanism acting on the Meddy, which was observed to decay over a two-year period.

Double-diffusively driven intrusions could turn out to be a primary horizontal mixing mechanism of the ocean. While "isopycnal" displacements do not require a change in potential energy, and thus should be readily accomplished, it appears that the internal wave climate provides rather modest lateral mixing rates (Young et al 1982). Though large-scale baroclinic instability is an active "stirring" mechanism, serving to increase mesoscale lateral gradients, it does not actually cause mixing (the destruction of gradients). Intrusions provide a key link in conveying heat and salt variance from the mesoscale to the microscale. The double-diffusive driving can be strong because the density ratio approaches one for both the finger and diffusive boundaries of the inversion. This releases energy in the isopycnal heat and salt gradients and appears to accomplish substantial lateral mixing. While we know rather little about lateral mixing processes, preliminary indications are that other finescale mechanisms are of lesser strength (Schmitt & Georgi 1982). Thus, there is reason to believe that double-diffusive intrusions are one of the most important lateral mixing agents in those regions with isopycnal gradients of temperature and salinity.

LARGE-SCALE IMPLICATIONS

Conditions favorable to salt fingering are very common in the main thermocline of the subtropical gyres. Ingham (1966) reports that in 90% of the Atlantic Ocean the main thermocline has a density ratio less than 2.3. At 24° N in the Atlantic, Schmitt (1990) finds that 95% of the upper kilometer is fingering favorable, with over 65% having a density ratio between 1.5 and 2.5. The Pacific is less conducive to salt fingering, but the Indian Ocean also harbors vast regions of low density ratio [65% has a thermocline $R_\rho < 4.5$, according to Ingham (1966)]. Given the widespread nature of fingering-favorable conditions in the world ocean, it is important to evaluate the effects of double diffusion on larger scale thermohaline structure.

As already discussed, evidence for active salt fingering in the ocean can be seen with a variety of measurement techniques. The agreement of spectra with theory, the increased occurrence of microstructure at low density ratio, the performance of spectral slope and kurtosis discriminators, the scaled ratio of χ to ε, the persistence and lateral property variations of thermohaline staircases, the finger-scale microstructure in the interfaces and the plume-like structure in the mixed layers, the occurrence of steps at low density ratio, and the behavior and microstructure of thermohaline intrusions, all indicate that salt fingering is an active ocean mixing process.

But is it an important process, given the failure of present models in describing the structural details of finger interfaces (Fleury & Lueck 1992) and the inadequacy of laboratory flux laws (Kunze 1987)? I would argue that it is, based on the structure of the large-scale temperature-salinity relation in the ocean. As noted by Ingham (1966) and Schmitt (1981), the form of the temperature-salinity relationship in most of the Central Waters of the various oceans is better fit by a curve of constant density ratio than a straight line (Figure 6a). Greengrove & Rennie (1991) find similar results for the South Atlantic. Schmitt (1981) showed how the dependence of salt finger mixing intensity on R_ρ and the greater transport of salt than heat could cause such structure. In this mechanism, variations away from a constant R_ρ profile are sites of salt flux convergence or divergence that tend to remove the R_ρ perturbations. A linearized analysis showed that R_ρ could "diffuse" vertically with an effective diffusivity well above that for heat or salt. Numerical examples of the nonlinear one-dimensional mechanism showed how salt fingers can rotate the slope of the T-S relation and induce curvature toward constant R_ρ, in distinct contrast to the straight line T-S relation given by ordinary turbulence, with equal mixing rates for T and S (Figure 6b). The suggested decrease in finger mixing rate as R_ρ exceeds 2 provides a rationale for the large portions of the Central Waters that have R_ρ close to 2. As an alternative, Stommel (1993) has proposed a lateral mixing mechanism for the surface mixed layer that also produces $R_\rho = 2$. However, the lateral gradients he requires would be strongly unstable to double-diffusive intrusions. Also, examination of density ratio distributions shows that it is generally less likely to have a value of 2 near the surface than deeper in the water column (Schmitt 1990). This suggests that the internal mixing and advection processes are most likely the dominant determinants of the density ratio.

In retrospect, the mixing rates chosen for the Schmitt (1981) model are too high, based on what was learned from C-SALT. However, our understanding of internal wave induced mixing has also led to a lowering of expected turbulence rates in the thermocline over the past decade. Gregg (1989) finds low levels of turbulent vertical mixing under typical internal wave conditions, giving $K_\rho < 0.1$ cm^2/s. The Schmitt (1981) mechanism will still function so long as the transfer of salt exceeds the transfer of heat and the mixing rate is a function of R_ρ. The model of Kunze (1990) shows a variation of mixing rate with R_ρ similar to what I proposed, though at a lower overall level. Progress in further understanding the role of salt fingers in large-scale balances may come from an analysis technique developed by Schmitt (1990). There, a relation is derived that expresses the potential for vertical shear to act on isopycnal gradients of temperature and salinity and thus modify the density ratio. The tendency of shear to rotate the slope of the T-S relation can be balanced by the differential vertical transport of heat and salt by fingers, but not by turbulence. The technique shows promise for the "constant R_ρ" regions of the subtropical gyres, where the absence of R_ρ gradients makes the R_ρ balance insensitive to the absolute velocity, and the vertical shear can be estimated from the thermal-wind relation.

Another approach has been taken by Gargett & Holloway (1992). They explored the influence of different diffusion rates for T and S in a numerical model of the ocean circulation. Dramatic modifications of the large-scale

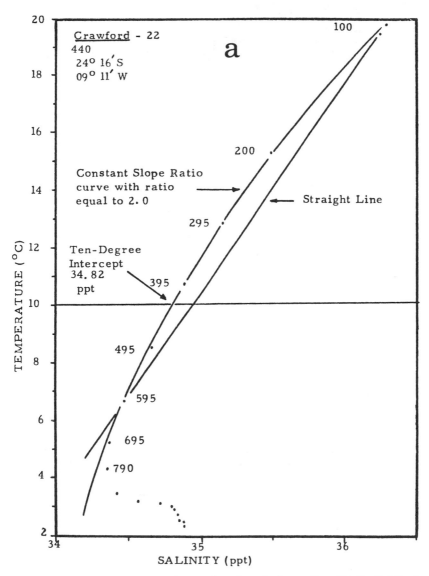

Figure 6 (*a*) Temperature-salinity diagram from the main thermocline of the South Atlantic, from Ingham (1966). The curve of constant $R_\rho = 2.0$ (his "slope ratio") is a better description of the data than a straight line fit for depths from 100 to 800 m.

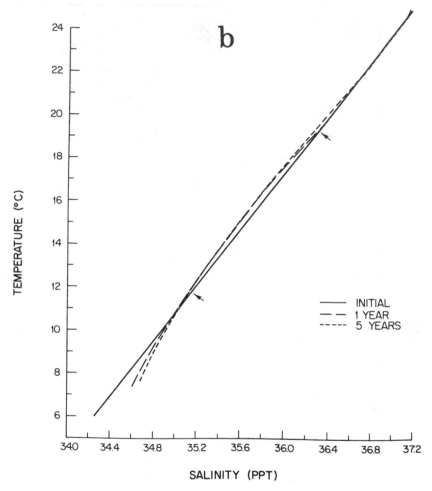

Figure 6 (*b*) Temperature-salinity diagram from the salt finger model of Schmitt (1981). The induction of curvature into an initially linear T-S relation over 1–5 years is caused by enhanced vertical transport of salt relative to heat in low R_ρ portions of the water column. This predicted effect of salt fingers eventually yields a constant R_ρ T-S curve, rather than the straight line expected from ordinary turbulent mixing. (Note: Constant R_ρ corresponds to a curve on a T-S diagram because of the nonlinear equation of state for seawater, primarily due to the increase in thermal expansion coefficient with temperature.)

meridional transports resulted because the vertical buoyancy-flux changes sign in the double-diffusive case relative to the constant diffusivity assumption, requiring different advective regimes to achieve a local density balance. Gargett & Holloway suggest that the ratio of the effective heat and salt diffusivities is more important than the specific value of the mixing rate. This is an encouraging result, since ratios are often more easily determined from the data (e.g. the horizontal density ratio observed in C-SALT) than are absolute fluxes. Much remains to be done with such modeling, including the use of diffusivities with the expected R_ρ dependence (Schmitt 1981, Kelley 1990), rather than a simple on/off function.

SUMMARY

Double diffusion has long graduated from the status of an "oceanographic curiosity." There is overwhelming evidence from fine- and microstructure studies that both salt fingers and diffusive convection are active ocean mixing processes. Much of the main thermocline of the mid- and low-latitude ocean is unstable to salt fingers, and double-diffusive intrusions are likely an important lateral mixing agent in some areas. There is also suggestive evidence from the character of the large-scale temperature-salinity relation in the Central Waters that fingers cause unequal mixing rates for heat and salt. Model results indicate that the presence of double-diffusive mixing has profound implications for the thermohaline circulation (and thus ocean climate as well). Further advancement of our understanding of the role of double diffusion in the ocean should have high priority. Improving observational capabilities will help to accomplish this goal, though field studies must be augmented by continuing improvement in laboratory, numerical, and theoretical models.

ACKNOWLEDGMENTS

Drs. Richard Lambert, Melvin Stern, and Sandy Williams provided advice, encouragement, and research opportunities that helped develop my initial understanding of double diffusion. Dan Georgi, Terry Joyce, Eric Kunze, Trevor McDougall, Barry Ruddick, and John Toole have been helpful contributors to that understanding. Eric Kunze provided a thorough review of the manuscript. The National Science Foundation and the Office of Naval Research have generously supported my work. This review was prepared under grant N00014-92-J-1323 from ONR. This is Contribution Number 8315 of the Woods Hole Oceanographic Institution.

282   SCHMITT

Literature Cited

Baines, P. G., Gill, A. E. 1969. On thermo-haline convection with linear gradients. *J. Fluid Mech.* 37: 289–306

Boyd, J. D. 1989. Properties of thermal stair-case off the northeast coast of South America, Spring and Fall 1985. *J. Geophys. Res.* 94: 8303–12

Boyd, J. D., Perkins, H. 1987. Charac-teristics of thermohaline steps off the northeast coast of South America, July 1983. *Deep-Sea Res.* 34(3): 337–64

Elliott, A. J., Howe, M. R., Tait, R. I. 1974. The lateral coherence of a system of thermo-haline layers in the deep ocean. *Deep-Sea Res.* 21: 95–107

Federov, K. N. 1988. Layer thicknesses and effective diffusivities in "diffusive" thermohaline convection in the ocean. See Nihoul & Jamart 1988, pp. 471–80

Fernando, H. J. S. 1987. The formation of layered structure when a stable salinity gradient is heated from below. *J. Fluid Mech.* 182: 425–42

Fluery, M., Lueck, R. G. 1991. Fluxes across a thermohaline staircase. *Deep-Sea Res.* 38(7): 745–47

Fluery, M., Lueck, R. G. 1992. Micro-structure in and around a double-diffusive interface. *J. Phys. Ocean.* 22: 701–18

Foster, T. D., Carmack, E. C. 1976. Tem-perature and salinity structure in the Wed-dell Sea. *J. Phys. Ocean.* 6: 36–44

Gargett, A. E., Holloway, G. 1992. Sen-sitivity of the GFDL ocean model to different diffusivities for heat and salt. *J. Phys. Ocean.* 22(10): 1158–77

Gargett, A. E., Schmitt, R. W. 1982. Obser-vations of salt fingers in the central waters of the eastern North Pacific. *J. Geophys. Res.* 87(C10): 8017–29

Garrett, C. 1982. On the parameterization of diapycnal fluxes due to double-diffusive intrusions. *J. Phys. Ocean.* 12: 952–59

Greengrove, C. L., Rennie, S. E. 1991. South Atlantic density ratio distribution. *Deep-Sea Res.* 38: 345–54 (Suppl. 1)

Gregg, M. C. 1980. The three-dimensional mapping of a small thermohaline intrusion. *J. Phys. Ocean.* 10: 1468–92

Gregg, M. C. 1989. Scaling turbulent dis-sipation in the thermocline. *J. Geophys. Res.* 94(C7): 9686–98

Gregg, M. C., McKenzie, J. H. 1979. Thermohaline intrusions lie across iso-pycnals. *Nature* 280: 310–11

Gregg, M. C., Sanford, T. B. 1987. Shear and turbulence in thermohaline staircases. *Deep-Sea Res.* 34: 1689–96

Hamilton, J. M., Lewis, M. R., Ruddick, B. R. 1989. Vertical fluxes of nitrate associ-ated with salt fingers in the world's oceans. *J. Geophys. Res.* 94(C2): 2137–45

Hamilton, J. M., Oakey, N. S., Kelley, D. E. 1993. Salt finger signatures in micro-structure measurements. *J. Geophys. Res.* 98(C2): 2453–60

Hebert, D., 1989. Estimates of salt finger fluxes. *Deep-Sea Res.* 35(12): 1887–1901

Hebert, D., Oakey, N., Ruddick, B. R. 1990. Evolution of a Mediterranean salt lens. *J. Phys. Ocean.* 20: 1468–83

Hoare, R. A. 1966. Problems of heat transfer in Lake Vanda, a density stratified Ant-arctic lake. *Nature* 210: 787–89

Holloway, G., Gargett, A. 1987. The infer-ence of salt fingering from towed micro-structure observations. *J. Geophys. Res.* 11: 1963–65

Huppert, H. E., Linden, P. F. 1979. On heat-ing a stable salinity gradient from below. *J. Fluid Mech.* 95: 431–64

Ingham, M. C. 1966. *The salinity extrema of the world ocean.* PhD dissertation. Oregon State Univ., Corvallis

Jevons, W. S. 1857. On the cirrous form of cloud. *London, Edinburgh, Dublin Philos. Mag. J. Sci.* Ser. 4, 14(90): 22–35

Johannessen, O. M., Lee, O. S. 1974. Thermohaline staircase structure in the Tyrrhenian Sea. *Deep-Sea Res.* 21: 629–39

Joyce, T. M. 1977. A note on the lateral mixing of water masses. *J. Phys. Ocean.* 7: 626–29

Joyce, T. M., Zenk, W., Toole, J. M. 1978. The anatomy of the Antarctic Polar Front in the Drake Passage. *J. Geophys. Res.* 83: 6093–6113

Kelley, D. E. 1984. Effective diffusivities within ocean thermohaline staircases. *J. Geophys. Res.* 89: 10,484–88

Kelley, D. E. 1988. Explaining effective diffusivities within diffusive staircases. See Nihoul & Jamart 1988, pp. 481–502

Kelley, D. E. 1990. Fluxes through diffusive staircases: a new formulation. *J. Geophys. Res.* 95: 3365–71

Kunze E. 1987. Limits on growing, finite length salt fingers: a Richardson number constraint. *J. Mar. Res.* 45: 533–56

Kunze, E. 1990. The evolution of salt fingers in inertial wave shear. *J. Mar. Res.* 48: 471–504

Kunze, E., Williams, A. J. III, Schmitt, R. W. 1987. Optical microstructure in the thermohaline staircase east of Barbados. *Deep-Sea Res.* 34(10): 1697–704

Lambert, R. B., Demenkow, J. W. 1972. On the vertical transport due to fingers in double diffusive convection. *J. Fluid Mech.* 54: 627–40

Lambert, R. B., Sturges, W. 1977. A thermohaline staircase and vertical mixing in the thermocline. *Deep-Sea Res.* 24: 211–22

Lee, J. H., Veronis, G. 1991. On the difference between tracer and geostrophic velocities obtained from C-SALT data. *Deep-Sea Res.* 38: 555–68

Levine, M. D., Paulson, C. A., Morison, J. H. 1987. Observations of internal gravity waves under the Arctic pack ice. *J. Geophys. Res.* 92: 779–82

Linden, P. F. 1974a. Salt fingers in a steady shear flow. *Geophys. Fluid Dyn.* 6: 1–27

Linden, P. F. 1974b. A note on the transport across a diffusive interface. *Deep-Sea Res.* 21: 283–87

Linden, P. F., Shirtcliffe, T. G. F. 1978. The diffusive interface in double-diffusive convection. *J. Fluid Mech.* 87: 417–32

Lueck, R. 1987. Microstructure measurements in a thermohaline staircase. *Deep-Sea Res.* 34(10): 1677–88

Mack, S. A. 1985. Two-dimensional measurements of ocean microstructure: the role of double diffusion. *J. Phys. Ocean.* 15: 1581–1604

Mack, S. A. 1989. Towed chain measurement of ocean microstructure. *J. Phys. Ocean.* 19(8): 1108–29

Mack, S. A., Schoeberlein, H. C. 1993. Discriminating salt fingering from turbulence-induced microstructure: analysis of towed temperature-conductivity chain data. *J. Phys. Ocean.* 23(9): 2073–106

Magnell, B. 1976. Salt fingers observed in the Mediterranean outflow region (34°N, 11°W) using a towed sensor. *J. Phys. Ocean.* 6: 511–23

Marmorino, G. O. 1987. Observations of small-scale mixing processes in the seasonal thermocline. Part I: Salt fingering. *J. Phys. Ocean.* 17: 1339–47

Marmorino, G. O. 1990. "Turbulent mixing" in a salt-finger staircase. *J. Geophys. Res.* 95: 12,983–94

Marmorino, G. O. 1991. Intrusions and diffusive interfaces in a salt fingering staircase. *Deep-Sea Res.* 38: 1431–54

Marmorino, G. O., Brown, W. K., Morris, W. D. 1987. Two-dimensional temperature structure in the C-SALT thermohaline staircase. *Deep-Sea Res.* 23(10): 1667–75

Marmorino, G. O., Caldwell, D. R. 1976. Heat and salt transport through a diffusive thermohaline interface. *Deep-Sea Res.* 23: 59–67

Marmorino, G. O., Greenewalt, D. 1988. Inferring the nature of microstructure signals. *J. Geophys. Res.* 93: 1219–25

Mazeika, P. A. 1974. Subsurface mixed layers in the northwest tropical Atlantic. *J. Phys. Ocean.* 4: 446–53

McDougall, T. J. 1981. Double-diffusive convection with a nonlinear equation of state. II. Laboratory experiments and their interpretation. *Prog. Ocean.* 10: 91–121

McDougall, T. J. 1985a. Double-diffusive interleaving. Part I: Linear stability analysis. *J. Phys. Ocean.* 15: 1532–41

McDougall, T. J. 1985b. Double-diffusive interleaving. Part II: Finite amplitude steady state interleaving. *J. Phys. Ocean.* 15: 1542–56

McDougall, T. J. 1986. Oceanic intrusions: some limitations of the Ruddick and Turner (1979) mechanism. *Deep-Sea Res.* 33: 1653–64

McDougall, T. J. 1988. Some implications of ocean mixing for ocean modelling. See Nihoul & Jamart 1988, pp. 21–36

McDougall, T. J. 1991. Interfacial advection in the thermohaline staircase east of Barbados. *Deep-Sea Res.* 38(3): 367–70

McDougall, T. J., Ruddick, B. R. 1992. The use of ocean microstructure to quantify both turbulent mixing and salt fingering. *Deep-Sea Res.* 39: 1931–52

McIntyre, M. E. 1970. Diffusive destabilization of the baroclinic vortex. *Geophys. Fluid Dyn.* 1: 19–57

Middleton, J. H., Foster, T. D. 1980. Fine-structure measurements in a temperature-compensated halocline. *J. Geophys. Res.* 85: 1107–22

Molcard, R., Tait, R. I. 1977. The steady state of the step structure in the Tyrrhenian Sea. In *A Voyage of Discovery*, George Deacon 70th Anniv. Vol., ed. M. V. Angel. *Deep-Sea Res.* 38: 221–33 (Suppl.)

Molcard, R., Williams, A. J. 1975. Deep-stepped structure in the Tyrrhenian Sea. *Mem. Soc. R. Sci. Liege* 6: 191–210

Muench, R. D., Fernando, H. J. S., Stegan, G. R. 1990. Temperature and salinity staircases in the northwestern Weddell Sea. *J. Phys. Ocean.* 20: 295–306

Neal, V. T., Neshyba, S., Denner, W. 1969. Thermal stratification in the Arctic Ocean. *Science* 166: 373–74

Neshyba, S., Neal, V. T., Denner, W. 1971. Temperature and conductivity measurements under Ice Island T-3. *J. Geophys. Res.* 76: 8107–20

Nield, D. A. 1967. The thermohaline Rayleigh-Jeffries problem. *J. Fluid Mech.* 29: 545–58

Nihoul, J., Jamart, B., eds. 1988. *Small-Scale Turbulence and Mixing in the Ocean*, Elsevier Oceanogr. Ser. Vol. 46. New York: Elsevier

Niino, H. 1986. A linear stability theory of double-diffusive horizontal intrusions in a temperature-salinity front. *J. Fluid Mech.* 171: 71–100

Oakey, N. S. 1985. Statistics of mixing parameters in the upper ocean during JASIN phase 2. *J. Phys. Ocean.* 10: 83–89

Osborn, T. R. 1980. Estimates of the local rate of vertical diffusion from dissipation measurements. *J. Phys. Ocean.* 10: 83–89

Osborn, T. R. 1991. Observations of the Salt Fountain. *Atmos.-Oceans* 29(2): 340–56

Osborn, T., Cox, C. S. 1972. Oceanic fine structure. *Geophys. Fluid Dyn.* 3: 321–45

Padman, L. 1991. Diffusive-convective staircases in the Arctic Ocean. In *Double-diffusion in Oceanography: Proc. of a Meeting.* Woods Hole Oceanogr. Inst. Tech. Rep. WHOI-91-200: 161–72

Padman, L., Dillon, T. M. 1987. Vertical fluxes through the Beaufort sea thermohaline staircase. *J. Geophys. Res.* 92: 10,799–806

Padman, L., Dillon, T. J. 1988. On the horizontal extent of the Canada Basin thermohaline steps. *J. Phys. Ocean.* 18: 1458–62

Padman, L., Dillon, T. M. 1989. Thermal microstructure and internal waves in the Canada Basin diffusive staircase. *Deep-Sea Res.* 36: 531–42

Posmentier, E. S., Hibbard, C. B. 1982. The role of tilt in double-diffusive interleaving. *J. Geophys. Res.* 87: 518–24

Posmentier, E. S., Kirwan, A. D. 1985. The role of double-diffusive interleaving in mesoscale dynamics: an hypothesis. *J. Mar. Res.* 43: 541–52

Proctor, M. R. E., Holyer, J. Y. 1986. Planform selection in salt fingers. *J. Fluid Mech.* 168: 241–53

Rayleigh, Lord 1883. Investigation of the character of the equilibrium of an incompressible heavy fluid of variable density. *Proc. London Math. Soc.* 14: 170–77

Richards, K. J. 1991. Double-diffusive interleaving at the equator. *J. Phys. Ocean.* 21(7): 933–38

Ruddick, B. 1984. The life of a thermohaline intrusion. *J. Mar. Res.* 42: 831–52

Ruddick, B. 1992. Intrusive mixing in a Mediterranean salt lens: intrusion slopes and dynamical mechanisms. *J. Phys. Ocean.* 22: 1274–85

Ruddick, B. R., Turner, J. S. 1979. The vertical length scale of double-diffusive intrusions. *Deep-Sea Res.* 26A: 903–13

Schmitt, R. W. 1979a. The growth rate of super-critical salt fingers. *Deep-Sea Res.* 26A: 23–40

Schmitt, R. W. 1979b. Flux measurements on salt fingers at an interface. *J. Mar. Res.* 37: 419–436

Schmitt, R. W. 1981. Form of the temperature-salinity relationship in the Central Water: evidence for double-diffusive mixing. *J. Phys. Ocean.* 11: 1015–26

Schmitt, R. W. 1983. The characteristics of

salt fingers in a variety of fluid systems, including stellar interiors, liquid metals, oceans and magmas. *Phys. Fluids* 26: 2373–77

Schmitt, R. W. 1988. Mixing in a thermohaline staircase. See Nihoul & Jamart 1988, pp. 435–52

Schmitt, R. W. 1990. On the density ratio balance in the Central Water. *J. Phys. Ocean.* 20(6): 900–6

Schmitt, R. W. 1993. Triangular and asymmetric salt fingers. *J. Phys. Ocean.* In press

Schmitt, R. W., Evans, D. L. 1978. An estimate of the vertical mixing due to salt fingers based on observations in the North Atlantic Central Water. *J. Geophys. Res.* 83: 2913–19

Schmitt, R. W., Georgi, D. T. 1982. Fine-structure and microstructure in the North Atlantic Current. *J. Mar. Res.* 40: 659–705 (Suppl.)

Schmitt, R. W., Lueck, R. G., Joyce, T. M. 1986. Fine- and micro-structure at the edge of a warm core ring. *Deep-Sea Res.* 33: 1665–89

Schmitt, R. W., Perkins, H. Boyd, J. D., Stalcup, M. C. 1987. C-SALT: an investigation of the thermohaline staircase in the western tropical North Atlantic. *Deep-Sea Res.* 34(10): 1697–1704

Schmitt, R. W., Toole, J. M. Koehler, R. L., Mellinger, E. C., Doherty, K. W. 1988. The development of a fine- and microstructure profiler. *J. Atmos. Ocean. Tech.* 5(4): 484–500

Shen, C. Y. 1989. The evolution of the double-diffusive instability: salt fingers. *Phys. Fluids* A1(5): 829–44

Shirtcliffe, T. G. L 1967. Thermosolutal convection: observation of an overstable mode. *Nature* 213: 489–90

Shirtcliffe, T. G. L., Turner, J. S. 1970. Observations of the cell structure of salt fingers. *J. Fluid Mech.* 41: 707–19

Stern, M. E. 1960. The "salt fountain" and thermohaline convection. *Tellus* 12: 172–75

Stern, M. E. 1967. Lateral mixing of water masses. *Deep-Sea Res.* 14: 747–53

Stern, M. E. 1975. *Ocean Circulation Physics.* New York: Academic

Stern, M. E., Turner, J. S. 1969. Salt fingers and convecting layers. *Deep-Sea Res.* 16: 497–511

Stommel, H. M. 1993. A conjectural regulating mechanism for determining the thermohaline structure of the oceanic mixed layer. *J. Phys. Ocean.* 23(1): 142–48

Stommel, H. M., Arons, A. B., Blanchard, D. 1956. An oceanographic curiosity: the perpetual salt fountain. *Deep-Sea Res.* 3: 152–53

Swallow, J. C., Crease, J. 1965. Hot salty water at the bottom of the Red Sea. *Nature* 205: 165–66

Tait, R. I., Howe, M. R. 1968. Some observation of thermo-haline stratification in the deep ocean. *Deep-Sea Res.* 15: 275–80

Taylor, J. 1991. Laboratory experiments on the formation of salt fingers after the decay of turbulence. *J. Geophys. Res.* 96(C7): 12,497–510

Toole, J. M. 1981a. Intrusion characteristics in the Antarctic Polar Front. *J. Phys. Ocean.* 11: 780 93

Toole, J. M. 1981b. Anomalous characteristics of equatorial thermohaline finestructure. *J. Phys. Ocean.* 11(6): 871–76

Toole, J. M., Georgi, D. T. 1981. On the dynamics and effects of double-diffusively driven intrusions. *Prog. Oceanogr.* 10: 123–45

Turner, J. S. 1965. The coupled turbulent transports of salt and heat across a sharp density interface. *Int. J. Heat Mass Transport* 8: 759–67

Turner, J. S. 1968. The behaviour of a stable salinity gradient heated from below. *J. Fluid Mech.* 33: 183–200

Turner, J. S. 1978. Double-diffusive intrusions into a density gradient. *J. Geophys. Res.* 83: 2887–2901

Turner, J. S. 1985. Multicomponent convection. *Annu. Rev. Fluid Mech.* 17: 11–44

Turner, J. S., Stommel, H. 1964. A new case of convection in the presence of combined vertical salinity and temperature gradients. *Proc. Natl. Acad. Sci.* 52: 49–53

Veronis, G. 1965. On finite amplitude instability in thermohaline convection. *J. Mar. Res.* 23: 1–17

Veronis, G. 1968. Effect of a stabilizing gradient of solute on thermal convection. *J. Fluid Mech.* 34: 315–36

Walin, G. 1964. Note on the stability of water stratified by both salt and heat. *Tellus* 16: 389–93

Williams, A. J. 1974. Salt fingers observed in the Mediterranean outflow. *Science* 185: 941–43

Williams, A. J. 1975. Images of ocean microstructure. *Deep-Sea Res.* 22: 811–29

Williams, A. J. 1981. The role of double-diffusion in a Gulf Stream frontal intrusion. *J. Geophys. Res.* 86: 1917–28

Yoshida, J., Nagashime, H., Niino, H. 1989. The behavior of double-diffusive intrusions in a rotating system. *J. Geophys. Res.* 94: 4923–37

Young, W. R., Rhines, P. B., Garrett, C. J. R. 1982. Shear flow dispersion, internal waves and horizontal mixing in the ocean. *J. Phys. Ocean.* 12: 515–27

Annu. Rev. Fluid Mech. 1994. 26 : 287–319

THE PHYSICS OF SUPERSONIC TURBULENT BOUNDARY LAYERS[1]

Eric F. Spina

Mechanical, Aerospace, and Manufacturing Engineering Department, Syracuse University, Syracuse, New York 13244

Alexander J. Smits

Mechanical and Aerospace Engineering Department, Princeton University, Princeton, New Jersey 08544

Stephen K. Robinson

Experimental Flow Physics Branch, NASA-Langley Research Center, Hampton, Virginia 23665

KEY WORDS: compressible flow, shear layers, distorted flow, hypersonics, flow structure

INTRODUCTION

When a vehicle travels at Mach numbers greater than one, a significant temperature gradient develops across the boundary layer due to the high levels of viscous dissipation near the wall. In fact, the static-temperature variation can be very large even in an adiabatic flow, resulting in a low-density, high-viscosity region near the wall. In turn, this leads to a skewed mass-flux profile, a thicker boundary layer, and a region in which viscous effects are somewhat more important than at an equivalent Reynolds number in subsonic flow. Intuitively, one would expect to see significant dynamical differences between subsonic and supersonic boundary layers. However, many of these differences can be explained by simply accounting for the fluid-property variations that accompany the temperature

---

[1] The US Government has the right to retain a nonexclusive, royalty-free license in and to any copyright covering this paper.

287

variation, as would be the case in a heated incompressible boundary layer. This suggests a rather passive role for the density differences in these flows, most clearly expressed by Morkovin's hypothesis (1962): The dynamics of a compressible boundary layer follow the incompressible pattern closely, as long as the fluctuating Mach number, M', remains small. Here M' is the r.m.s. perturbation of the instantaneous Mach number from its mean value, taking into account the fact that both the velocity and the sound speed vary with time. If the fluctuating Mach number approaches unity at any point, we would expect direct compressibility effects such as local "shocklets" and pressure fluctuations to become important. If we take $M' = 0.3$ as the point where compressibility effects become important for the turbulence behavior, we find that for zero-pressure-gradient adiabatic boundary layers at moderately high Reynolds numbers, this point will be reached with a freestream Mach number of about 4 or 5 (see Figure 1).

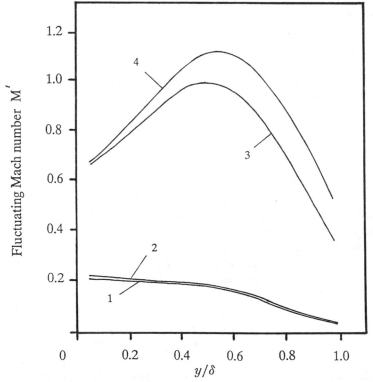

Figure 1   Fluctuating Mach number distributions. Flow 1: $M_e = 2.32$, $Re_\theta = 4700$, adiabatic wall (Elena & Lacharme 1988); Flow 2: $M_e = 2.87$, $Re_\theta = 80,000$, adiabatic wall (Spina & Smits 1987); Flow 3: $M_e = 7.2$, $Re_\theta = 7100$, $T_w/T_r = 0.2$ (Owen & Horstman 1972); Flow 4: $M_e = 9.4$, $Re_\theta = 40,000$, $T_w/T_r = 0.4$ (Laderman & Demetriades 1974).

Recently, some measurements in moderately supersonic boundary layers ($M_\infty < 5$) have indicated subtle differences in the instantaneous behavior of certain quantities and parameters as compared to incompressible flow. In particular, differences in length scales, the intermittency function, and the structure of the large-scale shear-stress motions may indicate that some alteration of turbulence dynamics occurs at a lower Mach number than previously believed.

More importantly, in the fifteen years since Bradshaw's review of compressible turbulent shear layers (1977), a significant number of new experiments have been performed to study the response of turbulent boundary layers in supersonic flow to perturbations such as the imposition of a pressure gradient, the development of longitudinal streamline curvature, and the interaction with a shock wave. These perturbations introduce phenomena that do not have an equivalent subsonic counterpart, and that cannot be explained in terms of fluid-property variations. For example, since the density is not just a function of pressure, vorticity can be produced through baroclinic torques. In addition, longitudinal pressure gradients lead to the compression or dilatation of vortex tubes, enhancing or reducing turbulent velocity and pressure fluctuations. When shock waves are present, separation occurs if the shock is strong enough (a phenomenon that can be understood from subsonic experience), but even in the absence of separation there exists a strong coupling between the shock and the turbulence, and the resultant distortions of the shock sheet have been widely observed. Understanding the shock motion and the resultant unsteady heat transfer and pressure loading is of great importance in many aerodynamic flows. Since the shock motion seems closely dependent on the incoming turbulence field, there is a clear need to understand the nature of the organized motions in the incoming boundary layer, particularly the large scales.

It is the aim of this article to review recent studies on the behavior of supersonic turbulent boundary layers. Particular attention is directed toward complex flows where pressure gradients and streamline curvature effects are important. The intent is to assess how far our understanding had advanced, and to provide some discussion of future research needs in this area. This effort was helped enormously by the availability of the catalogs of supersonic turbulence data compiled by Fernholz & Finley (1976, 1980, 1981) and Fernholz et al (1989), and the report by Settles & Dodson (1991). We restrict our attention to flows in two-dimensional geometries, emphasizing experiments that shed light on the turbulence evolution. We do not review the large body of work on the distortion of supersonic turbulent boundary layers by shock waves. This represents a major topic in itself, and the reader is referred to Adamson & Messiter

(1980) and Green (1970) for reviews of two-dimensional shock-wave/boundary-layer interactions; Settles & Dolling (1992), who consider the mean-flow behavior of swept interactions; and Dolling (1993), who discusses measurements of fluctuating pressure loads in these flows.

EQUATIONS OF COMPRESSIBLE TURBULENT BOUNDARY LAYERS

The dimensionless form of the energy equation reveals the source of the differences between incompressible and compressible boundary layers, whether laminar or turbulent:

$$\rho c_{\mathrm{p}} \frac{DT}{Dt} = (\gamma - 1) M_{\infty}^2 \frac{Dp}{Dt} + \frac{1}{Re\,Pr} \nabla \cdot (k \nabla T) + \frac{(\gamma - 1) M_{\infty}^2}{Re} \Phi,$$

where Re is the Reynolds number, Pr is the Prandtl number, ρ is the density, c_{p} is the specific heat at constant pressure, k is the thermal conductivity, T is the temperature, and γ is the ratio of specific heats. The freestream Mach number appears as a parameter for both the dissipation function, Φ, and the rate of work due to compression, Dp/Dt. As the Mach number increases, the positive-definite dissipation function serves as a dominant source of internal energy where the strain rates are high; that is, near the wall. The resulting static-temperature profile skews the wall-normal distribution of many mean quantities, and it has important implications for turbulence parameters. The effect of the compression work is also increased at high Mach number, and this has implications for supersonic flows subject to perturbations such as shock waves and curvature.

Detailed derivations of the equations for compressible turbulent boundary layers have been provided in kinematic variables by van Driest (1951), Schubauer & Tchen (1959), and Fernholz & Finley (1980). While it is well known that the inclusion of density as an instantaneous variable adds terms other than $\bar{\rho} \overline{u'v'}$ to the Reynolds-averaged boundary layer equations, the interpretation of these terms and their significance is not universally agreed upon. One of the reasons is that these terms do not appear in the mass-averaged (Favre-averaged) equations, as shown by, for example, Morkovin (1962), Favre (1965), and Rubesin & Rose (1973). A critical review of the equations of compressible turbulent flow and a discussion of the relative merits of the mass-averaged form is given by Lele (1994).

The Continuity Equation

The boundary-layer approximation applied to the Reynolds-averaged, stationary, two-dimensional continuity equation for compressible flow results in

$$\frac{\partial}{\partial x}(\bar{\rho}U) + \frac{\partial}{\partial y}(\bar{\rho}V) + \frac{\partial}{\partial y}(\overline{\rho'v'}) = 0.$$

The additional term in this equation, $\partial/\partial y(\overline{\rho'v'})$, acts as an apparent source/sink to the mean flow (Schubauer & Tchen 1959). A simple mixing-length argument indicates that $\overline{\rho'v'}$ is negative. Considering that the absolute magnitude of both ρ' and v' increases with y near the wall and decreases with increasing y in the outer part of the boundary layer, then this additional term acts as a mass-flux source in the inner layer and as a sink in the outer region of the boundary layer. Such an interpretation is consistent with the outward displacement of streamlines, but the presence of a source term in the continuity equation may indicate that the physics of the flowfield are not properly represented.

An alternative approach uses "Favre averaging," where the instantaneous variable is decomposed into the sum of a mass-weighted average, $\tilde{\alpha}$, and a fluctuation, α''. The use of mass-averaged variables leaves the continuity equation devoid of turbulent mass transport terms:

$$\frac{\partial}{\partial x}(\bar{\rho}\tilde{u}) + \frac{\partial}{\partial y}(\bar{\rho}\tilde{v}) = 0.$$

The x-Momentum Equation

If the continuity equation is multiplied by the streamwise velocity and added to the boundary-layer approximation of the x-momentum equation, and the resulting equation is Reynolds-averaged, the result is

$$\frac{\partial}{\partial x}(\bar{\rho}U^2) + \frac{\partial}{\partial y}(\bar{\rho}UV) = \frac{-d\bar{p}}{dx}$$

$$+ \frac{\partial}{\partial y}\left(\mu\frac{\partial U}{\partial y} - \bar{\rho}\overline{u'v'} - U\overline{\rho'v'} - V\overline{\rho'u'} - \overline{\rho'u'v'}\right). \quad (1)$$

Reynolds averaging is a process in which instantaneous variables are replaced by the sum of a mean and a fluctuating variable, and the entire equation is time averaged. Equation 1 is the most general form of the compressible boundary-layer equation. The usual practice is to neglect the triple-product term as being one order of magnitude smaller than the other terms, and to neglect $V\overline{\rho'u'}$ since it is smaller than $U\overline{\rho'v'}$ ($\overline{\rho'u'}$ and $\overline{\rho'v'}$ are assumed to be the same order and $V \ll U$). The resulting equation is

$$\frac{\partial}{\partial x}(\bar{\rho}U^2) + \frac{\partial}{\partial y}(\bar{\rho}UV) = \frac{-d\bar{p}}{dx} + \frac{\partial}{\partial y}\left(\mu\frac{\partial U}{\partial y} - \bar{\rho}\overline{u'v'} - U\overline{\rho'v'}\right). \quad (2)$$

Alternatively, the boundary-layer form of the compressible x-momentum equation can be written

$$\overline{\rho u}\frac{\partial}{\partial x}U+\overline{\rho v}\frac{\partial}{\partial y}U = \frac{-d\bar{p}}{dx}+\frac{\partial}{\partial y}\left(\mu\frac{\partial U}{\partial y}-\overline{\rho u'v'}\right),\tag{3}$$

where $\overline{\rho u}=\bar{\rho}U+\overline{\rho'u'}$ and $\overline{\rho v}=\bar{\rho}V+\overline{\rho'v'}$, and $\overline{\rho'u'}$ can usually be neglected. When the Favre-averaged form of the x-momentum equation is considered,

$$\frac{\partial}{\partial x}(\bar{\rho}\tilde{u}^2)+\frac{\partial}{\partial y}(\bar{\rho}\tilde{u}\tilde{v}) = \frac{-d\bar{p}}{dx}+\frac{\partial}{\partial y}(\bar{\tau}_{yx}-\overline{\rho u''v''}),\tag{4}$$

it is clear that three different forms of the equation exist, and some physical insight to the differences is necessary.

In Equation 2 the traditional Reynolds stress and another "apparent" stress, $-U\overline{\rho'v'}$, comprise the turbulent shear stress. Both correlations are negative (as evident from a mixing-length argument), and thus $U\overline{\rho'v'}$ enhances the incompressible Reynolds shear stress. Assuming small pressure fluctuations and using the Strong Reynolds Analogy (SRA) (Morkovin 1962), it is a simple matter to express the ratio of $U\overline{\rho'v'}$ to $\bar{\rho}\overline{u'v'}$ as $(\gamma-1)M^2$ (e.g. Spina et al 1991a). Of course, this expression is subject to the inaccuracies inherent in the SRA (see below), but it is a good approximation to at least $M=5$, and provides an order-of-magnitude comparison even at higher Mach numbers. This relation indicates how quickly $U\overline{\rho'v'}$ becomes important in the boundary layer. For a Mach 3 adiabatic-wall boundary layer with $Re_\theta = 80,000$, $(\gamma-1)M^2$ rises to 1.0 at approximately 0.05δ ($\sim 500y^+$), and asymptotes to 3.5 at the boundary-layer edge (Spina 1988). Since the value of the Mach number is small across much of the constant-stress layer, Schubauer & Tchen (1959) neglected the "second-order term" when developing a skin-friction theory, but this should not be considered a general result.

An accurate classification of $U\overline{\rho'v'}$ may be aided by a physical comparison to the traditional Reynolds shear stress. The term $\bar{\rho}\overline{u'v'}$ can be interpreted as the mean rate of transfer of turbulent x-momentum across a plane normal to the y-direction. In a similar vein, $U\overline{\rho'v'}$ can be interpreted as the mean rate of transfer of turbulent mass flux across the same plane. It appears, then, that $U\overline{\rho'v'}$ is not a "true" Reynolds stress, but a fictitious one because it does not involve the transport of turbulent momentum. An additional clue to the nature of $U\overline{\rho'v'}$ can be found by examining the turbulent kinetic energy (TKE) equation for a compressible boundary layer. This is much more complex than the incompressible TKE equation, with eight production terms, including one due to the Reynolds shear stress, $-\overline{\rho u'v'}(\partial U/\partial y)$, and one due to the "fictitious" stress, $-U\overline{\rho'v'}(\partial V/\partial x)$. A comparison between these two terms indicates that the production of turbulent kinetic energy due to the Reynolds shear stress is two orders of magnitude greater than that due to the term in question (in fact, there are three other terms that are an order of magnitude larger than $-U\overline{\rho'v'}(\partial V/\partial x)$. This indicates that $U\overline{\rho'v'}$ is less important than the other terms in determining the energy flow in a compressible boundary layer because it interacts with a considerably smaller mean strain.

If the convective terms are written as the product of the average instantaneous mass flux and a strain (Equation 3), the only additional term (in addition to those found in laminar flow) is the traditional Reynolds stress, $\overline{\rho u'v'}$. This form of the equation was advocated by Morkovin (1962) to isolate the turbulent momentum transport, and the new part of the convective terms represents the fact that there is no mean mass transfer between mean streamlines. Since $U\overline{\rho'v'}$ is a turbulent mass transport term, it is not surprising that this form of the equation is free from this term, and the interpretation of the equation is physically and intuitively attractive.

As for the x-momentum equation in Favre-averaged variables (Equation 4), the major drawback to such a representation is that $\bar{\tau}_{yx}$ is more complex than for incompressible boundary layers (Rubesin & Rose 1973). Expressing the instantaneous stress tensor in mass-weighted variables, expanding, and time-averaging results in:

$$\bar{\tau}_{ij} = \bar{\mu}\tilde{S}_{ij} + \bar{\mu}\overline{S''_{ij}},$$

where $S_{ij} = [(u_{i,j} + u_{j,i}) - \tfrac{2}{3}\delta_{ij}u_{k,k}]$. This expression contains additional terms that are not amenable to a simple physical interpretation, but the similarity of the Favre-averaged representation of the compressible momentum equation to that of the incompressible equation makes its use compelling, especially in computations.

The Energy Equation and the Strong Reynolds Analogy

The mean energy equation was developed in terms of the stagnation enthalpy by Young (1951) (see Howarth 1953) and Gaviglio (1987) in the forms corresponding to the Reynolds-averaged and Favre-averaged variables, respectively. In Reynolds-averaged variables, the boundary-layer approximation for the equation is:

$$\overline{\rho u}\frac{\partial \bar{H}}{\partial x} + \overline{\rho v}\frac{\partial \bar{H}}{\partial y} = \frac{\partial}{\partial y}\left[\frac{k}{c_p}\frac{\partial \bar{H}}{\partial y} + \mu\left(1 - \frac{1}{Pr}\right)\frac{\partial}{\partial y}\left(\frac{U^2}{2}\right) - \overline{\rho v'H'}\right], \tag{5}$$

where, neglecting higher-order terms, $\bar{H} = \bar{h} + \tfrac{1}{2}U^2$, and $H' = h' + Uu'$. As in the development of the mean x-momentum equation (Equation 3), there are no additional terms beyond those found in incompressible flow, although the convective terms are slightly altered, as noted by Morkovin (1962).

A useful relation for the reduction of experimental data and the comparison of compressible to incompressible results is the Strong Reynolds Analogy (first identified as such by Morkovin 1962, but primarily due to Young 1951). This analogy, leading to simplified solutions of the energy equation, is based upon the similarity between Equations 3 and 5 when $Pr = 1$ (or if molecular effects are negligible compared to turbulent processes) and the similarity of the boundary conditions for $\bar{T}_0$ and U, and T_0' and u'. For zero-pressure-gradient flow with heat transfer, the equations admit the solution

$$c_p(\bar{T}_0 - T_w) = Pr_w \frac{\bar{q}_w}{\bar{\tau}_w} U, \tag{6a}$$

$$c_p T_0' = Pr_w \frac{\bar{q}_w}{\bar{\tau}_w} u', \tag{6b}$$

where the heat transfer rate and shear stress at the wall enter through the boundary conditions. For adiabatic flows, it follows that

$$T_0' = 0, \tag{7a}$$

$$\frac{T'}{\bar{T}} = -(\gamma - 1)M^2 \frac{u'}{U} \quad \text{(to first-order)}, \tag{7b}$$

$$R_{uT} = -1, \tag{7c}$$

where R_{uT} is the correlation between u and T. The solution given by $\bar{T}_0 = T_w$ and Equation 7a satisfies the energy equation independently, and therefore may be applied for any pressure gradient (Gaviglio 1987).

Gaviglio notes that these relations (Equations 6a–7c) are so strict (i.e. they apply in an instantaneous sense) that they cannot be expected to hold exactly. Morkovin (1962) gives a "milder" form of the SRA that relates the r.m.s. of the static temperature fluctuations to that of the velocity fluctuations (also see Spina et al 1991a). Morkovin (1962) and Gaviglio (1987) tested the time-averaged form of the SRA and found that R_{uT} is not -1.0 but is closer to -0.8 or -0.9. Still, the high anticorrelation indicates that large-scale eddies moving away from the wall in a supersonic flow contain warmer, lower-speed fluid than the average values found at that distance from the wall. As for the instantaneous form of the SRA (Equation 7a), Morkovin & Phinney (1958), Kistler (1959), Dussauge & Gaviglio (1987), and Smith & Smits (1993) have shown that T_0' is not negligible, but that the results derived from such an assumption still represent very good approximations. The instantaneous form of the SRA has been validated to a freestream Mach number of 3 (Smith & Smits 1993), but the only limit to its first-order approximation may be the increasing importance of low-Reynolds-number effects near the wall at higher hypersonic Mach numbers (Morkovin 1962).

MEAN FLOW ENVIRONMENT

The increased influence of viscous dissipation in high-Mach-number flows is illustrated by the evolution of the mass-flux profile with Mach number. The elevated static temperature at the wall creates a low-density region that shifts the majority of the mass flux toward the outer part of the boundary layer, and the profile becomes more skewed as the freestream Mach number becomes larger. In addition to affecting the density profile, the mean static temperature variation also creates fluid-property gradients across the boundary layer. Typical wall-to-freestream ratios of some properties are provided in Table 1 for three different Mach numbers. Since the density of gases decreases and the viscosity increases with temperature, the ratio of wall to freestream dynamic viscosity can be dramatic. Of course, when the wall is heated or when the flow is perturbed so that normal pressure gradients exist, the gradients of ρ, μ, and k may be even more severe. As a result of the fluid-property gradients, the "local" Reynolds number, Re_y (a concept introduced by Morkovin 1962), is considerably smaller at a given height in the boundary layer than in an equivalent-Reynolds-number incompressible flow. The low Reynolds number effects usually found only very near the wall will encompass a larger portion of the boundary layer as M_∞ increases, and the influence of the viscous sublayer will increase.

Stagnation-Temperature Distribution

The stagnation-temperature profile must be known to calculate the velocity and to predict the turbulent Prandtl number. Measurements and theory

Table 1   The ratio of fluid properties across three different adiabatic boundary layers

M_∞	T_w/T_∞	ρ_w/ρ_∞	μ_w/μ_∞	v_w/v_∞
2.9 (air)[a]	3	0.33	1.4	4.2
4.5 (air)[b]	5	0.2	2.9	14
10.3 (He)[c]	33	0.03	9.6	320

[a] Spina & Smits 1987.
[b] Mabey et al 1974.
[c] Watson et al 1973.

often seem to conflict, however, and a truly representative stagnation-temperature profile is difficult to define, particularly at high Mach number. The measurement difficulty stems from the compromise that must be made between spatial resolution and accuracy when selecting a stagnation-temperature probe (see, for example, Fernholz & Finley 1980). Since approximately one half of the decrease in $\bar{T}_0$ to the wall-recovery value occurs in the inner layer of near-unity Prandtl-number gases (Morkovin 1962), this compromise limits the accuracy of the data.

As for theoretical stagnation-temperature distributions, Fernholz & Finley (1980) present and discuss many of the energy equation solutions commonly applied to supersonic turbulent boundary layers. They note that many of the relations are applied beyond their range of validity when used to benchmark experimental data. The two most widely discussed stagnation-temperature distributions in the literature are the "linear" and "quadratic" solutions. It has been commonly assumed that $(T_0 - T_w)/(T_{0_e} - T_w) \equiv \Theta = U/U_e$ is the proper distribution for flat-plate flows, while $\Theta = (U/U_e)^2$ is the appropriate tunnel/nozzle wall solution. It has been claimed that the quadratic nature of the measurements along tunnel walls is due to the upstream history of the flow (significant dT/dx and dp/dx) and the resultant local nonequilibrium. While Feller (1973), Bushnell et al (1969), and Beckwith (1970) offer convincing arguments for flow-history effects, there is little experimental evidence that the linear Crocco profile is the equilibrium stagnation-temperature distribution in supersonic, turbulent boundary layers.

The classic (linear) Crocco solution, $\Theta = U/U_e$, is derived from the energy and momentum equations for laminar flow with $Pr = 1$, zero pressure gradient, and an isothermal wall. The Crocco solution is extended to turbulent flows under the same conditions with the additional assumption of unity turbulent Prandtl number (Pr_t). However, it has been shown that Pr_t is less than 1.0 across the outer layer for both near-adiabatic walls (Meier & Rotta 1971) and cold walls (Owen et al 1975). Fernholz & Finley (1980) show that the origins of a quadratic profile for turbulent flow lie in a solution attributed to Walz by Voisinet & Lee (1973):

$$\Theta = \beta \frac{U}{U_e} + (1 - \beta)\left(\frac{U}{U_e}\right)^2,$$

where $\beta = (T_{aw} - T_w)/(T_{0_e} - T_w)$. The assumptions inherent in this solution are zero pressure gradient, isothermal wall, and "mixed" Prandtl number, $Pr_M \equiv c_p(\mu + \mu_t)/(k + k_t)$, constant and between 0.7 and 1.0. The linear profile therefore holds only for $\beta = 1$, that is, $T_{0_e} = T_{aw}$, and the purely quadratic profile holds only for a zero-pressure-gradient flow, with constrained Pr_M, and an adiabatic wall ($\beta = 0$). The range of validity of the quadratic relation is often extended improperly to flows with pressure gradients because of the similarity of the equation to one that is valid for laminar flow with pressure gradients. Perhaps due to the relaxed constraint on the Prandtl number (as compared to the linear solution), much of the stagnation-temperature data appears to be characterized by a quadratic trend (Bushnell et al 1969, Bertram & Neal 1965, Wallace 1969, Hopkins & Keener 1972, etc).

A critical shortcoming is the scarcity of near-wall T_0 measurements, which are critical for determination of the wall heat-transfer rate. The lack of data makes it impossible to determine whether these temperature-velocity relations, or even that provided by Bradshaw (1977) to represent the inner layer, accurately describe the near-wall behavior of the stagnation temperature. For flows with non-isothermal walls and significant pressure gradients the situation is much worse, however, as no theoretical temperature-velocity relations exist for these conditions. Much of the confusion surrounding stagnation-temperature distributions is due to comparison between data taken under these conditions and theoretical relations that are applied beyond their range of validity.

Mean-Velocity Scaling

When the mean velocity in a supersonic boundary layer is plotted as U/U_e vs y/δ, the profile appears qualitatively similar to that of an incompressible flow. When the velocity is replotted in classic inner- or outer-layer coordinates, however, the velocity does not follow the familiar incompressible scaling laws for these regions. But a modified scaling that accounts for the fluid-property variations correlates much of the existing compressible mean-velocity data with the "universal" incompressible distribution. This velocity scaling was first employed in the viscous sublayer and the logarithmic region by van Driest (1951). It was later extended to the wake region and to velocity-defect scaling by Maise & McDonald (1968), and to Coles' universal wall-wake scaling by Mathews et al (1970). The use of van Driest's "transform" has become an accepted standard. For example, a preliminary investigation of a supersonic boundary layer usually includes a transformation of the mean velocities to van Driest's generalized velocities. The data are then compared to the incompressible correlation to determine whether the boundary layer has transitioned and reached an equilibrium turbulent state. The application of the van Driest transform has become so common, and is usually so successful, that its physical significance and shortcomings are often overlooked.

Unlike Coles' (1962) compressible turbulent boundary-layer transformation, the van Driest (1951) approach bypasses the need for a point-to-point transformation to specify the flowfield. Rather, it is assumed that the laws of the wall/wake hold, and that the mixing-length hypothesis can be applied to determine the scaling. As part of the procedure, the velocity coordinate is stretched to account for ρ and μ distributions caused by viscous dissipation near the wall. The empirical validity of Morkovin's hypothesis (1962) offers some support for the concept behind the van Driest transform, simply by suggesting that multilayer scaling holds in compressible boundary layers; and despite the naive assumptions inherent to the mixing-length hypothesis, the underlying dimensional argument is sound as long as the length-scale distributions in supersonic boundary layers follow the same behavior as in subsonic flows. However, the absence of a firm foundation for this standard transformation indicates that there is a critical need for advances in the understanding of turbulence physics at supersonic speeds.

These warnings are not meant to imply that there is no place for the van Driest transformation in the analysis of boundary-layer data. In fact, experimental data taken over a wide Mach-number range, with various wall-heating conditions and modest pressure gradients, and transformed via van Driest show good agreement with incompressible data correlations (e.g. Kemp & Owen 1972, Laderman & Demetriades 1974, Owen et al 1975, Watson 1977). It is important to note what the limits of applicability appear to be, however. Other than strong pressure gradients, the primary constraint is imposed by the dependence of similarity on large values of the Reynolds number, implying universality and independence from upstream history. Fernholz & Finley (1980) observe that the low Re region that begins to dominate the inner layer at high Mach number may eventually cause the failure of the velocity scaling laws that the transformed data follow. Hopkins et al (1972) attribute the poor performance of van Driest at $M_e = 7.7$ to the low Reynolds number of the flow, $Re_\theta = 5000$. This can be compared to a successful application of van Driest at $M_e = 9.4$ and $Re_\theta \approx 37,000$ by Laderman & Demetriades. In their critical examination of mean-flow data, Fernholz & Finley (1980) offer a slightly more accurate variation of van Driest, basing the temperature distribution on a Prandtl number assumption ($0.7 \leq Pr_M \leq 1.0$) that is more realistic than van Driest's assumption of $Pr_t = Pr = 1$.

Skin Friction

Another result of the increased viscous dissipation in compressible boundary layers is a decrease in the skin-friction coefficient c_f with increasing Mach number (at fixed Re). The low density of the fluid near the wall indirectly results in a decrease in the slope of the nondimensionalized velocity profile relative to that for an equivalent-Reynolds-number incompressible boundary layer. Since density has a stronger dependence on temperature than viscosity does, the skin-friction coefficient decreases with Mach number (although the dimensional wall shear increases due to the increase in velocity). The general trends for hot and cold walls can be predicted from these considerations, with heated walls leading to lower c_f.

While the Howarth-Dorodnitsyn compressibility transformation provides an analytical solution for c_f in laminar boundary layers ($Pr = 1$), no such solution exists in turbulent compressible boundary layers. Instead, a variety of experimental correlations, transformations, and finite-difference solutions exist. Bradshaw (1977) critically reviewed the most widely-used skin-friction formulas and found that a variation of the formula known as van Driest II (1956) exhibited the best agreement with reliable zero-pressure-gradient data, with less than 10% error for $0.2 \leq T_w/T_{aw} \leq 1$.

Skin-friction measurements are more difficult to make and to interpret in perturbed supersonic flows (Fernholz & Finley 1980, Smith et al 1992). Floating-element gauges are susceptible to inaccuracies stemming from moments applied by streamwise pressure gradients. Reduction of Preston-tube data depends upon a variety of assumptions that are dubious in severely perturbed flows—including the existence of a logarithmic region that is larger than the Preston-tube diameter and the need for low turbulence levels (Hirt & Thomann 1986). Most schemes for reducing Preston-tube data rely on boundary-layer edge conditions (Hopkins & Keener 1966, for example), and this can introduce additional errors since the edge properties are often unrelated to the flow behavior near the wall in perturbed flows. The method of Bradshaw & Unsworth (1974) is preferred to reduce Preston-tube data since it uses wall variables exclusively, but it still holds exactly only for zero-pressure-gradient flows. Use of the Clauser method (1954) to determine the skin friction from the transformed velocity profile also relies upon the existence of a logarithmic region. In perturbed-flow applications with the Clauser method, the compressibility transformation of Carvin et al (1988) should be more reliable than that of van Driest because it does not have the additional requirement of a self-preserving boundary layer. In practice, for a wide variety of strongly perturbed flows, the differences between the Clauser-chart results obtained using the two transformations (and Preston-tube results) seem to be within about $\pm 15\%$ (Smith et al 1992). The laser interferometer skin friction meter (LISF) is a promising new technique that does not require assumptions about the character of the wall region to deduce the wall shear stress (Kim et al 1991), and can thus provide direct measurement of the skin friction in a perturbed flow.

MEAN TURBULENCE BEHAVIOR

Sandborn (1974) and Fernholz & Finley (1981) both critically reviewed turbulence measurements in supersonic boundary layers. While many data sets were acquired in the period between the two reviews, their conclusions were similar and they are still relevant. In particular, accurate, repeatable measurements of the Reynolds-stress tensor are needed over a wide Mach-number range. The most well-documented component is the longitudinal normal stress, which has been widely measured and properly scaled. But there have been so few systematic investigations of the effects of Reynolds number, wall heat transfer, and extra strain rates in supersonic flow that their influence on the turbulence field is not well known. The reason for the scarcity of measurements and their generally poor quality is simple: The measurement of turbulence quantities in supersonic boundary layers is exceedingly difficult, with the level of difficulty increasing with flow complexity and Mach number. There are significant measurement and data-reduction errors associated with every technique designed to measure fluctuating velocities in supersonic flow: thermal anemometry (see Smits & Dussauge 1989), laser-Doppler velocimetry (see Johnson 1989), and advanced laser-based techniques such as laser-induced fluorescence (see Logan 1987 and Miles & Nosenchuck 1989). What we know of the mean-turbulence field in zero-pressure-gradient flows will be reviewed here; that of perturbed flows will be discussed in a later section.

When the longitudinal velocity fluctuations are normalized by the shear velocity, $\overline{u'^2}/u_\tau^2$, there is a clear decrease in fluctuation level with increasing Mach number (see Kistler 1959 and Fernholz & Finley 1981). However, when the streamwise normal stress is normalized by the wall shear stress, the data exhibit some degree of similarity, particularly in the outer layer (see, for example, Dussauge & Gaviglio 1987). This formulation of the velocity fluctuations indicates the success of the scaling suggested by Morkovin (1962) to account for the mean density variation: $(\overline{u'^2}/u_\tau^2) \cdot [\bar{\rho}(y)/\rho_w] \equiv (\overline{\rho u'^2}/\tau_w)$. The streamwise normal stress distribution is in fair agreement with the incompressible results of Klebanoff (1955), except near the wall where reduced accuracy affects the supersonic measurements. Morkovin's scaling appears to be appropriate to at least Mach 5. Measurements by Owen et al (1975) at $M_\infty = 6.7$ and Laderman & Demetriades (1974) at $M_\infty = 9.4$ exhibit damped turbulent fluctuations, particularly near the wall. Since both of the hypersonic data sets are for cold-wall conditions, this may simply indicate the stabilizing effect of cooling.

Measurements of $\overline{v'^2}$ and $\overline{w'^2}$ are less common than those of $\overline{u'^2}$, the data exhibit more scatter, and the conclusions are therefore less certain. In contrast to the streamwise turbulence intensity, both distributions appear to increase slightly with increasing Mach number (Fernholz & Finley 1981). In this case, Morkovin's scaling does not collapse the data, and $\bar{\rho}\overline{v'^2}/\tau_w$ and $\bar{\rho}\overline{w'^2}/\tau_w$ show no real trend toward similarity. A difference with subsonic boundary layers is indicated in the approximate equality of the $\overline{w'^2}$ and $\overline{v'^2}$ distributions. However, the limited nature of the data precludes any physical interpretation of the reduced anisotropy in supersonic shear flows and how compressibility may be influencing this fundamental structure parameter.

Sandborn (1974) reviewed direct measurements and indirect evaluations of the zero-pressure-gradient Reynolds shear stress, $-\bar{\rho}\overline{u'v'}$. He constructed a "best fit" of normalized shear stress profiles (τ/τ_w) from integrated mean-flow data taken by a variety of researchers over a wide Mach-number range, $2.5 < M_\infty < 7.2$ (extended to Mach 10 by Watson 1978) for adiabatic and cold walls. The data indicate a near-universal shear-stress profile that is in excellent agreement with the incompressible measurements of Klebanoff (1955). The universality of τ/τ_w over such a wide Mach-number range is not surprising in light of the fixed constraints on the normalized shear stress at the wall and approaching the freestream. Even so, the only Reynolds shear stress measurements to agree with the "best fit" in 1974, and then only in the outer layer, were the LDV data of Rose & Johnson (1975). Subsequent Reynolds shear stress measurements by Mikulla & Horstman (1975), Kussoy et al (1978), Robinson (1983), Smits & Muck (1984), and Donovan & Spina (1992) (all using hot wires except Robinson) have exhibited modest agreement with Sandborn's best fit and the incompressible distribution. The agreement is limited to the outer layer, with tremendous scatter in the inner layer and most profiles not tending toward $\tau/\tau_w = 1$ near the wall. The data in the inner layer do not scale with yu_τ/v_w, probably because of the difficulties with the measurements.

At hypersonic Mach numbers, it is possible that the triple correlation $\overline{\rho'u'v'}$ may become comparable to the "incompressible" Reynolds shear stress, $\bar{\rho}\overline{u'v'}$, since $\rho'/\bar{\rho} \sim M^2u'/U$. Owen (1990) evaluated the various contributions to the "compressible" Reynolds shear stress at Mach 6 through simultaneous use of two-component LDV and a normal hot wire. His results indicate that $\overline{\rho'u'v'}$ is negligible compared to $\bar{\rho}\overline{u'v'}$. Even though density fluctuations increase with the square of the Mach number, it should be remembered that the main contribution to the Reynolds shear stress occurs in the region where $M(y)/M_\infty$ is small, so this "hypersonic effect" should only be important at very high freestream Mach number.

The stagnation-temperature fluctuation must be known to evaluate the turbulent heat-flux correlation, $-c_p\bar{\rho}\overline{v'T'}$. Kistler (1959) observed that $T'_{0_{rms}}/\bar{T}_0$ increased with Mach number, with maxima of 0.02 at $M_\infty = 1.72$

and 0.048 at $M_\infty = 4.67$. If Kistler's data are alternatively non-dimensionalized by either T_w (Fernholz & Finley 1981) or $T_r - T_e$ (Sandborn 1974), the Mach-number dependence appears to be eliminated, but similarity of the stagnation-temperature distributions is not achieved. Similar conclusions are reached from measurements by Morkovin & Phinney (1958) and Horstman & Owen (1972). The maximum level of stagnation-temperature fluctuations is about 6% (for $M < 7$), which is less than that of u'_{rms} and T'_{rms}, but not low enough to satisfy the strict Strong Reynolds Analogy (see Gaviglio 1987). In fact, the SRA can be used to estimate that $T'_{0_{rms}}$ is about 60% of T'_{rms} at Mach 3 (Smits et al 1989).

BOUNDARY-LAYER STRUCTURE

The eddy structure and internal dynamics of compressible turbulent boundary layers play an important role in many aerospace engineering applications. These include turbulent mixing for high-speed propulsion systems, tripping of hypersonic laminar boundary layers (for inlet efficiency), acoustic noise generation and propagation from high-speed engines, surface heat transfer on high-speed vehicles, performance optimization for low-observable configurations, and unsteadiness in shock/turbulent boundary layer interactions.

For incompressible boundary layers, turbulence is now known to consist of dynamically regenerating "structural" features such as low-speed streaks in the sublayer, outward ejections of low-speed fluid, wallward sweeps of high-speed fluid, and various forms of compact embedded vortices (Robinson 1991). The function of these fundamental elements of turbulent boundary layers is to draw energy from the mean flow to create turbulent kinetic energy at a rate that sustains the boundary-layer turbulence.

As Morkovin (1962) suggested, for moderate Mach numbers, the "essential dynamics" of compressible boundary layers are not expected to differ greatly from the incompressible case. For higher Mach numbers, however, significant differences are likely, especially for the inner regions of the boundary layer, but there are very few results to support any conjectures concerning hypersonic turbulent boundary-layer structure.

The current state of knowledge concerning compressible boundary layer structure is limited to large-scale motions in the outer region, and is derived largely from studies at Princeton University (Spina et al 1991a,b; Smits et al 1989; Spina & Smits 1987; Fernando & Smits 1990; Donovan et al 1994), and at NASA Ames (Robinson 1986) of flat-plate layers with freestream Mach numbers of approximately 3.0. These studies were preceded by a pioneering investigation by Horstman & Owen (1972), who made extensive two-point cross-correlation measurements with hot wires in a Mach 7.2 boundary layer. Most of the results available in the literature were obtained using hot-wire anemometry (with its attendant limitations), with some degree of corroboration by high-speed flow-visualization techniques (Cogne et al 1993, Smith et al 1989).

For moderate Mach numbers, the outer region of the boundary layer (beyond the logarithmic region) is dominated by the entrainment process rather than by turbulence production. Thus the available studies of supersonic turbulent boundary-layer structure are primarily relevant to the processes by which the boundary layer grows. In contrast, for subsonic turbulent boundary layers, most of the attention has focused upon the near-wall turbulence-production processes. In addition, while supersonic-structure experiments have been conducted at very high Reynolds number, the majority of subsonic-structure research has been at quite low Reynolds number. These mismatches in emphasis between subsonic and supersonic investigations have made comparisons inconclusive, at least for isolating effects of compressibility on turbulence physics.

One measure of the wallward extent of the entrainment process is the intermittency profile, which is often estimated with measurements of u' flatness. The measured intermittency profile in Mach 0, 3, and 7 boundary layers (Klebanoff 1955, Robinson 1986, Spina & Smits 1987, Horstman & Owen 1972) displays an apparent Mach-number dependence, wherein the onset of intermittency (corresponding to the rise in flatness factor) occurs nearer to the boundary-layer edge as the Mach number is increased. Since the cone of influence of a flow disturbance is inversely proportional to Mach number, the turbulent/nonturbulent interface at the boundary-layer edge may be confined to thinner layers as the Mach number increases. This interpretation is not fully supported by high-speed flow visualizations, however, so the data remain provocative. For example, dual-pulse Rayleigh scattering flow visualization by Cogne et al (1993) shows deep potential incursions into the turbulent eddies of a Mach 3 boundary layer (Figure 2) in patterns that are strikingly similar to visualizations of low-speed boundary layers.

For both incompressible low Reynolds number, and compressible high Reynolds number boundary layers, the most identifiable feature of the outer region is a downstream-sloping shear-layer interface between upstream high-speed fluid and downstream low-speed fluid. (Unfortunately, these structures have been labeled both "fronts" and "backs" in the literature.) These interfaces are three-dimensional shear layers which are believed to form the upstream side of the largest of the boundary-layer

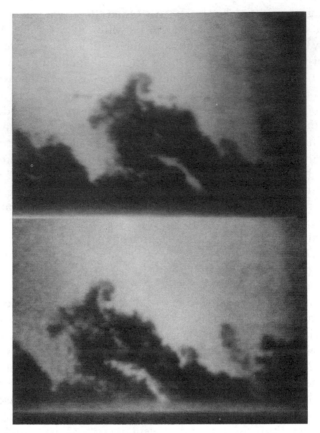

Figure 2   Double-pulsed Rayleigh images from a Mach 3 turbulent boundary layer; (*top*) time t, (*bottom*) time $t + 20$ μs. The flow is from right to left. (From Cogne et al 1993).

eddies, and remain coherent long enough to convect several boundary-layer thicknesses downstream. They are not inert, however, since Spina et al (1991a) have shown that 40% of the outer-layer Reynolds shear stress can be found in the neighborhood of these sloping interfaces (causality is not implied). The intense turbulence-production processes near the wall in the Mach 3 layer have not been investigated, but incompressible experience suggests that the large-scale sloping interfaces are not closely affiliated with near-wall regions of high Reynolds shear stress.

Sloping interfaces are easily detected with dual hot wires separated in y, using either traditional space-time correlations, or a variety of conditional sampling techniques. For Mach 3 turbulent boundary layers, the effect of compressibility on the large-scale outer structures has been found to be generally small, which may be expected since the fluctuating Mach number in the outer regions is unlikely to approach unity (Figure 1). However, some differences between subsonic and supersonic large-scale motions have been measured; these are summarized below (references are the Princeton and NASA Ames papers cited previously).

The average "structure angle" at which delta-scale interfaces lean downstream in a Mach 3 turbulent boundary layer ranges from 45° to 60° (standard deviation $\approx 20°$) across most of the boundary layer, with a decrease near the wall and an increase near the boundary layer edge. The measured value of the structure angle is strongly dependent on measurement technique, although the method in current favor employs two hot wires, separated by a fixed distance in y of 0.1 to 0.3δ, with both traversed across the layer. Structure angles measured using this technique in subsonic, low Reynolds number turbulent boundary layers are somewhat lower than those for Mach 3, high Reynolds number layers, but this seems more likely to be a Reynolds-number effect than a Mach-number effect.

Investigations of large-scale structure angles in moderately supersonic boundary layers following perturbations such as adverse pressure gradients and concave curvature have also been reported (Donovan et al 1994, Fernando & Smits 1990). In all cases, the perturbations resulted in somewhat increased structure angles, with stronger perturbations (tighter curvature) showing higher angles.

Hot-wire and flow visualizations show that the sloping delta-scale structures convect downstream at approximately 90% of the freestream velocity (slightly greater than for similar structures in low Reynolds number, incompressible turbulent boundary layers), and persist for at least 4 boundary-layer thicknesses (and probably much farther) downstream (Spina et al 1991b).

Outer-region space-time correlations suggest that the average spanwise extent of the largest eddies in the Mach 3 turbulent boundary layer is similar to that of subsonic turbulent boundary layers: approximately 1/2δ in the outer layer, decreasing near the wall. (Although mean and instantaneous results for the sloping interface structure are in good agreement, the average cross-correlations used to deduce spanwise extent probably suffer from "jitter" averaging, and the instantaneous extents may be larger.) The average streamwise scales of the largest eddies in the high Reynolds number, Mach 3 turbulent boundary layer are about twice those of low Reynolds number, subsonic turbulent boundary layers. This seems to be the most significant structural difference between the two flows yet found, although both Reynolds number and compressibility could be responsible.

Cross-wire measurements of both streamwise and wall-normal components of velocity have suggested additional apparent differences between Mach 3 and incompressible boundary-layer structure (Smits et al 1989). The shear correlation coefficient R_{uv} decreases significantly with distance from the wall for the Mach 3 case, but remains nearly constant between 0.1 and 0.8δ for the subsonic layer. Joint probability density distributions of the two velocity (or mass flux) components are also somewhat different between the two flows, with the supersonic case favoring vertical fluctuations in the mid-layer slightly more than the subsonic case.

Since the influence of compressibility on the large-scale turbulent boundary-layer motions seems to be subtle, explanations for the observed differences between low- and high-speed boundary-layer structure are mostly speculative. Density-gradient effects are known to play a significant role in turbulent shear layers, but these effects are most likely to influence the near-wall region of the wall-layer, out of reach of standard measurement techniques. Parallels have also been drawn between the 45° slope of the interfacial structures in supersonic boundary layers and that of the hypothesized hairpin-vortex structure for incompressible boundary layers. Little evidence exists to support either side of this comparison, however. More conclusive results concerning compressibility effects on large-scale structure require higher Mach number investigations.

For boundary layers with freestream Mach numbers above 5, the near-wall region is more likely to show significant departures from known incompressible structure. The viscous sublayer for hypersonic boundary layers is likely to be much more quiescent than for incompressible flows (although pressure fluctuations will be imposed from above), and may not display the familiar streaky structure. Since the mass flux near the wall is very low for high Mach numbers, the buffer region may not be the dominant region for turbulence production, as in subsonic boundary layers. (Recall that hypersonic laminar boundary layers undergo transition by disturbances spreading inward from the outer layer). Investigation of these speculations awaits the development and application of nonintrusive measurement techniques to the near-wall regions of hypersonic boundary layers.

In summary, the large-scale, outer-region structure is not dramatically affected by compressibility for flat-plate and mildly perturbed boundary layers up to Mach 3. Beyond that conclusion, very little has been learned about the influences of compressibility on boundary-layer structure; significant progress will require significantly improved instrumentation (particularly optical techniques). Especially important is the differentiation of the influences of Mach and Reynolds numbers on turbulent boundary-layer structure.

EFFECT OF EXTRA STRAIN RATES

If properly scaled, many characteristics of canonical supersonic turbulent boundary layers are remarkably similar to those of comparable incompressible flows. This is not always the case when "extra strain rates" are applied in addition to the principal strain rate, $\partial U/\partial y$ (Bradshaw 1974). In supersonic flow, significant extra strain rates result from perturbations such as shock and expansion waves, strong pressure gradients, and streamline curvature. Extra strain rates in supersonic boundary layers can lead to counterintuitive mean-flow behavior and can dramatically alter the turbulence field. Their effect can be up to an order of magnitude greater than that predicted by conventional means, and their presence in many flows of practical interest makes an improved understanding of these perturbations particularly important. Indeed, most of the progress made in supersonic turbulent boundary-layer research since Bradshaw's review (1977) has been in an improved understanding of distorted flows (see, for example, Fernholz et al 1989).

Adverse Pressure Gradients

When a streamwise pressure gradient is imposed on a supersonic boundary layer, the flow is distorted both by the direct effects of the pressure gradient and by bulk compression, $-\nabla \cdot \mathbf{V}$. While few turbulence measurements have been made in such flows, the mean-flow behavior is comparatively well documented. When the sonic region of the boundary layer is small compared to the boundary-layer thickness (in practice, when M_∞ is greater than about 1.8), the adverse pressure gradient causes the wall shear stress to increase and the boundary-layer thickness to decrease. This counterintuitive behavior occurs because of the unique character of supersonic flow: When the pressure rises the density increases more rapidly than the velocity decreases. This results in a compression of the streamtubes in the supersonic part of the flow and thus a decrease in boundary-layer thickness. The compression of the streamtubes also increases all components of the vorticity and amplifies the turbulence intensities (Green showed that bulk compression is destabilizing through an argument based on conservation of angular momentum; see Bradshaw 1974).

The effects of an adverse pressure gradient (APG) can only be isolated in a reflected-wave flow, in which a freestream body imposes a pressure gradient on a flat-plate boundary layer. Investigations of this type of APG flow were performed by Peake et al (1971), Waltrup & Schetz (1973), Lewis et al (1972), Thomas (1974), Kussoy et al (1978), Zwarts (1970), Fernando & Smits (1990), and Smith (1993). The initial conditions of these flows (Mach number, Reynolds number, upstream history) vary greatly, as do

Table 2a Property trends in the "Group I" ($8°$ of turning, $p_2/p_1 \approx 2$) boundary-layer-perturbation studies of Smits and his coworkers[a]

	Ramp	Curved wall	Curved wall	Flat-plate APG
R_c	0	$10\delta_0$	$50\delta_0$	∞
Interaction length	0	$1.4\delta_0$	$7\delta_0$	$7\delta_0$
I_θ	0.14	0.14	0.14	≤ 0.05
I_p	0.46	0.46	0.46	0.46
τ_w	$+60\%$	$+40\%$	$+55\%$	$+45\%$
Appearance of dip in log-law	$2.5\delta_0$	$2\delta_0$	$9\delta_0$	$11.5\delta_0$
$\bar{\rho}\overline{u'^2}$	$+350\%$	$+240\%$	$+320\%$	$+200\%$
$\bar{\rho}\overline{u'v'}$	$+390\%$	$+390\%$	$+200\%$	$+220\%$
a_1'' $(-\overline{u'v'}/\overline{u'^2})$	$+70\%$	$\downarrow$ near wall	$\downarrow$ near wall	Approx. constant

[a] The quoted amplification percentages are peak values (although most occur near $y/\delta = 0.4$).

the strengths of the applied adverse pressure gradients, the lengths over which the APG is applied, and the lengths over which the flow relaxes. Nevertheless, some general conclusions can be made. If an appropriate velocity transformation (van Driest 1951 or Carvin et al 1988) is used in moderate APGs, the pressure gradient alters the mean-velocity scaling in a similar manner as observed for incompressible boundary layers. Specifically, the law of the wall holds, the strength of the wake is increased over the zero pressure gradient (ZPG) case, and the outer law shows a higher negative slope than in a ZPG flow. For all flows where the Mach number is outside the transonic range (in practice, $M_\infty > 1.8$), the wall shear stress increases monotonically through the region of applied pressure gradient, although c_f remains nearly constant in most cases because of the increase in $\rho_e U_e^2$. There is a clear lag in the boundary-layer response, as the wall stress reaches a maximum after the removal of the APG and only then relaxes toward an equilibrium value. The data also clearly indicate a decrease in δ, although the momentum thickness still increases due to entrainment, as indicated by the mean streamline patterns plotted by Fernando & Smits (1990) and Smith (1993).

The only turbulence measurements in such flows are by Fernando & Smits (1990) and Smith (1993), and they indicate that the Reynolds stress components are strongly amplified (Tables 2a,b). The crossed-wire data of Fernando & Smits show that the turbulence stress ratios (whether $\overline{u'^2}/\overline{v'^2}$, R_{uv}, or a_1'') change less than the stresses themselves through the interaction. This evidence that the nature of the turbulent motions is

Table 2b Property trends in the "Group II" ($16°$ of turning, $p_2/p_1 \approx 3$) boundary-layer-perturbation studies of Smits and his coworkers

	Ramp	Curved wall	Curved wall	Flate-plate APG
R_c	0	$12\delta_0$	$50\delta_0$	∞
Interaction length	0	$3.5\delta_0$	$15\delta_0$	$3.5\delta_0$
I_θ	0.28	0.28	0.28	≤ 0.09
I_p	0.78	0.78	0.78	0.78
τ_w	$+115\%^b$	$+130\%^b$	$+105\%$	$+70\%^b$
Appearance of dip in log-law	$1.5\delta_0$	$4.5\delta_0$	$7.5\delta_0$	c
$\bar{\rho u'^2}$	$+1700\%$	$+900\%$	—	$+600\%$
$\bar{\rho u'v'}$	$+650\%$	$+700\%$	—	—
a_1'' $(-\overline{u'v'}/\overline{u'^2})$	↓ near wall ↑ in outer layer	↑ in outer layer	—	—

[b] Still increasing at last measurement location.
[c] Limited streamwise extent may prevent detection of dip.

essentially unchanged is supported by a series of detailed space-time correlation measurements showing little effect on the large-scale organized structures.

Most of the investigations have shown that equilibrium is not recovered uniformly across the boundary layer after the removal of the extra strain rate. For example, the log law recovers more quickly than the defect law and the Reynolds stress distributions exhibit initial relaxation near the wall. A slower recovery of the outer layer is not unexpected, since the relative effect of the extra strain rate is largest there due to the lower value of the principal strain rate away from the wall (Smits & Wood 1985). The nonuniformity of the boundary-layer response was also observed by Fernando & Smits in the form of a dip below the log law during the latter part of the pressure rise. This is a common feature of subsonic flows subjected to concave curvature, and it appears to indicate that the turbulence length scales increase faster with distance from the wall than in an equilibrium boundary layer (Smits et al 1979). The spectra of the longitudinal mass-flux fluctuations exhibit a decrease in the frequency of the most energetic motions, also signaling a decrease in length scale. The alteration of length scale is first noticeable downstream of the point where the pressure reaches a maximum, but it seems that the dip is due to the streamline curvature and not due to the APG nor the bulk compression. The dip in the log law has not been observed either in subsonic APG flows or in any other supersonic APG boundary layers. The reason for this may be either the high Reynolds number and large streamwise extent of this study, or the small degree of concave curvature of the streamlines in this flow (about 3° where the dip occurs). Smith (1993) measured about a 5° streamline curvature in a stronger pressure gradient and did not observe the dip, but the streamwise extent of his study was much less than that of the weaker case.

The only investigation that allows comment on the temperature-velocity correlation in an APG flow over an adiabatic wall is that of Waltrup & Schetz (1973). Their data (plotted in Fernholz & Finley 1980) indicate that the temperature-velocity relation is well represented by a quadratic variation. This relation is strictly true only for adiabatic, ZPG flow, but the moderate nature of the APG in the study by Waltrup & Schetz is probably why this relation can be extended beyond its exact limits of applicability. No temperature-velocity relation exists for nonadiabatic supersonic turbulent boundary layers subjected to a pressure gradient, and the only existing data from such an APG flow are of questionable quality (Gran et al 1974). This helps to illustrate that APG flows need further study, with emphasis on an increased range of wall-temperature ratios, higher and lower freestream Mach numbers (greater than Mach 5 and lower than Mach 2), stronger pressure gradients ($\beta \equiv \delta^*/\tau_w \, \partial p/\partial x > 5$), and turbulence measurements.

Favorable Pressure Gradients

The application of a favorable pressure gradient (FPG) to a supersonic flow results in bulk dilatation and the expansion of supersonic streamtubes. This stabilizing extra strain rate serves to decrease the turbulence intensities and may even relaminarize part of the boundary layer if the pressure gradient is strong enough. Both nozzle-wall flows and expansion corners introduce multiple extra strain rates, so a review of "pure" FPG effects is limited to a few studies done on flat-wall boundary layers with an imposed pressure gradient. Several of these investigations are of nonadiabatic boundary layers, but the data are still too limited to offer much insight into the temperature-velocity correlation.

Experimental investigations of supersonic FPG boundary layers have been performed by Morkovin (1955), Hill (1959), Perry & East (1968), Lewis et al (1972), Voisinet & Lee (1973), and Thomas (1974), but none of the measurements include turbulence data. While their conclusions are generally consistent, some concerns exist regarding the quality of the data. Despite these concerns, the qualitative effects of favorable pressure gradients on the mean-flow behavior of supersonic turbulent boundary layers is clear. As expected (by reasoning parallel to that applied in the APG case), bulk dilatation decreases the wall shear stress and increases the boundary-layer thickness. Lewis et al observed that the boundary-layer edge approximately coincides with a mean streamline; this is consistent with reduced turbulent entrainment resulting from decreased turbulence intensities. The bulk dilatation also affects the inner and outer velocity scaling in an opposite manner than bulk compression: The logarithmic region is extended, the strength of the wake decreases dramatically, and the defect law exhibits a smaller negative slope than in a zero pressure gradient. There is also evidence that the boundary-layer response to the FPG is slower in the outer layer than in the wall region.

Voisinet & Lee (1973) examined a Mach 3.8 FPG boundary layer along an isothermal wall with high cooling rates $(T_w/T_r \approx 0.25)$. The temperature-velocity correlation exhibits linear behavior, and is very similar to data taken by the same researchers in a ZPG flow along a cold, isothermal wall (Voisinet & Lee 1972). Fernholz & Finley speculate that the cold wall has such a dominant influence in this flow that the rather modest FPG is a second-order perturbation to the temperature distribution. In contrast, the lower Reynolds number data of Perry & East $(M_{\infty,\text{ref}} = 10.2$–$11.6, T_w/T_r = 0.29)$ indicate quadratic behavior for the temperature-velocity correlation. The difference in the temperature-velocity relation in these two experiments may be caused by the disparate Mach numbers, the order-of-magnitude difference in Reynolds number, or the stronger pressure gradient of Perry & East.

Streamline Curvature

Streamline curvature, $\partial V/\partial x$, is an important extra strain rate in flow over curved surfaces as well as that over sharp compression and expansion corners. Bulk compression and bulk dilatation are additional extra strain rates in these flows through the effect of the pressure gradient and, for the sharp-corner flows, through the concentrated wave structure of the flow. Another important element of the flow field is the pressure gradient created normal to the curved streamlines: $\partial p/\partial n \approx -\rho U^2/R = -\gamma M^2 p/R$. Clearly, the pressure difference across the boundary layer can be significant even for a large radius of curvature (R_c) if the Mach number is high enough: $\Delta p \approx \delta(\partial p/\partial n) \approx \gamma M^2(\delta/R)p$.

Concave curvature is destabilizing to the flow in the sense that it enhances turbulent mixing, and convex curvature is stabilizing. This behavior is exhibited in all curvature studies, and a good example in supersonic flow is the work of Thomann (1968) at Mach 2.5. He found that 20° of concave curvature (with the streamwise pressure gradient eliminated) increased the mean heat transfer rate by 20%, whereas isolated convex streamline curvature decreased the mean heat transfer rate by about the same amount. A classic manifestation of the destabilizing effect of concave curvature in subsonic flow is the generation of Taylor-Görtler-type (T-G) vortices, even when the flow is turbulent (Bradshaw (1973). No evidence of steady T-G vortices was found by Sturek & Danberg (1972a,b), Laderman (1980), or Jayaram et al (1987). The techniques used to search for the existence of T-G vortices would not detect unsteady longitudinal roll-cells, however. Jayaram et al suggested that the nonlinear interaction of concave curvature and compression, amplifying different components of vorticity, may prevent roll-cell formation, at least in a steady array.

Studies that focused on the behavior of the mean flow when subjected to concave curvature include those by McLafferty & Barber (1962), Hoydysh & Zakkay (1969), Sturek & Danberg (1972a,b), Laderman (1980), and Chou & Childs (1983). None of these investigations allow the influence of streamline curvature to be isolated, however. Only Thomann (1968) and Smits and his co-workers structured their experiments to provide insight into the separate effects of concave streamline curvature and bulk compression. Taylor & Smits (1984), Smits & Muck (1987), Jayaram et al (1987), Fernando & Smits (1990), Smith (1993), and Donovan et al (1994) studied a variety of extra strain rates applied to the same incoming boundary layer (Mach 2.9, $Re_\theta \approx 81,000$). They applied two groups of flow perturbations, characterized by pressure rises equivalent to $8°$ and $16°$ of curvature. Each group consists of two models with concave curvature (radii of curvature of $50\delta_0$ and about $10\delta_0$), a compression corner, and a flat plate with externally imposed pressure gradient. The parametric nature of these investigations provides some insight into the separate effects of concave streamline curvature and bulk compression. The perturbation characteristics and significant results are summarized in Tables 2a and b.[1]

One way to evaluate the relative influence of extra strain rates is to consider the integrated effect, on the basis that for a rapid perturbation the overall turning (for example) is likely to be more important than the rate at which the turning takes place (Smits et al 1979). Hence we compute the total impulse: $I = \int e\, dt$, where e is the extra strain rate. For example, the impulse to the $8°$ curved-wall flow is 0.46 from bulk compression (I_p) and 0.14 from streamline curvature (I_θ), whereas the companion flat-wall/APG case (Fernando & Smits) only has an impulse from bulk compression (0.46). An examination of Table 2 leads to the general conclusion that concave streamline curvature enhances the destabilizing effects of bulk compression. Furthermore, it appears that the extra strain rates do not interact linearly, and the effect of curvature is not uniform or consistent

[1] The figures quoted in Tables 2a,b are approximate values only. Each of the measurements has an appreciable error band associated with it, and the original papers should be consulted for detailed values.

from one property to another. Clearly, much more data must be acquired before the relative effects of the extra strain rates are known with any certainty. This is further confirmed by the APG/curvature experiments of Thomann (1968). An applied pressure gradient ($I_p = 0.86$) increased the mean heat transfer rate by 5% over that of the ZPG flow, whereas a curved model ($I_p = 0.86$ and $I_\theta = 0.35$) increased it by 24%. This disproportionate influence of streamline curvature illustrates that the importance of the individual extra strain rates is not determined simply by summing the individual impulses. However, the relative influence of bulk compression can be expected to increase at high Mach number since $I_p \sim \ln M^2$ and $I_\theta \sim \theta$.

Streamline curvature appears to have a large influence on the mean-velocity scaling, as observed by Donovan et al (1994) and Jayaram et al (1987). Similar to the observations in the flat-plate/APG flow, the velocity profiles on the curved walls show an increase in strength of the wake. However, the dominant feature of the inner-layer scaling is the dip below the log law, indicating a distortion of the length scales relative to those in an equilibrium boundary layer. The magnitude of the dip (Jayaram et al 1987, Donovan et al 1994) is considerably greater than that observed in the APG flow of Fernando & Smits, indicating the influence of streamline curvature. The decrease in the frequency corresponding to the peak of the power spectra is also more pronounced in the curved flows than in the applied pressure gradient case.

The rate of curvature has an uncertain effect on both mean and turbulence quantities. In Group I, for example, the rapid curvature has an effect on the Reynolds shear stress that is similar to that of the ramp flow and greater than that of the gradual curvature. However, the effect of perturbation rate appears to be just the opposite on the behavior of the longitudinal Reynolds stress and the wall shear stress.

The behavior of supersonic boundary layers subject to convex curvature has not been studied extensively. Most of the available data sets are from nozzle walls or are not complete. The nozzle flows suffer from low Reynolds number effects and severe nonequilibrium due to the strong pressure gradients (see, for example, Beckwith et al 1971 or Kemp & Owen 1972). Thomann (1968) isolated the effect of convex curvature by eliminating the streamwise pressure gradient, but only measurements of the wall heat transfer rate were made. No data have been taken on smooth convex corners in supersonic flow, leaving the expansion-corner studies of Dussauge & Gaviglio (1987) and Smith & Smits (1991) (also see Smith 1993) as the only flows that provide insight to the effect of convex streamline curvature.

Dussauge & Gaviglio (1987) compared measurements and rapid distortion theory for a Mach 1.7 flow over a 12° expansion corner, and Smith (1993) did the same for a Mach 2.9 flow over a 20° corner. The shape of the log law and the outer law of both perturbed flows closely mirror that of a flat-wall flow subjected to an FPG (such as that of Lewis et al 1972). The perturbations are rapid, however, as shown by a comparison of the "time of flight" for a fluid element to the turbulence time scales. As such, the flows relax toward an equilibrium state downstream of the expansion (about $10\delta_0$ in Dussauge & Gaviglio's flow). The sublayer appears to thicken immediately downstream of the corner, and this could indicate the start of a relaminarization of the boundary layer. Such a possibility is given credence by the dramatic reduction in streamwise Reynolds stress (85% in Smith's 20° expansion). However, both studies actually provide more insight into the effect of a rapid application of bulk dilatation than the influence of streamline curvature: The expansion of Dussauge & Gaviglio results in a value of streamline curvature that is about one-third the upstream principal strain rate, whereas the bulk dilatation is five times that strong. A comparison with computations indeed indicates that bulk dilatation plays the dominant role in the distortion of the boundary layer. It is further noted that the relative influence of bulk dilatation will only become greater as the Mach number increases because the total impulse due to this extra strain rate is proportional to $\ln M^2$.

OUTLOOK FOR ADVANCEMENT

In zero-pressure-gradient supersonic flows with moderate Mach number, it appears that the direct effects of compressibility on wall turbulence are rather small: The most notable differences between subsonic and supersonic boundary layers may be attributed to the variation in fluid properties across the layer. This is not very surprising since the fluctuating Mach number ($M' = M - \bar{M}$) for moderately supersonic flows is considerably less than one, as illustrated in Figure 1. However, a more detailed inspection of the turbulence properties reveals certain characteristics that cannot be collapsed by a simple density scaling. For example, the intermittency profile is fuller than the corresponding subsonic profile, the fractal dimension of the density interface appears to decrease with Mach number, the shear correlation coefficient R_{uv} decreases with distance from the wall instead of remaining approximately constant, the streamwise length scales are significantly smaller, and there is an order-of-magnitude decrease in the rate of decay of the large-scale motions as the Mach number increases from low-subsonic to supersonic values (Smits et al 1989).

How can we explain these differences? Part of the answer may lie in understanding the role of Reynolds number more clearly, but understanding the effects of fluid-property variations may be more important. In that respect, a direct numerical simulation of a strongly heated, incompressible turbulent boundary layer in the absence of buoyancy effects would be particularly valuable. Experimentally, we urgently need detailed turbulence data at higher Mach numbers. We are seeing subtle differences at supersonic speeds that may signal the onset of direct compressibility effects such as the increased importance of pressure fluctuations and pressure-velocity correlations. These effects will become more obvious at hypersonic Mach numbers, and such studies would contribute to our understanding of the supersonic behavior.

While few specifics are known, the turbulence physics become more complex as the Mach number increases beyond about five. For example, the Strong Reynolds Analogy and Morkovin's hypothesis (1962) are staples of boundary-layer analysis at moderate Mach number. However, an upper Mach-number limit must exist on the applicability of these simplifying assumptions, if only because there is a limit on the magnitude of temperature fluctuations. Indeed, the change in magnitude of the fluctuating Mach-number distribution as the flow enters the hypersonic range (see Figure 1) points to the possibility of a dramatic alteration of turbulence dynamics due to compressibility effects around Mach 5 (in comparison, the Mach number of the fluctuations, $u'_{rms}/\bar{a}$, is less than 0.3 even for the Mach 7.2 and 9.4 flows). Unlike the distribution of $u'_{rms}/\bar{a}$, the fluctuating Mach number develops a peak near the middle of the boundary layer where both the velocity and temperature fluctuations are important. This behavior, when considered together with the large gradients in density and viscosity near the wall, also leads to the conclusion that there may be substantial differences in turbulence dynamics at high Mach number.

At the same time, the near-wall gradients in density and viscosity are strongly dependent on heat transfer, and therefore the thickness of the sublayer will depend on Mach number, Reynolds number, and wall temperature. This leads to the issue of how the viscous instability of the sublayer changes when fluid properties vary with distance from the wall (see Morkovin 1992). Do fluid-property variations just change the effective Reynolds number? Since the local Reynolds number increases away from an adiabatic wall faster in supersonic flow than in incompressible flow, we would expect the flow to become less stable as we move away from the wall at a rate that is faster than in an incompressible flow at the same friction velocity. What is the proper basis of comparison between compressible and incompressible boundary layers in the near-wall region and in the outer region? Given that the kinematic viscosity changes dramatically across the layer, how do the dissipation rates vary?

For perturbed flows, where pressure gradients, streamline curvature, and divergence are important, the answers to these questions become even more important. There is clearly a need for detailed studies to document the response of boundary-layer turbulence to these effects, separately and in combination, but more importantly there is a need to combine these studies with numerical experiments of the kind that are now being performed for incompressible flows. Direct numerical simulations and large-eddy simulations for compressible wall-bounded flows are now possible, and they will provide crucial data for the development of physical models and, eventually, turbulence models for complex supersonic flows. At the same time, the development and application of advanced diagnostics such as Rayleigh scattering, filtered Rayleigh scattering, methods for tagging molecules, and high-speed multiple-exposure imaging is transforming the depth and breadth of experimental data available. These new techniques are particularly valuable in studies where three-dimensional data are needed, as in the study of turbulence structure or in the investigation of flows which are three dimensional in the mean. In terms of new computational and experimental tools, the study of high-speed turbulent flows is at the beginning of a new era, and we expect that it will provide many exciting rewards for the workers in the field.

Literature Cited

Adamson, T. C., Messiter, A. F. 1980. Analysis of two-dimensional interactions between shock waves and boundary layers. *Annu. Rev. Fluid Mech.* 12: 103–38

Beckwith, I. E. 1970. Recent advances in research on compressible turbulent boundary layers. *NASA Spec. Publ. SP-228*, pp. 355–416

Beckwith, I. E., Harvey, W. D., Clark, F. L. 1971. Comparisons of turbulent boundary layer measurements at Mach number 19.5 with theory and an assessment of probe errors. *NASA Tech. Note D-6192*

Bertram, M. H., Neal, L. 1965. Recent experiments in hypersonic turbulent boundary layers. *NASA Tech. Mem. X-56335*

Bradshaw, P. 1973. Effects of streamline curvature on turbulent flow. *AGARDograph-AG-169*

Bradshaw, P. 1974. The effect of mean compression or dilatation on the turbulence structure of supersonic boundary layers. *J. Fluid Mech.* 63: 449–64

Bradshaw, P. 1977. Compressible turbulent shear layers. *Annu. Rev. Fluid Mech.* 9: 33–54

Bradshaw, P., Unsworth, K. 1974. Comment on "Evaluation of Preston tube calibration equations in supersonic flow." *AIAA J.* 12: 1293–96

Bushnell, D. M., Johnson, C. B., Harvey, W. D., Feller, W. V. 1969. Comparison of prediction methods and studies of relaxation in hypersonic turbulent nozzle-wall boundary layers. *NASA Tech. Note D-5433*

Carvin, C., Debieve, J. F., Smits, A. J. 1988. The near-wall temperature profile of turbulent boundary layers. *AIAA Pap. 88-0136*

Chou, J. H., Childs, M. E. 1983. An experimental study of surface curvature effects on a supersonic turbulent boundary layer. *AIAA Pap. 83-1672*

Clauser, F. H. 1954. Turbulent boundary layers in adverse pressure gradients. *J. Aeronaut. Sci.* 21: 91–108

Cogne, S., Forkey, J., Miles, R. B., Smits, A. J. 1993. The evolution of large-scale structures in a supersonic turbulent boundary layer. *Proc. of the Symp. on Transitional and Turbulent Compressible Flows*, ASME Fluids Engrg. Div.

Coles, D. E. 1962. The turbulent boundary layer in a compressible fluid. *Rand Corp. Rep. R-403-PR*, Santa Monica, Calif.

Dolling, D. S. 1993. Fluctuating loads in shock wave/turbulent boundary layer interaction: tutorial and update. *AIAA Pap.* 93-0284

Donovan, J. F., Spina, E. F. 1992. An improved analysis method for crossed wire signals obtained in supersonic flow. *Exp. Fluids* 12: 359–68

Donovan, J. F., Spina, E. F., Smits, A. J. 1994. The structure of a supersonic turbulent boundary layer subjected to concave surface curvature. *J. Fluid Mech.* In press

Dussauge, J.-P., Gaviglio, J. 1987. The rapid expansion of a supersonic turbulent flow: role of bulk dilatation. *J. Fluid Mech.* 174: 81–112

Elena, M., Lacharme, J.-P. 1988. Experimental study of a supersonic turbulent boundary layer using a laser doppler anemometer. *J. Mech. Theor. Appl.* 7: 175–90

Favre, A. 1965. Equation des gaz turbulents compressibles. *J. Mec.* 4: 361–421

Feller, W. V. 1973. Effects of upstream wall temperatures on hypersonic tunnel wall boundary-layer profile measurements. *AIAA J.* 11: 556–58

Fernando, E. M., Smits, A. J. 1990. A supersonic turbulent boundary layer in an adverse pressure gradient. *J. Fluid Mech.* 211: 285–307

Fernholz, H. H., Finley, P. J. 1976. A critical compilation of compressible turbulent boundary layer data. *AGARDograph-AG-223*

Fernholz, H. H., Finley, P. J. 1980. A critical commentary on mean flow data for two-dimensional compressible turbulent boundary layers. *AGARDograph-AG-253*

Fernholz, H. H., Finley, P. J. 1981. A further compilation of compressible turbulent boundary layer data with a survey of turbulence data. *AGARDograph-AG-263*

Fernholz, H. H., Smits, A. J., Dussauge, J.-P., Finley, P. J. 1989. A survey of measurements and measuring techniques in rapidly distorted compressible turbulent boundary layers. *AGARDograph-AG-315*

Gaviglio, J. 1987. Reynolds analogies and experimental study of heat transfer in the supersonic boundary layer. *Int. J. Heat Mass Transfer* 30: 911–26

Gran, R. L., Lewis, J. E., Kubota, T. 1974. The effect of wall cooling on a compressible turbulent boundary layer. *J. Fluid Mech.* 66: 507–28

Green, J. E. 1970. Interaction between shock waves and turbulent boundary layers. *Prog. Aeronaut. Sci.* 11: 235–340

Hill, F. K. 1959. Turbulent boundary layer measurements at Mach numbers from 8 to 10. *Phys. Fluids* 2: 668–80

Hirt, F., Thomann, H. 1986. Measurements of wall shear stress in turbulent boundary layers subject to strong pressure gradients. *J. Fluid Mech.* 171: 547–63

Hopkins, E. J., Keener, E. R. 1966. Study of surface Pitots for measuring turbulent skin friction at supersonic Mach numbers—adiabatic wall. *NASA Tech. Note* D-3478

Hopkins, E. J., Keener, E. R. 1972. Pressure-gradient effects on hypersonic turbulent skin-friction and boundary-layer profiles. *AIAA J.* 10: 1141–42

Hopkins, E. J., Keener, E. R., Polek, T. E., Dwyer, H. A. 1972. Hypersonic turbulent skin-friction and boundary-layer profiles on nonadiabatic flat plates. *AIAA J.* 10: 40–48

Horstman, C. C., Owen, F. K. 1972. Turbulent properties of a compressible boundary layer. *AIAA J.* 10: 1418–24

Howarth, L., ed. 1953. *Modern Developments in Fluid Dynamics, High Speed Flow.* Oxford: Clarendon

Hoydysh, W. G., Zakkay, V. 1969. An experimental investigation of hypersonic turbulent boundary layers in adverse pressure gradient. *AIAA J.* 7: 105–16

Jayaram, M., Taylor, M. W., Smits, A. J. 1987. The response of a compressible turbulent boundary layer to short regions of concave surface curvature. *J. Fluid Mech.* 175: 343–62

Johnson, D. A. 1989. Laser doppler anemometry. In *AGARDograph-AG-315*, Chap. 6

Kemp, J. H., Owen, F. K. 1972. Nozzle wall boundary layers at Mach numbers 20 to 47. *AIAA J.* 10: 872–79

Kim, K.S., Lee, Y., Settles, G. S. 1991. Laser interferometry/Preston tube skin-friction comparison in a shock/boundary layer interaction. *AIAA J.* 29: 1007–9

Kistler, A. L. 1959. Fluctuation measurements in a supersonic turbulent boundary layer. *Phys. Fluids* 2: 290–96

Klebanoff, P. S. 1955. Characteristics of turbulence in a boundary layer with zero pressure gradient. *NACA Tech. Rep.* 1247

Kussoy, M. I., Horstman, C. C., Acharya, M. 1978. An experimental documentation of pressure gradient and Reynolds number effects on compressible turbulent boundary layers. *NASA Tech. Mem.* 78488

Laderman, A. J. 1980. Adverse pressure gradient effects on supersonic boundary-layer turbulence. *AIAA J.* 18: 1186–95

Laderman, A. J., Demetriades, A. 1974. Mean and fluctuating flow measurements in the hypersonic boundary layer over a cooled wall. *J. Fluid Mech.* 63: 121–44

Lele, S. K. 1994. Compressibility effects on turbulence. *Annu. Rev. Fluid Mech.* 26: 211–54

Lewis, J. E., Gran, R. L., Kubota, T. 1972.

An experiment on the adiabatic compressible turbulent boundary layer in adverse and favourable pressure gradients. *J. Fluid Mech.* 51: 657–72

Logan, P. 1987. *Studies of supersonic turbulence and hot wire response using laser-induced fluorescence.* PhD thesis. Stanford Univ. 154 pp.

Mabey, D. G., Meier, H. U., Sawyer, W. G. 1974. Experimental and theoretical studies of the boundary layer on a flat plate at Mach numbers from 2.5 to 4.5. *RAE Tech. Rep.* 74127

Maise, G., McDonald, H. 1968. Mixing length and kinematic eddy viscosity in a compressible boundary layer. *AIAA J.* 6: 73–80

Mathews, D. C., Childs, M. E., Paynter, G. C. 1970. Use of Coles' universal wake function for compressible turbulent boundary layers. *J. Aircraft* 7: 137–40

McLafferty, G. H., Barber, R. E. 1962. The effect of adverse pressure gradients on the characteristics of turbulent boundary layers in supersonic streams. *J. Aeronaut. Sci.* 29: 1–10

Meier, H. U., Rotta, J. C. 1971. Temperature distributions in supersonic turbulent boundary layers. *AIAA J.* 9: 2149–56

Mikulla, V., Horstman, C. C. 1975. Turbulence stress measurements in a non-adiabatic hypersonic boundary layer. *AIAA J.* 13: 1607–13

Miles, R. B., Nosenchuck, D. M. 1989. Three-dimensional quantitative flow diagnostics. *Lect. Notes Eng., Advances in Fluid Mechanics Measurements* 45: 33–107

Morkovin, M. V. 1955. Effects of high acceleration on a turbulent supersonic shear layer. *Proc. of the 1955 Heat Transfer and Fluid Mech. Inst.* Stanford: Stanford Univ. Press

Morkovin, M. V. 1962. Effects of compressibility on turbulent flows. In *Mechanique de la Turbulence*, ed. A. Favre, pp. 367–80. Paris: CNRS

Morkovin, M. V. 1992. Mach number effects on free and wall turbulent structures in light of instability flow interactions. In *Studies in Turbulence*, ed. T. B. Gatski, S. Sarkar, C. G. Speziale, pp. 269–84. Berlin: Springer-Verlag

Morkovin, M. V., Phinney, R. E. 1958. Extended applications of hot wire anemometry to high-speed turbulent boundary layers. *AFOSR TN-58-469.* Baltimore: Johns Hopkins Univ.

Owen, F. K. 1990. Turbulence and shear stress measurements in hypersonic flow. *AIAA Pap.* 90-1394

Owen, F. K., Horstman, C. C. 1972. On the structure of hypersonic turbulent boundary layers. *J. Fluid Mech.* 53: 611–36

Owen, F. K., Horstman, C. C., Kussoy, M. I. 1975. Mean and fluctuating flow measurements of a fully-developed, non-adiabatic, hypersonic boundary layer. *J. Fluid Mech.* 70: 393–413

Peake, D. J., Brakman, G., Romeskie, J. M. 1971. Comparisons between some high Reynolds number turbulent boundary layer experiments at Mach 4 and various recent calculation procedures. *AGARD CP*-93-71, Pap. 11

Perry, J. H., East, R. A. 1968. Experimental measurements of cold wall turbulent hypersonic boundary layers. *AGARD CP No.* 30

Robinson, S. K. 1983. Hot-wire and laser doppler anemometer measurements in a supersonic boundary layer. *AIAA Pap.* 83-1723

Robinson, S. K. 1986. Space-time correlation measurements in a compressible boundary layer. *AIAA Pap.* 86-1130

Robinson, S. K. 1991. Coherent motions in the turbulent boundary layer. *Annu. Rev. Fluid Mech.* 23: 601–39

Rose, W. C., Johnson, D. A. 1975. Turbulence in a shock-wave boundary-layer interaction. *AIAA J.* 13: 884–89

Rubesin, M. W., Rose, W. C. 1973. The turbulent mean-flow, Reynolds-stress, and heat-flux equations in mass-averaged dependent variables. *NASA Tech. Mem. X*-62248

Sandborn, V. A. 1974. A review of turbulence measurements in compressible flow. *NASA Tech. Rep. X*-62337

Schubauer, G. B., Tchen, C. M. 1959. Turbulent Flow. In *Turbulent Flows and Heat Transfer*, ed. C. C. Lin, pp. 75–195; Vol. V of *High-Speed Aerodynamics and Jet Propulsion.* Princeton: Princeton Univ. Press

Settles, G. S., Dodson, L. J. 1991. Hypersonic shock/boundary layer interaction database. *NASA Contractor Rep.* 177577

Settles, G. S., Dolling, D. S. 1992. Swept shock wave boundary layer interaction. *AIAA Prog. Ser. Vol. on Tactical Missile Aerodynamics* 141: 505-74

Smith, D. R. 1993. *The effects of successive distortions on a turbulent boundary layer in a supersonic flow.* PhD thesis. Princeton Univ. 217 pp.

Smith, D. R., Fernando, E. M., Donovan, J. F., Smits, A. J. 1992. Conventional skin friction measurement techniques for strongly perturbed supersonic turbulent boundary layers. *Eur. J. Mech., B/Fluids* 6: 719–40

Smith, D. R., Smits, A. J. 1991. The rapid expansion of a turbulent boundary layer in a supersonic flow. *Theor. Comput. Fluid Dyn.* 2: 319–28

Smith, D. R., Smits, A. J. 1993. The simultaneous measurement of velocity and temperature fluctuations in the boundary layer of a supersonic flow. *Exp. Thermal Fluid Sci.* In press

Smith, M. W., Smits, A. J., Miles, R. B. 1989. Compressible boundary-layer density cross sections by UV Rayleigh scattering. *Opt. Lett* 14: 916–18

Smits, A. J., Dussauge, J.-P. 1989. Hot-wire anemometry in supersonic flow. In *AGARDograph-AG*-315, Chap. 5

Smits, A. J., Muck, K.-C. 1984. Constant-temperature hot-wire anemometer practice in supersonic flows. Part 2: The inclined wire. *Exp. Fluids* 2: 33–41

Smits, A. J., Muck, K.-C. 1987. Experimental study of three shock wave/turbulent boundary layer interactions. *J. Fluid Mech.* 182: 291–314

Smits, A. J., Spina, E. F., Alving, A. E., Smith., R. W., Fernando, E. M., Donovan, J. F. 1989. A comparison of the turbulence structure of subsonic and supersonic boundary layers. *Phys. Fluids A* 1: 1865–75

Smits, A. J., Wood, D. H. 1985. The response of turbulent boundary layers to sudden perturbations. *Annu. Rev. Fluid Mech.* 17: 321–58

Smits, A. J., Young, S. T. B., Bradshaw, P. 1979. The effect of short regions of high surface curvature on turbulent boundary layers. *J. Fluid Mech.* 94: 209–42

Spina, E. F. 1988. *Organized structures in a supersonic turbulent boundary layer.* PhD thesis. Princeton Univ. 233 pp.

Spina, E. F., Donovan, J. F., Smits, A. J. 1991a. On the structure of high-Reynolds-number supersonic turbulent boundary layers. *J. Fluid Mech.* 222: 293–327

Spina, E. F., Donovan, J. F., Smits, A. J. 1991b. Convection velocity in supersonic turbulent boundary layers. *Phys. Fluids A* 3: 3124–26

Spina, E. F., Smits, A. J. 1987. Organized structures in a compressible turbulent boundary layer. *J. Fluid Mech.* 182: 85–109

Sturek, W. B., Danberg, J. E. 1972a. Supersonic turbulent boundary layer in adverse pressure gradient. Part I: The experiment. *AIAA J.* 10: 475–80

Sturek, W. B., Danberg, J. E. 1972b. Super-sonic turbulent boundary layer in adverse pressure gradient. Part II: Data analysis. *AIAA J.* 10: 630–35

Taylor, M., Smits, A. J. 1984. The effects of a short region of concave curvature on a supersonic turbulent boundary layer. *AIAA Pap.* 84-0169

Thomann, H. 1968. Effect of streamwise wall curvature on heat transfer in a turbulent boundary layer. *J. Fluid Mech.* 33: 283–92

Thomas, C. D. 1974. Compressible turbulent boundary layers with combined air injection and pressure gradient. *Aeronaut. Res. Council Rep. Memo.* 3779

van Driest, E. R. 1951. Turbulent boundary layer in compressible fluids. *J. Aeronaut. Sci.* 18: 145–60

van Driest, E. R. 1956. On turbulent flow near a wall. *J. Aeronaut. Sci.* 23: 1007–11, 1036

Voisinet, R. L. P., Lee, R. E. 1972. Measurements of a Mach 4.9 zero pressure gradient boundary layer with heat transfer. *Nav. Ord. Lab. Tech. Rep.* 72-232

Voisinet, R. L. P., Lee, R. E. 1973. Measurements of a supersonic favorable-pressure-gradient turbulent boundary layer with heat transfer, Part I. *Nav. Ord. Lab. Tech. Rep.* 73-224

Wallace, J. E. 1969. Hypersonic turbulent boundary-layer measurements using an electron beam. *AIAA J.* 7: 757–59

Waltrup, P. J., Schetz, J. A. 1973. Supersonic turbulent boundary layer subjected to adverse pressure gradients. *AIAA J.* 11: 50–57

Watson, R. D. 1977. Wall cooling effects on hypersonic transitional/turbulent boundary layers at high Reynolds numbers. *AIAA J.* 15: 1455–61

Watson, R. D. 1978. Characteristics of Mach 10 transitional and turbulent boundary layers. *NASA TP*-1243

Watson, R. D., Harris, J. E., Anders, J. B. 1973. Measurements in a transitional/turbulent Mach 10 boundary layer at high Reynolds number. *AIAA Pap.* 73-165

Young, A. D. 1951. *The equations of motion and energy and the velocity profile of a turbulent boundary layer in a compressible fluid.* College of Aeronaut., Cranfield, Rep. No. 42

Zwarts, F. 1970. *Compressible turbulent boundary layers.* PhD thesis. McGill Univ.

Annu. Rev. Fluid Mech. 1994. 26 : 321–52

PREMIXED COMBUSTION AND GASDYNAMICS

P. Clavin

Laboratoire de Recherche en Combustion, URA 1117 CNRS/Université Aix-Marseille I, S. 252 Centre St.-Jérôme, 13397 Marseille cedex 20, France

KEY WORDS: laminar and turbulent flames, flammability limits, quasi-isobaric ignition, sound generation, acoustic instabilities of combustion

1. INTRODUCTION

Combustion science is a fascinating field of nonequilibrium phenomena in complex systems. Although combustion processes are very diverse in nature, nevertheless, they have two fundamental characteristics: heat release by excess energy of chemical bonds and a strong temperature dependence of the chemical reaction rate. The basic laws of combustion may be obtained analytically by taking this thermal nonlinearity to its extreme limit. Yakov Borisovich Zeldovich is at the origin of this remarkable asymptotic method and one is readily convinced by reading Volume I of his selected works (Ostriker 1992), that nobody has contributed more than him to the understanding of combustion theory.

The present paper will be restricted to combustion phenomena in premixed gases; detonations are not included. The theory of the thermal propagation mechanism of flames is briefly recalled in Section 2. New results on the intriguing cool flame phenomena are also presented in this section. Results obtained during this past decade on dynamics of flame fronts are outlined in Section 3. Flammability limits and ignition problems are considered in Sections 4 and 5. Section 6 presents the current status of turbulent flame theory. The rest of this paper is devoted to compressible effects. Recent theoretical results on sound generation by turbulent flames are presented in Section 7. The last section is devoted to acoustic instabilities of combustion.

Combustion in gases is governed by the following set of equations: mass and momentum conservation (inviscid approximation), equation of state (ideal-gas law), and energy and species conservation,

0066–4189/94/0115–0321$05.00

$$\partial\rho/\partial t+\nabla\cdot(\rho\mathbf{v}) = 0, \qquad \rho D\mathbf{v}/Dt = -\nabla p, \qquad p = (C_p-C_v)\rho T, \quad \text{(1a,b,c)}$$

$$\rho C_p DT/Dt = Dp/Dt+\nabla\cdot(\lambda\nabla T)+\dot{Q}, \qquad \dot{Q} = \sum_j Q^{(j)}\dot{W}^{(j)}, \qquad \text{(1d,e)}$$

$$\rho DY_i/Dt = \nabla\cdot(\rho D_i\nabla Y_i)+\sum_j v_i^{(j)}\dot{W}^{(j)}, \qquad \text{(1f)}$$

where ρ is density, T is temperature, $\mathbf{v}$ is velocity, C_p and C_v are the specific heats, λ is the thermal conductivity, $\dot{Q}$ is the rate of heat release per unit volume, $\dot{W}^{(j)}$ is the reaction rate of the jth reaction, Y_i is the mass fraction of species i, and $v_i^{(j)}$ and $Q^{(j)}$ are the changes of mass of species i and enthalpy by reaction j. Fick's laws, which are valid with a good accuracy when an abundant inert gas is present, and Fourier's law have been used for simplicity. D_i is the binary diffusion coefficient of species i in the species present in excess. The reaction rate of fuel-oxidant mixtures is very sensitive to temperature. Under normal conditions ($p \approx 1$ atm, $T_u \approx 300$ K,) reactive mixtures of common use are frozen at an initial composition far from chemical equilibrium, $\dot{W}^{(j)} \approx 0$. But the reaction time at the burned gas temperature $T_b \approx 1500\text{--}3000$ K, is as short as τ_r $(T_b) \approx 10^{-6}$ s. Such a strong temperature dependence is due to large activation energies, $E \approx 20\text{--}50$ Kcal/mol, $E/RT_b \approx 10$. This is well described by a one-step model for the exothermic and irreversible decomposition of a reactant R into a product P governed by an Arrhenius law

$$R \rightarrow P+Q, \qquad \dot{W} = \rho_b Y/\tau_r(T), \qquad T_b-T_u = QY_u/C_p, \qquad \text{(2a,b,c)}$$

where Y is the mass fraction of R, Y_u its initial value, and T_u and T_b are the temperature of the unburned and burned gases, respectively. The characteristic reaction time, $\tau_r(T) = \tau_c\exp(E/RT)$, is much larger than the elastic collision time τ_c, $\tau_c(T_b) \approx 3.0\times10^{-10}$ s, and varies strongly with temperature, $\tau_r(T_b)/\tau_r(T_u) \approx 10^{-18}$ typically. When the reduced temperature $\theta = (T-T_u)/(T_b-T_u)$ and the Zeldovich number $\beta \equiv aE/RT_b$ are introduced [$a \equiv (T_b-T_u)/T_b \approx 0.85$, $\beta \approx 8$], the reduced reaction rate takes the approximate form $\tau_r(T_b)/\tau_r(T) \approx \exp[-\beta(1-\theta)]$ valid for $\beta \gg 1$ and $(1-\theta) = O(1/\beta)$. The relative reaction rate becomes negligibly small as soon as the temperature decreases slightly below the burned gas temperature, and the reaction rate is concentrated at high temperature where $(T_b-T)/(T_b-T_u) \equiv (1-\theta) = O(1/\beta)$. This stands at the root of the pioneering work of Zeldovich & Frank-Kamenetskii (ZFK, 1938) which is at the origin of the modern development of combustion theory. To summarize, the validity of a hydrodynamical description (1a–f) relies on the inequality $E \gg RT$ which also ensures that the reactive mixture is frozen at normal conditions, far from the equilibrium composition.

2. LAMINAR FLAME PROPAGATION

Propagation of the combustion into a fresh mixture from a hot spot may result from diffusion of the heat released by the reaction. This is the basic mechanism in flames.

2.1 *Quasi-Isobaric Approximation*

A balance of the rates of heat release and heat diffusion yields an approximate value of the flame velocity: $U_L \approx [D_T/\tau_r(T_b)]^{1/2}$, where $D_T \equiv \lambda/\rho C_p$ is the thermal diffusivity. According to the kinetic theory of gases one has $D_T \approx c^2\tau_c$, where c is the sound speed, and the Mach number $M_L \equiv U_L/c$ with U_L estimated above, is a small term when $E \gg RT_b$, $M_L \approx [\tau_c/\tau_r(T_b)]^{1/2} \approx \exp(-E/2RT_b)$. For typical flow velocity variations $\delta u \approx U_L$, hydrodynamical and compressible acoustic effects yield $\delta p/p = O(M_L^2)$ and $\delta p/p = O(M_L)$, respectively. Thus, pressure variations across the flame may be neglected in (1c), $\rho T = Ct$, as well as in the energy equation (1d) which reduces to a reaction-diffusion equation similar to (1f). According to (1a), the mass flux $m = \rho u$ is constant across a plane flame propagating at a constant velocity U_L. Then, species and energy equations (1d–f), written in the referential frame of the flame in terms of $m \equiv \rho_u U_L = \rho_b U_b$ with $\rho D/Dt \to m d/dx$, form a closed system of two nonlinear ordinary differential equations,

$$mC_p \, dT/dx - \lambda \, d^2T/dx^2 = \sum_j Q^{(j)}\dot{W}^{(j)}, \tag{3a}$$

$$m \, dY_i/dx - \rho D_i \, d^2Y/dx^2 = \sum_j v_i^{(j)}\dot{W}^{(j)}, \tag{3b}$$

where λ and ρD_i have been assumed constant for simplicity and where m is an eigenvalue determining the flame velocity U_L, obtained by imposing upstream and downstream boundary conditions; $x = -\infty$ corresponds to the initial unburned state $Y_i = Y_{iu}$ and $T = T_u$ with $W_u^{(j)} \approx 0$; $x = +\infty$ represents burned gases at equilibrium. Numerical solutions of such systems of equations are carried out routinely for hydrocarbon-air mixtures involving complex chemistry—typically 350 reactions and 60 species or more. As shown later in particular examples, most of the basic phenomena may be described and understood by reduced schemes, typically 3 or 4 reactions and 2 or 3 intermediate species, which are derived more or less systematically from real chemistry (see Smooke 1992, Peters & Rogg 1993).

2.2 *Thermal Propagation Mechanism*

The flame propagation problem (3a,b) belongs to the wider class of reaction-diffusion waves. The problem is mathematically well posed when the initial conditions at $x = -\infty$ correspond to a steady state, $W_u^{(j)} = 0$. Usually, the mathematical structure of a traveling wave solution depends on the stability property of this state: metastable, unstable, or excitable. This is not the case for ordinary flames whose reaction rates are highly temperature sensitive. The underlying physics may be understood in the framework of a simple model. When the Lewis number—the ratio of the thermal diffusivity to the molecular diffusivity of the limiting reactant, $Le \equiv D_T/D$—is unity, the mass fraction and temperature profiles of model (2a) are similar, $C_p(T - T_u) + Q(Y - Y_u) = 0$. System (3a,b) reduces to a single reaction-diffusion equation which can be written in a dimensionless form as:

$$\mu \, d\theta/d\xi - d^2\theta/d\xi^2 = \omega(\theta) \quad \text{with} \quad \xi = -\infty: \theta = 0, \quad \xi = +\infty: \theta = 1$$

$$(4a,b)$$

where $\xi \equiv x/d_L$, $\mu \equiv m\tau_L/\rho_b d_L$, $\omega(\theta) = (\beta^2/2)(1 - \theta)\exp[-\beta(1 - \theta)]$, and where the length and time scales are $d_L \equiv (\tau_L D_T \rho/\rho_b)^{1/2}$ and $\tau_L \equiv \beta^2\tau_r(T_b)/2$. The factor $\beta^2/2$ is introduced for convenience by anticipating the ZFK result, written in this dimensionless form as: $\mu = \mu_{ZFK} = 1 + O(1/\beta)$ with $\beta \gg 1$. This result which is derived below, yields flame velocities, transit times, and flame thickness which agree with experimental data for ordinary flames:

$$U_L = (\rho_b/\rho_u)\sqrt{D_{Tb}/\tau_L}, \qquad \tau_L = (\beta^2/2)\tau_c \exp(E/RT_b), \qquad d_L \equiv U_L\tau_L,$$

$$(4c,d,e)$$

yielding $U_L \approx 37$ cm/s, $d_L \approx 1.2 \times 10^{-2}$ cm, $\tau_L \approx 3 \times 10^{-4}$ s for $\beta = 10$, $D_{Tb} \approx 3$ cm²/s, $\tau_r(T_b) \approx 10^{-6}$ s.

When $\beta \gg 1$, the reaction rate is located in a thin layer of thickness d_L/β, much smaller than the total flame thickness d_L and in which the temperature is close to T_b, $\theta = 1 + O(1/\beta)$ (Figure 1). Equation (4a) reduces to $d^2\theta/d\xi^2 \approx -\omega(\theta)$ in this layer (arbitrarily located at $\xi = 0$) in which the heat flux is

$$d\theta/d\xi \approx \left[2\int_\theta^1 \omega(\theta')\,d\theta'\right]^{1/2}.$$

The heat flux leaving the reaction layer which warms the upstream frozen gases is obtained by noticing that the integral converges quickly when θ decreases slightly from unity,

$$2\int_\theta^1 \omega(\theta')\,d\theta' = -\int_{-\beta(1-\theta)}^0 \exp(\Theta)\,\Theta\,d\Theta \to 1$$

when $\beta(1 - \theta) \to \infty$, a limit which is reached for small values of $1 - \theta$ when

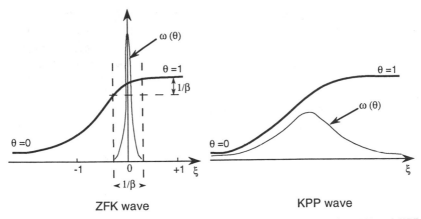

Figure 1   Sketch of typical profiles across reaction-diffusion waves in the ZFK and KPP regimes.

β is sufficiently large. The final result is obtained by matching this inner value of the heat flux with its outer expression $d\theta/d\xi|_{\xi=0-}$ computed from the solution $\theta = \exp \mu\xi$ of the preheated zone at $\xi < 0$ in which $\omega \approx 0$ and $\mu \, d\theta/d\xi - d\theta/d\xi^2 \approx 0$. This gives

$$\mu_{ZFK} \equiv \left[2 \int_0^1 \omega(\theta) \, d\theta \right]^{1/2} = 1 + O(1/\beta)$$

leading to result (4c–e). The initial state $\theta = 0$ of (4a,b) is not, strictly speaking, a steady state. The initial reaction rate $\omega(\theta = 0) = \beta^2/2 \exp(-\beta)$ is a transcendentally small term when $\beta \gg 1$, and (4c) is indeed the solution of a two time-scale problem (cold boundary difficulty).

Historically problem (4a,b) was first studied by Kolmogorov et al (KPP 1937) and Fisher (1937) in the context of population genetics, for an unstable initial state, $\omega(\theta = 0) = 0$, $\omega'_u \equiv d\omega/d\theta|_{\theta=0} > 0$ and for a convex and positive function $\omega(\theta)$ in $\theta \in {]}0, 1{[}$, with a typical form $\theta(1-\theta)$. In this case, when attention is restricted to traveling wave solutions satisfying $\theta(\xi) \geq 0$, there is a continuous set of possible reduced velocities μ bounded from below by the KPP value, $\mu_{KPP} = 2\omega'^{1/2}_u$, corresponding to a dimensional wave velocity $U_{KPP} = 2(D/\tau_u)^{1/2}$ where τ_u^{-1} is the linear growth rate at initial conditions, $\theta = 0$. This result is indeed not restricted to a convex function $\omega(\theta)$, but it differs drastically from the ZFK result for flames where the propagation is controlled by the reaction rate $\tau_r^{-1}(T_b)$ at the burned temperature, $\theta = 1$. Moreover, the KPP solution is selected in the long time limit from all initial conditions well localized in space, i.e. $\theta \neq 0$

only in a bounded ξ-domain. Following Zeldovich et al 1985 (p. 298) a linearization around the initial state, $\theta = 0$, put the unsteady version of (4a) in the form $\partial\theta/\partial\tau - \partial^2\theta/\partial\xi^2 = \theta\omega'_u$ with a Green's function $\theta = (4\pi\tau)^{-1/2}\exp[-(\xi^2/4\tau)+\omega'_u\tau]$ [i.e. a solution associated with the initial condition $\theta = \delta(\xi)$ at $\tau = 0$]. The velocity of points $\ln\theta = $ const. of the Green's function, tends to μ_{KPP} when $\tau \to \infty$. This clearly exhibits the nature of KPP solutions as waves slaved by the unstable initial state at $\xi = -\infty$ ($\theta \approx 0$). The KPP result is indeed no longer valid when the production rate $\omega(\theta)$ is sufficiently nonlinear, and a transition appears when $\omega(\theta)$ is modified continuously from a weakly nonlinear form as $\theta(1-\theta)$, to a strongly nonlinear one as $(1-\theta)\exp[-\beta(1-\theta)]$. This problem has been investigated in the phase space of (4a,b) by Zeldovich et al (1979) and later by Clavin & Liñan (1984). By using a one parameter family, as for example $\omega(\theta) = (\beta^2/2)(1-\theta)\{\exp[\beta(\theta-1)]-\exp[-\beta]\}$ and by varying β from zero to large values, it is shown that the KPP solution is valid throughout the range $0 < \beta < \beta^*$. A transition is observed at a critical value $\beta = \beta^*$ ($\beta^* = 3.04$ in this example) at which the trajectory of the marginal solution (lower bound of μ) changes nature in phase space. As a result, the mathematical structure of the marginal solution, which is associated with the lower bound eigenvalue μ, becomes different from all the others in the spectrum and takes the form of a ZFK solution when β is further increased. It has also been shown rigorously that the marginal solution is always selected as $\tau \to \infty$ for any disturbance that is initially of bounded support (Aronson & Weinberger 1978). The results presented in this section exhibit the special character of flame propagation, due to the strong nonlinear temperature dependence.

2.2 *Diffusion of Active Species and Cool Flame Propagation*

Model (2a–c) describes well the thermal propagation mechanism but cannot respresent all the phenomena of real flames. The propagation mechanism is influenced by a complex kinetic scheme. Chain reactions involving active intermediate species such as radicals H, OH, etc are essential in combustion kinetics and control the limits of explosion (see Zeldovich et al 1985; pp. 39–46). An autocatalytic reaction such as $X + R \to 2X$ with an X production rate proportional to $Y_X Y_R$, may sustain a traveling wave propagating into a medium R, $Y_X = 0$, $Y_R = Y_{Ru}$, through diffusion of X radicals only. Such isothermal waves are observed in aqueous arsenous acid-iodate systems, studied by Harrison & Showalter (1986), and used recently by Ronney (1992) in turbulent flows. In the simplified case $D_X = D_R$, this wave is a KPP solution of (4a,b) with $\theta \to \varphi$, $\omega(\varphi) = \dot\varphi(1-\varphi)$, where $\varphi \equiv Y_X/Y_{Ru}$ and $(1-\varphi) \equiv Y_R/Y_{Ru}$. But the thermal propagation mechanism becomes dominant as soon as the reaction is exothermic with a sufficiently large activation energy E. A transition, similar to the KPP-ZFK one described above, appears when E/RT_b increases from 0 to large values yielding a flame velocity that differs from (4c) only through numerical factors as $(D_X/D_T)^n$. Analytical results of flame structures have been obtained for ozone, methane-air (see references in Williams 1992), and halogen-hydrogen (Liñan & Clavin 1987).

Chain reactions and species diffusion play dominant roles in *cool flames*—exotic phenomena observed in most hydrocarbon-air mixtures at initial temperatures in the range 500–800 K during the induction period preceding the thermal runaway. Cool flames correspond to incomplete combustion with a weak exothermicity, $\Delta T \approx 50$–200 K. They yield traveling waves which may occur in succession and which are materialized by a faint bluish luminescence characteristic of excited formaldehyde radicals.

A theory has been developed for the combustion of carbon disulfide in oxygen (see Zeldovich et al 1985, pp. 123–29 and 415–28). These cool flames have been represented by isothermal waves propagating into a metastable state ($Y_X = 0$: $\omega'_u < 0$), sustained by the competition of a chain branching process $X + X + R \rightarrow 3X + P_1$ and a volume loss of active centers $X \rightarrow P_2$, where P_1 and P_2 are inert products. When $D_X = D_R$ the problem reduces to (4a,b) with $0 \rightarrow \varphi$, $\omega = \varphi(1 - \varphi)(\varphi - \varphi^*)$, where φ^* is between 0 and $1/2$; this problem has a unique solution $\mu = 2^{1/2}(1/2 - \varphi^*)$, $d\varphi/d\xi = 2^{-1/2}\varphi(1 - \varphi)$.

However, this description is not valid for hydrocarbon combustion in which cool flames are associated with an exothermic autocatalytic process that is self quenched by the temperature increase. Such a negative temperature coefficient of the reaction rate, appearing in the temperature range 700–900 K, is rather unique in combustion and is considered as the main characteristic of hydrocarbon mixtures exhibiting cool flames. The temperature rise in cool flames is of kinetic significance but for a reason that is just the opposite to ordinary flames. This chemistry of combustion at moderate temperature is very complex, but a reduced scheme proposed initially by Yang & Gray (1969), represents all these characteristics and has been justified for acetaldehyde oxidation. In their scheme, a weakly exothermic chain branching reaction $X + R \rightarrow 2X + Q$, with a moderate activation energy E competes with two chain-breaking reactions. The first one, $X \rightarrow P_1 + Q_1$, is very exothermic and has a high activation energy $E_1 > E$. Since its X consumption rate strongly increases with temperature, this chain breaking is dominant at high temperature, $T > T_c$, and quenches the reaction above a cross-over temperature $T_c \approx 700$ K, defined by a balance between production and destruction rates of X radicals by chain branching and chain breaking. The second chain breaking $X \rightarrow P_2$ is athermal, $Q_2 = 0$, with a zero activation energy $E_2 = 0$. Its X consumption rate is higher than the X production rate at lower temperature $T < T_i$ and quenches the reaction below an ignition temperature, $T_i \approx 500 \, \text{K} < T_c$, defined by a similar balance. The limiting process is not the consumption of the reactant R but the self-quenching at high temperature $(T > T_c)$. The final temperature is an unknown which cannot be obtained by a simple energy balance such as (2c). The cool flame structure is described by a system of two reaction-diffusion equations for X and T. The analysis of such waves was carried out in terms of the initial temperature T_u by Nicoli et al (1990) for $D_X \neq D_T$. The burned gas temperature T_b is slightly higher than T_c and does not vary much with T_u. The initial state $Y_X = 0$ is stable against Y_X fluctuations for $T_u < T_i$ and unstable for $T_i < T_u < T_c$. Cool flame propagation is possible for $T_u < T_c$ but there is a critical temperature T_o in the unstable domain $T_i < T_o < T_c$, denoting a change in the propagation regime. When $T_o < T_u < T_c$, the propagation is sustained by diffusion of the X chain carrier and the wave is of the KPP type with a velocity $U = 2(D_X/\tau_u)^{1/2}$, where $1/\tau_u$ is the X linear growth rate at T_u, varying from zero at $T_u = T_i$ to zero at $T_u = T_c$ with a maximum between T_o and T_c. At $T_u = T_o$ a transition appears similar to the KPP-ZFK one. For $T_u < T_o$, the flame velocity decreases with T_u but is no longer zero at $T_u = T_i$ and the propagation is now controlled by heat conduction. Small heat losses cooling the burned gases below T_c are responsible for successive traveling waves.

3.   DYNAMICS OF LAMINAR FLAMES

Our understanding of laminar flame dynamics has developed mainly since 1977, and has been already reviewed by Sivashinsky (1983), Clavin (1985), and Williams (1985, 1992). Only some key points are recalled here. In addition to reaction-diffusion phenomena, gas expansion induces strong hydrodynamical effects on nonplanar flames. As a result of variable-density effects, the energy and species conservation equations (1d–f) are coupled with fluid mechanics (1a–c) in wrinkled flames, even in the quasi-isobaric approximation. The one-step model (2a) conveniently describes such phenomena, and when $Le \neq 1$, energy and species conservation reduces to two coupled equations for θ and the nondimensional mass fraction $\psi \equiv Y/Y_u$. They form a closed set only in the planar case for which an unsteady version of 3(a,b) is obtained by introducing a mass-weighted coordinate (e.g. Clavin et al 1990), to give in a dimensionless form

$$\frac{\partial \theta}{\partial \tau} - \frac{\partial^2}{\partial \xi^2} \theta = \omega(\theta, \psi), \qquad \frac{\partial \psi}{\partial \tau} - \frac{1}{Le} \frac{\partial^2}{\partial \xi^2} \psi = -\omega(\theta, \psi) \qquad \text{(5a,b)}$$

$$\omega(\theta,\psi) \equiv [\beta^{n+1}/(n+1!\,Le^n)]\psi^n \exp[-\beta(1-\theta)] \tag{5c}$$

with boundary conditions,

$$\xi = -\infty: \theta = 0,\ \psi = 1 \quad \text{and} \quad \xi = +\infty: \theta = 1,\ \psi = 0. \tag{5d}$$

Here $n \neq 1$ is the order of the reaction. Equations (4a,b) are a particular case of (5a,b): $\partial/\partial\tau \to M\partial/\partial\xi$, $n = 1$, $Le = 1 \Rightarrow \psi = 1-\theta$.

For planar flames, the effects of $n \neq 1$ and $Le \neq 1$ are simply to change the factor $\beta^2/2$ of τ_L into $\beta^{n+1}/(n+1)!\,Le^n$ in (4c,d). But due to the temperature sensitivity, a difference of the diffusion coefficients of heat conduction and molecular diffusion, $Le \neq 1$, gives rise to important unsteady phenomena. In the limit $\beta \gg 1$, the reaction layer of model (2a–c) is considered as a "sheet" with an internal structure locally planar and quasi-steady. The jump conditions for θ and ψ are obtained by an analysis similar to that of ZFK (1938) but with $\theta_f \neq 1$ and $Le \neq 1$,

$$\partial\theta/\partial\xi|_{\xi=\alpha^-} = \exp[\beta(\theta_f-1)/2], \qquad [\partial\theta/\partial\xi + (1/Le)\partial\psi/\partial\xi]_{\xi=\alpha^-}^{\xi=\alpha^+} = 0. \tag{5e,f}$$

Here ξ is the coordinate normal to the front in the referential frame moving with the flame, $\xi = \alpha$ is the position of the reaction layer, and $\xi = \alpha^-$ and $\xi = \alpha^+$ denote the unburned and burned gas sides. Equation (5e) is a kinetic condition; jump condition (5f) results from the energy balance across the reaction layer. Conditions (5e,f) are valid in the limit $\beta \to \infty$, and unsteady problems (5a–d) reduce to free boundary problems where θ and ψ satisfy linear equations (5a,b) with $\omega = 0$, and boundary conditions (5d–f). This also applies to wrinkled flames but with more complex equations in the external zones.

As a result of the temperature sensitivity in (5e), small variations of flame temperature θ_f—$\beta(\theta_f-1) = O(1)$ when $\beta \to \infty$—produce strong effects on flame dynamics. As θ_f is controlled by a competition between heat conduction and reactant diffusion in the preheated zone, a small unbalance has a large influence. Thus, distinguished limits, such as $\beta(Le-1) = O(1)$ when $\beta \to \infty$, are the key points in the analysis of dynamics of wrinkled flames.

Transitions between different regimes appear in the domain of parameters $\beta(Le-1) = O(1)$ which is relevant for gaseous mixtures where Le is close to unity. If the free boundary problem is solved directly with β considered as a parameter of order unity in (5e,f) (e.g. Sivashinsky 1977), then $(Le-1)$ and β appear in the end results only through $\beta(Le-1)$, showing a posteriori the distinguished limit which has to be considered.

Such an approach was also used for solid combustion (gasless), outside its domain of validity for $Le \to \infty$ and finite β (e.g. Sivashinsky 1981). Comparisons with direct numerical simulations of (5a–c), as those of

Clavin et al (1986) for the retrocombustion of coal, exhibit a good agreement, showing that the dominant phenomenon of temperature sensitivity, is well represented by this phenomenological approach. In analytical studies of more complex cases, such as diffusive-thermal instabilities of nonadiabatic flames (Joulin & Clavin 1979) or acoustic instability of flames in sprays (Clavin & Sun 1991), it is necessary to use disginguished limits right from the very beginning.

Flame instabilities have been studied in great detail by this method with model (2a–c) (e.g. Sivashinsky 1983, Clavin 1985). Viscous effects do not play any significant role. Gas expansion produces a natural hydrodynamic instability of flames at all wavelengths. This Darrieus-Landau instability is responsible for the smooth curved front of flames propagating in tubes. Moreover, additional diffusive-thermal instabilities appear both above a positive critical value of $\beta(Le-1)$ and below a negative one (Zeldovich et al 1985, p. 324).

A nonlinear equation describing the evolution of a flame front succumbing to the Darrieus-Landau instability has been obtained by Sivashinsky (1977) in the limit of a small gas expansion (Sivashinsky 1977). Analytical and numerical investigations of such an equation have been carried out recently by Joulin (1989) and Denet (1993) to study the dynamics of curved fronts, a topic reviewed by Pelcé (1988).

Detailed analyses, coupling diffusive-thermal mechanisms and variable-density effects were carried out by Pelcé & Clavin (1982) and Searby & Clavin (1986) to study the dynamical properties of a flame propagating downward in a quiescent gas and also in a weakly turbulent flow. Most planar flames propagating downwards with $U_L < 12$ cm/s are predicted to be stable. As qualitatively explained by Markstein (1964), diffusive-thermal phenomena and buoyancy stabilize, respectively, the short and long wavelengths. Experiments of Searby et al (1983) and Searby & Quinard (1990) show good quantitative agreement with the theory. When the wavelengths of wrinkling are much larger than the flame thickness, the flame may be considered as an hydrodynamic discontinuity.

Following the multiple scale method developed by Clavin & Williams (1982) for wrinkled flames, a geometrical model of the flame front has been obtained independently by Matalon & Matkowsky (1982) and Clavin & Joulin (1983) (see also Matalon 1983), in which the normal burning velocity U_n is locally modified by curvature of the front and inhomogeneities of the flow in a form $U_n/U_L - 1 = M\tau_L\sigma^{-1}d\sigma/dt$, where $\sigma^{-1}d\sigma/dt$ is the total stretch rate of an elementary area of flame surface by these two effects. Theoretical expressions of Markstein number M were obtained by Clavin & Williams (1982) for a simple model (2a–c) and also for more realistic cases (see references in Clavin 1985 and Nicoli & Clavin 1987).

The values measured by Searby & Quinard (1990) are in good agreement with theory. However, it must be kept in mind that for such a local law accounting for a nonzero flame thickness, M depends on the way the flame surface is defined and which flow (burned/unburned) is considered to calculate the stretch rate (Frankel & Sivashinsky 1983, Clavin & Joulin 1983). But the resulting dynamics are independent of these choices.

4. FLAMMABILITY LIMITS

Flammability limits are defined by critical compositions beyond which planar flame propagation is no longer possible. Chemical kinetics play an important role in these phenomena. But, a sharp transition exists only when kinetic effects are coupled with small external perturbations which are always present, such as heat losses. This problem is well understood for H_2-O_2 flames and similar mechanisms are expected to hold for hydrocarbon flames. The simplest model, representative of the H_2 rich flammability limits in H_2-O_2 mixtures, is presented here. Its derivation may be found in Clavin (1985).

Among active species, H radicals play a central role. The H production by chain branching

$$R + X \to 2X + Q_1 \tag{6a}$$

competes with two trimolecular chain terminations

$$R + X + M \to P + M + Q_2, \qquad X + X + M \to P_2 + M + Q_3, \tag{6b,c}$$

where X and R represent respectively H and the limiting component (O_2 for H_2-rich mixtures), M is a third body, and P_1 and P_2 are inert products. The three reactions are exothermic; the chain branching one (6a) has a high activation energy E and its rate increases with temperature; the two others (6b,c) with zero activation energy are not influenced by temperature variations. The overall X production rate by (6a,b) is $Y_R Y_X k(T)$ with $k(T) \equiv [B_1 \exp(E/RT) - B_2]$, where B_1 and B_2 are prefactors of reactions (6a,b). A crossover temperature T^* is defined by $\exp(E/RT^*) = B_2/B_1$ ($T^* \approx 960$ K at $p = 1$ atm). For initial conditions (Y_{Ru}, T_u) such that the flame temperature T_b would be below T^*, the reaction cannot proceed through (6a–c) because since $Y_X = 0$ is stable at all temperatures in the range $[T_u, T_b]$, $k(T) < 0$, R cannot be consumed and heat cannot be released. In opposite situations, $T_b > T^*$, $k(T_b) > 0$, a wave propagates as a ZFK solution of (6a–c) because heat release is conditioned by the X production rate which is very sensitive to temperature through $k(T)$.

Analytical studies of this model were carried out by Joulin et al (1985). Condition $T_b = T^*$ corresponds to a H_2 mole fraction of 0.96 in a H_2-rich mixture of H_2-O_2 at $p = 1$ atm. The flame velocity U_L obtained from (6a–c) decreases toward zero when the initial composition approaches its critical value. The planar flame velocity obtained from direct numerical simulations with a complete kinetic scheme by He & Clavin (1993) does not vanish at the critical conditions of (6a–c), but decreases toward very small values, less than 1 mm/s. Even slower ($U_L < 0.1$ mm/s) and thicker ($d_L > 10$ cm) flames are obtained beyond the flammability limits of (6a–c) which correspond merely to a change in the propagation regime. As for the transition between the second and third explosion limits (see Williams 1985, p. 574 and Zeldovich et al 1985, p. 38 and p. 45), a secondary slow burning path, involving the reactivity of the HO_2 radical, takes over from (6a–c). As explained later, such slowly propagating planar flames cannot be observed in experiments. A useful two-equation model which represents well the numerical results of He & Clavin (1993) near the limits may be obtained. When the limit is approached, $T_b \to T^*$, the H production rate by (6a,b) decreases to zero, $k \to 0$, and the H recombination (6c) with a constant prefactor, $B_3 \approx Ct$, maintains the mass fraction of H at a small value and may justify the quasi-steady state approximation for H radicals: $Y_X Y_R k - Y_X^2 B_3 \approx 0$, i.e. $Y_X \approx Y_R k / B_3$ for $T > T^*$ and $Y_X = 0$ for $T < T^*$, where $k < 0$. Then, since Y_X is expressed in terms of Y_R, the flame problem (6a–c) only requires solving two equations (3a,b) for T and Y_R. This approximation has proven valid for a sufficiently large prefactor B_3 which is not attained in real H_2-O_2 mixtures. However, the analytical results of the two-equations model, when compared with numerical results based on a full scheme, show only small quantitative differences. The resulting reduced model is similar to (5a,b) but with a cut-off of the reaction rate $\omega(\theta, \psi)$ at high temperature $T^* \approx 960$ K below which the reaction rate vanishes

$$\theta \geq \theta^*: \ \omega(\theta, \psi) = (\beta^3/4Le^2)\psi^2 \{\exp[\beta(\theta - 1)] - \exp[-\varepsilon]\},$$

$$\theta \leq \theta^*: \ \omega \equiv 0 \tag{7a,b}$$

with $\theta^* \equiv (T^* - T_u)/(T_b - T_u) \equiv 1 - \varepsilon/\beta$ and where $\varepsilon \equiv \beta(T_b - T^*)/(T_b - T_u)$ is a new parameter which characterizes the distance from the flammability limit at $\varepsilon = 0$. Propagation of plane flames is only possible for $\varepsilon > 0$ and the standard results of model (5a–c) are recovered when $\varepsilon \to +\infty$.

Model (7a,b) may be analyzed by a ZFK method for $\beta \gg 1$ and $\varepsilon = O(1)$. As a result, the flame velocity decreases to zero when $\varepsilon \to 0^+$: $U_L(\varepsilon)/U_L(\infty) \approx \varepsilon^2$, and the flame thickness diverges: $d_L^{(\varepsilon)}/d_L^{(\infty)} \approx \varepsilon^{-2}$. Temperature sensitivity increases at flammability limits: $U_L \approx (T_b - T^*)^2 \Rightarrow d \ln U_L/d \ln T_b \approx 2T^*/(T_b - T^*)$. Kinetic induced dynamical effects appear for $\beta^* \gg 1$ when $\varepsilon = O(1)$, i.e. $(T_b - T^*)/(T_b - T_u) = O(1/\beta^*)$. They become dominant when $\varepsilon \to 0^+$ and vanish for $\varepsilon = O(\beta^*)$. They are located in a narrow domain around the flammability limit. For small ε, the net effect on flame dynamics is only to change β into $4\beta^*/\varepsilon$ in the final results: $\beta(Le-1) \to 4\beta^*(Le-1)/\varepsilon$. Preferential diffusion mechanisms, $D_i \neq D_T$, are thus reinforced and multidimensional effects are even more important. This is the case for the diffusive-thermal instabilities. The corresponding stability domain of planar flames shrinks to zero when $\varepsilon \to 0^+$, $Le-1 = O(1/\beta) \Rightarrow Le-1 = O(\varepsilon/\beta^*) \to 0$. For similar reasons, flame quenching which is produced by small heat losses with a relative order of magnitude $1/\beta$, is also reinforced when $\varepsilon \to 0^+$. This explains why planar flames cannot be observed experimentally in the neighborhood of flammability limits.

5. QUASI-ISOBARIC IGNITION PROBLEMS

Ignition is an important problem for spark-ignition engines and cryogenic liquid rocket engines where difficulties appear both in lean hydrocarbon-air mixtures and rich hydrogen-oxygen mixtures. These mixtures correspond to abundant species much smaller than the limiting species, $Le > 1$. In H_2-O_2 systems, $Le = 2$ for H_2-rich and $Le - 0.3$ for H_2-lean mixtures respectively. Competition between heat conduction and molecular diffusion of reactants is an essential mechanism in ignition problems.

The experiments of Ronney (1988, 1990) on near-limit flames at microgravity have clearly pointed out the distinction between ignition limits and flammability limits. The numerical studies of ignition of H_2-O_2 mixtures by a hot pocket of burned gases carried out by He & Clavin (1993) in a spherical geometry and under adiabatic conditions, are also instructive. The critical size of the initial pocket below which ignition fails, is determined as a function of the equivalence ratio. In H_2-rich mixtures, the critical radius is much larger than the flame thickness and diverges at a H_2 mole fraction 0.945 before the planar flammability limit at 0.96. Flames cannot be ignited when the H_2 mole fraction is in the range [0.945, 0.96] where the burning velocity of planar flames varies from 50 cm/s to a few mm/s. In H_2-lean mixtures the situation is quite different. The critical size is smaller than the flame thickness. Beyond the flammability limit, in a region where planar flames cannot propagate, small hot pockets produce ignition and spherical flames grow from a small radius up to a large one—typically half the box size, at which quenching occurs. These numerical results are in agreement with Ronney's experiments. A high-energy spark (up to 10 J) is required for ignition in the neighborhood of the H_2-rich limit. Lean mixtures are easily ignited with only a few millijoules but, as expected from the theory of the diffusive-thermal instability at small Lewis number, cellular structures appear on the expanding spherical flame front. Moreover, in some lean conditions, small quasi-steady "flame balls" are observed.

A solution for spherical and steady premixed flames in motionless gas was obtained by Zeldovich with the standard model (2a–c); see Zeldovich et al (1985, p. 327). Equations to be solved are

$$\frac{1}{\xi^2}\frac{\partial}{\partial\xi}\left(\xi^2\frac{\partial}{\partial\xi}\theta\right) = \omega(\theta,\psi), \qquad \frac{1}{Le}\frac{1}{\xi^2}\frac{\partial}{\partial\xi}\left(\xi^2\frac{\partial}{\partial\xi}\psi\right) = -\omega(\theta,\psi) \qquad \text{(8a,b)}$$

and (5c), where ξ is the dimensionless radial coordinate, $\xi = +\infty$: $\theta = 0$, $\psi = 1$; $\xi = 0$: $\partial\theta/\partial\xi = 0$, $\psi = 0$. The flame temperature T_f depends on the diffusive properties, $\theta_f = 1/Le$. The flame radius r_f is obtained when jump conditions (5e,f) are applied to the solutions of (8a,b) for $\xi > \alpha$ where $\omega = 0$ and for $\xi < \alpha$ where $\psi = 0$, $\theta = \theta_f$, to give $\alpha \equiv r_f/d_L = \exp[(\beta/2)(1-1/Le)]$. As in nucleation theory, this solution is unstable and yields the critical radius in ignition problems by hot pockets of burned gases. Notice that $\theta_f < 1$ ($T_f < T_b$) and $r_f/d_L > 1$ for $Le > 1$ but $\theta_f > 1$ and $r_f/d_L < 1$ for $Le < 1$. Because the spherical reacting sheet is cooled more efficiently with a stronger thermal conductivity, it is more difficult to ignite systems with $Le > 1$.

Stability analysis in the presence of external perturbations, such as heat loss, have been carried out to explain the flame balls observed by Ronney (see Buckmaster 1993). The unsteady problem of point source initiation was studied earlier by G. Joulin (1985). The kinetically induced effects that appear near the flammability limits have not been considered in these studies. They are well described by the minimal model (7a,b), (8a,b) as recently shown by He & Clavin (1993). If β is replaced by $4\beta^*/\varepsilon$ in the result r_f/d_L obtained by Zeldovich within the standard model, one simply gets a divergence at the flammability limit when $Le > 1$. The analytical results obtained with (7a,b) are different: The critical radius diverges when ε approaches from above a nonzero value $\varepsilon^* \equiv \beta^*(1-1/Le)$ since $r_f/d_L \sim \varepsilon^{*2}(\varepsilon-\varepsilon^*)^{-2}$ as $\varepsilon \to \varepsilon^*$. For $Le \neq 1$, this limiting value ε^* is different from the planar flammability limits corresponding to $\varepsilon = 0$. The divergence of r_f/d_L corresponds to a flame temperature T_f approaching the crossover temperature T^* ($T_f \to T^*$ from above), and the difference from the planar flammability limits ($\varepsilon^* \neq 0$) is due to the fact that T_f is not equal to the adiabatic flame temperature T_b when $Le \neq 1$. As a consequence, when $Le > 1$, since T_f is smaller than T_b, flammable mixtures corresponding to $0 < \varepsilon < \varepsilon^*$ cannot be ignited by a hotspot of burned gases. In the opposite situation, $Le < 1$, nonsteady spherical flames are easily ignited in nonflammable mixtures corresponding to $\varepsilon^* < \varepsilon < 0$, in accordance with the results of direct numerical simulation of He & Clavin (1993).

6.   TURBULENT FLAME PROPAGATION

Despite a considerable amount of work devoted to turbulent combustion modeling (e.g. Borghi 1988), turbulent flame propagation is in an immature scientific state. Except for weak turbulence intensities, no sound theory is available and experimental data on turbulent burning velocities show too much scatter. Two different regimes are well identified (e.g. Peters 1986): the wrinkled flamelet regime and the corrugated flamelet regime. Both correspond to turbulent flows of fresh mixture with scales larger than the laminar flame thickness d_L. The wrinkled flamelet regime is characterized by a small turbulence intensity, $u' < U_L$; for the corrugated flamelet regime, turbulence is more prevalent and $u' > U_L$.

6.1   *Wrinkled Flamelet Regime*

The hydrodynamic flow induced by the thermal expansion of gases cannot be neglected in this regime. The case of a stable flame in a weakly turbulent flow has been solved analytically as a linear response problem by the method presented in Section 3. The results are in good agreement with experiments (Searby et al 1983, 1986, 1990). Unstable flames are more complex. Diffusive-thermal instabilities may lead to self-turbulizing fronts of cellular flames propagating in quiescent flows, as shown by solutions of the Kuramoto-Sivashinsky equation obtained within the constant density model (e.g. Sivashinsky 1983). Moreover, turbulence induced cells appear on fronts which are subject to hydrodynamical instability, as is observed for spherical flames expanding in a turbulent flow (see Strehlow 1984, p. 432). Inspired by the theory of Zeldovich et al (1980) on stability of curved fronts, a relation between the mean cell size and turbulence properties has been proposed (Clavin 1988). Solutions of the Sivashinsky equation subject to an external noise, provide a more detailed insight into this problem of turbulence selected patterns (Cambray & Joulin 1993).

6.2   *Corrugated Flamelet Regime*

This regime is more difficult to analyze. The turbulent flame front is something like a fractal object (see Figure 2). Knowledge of the geometrical and statistical properties of the flame front is required. The experimental situation is improving quickly. By contrast, on the theoretical side, one can do little better than to use scaling-Prandtl type arguments (but see Pocheau 1993a,b). Results have been obtained only for an idealized limit in which the turbulent Reynolds number is sufficiently large that Kolmogorov

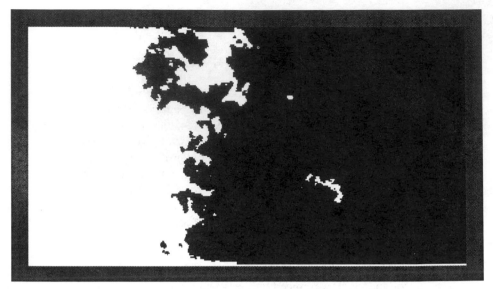

Figure 2 Cross-section of a premixed turbulent flame propagating in a closed vessel. Propane-air mixture, equivalence ratio = 0.85, $u'/U_L = 1.44$ see Pocheau (1933b). (Courtesy of A. Pocheau.)

scaling applies between velocities u_i', length scales l_i, and turnover time $\tau_i \equiv l_i/u_i'$, $u_i' \approx (\varepsilon l_i)^{1/3}$, where ε is the energy viscous dissipation rate per unit mass.

Combustion complicates conventional turbulence theory by creating something akin to a two-phase system consisting of burnt and unburnt gases with discontinuous density and velocity across the interface. Clavin & Siggia (1991) have argued for no qualitative change in the Kolmogorov picture. General conclusions are obtained only in the limiting case,

$$u_K' \ll U_L \ll u_I', \qquad (10a)$$

where u_K' and u_I' are orders of magnitude of the fluctuating turbulent velocities at the viscous Kolmogorov scale and at the integral scale respectively. The inequality $u_K' \ll U_L$ ensures that the local flame structure is not modified from the laminar and planar one. This follows from the approximate equality of the kinematic viscosity v and the thermal diffusivity D_T valid in gases. Introducing the Kolmogorov length scale l_K, one has by definition $l_K u_K' \approx v$ and $d_L U_L \approx D_T$; see (4c–e). Then $v \approx D_T \Rightarrow d_L/l_K \approx u_K'/U_L$ and $u_K' \ll U_L \Rightarrow d_L \ll l_K$, $\tau_L \ll \tau_K \equiv l_K/u_K'$. Fluctuations in the flows are thus, restricted to length and time scales much larger than the thickness and the transit time of the planar flame.

Moreover, following Peters (1986), one introduces a new length scale l_G, $l_K \ll l_G$, called the Gibson scale and defined by $U_L \approx (\varepsilon l_G)^{1/3}$ with $\varepsilon \equiv u_K'^3/l_K = u_1'^3/l_1$. The Gibson scale l_G corresponds to the length below which the turbulent flow is not able to further wrinkle the flame front because the corresponding eddies have an interaction time with the flame which is smaller than the turnover time, $u_j' < U_L \Rightarrow l_j/U_L < \tau_j$. Thus, the flame front can be substantially corrugated by the flow only for scales larger than l_G, where its temporal dynamics are those of the turbulence because $U_L \ll u_i'$. The inequality $U_L \ll u_1'$, valid for large Reynolds number, ensures there is a large range of such scales to make meaningful self-similar properties. As a result, the turbulent combustion is fully controlled by the turbulent flow. All the molecular properties of chemistry and transport mechanisms disappear from the final results in the limiting case (10a): The turbulent flame velocity U_{tur} is the turbulence intensity u_1', the thickness of the turbulent flame brush d_{tur} is the integral scale l_1, and, in the range between l_G and l_1, the flame front has the same fractal dimension $D_f = 7/3$ (Kerstein 1988) as a passive scalar surface,

$$U_{tur} \approx u_1', \qquad d_{tur} \approx l_1, \qquad D_f = 7.3. \tag{10b,c,d}$$

An essential reason for this result is the equality between the time necessary for the combustion to go to completion within each active eddy and the turnover time, as first proposed by Kerstein (1988). Imagine a series of length scales $l_i = l_1/2^i$. A front that is being folded on a scale l_i will be simultaneously and more rapidly corrugated on smaller scales l_{i+1} and so on down to l_G. The surface area in a typical volume $l_i^3 > l_G^3$, increases exponentially due to folding at a rate $\Sigma_{j>i}\tau_j^{-1} \approx \tau_G^{-1} \gg \tau_i^{-1}$, valid for a typical Kolmogorov scaling, where $\tau_G \equiv l_G/U_L$ corresponds also to the characteristic time for combustion at the Gibson scale to be complete. Since $\tau_G \ll \tau_i$, the combustion at length scale l_i is limited to the turnover time τ_i. Moreover, because $u_i' > u_{i+1}'$, the largest eddies are the fastest to contaminate the fresh mixture of flamelets and thus, one expects the region of active combustion to move with velocity u_1' provided that the flame thickness remained bounded. This is the case in fully developed turbulence where the time to completely burn an eddy of size l_i is the turnover time τ_i, valid at all scales in the range between l_G and l_1.

The maintenance of the fractal flame surface under stationary conditions is more transparent than for a passive surface. Wrinkles of size l_i on the flame front are produced only by vortices of the same size, and they only have time to acquire an amplitude l_i leading to the hierarchy of folds characterizing a fractal object. If we work with a resolution $l_i > l_G$, the surface area S_i measured for a typical burning volume l_1^3 is $S_i/l_1^2 \approx (l_1/l_i)^{D_f-2}$ and the total flame area is $S/l_1^2 \approx (l_1/l_G)^{D_f-2} \approx (u_1'/U_L)^{3(D_f-2)}$.

Then, mass conservation, $l_I^2 U_{tur} = S U_L$, implies $U_{tur}/U_L \approx (u_I'/U_L)^{3(D_f - 2)}$ (Peters 1986) and $U_{tur} \approx u_I' \Rightarrow D_f = 2 + 1/3$. Even if the laminar flame velocity disappears from the final result in the limiting case (10a), $U_L \neq 0$ is essential for the validity of (10b–d). The multiplicity of length scales in turbulence is also necessary for the validity of (10b–d). The results would be different for a chaotic flow with a single length scale l_I satisfying $U_L \ll u_I'$. More details may be found in Clavin & Siggia (1991). Experimental data on flames freely propagating in a uniform turbulent flow satisfying (10a) are missing. As shown by the Kolmogorov scaling, $u_I'/u_K' = Re^{1/4}$, $l_I/l_K = Re^{3/4}$, major experimental difficulties are the large integral scale and high Reynolds number of turbulent reactive flows satisfying (10a). On the theoretical side, a simplified problem is provided by the kinematic equation for a front propagating with a normal velocity U_n relative to the fresh mixture, $\partial G/\partial t + \mathbf{v} \cdot \nabla G = U_n |\nabla G|$, where $G(\mathbf{r}, t) = Ct$ is the equation of the front and where the turbulent velocity field $\mathbf{v}(\mathbf{r}, t)$ is prescribed (constant density approximation) and verifies (10a). No solution is known with either $U_n = U_L = Ct$ or the geometrical model mentioned in Section 3.

Results (10b,c) are in agreement with the phenomenological *Eddy-Break-Up* model of Spalding (1977), used extensively by engineers. In this model, the turbulent flame is represented by a KPP wave which is a solution of (4a,b). Here θ is now the reduced mean temperature and $\omega(\theta) = \theta(1 - \theta)$ is the reduced time-average reaction rate. The characteristic turbulent reaction time is proportional to τ_I and the diffusion coefficient is proportional to $u_I' l_I$ (mixing-length model), yielding $U_{tur} \approx (u_I' l_I/\tau_I)^{1/2} \approx u_I'$. Unfortunately conditions (10a) are not fulfilled in many applied systems where, unlike (10b–d), the combustion regime varies with the equivalence ratio. When the inequality $u_K' \ll U_L$ is not fulfilled, the local laminar flame structure is modified. Intermittent quenching and ignition of flamelets decrease the turbulent combustion rate at small scales. This may increase the thickness of the turbulent flame brush. It may also prevent the largest eddies from efficiently propagating the combustion at a constant velocity u_I'. Then, chemistry and molecular transport may influence this turbulent combustion regime. Despite many modeling efforts, no pertinent model is presently available and a better physical insight is necessary to address this challenging and important problem.

7. COMPRESSIBLE EFFECTS, SOUND GENERATION

Acoustics and combustion are coupled through the unsteady pressure term Dp/Dt in (1d) together with (1a–c). Neglecting heat conduction, elimination of T between (1c) and (1d) yields a nonadiabatic pressure-density relation in the form (11a),

$$Dp/DT = c^2 D\rho/Dt + \dot{Q}(C_p - C_v)/C_v, \qquad \partial^2 p/\partial t^2 - c^2 \nabla^2 p = (\partial \dot{q}/\partial t),$$

$$\dot{q} \equiv \dot{Q}(C_p - C_v)/C_v. \tag{11a,b,c}$$

Equation (11b) is the linear acoustic wave equation which is obtained from (1a–c) and (11a) when inhomogeneities of the unperturbed flow and of sound speed are omitted to simplify the presentation. Consider a geometry in which a fire ball is burning in free surroundings, fed by turbulent jets of premixed gas. According to (11b,c), combustion—or more precisely fluctuations in the rate of heat release—acts as a time-dependent volume source and displays a monopole sound distribution to lowest order in a Mach number expansion. In the limit where the wavelength is much larger than the size of the burning region, the acoustic pressure fluctuations are easily computed from (11b) in terms of time derivative of the total heat-release rate $\dot{Q}_{tot} \equiv \iiint \dot{Q} d^3\mathbf{r}$ (e.g. Strahle 1985). When the mass rate of fresh gas injected is kept constant in time and when the turbulent combustion is represented by the corrugated flamelet regimes, the rate of total heat release is related to the flame surface area S by $\dot{Q}_{tot} = C_p(T_b - T_u)\rho_u U_L S$. Then, the resulting sound intensity I, is expressed in terms of the time derivative of the flame surface area as (12a)

$$I = \frac{1}{4\pi c}\left(\frac{1}{\rho_b} - \frac{1}{\rho_u}\right)^2 \rho_b (\rho_u U_L)^2 \langle (dS/dt)^2 \rangle, \qquad \frac{\langle (dS/dt)^2 \rangle}{\langle S \rangle^2} = \sum_i \frac{l_i^2 \omega_i^2}{\langle S_i \rangle}, \tag{12a,b}$$

where angular brackets denote time averages. Note that this result corresponds to sound emitted by a time-dependent volume source $V(t)$ with $dV/dt = \rho_u U_L S(1/\rho_b - 1/\rho_u)$. Thus, the power spectrum of combustion noise may be related to the dynamics of the turbulent flame. Clavin & Siggia (1991) have done this in the corrugated flamelet regime from an analysis of the dynamics of the fractal surface. The result is shown in (12b) where ω_i is the characteristic frequency of fluctuations at scale l_i. Note that $\langle S \rangle$ and $\langle S_i \rangle$ are both extensive quantities. The sum in (12b) converges rapidly in i since $\omega_i^2 \approx \varepsilon^{2/3} l_i^{-4/3}$ and $\langle S_i \rangle / l_i^2 \approx (l_1/l_i)^{D_f}$ in one integral volume l_1^3. Thus, as first pointed out by Strahle (1985), the sound intensity scales with the volume of the brush and is predominant at low frequencies. According to (10d) and (12a,b), the acoustic power falls off with frequency as a $-5/2$ power law, $dI(\omega) \approx \omega^{-5/2} d\omega$, valid for a Kolmogorov scaling and in the frequency range between the inverse of the turnover time at integral scale, τ_I^{-1}, and the inverse of the transit time in a laminar flame, τ_L^{-1}.

The relations of the acoustic intensity to the surface fluctuations, as expressed by (12a,b), are more general. Equation (12a) is valid within the corrugated flamelet regime (10a), without detailed assumptions about the flow which, in particular, may be sustained by gas expansion. Equation (12b) is true, even as an expression for the acoustic power per frequency band in the frequency range $[\tau_I^{-1}, \tau_L^{-1}]$, provided one inserts the correct ω_i for the flow and the value of $\langle S_i \rangle$ measured by tomography techniques. At the limit of the flamelet regime ($u'_K \approx U_L$), singularities of dS/dt are produced by local quenching and ignitions. But even stronger singularities may be produced in the corrugated flamelet regime, at the final stage of combustion of thin tongues of fresh gas which are engulfed in coherent structures. The effects on the combustion noise of such singularities are not included in (12b). They increase the power spectrum at high frequencies ($\omega > \tau_L^{-1}$). Sound emission by nonreactive low speed vortical flows requires delicate experiments and subtle theory to elucidate quantitatively (Lighthill 1952, Kambe 1986). However, turbulent combustion displays a distribution of monopole sources and the emitted sound intensity is much stronger.

The order of magnitude of the acoustic energy is easily estimated from (12a,b). Mass conservation $l_I^2 U_{\text{tur}} = \langle S \rangle U_L$ and (10b) yields $\langle S \rangle \approx l_I^2 u'_I / U_L$ for one integral volume. Moreover, (12b) $\Rightarrow \langle (dS/dt)^2 \rangle \approx \langle S \rangle^2 (u'_I / l_I)^2 \approx \varepsilon u'_I l_I^3 / U_L^2$ where ε is the viscous dissipation rate $\varepsilon \equiv u'^3_I / l_I$. According to (12a), one has $I \approx \rho_u U_L^2 c^{-1} \langle (dS/dt)^2 \rangle$. Then, the ratio η of the radiated acoustic energy to the dissipated turbulent energy, $\rho_u \varepsilon l_I^3$ (per unit time), has the same order of magnitude as the turbulent Mach number M_I: $\eta = O(M_I)$ with $M_I \equiv u'_I / c$. This has to be compared with the case of a nonreactive turbulent jet where, according to the theory of Lighthill (1952), the same ratio is $\eta = O(M_I^5)$ (see Landau & Lifshitz 1989, p. 416). Thus, for $M_I \approx 10^{-2}$, the acoustic power is approximately 10^8 times stronger with than without combustion for the same nonreactive turbulent jet.

8. ACOUSTIC INSTABILITIES

Combustion-induced acoustic instabilities are frequently observed in combustion chambers. They have been encountered during the development of many continuous-flow combustors such as rocket engines. Similar phenomena have been observed very early, in 1883, by Mallard & Le Chatelier for flames propagating in tubes. The rate of heat release $\dot{Q}$ is usually weakly modified by acoustic waves, and, according to the approximate Equation (11b), a feedback mechanism may trigger standing acoustic waves of the cavity.

8.1 Combustion Systems with a Distributed Heat Release

This is typically the case in liquid propellant rocket engines. In linear approximation, the time Fourier transforms ($\widetilde{\delta q}(\omega)$, $\widetilde{\delta p}(\omega)$) of heat release rate and pressure fluctuations [$\delta \dot{q}(t)$, $\delta p(t)$], may be expressed locally as $\widetilde{\delta q}(\omega) = a(\omega)\widetilde{\delta p}(\omega)/\tau_i$, where τ_i is a characteristic coupling time and $a(\omega)$ is a dimensionless complex function describing the frequency dependence of the linear response. They both depend on the details of the physical mechanism involved in the combustion response. (We will return later to this problem and also make order of magnitude estimates.) As a rule, this coupling is weak, $\tau_a/\tau_i \ll 1$ where $\tau_a \equiv L/c$ is a characteristic acoustic time, L being a characteristic dimension of the combustion chamber. For liquid propellants, vaporization may also be an important phenomenon. In sprays, vaporization of droplets induces mass sources per unit volume, $\dot{R}$, in the rhs of the gas mass conservation equation (1a). Such mass sources modify the rhs of (11b) only through a new term $c^2(\partial\dot{R}/\partial t)$ which is added to $(\partial\dot{Q}/\partial t)(C_p - C_v)/C_v$. From now, such phenomena are supposed to be included in $\dot{q}$.

Consider first the ideal case of a well stirred reactor in which heat release is distributed homogeneously throughout the system and where, as a consequence, the transfer function $a(\omega)$ does not depend on the space coordinate $\mathbf{r}$. Then, model equation (11b) describes damped or amplified standing acoustic waves, depending on the sign of $\mathrm{Re}\{a(\omega)\}$, in agreement with Rayleigh's criterion stated in 1878: Instability occurs when fluctuations of heat release are positively correlated with pressure fluctuations, $\mathrm{Re}\{a(\omega_0)\} > 0$, where ω_0 is the unperturbed acoustic frequency, $\omega_0\tau_a = O(1)$. The linear growth (or damping) rate and the frequency shift are easily obtained from (11b), to give respectively, $(2\tau_i)^{-1}\mathrm{Re}\{a(\omega)_0\}$ and $(2\tau_i)^{-1}\mathrm{Im}\{a(\omega_0)\}$, where, by definition, $\tau_i^{-1} = O(\delta\dot{q}/\delta p)$ and scaling is such that $|a(\omega_0)| = O(1)$. A full stability analysis requires an investigation of all the damping mechanisms: losses through the choked end nozzle, viscous and heat transfer at the walls across the acoustic boundary layers, particle damping, etc (see Williams 1985, pp. 305–14).

Inhomogeneities of the mean flow, which are neglected in (11b), may also be of importance. Attention will be focused here on possible amplification mechanisms only. In most combustion-systems that can exhibit combustion-induced acoustic instabilities, heat release is not distributed homogeneously and the transfer function has a space dependence $\mathbf{r}$, $a(\omega_0, \mathbf{r})$. In such a case, the linear growth or damping rate of an acoustic mode is obtained from (11b) by a multiple scale analysis in a weak coupling limit, to give $(2\tau_i)^{-1}\iiint \mathrm{Re}[a(\omega_0, \mathbf{r})]|\Pi_0(\mathbf{r})|^2 d^3\mathbf{r}$, where $\Pi_0(\mathbf{r})$ is the normalized unperturbed eigenmode of the acoustic pressure, $\iiint |\Pi_0(\mathbf{r})|^2 d^3\mathbf{r} =$

1 (e.g. Clavin et al 1993). This shows how the positions in the chamber of maxima of the heat release rate are important for controlling acoustic instabilities. These results correspond to the balance of acoustic energy ε per unit volume obtained from (11b), $\partial \varepsilon / \partial t + \nabla \cdot \mathbf{J}_\varepsilon = (\delta \dot{Q} \delta p / p)(C_p - C_v)/C_p$ (see Culick 1988 and Candel 1992).

A large mass rate of reactants injected into a combustion chamber produces a turbulent reacting flow. Turbulence-induced fluctuations of $a(\omega_o, \mathbf{r})$ yield fluctuations of the linear growth or damping rate of acoustic modes on a time scale typically longer than the acoustic time τ_a. In the neighborhood of the stability limits where the linear rate is zero by definition, these fluctuations may produce an erratic evolution of the amplitude of the oscillatory pressure time-traces with random bursts of strong amplitude oscillations, as sometimes observed during development of liquid-propellant rocket engines. This nonlinear problem has recently been studied in the framework of a Fokker-Planck equation by Clavin et al (1993).

An evaluation in order of magnitude of $\tau_i^{-1} = O(\delta \dot{q}/\delta p)$ may be obtained for simple coupling mechanisms. Pressure dependence through prefactors of reaction rates yields $\delta \dot{Q}/\dot{Q} \approx \delta p/p$. Consider for simplicity a well stirred reactor with an elongated shape of section Φ and length L, where fresh mixture at temperature T_u is injected at one end in the axial direction, and burned gas at T_b flows out at the other end. $\dot{Q}$ may be estimated as $\dot{Q}\Phi L \approx \rho U \Phi C_p (T_b - T_u)$, where U is the axial velocity of the flow. With $p = (C_p - C_v)\rho T_b$, one gets $\dot{q}/p \approx \tau_{res}^{-1}$, where $\tau_{res} \equiv L/U$ is the residence time. Thus, $\tau_i^{-1} \approx \tau_{res}^{-1} \approx M\tau_a^{-1}$, where $\tau_a = L/c$ is the period of axial acoustic modes and where $M = U/c$ is the Mach number of the reacting flow—a small number, 10^{-3} to 10^{-2} in ordinary subsonic combustion chambers. But chemical reaction rates are more sensitive to temperature than to pressure, $\delta \dot{Q}/\dot{Q} \approx \beta \delta T/T$ with $\beta \approx 10$. Adiabatic compression in acoustic waves yields $\delta T/T \approx \delta p/p$, and, according to the above estimate, one obtains $\tau_i^{-1} \approx \beta M \tau_a^{-1}$. Since βM is also a small number, typically 10^{-2}, the weak coupling approximation is still valid. Such small effects are sufficient to produce instabilities because damping rates by losses are also weak, as for example the one associated with a choked end nozzle, $M\tau_a^{-1}$ (Crocco & Cheng 1956, pp. 168–87).

Regardless of the final amplitude, linear instabilities due to acoustic-combustion coupling are often controlled by a fine imbalance of energy rates which are all much smaller than the heat release rate. This explains why active control may be an efficient mechanism as shown by Poinsot et al (1992); for a review of this topic, see Candel (1992). Combustion-acoustic coupling may also be produced by unsteady geometrical modifications in the position and extension of reaction zones, associated with fluctuations of the flow velocity (see Section 8.3). The linear growth rate is more difficult to evaluate in this case. A strong coupling could be obtained in principle through a pure velocity mechanism defined by $\delta\dot{Q}/\dot{Q} \approx \delta u/U$. Since velocity and pressure fluctuations in acoustic waves are related by $\delta p \approx \rho c \delta u$, this would lead to $\delta\dot{Q}/\delta p \approx \dot{Q}/\rho c U \approx M^{-1}\dot{q}/p$, where $p \approx \rho c^2$ has been used, to give $\tau_i^{-1} \approx \tau_a^{-1}$. Such a strong instability is not usually observed. Some mechanisms proceed through gas acceleration $\delta u/\tau_a$ and inertia with a relaxation time $\tau_{relax} < \tau_a$ as in biphasic mixtures with a viscous friction of small fuel particles. In this case, one has typically $\delta\dot{q}/\dot{q} \approx (\tau_{relax}/\tau_a)\delta u/U$ yielding a small coupling $\tau_i^{-1} \approx (\tau_{relax}/\tau_a)\tau_a^{-1}$. Examples are given below for flames.

Technical difficulties for computing $a(\omega, \mathbf{r})$ in liquid propellant rocket motors, also result from the nonlocal character of the coupling mechanisms. The reaction rate of a fluid element could, in principle, depend on its Lagrangian history. For example, pulverization modifications at the injectors by acoustic waves may influence the reaction rate downstream, after a time lag associated with the velocity field and the mixing process. Similar problems occur for combustion instabilities associated with vortex shedding. A general "time-lag" formalism was developed very early by Crocco & Cheng (1956) but each case must be studied separately. Moreover, linear approaches are not sufficient for subcritical bifurcations which may occur in practical systems as suggested by tests in which acoustical instabilities are triggered by exploding powder charges.

8.2   Acoustic and Flame Structure Coupling: Homogeneous Solid Propellants

Combustion of solid propellants takes place in a thin region adjacent to the solid surface. The acoustic coupling is better expressed in this case by an admittance function Z relating the fluctuations of the burned gas velocity normal to the surface, δU_b, to the acoustic pressure fluctuations, δp: $\widetilde{\delta U_b}(\omega) = Z(\omega)\widetilde{\delta p}(\omega)/\rho c$, where U_b is called the burning velocity. Note that $|Z| = O(\delta U_b/\delta u)$, where δu is the acoustic velocity. Orders of magnitude may be easily obtained for axial oscillations in an end burning elongated rocket. In such a one-dimensional case, the boundary condition at one end of the tube is given by the burning velocity of the solid propellant which has a burning surface perpendicular to the axis. The time derivative of acoustic energy per unit cross-sectional area, $\mathsf{E} = O(\tau_a \delta u \delta p)$, due to combustion coupling is approximately $d\mathsf{E}/dt \approx \delta U_b \delta p \Rightarrow (1/\mathsf{E})d\mathsf{E}/dt \approx \tau_a^{-1}\delta U_b/\delta u$, and the order of magnitude of the corresponding linear rate, $\tau_i^{-1} \equiv (1/\mathsf{E})d\mathsf{E}/dt$, is $\tau_i^{-1}/\tau_a^{-1} = O(|Z|)$. For a pressure coupling, $\delta U_b/U_b \approx \delta p/p \Rightarrow |Z| = O(M_b)$, one obtains a weak coupling $\tau_i^{-1}/\tau_a^{-1} = O(M_b)$, where $M_b = U_b/c \ll 1$ is the burning Mach number.

For a homogeneous solid propellant, the flame structure is quasi-planar and its linear response to acoustic oscillations can be investigated analytically. Until recently, the only analytical results available for the admittance function $Z(\omega)$ were those of Zeldovich (1942) and Novozhilov (1967) obtained in the framework of a quasi-steady state approximation of the gas phase, referred to as the QS approximation and ZN theory. For a small gas-to-solid density ratio, the transit time in the gas phase, τ_L, is much smaller than the response time of heat transfer in the condensed phase. At low frequency, $\omega\tau_L \ll 1$, and for stable dynamics in the gas phase, the combustion response is thus slaved by heat transfer in the solid and the QS assumption is valid. The ZN combustion response vanishes with the pressure exponents n of the prefactor in the Arrhenius laws controlling the pyrolisis at the interface and the reaction rate in gases, to give $|Z| = O(nM_b)$. No temperature coupling exists in the QS approximation.

A general analysis taking into account the non-steady effects in the gas phase has been recently carried out by Clavin & Lazimi (1992) following a method similar to that described in Section 3. The ZN result is recovered with a good approximation only for $\omega\tau_L < 0.05$, but a qualitatively different result is obtained for $\omega\tau_L > 1$ (see Figure 7 of Clavin & Lazimi 1992)—a case frequently encountered in practical systems (500 to 5000 HZ). There, contrary to the ZN result, the combustion-acoustic coupling is unstable due to acoustic-driven oscillations of the flame temperature of the thin reaction layer in the gas phase. Temperature oscillations in the bulk of the gas are damped in a thermal layer at the solid-gas interface whose thickness, $l_{therm} \approx U_b^3/\omega^2 D_T \approx d_L/(\tau_L\omega)^2$, increases when the frequency decreases. In the QS approximation, $\omega\tau_L \ll 1 \Rightarrow l_{therm} \gg d_L$, the reaction layer does not feel the acoustic-driven temperature oscillations in the bulk of the gas and the ZN result is valid with $|Z| = O(nM_b)$. In the opposite case, $\omega\tau_L > 1 \Rightarrow l_{therm} < d_L$, the temperature coupling is dominant and yields a stronger oscillatory instability $|Z| = O(\beta M_b)$.

Similar analytical studies of unsteady coupling between acoustic and flame structure have also been carried out to describe the vibratory instability of planar flames propagating in tubes. The flame is modeled as a moving hydrodynamical discontinuity of the axial flow velocity and of its acoustic fluctuations. A Dirac transfer function $a(\omega, \mathbf{r})$ (defined in Section 8.1) would be sufficient for a qualitative description, but a detailed analysis must take into account not only the difference of temperature and sound speed in burned and unburned gases but also the acoustic frequency variation when the flame propagates. It is convenient to introduce an admittance function for the acoustic velocity jump across the flame, resulting from modifications of the combustion rate, $\widetilde{\delta u}(\omega)|_-^+ = Z(\omega)\widetilde{\delta p}(\omega)/\rho_u c_u$. Then the full acoustic problem can be solved with a multiple time scale analysis by using this jump condition. The order of magnitude of the linear rate is still $\tau_i^{-1}/\tau_a^{-1} = O(|Z|)$. As for a solid propellant, the admittance function $Z(\omega)$ is obtained from a nonsteady analysis of flame structure. The analytical results of Clavin et al (1990) for premixed gas flames, show that a temperature coupling, $|Z| = O(\beta M)$, can be sufficient to overcome the damping mechanisms in the boundary layers at the tube walls and the acoustic loss by sound radiation from the open end of the tube. But, except in conditions close to diffusive-thermal instability limits, the resulting acoustic instability does not yield very large final amplifications of initial perturbations. It is not yet clear whether or not this mechanism is involved in the observed vibratory instabilities (see below). A much stronger one-dimensional acoustic instability is found in the analytical results of Clavin & Sun (1991) for planar flames propagating in sprays or particle laden gases. Inertia and drag of particles whose velocity lags behind the velocity fluctuations in the gas phase of the fresh mixture, are responsible for acoustic-driven oscillations of the combustion rate. Detailed results have been obtained for sufficiently small particles. The order of magnitude of the acoustic instability growth rate is $|Z| = O(\beta \tau_{\text{relax}}/\tau_{\text{a}})$, with a Stokes relaxation time defined as $\tau_{\text{relax}} \approx (\rho_1/\rho)r^2/\nu$ where r and ρ_1 are the radius and density of particles, respectively. For particles of 10 μm radius in a tube one meter long, the growth rate is 10^2–10^3 times stronger than for planar flames propagating in gases. Strong vibratory instabilities have been observed in lycopodium-air mixtures by Berlad et al (1990) but it is not clear that multidimensional effects as those described below, are absent in this experiment. Preliminary experiments of Searby on sprays confirm these predictions (see end of Section 8.3).

8.3   Vibratory Instability of Flames Propagating in Tubes

The kinematic study of an anchored V-shaped flame by Boyer & Quinard (1990), shows how homogeneous oscillations δu of a uniform flow velocity induce amplitude waves on the flame front with a typical wavelength λ equal to the tangential mean velocity times the period of oscillations, $\lambda = O(U/\omega)$, and with an amplitude of the same order of magnitude as the displacement $\delta u/\omega$ (see Figure 3). When modifications to flame structure are neglected, the heat release rate is constant per unit flame surface area and fluctuations of the total flame surface area δS induce fluctuations of the heat release rate. Thus, the coupling of acoustics with the dynamics and geometry of flame fronts can be a very efficient instability mechanism. But when the unperturbed front is planar, δS is a second-order quantity, $\delta S/S = O[(\delta u/U)^2]$, and no linear acoustic instability is expected through this mechanism.

The vibratory instability of flames propagating in tubes was studied a long time ago by Markstein (1964). New developments have been obtained

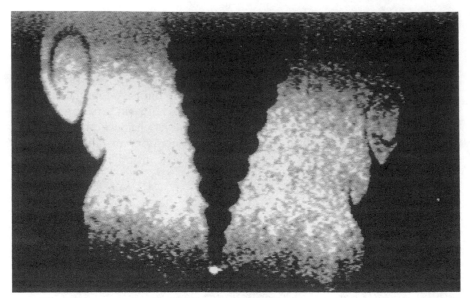

Figure 3 Tomographic cut through a "V" shaped lean propane-air flame (equivalence ratio $\phi = 0.8$, mean flow speed $U = 3$ m/s) anchored on a rod 2 mm diameter. The flame is excited by a planar acoustic wave of frequency $f = 330$ Hz with an acoustic amplitude of 5 cm/s. The mean wavelength of the structure is given by $\lambda = U/f$. (Courtesy of J. Quinard.)

by Searby (1992) with his recent experimental study of propane-air flames in a vertical tube 120 cm long, open at the top and closed at the bottom. Searby's experiment mainly concerns lean mixtures without spontaneous cellular structures of a diffusive-thermal nature. The flame is ignited at the top and initially takes the form of a smooth curved front, resulting from the Darrieus-Landau hydrodynamical instability. This curved front (shown in Figure 4a) propagates at about twice the laminar flame velocity (planar front) as would be expected from the increase of flame surface area. The laminar flame velocity is a sensitive parameter in Searby's experiments. For sufficiently lean mixtures with a laminar flame velocity below 16 cm/s, the flame propagates from top to bottom producing no sound. For faster flames, three qualitatively different stages are observed. The first stage corresponds to a primary acoustic instability of the 1/4 wavelength mode (≈ 130 Hz), which occurs when the flame reaches about the center of the tube, with a linear growth rate approximately equal to 10 s^{-1} (see Figure 5). This rate is fifty times stronger than the one computed by Clavin et al (1990) for a pressure coupling, and the coupling mechanism is more likely associated with the hydrodynamical response of a curved flame front submitted to an oscillatory acceleration. Such a mechanism was inves-

(a)   (b)

(c)   (d)

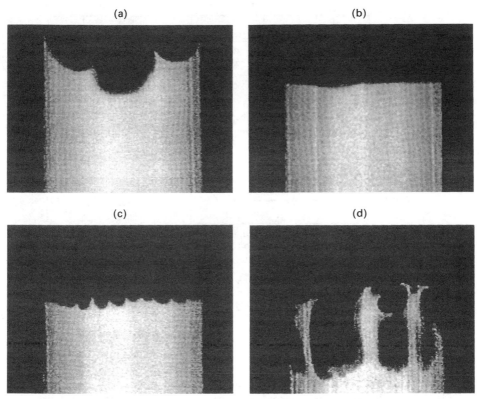

Figure 4 High speed tomographic cuts through a premixed flame at different stages in an acoustic instability. Propane-air flame in a tube 1.2 m long and 10 cm diameter, open at burnt gas extremity. Equivalence ratio = 0.82. (*a*) Curved flame at onset of primary instability. (*b*) Quasi-flat flame during saturation of primary instability. (*c*) Small cells on nearly flat flame at onset of secondary acoustic instability. (*d*) High amplitude cells during development of secondary acoustic instability. (Courtesy of G. Searby.) See Searby (1992) and Searby & Rochwerger (1991).

tigated analytically by Pelcé & Rochwerger (1992) for the vibratory instability of cellular flames but under the approximation of small amplitude cellular structures. Orders of magnitude of linear growth rate obtained by extrapolating these theoretical results to the finite amplitude of the curved flame front, are surprisingly in good agreement with the experimental data of Searby (1992). For intermediate flame velocities, 16 cm/s $< U_L < 25$ cm/s, the primary instability saturates after about 0.25 s when the acoustic pressure reaches about 0.005 bar (see Figure 5). This sudden saturation is accompanied by a transition, occurring over a few acoustical cycles, from a curved front to a quasi-planar flame with a position oscillating at the

348   CLAVIN

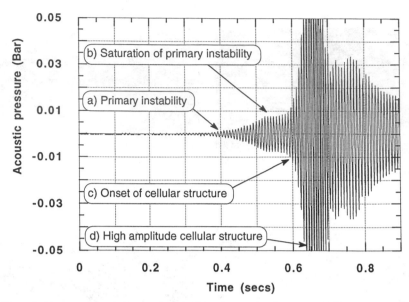

Figure 5 Pressure record of acoustic instability of a premixed propane-air flame propagating in a tube. Equivalence ratio = 0.82. (Courtesy of G. Searby.)

acoustic frequency and propagating at the laminar flame speed (see Figure 4b). This phenomenon, referred to here as the second stage, was first observed by Kaskan (1953). The sound intensity continues to increase but at a smaller rate, about 0.6 s^{-1}, quite compatible with a pressure coupling mechanism described by Clavin et al (1991). For faster flames, 25 cm/s $< U_L <$ 40 cm/s, a third stage appears at the beginning of the second half of the tube where a violent secondary instability occurs on the planar front. Small pulsating cellular structures, with a few millimeters wavelength, develop on the flame and increase very rapidly in amplitude (see Figure 4c,d). The average propagation velocity can reach 170 cm/s (Searby 1992). At the same time the sound level increases with a linear growth rate about 30 s^{-1} and reaches an intensity of 0.05–0.2 bar, depending on the flame velocity (see Figure 5). The period of oscillation of these structures is 14 ms—just twice the acoustic period, the characteristic signature of a parametric forcing as first noticed by Markstein (1953).

As explained qualitatively by Markstein (1964), the origin of the second instability is similar to that of the Faraday instability resulting from a periodic acceleration of an interface separating two regions of different density, but with a nonzero mass flux crossing this interface. Searby &

Rochwerger (1991) have recently carried out an original experimental study of this phenomenon for freely propagating flames, kept stationary in the laboratory frame and interacting with a standing acoustic field generated by a loudspeaker. They have successfully compared their results with a theoretical analysis, based on an extension of the linear evolution equation obtained by Pelcé & Clavin (1982) for flame fronts propagating downwards, in which acoustic acceleration is added to the acceleration due to gravity. For each wavelength, the amplitude of wrinkling of the front satisfies a Mathieu equation representing a parametrically-driven damped oscillator. Two amplitude thresholds of the acoustic acceleration are exhibited. Above the first one, the natural Darrieus-Landau hydro-dynamic instability is suppressed and the oscillating acceleration restores the stability of planar flame. Above the second one a parametric cellular instability is triggered. Thus, the three stages reported in Searby's experiment of self-excited vibratory instabilities of flames propagating in tubes, are explained when the first amplitude threshold is below the second one. But a theoretical analysis of the full problem describing the strong growth rate in the third stage is still missing.

A recent analysis of Joulin (1993) shows that the flame response at high frequencies is not controlled by the same Markstein number M as at low frequencies and much theoretical works needs to be done in that direction. The two thresholds mentioned above depend on frequency, laminar flame velocity U_L, and Markstein number M. When the two domains of instability overlap, the parametric cellular instability must appear directly on a curved flame front and the third stage may occur from the first one without going through a planar flame front. This has been observed by Searby (1992) close to the stoichiometry ($U_L = 42$ cm/s) where a strong secondary instability develops without any observation of the relative stabilization of sound level characteristics of the second stage.

A surprising new phenomenon is also observed by Searby in this experiment. When the intensity reaches ± 0.25 bar, a sudden breakdown of the parametric cellular structure occurs (10^{-1} s after its birth) degenerating into an incoherent highly turbulent motion with a very rapid drop in the sound level at a rate of about 60 s^{-1}, much higher than the laminar decay rate of 5 s^{-1} (see Figure 5). It is as if the coherent energy of the acoustic modes is suddenly transferred into a turbulent motion. The turbulent flame propagates through the last 50 cm during 10^{-1} s with a flame velocity increasing from approximately 1 m/s up to 7.5 m/s. This new phenomenon requires further work to elucidate qualitatively.

Searby has also recently developed the same experiment in sprays. The growth rate of the primary instability is 10^3 times stronger than in gases and is in quantitative agreement with theoretical results of Clavin & Sun

(1991, 1993). But, on the other hand, fuel droplets introduce a saturation mechanism which decreases the strength of the secondary instability mechanism. Much work remains to be done on this problem.

Literature Cited

Aronson, D. G., Weinberger, H. F. 1978. Multidimensional nonlinear diffusion arising in population genetics. *Adv. Math.* 30: 33–76

Berlad, A. L., Ross, H., Facca, L., Tangisala, V. 1990. Particle cloud flames in acoustic fields. *Combust. Flame* 82: 448–50

Borghi, R. 1988. Turbulent combustion modelling. *Prog. Energy Combust. Sci.* 14: 245–92

Boyer, L., Quinard, J. 1990. On the dynamics of anchored flames. *Combust. Flame* 82: 51–65

Buckmaster, J. 1993. The structure and stability of laminar flames. *Annu. Rev. Fluid Mech.* 25: 21–53

Cambray, P., Joulin, G. 1993. On the length-scales of weakly-forced unstable premixed flame. *Combust. Sci. Technol.* To appear

Candel, S. M. 1992. Combustion instabilities coupled by pressure waves and their active control. *24th Symp. on Combustion*, pp. 1277–96. Combust. Inst.

Clavin, P. 1985. Dynamic behavior of premixed flame fronts in laminar and turbulent flows. *Prog. Energy Combust. Sci.* 11: 1–59

Clavin, P. 1988. Theory of flames in disorder and mixing. *NATO ASI Ser. E* 152: 293–315

Clavin, P., Denet, B., Monteiller, J., Pelcé, P. 1986. Study of reaction-diffusioon waves representative of reverse combustion in porous combustible media. *J. Mec. Theor. Appl.* Spec. No.: 173–92

Clavin, P., Joulin, G. 1983. Premixed flames in large and high intensity turbulent flow. *J. Phys. (Paris) Lett.* 44: L1–12

Clavin, P., Kim, J. S., Williams, F. A., 1993. Turbulence-induced noise effects on high-frequency combustion instatilities. *Combust. Sci. Technol.* To appear

Clavin, P., Lazimi, D. 1992. Theoretical analysis of oscillatory burning of homogeneous solid propellant including non-steady gas phase effects. *Combust. Sci. Technol.* 83: 1–32

Clavin, P., Liñan, A. 1984. Theory of gaseous combustion. *NATO ASI Ser. B* 116: 291–338

Clavin, P., Pelcé, P., Hé, L. 1990. One-dimensional vibratory instability of planar flames propagating in tubes. *J. Fluid Mech.* 216: 299–322

Clavin, P., Siggia, E. D. 1991. Turbulent premixed flames and sound generation. *Combust. Sci. Technol.* 78: 147–55

Clavin, P., Sun, J. 1991. Theory of acoustic instabilities of planar flames propagating in sprays or particle-laden gases. *Combust. Sci. Technol.* 78: 265–88

Clavin, P., Sun, J. 1993. Effect of a non-equilibrium between vapor and particles on the acoustic instability of biphasic flames. In prep.

Clavin, P., Williams, F. A. 1982. Effects of molecular diffusion and of thermal expansion on the structure and dynamics of premixed flames in turbulent flows of large scale and low intensity. *J. Fluid Mech.* 116: 251–82

Crocco, L., Cheng, S. I. 1956. Theory of combustion instability in liquid propellant rocket motors. *AGARD graphs No. 8.* London: Butterworth Sci.

Culick, F. E. C. 1988. *Combustion instabilities in liqui-fueled propulsion systems—an overview.* Presented at AGARD 72 B PEP Meet.

Denet, B. 1993. Intrinsic instabilities of curved premixed flames. *J. Eur. Phys. Lett.* 21(3): 299–304

Fisher, R. A. 1937. The wave of advance of advantageous genes. *Ann. Eugenics* 7: 355–69

Frankel, M. L., Sivashinsky, G. I. 1983. On effects due to thermal expansion and Lewis number in spherical flame propagation. *Combust. Sci. Technol.* 31: 131–38

Harrison, J., Showalter, K. 1986. Propagating acidity fronts in the iodate-arsenous acid reaction. *J. Phys. Chem.* 90: 225–26

Hé, L., Clavin, P. 1993. Premixed hydrogen-oxygen flames. Part I & II. *Combust. Flame* 93: 391–420

Joulin, G. 1985. Point-source initiation of lean spherical flames of light reactants: an asymptotic theory. *Combust. Sci. Technol.* 43: 99–113

Joulin, G. 1989. On the hydrodynamic stability of curved premixed flames. *J. Phys. (Paris)* 50: 1069–82

Joulin, G. 1993. On the response of premixed flames to time dependent stretch and curvature. *Combust. Sci. Technol.* To appear

Joulin, G., Clavin, P. 1979. Linear stability

analysis of non-adiabatic flames. *Combust. Flame* 35: 139–53

Joulin, G., Liñan, A., Ludford, G. S. S., Peters, N., Schmidt, E. 1985. Flame with chain-branching/chain-breaking kinetics. *SIAM J. Appl. Math.* 45(3): 420–34

Kambe, T. 1986. Acoustic emission by vortex motion. *J. Fluid Mech.* 173: 643

Kaskan, W., E. 1953. An investigation of vibrating flames. *Fourth Symp. on Combustion*, pp. 575–91. Baltimore: Williams & Wilkins

Kerstein, A. 1988. Fractal dimension of turbulent premixed flames. *Combust. Sci. Technol.* 60: 441–45

Kolmogorov, A., Petrovsky, I., Piscounoff, N. 1937. Etude de l'équation de la diffusion avec croissance de la quantité de matière et de son application à un problème biologique. *Bull. Univ. Moscou* A1: 1–25

Landau, L., Lifshitz, E. 1989. *Mecanique des Fluides*. 2nd Ed. MIR

Lighthill, M. J. 1952. On sound generated aerodynamically. *Proc. R. Soc. London Ser. A* 211: 564

Liñan, A., Clavin, P. 1987. Premixed flames with non-branching chain reactions. *Combust. Flame* 68: 69–71

Mallard, E. E., Le Chatelier, H. 1883. Recherches expérimentales et théoriques sur la combustion des melanges gazeue explosifs. *Ann. Mines* 8(4): 274–376

Markstein, G. H., 1953, Instability phenomena in combustion waves. *Fourth Symp. on Combustion*, pp. 44–59. Baltimore: Williams & Wilkins

Markstein, G. H. 1964. *Non-Steady Flame Propagation*. Oxford: Pergamon

Matalon, M. 1983. On flame stretch. *Combust. Sci. Technol.* 31: 169–81

Matalon, M., Matkowsky, B. 1982. Flame as gas dynamic discontinuities. *J. Fluid Mech.* 124: 239–59

Nicoli, C., Clavin, P. 1987. Effect of variable heat loss intensities on the dynamics of a premixed flame front. *Combust. Flame* 68: 69–71

Nicoli, C., Clavin, P., Liñan, A. 1990. Travelling waves in the cool flame regime. In *Spatial Inhomogeneities and Transient Behaviour in Chemical Systems*, pp. 317–34, ed. P. Gray, G. Nicoles. F. Baras, P. Borckmaus, S. K. Scott. Manchester: Manchester Univ. Press

Novozhilov, B. V. 1967. Non-steady burning of powder having variable surface temperature. *J. Appl. Mech. Tech. Phys.* 8: 37–43 (In Russian)

Ostriker, J. P., ed. 1992. *Selected Works of Yakov Borisovich Zeldovich*. Vol. I. Princeton: Princeton Univ. Press

Pelcé, P. 1988. *Dynamics of Curved Fronts*. London: Academic

Pelcé, P., Clavin, P. 1982. Influence of hydrodynamics and diffusion upon the stability limits of laminar premixed flames. *J. Fluid Mech.* 124: 219–37

Pelcé, P., Rochwerger, D. 1992. Vibratory instability of cellular flames propagating in tubes. *J. Fluid Mech.* 239: 293–307

Peters, N. 1986. Laminar flamelet concepts in turbulent combustion. *Twenty first Symp. on Combustion*, pp. 1231–50. Combust. Inst.

Peters, N., Rogg, B. 1993. *Reduced Kinetic Mechanisms for Application in Combustion Systems*. New York: Springer-Verlag

Pocheau A. 1993a. Scale invariance in turbulent combustion. *Phys. Rev.* To appear

Pocheau A. 1993b. Intrinsic velocity of a turbulent premixed flame propagating in a closed vessel. *Combust. Sci. Technol.* To appear

Poinsot, T., Yip, B., Veynante, D., Trouvé, A., Sarmaniego, J. M., Candel, S. 1992. Active control: an investigation method for combustion instabilities. *J. Phys. (Paris)* 2: 1331–57

Rayleigh, J. W. S. 1877. *The Theory of Sound*. 2nd ed. 1945. Mineola, Minn: Dover 1945

Ronney, P. D. 1988. Effects of chemistry and transport properties on near-limit flames at microgravity. *Combust. Sci. Technol.* 59: 123–44

Ronney, P. D. 1990. Near-limit flame structures at low Lewis number. *Combust. Flame* 82: 1–14

Searby, G. 1992. Acoustic instability in premixed flames. *Comb. Sci. Technol.* 81: 221–31

Searby, G., Clavin, P. 1986. Weakly turbulent wrinkled flames in premixed gases. *Combust. Sci. Tech.* 46: 167–94

Searby, G., Quinard, J. 1990. Direct and indirect measurements of Markstein numbers of premixed flames. *Combust. Flame* 82(3–4): 298–311

Searby, G., Rochwerger, D. 1991. A parametric acoustic instability in premixed flames. *J. Fluid Mech.* 231: 529–43

Searby, G., Sabathier, F., Clavin P., Boyer, L. 1983. Hydrodynamical coupling between the motion of flame front and upstream gas flow. *Phys. Rev. Lett.* 51(6): 1450–53

Sivashinsky, G. I. 1977. Nonlinear analysis of hydrodynamic in laminar flames. *Acta Astron.* 4: 1177–206

Sivashinsky, G. I. 1981. On spinning propagation of combustion waves. *SIAM J. Appl. Math.* 40: 432–38

Sivashinsky, G. I. 1983. Instabilities, pattern formation, and turbulence in flames. *Annu. Rev. Fluid Mech.* 15: 179–99

Smooke, M. D. 1992. *Reduced Kinetic Mechanisms and Asymptotic Approximations for Methane—Air Flames. Lect. Notes Phys.* 384. New York: Springer-Verlag

Spalding, D. B. 1977. Development of the Eddy-Break-Up model of turbulent combustion. *Sixteenth Symp. on Combustion*, pp. 1657–63. Combust. Inst.

Strahle, W. C. 1985. A modern theory of combustion noise. In *Recent Advances in the Aerospace Sciences*, pp. 103–14, ed. C. Casci. New York: Plenum

Strehlow, R. A. 1984. *Combustion Fundamentals*. New York: McGraw-Hill. 2nd. ed.

Williams, F. A. 1985. *Combustion Theory*. Redwood City, Calif.: Benjamin/Cummings

Williams, F. A. 1992. The role of theory in combustion sciences. *24th Symp. on Combustion*, pp. 1–17. Combust. Inst.

Yang, C. H., Gray, B. F. 1969. On the slow oxidation of hydrocarbon and cool flames. *J. Phys. Chem.* 78: 3395–3406

Zeldovich, Y., B. 1942. On the theory of combustion of powders and explosives. *Z. Eksp. Teor. Fiz.* 12: 498–524. See also Ostriker 1992, pp. 330–63

Zeldovich, Y. B., Aldushin, A. P., Khudyaev, S. I. 1979. See Ostriker 1992, pp. 328–29

Zeldovich, Ya., B., Barenblatt, G. I., Librovich, V. B., Makhvildaze, G. M. 1985. *The Mathematical Theory of Combustion and Explosions*. New York: Consult. Bur.

Zeldovich, Y., B., Frank-Kamenetsky, D. A. 1938. A theory of thermal propagation of flame. *Acta Physicochim. URSS* 9: 341–50. See also Ostriker 1992, pp. 262–70

Zeldovich, Y. B., Istratov, A. G., Kidin, N. I., Librovich, V. B. 1980. Flames propagation in tubes. *Combust. Sci. Technol.* 24: 1–13

Annu. Rev. Fluid Mech. 1994. 26 : 353–78

CLIMATE DYNAMICS AND GLOBAL CHANGE

R. S. Lindzen

Center for Meteorology and Physical Oceanography, MIT, Cambridge, Massachusetts 02139

KEY WORDS: climate variability, greenhouse effect, glacial cycles, carbon dioxide, water vapor

1. INTRODUCTION

The question of global climate change has been a major item on the political agenda for several years now. Politically, the main concern has been the impact of anthropogenic increases in minor greenhouse gases (the major greenhouse gas is water vapor). The question is also an interesting scientific question. Restricting ourselves to issues of climate, the answer requires that we be able to answer at least two far more fundamental questions, both involving strong fluid mechanical components:

1. What determines the mean temperature of the Earth?; and
2. What determines the equator-pole temperature distribution of the Earth's surface?

The popular literature lays stress on the first question, but the two are intimately related, and there are reasons for considering the second question to be the more fundamental. When we discuss climate observations, we will see that climate changes in the past history of the Earth were primarily associated with almost unchanged equatorial temperatures and major changes in the equator-pole distribution. There are good reasons to view changes in the mean temperature of the Earth as residual terms arising from the change in the equator-pole distribution. Section 2 of this review quickly summarizes observations of climate. We discuss not only the temperature trends of the past century, but also earlier climate. In

353

0066–4189/94/0115–0353$05.00

connection with the discussion of glaciation cycles, we also introduce the Milankovitch hypothesis, which attempts to relate these cycles to variations in the Earth's orbital parameters. Section 3 reviews our current understanding of the two fundamental questions. The answer to both questions depends on the heat budget of the Earth—both its radiative and dynamic components. The radiative contributions depend on the radiatively active constituents of the atmosphere—mainly water vapor and cloud cover. In the context of the present political debate, the focus is on the minor greenhouse gases (primarily CO_2), and the behavior of water vapor and clouds is subsumed under the title "feedbacks." In present models, water vapor automatically increases with warming, and constitutes the major positive feedback. Without this feedback, no current model would produce *equilibrium* warming due to a doubling of CO_2 in excess of about 1.5°C regardless of other model feedbacks. This brings us to two additional questions: Namely, what determines the density of water vapor in the atmosphere?, and how is the time-dependent response of the atmosphere to radiative perturbations related to the equilibrium response? Sections 4 and 5 deal with these two questions.

The time-dependent behavior of the climate is highly contingent on the presence of oceans. This is true not only for temperature, but also for CO_2. The behavior of CO_2 strongly involves chemistry and largely transcends the scope of this review. However, we briefly discuss this issue in Section 6. Section 7 summarizes our discussion, emphasizing relatively simple and focused approaches to the question of how we might expect the climate to respond to increased emissions of CO_2. At present, we note that there is no basis in the data for current fears, and model predictions result from physically inadequate model features. We suggest reasons for expecting small warming to result from expected increases in CO_2.

2. OBSERVATIONS OF CLIMATE

The following is only a cursory treatment of the observed climate, focusing only on those aspects essential to subsequent discussion. Houghton et al (1990, 1992), Crowley & North (1991), Imbrie & Imbrie (1980), Balling (1992), and Peixoto & Oort (1992) provide much more material for those interested.

The definition of climate variations is not without ambiguities. We shall take climate variability to refer to changes on time scales of a year or longer. In this paper, moreover, we will restrict ourselves to surface climate on a global scale. For simplicity, most global change studies have focused on the globally and annually averaged temperature. The behavior of this quantity since 1860 is illustrated in Figure 1. Figure 1 is based primarily on temperature records over land. Attempts have been made to use primarily records from areas minimally affected by urbanization, though urban heat island effects may have introduced errors on the order of 0.1°C. Ocean data are limited, and such data as are available have been "corrected" substantially (several tenths of a degree). In forming global averages, data have been interpolated over a regular grid enabling one to take area-weighted averages. Given the preponderance of data from Northern Hemisphere land stations, this means that large areas are based on minimal observation. What Figure 1 shows is a global temperature that rose noticeably between 1915 and 1940, remained relatively steady until the early 1970s, and rose again in the late 1970s. The change over the past century is estimated to be 0.45°C ± 0.15°C. The temperature increase in the late 1970s is largely a Southern Hemisphere phenomenon. In the Northern Hemisphere, there was a decline in temperature from the mid 1950s until the early 1970s. A more complete discussion and extensive references may be found in Houghton et al (1990) and Balling (1992). It is clear that there has been a leveling of temperatures in the 1980s, and although these are referred to as record-breaking years, they are not appreciably warmer than the record-breaking years of the 1940s. Recently, satellite data have been used to obtain global average temperature (Spencer & Christy 1990). Such data are available since 1979. The data tend to be representative of the whole troposphere rather than the surface. According to all existing models, climate response to additional greenhouse gases should affect the entire troposphere, so that the satellite data might, in some respects, be preferable to surface data. On the whole, satellite data correlate well with surface data when the latter are plentiful, and relatively poorly when surface data are sparse. Trenberth et al (1992) have recently reviewed this situation, concluding that surface data from the late 19th century may be less reliable than claimed. However, even without this caveat, the Intergovernmental Panel on Climate Change (Houghton et al 1990) notes that the surface record depicts nothing that can be distinguished from natural variability.

Proxy records exist that offer some suggestion of how climate varied prior to the instrumental record. However, these records are usually insufficient for global averages. There is some evidence for a "little ice age" in the 18th Century, as well as a medieval optimum when temperatures were significantly warmer than at present (Crowley & North 1991). Budyko & Izrael (1991) in reviewing past climates notably different from the present observed that these climates differed from the present not only in mean temperature but in temperature distribution with latitude. They argue for a universal distribution in latitude shown in Figure 2. This distribution is characterized by very small changes near the equator, and

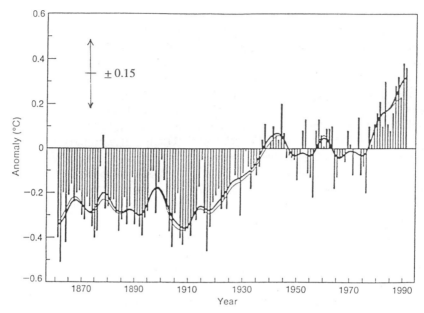

Figure 1 Globally averaged surface temperature record since 1860. (From Houghton et al 1992. Light line from Houghton et al 1990.)

major changes in the equator-to-pole temperature difference. This hypothesized "universal" curve poses two major questions:

1. Since the change of equator-to-pole temperature difference must indicate a change in the heat flux from the tropics to higher latitudes, why are the variations at low and high latitudes not out of phase with each other?
2. What is preventing significant variation of equatorial temperatures? The crucial point here is that the changing equator-to-pole temperature differences would appear to call for profound changes in the heat flux out of the tropics, which for the tropics represents a large change in thermal forcing.

In this connection, it should be noted that there are suggestions (Barron 1987) that during the very warm climate of the Eocene (~ 50 million years ago) the equator may have been colder than at present. Also, in connection with the "warming trend" of the past century, the pronounced latitude variation of Figure 2 was absent, and in the warming episode of the late 1970s, tropical warming exceeded polar warming, which may even have amounted to cooling.

Temperature scaled by global mean change

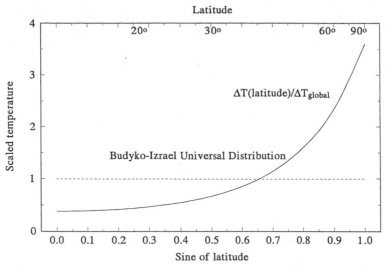

Figure 2   Universal latitude variation of climate change. (After Budyko & Izrael 1991.)

The past million years or so have also manifested an additional striking aspect of climate: cycles of major glaciation and deglaciation. The cycles are determined from the study of δO^{18} in ice cores. δO^{18} is primarily indicative of ice volume. While there are problems in dating different levels in such cores, Figure 3 gives a widely accepted time history from such cores. Figure 4 shows a power spectrum of this time series. The 100,000 year component clearly dominates, but significant peaks are claimed near

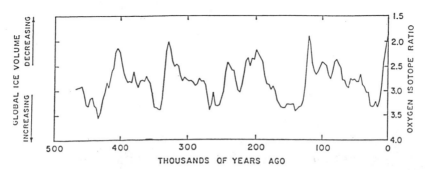

Figure 3 δO^{18} as a function of time over the past 700,000 years. (From Imbrie & Imbrie 1980.)

40,000 and 20,000 years. It should be noted that there remain arguments concerning these peaks based on both core dating (Winograd et al 1992) and analysis method (Evans & Freeland 1977). The general hypothesis for this periodic behavior is that the climate has been forced by the orbital variations of the Earth; this is referred to as the Milankovitch mechanism. The orbital variations consist of variations in the obliquity (tilt) of the Earth's rotation axis, the precession of the equinoxes along the Earth's elliptic orbit, and the changes in eccentricity of the orbit. These variations are schematically illustrated in Figure 5. The obliquity variations are characterized by periods of about 40,000 years, the precession is associated with periods of about 20,000 years, and the eccentricity is associated with periods of 100,000 years and 400,000 years. The precessional cycle, moreover, is strongly modulated by the eccentricity cycle (Berger 1978). The relevant periods are indicated in Figure 4. Both the 24,000 yr and 19,000 yr peaks correspond to the precessional cycle. Imbrie & Imbrie (1980) provide a readable treatment of this phenomenon. From the point of view of climate change, there are several aspects of the glaciation cycles

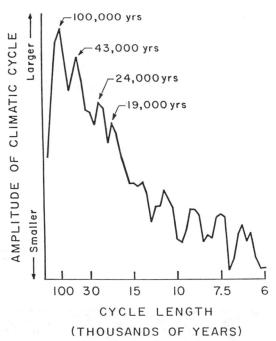

Figure 4 Normalized power spectrum of time series shown in Figure 3. (From Imbrie & Imbrie 1980.)

that deserve comment. First, the change in annually and globally averaged insolation associated with orbital variations is very small ($\leqslant 1\%$). On the other hand, orbital changes lead to substantial changes in the geographical distribution of insolation. Milankovitch (1930) stressed the importance of summer insolation at high latitudes for the melting of winter snow accumulation. More recently, Lindzen & Pan (1993) have noted that orbital variations can greatly influence the intensity of the Hadley circulation, a basic component of planetary heat transport. Relatively uniform changes in heating, it should be recalled, will have little effect on heat transport.

3. BASIC PHYSICS OF GLOBAL CLIMATE

Global Mean Temperature

The most common but, as we shall see, severely incomplete approach to global mean temperatures is to consider a one-dimensional radiative convective model with solar insolation characteristic of some "mean" latitude at equinoxes. An example of such an approach is illustrated in Figure 6, taken from Möller & Manabe (1961). Several vertical profiles of temperature are shown: pure radiative equilibrium, with and without the infrared properties of clouds, and radiative-convective equilibrium with the infrared properties of clouds included. (The visible reflectivity of clouds is included in all the calculations.) In most popular depictions of the

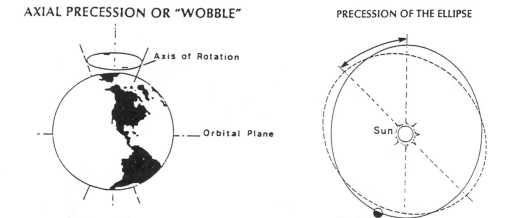

Figure 5 Schematic illustration of orbital parameters involved in Milankovitch mechanism. (After Crowley & North 1991.)

PRECESSION OF THE EQUINOXES

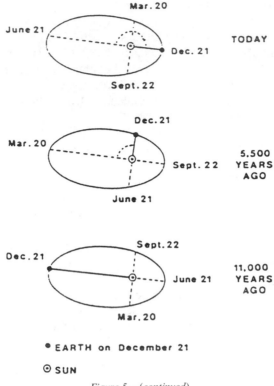

Figure 5—(continued)

greenhouse effect, it is noted that in the absence of greenhouse gases, the Earth's mean temperature would be 255 K, and that the presence of infrared absorbing gases elevates this to 288 K. In order to illustrate this, only radiative heat transfer is included in the schematic illustrations of the effect (Houghton et al 1990, 1992); this lends an artificial inevitability to the picture. Several points should be made concerning this picture:

1. The most important greenhouse gas is water vapor, and the next most important greenhouse substance consists in clouds; CO_2 is a distant third (Goody & Yung 1989).
2. In considering an atmosphere without greenhouse substances (in order to get 255 K), clouds are retained for their visible reflectivity while ignored for their infrared properties. More logically, one might assume

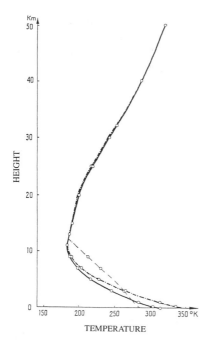

Km
50

40

HEIGHT
30

20

10

0
150   200   250   300   350 °K

TEMPERATURE

Figure 6   Pure radiative equilibrium with infrared effects of clouds included (*thick dashed-dotted curve*) and without clouds (*solid curve*); radiative-convective equilibrium (*thin dashed curve*). (After Möller & Manabe 1961.)

that the elimination of water would also lead to the absence of clouds, leading to a temperature of about 274 K rather than 255 K.

3. Pure radiative heat transfer leads to a surface temperature of about 350 K rather than 288 K. The latter temperature is only achieved by including a convective adjustment that consists simply in adjusting the vertical temperature gradient so as to avoid convective instability while maintaining a consistent radiative heat flux. [It should be noted that this is a crude and inadequate approach to the treatment of convection; however, the development of better approaches is still a matter of active research (Arakawa & Schubert 1974, Lindzen 1988, Geleyn et al 1982, Emanuel 1991).] The greenhouse effect can be measured in terms of the change in T^4 at the surface necessitated by the presence of infrared absorbing gases. From this perspective, the presence of convection diminishes the purely radiative greenhouse effect by 75%. The reason is that the surface of the Earth does not cool primarily by radiation. Rather, convection carries heat away from the surface, bypassing much of the greenhouse gases, and depositing heat at higher levels where there is less greenhouse gas to inhibit cooling to space.

4. Water vapor decreases much more rapidly with height than does mean air density. Crudely speaking, the scale height for water vapor is 2–3

km compared with 7 km for air. As a result of the convection in item 3 above, water vapor near the surface contributes little to greenhouse warming. A molecule of water at 10 km altitude is comparable in importance to 1000 molecules at 2–3 km, and far more important than 5000 molecules at the surface (Arking 1993). In fact, water vapor decreases rapidly not only with increasing altitude but also with increasing latitude. Thus, the mean temperature of the Earth will depend not only on vertical transport of heat but also on meridional transport of heat. In attempting to calculate the mean temperature of the Earth by means of one-dimensional models one is assuming that one can find a latitude where the divergence of the dynamic heat flux is zero. However, in the absence of knowledge of the horizontal transport the choice of such a latitude is no more than a tuning parameter.

The situation summarized in items 3 and 4 above is schematically illustrated in Figure 7.

The Equator-to-Pole Temperature Distribution

The current annually averaged equator-to-pole temperature difference is about 40°C. In the absence of dynamic transport this quantity would be about 100°C (Lindzen 1990). The difference is even more striking for the winter hemisphere where the polar regions do not receive any sunlight.

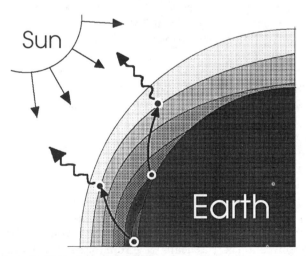

Figure 7 Schematic illustration of greenhouse effect with dynamic heat transfer. Infrared capacity is greatest at the ground over the tropics, and diminishes as one goes poleward. Air currents bodily carry heat to regions of diminished infrared opacity where the heat is radiated to space—balancing absorbed sunlight. Lighter shading schematically represents reduced opacity due to diminishing water vapor density.

Interestingly, there is currently no simple theory that quantitatively predicts the equator-to-pole temperature, despite the fact that it is this quantity that seems most relevant to climate change. As noted above, moreover, a knowledge of horizontal transport is also essential to calculating the global mean temperature. Current large-scale numerical models have difficulties here: both in the prediction of eddies (Stone & Risbey 1990) and in the prediction of polar temperatures (Boer et al 1992).

The usual picture is that heat is transported within the tropics by a large-scale cellular flow known as the Hadley circulation, and from 30° to the poles by baroclinically unstable eddies (Lorenz 1967). While this picture is roughly correct, it has a profound seasonal character. Except for a brief period when the zonally averaged surface temperature maximum is exactly at the equator, the Hadley circulation consists in a single cell with ascent in the summer hemisphere and descent in the winter hemisphere (Oort & Rasmussen 1970, Lindzen & Hou 1988). The descending branch typically extends 30° into the winter hemisphere and is associated with strong lateral gradients in potential vorticity (Hou & Lindzen 1992), and such gradients are generally associated with eddy instability. Indeed, eddy heat transport is much larger in the winter than in the summer (Oort 1983). The heat transport situation is schematically illustrated in Figure 8. The above only refers to the atmosphere; heat transport in the ocean is much less well understood, though it appears to be significant (Carrissimo et al 1985). Here we focus on the atmospheric transport for several reasons: the shallow ocean circulation is wind driven and is not a direct response to heating gradients; and the deep thermohaline circulation in the ocean is slow compared to the time it takes for the surface of the ocean to equilibrate thermally with the air above and the radiative forcing.

There is, in fact, reason to believe that the mean meridional temperature distribution is largely determined by the atmospheric transport. As noted above, heat transport between the tropics and high latitudes is carried by eddies that are believed to arise from baroclinic instability (Eady 1949, Charney 1947). The possibility that these eddies act to neutralize the basic state has long been suggested (Pocinki 1955, Stone 1978, Lindzen & Farrell 1980). However, the neutral states considered were based on the Charney-Stern condition (Charney & Stern 1962)—an extension of the Rayleigh inflection point condition to rotating stratified fluids (Lindzen 1990), and these states differed from the observed state rather profoundly. Recently, however, a new neutral state has been found (Lindzen 1993a) which is remarkably compatible with the observed state (Sun & Lindzen 1993c). The match between this state and the Hadley regime appears to depend on the intensity of the Hadley circulation (Hou 1993). The intensity of the Hadley circulation depends, in turn, on the displacement of the summer

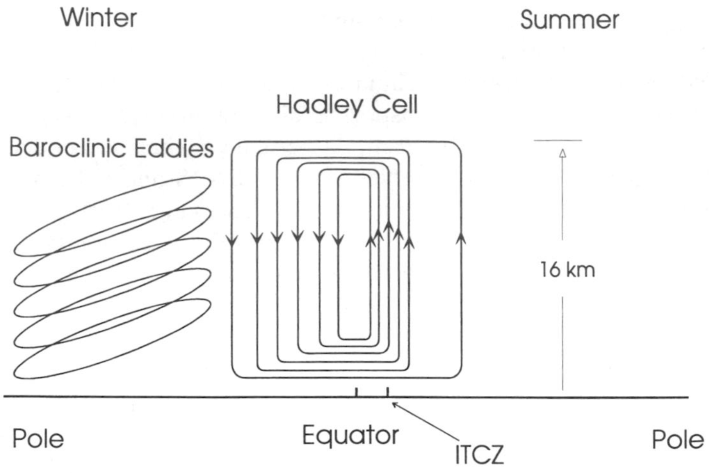

Winter   Summer

Hadley Cell

Baroclinic Eddies

16 km

Pole   Equator   Pole
                                    ITCZ

Figure 8 Schematic illustration of major dynamic mechanisms for meridional heat transfer in the atmosphere.

surface temperature maximum from the equator (Lindzen & Hou 1988) and on the sharpness of the maximum (Hou & Lindzen 1992). The former depends pronouncedly on the orbital variation (Lindzen & Pan 1993) and provides a possible physical basis for the Milankovitch mechanism. It thus appears that atmospheric processes alone may largely determine the gross temperature structure of the atmosphere (including the surface). This does not preclude an oceanic contribution to the heat flux—as long as it is not so great as to completely preclude atmospheric eddies.

4. CLIMATE SENSITIVITY AND FEEDBACKS

Current climate concerns focus on the response of the global mean temperature to increasing atmospheric CO_2. Sensitivity is defined, for convenience, as the equilibrium response of the global mean temperature to a doubling of CO_2. Such a definition is meaningful for the problem at hand, but in no way suggests that the sensitivity, so defined, is relevant to past climate change. As we have already noted, past climate change appears to have been primarily associated with changes in the equator-to-pole heat flux, and gross forcing such as that provided by increasing CO_2 does not significantly affect such fluxes in any way currently identified. Nevertheless, current large-scale numerical simulations of climate commonly suggest significant climate response to a doubling of CO_2, even in the tropics (Houghton et al 1990, 1992); there is no evidence in these models of any tropical stabilization. While these results immediately suggest a certain questionability concerning the models, it is of interest to examine the model response in terms of the physics they contain. It should first be noted that the expected globally averaged warming from a doubling of CO_2 alone without any feedbacks is 0.5–1.2°C (Lindzen 1993b). Model predictions of values from 1.5–4.5°C (Houghton et al 1990, 1992) depend on positive feedbacks within the models. One may write the globally averaged equilibrium warming for a doubling of CO_2 as

$$\Delta T_{2 \times CO_2} = \text{gain} \times \Delta T_{ng}, \tag{1}$$

where ΔT_{ng} is the response to a doubling of CO_2 in the absence of feedbacks. Gain is related to feedback by the expression

$$\text{gain} = \frac{1}{1-f}, \tag{2}$$

where f is the feedback factor. To the extent that feedbacks from different processes are independent, their contributions to the feedback factor are additive: i.e.

$$f = \sum_i f_i. \tag{3}$$

Note that gains from various processes are not additive. Crude analyses have been made of the physical origins of feedbacks in various models.

Results for several commonly cited models are given in Table 1. As we see from Table 1, the largest feedback is from water vapor, and arises because in all models, water vapor at all levels increases with increasing surface temperature. Recall that it is upper-level (above 2–3 km) water vapor that is of primary importance for greenhouse warming. (Note that water vapor and lapse rate tend to be lumped together because lapse rate changes accompany water vapor changes in current models, and the lapse

Table 1   Feedback factors, f_i, for GISS[a] and GFDL[b]

Process	Model	
	GISS	GFDL
Water vapor/lapse rate	0.40	0.43
Cloud	0.22	0.11
Snow/albedo	0.09	0.16
TOTAL	0.71	0.70
gain $[= 1/(1-f)]$	3.44	3.33

[a] Goddard Institute for Space Studies.
[b] Geophysical Fluid Dynamics Laboratory at Princeton.

rate, itself, influences the surface temperature.) The next feedback arises from clouds. In most models, surface warming is accompanied by increasing cloud cover. It may seem surprising that this leads to a positive feedback, but in these models, the infrared properties of the clouds outweigh their visible reflectivity. The remaining feedback is from snow/albedo. This refers to the fact that in the models, increased surface temperature is associated with reduced snow cover, which in turn leads to reduced visible reflectivity. There is, in fact, substantial uncertainty over the last two feedback mechanisms (Cess et al 1990, Cess et al 1991). The cloud feedback may very well be negative rather than positive (Mitchell et al 1989, Ramanathan & Collins 1991). Even the snow/albedo feedback is subject to doubt because of such factors as winter-night, high-latitude cloud cover, etc. The magnitude of this feedback in existing models varies greatly from model to model. Oddly enough, there is a tendency to regard the water vapor/lapse rate feedback as well determined because most models behave similarly despite the fact that the physics relevant to the upper level water vapor is absent in these models. Upper level water vapor in these models appears to be determined by diffusion from below which, as we show later, is impossible. For the present, we should note that the water vapor feedback alone would only produce a gain of about 1.67, but its presence is essential to the total gain. Without it, no model would produce a ΔT_{2CO_2} greater than 1.7°C regardless of the presence of other feedbacks, and under the assumption that ΔT_{ng} has the large value of 1.2°C.

For our purposes, the central fact about water vapor is the Clausius-Clapeyron relation for the saturation vapor pressure. This quantity drops sharply with temperature as illustrated in Figure 9. A parcel of air suffers compressional heating as it descends, and adiabatic expansion and cooling as it rises (both at a rate $g/c_p = 9.8$°C/km). Thus a rising parcel that starts at 80% relative humidity (i.e. 80% of saturation) will saturate within a few hundred meters, while a saturated parcel at 12 km will have its relative humidity decrease exponentially (with a scale height of 2–3 km) with depth as it sinks. This situation is schematically illustrated in Figure 10. If the source of water vapor in the upper troposphere were water vapor deposited aloft by deep cumulus convection and carried downward, we would expect rapidly decreasing relative humidity with depth. Instead, relative humidity is relatively constant with depth. If the source were upward transport from below, we would expect widespread deep cloud up to the level being supplied. This too is not observed. Rather, it appears that the source of upper troposphere water vapor is the evaporation of precipitation originating in ice crystals near the tropopause (Smith 1992, Sun & Lindzen 1993b, Betts 1990). Such a process is free of the problems associated with

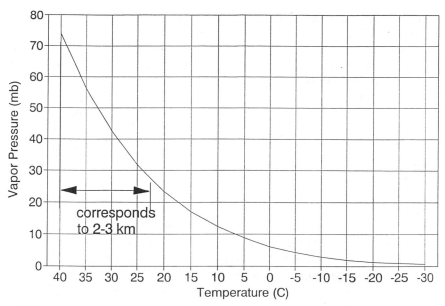

Figure 9 Saturation water vapor pressure (over pure liquid water) as a function of temperature.

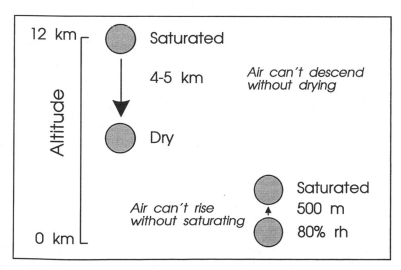

Figure 10 Schematic illustration of consequences of simple vertical transport of water vapor.

the direct transport of water vapor. In the tropics, at least, these crystals appear to have been directly detrained from cumulonimbus towers. What we need to know is how the amount of detrained ice is determined. The simplest models of cumulus convection (Arakawa & Schubert 1974, Lindzen 1988) assumed that as air rises in a cumulonimbus tower, all condensate rains out, leaving just saturated air to be detrained at the cloud top. At the very least, this requires slow ascent. With rapid ascent, at least some of the condensate will be lofted by the updraft to above the freezing level (ca 5 km) and detrained in the form of ice. The speed of the updrafts in cumulonimbus towers will be determined by the convective available potential energy (CAPE) of the ambient air. This simply refers to the integral of the buoyancy force over the depth of the cloud (Bolton 1980, Williams et al 1992). It seems reasonable to assume that the amount of detrained ice will depend on CAPE, and that the nature of the water vapor feedback will depend on how CAPE responds to warming.[1]

In this connection, it is important to note that the dependence of CAPE on surface temperature within a given climate regime may differ from the dependence of CAPE on global (or tropical) mean temperature in different climate regimes. The point is simply that (at least within the tropics), the large-scale circulation acts to eliminate horizontal temperature gradients above the trade wind boundary layer. However, within the turbulent boundary layer the horizontal temperature variation is less well mixed; warmer surface temperatures are therefore associated with greater CAPE. However, the possibility remains that in different climates, the contribution to CAPE from above the boundary layer may be different, and that warmer climates (as opposed to warmer local surface temperatures) may be associated with reduced overall CAPE (at least in the tropics). This was argued by Sun & Lindzen (1993a). Observationally, Oort (1993), analyzing routine radiosonde data for globally averaged temperature and specific humidity at various levels in the atmosphere, did find that the global warming of the late 1970s was indeed accompanied by reduced specific humidity above 700 mb. Indeed the observed reduction was such to produce a water vapor feedback contribution to f (in Equation 3) of -6 as opposed to $+0.4$ as is found in current models (Sun & Lindzen 1993a). Unfortunately, changes in sensors used to measure humidity make the water vapor time series unreliable (Elliot & Gaffen 1991).

Interestingly, however, the inferred value of $f = -6$ would be sufficient

---

[1] It should be noted that the supposition in a number of papers (Raval & Ramanathan 1989, Rind et al 1991) that upper level water vapor is determined simply by surface temperature immediately below is at variance with both the above physics, and the fact that over 99.9% of the air in the tropics is descending, not rising from the surface (Sarachik 1985).

to explain the stability of tropical temperatures despite changes in equator-to-pole heat fluxes equivalent to changes in surface fluxes of 10s of watts/m^2. It should be emphasized that the truly remarkable fact about tropical stability is its existence in the presence of very large changes in tropics-to-high latitude heat fluxes. These changes are far larger than those envisaged as being due to changing CO_2, solar radiation, etc.

5. EQUILIBRIA VS TIME-DEPENDENT RESPONSE TO CLIMATE PERTURBATIONS

The above discussion was framed in terms of the equilibrium response to climate perturbations. In point of fact, most of the Earth is ocean covered, and the heat capacity of the ocean will delay the response of the climate system as a whole. As noted by Hansen et al (1985), the delay will depend on the rate at which heat is transported downward in the ocean (the more rapidly heat is transported, the longer the delay), and the feedbacks in the atmosphere. The stronger the atmospheric feedbacks are, the longer the delay since, as we shall see, stronger atmospheric feedbacks imply weaker coupling to the sea surface. The simplest approach to ocean delay has been to use simple box-diffusion models for communication between the atmosphere and the ocean. A typical geometry is shown in Figure 11. The relevant equations for this geometry are:

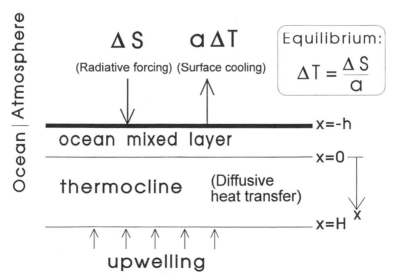

Figure 11 Geometry of simple box-diffusion model for ocean heat absorption. (From Lindzen 1993b.)

$$\rho c h \frac{\partial \Delta T_1}{\partial t} = \Delta S - a \Delta T_1 - \kappa \frac{\partial T_2}{\partial x}\bigg|_{x=0}, \tag{4a}$$

$$\frac{\partial T_2}{\partial x} = \kappa \frac{\partial^2 T_2}{\partial x^2} \quad \text{for} \quad x > 0, \tag{4b}$$

$$T_2 = T_1 \quad \text{at} \quad x = 0, \tag{4c}$$

$$\frac{\partial T_2}{\partial x} = 0 \quad \text{at} \quad x = H, \tag{4d}$$

where T_1 is the mixed layer temperature, and $T_2(x)$ is the temperature below the mixed layer, ρ is the density of water, c is the heat capacity of water, and h is the thickness of the mixed layer (taken to be 70 m). H is the depth of the thermocline below which upwelling inhibits downward heat diffusion. κ, the eddy heat diffusion, is typically taken to be 1.5 cm^2 sec^{-1}. ΔS represents a radiative perturbation, while $a \Delta T_1$ represents the response of the surface temperature. If $a = a_0$ in the absence of gain, then, in the presence of gain, $a = a_0/g$, and $g = 1/(1-f)$. As already noted the larger g is, the weaker the coupling in Equation 4a.

Equations 4 above are relatively standard equations for diffusive heat transfer. The response of the surface temperature to perturbed forcing is not simply exponential in time, but rather involves a continuously increasing time scale as heat penetrates to greater depths (Lindzen 1993). However, for purposes of discussion, it is convenient to define a characteristic time as that time over which the response to impulsive forcing reaches to within $1/e$ of its equilibrium value. A plot of how this characteristic time varies with gain is given in Figure 12. We see that the depen-

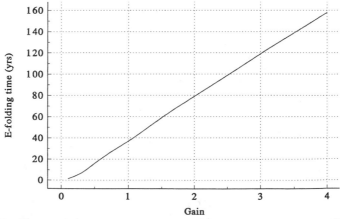

Figure 12 Characteristic ocean delay time as a function of climate system gain. (From Lindzen 1993b.)

dence is almost linear away from very small gains. This relation between climate sensitivity (i.e. gain) and ocean delay allows a more sophisticated approach to assessing both the past record of climate and projections for the future. In Figure 13, we show expected warming on the basis of various climate sensitivities for the IPCC (Intergovernmental Panel on Climate Change) business-as-usual emissions scenario (which involves quadrupling effective CO_2 by 2100). Although differing sensitivities lead to different results by 2100, the fact that higher sensitivities are associated with longer delays leads to the disappointing result that the record of the past century (when effective CO_2 increased by almost 50%) is broadly consistent with virtually any sensitivity (assuming natural variability on the order of 0.5°C). On the other hand, if we assume a gain of 3.33, and vary κ in order to obtain different delay times, we get the results shown in Figure 14. These results make it clear that such high gain is broadly consistent with the observed record only if the characteristic delay time is greater than 100 yrs.

A matter of some interest is whether one can directly measure delay time. For example, large volcanic eruptions provide a significant change in albedo for a year or so following eruption (Oliver 1976). The usual picture is that it takes about three months for volcanic emissions to spread around the world forming a sulfate aerosol layer which then decays with a time scale of about one year. The response of the system described by Equations 4 to such forcing (for different choices of gain) is shown in Figure 15. Given the numerous uncertainties associated with volcanic forcing, the responses during the first three years following an eruption do

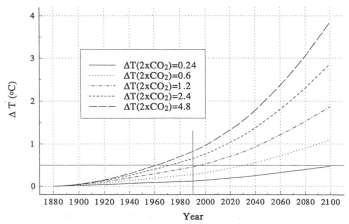

Figure 13 Change in global temperature expected for IPCC business-as-usual emissions scenario and various climate sensitivities. (From Lindzen 1993b.)

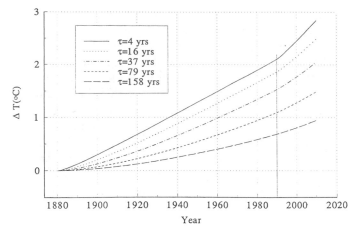

Figure 14   Behavior of global temperature for IPCC business-as-usual emissions scenario, $\Delta T_{2CO_2} = 4°C$, and various choices of ocean delay. (From Lindzen 1993b.)

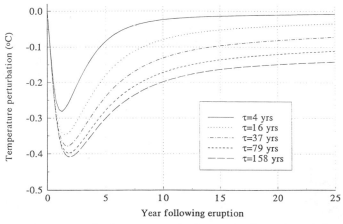

Figure 15   Response to Krakatoa type volcanic eruption for various choices of climate system gain. (From Lindzen 1993b.)

not significantly distinguish one gain from another.[2] The only possible distinction that can be noted in Figure 15 is that the response for strong negative feedback (gain < 1) tends to peak a year following eruption, while the response for positive feedback (gain > 1) tends to peak two years after

---

[2] The claim (Hansen et al 1992) that the prediction of cooling following Pinatubo forms a test of a model, ignores the fact that such a prediction fails to distinguish between models that predict $\Delta T_{2CO_2} = 5°C$ and those that predict $\Delta T_{2CO_2} = 0.25°C$.

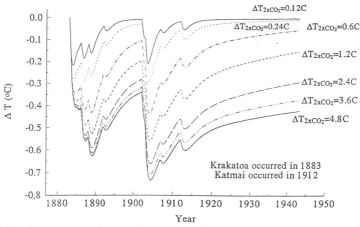

Figure 16   Response to series of volcanoes between Krakatoa in 1883 and Katmai in 1912. (From Lindzen 1993b.)

eruption. After 5–10 years, models with high sensitivity retain much more of the volcanic cooling than do models with low sensitivity. However, for individual volcanoes, the amount in either case is smaller than normal interannual variability. The situation is potentially better if one has a sequence of relatively closely spaced major volcanic eruptions, as was the case between Krakatoa in 1883 and Katmai in 1912 (Oliver 1976). Here, one might expect that for high sensitivity, the response of each volcano will add to the tail of the preceding volcano leading to a pronounced cooling trend. For low sensitivity, however, each volcano will produce an almost independent blip. The situation is illustrated in Figure 16. We see that high sensitivity leads to a pronounced cooling trend from 1883 to 1912 followed by persistent low temperatures (there were no further major volcanoes for about 50 years). Looking at Figure 1, we see only cooling blips corresponding to low sensitivity followed by pronounced warming between 1919 and 1940. It could of course be argued that the warming trend had begun earlier and had been cancelled by the volcanic cooling. Recall, however, that high sensitivity is associated with long delay so that the posited warming trend would have required forcing far greater than could have been accounted for by increasing CO_2 or any other known source.

6.   REMARKS ON INCREASING CO_2

CO_2 in the atmosphere has increased from about 280 ppmv in 1800 to about 355 ppmv today. This represents an increase in excess of 25%. In addition, other minor greenhouse gases have also increased leading to an effective increase of about 50% in greenhouse gases. There is concern that continuing increases will lead to significant global warming. The focus of the present review has been on the basic physics of climate, and implications for climatic response to specified effective CO_2 increases. However, a few words are in order on how these increases have been specified. The reader may have noticed that the common measure of climate sensitivity is the equilibrium response to a doubling of CO_2, while the IPCC projections are based on the transient response to a quadrupling of effective CO_2 by 2100. As was noted in Section 5, high sensitivity is inevitably accompanied by long ocean delays. We have also seen that if the delays were shorter, then the record of the past century would have been inconsistent with high sensitivities. Given the need for long delays, the response to a mere doubling of CO_2 by 2100 would have been far smaller than the equilibrium responses; quadrupling was necessary for the transient response to be comparable by 2100 to the equilibrium response for a doubling of CO_2. Such a scenario was labeled by Houghton et al (1990) as the "business as usual scenario." However, it was, in fact a scenario designed to double effective CO_2 by 2030 and quadruple it later in the century. In order to arrive at such a scenario, it was necessary to project substantial increases in population, higher standards of living in the currently less developed world, increased reliance on coal, restrictions on nuclear power, etc. Recognition of the vast uncertainty of all projections over such long periods led to the presentation of a broad range of possibilities in Houghton et al (1992). It became clear that the main determinant of emissions would be population and economic growth in the currently less developed countries, and that emission controls in the currently developed countries was of relatively small long-term importance. In addition to socio-economic uncertainties, there are significant geochemical uncertainties in translating emissions into atmospheric CO_2.

The complexity and uncertainty of the chemistry are substantial (Heimann 1992). Here, we simply wish to consider some broad aspects of the problem. Measurements of ice cores suggest that preindustrial levels of CO_2 were approximately 280 ppmv for at least hundreds of years. This is suggestive of an equilibrating process. The simplest representation of such a process is

$$\frac{dn_{CO_2}}{dt} = a(\bar{n}_{CO_2} - n_{CO_2}) + S_{CO_2}, \tag{5}$$

where n refers to density, the overbar to an equilibrium value, a^{-1} to an equilibration time, and S to a source. Equation 5 is of course largely schematic, but it provides a framework for discussion. Clearly, an S that remains constant, will simply lead eventually to a new equilibrium with none of the emissions contributing to additional atmospheric CO_2. However, from 1800 to 1973, CO_2 sources are believed to have been exponentially increasing with an e-folding time, τ, of about 45 years. Such a system cannot be in equilibrium; the degree of disequilibrium depends on the relative magnitude of τ and a^{-1}. If a^{-1} is much shorter than τ then the system will always be near equilibrium and very little of the increased emission will appear in the atmosphere. If, on the other hand, a^{-1} is much longer than τ, then most of S will appear in the atmosphere. In point of fact, about half the CO_2 provided by S has remained in the atmosphere, suggesting that a^{-1} and τ are comparable. One remarkable aspect of all the IPCC scenarios is that they all call for τ to be much larger than 45 years during the next century, and at the same time rather inconsistently have the fraction of CO_2 remaining in the atmosphere increasing as well. It would appear that one is exaggerating the atmospheric consequences of the uncertain emissions.

Of course, the questions posed in this review concerning how climate behaves are of immense importance regardless of the specific behavior of CO_2. However, there can be little question that without increasing minor greenhouse gases the political import of the question diminishes significantly.

7. SUMMARY AND REMARKS

This review has stressed the basic questions in climate that transcend the specific concern for the role of minor greenhouse gases: Namely, what determines the mean temperature of the Earth and its distribution— specifically the distribution with latitude? Major climate changes in the past have been characterized by large changes in the equator-to-pole temperature difference (ranging from about 19°C to 60°C; the present annually averaged value is about 40°C), and relatively constant equatorial temperatures (within about 2°C of the present). Changes in the mean temperature of the Earth appear to have been a by-product of these changes rather than a cause. The cause(s) for the near constancy of equatorial temperatures constitutes an additional major question. The stability of equatorial temperatures is particularly remarkable when one considers that the changes in heat flux out of the tropics implicit in the changes in equator-to-pole temperature difference are likely to have been far larger than any proposed external radiative forcing. Indeed, given the stability of the tropics, it is hard to see how gross external radiative forcing, which does not affect meridional heat fluxes in any evident way, can substantially alter the mean temperature of the Earth.

Indeed, despite an increase of effective CO_2 of about 50% over the past century, the data do not display any change that can be distinguished from normal climate variability. If the ocean delay is on the order of 160 years, then the observed warming over the past century of $0.45°C \pm 0.15°C$ is "broadly" consistent with any ΔT_{2CO_2} between 0 and 5°C. However, if the delay times should prove shorter, then the larger values become progressively incompatible with the data. As we have noted, such data as one has from the sequence of volcanoes between Krakatoa (1883) and Katmai (1912) are suggestive of small delays and negative feedbacks.

It has been commonly suggested that it will be decades before anthropogenic warming will be identified in the data, and that it may take a similar length of time before large-scale computer simulations of climate become dependable. This may be true at some level. However, a distinctly more optimistic view is appropriate at another level. For example, our present fears of large warming are based on specific aspects of current models: most notably their treatment of upper level water vapor. Improved understanding of the water vapor budget based on both theory and thoughtful observational analyses (both of which can be achieved in a much shorter time) should enable us to determine whether current predictions have any substantial foundation. To be sure, there may exist hitherto unknown processes that could still lead to warming, but unknown processes offer little policy guidance. We have also noted that ocean delay offers a direct measure of feedback. It should be possible to explore existing data concerning this matter. Finally, our increasing knowledge of the Earth's past climate provides a valuable test-bed for our quantitative understanding of climate. Such understanding is not limited to the output of large-scale simulations.

ACKNOWLEDGEMENTS

The preparation of this review was supported by NASA under grant NAGW-525, and by the NSF under grant ATM-914441.

Literature Cited

Arakawa, A., Schubert, W. H. 1974. Interaction of a cumulus cloud ensemble with the large scale environment, Part I. *J. Atmos. Sci.* 31: 674–701

Arking, A. 1993. Water vapor and lapse rate feedback: insight from one-dimensional climate model. *J. Clim.* In press

Balling, R. C. Jr. 1992. *The Heated Debate.* San Francisco: Pacific Res. Inst. 195 pp.

Barron, E. J. 1987. Eocene equator-to-pole surface ocean temperatures: A significant climate problem? *Paleoceanogr.* 2: 729–39

Berger, A. L. 1978. Long-term variations of

daily insolation and Quaternary climatic changes. *J. Atmos. Sci.* 35: 2362–67

Betts, A. K. 1990. Greenhouse warming and the tropical water vapor budget. *Bull. Am. Meteorol. Soc.* 71: 1465–67

Boer, G. J., Arpe, K., Blackburn, M., Déqué, M., Gates, W. L., et al. 1992. Some results from an intercomparison of the climates simulated by 14 atmospheric general circulation models. *J. Geophys. Res.* 97: 12,771–86

Bolton, D. 1980. The computation of equivalent potential temperature. *Mon. Weather Rev.* 108: 1046–53

Budyko, M. I., Izrael, Y. A. 1991. In *Anthropogenic Climate Change*, ed. M. I. Budyko, Y. A. Izrael, pp. 277–318. Tucson: Univ. Ariz. Press

Carissimo, B. C., Oort, A. H., Vonder Haar, T. H. 1985. Estimating the meridional energy transports in the atmosphere and ocean. *J. Phys. Oceanogr.* 15: 82–91

Cess, R. D., Potter, G. L., Blanchet, J. P., Boer, G. J., Del Genio, A. D., et al. 1990. Intercomparison and interpretation of climate feedback processes in 19 atmospheric general circulation models. *J. Geophys. Res.* 95: 16,601–15

Cess, R. D., Potter, G. L., Zhang, M.-H., Blanchet, J.-P., Chalita, S., et al. 1991. Interpretation of snow-climate feedback as produced by 17 general circulation models. *Science* 253: 888–92

Charney, J. G. 1947. The dynamics of long waves in a baroclinic westerly current. *J. Meteorol.* 4: 135–63

Charney, J. G., Stern, M. E. 1962. On the instability of internal baroclinic jets in a rotating atmosphere. *J. Atmos. Sci.* 19: 159–72

Crowley, T. J., North, G. R. 1991. *Paleoclimatology*. New York: Oxford Univ. Press. 339 pp.

Eady, E. T. 1949. Long waves and cyclone waves. *Tellus* 1: 33–52

Elliot, W. P., Gaffen, D. J. 1991. On the utility of radiosonde humidity archives for climate studies. *Bull. Am. Meteorol. Soc.* 72: 1507–20

Emanuel, K. A. 1991. A scheme for representing cumulus convection in large-scale models. *J. Atmos. Sci.* 48: 2313–35

Evans, D. L., Freeland, H. J. 1977. Variations in the earth's orbit: Pacemaker of the ice ages? *Science* 198: 528–30

Geleyn, J.-F., Girard, C., Louis, J.-F. 1982. A simple parameterization of moist convection for large-scale atmospheric models. *Beitr. Phys. Atmosph.* 55: 325–34

Goody, R. M., Yung, Y. L. 1989. *Atmospheric Radiation*. New York: Oxford Univ. Press. 519 pp.

Hansen, J., Lacis, A., Ruedy, R., Sato, M. 1992. Potential climate impact of Mount Pinatubo eruption. *Geophys. Res. Lett.* 19: 215–18

Hansen, J., Russell, G., Lacis, A., Fung, I., Rind, D. 1985. Climate response times: dependence on climate sensitivity and ocean mixing. *Science* 229: 857–59

Heimann, M. 1991. *Modelling the global carbon cycle*. Presented at First Demetra Meeting on Climate Variability and Global Change, Chianciano Therme, Italy, Oct. 28–Nov. 3, 1991. Proceedings of Meeting in preparation

Hou, A. Y. 1993. The influence of tropical heating displacements on the extratropical climate. *J. Atmos. Sci.* In press

Hou, A. Y., Lindzen, R. S. 1992. Intensification of the Hadley circulation due to concentrated heating. *J. Atmos. Sci.* 49: 1233–41

Houghton, J. T., Callander, B. A., Varney, S. K. 1992. Update to *Climate Change 1992, Suppl. Rep. IPCC Sci. Assessment*. Cambridge: Cambridge Univ. Press. 200 pp.

Houghton, J. T., Jenkins, G. J., Ephraums J. J., eds., 1990. *Climate Change. The IPCC Scientific Assessment*. Cambridge: Cambridge Univ. Press. 365 pp.

Imbrie, J., Imbrie, K. P. 1980. *Ice Ages, Solving the Mystery*. Hillside, NJ: Enslow. 213 pp.

Lindzen, R. S. 1988. Some remarks on cumulus parameterization. *Pageoph* 126: 123–35

Lindzen, R. S. 1990. *Dynamics in Atmospheric Physics*. New York: Cambridge Univ. Press. 310 pp.

Lindzen, R. S. 1993a. Baroclinic neutrality and the tropopause. *J. Atmos. Sci.* 50: 1148–51

Lindzen, R. S. 1993b. Constraining possibilities versus signal detection. In *Review of Climate Variability on the Decade to Century Time Scale*, National Research Council. To appear

Lindzen, R. S., Farrell, B. 1980. The role of polar regions in global climate, and the parameterization of global heat transport. *Mon. Weather Rev.* 108: 2064–79

Lindzen, R. S., Hou, A. Y. 1988. Hadley circulations for zonally averaged heating centered off the equator. *J. Atmos. Sci.* 45: 2416–27

Lindzen, R. S., Pan, W. 1993. A note on orbital control of equator-pole heat fluxes. *Clim. Dyn.* In press

Lorenz, E. N. 1967. *The Nature and Theory of the General Circulation of the Atmosphere*. Geneva: World Meteorol. Organ. 161 pp.

Milankovitch, M. 1930. *Mathematische Klimalehre und Astronomische Theorie der*

Klimaschwankungen. Berlin: Gebrüder Borntraeger 176 pp.

Mitchell, J. F. B., Senior, C. A., Ingram, W. J. 1989. CO_2 and climate: a missing feedback? *Nature* 341: 132–34

Möller, F., Manabe, S. 1961. Über das Strahlungsgleichgewicht der Atmosphäre. *Z. Meteorol.* 15: 3–8

Oliver, R. C. 1976. On the response of hemispheric mean temperature to stratospheric dust: an empirical approach. *J. Appl. Meteorol.* 15: 933–50

Oort, A. H. 1983. Global atmospheric circulation statistics, 1958–1973. *NOAA Prof. Pap.* 14. Rockville, Md.: NOAA, U.S. Dept. Commerce. 180 pp.

Oort, A. H. 1993. Observed humidity trends in the atmosphere. In *Proc. 17th Annu. Climate Dynamics Workshop.* To appear

Oort A. H., Rasmussen, E. M. 1970. On the annual variation of the monthly mean meridional circulation. *Mon. Weather Rev.* 98: 423–42

Peixoto, J. P., Oort, A. H. 1992. *Physics of Climate.* New York: Am. Inst. Phys. 520 pp.

Pocinki, L., 1955. Stability of a simple baroclinic flow with horizontal shear. *A. F. Cambridge Res. Center, Res. Pap. No.* 38. 78 pp.

Ramanathan. V., Collins, W. 1991. Thermodynamic regulation of ocean warming by cirrus clouds deduced from the 1987 El Niño. *Nature* 351: 27–32

Raval, A., Ramanathan, V. 1989. Observational determination of the greenhouse effect. *Nature* 342: 758–61

Rind, D., Chiou, E. W., Chu, W., Larsen, J., Oltmans, S., et al. 1991. Positive water vapor feedback in climate models confirmed by satellite data. *Nature* 349: 500–3

Sarachik, E. S. 1985. A simple theory for the vertical structure of the tropical atmosphere. *Pageoph* 123: 261–71

Smith, R. B. 1992. Deuterium in North Atlantic storm tops. *J. Atmos. Sci.* 49: 2041–57

Spencer, R. W., Christy, J. R. 1990. Precise monitoring of global temperature trends from satellites. *Science* 247: 1558–62

Stone, P. H. 1978. Baroclinic adjustment. *J. Atmos. Sci.* 35: 561–71

Stone, P. H., Risbey, J. 1990. On the limitations of general circulation models. *Geophys. Res. Lett.* 17: 2173–76

Sun, D.-Z., Lindzen, R. S. 1993a. Negative feedback of water vapor inferred from the mountain snowline record. *Ann. Geophys.* 11: 204–15

Sun, D.-Z., Lindzen, R. S. 1993b. On the distribution of tropical tropospheric water vapor. *J. Atmos. Sci.* 50: 1643–60

Sun, D.-Z., Lindzen, R. S. 1993c. An EPV view of the zonal mean distribution of temperature and wind in the extra-tropical troposphere. *J. Atmos. Sci.* Submitted

Trenberth, K. E., Christy, J. R., Hurrell, J. W. 1992. Monitoring global monthly mean surface temperatures. *J. Clim.* 12: 1405–23

Williams, E. R., Rutledge, S. A., Geotis, S. G., Renno, N., Rasmussen, E., et al. 1992. A radar and electrical study of tropical "hot towers." *J. Atmos. Sci.* 49: 1386–95

Winograd, I. J., Coplen, T. B., Landwehr, J. M., Riggs, A. C., Ludwig, K. R., et al. 1992. Continuous 500,000-year climate record from vein calcite in Devils Hole, Nevada. *Science* 258: 255–60

Annu. Rev. Fluid Mech. 1994. 26 : 379–409

GÖRTLER VORTICES

William S. Saric

Mechanical and Aerospace Engineering, Arizona State University, Tempe, Arizona 85287-6106

KEY WORDS: centrifugal instabilities, boundary layers on concave walls, stream-wise vortices, curved flows, nonparallel stability

1.   INTRODUCTION

The origins of transition to turbulence in a bounded shear flow within a low disturbance environment are typically found in the instabilities of the basic state. There are, of course, a number of mechanisms that can lead to breakdown of laminar boundary-layer flow and many have been reviewed in these Volumes, e.g. inflectional profiles and viscous instabilities (Reshotko 1976, Bayly et al 1988), secondary instabilities (Herbert 1988), and crossflow (Reed & Saric 1989). This review examines boundary-layer instabilities induced by wall curvature. In this case the instability is in the form of steady, streamwise-oriented, counter-rotating vortices, commonly called Görtler vortices.

1.1   Basic Ideas

The shear flow over a concave surface is subject to a centrifugal instability whose inviscid mechanism was first given by Rayleigh (1916). For a circular geometry (r, θ, z) with basic-state velocity vector, $\mathbf{V} = (U, V, W)$, defined as

$$U = 0, \quad V = V(r), \quad W = 0, \tag{1}$$

where U, V, and W represent the radial, tangential, and axial velocity components, Rayleigh showed that the necessary and sufficient condition for the existence of an inviscid axisymmetric instability is:

$$d(\Gamma^2)/dr < 0, \quad \text{anywhere in the flow}, \tag{2}$$

where Γ is the circulation defined as $\Gamma = rV$. This particular Rayleigh

379

criterion for instability is called the *Rayleigh circulation criterion.* The physical description of this well-known mechanism and the necessary mathematical derivation are given by Drazin & Reid (1981) and Stuart (1986). Essentially, the inability of the local pressure gradient, under conditions of (2), to restrain an excess in angular momentum of a particle undergoing an outward virtual displacement, leads to the instability.

Examples of the Rayleigh circulation criterion are shown in Figure 1. If, in an inviscid circular flow, $|rV|$ decreases with an increase in r, as shown in Figures 1*b* and 1*c*, the flow will be unstable. On the other hand, the flows of Figures 1*a* and 1*d* are stable since $|rV|$ increases with r. Bayley (1987) showed that this behavior holds for noncircular closed streamlines

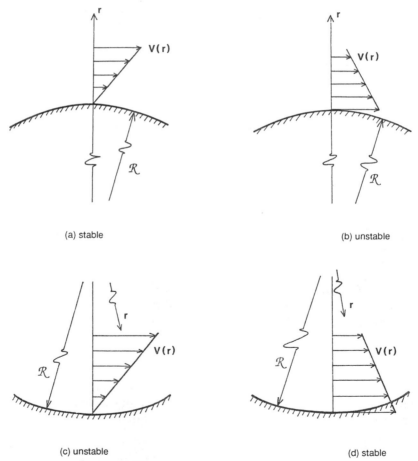

(a) stable   (b) unstable

(c) unstable   (d) stable

Figure 1   Illustrations of velocity distributions that lead to centrifugal instabilities.

and Floryan (1986) showed this for boundary-layer type profiles. Floryan also points out that this behavior need not be restricted to monotonic profiles.

With viscosity present, the Rayleigh circulation criterion is only a necessary condition for a centrifugal instability since one expects stability in some small Reynolds number limit. Taylor (1923) used Rayleigh's result in his classic work on the stability of viscous circular Couette flow. In the case of the inner cylinder rotating and the outer cylinder fixed (as in Figure 1*b*), Equation (2) is satisfied. This geometry (tangential flow in the annulus between rotating cylinders) defines the *Taylor instability*. When a critical value of the inner cylinder speed is reached, the instability is initially observed in the form of counter-rotating toroidal vortices usually called *Taylor vortices*. This is a *closed* system of a *parallel* flow which has been studied quite extensively (e.g. Coles 1965, Drazin & Reid 1981, Stuart 1986).

When a fully-developed channel flow is curved, Rayleigh's circulation criterion is satisfied along the outer race of the channel (as in Figure 1*c*) and hence, this region of the flow is unstable. This is called the *Dean instability* (Dean 1928) and the instability is in the form of *Dean vortices*. This is an *open* system with a parallel basic state which is described by Drazin & Reid (1981).

1.2   Görtler Vortices

In a boundary-layer flow over a concave surface, the radial direction opposes the velocity gradient as shown in Figure 1*c*. At the center of curvature, $rU_\infty = 0$, and at the wall, $r_w U = 0$. Away from the center of curvature, where $U = U_\infty$, rU_∞ increases with r. Therefore a maximum of rU is present in the flow, and a region where $d|rU|/dr < 0$ must exist. The first application of this idea to boundary layers was presented by Görtler (1941) who showed solutions of the disturbance equations to be in the form of streamwise-oriented, counter-rotating vortices. Hence, the instability of viscous boundary-layer flows over concave surfaces bears the name of *Görtler* and the disturbance flow is generally referred to as *Görtler vortices*. This is an open system with a *weakly nonparallel* basic state. It should be noted that this instability is not limited to concave geometries and that the Görtler instability is operative for a wall jet over a convex surface (Floryan 1986, 1989; Herron 1991).

All three of these instabilities share the same physical mechanism given by Rayleigh. Their basic differences are whether they are closed (Taylor) or open (Görtler and Dean) systems and whether they are parallel (Taylor and Dean) or boundary-layer (Görtler) flows. The distinction between these instabilities has not always been clear and hence, one is likely to see the name of Taylor associated with all three of these types of flows. However, the presentation that follows will show that the spatial, non-parallel, open-flow characteristics of the Görtler problem make it quite unique. It shares neither the complex and interesting bifurcations of the Taylor problem (Coles 1965) nor the mathematical simplicity of the Dean problem (Drazin & Reed 1981). Very little of the vast literature on the Taylor problem is applicable to the understanding of the growth of the vortices in an open, nonparallel system. Therefore we reserve the appellation of Görtler to this latter problem and concentrate the review in this area. The reader may look to Stuart (1986), Koschmieder (1992), and Tagg (1992) for reviews of the Taylor problem and to Guo & Finlay (1993) for a current review of the Dean problem.

The Görtler instability is an important boundary-layer instability that, under some conditions, leads the flow through a transition to turbulence. It is known that a Görtler instability can cause transition on the wall of a supersonic nozzle in a boundary layer that would be otherwise laminar (Beckwith et al 1985, Chen et al 1985). Moreover, the Görtler vortex structure exists in a turbulent boundary layer over a concave surface such as turbine-compressor blades (e.g. see Floryan 1991 for a review). This instability is visualized, for example, with surface striations on the reentry vehicles in the Smithsonian Air & Space Museum where differential surface ablation caused locally concave surfaces. On the other hand, recent experiments (Swearingen & Blackwelder 1987; Peerhossaini & Wesfreid 1988a,b) show that the breakdown to turbulence in the presence of Görtler vortices is typically through a strong secondary instability caused by distortion of the steady velocity profile. This leads to arguments regarding the *linear* nature of this instability. In addition, the spanwise modulations of the steady flow caused by the Görtler vortices can also destabilize Tollmien-Schlichting waves (Nayfeh & Al-Maaitah 1987, Hall & Seddougui 1989, Malik & Hussaini 1990). These brief comments serve to illustrate that this is a rich area of study.

Before completing the review of the literature, it is important to bring the reader up to date with regard to the current mathematical formulation of the problem since it is critical to understanding the literature. This is carried out in Section 2 with the idea to clear up some confusion that has existed in solution techniques. In the coverage of experimental results in Section 3, the mean flow distortion effects are described. This puts into perspective the difficulty of using linear stability theory for the Görtler problem. Section 5 describes nonlinear work and a few applications where the linear instability analysis is a useful evaluation tool. It also concentrates on secondary instabilities which is the most recent area of research activity. This review is aided by recent worthwhile and complementary reviews by Hall (1990) and Floryan (1991) which cover some different ground than is discussed here.

2.   PROBLEM FORMULATION AND LINEAR STABILITY ANALYSIS

The Görtler instability in a 2-D flow, occurs in the form of steady, streamwise-oriented, counter-rotating vortices as shown in Figure 2. In this figure, $\mathscr{R}$ identifies the wall radius of curvature, λ is the spanwise wavelength of the disturbance, and U_∞ is the edge velocity. The disturbance velocities associated with the steady vortex grow in the stream direction while holding their spanwise wavelength fixed. The experimental evidence for such a picture, in either forced or natural environments, for the early development of the instability is overwhelming (Tani 1949, 1962; Aihara 1962; Bippes 1972; Yurchenko 1981; Ito 1980). This leads to the fundamental assump-

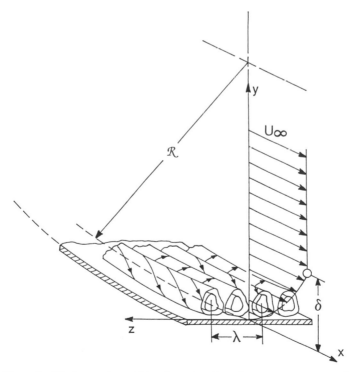

Figure 2   Görtler vortices within a boundary-layer flow over a concave wall.

tion, made in the linear regime, that the vortex structure is spanwise periodic and can be given by a single Fourier mode.

With weak curvature, the vortices grow weakly in the stream direction. Since there is no $O(1)$ streamwise variation of the instability due to a traveling wave, it is not possible to separate the weak growth of the boundary layer from the growth of the disturbance. This leads to a failure of the normal-mode approach and is the heart of the Görtler problem. Therefore, it is important in this case to first develop the disturbance equations for the Görtler instability which include the effects of nonparallel growth of the boundary layer.

2.1 *Disturbance Equations*

An incompressible flow is assumed and by following Floryan & Saric (1982), a small viscous parameter $\varepsilon = 1/R = (v/U_\infty L)^{1/2}$ is introduced and the dimensionless coordinates (ξ, y, z) are defined as

$$(\xi, y, z) = (\varepsilon x^*/L, y^*/L, z^*/L), \tag{3}$$

with dimensionless, weakly-nonparallel, basic-state velocity field

$$\mathbf{V} = [U(\xi, y), \varepsilon V(\xi, y)]. \tag{4}$$

Here, x denotes the stream direction, y denotes the wall-normal direction, and z is the span direction. L is a reference length, * denotes dimensional variables, and v is the kinematic viscosity. When $L = x^*$, R is the typical boundary-layer Reynolds number based on the length scale $\delta_r = (vx^*/U_\infty)^{1/2}$. When considering the *weak* nonparallel nature of the basic state, it is more convenient to express the stream direction as a slow scale (in this case $\xi = \varepsilon x^*/L$) rather than express the wall-normal direction as a fast scale.

The stability equations are obtained by superposing small, steady, spanwise periodic disturbances onto the basic state in the following form:

$$u^*/U_\infty = U(\xi, y) + u(\xi, y) \cos \beta z, \tag{5a}$$

$$v^*/U_\infty = \varepsilon V(\xi, y) + \varepsilon v(\xi, y) \cos \beta z, \tag{5b}$$

$$w^*/U_\infty = \varepsilon w(\xi, y) \sin \beta z, \tag{5c}$$

$$p^*/\rho U_\infty^2 = P + \varepsilon^2 p(\xi, y) \cos \beta z, \tag{5d}$$

where the dimensionless spanwise wave number[1] is defined as $\beta = 2\pi\varepsilon L/\lambda$.

---

[1] The historical notation uses α for spanwise wavenumber and β for growth rate. However, when considering interactions and secondary instabilities with other wave forms, the more common usage of α for chordwise wavenumber, β for spanwise wavenumber, and σ for growth rate is less confusing (e.g. Yu & Liu 1991).

An important feature of steady streamwise vortices within a shear layer is the convection of streamwise momentum in the normal and spanwise directions by very weak velocity components of the vorticity. This produces large changes in the mean velocity profiles. Therefore, in Equation (5), the streamwise disturbance velocity is scaled differently from v and w in order to account for $O(1)$ changes in u^* with $O(\varepsilon)$ changes in v^* and w^* because of the convection of streamwise momentum (Floryan & Saric 1982). This ordering was first suggested by DiPrima & Stuart (1972, 1975).

The terms in Equation (5) are substituted into the Navier-Stokes equations written in *optimal* curvilinear coordinates. This optimal coordinate system, based on the inviscid streamlines and potential lines of a flow over a curved surface, are surface-oriented for the inner flow and Cartesian for the outer flow and permit uniformly-valid matching (Floryan & Saric 1982). In addition to the viscous scale, ε, a small curvature scale, $\kappa = \varepsilon L/\mathscr{R}$ is introduced. The equations are linearized and the metrics are expanded in powers of ε and κ. Terms of higher order than the first in curvature and Reynolds number are disregarded to obtain the disturbance equations:

$$u_\xi + v_y + \beta w = 0, \tag{6}$$

$$Uu_\xi + U_\xi u + Vu_y + U_y v - u_{yy} + \beta^2 u = 0, \tag{7}$$

$$Uv_\xi + V_\xi u + Vv_y + V_y v + 2G^2 Uu + p_y - v_{yy} + \beta^2 v = 0, \tag{8}$$

$$Uw_\xi + Vw_y - \beta p - w_{yy} + \beta^2 w = 0, \tag{9}$$

where the Görtler number is

$$G = (\kappa/\varepsilon)^{1/2} \tag{10}$$

and κ is the dimensionless wall curvature, $\kappa = \varepsilon L/\mathscr{R}$. Floryan & Saric (1982) showed that Equations (6)–(9) are the result of the formal limits:

$$\varepsilon \to 0, \quad \kappa \to 0, \quad \text{with } G \text{ fixed.}[2] \tag{11}$$

If, for example, L is taken to be the streamwise position, x^*, then $\varepsilon L = \delta_r = (vx^*/U_\infty)^{1/2}$ and the Görtler number is written as $G = [(U_\infty \delta_r/v)(\delta_r/\mathscr{R})]^{1/2}$. The reader will occasionally find the momentum thickness in place of δ_r in some papers.

Finally, the boundary conditions are specified as

---

[2] In order to avoid confusion between the Reynolds number, R, and the radius of curvature, $\mathscr{R}$, the viscous scale will always be given by ε and the curvature scale by κ.

$$u = v = w = 0 \quad \text{at} \quad y = 0, \, y \to \infty. \tag{12}$$

2.2 Regions of Validity of the Disturbance Equations

It is important to discuss the different aspects of the limit process stated in (11) on the formulation of the basic state. The $\varepsilon \to 0$ limit gives Prandtl's boundary-layer equations and in the absence of streamwise pressure gradients, the velocity profile of Equation (4) is that of the Blasius boundary layer. Thus to zeroth order, the boundary layer is not affected by the curvature. The $\kappa \to 0$ limit reduces curvature effects (the wall-normal pressure gradient due to rotation) to a body-force-like term in the y-momentum equation analogous to the Boussinesq approximation in Bénard convection or the small-gap Taylor instability. This is essentially what Görtler (1941) did on an ad hoc basis for a parallel boundary layer.

Thus, Equations (6)–(9) represent the generally accepted zeroth-order statement of the Görtler problem. Floryan & Saric (1982) showed that any attempt to consider higher-order curvature terms (e.g. Schultz-Grunow & Behbahani 1973) must include higher-order boundary-layer theory in the basic state. At this next level of approximation, the disturbance equations contain the Reynolds number, $1/\varepsilon$, and another small (but different) curvature scale. Thus, Equations (6)–(9) break down in the $G \to \infty$ limit. These equations are also suspect in the $\beta \to 0$ limit. The disturbance-velocity scaling of Equation (5) is also essential. Smith (1955) looked at the correct approximation in the curvature terms but because he had identical scaling on all disturbance quantities, the Reynolds number explicitly appeared in his disturbance equations. Herbert (1976), Floryan & Saric (1982), and Ragab & Nayfeh (1981) review the many different basic-state approximations and explain why so many different neutral curves have appeared. Therefore, with just a few exceptions, the linear analysis literature prior to 1980 is not reviewed in detail.

The historical approach (prior to Hall 1983) to solving Equations (6)–(9) utilizes a separation-of-variables method (normal-modes). Unfortunately, these equations are not susceptible to a normal-mode solution because the coefficients depend on the stream coordinate ξ. Indeed, the equations are parabolic in (ξ, y) and are not separable. Thus, the boundary conditions given in Equation (12) must be supplemented by spatial initial conditions given as

$$[u(\xi, y), v(\xi, y), w(\xi, y)] = [u_0(y), v_0(y), w_0(y)] \quad \text{at} \quad \xi = x_0. \tag{13}$$

Hall (1983) shows that the numerical integration of the parabolic stability equations is the only appropriate method for analyzing the Görtler instability for $\beta = O(1)$. Whereas Hall (1982a) shows that the separation-of-variables solution is valid only in the asymptotic limit of $\beta \gg 1$, Floryan

(1991) argues that it can be interpreted as a *local* analysis at all wave numbers. A comparison of the numerical integration of the parabolic stability equations and the separation-of-variables solution (the normal-mode or eigenvalue or local analysis) is given by Day et al (1990). They support Hall's conclusions and give guidelines regarding initial conditions and the interpretation of the local analysis.

Since a substantial literature already exists on the normal-mode solutions and researchers continue to use this technique, a comparison of the two techniques follows.

2.3   Normal-Mode Solution

Briefly, the normal-mode solution or *local solution* is written by letting the disturbance quantities in Equations (6)–(9) take the form

$$\Phi_i(\xi, y) = \phi_i(y) \exp(\sigma \xi), \tag{14}$$

where $\Phi_i = (u, v, w, p)$ and the spatial growth rate, σ, are assumed real. The resulting equations form a sixth-order system of homogeneous, ordinary differential equations forming a real eigenvalue problem, $F(G, \beta, \sigma) = 0$, with the parameters β, σ, and G.

Typical solutions of $G = f(\beta, \sigma)$ are shown in Figures 3 and 4. Figure 3 presents the calculated eigenvalues and data in the usual sense. The experimental points group themselves according to their initial wavelength since the dimensional wavelength is conserved. Because the wavenumber, β, is made dimensionless with the scale εL, β varies with G. A more physically meaningful way to present the data is to define a dimensionless wavelength parameter, $\Lambda = (U_\infty \lambda / v)(\lambda / \mathcal{R})^{1/2}$, which is constant in the stream direction and plot $G = f(\beta, \Lambda)$ (Figure 4). The eigenvalue problem admits a neutral stability curve that flattens in the limit of small wavenumbers, i.e. $dG/d\beta|_{\sigma=0} \to 0^+$ as $\beta \to 0$ ($\Lambda \to \infty$). Physical arguments and experimental evidence (Bippes 1972) suggest that $dG/d\beta|_{\sigma=0} < 0$ as $\beta \to 0$. This is interpreted as a failure of the zeroth-order approximation of Equations (6)–(9) in that higher-order curvature terms must be included in the limit $\beta \to 0$.

Bippes (1972) took some care to examine the most likely vortex that would appear in his experiments and it is encouraging that the most amplified spanwise wavelength indicated from the eigenvalue problem corresponds to the observations of Bippes shown in Figures 3 and 4. The Tani & Sakagama (1962) experiments forced different spanwise wavelengths as initial conditions and verify the constant-Λ assumption. However, neither the normal-mode solution nor the numerical solution of Equations (6)–(9) predict the disturbance velocity distributions of Bippes (1972) within acceptable accuracy. (See Section 3.2.) Moreover, for Blasius

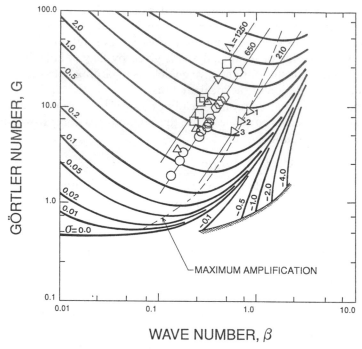

WAVE NUMBER, β

Figure 3 Curves of constant amplification rate as a function of Görtler number, $G = (U_\infty \delta_r / \nu)(\delta_r / \mathscr{R})^{1/2}$, and wavenumber, β, for the Blasius boundary layer. Comparison of theory with experiments. Experimental points, $\bigcirc$, $\square$, O, $\triangle$, and ∇ from Tani & Sakagami (1962); experimental point $\triangleright$ from Bippes (1972). (From Floryan & Saric 1982.)

flow, the eigenvalue problem does not admit solutions for highly-damped vortices, zero curvature, or convex curvature (Floryan & Saric 1982).

2.4   *Marching Solution of the Disturbance Equations*

Because of the inseparable nature of the disturbance equations, Hall (1983) discarded the normal-mode approach and used a marching solution for solving the disturbance equations. Since no streamwise derivative appears on the pressure in (6)–(9), p and w are eliminated by cross-differentiation resulting in coupled second-order and fourth-order equations in u and v. Initial conditions in the form of (13) accompany the boundary conditions (12) where $y \to \infty$ is taken as the last two grid points in y, far enough away from the wall as to not affect the solution. The length scale is taken to be x_0, the streamwise location of the initial conditions. The equations are easily solved using an implicit scheme.

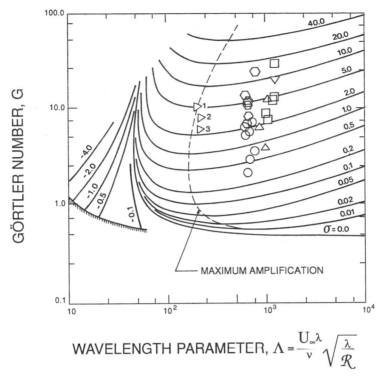

$$\text{WAVELENGTH PARAMETER, } \Lambda = \frac{U_\infty \lambda}{\nu} \sqrt{\frac{\lambda}{\mathcal{R}}}$$

Figure 4 Curves of constant amplification rate as a function of Görtler number, $G = (U_\infty \delta_r / \nu)(\delta_r / \mathcal{R})^{1/2}$, for the Blasius boundary layer. (From Floryan & Saric 1982.)

As with any nonparallel (or two-dimensional) basic state, the measure of stability depends on which disturbance quantity is chosen to be tracked (Gaster 1984, Saric & Nayfeh 1977). Following Day et al (1990), the following definitions are introduced:

$$\sigma_1 = (dA_1/d\xi)/A_1, \quad A_1(\xi) = \max_y [u(\xi, y)], \tag{15}$$

$$\sigma_2 = (dA_2/d\xi)/A_2, \quad A_2(\xi) = \int (u^2 + v^2 + w^2)\, dy, \tag{16}$$

$$\sigma_3 = (dA_3/d\xi)/A_3, \quad A_3(\xi) = \int (u')\, dy, \tag{17}$$

where σ_1 is used to compare with the normal-mode analysis and uses the length scale x^* for that reason, whereas σ_2 and σ_3 use x_0 (the initial x^*) as the length scale. Because of the two-dimensionality of the basic state, an integral measure, such as σ_2 or σ_3, is more appropriate. Although Hall (1983) used σ_2, if this function is to be interpreted as disturbance energy, v and w should be renormalized as εv and εw in order to regain the correct physical scaling with respect to u. Thus, σ_3 is more closely the correct form for disturbance energy for large Reynolds numbers.

Hall (1983) showed that the existence of a neutral point, $\sigma_i = 0$, strongly depends on the location and shape of the initial conditions that are specified in (13). For all of his initial disturbance velocities, he showed initial stability. Thus, the neutral point can be placed anywhere by simply changing x_0. The nonuniqueness of this situation is not as bad as it sounds for two reasons. First, if one chooses arbitrary initial disturbances that are quite unusual, the parabolic nature of the equations will cause numerical relaxation of the initial conditions to distributions that are consistent with the equations. This relaxation will always be interpreted as decay by any stability measure. Second, the real issue in open system flows is *receptivity*, i.e. the mechanisms by which freestream disturbances enter the boundary layer and provide initial conditions for unstable disturbances. Receptivity is very apparent in the experiments of Bippes (1972) who systematically changed upstream conditions and observed the results.

The marching solutions of the disturbance equations show the sensitivity to initial conditions and, most importantly, provide the means for assessing receptivity issues in Görtler flows that are not possible with eigenvalue methods. If one was interested in integrated growth for stability-control calculations (i.e. suction and blowing, heating and cooling, etc) the marching solution is as easy as any other method. Moreover, the marching solution can calculate vortex stability characteristics for zero- and convex-curvature flows. What it cannot do, of course, is provide the somewhat universal charts like Figures 3 and 4 that are of use to the designer or someone who is unwilling to do the calculations. This is discussed in the next section.

2.5   Evaluation of Linear Stability Analyses

Day et al (1990) and Kalburgi et al (1988a,b) did direct comparisons between the marching solution and the normal-mode solution. In order to resolve the question of initial conditions, they chose to use the eigenfunctions from the normal-mode solution. Figure 5 is a typical result from Day et al (1990) that uses the definitions of Equations (15)–(17) for growth rates. The three marching curves all start at the first neutral point and have the same initial conditions. The circles show the eigenvalues of the normal-mode solution. The details are in their paper; here it is only necessary to point out a critical feature. When the marching solution is initiated from each of the eigenvalues of the normal-mode solution, the growth rate quickly tracks to the σ_1 curve as if the σ_1 curve were a universal curve for that growth-rate measure. However, it should be mentioned that Day et al show that the local analysis cannot predict the second neutral point (where $\partial/\partial\xi = 0$ for the disturbance state) even if the eigenfunctions from the local analysis are used as initial conditions. Thus, the local analysis is not formally justified even in the case of neutral stability. The other important result is that the normal-mode solution appears to give a maximum growth rate at any streamwise location. The differences in growth rates are modest, so that the designer who uses the charts of Figures 3 and 4 would have a conservative estimate.

Day et al (1990) support the conclusion of Hall (1983) that the lower neutral curve cannot be predicted without some information regarding the initial conditions. The normal-mode solutions are not helpful in this case since they cannot handle strongly decaying vortices and their upstream implementation is limited. However, overall evaluation of disturbance growth such as the use of an amplification factor, $N = \ln [A(x_1)/A(x_o)]$, does not depend strongly on the location of the neutral curve.

Hall (1982b) shows that one can use separation of variables in Equations (6)–(9) in the limit of $\beta \gg 1$. He argues that since β is always increasing in the stream direction, as one can see from the track of the experimental

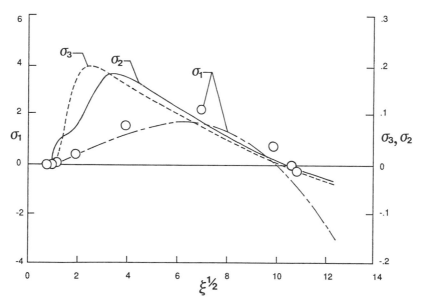

Figure 5 Growth rates β, α, and σ for the run of Figures 3 and 4. The circles show the values of β obtained from local analysis. (From Day et al 1990.)

points in Figure 3, the vortices always reach a point where β is large, hence the approximation is justified. Thus, the normal-mode approach is called the *large wavenumber limit* and a large number of papers have appeared using this reasoning. However, the discussion of Section 3.2 shows that the large wavenumber limit can only be used if the vortices are short-fetched—i.e. if profile distortion effects have not already taken place—a rather critical restriction.

3. EXPERIMENTS AND NONLINEAR SATURATION

3.1 *Early Work*

Liepmann (1943, 1945) conducted boundary-layer transition experiments on concave walls and showed that concavity lowered the transition Reynolds number. He was able to demonstrate the importance of the Görtler number in correlating his results and thus attributed the earlier transition to a Görtler instability. Liepmann also clarified some ambiguity in earlier work by Clauser & Clauser (1937). Gregory & Walker (1950) visualized surface streaks that could be identified with Görtler vortices; the first detailed measurements were conducted by Tani (1949, 1962) and Tani & Sakagami (1962). The Tani experiments used an upstream airfoil to fix the Görtler vortex wavelength and thus showed: (*a*) that the vortex wavelength, Λ, remains fixed as shown in Figures 3 and 4; (*b*) the sensitivity of Λ to initial conditions; and (*c*) that the stationary vortex produces a strong spanwise modulation of the shear stress which produces the wall streaks in the visualization. Aihara (1962) and Wortmann (1964a,b) provided additional qualitative data through flow visualization.

It remained for Bippes (1972) and Bippes & Görtler (1972) to give detailed measurements of disturbance velocity distributions within the boundary layer and to attempt to define a neutral stability point (see Figures 3 and 4). These experiments were conducted in a towing tank and thus, are the only really low-freestream-disturbance experiments to which one may refer. Bippes found that in the lowest freestream turbulence conditions, it was extremely difficult to visualize the vortex structure and to identify the characteristic wavelength. He found it necessary to put a turbulence-generating screen ahead of the model or to place small heating wires along the span. In fact, he measured the most dominant wavelength by use of a screen and that measurement determined the spacing of the heated wires in the absence of the screen. This brings us back to the discussion of Sections 2.2–2.4: the importance of initial conditions given by Equation (13), the issue of receptivity, and the need to do careful experiments. Bippes also visualized the unsteadiness of the secondary instability that leads to breakdown. Moreover, in the use of surface visualization and smoke, one must already have a large disturbance in order for the technique to work. Thus nonlinearity and saturation become important issues and explain to some extent the failure to date of any linear theory to predict the disturbance velocity profiles. This is discussed further in Section 3.2.

Experiments by Mangalam et al (1985) used flow visualization and a laser-Doppler velocimeter to determine the spanwise wavelength of the vortices at different Görtler numbers in a low-turbulence tunnel. They found that the observed wavelengths corresponded in a general sense to the most amplified wavelengths predicted by linear theory—in agreement, more or less, with Bippes (1972) and Babenko & Yurchenko (1980). However, since the Görtler instability is strongly initial-condition dependent, this agreement is fortuitous.

3.2 Distortion of the Mean Flow: The Role of the Stationary Vortex Structure

Since the experimental and theoretical developments in the 1980s progressed in a inchoate manner, it is appropriate to describe the physical aspects of the vortex growth before continuing with the review. The significant feature of a stationary, streamwise-oriented vortex in a spatially developing flow is the convection of streamwise momentum normal to the wall. Figure 6 is a sketch of such behavior and Figure 6a shows the orientation of the vortex motion. At the $z = 0$ position, the combined action of the two vortices produces an upwelling of the flow. At the $\pm \pi$ position, there is a downwelling. If the low-momentum fluid is identified by the shaded area of Figure 6b as an initial condition, the upwelling at $z = 0$ raises the low-momentum fluid and reduces the shear. On the other hand, the downwelling at $\pm \pi$ decreases the region of low-momentum fluid and increases the shear. As the motion continues, a mushroom-shaped distribution is formed as shown in Figure 6d. This distribution is verified experimentally by Peerhossaini (1987) in the photograph shown in Figure 7a, where the light-colored fluid is visualized by laser-fluorescent dye initially placed in the wall region and hence is a low-momentum tracer. Figure 7b shows a typical nonlinear calculation [performed in this case by Benmalek (1993); but Liu & Sabry (1990), Sabry & Liu (1991), Lee & Liu (1992), Liu & Domaradzki (1993), and Guo & Finlay (1993) have made similar calculations]. One can clearly see the large distortion due to the strong changes in streamwise momentum. The same phenomenon also exists in other flows with a stationary streamwise vortex structure, such as the crossflow instability on a swept wing (Kohama et al 1990) and the curved channel problem (Guo & Finlay 1993).

The consequence of this behavior is shown in Figure 8, which shows the

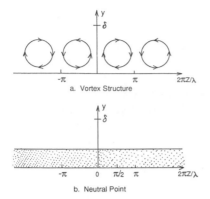

a. Vortex Structure

b. Neutral Point

Figure 6 Evolution of the profile distortion.

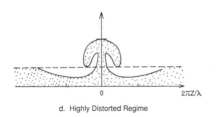

c. Linear Regime

d. Highly Distorted Regime

velocity profile, calculated for the different streamwise locations, and the development of the highly inflectional velocity profiles that would give rise to a Rayleigh instability. The computations are for the nonlinear, parabolized disturbance equations (Benmalek 1993) and should be considered as generic features that are characteristic of these distorted profiles. Figure 9 is a comparable development of the spanwise gradients of the velocity profiles. These profiles are subject to a Kelvin-Helmholtz instability. It is apparent that the spanwise gradients are as large as the wall-normal gradients and that flows such as this are subject to strong secondary instabilities.

The other feature of this nonlinear profile distortion is saturation. At some streamwise location, the disturbance energy saturates as shown in the generic Figure 10, which shows the trajectories of Equation (17) for all of the Fourier modes used in the calculation (Benmalek 1993). Saturation

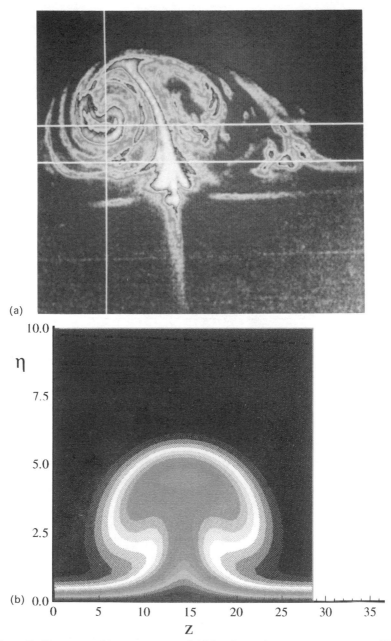

Figure 7 Transport of low-momentum fluid by the stationary vortex structure. (From Peerhossaini 1987, Benmalek 1993.) (*a*) Dye transport experiment. (*b*) Computation with velocity contours at $\xi = 3.37$ ($x = 11.36$).

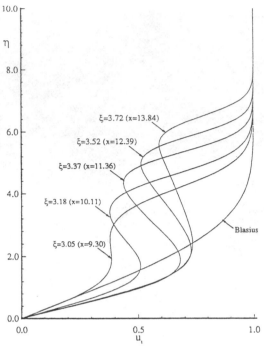

Figure 8 An example of distortion of a streamwise velocity profile with different x and $z = 0$. (From Benmalek 1993.)

occurs at a modest value of the Görtler number, and brings into question linear analyses with large Görtler number.

The first measurements of the strong distortion of the mean flow and the description of the nonlinear process are found in Aihara (1979), Yurchenko et al (1980), Yurchenko (1981), and Ito (1980). Yurchenko also noted the importance of the spanwise gradient of the streamwise flow, $\partial u/\partial z$, which is as large as the wall-normal gradient, $\partial u/\partial y$. Aihara & Koyama (1981) later identified the breakdown of the vortex structure as a secondary instability due to a horseshoe-vortex structure. This type of work continued in a succession of short papers by Aihara & Sonoda (1981), Aihara & Koyama (1982), and Aihara et al (1985).

In a series of papers, Ito (1980, 1985, 1987, 1988) addressed the details of the breakdown process with LDV and smoke. These experiments are reviewed in some detail by Floryan (1991) so only the highlights are given here. Ito showed measurements of the strongly distorted velocity profiles and was the first to visualize the "mushroom-shaped" streaklines that

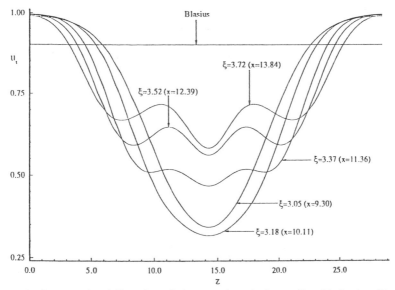

Figure 9 An example of distortion of the spanwise velocity profile with fixed *y*. (From Benmalek 1993.)

occur when low-momentum fluid from the wall is convected into the boundary layer.

Swearingen & Blackwelder (1986, 1987), using a specially designed low-turbulence wind tunnel, carried out detailed measurements in the neighborhood of vortex breakdown and tied together a data base for suitable analytical comparison. They identified two types of secondary instabilities—wavy structures resembling either a sinuous mode or a horseshoe vortex mode—and made the detailed measurements to identify the origins of the mechanisms by careful mappings of u'_{rms}. They found that the sinuous mode appears more frequently than the horseshoe mode [this was qualitatively described by Bippes (1972)], and is associated with a Kelvin-Helmholtz instability originating at extrema in $\partial u/\partial z$. The horseshoe mode, sometimes called the varicose mode, is due to a Rayleigh instability that is correlated with extrema of $\partial u/\partial y$ [qualitatively described by Aihara & Koyama (1981)]. This body of work was extended by Myose & Blackwelder (1991) who readdress the issue of wavelength selection and demonstrate the sensitivity of the initial conditions of Equation (13). Bippes (1990) also reconsiders the wavelength selection mechanism and the flow quality of wind tunnel and water tunnels.

Peerhossaini (1987) and Peerhossaini & Wesfreid (1988a,b) conducted

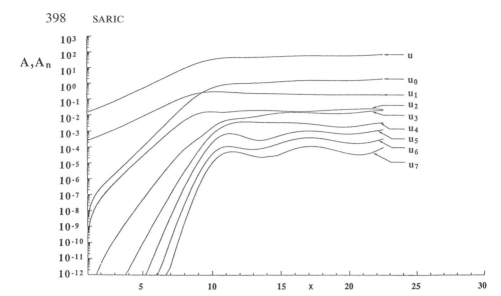

Figure 10   An example of energy saturation of the vortices. Amplitudes of the disturbance and their Fourier components in the streamwise direction where

$$A = \int_0^{2\pi/\alpha} \int_0^\infty u^2 \, dy \, dz \quad \text{and} \quad A_n = \int_0^\infty |u_n|^2 \, dy.$$

extensive flow visualization studies in a water channel that produced amongst other things, the photograph of Figure 7. They systematically documented the unsteady nature of the breakdown process and the interaction of vortex pairs in the unsteady region of breakdown.

Petitjeans (1992) conducted a series of experiments in a water channel and produced another set of photographs of the mushroom shaped streaklines of Figure 7. He measured disturbance velocity distributions but did not obtain agreement with linear theory because he was near saturation where linear theory should fail. His was the first effort to systematically document the saturation phenomena at different wavelengths. Although he demonstrated its existence, his saturation measurements were inconclusive because of possible random input from the upstream disturbance environment.

3.3   Other Experiments

The role of freestream disturbances was examined by Kottke (1986, 1988) and Kottke & Mpourdis (1987) using flow visualization techniques. Görtler vortices in high turbulence environments characteristic of turbine blades are studied by Crane & Winoto (1979), Crane et al (1986, 1988), Winoto et al (1979), Winoto & Crane (1980), Sabzvari & Crane (1985), and Leoutsakos & Crane (1990). These are reviewed by Floryan (1991). Görtler vortices in turbulent boundary layers and in boundary layers with heating and suction are also reviewed by Floryan (1991).

The impulsive spin-down to rest of a fluid inside a rotating cylinder produces an unsteady Görtler instability within the fluid near the cylinder wall. Because it has the mathematical simplicity of the Taylor problem, this application has been addressed by many; the most recent works are by Mathis & Neitzel (1985) and Kohuth & Neitzel (1988). The fact that it is a closed, parallel system makes for much better agreement between experiment and analysis.

4.   OTHER LINEAR ANALYSES

The lessons of Sections 2 and 3 are that the analytical road to success has many hazards. The separation-of-variable technique does not work in any real sense for $\beta = O(1)$. Moreover, any linear stability approach may fail because it examines the stability of a basic state that does not exist after the onset of the instability. The discussion of Section 3.2 and the weight of the experimental evidence regarding profile distortion effects would seem to obviate all but the short-fetched vortices. Approximations such as the large spanwise wavenumber limit would tend to push the instability to higher Görtler numbers and hence longer fetch. Higher Görtler numbers would also imply larger curvature, so the loss of $G = O(1)$ results in higher-order curvature effects at the level of the basic state. Therefore one must chose the right problem before embarking on a linear analysis. Some examples follow.

4.1   Swept-Wing Flows and Stagnation Flows

Hall (1985) has shown that beyond a very small sweep angle, the principal instability on a swept wing with concave curvature is a crossflow instability and not a Görtler instability. This analysis reduces the concern about Görtler problems in swept-wing flows. Bassom & Hall (1991) extend these ideas to the receptivity question. Experiments are now underway (Peerhossaini et al 1990), whose preliminary results verify the suppression of Görtler vortices in the presence of crossflow.

Another issue that frequently comes up in the context of Görtler vortices concerns stagnation flows. The Rayleigh circulation criterion appears to be satisfied near the stagnation region of bluff bodies and a Görtler instability has been a suspect regarding the origin of streamwise vorticity in a boundary layer such as observed by Klebanoff et al (1962). Stuart (1984) has shown conclusively that an instability does not exist.

4.2   *Compressibility Effects*

A recent application for information regarding transition induced by Görtler vortices in a compressible flow is in the design of "quiet" supersonic wind tunnels. The major source of freestream noise is radiation from the turbulent boundary layers on the nozzle. Since the accelerating pressure gradient generally stabilizes Tollmien-Schlichting instabilities, transition is induced by Görtler vortices on the concave region of the nozzle (Beckwith et al 1985, Chen et al 1985). Efforts to calculate the growth of the instability with compressibility are given by Hämmerlin (1961), Kobayashi & Kohama (1977), El Hady & Verma (1983a,b), Tumin & Chernov (1988), and Jallade et al (1990) using the separation-of-variables technique without the large spanwise wavenumber limit. Hall & Malik (1989) analyze the role of compressibility within the large wavenumber limit. As mentioned earlier, these results will have limited applicability. Recently Dando & Seddougui (1991) and Dando (1992) completed analyses for two-dimensional and three-dimensional boundary layers, respectively in the large Görtler number limit (inviscid flow). Both papers use unit Prandtl number and Chapman viscosity law approximations that have been shown to be grossly incorrect for other types of instabilities in high-speed flows. In the absence of an experiment, there is no telling whether these approximations are valid for Görtler vortices.

Spall & Malik (1989) integrate the linearized parabolic stability equations and show that compressibility is stabilizing but the stabilization is reduced as hypersonic speeds are approached. Hall & Fu (1989) and Fu et al (1990) assess the role of Sutherland's viscosity law and real-gas effects on the linear stability of Görtler vortices. Fu & Hall (1991a) then extend this to nonlinear effects; Fu & Hall (1991b) consider secondary instabilities. These last four papers have jumped so far ahead of the experimental capability and present knowledge base that it is difficult to assess their value. We know that the approximations made on the mean flow would not be appropriate for streamwise traveling instabilities. On the other hand, they may offer some insight into operative mechanisms that a well-established experiment could examine.

4.3   *Wavy Walls*

It may be worthwhile to describe one type of calculation of Equations (6)–(9) that could be useful. Saric & Benmalek (1991) examine the problem of the Görtler instability over wavy walls, i.e. when the wall curvature is periodic and can be convex as well as concave. The problem is solved by the direct integration of the parabolized disturbance equations, Equations (6)–(9). In spite of any presumed advantages to the normal-mode solution or a large-β solution (e.g. Jallade 1990), it is well known that zero or convex curvature cannot be accounted for by these methods because the discrete spectrum of eigenvalues does not exist.

The wavy-wall problem is of interest because of stability and transition studies over surfaces that unintentionally may be wavy and because of the *Görtler-Witting* mechanism which is still being discussed (e.g. Lessen & Koh 1985). With wavy walls, the question always arises regarding the nature of stabilization in the convex portion of the wall and whether it is sufficient to overcome the destabilization of the concave region. No ad hoc interpretation of the local analysis can correctly address this problem. The basis of the Görtler-Witting mechanism lies in the conjecture that large amplitude T-S waves locally induce concave curvature in the streamlines and hence a Görtler instability.

Saric & Benmalek (1991) show that convex curvature has an extraordinary stabilizing influence on the Görtler vortex, and they give examples of wavy-wall computations where the net result is stabilizing. The result of the computations show that an oppositely rotating vortex pair is generated in the convex region giving disturbance velocity profiles that resemble higher eigenmode distributions (Herbert 1979, Floryan 1985). They conclude that the stabilizing effects of a convex surface make it unlikely that the boundary layer over a wavy surface is subject to a strong Görtler instability. Moreover, since the Görtler-Witting mechanism has concave/convex curvature in the middle of the boundary layer and the maximum source term for the Görtler instability (dU/dy) is at the wall, Görtler-Witting is not an important instability. This has been further verified with the nonlinear parabolized equations (Benmalek 1993).

4.4   *Rotation Effects*

Two papers address rotation effects on Görtler vortices. Both Aouidef et al (1992) and Zebib & Bottaro (1993) consider a concave wall subject to uniform rotation around the spanwise axis of the curved plate. This configuration has an immediate application to Görtler vortices on rotating turbine blades. Both are preliminary papers since they use the separation-of-variables technique without considering the large spanwise wavenumber limit. They intend to show that the effect of rotation can be stabilizing or destabilizing depending on the sign of the rotation. In a qualitative sense, these results agree (except that Zebib & Bottaro have negative rotation about a left-handed coordinate system giving the same sense of rotation as Aouidef et al).

4.5   *Receptivity*

Receptivity is defined as the mechanism by which freestream disturbances enter the boundary layer and create the initial conditions for unstable waves. This can take place by the interaction of a sound wave with a geometrical discontinuity or in the present case, anything that sheds a streamwise vortex. Receptivity is a critical issue for the Görtler problem since the problem is truly initial-condition dependent. For this reason, there should be considerable growth in future research activity in this area. Denier et al (1991) address exactly this issue by considering the vortex motion induced by wall roughness. Although their work is still under development and the critical roughness height has not yet been established, it does show promise.

5. NONLINEAR COMPUTATIONS AND SECONDARY INSTABILITIES

5.1 *Nonlinear Computations*

The Görtler vortex problem is really a nonlinear, nonparallel instability in which the basic state cannot be decoupled from the disturbance state or the initial conditions. In fact, the only reliable assumptions about the Görtler problem are that it is spanwise periodic and initially stationary. Prior to the onset of the secondary instability, the experiment teaches us that (*a*) there is significant profile distortion from Blasius flow (Figures 8, 9); (*b*) saturation will occur (Figure 10); and (*c*) the low-momentum streaks form a mushroom-shaped cross section (Figure 7). We use these events to evaluate the recent contributions. Prior to 1988, no nonlinear theorries could predict these events, so we begin with Hall (1988) who solved the nonlinear parabolized equations in a spatial formulation and was able to calculate the distortion of the mean flow. He chose to have the curvature vary with streamwise position and hence made it difficult to compare the flow to anything physical. He did not observe saturation of the mean flow distortion. At low initial amplitudes, he calculated growth followed by decay. At higher initial amplitudes he calculated continuous growth. Later, Hall (1990) performed constant-curvature calculations, which he compared with data from Swearingen & Blackwelder (1987) and was able to obtain qualitative agreement.

Sabry & Liu (1988, 1991) did a temporal calculation of the spatially developing vortex and used a constant convective velocity to convert from temporal to spatial behavior. The constant convective velocity plays the role of a free parameter and one can adjust it to optimize comparison with experiment. Whereas this is intrinsically incorrect, it provides a good comparison with experiment in calculating distorted velocity profiles, saturation, and the low-momentum mushroom. Liu & Domaradzki (1993) used the same idea of a convective velocity in a temporal simulation and they also were able to achieve good agreement with experiment.

Lee & Liu (1992) did the nonlinear spatial computations of the parabolized Navier-Stokes equations as did Guo & Finlay (1993b) and Benmalek (1993). Whereas some numerical details differ between these computations, they are stable computational models of the steady spatial evolution of the Görtler vortex structure. Both Lee & Liu (1992) and Guo & Finlay (1993) give successful comparisons with the data of Swearingen & Blackwelder (1987). Lee & Liu go one step further and show that the initial conditions typically used by Hall (1983, 1988) do not give as good a comparison with experiment as compared with initial conditions chosen as eigen solutions to the separation of variable solution (e.g. Day et al 1990). This result is surprising but nevertheless emphasizes the sensitivity to initial conditions.

Liu & Sabry (1990), Liu (1991), and Liu & Lee (1993) set out to answer the question of the analogy between the transport of concentration, heat, and momentum as it applies to the visualization techniques that are used in air and water. Their work is of interest because it gives a practical application for the nonlinear computations and provides some insight regarding what is used in the laboratory and what is modeled on the computer.

5.2 Secondary Instabilities

It appears that the breakdown process of Görtler vortices is through a secondary instability that occurs when the vortex motion begins to saturate while distorting the mean flow. Subsequently, unsteady fluctuations appear which lead to transition to turbulence. Although the existence of a secondary instability can be deduced from the experiments of Bippes, Yurchenko, Aihara, and Ito, it was really the work of Swearingen & Blackwelder (1986, 1987) that provided the concrete evidence and the data base for comparisons with theory and computations.

The presence of highly inflectional profiles in a shear layer is a harbinger of secondary instabilities that lead to transition and Figures 8 and 9 show that the necessary large gradients exist in both the spanwise and wall-normal directions because of steady-state profile distortion. Swearingen & Blackwelder were able to locate the source of the breakdown and thus identify which mechanism was causing breakdown in either the sinuous or varicose modes. The sinuous mode is characterized by having the disturbances u, v, and p asymmetric and w symmetric about $z = 0$ (Figure 6) and is associated with the spanwise gradient $\partial U/\partial z$. The varicose mode has u, v, and p symmetric and w asymmetric about $z = 0$ and is associated with the wall-normal gradient $\partial U/\partial y$.

Hall & Seddougui (1989), Sabry et al (1990), Bassom & Seddougui (1990), Park (1990), Hall & Horseman (1991), Yu & Liu (1991, 1993), and Guo & Finlay (1993) all attack the inflection-point instabilities with predictable qualitative results, i.e. both the wall-normal and spanwise inflection points are the source instabilities with the corresponding symmetries. Whereas the details are beyond the scope of this review, some general comments can be given. The Hall & Seddougui (1989) and Bassom & Seddougui (1990) papers are in the large Görtler number and large wavenumber limit which may not be applicable in this case since the observed breakdowns have $O(1)$ wavenumber. The work of Hall & Horseman (1991) is an inviscid formulation that is able to describe some but not all of the important features of the secondary instability. The apparently most successful effort is the recent work by Yu & Liu (1993). They include the effects of viscosity in a linear stability analysis of the nonlinearly deformed basic state. An energy balance is performed and they conclude that the sinuous mode would prevail over the varicose modes (in agreement with experiment). They give details of the instability process and, in particular, illustrate the importance of viscous effects. Moreover, they correctly locate the origin of the unstable fluctuations within the boundary layer. It now remains to collect significantly more experimental data. In particular, details of the dominant frequencies in the breakdown process will permit a critical assessment of the analytical work.

6.   CONCLUSIONS

In the case of Görtler vortices, the most interesting development in the past five years has been the recognition that the mean velocity distortion is the key to the secondary instabilities that lead to transition. The usual linear stability analysis is restricted to a very small region of the flow that is typically upstream of most experimental observations. Much of the heavy weather developed over the use of normal-modes in a linear analysis loses its impact since most of what has been done is simply not applicable. The Görtler vortex motion (as well as that of Dean vortices and crossflow vortices) produces a situation in spatially developing flows that is unlike other stability phenomena. The disturbance motion is inseparable in three dimensions from the basic-state motion. It seems as though all interesting phenomena associated with Görtler vortices share this three-dimensional inseparability.

   After the recent Görtler-vortex workshop (Peerhossaini & Wesfreid 1990) and symposium (Laure 1993) it has become clear that the theory is far ahead of the experiments and that the most important missing ingredient is an experiment in the linear range, with modern diagnostics, in a low disturbance environment, e.g. a modern version of Bippes (1972). The fact that the disturbances are governed by Equations (6)–(9) with initial conditions given by Equation (13) makes it difficult to have two experiments agree unless the initial conditions are controlled. On the other hand, the experiments of Swearingen & Blackwelder (1987), described in Section 3, are an example of a well-conducted experiment concentrating on the secondary instability.

ACKNOWLEDGMENTS

The author prepared most of this manuscript while a Professor of Aeronautics and Astronautics at Tohoku University, Sendai, Japan during the 1991–1992 Academic Year. He would like to acknowledge the support of Tohoku University and the encouragement of Professors Kobayashi, Kohama, and Fukunishi during this time. The author would also like to thank Dr. H. Reed for her helpful comments and suggestions during the course of this work. The research associated with this paper was supported by the Air Force Office of Scientific Research Grant AFOSR-90-0234 and NASA-Langley Research Center Fellowship NAG1-1111. The author would like to thank Dr. Hassan Peerhossaini for sharing his photograph.

Literature Cited

Aihara, Y. 1962. Transition in an incompressible boundary layer along a concave wall. *Bull. Aerosp. Res. Inst., Univ. Tokyo* 3: 195–240
Aihara, Y. 1979. Görtler vortices in the nonlinear region. In *Theoretical and Experimental Fluid Mechanics*, ed. U. Mller, K. G. Roesner, B. Schmidt, pp. 331–38. Berlin: Springer-Verlag
Aihara, Y., Koyama, H. 1981. Secondary instability of Görtler vortices: formation of periodic three-dimensional coherent structure. *Trans. Jpn. Soc. Aeronaut. Space Sci.* 24: 78–94
Aihara, Y., Koyama, H. 1982. Nonlinear development and secondary instability of Görtler vortices. In *Stability in the Mechanics of Continua*, ed. F. H. Schroeder, pp. 345–54. Berlin: Springer-Verlag
Aihara, Y., Sonoda, T. 1981. Effects of pressure gradient on the secondary instability of Görtler vortices. *AIAA Pap. No.* 81-0197
Aihara, Y., Tomita, Y., Ito, A. 1985. Generation, development and distortion of longitudinal vortices in boundary layers along concave and flat plates. In *Laminar-Turbulent Transition*, ed. V. V. Kozlov, pp. 447–54. New York: Springer-Verlag
Aouidef, A., Wesfreid, J. E., Mutabazi, I. 1992. Coriolis effects on Görtler vortices

in a boundary-layer flow on a concave wall. *AIAA J.* 30(11): 2779–82
Babenko, V. V., Yurchenko, N. H. 1980. Experimental investigation of Görtler instability on rigid and elastic flat plates. *Gidromekhanika* 41: 103–8 (In Russian)
Bassom, A. P., Hall, P. 1991. Vortex instabilities in three-dimensional boundary layers: the relationship between Görtler and crossflow vortices. *J. Fluid Mech.* 232: 647–80
Bassom, A. P., Seddougui, S. O. 1990. The onset of three-dimensionality and time-dependence in Görtler vortices: neutrally stable wavy modes. *J. Fluid Mech.* 220: 664–72
Bayly, B. J. 1987. Three-dimensional centrifugal-type instabilities in inviscid two-dimensional flows. *Phys. Fluids* 31: 56–64
Bayly, B. J., Orszag, S. A., Herbert, Th. 1988. Instability mechanisms in shear flow transition. *Annu. Rev. Fluid Mech.* 20: 359–91
Beckwith, I. E., Malik, M. R., Chen, F. J., Bushnell, D. M. 1985. Effects of nozzle design parameters on the extent of quiet test flow at Mach 3.5. In *Laminar-Turbulent Transition*, ed. V. V. Kozlov, pp. 589–600. New York: Springer-Verlag
Benmalek, A. 1993. *Nonlinear development of Görtler vortices over variable curvature walls*. PhD thesis. Ariz. State Univ.

Bippes, H. 1972. Experimentelle Untersuchung des laminar-turbulenten Umschlags an einer parallel angestrmten konkaven wand. *Sitzungsberichte der Heidelberger Akademie der Wissenschaften Mathematischnaturwissenchaftliche Klasse* 3: 103–80 (also *NASA TM*-75243, 1978)

Bippes, H., Görtler, H. 1972. Dreidimensionale Strungen in der Grenzschicht an einer konkaven Wand. *Acta Mech.* 14: 251–67

Chen, F. J., Malik, M. R., Beckwith, I. E. 1985. Instabilities and transition in the wall boundary layers of low-disturbance supersonic nozzles. *AIAA Pap. No.* 85-1573

Clauser, M., Clauser, F. 1937. The effect of curvature on the transition from laminar to turbulent boundary layer. *NACA Rep.* 613

Coles, D. 1965. Transition in circular Couette flow. *J. Fluid Mech.* 21: 385–425

Crane, R. I., Leoutsakos, G., Sabzvari, J. 1986. Transition in pressure-surface boundary layers. *ASME Pap. No.* 7, 31*st Intl. Gas Turb. Conf. and Exhibit*, Düsseldorf

Crane, R. I., Sabzvari, J. 1988. Heat transfer visualization and measurement in unstable concave-wall laminar boundary layers. *ASME* 88-*GT*-36

Crane, R. I., Winoto, S. H. 1979. Longitudinal vortices in a concave surface boundary layer. *AGARD CP No.* 271

Dando, A. 1992. The inviscid compressible Görtler problem in three-dimensional boundary layers. *Theoret. Comput. Fluid Dyn.* 3: 253–65

Dando, A., Seddougui, S. O. 1991. The inviscid compressible Görtler problem. *ICASE Rep. ICASE*-91-54

Day, H. P., Herbert, T., Saric, W. S. 1990. Comparing local and marching analyses of Görtler instability. *AIAA J.* 28(6): 1010–15

Dean, W. R. 1928. Fluid motion in a curved channel. *Proc. R. Soc. London Ser. A* 15: 623–31

Denier, J. P., Hall, P., Seddougui, S. O. 1991. On the receptivity problem for Görtler vortices: vortex motions induced by wall roughness. *Phil. Trans. R. Soc. London Ser. A* 335: 51–85

DiPrima, R. C., Stuart, J. T. 1972. Nonlocal effects in the stability of flow between eccentric rotating cylinders. *J. Fluid Mech.* 54: 393–415

DiPrima, R. C., Stuart, J. T. 1975. The nonlinear calculation of Taylor vortex flow between eccentric rotating cylinders. *J. Fluid Mech.* 67: 85–111

Drazin, P. G., Reid, W. H. 1981. *Hydro-*

dynamic Stability. Cambridge: Cambridge Univ. Press

El Hady, N. M., Verma, A. K. 1983a. Growth of Görtler vortices in compressible boundary layers along curved surfaces. *J. Eng. Appl. Sci.* 2: 213–38

El Hady, N. M., Verma, A. K. 1983b. Görtler instability of compressible boundary layers. *AIAA J.* 22: 1354–55

Floryan, J. M. 1985. The second mode of the Görtler instability of boundary layers. *AIAA J.* 23: 1828–30

Floryan, J. M. 1986. Görtler Instability of boundary layers over concave and convex walls. *Phys. Fluids* 29(8): 2380–87

Floryan, J. M. 1989. Görtler instability of wall jets. *AIAA J.* 27: 112–14

Floryan, J. M. 1991. On the Görtler instability of boundary layers. *Prog. Aerosp. Sci.* 28: 235–71

Floryan, J. M., Saric, W. S. 1982. Stability of Görtler vortices in boundary layers. *AIAA J.* 20(3): 316–24

Fu, Y. B., Hall, P. 1991a. Nonlinear development and secondary instability of Görtler vortices in hypersonic flows. *ICASE Rep. ICASE*-91-39

Fu, Y. B., Hall, P. 1991b. Effects of Görtler vortices, wall cooling and gas dissociation on the Rayleigh Instability in a hypersonic boundary layer. *ICASE Rep. ICASE*-91-87

Fu, Y. B., Hall, P., Blackaby, N. D. 1990. On the Görtler instability in hypersonic flows: Sutherland's law fluids and real gas effects. *ICASE Rep. ICASE*-90-85

Gaster, M. 1974. On the effects of boundary-layer growth on flow stability. *J. Fluid Mech.* 66: 465–80

Görtler, H. 1941. Instabilita–umt laminarer Grenzchichten an Konkaven Wänden gegenber gewissen dreidimensionalen Störungen. *ZAMM* 21: 250–52; also *NACA Rep.* 1375 (1954)

Gregory, N., Walker, W. S. 1950. The effect on transition of isolated surface excrescences in the boundary layer. *ARC Tech. Rep.* 13, 436; published as Aeronaut. Res. Council R&M 2779 (1956) 1–10

Guo, Y., Finlay, W. H. 1993. Wavenumber selection and irregularity of spatially developing nonlinear Dean & Görtler vortices. *J. Fluid Mech.* To appear

Hall, P. 1982a. Taylor-Görtler vortices in fully developed or boundary-layer flows: linear theory. *J. Fluid Mech.* 124: 475–94

Hall, P. 1982b. On the nonlinear evolution of Görtler vortices in non-parallel boundary layers. *IMA J. Appl. Math* 29: 173–96

Hall, P. 1983. The linear development of Görtler vortices in growing boundary layers. *J. Fluid Mech.* 130: 41–58

Hall, P. 1985. The Görtler vortex instability

mechanism in three-dimensional boundary layers. *Proc. R. Soc. London Ser. A* 399: 135–52

Hall, P. 1988. The nonlinear development of Görtler vortices in growing boundary layers. *J. Fluid Mech.* 193: 243–66

Hall, P. 1990. Görtler vortices in growing boundary layers: The leading edge receptivity problem, linear growth, and nonlinear breakdown stage. *Mathematika* 37: 151–89

Hall, P., Fu, Y. B. 1989. On the Görtler vortex instability mechanism at hypersonic speeds. *Theoret. Comput. Fluid Dyn.* 1: 125–34

Hall, P., Horseman, N. J. 1991. The linear inviscid secondary instability of longitudinal vortex structures in boundary layers. *J. Fluid Mech.* 232: 357–75

Hall, P., Malik, M. R. 1989. The growth of Görtler vortices in compressible boundary layers. *J. Eng. Math.* 23: 239–45

Hall, P., Seddougui, S. 1989. On the onset of three-dimensionality and time dependence in the Görtler vortex problem. *J. Fluid Mech.* 204: 405–20

Hämmerlin, G. 1961 Uber die Stabilität einer kompressiblen Störmung längs einer konkaven Wand bei verschiedenen Wandtemperaturverhältnissen. *Deutsche Versuchsanstalt für Luftfahrt,* Bericht 176

Herbert, Th. 1976. On the stability of the boundary layer along a concave wall. *Arch. Mech. Stosowanej* 28: 1039–55

Herbert, T. 1979. Higher eigenstates of Görtler vortices. In *Theoretical and Experimental Fluid Mechanics*, ed. U. Müller, K. G. Roesner, B. Schmidt, pp. 322–30. Berlin: Springer-Verlag

Herbert, Th. 1988. Secondary instability of boundary layers. *Annu. Rev. Fluid Mech.* 20: 487–526

Herron, I. H. 1991. Stability criteria for flow along a convex wall. *Phys. Fluids A* 3(7): 1825–27

Ito, A. 1980. The generation and breakdown of longitudinal vortices along a concave wall. *J. Jpn. Soc. Aeronaut. Space Sci.* 28: 327–33 (In Japanese)

Ito, A. 1985. Breakdown structure of longitudinal vortices along a concave wall. *Trans. Jpn. Soc. Aeronaut. Space Sci.* 33: 166–73

Ito, A. 1987. Visualization of boundary layer transition along a concave wall. *Proc. 4th Intl. Symp. Flow Visualization*, pp. 339–44. Washington, DC: Hemisphere

Ito, A. 1988. Breakdown structure of longitudinal vortices along a concave wall: on the relation of horseshoe-type vortices and fluctuating flows. *J. Jpn. Soc. Aeronaut. Space Sci.* 36: 274–79

Jallade, S. 1990. Effects of streamwise curvature variations on Görtler vortices. In *Instability and Transition, II*, ed. M. Y. Hussaini, R. G. Voigt, pp. 63–78. New York: Springer-Verlag

Jallade, S., Arnal, D., Ha Minh, H. 1990. Theoretical study of Görtler vortices: linear stability approach. In *Laminar-Turbulent Transition*, ed. D. Arnal, R. Michel, pp. 563–72. Berlin: Springer-Verlag

Kalburgi, V., Mangalam, S. M., Dagenhart, J. R. 1988a. A comparative study of theoretical methods on Görtler instability. *AIAA Pap. No.* 88-0407

Kalburgi, V., Mangalam, S. M., Dagenhart, J. R. 1988b. Görtler instability on an airfoil: comparison of marching solution with experimental observations. *AGARD CP No.* 438, Pap. No. 8

Klebanoff, P. S., Tidstrom, K. D., Sargent, L. M. 1962. Three-dimensional nature of boundary layer instability. *J. Fluid Mech.* 12: 1–34

Kobayashi, R., Kohama, Y. 1977. Taylor-Görtler instability of compressible boundary layers. *AIAA J.* 15: 1723–27

Kohama, Y., Saric, W. S., Hoos, J. A. 1990. A high-frequency secondary instability of crossflow vortices that leads to transition. *Proc. R. Aeronaut. Soc.: Boundary-Layer Transition and Control*, Cambridge

Kohuth, K. R., Neitzel, G. P. 1988. Experiments on the stability of an impulsively-initiated circular Couette flow. *Exp. Fluids* 6: 199–208

Koschmieder, L. 1992. *Benard Cells and Taylor Vortices*. Cambridge: Cambridge Univ. Press

Kottke, V. 1986. Taylor-Görtler vortices and their effect on heat and mass transfer. *Proc. Intl. Heat Trans. Conf., 8th*, pp. 1139–44. San Francisco

Kottke, V. 1988. On the instability of laminar boundary layers along concave walls towards Görtler vortices. In *Propagation in Systems Far From Equilibrium*, ed. J. E. Wesfreid, H. R. Brand, D. Manneville, G. Albinet, N. Boccara, et al. Berlin: Springer-Verlag

Kottke, V., Mpourdis, B. 1987. On the existence of Taylor-Görtler vortices on concave walls. *Proc. 4th Intl. Symp. Flow Vis.*, pp. 475–80. Washington: Hemisphere

Laure, P. 1993. Workshop on: Spatio-temporal properties of centrifugal instabilities. Nice: INLN

Lee, K., Liu, J. T. C. 1992. On the growth of mushroomlike structures in nonlinear spatially developing Görtler vortex flow. *Phys. Fluids A* 4(1): 95–103

Leoutsakos, G., Crane, R. I. 1990. Three-dimensional boundary layer transition on a concave surface. *Intl. J. Heat Fluid Flow* 11(1): 2–9

Lesson, M., Koh, P. H. 1985. Instability and turbulent bursting in the boundary layer. In *Laminar-Turbulent Transition*, ed. V. V. Kozlov, pp. 39–52. New York: Springer-Verlag

Liepmann, H. W., 1943. Investigations on laminar boundary-layer stability and transition on curved boundaries. *NACA Rep. W-107*

Liepmann, H. W. 1945. Investigation of boundary-layer transition on concave walls. *NACA Rep. W-87*

Liu, J. T. C. 1991. On scalar transport in nonlinearly developing Görtler vortex flow. *Geophys. Astrophys. Fluid Dyn.* 58: 133–45

Liu, J. T. C., Lee, K. 1993. Heat transfer under longitudinal vortices arising from concave surface curvature. *Trans. Polish Acad. Sci.—Fluid Flow Machinery*, pp. 1–18

Liu, J. T. C., Sabry , A. S. 1990. Concentration and heat transfer in nonlinear Görtler vortex flow and the analogy with longitudinal momentum transfer. *Proc. R. Soc. London Ser. A* 432: 1–12

Liu, W., Domaradzki, J. 1993. Direct numerical simulation of transition to turbulence in Görtler flow. *J. Fluid Mech.* 246: 267; also *AIAA Pap. No.* 90-0114

Malik, M. R., Hussaini, M. Y. 1990. Numerical simulation of interactions between Görtler vortices and Tollmien-Schlichting waves. *J. Fluid Mech.* 210: 183–89

Mangalam, S. M., Dagenhart, J. R., Hepner, T. E., Meyers, J. F. 1985. The Görtler instability of an airfoil. *AIAA Pap. No.* 85-0491

Mathis, D. M., Neitzel, G. P. 1985. Experiments on impulsive spin-down to rest. *Phys. Fluids* 28(2): 449

Myose, R., Blackwelder, R. F. 1991. Controlling the spacing of streamwise vortices on concave walls. *AIAA J.* 29(11): 1901–5

Nayfeh, A. H., Al-Maaitah, A. 1987. Influence of streamwise vortices on Tollmien-Schlichting waves. *AIAA Pap. No.* 87-1206

Park, D. S. 1990. *The primary and secondary instabilities of Görtler flow.* PhD thesis. Univ. South. Calif.

Peerhossaini, H. 1987. *L'Instabilite d'une couche limite sur une paroi concave (les tourbilons de Görtler).* Thèse de Doctorat. Univ. Pierre et Marie Curie, Paris

Peerhossaini, H., Bippes, H., Steinbach, D. 1990. Modèle pour l'étude expérimentale des effets de la courbure sur la transition de la couche limite sur une aile en flèche: résults préliminaires. *Rech. Aerosp.* 6: 15–21

Peerhossaini, H., Wesfreid, J. E. 1988a. On the inner structure of streamwise Görtler rolls. *Intl. J. Heat Fluid Flow* 9(1): 12–18

Peerhossaini, H., Wesfreid, J. E. 1988b. Experimental Study of the Taylor-Görtler Instability. In *Propagation in Systems Far From Equilibrium*, ed. J. E. Wesfreid, H. R. Brand, D. Manneville, G. Albinet, N. Boccara, et al, pp. 399–412. Berlin: Springer-Verlag

Peerhossaini, H., Wesfreid, J. E. 1990. Colloquium on Görtler vortex flows: Synopsis of contributions. *EUROMECH* 261. Nantes: ISITEM

Petitjeans, P. 1992. *Etude expérimentale des instabilités de couche limites sur des parois concaves: Instabilité de Görtler.* Thèse de Doctorat. Univ. Paris VI

Ragab, S. A., Nayfeh, A. H. 1981. Görtler instability. *Phys. Fluids* 24(8): 1405–17

Rayleigh, Lord. 1916. On the dynamics of revolving fluids. *Scientific Papers* 6: 447–53. Cambridge: Cambridge Univ. Press

Reed, H. L., Saric, W. S. 1989. Stability of three-dimensional boundary layers. *Annu. Rev. Fluid Mech.* 21: 235–84

Reshotko, E. 1976. Boundary-layer stability and transition. *Annu. Rev. Fluid Mech.* 8: 311–49

Sabry, A. S., Liu, J. T. C. 1988. Nonlinear Development of Görtler vortices and the generation of high shear layers in the boundary layer. In *Appl. Math., Fluid Mech., Astrophys.*, ed. D. J. Benney, F. H. Shu, C. Yuan, pp. 175–83. Singapore: World Sci.

Sabry, A. S., Liu, J. T. C. 1991. Longitudinal vorticity elements in boundary layers: nonlinear development form initial Görtler vortices as a prototype problem. *J. Fluid Mech.* 231: 615–63

Sabry, A. S., Yu, X., Liu, J. T. C. 1990. Secondary instabilities of three-dimensional inflectional velocity profiles resulting from longitudinal vorticity elements in boundary layers. In *Laminar-Turbulent Transition*, ed. D. Arnal, E. Michel, pp. 441–51. Berlin: Springer-Verlag

Sabzvari, J., Crane, R. I. 1985. Effect of Görtler vortices on transitional boundary layers. In *Three Dimensional Flow Phenomena in Fluid Machinery, Proc. ASME FED* 32: 113–19

Saric, W. S., Benmalek, A. 1991. Görtler vortices with periodic curvature. In *Boundary Layer Stability and Transition to Turbulence, ASME-FED* 114: 37–42

Saric, W. S., Nayfeh, A. H. 1977. Non-parallel stability of boundary-layer flows with pressure gradients and suction. *AGARD Rep. No.* 224, Pap. No. 6

Schultz-Grunow, F., Behbahani, D. 1973. Boundary-layer stability on longitudinally curved walls. *ZAMP* 24: 499–506

Smith, A. M. O. 1955. On the growth of Taylor-Görtler vortices along highly concave walls. *Quart. J. Math* 13: 233–62

Spall, R. E., Malik, M. R. 1989. Görtler vortices in supersonic and hypersonic boundary layers. *Phys. Fluids A* 1: 1822–35

Stuart, J. T. 1984. Instability of laminar flows, nonlinear growth of fluctuations and transition to turbulence. In *Turbulence and Chaotic Phenomena in Fluids*, ed. T. Tatsumi, pp. 17–26. Amsterdam: North-Holland

Stuart, J. T. 1986. Taylor-vortex flow: a dynamical system. *SIAM Rev.* 28: 315–42

Swearingen, J. D., Blackwelder, R. F. 1986. Spacing of streamwise vortices on concave walls. *AIAA J.* 24: 1706–9

Swearingen, J. D., Blackwelder, R. F. 1987. The growth and breakdown of streamwise vortices in the presence of a wall. *J. Fluid Mech.* 182: 255–90

Tagg, R. 1992. A guide to literature related to the Taylor-Couette problem. In *Ordered and Turbulent Patterns in Taylor-Couette Flow, Proc. NATO Adv. Res. Workshop*, ed. C. D. Andereck, F. Hayot. New York: Plenum

Tani, I. 1949. Some effects of surface curvature on the velocity distribution in a laminar boundary. *J. Jpn. Soc. Mech. Eng.* 52: 476–77 (In Japanese); also 57: 596–98 (1954) (In Japanese)

Tani, I. 1962. Production of longitudinal vortices in the boundary layer along a curved wall. *J. Geophys. Res.* 67(8): 3075–80

Tani, I., Sakagami, J. 1962. Boundary-layer instability at subsonic speeds. *Proc. Intl. Council Aerosp. Sci., 3rd Cong.*, pp. 391–403. Washington: Spartan

Taylor, G. I. 1923. Stability of a viscous liquid contained between two rotating cylinders. *Phil. Trans. R. Soc. London Ser. A* 223: 289–343

Tumin, A. M., Chernov, Yu. P. 1988. Asymptotic analysis of flow instability in a compressible boundary layer on curved surface. *Zh. Prikl. Mekh. Tekh. Fiz.* 3: 84–89

Winoto, S. H., Crane, R. I. 1980. Vortex structure in laminar boundary layers on a concave wall. *Intl. J. Heat Fluid Flow* 2: 221–31

Winoto, S. H., Durao, D. F. G., Crane, R. I. 1979. Measurements within Görtler vortices. *J. Fluids Engrg.* 100: 517–20

Wortmann, F. X. 1964a. Experimentelle Untersuchungen laminarer Grenzschichten bei instabilier Schichtung. *Proc. Intl. Congr. Appl. Mech. XI*, 815–25

Wortmann, F. X. 1964b. Experimental investigations of vortex occurrence at transition in unstable laminar boundary layers. *AFOSR Rep.* 64-1280, AF 61 (052)-220

Yu, X., Liu, J. T. C. 1991. The secondary instability in Görtler flow. *Phys. Fluids A* 3(8): 1825–27

Yu, X., Liu, J. T. C. 1993. On the mechanism of sinuous and varicose modes in three-dimensional viscous secondary instability of nonlinear Görtler rolls. *Phys. Fluids A* Submitted

Yurchenko, N. F. 1981. Method of experimental investigation of a system of streamwise vortices in a boundary layer. *Inz.-Fiz. Zh.* XLI(6): 996–1002 (In Russian)

Yurchenko, N. F., Babenko, V. V., Kozlov, L. F. 1980. Experimental investigation of the Görtler instability in boundary layers. In *Stratified and Turbulent Flows* 50–9. Kiev: Naukova Dumka (In Russian)

Zebib, A., Bottaro, A. 1993. Görtler vortices with system rotation: linear theory. *Phys. Fluids A* 5(4)

—

Annu. Rev. Fluid Mech. 1994. 26:411–82

PHYSICAL MECHANISMS OF LAMINAR-BOUNDARY-LAYER TRANSITION

Yury S. Kachanov

Institute of Theoretical and Applied Mechanics, Novosibirsk,
630090 Russia

KEY WORDS: onset of turbulence, resonant wave interactions, K-regime,
N-regime, solitons

1.   INTRODUCTION

The problem of turbulence onset in shear flows has attracted the attention
of investigators for more than a century. Despite its complexity, interest in
the laminar-turbulent transition has increased during the past few decades
owing to its importance in both fundamental and applied aspects of fluid
mechanics.

   The physical mechanisms of the transition phenomenon depend essentially on the specific type of flow and the character of environmental
disturbances. For boundary-layer flows two main classes of transition are
known (Morkovin 1968, 1984; Morkovin & Reshotko 1990). The first of
them is connected with boundary-layer instabilities (described initially
by linear stability theories), amplification, and interaction of different
instability modes resulting in the laminar flow breakdown. This class is
usually observed when environmental disturbances are rather small. The
second class of transition, usually called *bypass*, is connected with "direct"
nonlinear laminar-flow breakdown under the influence of external disturbances. This is observed when high enough levels of environmental
perturbations (free-stream disturbances, surface roughness, etc) are present.

   This article focuses on the first class of transition because of its fundamental and practical importance in problems involving moving vehicles

411

0066–4189/94/0115–0411$05.00

in air and water. We concentrate on experimental studies of the transition phenomenon in boundary layers of incompressible fluids. Theoretical studies in this field are discussed in the recent books by Craik (1985) and Zhigulyov & Tumin (1987) and reviews by Nayfeh (1987a), Herbert (1988), Reed & Saric (1989), Fasel (1990), Kleiser & Zang (1991), and others. Section 2 is devoted to a brief review of three main aspects of the transition process: receptivity, linear stability, and nonlinear breakdown. Nonlinear mechanisms of the breakdown process are discussed in Sections 3 to 7. Section 3 introduces two main types of boundary-layer breakdown which were found experimentally from the 1960s to 1980s, namely: K- and N-regimes of transition. The physical nature of the latter is discussed in Section 4. Sections 5 and 6 are devoted to the more complicated case— the K-regime. A brief review of experimental studies and theoretical ideas concerning the K-breakdown is presented in Section 5; Section 6 is devoted to a discussion of various aspects of this type of transition and to its main physical mechanisms, as well as to their application to understanding the physical nature of developed wall turbulence. Finally, in Section 7 some important nonlinear aspects of boundary-layer transition are discussed.

2.   ASPECTS OF THE TURBULENCE ONSET PROBLEM

Progress in hydrodynamic stability theory and turbulence-onset studies has led to the understanding that transition starts long before the pronounced phenomena of breakdown are seen. The onset of turbulence in the boundary layer comprises three main stages, schematically depicted in Figure 1 for the relatively simple case of a flat plate. These stages correspond to the

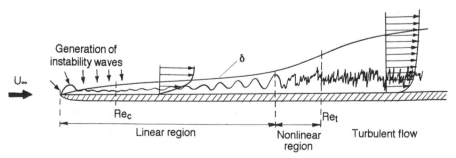

Figure 1   Qualitative sketch of the process of turbulence onset in a boundary layer (after Kachanov et al 1982).

three main aspects of the problem that are studied both experimentally and theoretically, namely: (a) receptivity, (b) linear stability, and (c) nonlinear breakdown.

2.1 Receptivity

During the first stage, in the region of relatively low local Reynolds number, instability waves (i.e. boundary-layer eigen oscillations, usually called Tollmien-Schlichting waves) are generated. This region extends from the vicinity of the model nose up to the vicinity of the first branch of the neutral stability curve, i.e. to the place where instability waves can begin to be amplified. The problem of generating these waves by perturbations (which include acoustic, vortical, temperature, and vibrational fluctuations) is referred to as the problem of boundary-layer receptivity to external disturbances. This aspect of the transition process was clearly formulated for the first time by Morkovin (1968) as the problem of transformation of external disturbances into eigen boundary-layer oscillations. (However, the idea that the Reynolds number of the pipe-flow transition has to increase when the amplitudes of disturbances in the incoming flow attenuate, had been suggested by O. Reynolds as long as a century ago and corroborated later in 1905.) The importance of the receptivity problem for understanding the transition phenomenon was also accentuated in reviews by Loehrke et al (1975) and Reshotko (1976). First successes in its solution had been achieved in the 1970s in both experimental and theoretical fields.

In particular, the important role of a model vibration in the acoustically excited boundary layer for the receptivity problem was shown experimentally and studied by Kachanov et al (1975a). Some aspects of acoustic receptivity were also investigated by Shapiro (1977). One of the mechanisms of transformation of free-stream vortices into Tollmien-Schlichting waves (the leading edge mechanism) was studied by Kachanov et al (1978a). Transformation of acoustic waves into eigen disturbances of the boundary layer in the vicinity of a small localized roughness element was investigated experimentally by Aizin & Polyakov (1979). (For a review of these Russian papers, see Nishioka & Morkovin 1986.)

The first theoretical studies of the receptivity problem also appeared in the 1970s. These were devoted to investigating the boundary-layer receptivity to free-stream vortices (Rogler & Reshotko 1975, Rogler 1977, Maksimov 1979) and to acoustic waves (Mangur 1977, Tam 1978, Maksimov 1979, Murdock 1980). In combined experimental and theoretical work by Kachanov et al (1979), the results of Maksimov's calculations were compared with experiment for the case of leading-edge receptivity to both acoustic-type and vortex disturbances of the free stream. The role of leading edge vibrations was also discussed. This period of direct studies of the receptivity problem has been summarized in the monograph by Kachanov et al (1982) (see also the review by Nishioka & Morkovin 1986).

The subsequent rapid development of receptivity investigations is detailed in the book by Zhigulyov & Tumin (1987) and in a number of recent reviews by Nishioka & Morkovin (1986), Goldstein & Hultgren (1989), Kerschen (1989), Kerschen et al (1990), Kozlov & Ryzhov (1990), Morkovin & Reshotko (1990), and others.

2.2 Linear Stability

The second stage of transition (also shown in Figure 1) corresponds to the propagation of small-amplitude instability waves down the boundary layer, which are either amplified, if the flow is unstable to them, or attenuated. This stage is described by linear hydrodynamic stability theory. Since the linear region is usually quite extensive and the phenomena that take place in it are simply described, linear stability theory, together with receptivity theory, are used for most engineering calculations (see, for example, Mack 1975, Hefner & Bushnell 1979, Kachanov et al 1982, Bushnell & Malik 1987).

Linear stability theory, especially of two-dimensional (2D) flows, is the most developed branch of the transition problem. For the case of 2D disturbances, this theory has been developed in numerous studies starting with Tollmien (1929) (see also Schlichting 1979, Lin 1955, and others) and carefully substantiated experimentally, starting with the classical work by Schubauer & Skramstad (1947) and later in works of Wortmann (1955), Ross et al (1970), Kachanov et al (1974a, 1975b) (see also Saric & Nayfeh 1977), Strazisar et al (1977), Kozlov & Babenko (1978), Klingmann et al (1993), and others. The influence of different flow and surface conditions on boundary-layer stability had already been predicted by very early works devoted to stability theory. For example, Lin (1955) showed that a negative streamwise pressure gradient stabilizes the boundary layer and vice versa; wall cooling in air flow or heating in water flow increases boundary-layer stability and vice versa; and the presence of an inflection point in the velocity profile plays a destabilizing role. However, the first experimental studies of the stability of more complicated flows began much later, although Schubauer & Skramstad (1947) corroborated the stabilizing influence of a negative streamwise pressure gradient. [Studies on the influence of a constant adverse pressure gradient on the stability of a flat-plate boundary layer were later repeated in much more detail by Wubben et al (1990) and Wubben (1990).]

The influence of free-stream unsteadiness on boundary-layer stability was studied experimentally in the 1960s and 1970s (see reviews by Loehrke et al 1975 and Kachanov et al 1982). The influence of surface cooling in air flow (Kachanov et al 1974a) and heating in water flow (Barker & Jennings 1977) on boundary-layer stability had been studied experimentally and compared with theoretical results by Levchenko et al (1975), Gaponov & Maslov (1971), and Lowell & Reshotko (1974). The influence of wall waviness (Kachanov et al 1974b) and suction (Kozlov et al 1978) on the stability characteristics of the boundary layer also had been investigated. Of course, many of these experiments were not exhaustive and were repeated in later works.

Many issues are connected with the influence of nonparallel effects on boundary-layer stability (see reviews by Fasel & Konzelmann 1990 and Bertolotti et al 1992). Experimentally this question was studied by Ross et al (1970) and Kachanov et al (1975b); the results of the latter work were compared with nonparallel stability theory by Saric & Nayfeh (1977). In the case of the flat-plate boundary layer without a pressure gradient, nonparallel effects were shown to strongly influence the amplification rates of 2D disturbances. [Fasel & Konzelmann (1990) and Bertolotti et al (1992) criticized this conclusion; their studies show the influence to be rather weak.] Stewart & Smith (1987) showed that nonparallel effects can influence the 3D stability of nonparallel flows, for example in a boundary layer with a strong adverse pressure gradient and separation. One of the most important conclusions concerning the influence of nonparallelism of the boundary layer on stability is that there is no universal criterion of stability for real flows (see Fasel & Konzelmann 1990). Whether the boundary layer is stable or not (and to what extent), with respect to a particular disturbance mode, depends on the method of determination of the stability criterion.

The first attempts to experimentally investigate boundary-layer stability with respect to 3D disturbances were undertaken in the 1960s by Vasudeva (1967), who used a localized source of instability waves. Later, Gaster & Grant (1975) obtained rich information about the development of 3D packets of instability waves in a Blasius boundary layer. Frequency and frequency-wavenumber Fourier analysis permitted them to obtain amplification rates of different 3D instability modes.

Interest in 3D linear disturbances, developing in 2D boundary layers, was initially lacking because of Squire's (1933) well-known theorem which says that a 2D boundary layer (at subsonic speeds) is usually more stable to 3D instability waves than to 2D ones. However, the properties of 3D modes, especially their dispersion characteristics (i.e. the dependencies of their phase velocities on frequency, spanwise wavenumber, and propagation angle), are very important for weak-nonlinear wave interactions. This explains why 3D instability theory began to attract attention in the

1970s, when weak-nonlinear theories were being developed (see below and Sections 4.1 and 5.1).

Gilyov et al (1981, 1983) studied experimentally the stability and dispersion characteristics of 3D waves propagating at angles of 0 to 80° to the flow direction for a flat-plate boundary layer. Their results were obtained by means of spanwise Fourier transforms of *spatial* wave packets (harmonic in time) generated by a point source. Similar results were obtained later by Schneider (1989) in water flow for waves inclined 0–30° from the flow direction which were harmonic in space and time. The development of 3D instability waves in a Blasius boundary layer was also studied experimentally by Cohen et al (1991) and Boiko et al (1991). In particular, it was shown by Gilyov et al (1981, 1983) (and later by Schneider 1989 and Boiko et al 1991) that downstream phase velocities $cx = \omega/\alpha_r$ of 3D instability waves grow with their propagation angle $\theta = \tan^{-1}(\beta/\alpha_r)$ (where ω is a disturbance frequency, α_r is a streamwise wavenumber, and β is a spanwise wavenumber). Larger angles θ led to larger growth rates of cx with θ. Eigenfunctions of 3D instability waves, propagating in the flat-plate boundary layer (i.e. normal-to-wall distributions of disturbance amplitude and phase), were also measured for the first time by Gilyov et al (1981, 1983) and later by Cohen et al (1991).

Kachanov & Michalke (1993) recently made the first direct quantitative comparison of theoretical and experimental data on the dispersion and stability characteristics of 3D instability waves, propagating in a flat-plate boundary layer. The dispersion characteristics (such as dependences of the streamwise wavenumber and downstream phase velocity on the propagation angle, frequency parameter, and Reynolds number) measured in experiments by Gilyov et al (1981, 1983) were shown to be in a very good quantitative agreement with the linear parallel spatial stability theory of A. Michalke and also with other available experimental results (Ross et al 1970, Schneider 1989). However, as expected, the amplification rates were found to display mainly qualitative agreement only. Kachanov & Michalke (1993) concluded that the problem of correctly predicting amplification rates for 3D instability modes needs much more precise investigation in both theory and experiment; in particular, the criteria of instability have to be determined more accurately, similar to what has been done for 2D disturbances (see Fasel & Konzelmann 1990 and Bertolotti et al 1992).

The dispersion characteristics of 3D modes and their eigenfunctions were later used for a weak-nonlinear description of late stages of laminar flow breakdown (see, for example, Kachanov & Levchenko 1982, Kachanov 1987, and Section 4.3). Development of linear spatial wave packets as a whole (i.e. without spanwise Fourier transforms) was also studied by

Gilyov et al (1981), and propagation of spatial-temporal packets has recently been investigated by Seifert (1990) and Cohen et al (1991).

Experimental studies of more complicated 2D boundary-layer flows were developed simultaneously. The 2D stability of a boundary layer (on an airfoil) with streamwise pressure gradient was studied experimentally by Dovgal et al (1981) and in a number of subsequent studies (see reviews by Dovgal et al 1987 and Dovgal & Kozlov 1990). The 3D stability of the boundary layer on an airfoil (including regions of negative and positive pressure gradient and the separation zone) was investigated experimentally by Gilyov et al (1988). Very similar studies were carried out by Dovgal et al (1988) and Boiko et al (1991) for laminar boundary layers with local separation zones, and by Kachanov & Tararykin (1987) for a Blasius boundary layer disturbed by steady streamwise vortices. In addition to amplification rates, the dispersion characteristics of 3D instability waves were also studied in all of these papers. In particular, Gilyov et al (1988) found that an adverse pressure gradient results in an almost complete disappearance of spatial dispersion (i.e. dependence of cx and α_r on θ or β) for 3D instability waves. The same result had been obtained by Kachanov & Tararykin (1987) and Boiko et al (1991), where it was observed that the appearance of local velocity profiles with inflection points led to a rapid weakening of the cx and α_r dependences on propagation angle θ. This phenomenon is probably connected with inviscid instability modes which predominate in flows with inflectional velocity profiles.

Experimental data of Gilyov et al (1988), Dovgal et al (1988), and Boiko et al (1991) on boundary-layer stability near a point of separation with respect to both 2D and 3D disturbances are in rather good agreement with calculations by A. Michalke (1990, and 1990 private communication) for inviscid and viscid instability within a flat-parallel approach. However, some observed peculiarities of 3D disturbance amplification probably cannot be described by parallel stability theory. The results of experiments by Gilyov et al (1988) corroborated the theoretical conclusion (Stewart & Smith 1987) that nonparallel effects can be important for calculating the 3D instability of 2D boundary layers with adverse pressure gradients (including separation), and can explain the more rapid amplification observed for 3D modes, compared with 2D ones.

The stability of 3D boundary layers represents a much more complicated case. Its theoretical investigation was carried out only recently, mainly starting during the 1980s (e.g. Arnal et al 1984; Reed 1984, 1985; Dallmann & Bieler 1987; Reed et al 1990; Itoh 1990; see also reviews by Zhigulyov & Tumin 1987 and Reed & Saric 1989). Experimental studies were initially devoted to observations of transition but not stability of 3D boundary layers. The first experiments for which some stability characteristics of swept-wing boundary layers were obtained were carried out by Michel et al (1985) and Saric & Yeates (1985). Subsequent development of experimental studies in this field is discussed in detail in reviews by Reed & Saric (1989) and Bippes (1990, 1991). In particular, very interesting results were obtained by Bippes and co-workers in a set of experiments on the investigation of traveling instability waves under "natural" conditions. Simultaneously, additional experimental studies (combined with theoretical calculations by Fyodorov) were undertaken by Kachanov et al (1989, 1990) and Kachanov & Tararykin (1990). In these studies a complete set of swept-wing boundary-layer stability characteristics to 3D steady disturbances (including eigenfunctions as well as amplification rates and wavevector orientations as functions of spanwise wavenumber) was obtained with the help of controlled excitation of the flow. Comparison of experimental data with theory demonstrated rather good agreement. All studies showed that 3D boundary layers are essentially unstable to steady (zero-frequency) disturbances which, together with traveling instability waves, play a very important role in the transition process. These steady disturbances have the form of counter-rotating vortices oriented nearly along the streamwise direction. (It is interesting to note that the superposition of these counter-rotating vortices with an undisturbed 3D mean flow gives a total mean flow with corotating vortices. This fact often led to confusion and misunderstandings.) Experimental investigations of the stability of 3D boundary layers to traveling instability waves have recently begun (see for example Bippes 1991 and Deyhle et al 1993).

2.3   Nonlinear Breakdown

When the amplitudes of instability waves reach considerable values (of order 1–2% of the free-stream velocity) the flow enters a phase of nonlinear breakdown, randomization, and a final transition into a turbulent state (see Figure 1). This breakdown phase is usually not very extensive, but just here the flow is transformed from a deterministic, regular, often two-dimensional laminar flow into a stochastic and at the same time ordered, three-dimensional, yet mysterious turbulent one.

Although the region of nonlinear breakdown had been studied for nearly forty years, many aspects remained a mystery. Recent progress is associated with: (a) the discovery of the decisive role played by resonant phenomena which occur in the process of transition and (b) the detection and description of coherent structures/solitons (CS-solitons) in the transitional boundary layer. Both ideas arose from a series of experimental and theoretical works. The former of these ideas led us to believe that the process of laminar-boundary-layer breakdown was a resonance phenomenon. The latter allows us to advance far downstream up to the strong-nonlinear region of the developed disturbance and thereby (hopefully) helps us to understand the very late stages of the laminar-turbulent transition and perhaps the structure of developed wall turbulence.

Studies in this field are now in full swing. Many questions are not yet clearly answered and many hypotheses are still to be proved. This article reviews the most important *experimental* studies and discusses the concepts and ideas related to the main physical mechanisms of nonlinear laminar-boundary-layer breakdown. Of course, most of these ideas are closely connected with theoretical work, to which we occasionally refer. Mainly, we concentrate on the simplest type of boundary-layer flow: flow over a flat plate. As we shall see, flat-plate boundary-layer transition is actually not simple at all, and many of its features are also observed in transitions of other boundary layers and even other flows (for example in plane channel flow).

3.   MAIN TYPES OF BOUNDARY-LAYER TRANSITION

The first thorough physical study of the laminar-flow breakdown in the boundary layer was carried out by Schubauer & Skramstad (1947). This work served as an experimental basis for the concept of hydrodynamic instability and provided the first details of nonlinear mechanisms of transition. At the end of the 1950s fundamental experiments (later to also become classical) were conducted by Klebanoff et al (1962); these laid the foundation for many ideas about the nature of laminar boundary layer breakdown. In these experiments controlled disturbances simulated the "natural" ones observed earlier by Schubauer & Klebanoff (1956) and by Klebanoff & Tidstrom (1959). In the work of Klebanoff et al (1962), the mechanisms of laminar flow breakdown were studied in detail and the applicability of a series of theoretical models and hypotheses was critically evaluated. New important features of the nonlinear breakdown were identified. Almost simultaneously, Kovasznay et al (1962) carried out a detailed investigation of three-dimensional flow velocity and vorticity in the same regime of transition. Hama & Nutant (1963) complemented hot-wire measurements with visual observations carried out with the help of a hydrogen-bubble flow-visualization technique. Tani & Komoda (1962) and Komoda (1967) studied the influence of spanwise mean-flow modulation on the process of boundary-layer breakdown.

3.1   K-*Regime*

Detailed flow-field hot-wire measurements carried out in the transitional boundary layer in a set of experiments (mainly by Schubauer & Klebanoff

1956, Klebanoff & Tidstrom 1959, Klebanoff et al 1962) provided subsequent investigators with minute fundamental knowledge of the main features of nonlinear boundary-layer breakdown. In experiments by Klebanoff et al (1962) (as well as in many subsequent studies) the investigation of transition phenomena was carried out under controlled conditions, i.e. with controlled excitation of steady and traveling initial disturbances which simulated the so-called natural ones observed in uncontrolled or partially controlled experiments (see, for example, Schubauer & Klebanoff 1956, Klebanoff & Tidstrom 1959). The experimental simulation of transition, starting with the classical studies by Schubauer & Skramstad (1947), is widely used by investigators and represents a very effective method for the investigation of this complicated phenomenon.

In the case of Klebanoff et al's (1962) experiment (Figure 2) a vibrating ribbon technique was used to introduce an almost flat fundamental instability wave, harmonic in time, which simulated the quasi-sinusoidal Tollmien-Schlichting wave usually observed in "natural" transition in the region

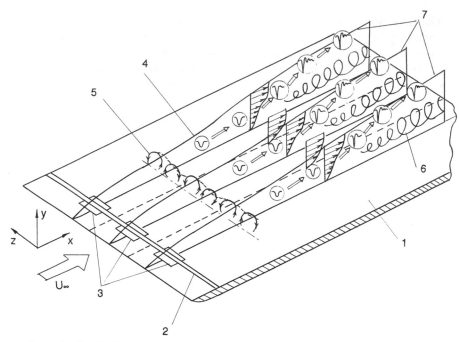

Figure 2 Sketch of experiments by Klebanoff et al (1962) and others on nonlinear breakdown of the boundary layer in the K-regime. 1: plate, 2: vibrating ribbon, 3: spacers, 4: boundary-layer edge, 5: pulsating streamwise vortices, 6: local flow randomization, 7: "peaks."

just before the flow breakdown. Strips of thin cellophane tape were placed on the plate surface beneath the vibrating ribbon (Figure 2) to simulate weak "natural" uncontrolled spanwise modulation of the mean flow that usually occurs in all real experimental situations (including experiments by Schubauer 1957 and Klebanoff & Tidstrom 1959). Interaction of the stationary and traveling disturbances resulted in a strong downstream amplification of the spanwise disturbance modulation (Figure 3) with streamwise vortices forming (Figure 2). The maxima (*peaks*) in the span-wise distributions of the disturbance amplitude are positions where rapid nonlinear disturbance amplification and fast laminar-flow breakdown and randomization take place (see qualitative sketch of this process in Figure 2).

In particular, Klebanoff et al (1962) discovered that breakdown starts with the appearance on the streamwise-velocity oscilloscope traces of powerful (up to 40%), high frequency flashes of disturbances called *spikes* which doubled, tripled, etc downstream (Figure 4). It was concluded that these spikes were responsible for the final laminar-boundary-layer break-down and flow randomization. The appearance of the spikes was attributed to a local high-frequency secondary instability of the primary wave (in the spirit of the work by Betchov 1960), which originated as a result of a *high-shear layer* in the instantaneous y-profiles of the flow velocity (the typical orientation of the coordinate axis is shown in Figure 2). The main global

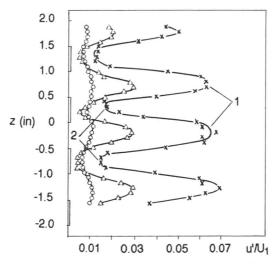

Figure 3   Downstream growth of spanwise modulation of rms amplitude of velocity dis-turbances in the K-breakdown (after Klebanoff et al 1962). 1: "peaks," 2: "valleys."

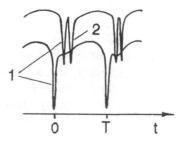

Figure 4 Typical single and double spikes (after Klebanoff et al 1962). 1: 1st spike, 2: 2nd spike, T: fundamental period.

conclusion drawn by Klebanoff et al (1962) was that the boundary-layer breakdown is of an essentially three-dimensional nature.

Up to the mid-1970s, the general opinion was that the succession of nonlinear phenomena, discovered in the experiments by Klebanoff et al (1962) (and in other studies, in particular quoted above) were a fundamental property of boundary layer flow. The results obtained in those works served as a basis for the majority of further experimental and theoretical studies. Later, in the end of the 1970s (see Herbert & Morkovin 1980), this type of laminar-turbulent breakdown was named the *K-regime* of transition (for Klebanoff).

3.2 N-*Regime*

The first indication that at least one more type of boundary-layer breakdown existed, appeared in experiments by Knapp & Roache (1968) who used a smoke visualization technique to study axisymmetric-boundary-layer transition. They observed Λ-shaped smoke accumulations which either *aligned in rows* or had a *staggered* order. The aligned order had been observed much earlier (probably for the first time) by Hama et al (1957), who introduced the term Λ-*vortex*, and later by Hama & Nutant (1963) and Hama (1963). However, the staggered order of Λ-vortices was not observed before Knapp & Roache (1968). Unfortunately, as pointed out by Herbert (1988), "the qualitative difference of the staggered pattern from the observations of Klebanoff et al (1962) was not recognized" by them. This was not surprising because Knapp & Roache (1968) had rather restricted (mainly visual) information and no theoretical basis for the description of the (resonant) interaction of 3D wave disturbances which either appeared some years later (Craik 1971) or was unknown to them (Maseev 1968, see also Herbert 1988). As was understood later, the staggered order of Λ-patterns was connected with the generation of 3D subharmonic disturbances (i.e. with frequency $\omega_1/2$, where ω_1 is the fundamental frequency) which had never been observed before in boundary-layer transition in the K-regime. [Note, nevertheless, that the staggered order of Λ-patterns was also observed almost ten years before the experiments of Knapp & Roach (1968) by Hama (1959) who observed a boundary layer for water flowing past a ring affixed to the nose of a 30° cone. However, it still remains unclear whether the staggered order of structures is also connected with the subharmonic resonance and N-type of breakdown.]

In work by Kachanov et al (1977), carried out in Novosibirsk, experimental data were obtained which indicated the existence of a new type of laminar-flow breakdown in the boundary layer. In 1980, when the existence of this new regime had been corroborated by two independent experiments (see below), Levchenko proposed to name this regime the N-type of transition (for *new* type, or for *Novosibirsk*). Another name often used for this regime is *subharmonic* because of the decisive role of subharmonic resonances in this transition (see Section 4). However, the latter term is rather inaccurate because subharmonic-type resonances were also found to be important in the K-regime of transition, as well as in other types of transitional flows (see also Sections 6.1, 6.5, and 7.1).

In the newly discovered transition regime no spikes, typical for the K-breakdown, were observed. N-breakdown started with low-frequency "vibration" of the fundamental wave observed on the oscilloscope screen. This "vibration" was amplified downstream until the signal became completely randomized. From the spectral viewpoint this type of transition was characterized by: (*a*) the gradual (very weak) growth of deterministic harmonics of the fundamental wave (with frequencies $n\omega_1$, where $n = 1, 3, 5, \ldots$), (*b*) the appearance of a broad spectrum of low-frequency 3D continuous spectrum oscillations, including a subharmonic (with frequency $\omega_{1/2} = \omega_1/2$), (*c*) their subsequent rapid amplification, and (*d*) the generation of 3D quasi-random (with continuous spectrum) disturbances near the frequencies $3\omega_1/2$, $5\omega_1/2$, etc. (See Figure 5). It was noted by Kachanov et al (1977) that the onset of the low-frequency quasi-subharmonic oscillations always (in all five regimes studied) played an important role in beginning the rapid development of the three-dimensionality, stochastisity, and final breakdown of the laminar flow.

As was shown later (see Section 4.3), the N-regime of transition is characterized by a staggered order of Λ-vortices which appeared as the result of amplification of 3D subharmonic disturbances. Therefore, we now realize that Knapp & Roache (1968) saw the N-regime of transition when observing the staggered patterns in their experiments.

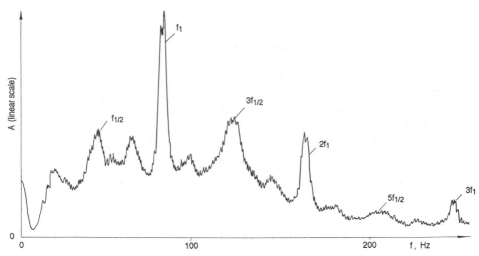

Figure 5 Typical amplitude spectra of velocity disturbances observed in the *N*-transition (Kachanov et al 1977).

4.   NATURE OF *N*-BREAKDOWN

4.1   *Early Studies of Resonant Wave Interactions*

The development of ideas on the important role of resonant wave interactions in the transition process in shear flows started many years ago with theoretical works by Raetz (1959) and Kelly (1967). Maseev (1968) was probably the first to develop a theory of secondary (nonlocal) instabilities of boundary-layer disturbances (see also Herbert 1988). Craik (1971) developed a theory of resonant three-wave interaction for the case of boundary-layer flow based on a weak-nonlinear approach (see also Craik 1985). Craik's triad consisted of one flat instability wave and two oblique subharmonic waves propagating at angles of the same value but of opposite sign. This weak-nonlinear branch of resonant theories was continued in the 1970s by Volodin & Zelman (1978). Herbert (1975) continued the study within the framework of the Floquet theory of secondary instability, begun by Kelly (1967).

However, for several years the resonant amplification of subharmonic modes, predicted by Craik and Volodin & Zelman, was not observed experimentally. As a result, many researchers began to doubt whether the resonant triads could play a significant role in boundary-layer transition. The nonparallelism of this flow and a continuously changing local Reynolds number seemed to break up this fragile resonance, the existence conditions for which were strongly influenced by the presence of dispersion of the boundary-layer waves.

The first observations of amplified subharmonic spectral modes in boundary-layer transition (Kachanov et al 1977) "revived interest in Craik's model" (Herbert 1988). The presence of Craik-type resonances were verified by Saric et al (1981), Thomas & Saric (1981), and Kachanov & Levchenko (1982). Nayfeh (1987a) noted in his review that "to explain the occurrence of the subharmonic in the experiments of Kachanov et al (1977), Nayfeh & Bozatli (1979a) used the method of multiple scales to analyze the nonparallel two-dimensional secondary instability of a primary two-dimensional T-S wave." They found that the threshold amplitude of the fundamental instability wave needed to excite the 2D subharmonic disturbance is very large—on the order of 29% of the mean flow! It meant that the subharmonic resonance with 2D waves cannot, in practice, amplify subharmonics in the flat-plate boundary layer. Therefore all subsequent theoretical approaches, devoted to mechanisms of subharmonic amplification in the Blasius boundary layer, dealt only with interactions of 3D subharmonics; just as had been done by Craik (1971).

The first direct attempt to find a resonant interaction between fundamental and subharmonic instability waves in the boundary layer was undertaken by Saric & Reynolds (1980). They concluded that "nonlinear interaction generating a subharmonic wave $F_1/2$ was not found to be present up to transition except when $F_1/2$ was introduced by the ribbon." Unfortunately the data illustrating this observations were not presented by Saric & Reynolds (1980). However the generation of harmonics ($3\omega_1/2$, etc) was noted in the case of artificial excitation of the subharmonic wave [the same as found in work by Kachanov et al (1977) without subharmonic excitation, see Figure 5].

4.2 Experimental Detection of Subharmonic Resonance and Its Decisive Role in N-Breakdown

In hot-wire experiments by Kachanov & Levchenko (1982) the physical nature of the N-regime of boundary-layer transition had been investigated, and the properties of subharmonic-type (mainly parametric) resonant instability-wave interactions were studied and documented in detail. These experiments were undertaken in 1980, a few years after the work of Kachanov et al (1977). The same regime of laminar flow breakdown was reproduced. The experiments consisted of two main parts. In (a), "natural" subharmonic disturbances were used; these were selected and amplified by resonance from background fluctuations. Part (b) incorporated an artificial excitation of small quasi-subharmonic priming disturbances of frequencies $\omega' = \omega_{1/2} + \Delta\omega$ and uncontrolled spatial spectrum (where the detuning $\Delta\omega$

was varied). Experiment (*a*) mainly permitted the detection of the resonant nature of the *N*-breakdown, while experiment (*b*) allowed the properties of the resonances to be studied.

In the *N*-regime of transition, which is usually observed when the initial fundamental-wave amplitude is not very large, the onset of three-dimensionality and stochasticity starts with the formation, and subsequent rapid development, of a broad packet of low-frequency spectral disturbances (Figure 6). This phenomenon was observed by Kachanov et al (1977) in five different regimes of observation for three frequencies and three initial amplitudes of the fundamental wave. Careful investigation of these broad packets of low-frequency disturbances showed that they represent quasi-subharmonic fluctuations with almost constant frequency equal to $\omega_{1/2}$, but with a random time-dependent amplitude and a constant phase, which jumps by π when the amplitude crosses its zero value (see Figure 7 *left*). Points of the corresponding phase trajectory in a plane of "instantaneous" complex subharmonic amplitudes $B_{1/2}$ are also shown in Figure 7 (*right*). It

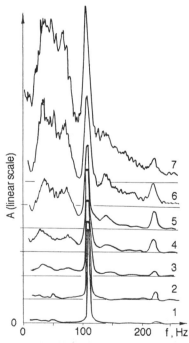

Figure 6 Downstream amplification of quasi-subharmonic continuous-spectrum disturbances observed in the *N*-transition (Kachanov & Levchenko 1982). 1, 2, ... 7 correspond to $x = 300, 480, 600, 640, 680, 720, 760$ mm.

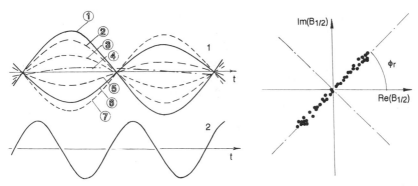

Figure 7 Properties of oscilloscope traces obtained at 7 different times (*left*) and points of phase trajectory (*right*) of quasi-subharmonic disturbances amplified in the *N*-transition (Kachanov & Levchenko 1982). $x = 640$ mm.

was also found by Kachanov & Levchenko (1982) that the amplified quasi-subharmonic consists of a pair of 3D instability waves inclined at $\theta_{1/2} \approx \pm 63°$ to the flow direction. The conditions for phase synchronism (Raetz 1959, Craik 1971)—$\omega_1 = 2\omega_{1/2}$ and $\alpha_1 = 2\alpha_{1/2}$—necessary for the resonant amplification of these subharmonics were also shown to be satisfied with high accuracy. These and other data obtained in the work of Kachanov & Levchenko (1982) permitted us to conclude that the broad continuous spectrum of low-frequency fluctuations as a whole (see Figure 6) is amplified in the *N*-regime as the result of subharmonic (parametric) resonant interaction of quasi-subharmonic 3D disturbances with the 2D fundamental wave. This conclusion was supported by comparison of the experimental data with predictions of the theory of Zelman & Maslennikova (1982), within the framework of weak-nonlinear theory. Subsequent careful experimental and theoretical studies fully corroborated this conclusion (see the next section).

4.3   *Properties of Subharmonic Resonances*

Experimental study of subharmonic resonance properties (Kachanov & Levchenko 1982) showed the resonance to be very wide in the frequency spectrum. It can amplify even quasi-subharmonic waves with frequencies $\omega_{1/2} \pm \Delta\omega$ with detunings $\Delta\omega$ close to half the subharmonic frequency. This result was corroborated by subsequent experiments (see, for example, Thomas 1987 and Corke 1990) and theoretical studies (Santos & Herbert 1986, Herbert 1988, Zelman & Maslennikova 1993). A large width of subharmonic resonances in the wavenumber spectrum was also shown theoretically by Zelman & Maslennikova (1982). Their theory also corroborated one more feature of the parametric amplification found experimentally by Kachanov & Levchenko (1982): the local symmetrization of resonantly amplified detuned modes in the frequency spectrum (when the form of the low-frequency part becomes almost symmetric relative to the subharmonic frequency, as seen in Figure 5). This property was explained by Kachanov & Levchenko (1982) within the framework of a quasi-stationary analysis of the behavior of amplitudes and phases of detuned subharmonics amplified by parametric resonance.

Figure 8 presents a comparison of measured amplitude and phase y-profiles of the resonantly amplified subharmonics with theoretical results obtained within the framework of different approaches: Floquet theory (Herbert 1984), weak-nonlinear theory (Zelman & Maslennikova 1989, 1990, 1993), and direct numerical solution of Navier-Stokes equations (Fasel et al 1987). Very good agreement between all the results is seen. These profiles are also very close to (or coincide with) the eigenfunctions of the corresponding 3D linear instability waves which were measured by Gilyov et al (1981, 1983) (see also Figure 12 in Kachanov 1987). This fact is attributed to the well-known *linear* character of subharmonic resonances at their parametric stage, i.e. when the disturbance amplitudes are rather small and there is no back influence of subharmonics on the fundamental wave. It is very interesting to note that the parametric stage of the subharmonic resonance, observed in experiments by Kachanov & Levchenko (1982), extended up to a stage when the subharmonic amplitude reached values of (at least) 2.5 times greater than those of fundamental wave! This phenomenon was explained later in theoretical works by Orszag & Patera (1983), Croswell (1985), Sinder et al (1987), and Spalart & Yang (1987)

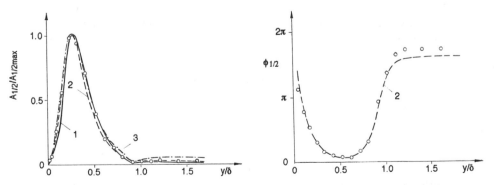

Figure 8   Measured (*points*) and calculated (*curves*) amplitude (*left*) and phase (*right*) y-profiles of the subharmonic mode amplified resonantly in the *N*-transition. Experiment by Kachanov & Levchenko (1982). Theories—1: Herbert (1984), 2: Fasel et al (1987), 3: Zelman & Maslennikova (1989, 1990, 1993).

(see also review by Herbert 1988) by the *catalytic* role of the fundamental wave in energy transfer directly from the mean flow to the amplified subharmonics. In particular Orszag & Patera (1983) showed that the direct energy flux from the 2D fundamental wave to 3D subharmonics is one order of magnitude less than that from the mean flow to the subharmonics (initiated catalytically by the fundamental wave). Direct fundamental-subharmonic energy exchange is only able to compensate for viscous dissipation of the subharmonics. These studies substantiated the possibility of using Floquet theory for describing the N-breakdown up to rather late stages of disturbance development. Of course, the subharmonic resonances become essentially nonlinear when the disturbance amplitudes exceed definite values. This nonlinear behavior, studied recently, is discussed in Section 4.4.

As shown by Volodin & Zelman (1978), the amplitudes of the subharmonics have a double exponential growth at resonance—a property well-known for parametric resonances in oscillatory systems. Figure 9 shows amplification curves for a number of main modes participating in the resonance, measured by Kachanov & Levchenko (1982) and calculated by Herbert (1984) and Fasel et al (1987). Again, rather good quantitative agreement between theory and experiment is observed for both the fundamental and subharmonic modes as well as for their harmonics.

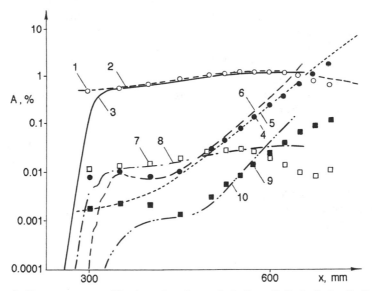

Figure 9 Downstream amplification of modes: $\omega_1(1, 2, 3)$, $\omega_1/2$ (4, 5, 6), $2\omega_1(7, 8)$, and $3\omega_1/2$ (9, 10) in the N-transition. Points are from the experiment by Kachanov & Levchenko (1982). Curves are from theory—2, 5: Herbert (1984); 3, 6, 8, 10: Fasel et al (1987).

For the first time the generation of higher subharmonic modes with frequencies $n\omega_1/2$ ($n = 3$, 5, 7) was observed in the boundary layer by Kachanov et al (1977) (Figure 5). Their amplification curves (see Figure 9), y-profiles, and spanwise distributions were studied by Kachanov & Levchenko (1982) and Corke & Mangano (1989). In all these works the amplification of these higher subharmonic modes was attributed to a combination (sum and difference) nonlinear interaction of the subharmonic mode with the fundamental one and its harmonics $m\omega_1$ ($m = 2$, 3, 4, ...). However, Herbert (1988) later showed that "combination resonance is governed by a linear Floquet system" which means that "occurrence of spectral peaks near odd multiples of $f_1/2$ in the experiments is not caused by nonlinearity"!

In the smoke-wire flow-visualization experiments by Thomas & Saric (1981) and Saric et al (1981) the staggered order of Λ-vortices observed in controlled boundary-layer transition at rather low initial amplitude of the fundamental wave were associated for the first time with the subharmonic hot-wire signals. The subsequent more detailed combined visual and hot-wire measurements by Saric et al (1984) permitted the authors to finally identify the K-regime of breakdown with the aligned order of Λ-patterns and the N-regime with the staggered order of patterns. Discussion over what type of resonant interaction (H-type for Herbert's theory or C-type for Craik's one) results in subharmonic amplification, continued for several years (see, for example, Herbert 1988 and Zelman & Maslennikova 1989, 1993; also Section 7.1). Today, we can conclude that both Floquet and weak-nonlinear theories demonstrate very good quantitative agreement with experiment, and seem to be accurate until certain stages of disturbance development. The results of direct numerical simulation are also in very good agreement with the observations (see the reviews by Fasel 1990 and Kleiser & Zang 1991).

Another important issue was connected with the propagation angles $\theta_{1/2}$ of the most amplified subharmonic disturbances. In the experiments by Kachanov & Levchenko (1982) $\theta_{1/2} \approx 63$–$64°$, but a weak-nonlinear parallel theory (see Volodin & Zelman 1978) initially predicted angles close to $50°$. However, subsequent studies solved this problem (see Saric & Thomas 1984; Saric et al 1984; Herbert 1984; Zelman & Maslennikova 1989, 1993). Figure 10 (*top*) shows the dependences of subharmonic amplification rates σ on spanwise-wavenumber parameter $b = \beta \times 10^3/Re$ and on the ratio of the spanwise (β) and streamwise (α) wavenumbers for different fundamental amplitudes, calculated by Zelman & Maslennikova (1989, 1993) within a weak-nonlinear approach. It is seen that the angles $\theta_{1/2} = \tan^{-1}(\beta/\alpha)$ of the most amplified subharmonics (*dotted curve*) increase together with the fundamental amplitude. A similar earlier result had been obtained by

Herbert (1984) within Floquet theory. Figure 10 (*bottom*) demonstrates dependencies of β/α of the most amplified subharmonics on the fundamental amplitude A_1 (determined at the point of the second branch of the neutral stability curve) calculated within the framework of weak-nonlinear theory (Zelman & Maslennikova 1989, 1993). The points represent corresponding experimental data (Kachanov & Levchenko 1982, Saric & Thomas 1984 and Saric et al 1984). Almost complete coincidence of theory and experiment is observed.

4.4   *Subsequent Studies of* N-*Breakdown*

Subsequent experimental and theoretical studies of the N-regime of boundary-layer transition proceeded in two main directions: (*a*) investigations of late (more nonlinear) stages of breakdown and (*b*) studies of more complicated cases such as compressible and three-dimensional boundary layers, the influence of suction, pressure gradients, and roughness elements, etc. The threshold character of the beginning of N-breakdown has also been studied recently in theoretical investigations.

The first experimental study of late stages of the N-regime of transition in controlled excitation of the subharmonic mode was carried out by Saric et al (1984). Some of the conclusions of the experiments by Kachanov & Levchenko (1982) were confirmed and clear correlation of visually observed Λ-patterns and hot-wire data was obtained. It was also shown how an N-type of breakdown can be transformed into a K-type and vice versa depending on the initial conditions. At high enough initial amplitude of the fundamental wave or low initial amplitude of the subharmonic priming disturbance, the K-regime of transition was observed. The N-regime appeared in the opposite situation.

Detailed experiments by Corke & Mangano (1989) (see also Corke 1990) were devoted to subsequent studies of the N-regime of boundary-layer transition in controlled conditions. In contrast to the previous investigations the subharmonic priming disturbances had a controlled spanwise spatial spectrum (not simply a controlled frequency and phase as before). Corke & Mangano (1989) confirmed all main results obtained earlier and studied late stages of the breakdown process. In particular they investigated spatial structure of modes $n\omega_1/2$ ($n = 3, 5, \ldots$) and their phase locking with the subharmonic mode. It was also found that growth of the subharmonic waves eventually saturates, after which time decay occurs. Amplification of broad-band low-frequency disturbances, which were coherent with subharmonic and fundamental modes, was observed. Subsequent sum and difference interaction led to a gradual filling of the spectrum. These observations were consistent with the previous ones and sup-

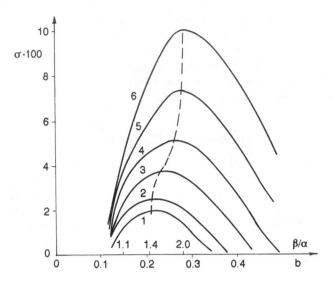

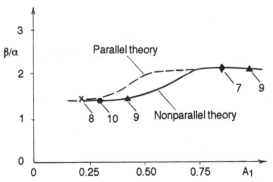

Figure 10 Amplification rates σ of 3D subharmonic modes at various amplitudes A_1 of the fundamental mode (*top*) and dependence of β/α for the most amplified subharmonics of A_1 (*bottom*) (after Zelman & Maslennikova 1989, 1993). 1, 2, ... 6: $A_1 = 0.14, 0.21, 0.28, 0.40, 0.53, 0.72\%$. Experimental points: 7—Kachanov & Levchenko (1982); 8, 9—Saric et al (1984); 10—Saric & Thomas (1984).

plemented them by providing more details of the 3D structure of the process.

In subsequent experiments reported by Corke (1990) the late stages of the N-breakdown were also studied for the case of detuned subharmonic resonance. In this work the generation of modes $n\omega_1/2 \pm \Delta\omega$ (where $n = 1$, $3, 5, \ldots$) and local symmetrization of the disturbance spectrum relative to the subharmonic frequency, and frequencies of modes $n\omega_1/2$ (found by Kachanov & Levchenko 1982), were also observed. Strong generation of these and other combination modes (of type $m\omega_1 \pm n\omega'$, where $2\omega'$ was close to ω_1 but not equal to it) was found further downstream. The properties of these combination modes were studied. These observations were very close to those obtained earlier by Kachanov et al (1978b) and Saric & Reynolds (1980) and complement them for the case of large detunings of frequencies ω_1 and ω'. [Kachanov et al (1978b) studied $\omega' \approx (0.73-0.85)\omega_1$, Saric & Reynolds (1980) $\omega' \approx (0.62-0.74)\omega_1$, and Corke (1990) $\omega' \approx (0.44-0.50)\omega_1$.] Although the focus of the Kachanov et al (1978b) study was on the combination instability-wave interaction and not the subharmonic resonance (which had not been discovered experimentally at that time), the results of those (and Saric & Reynolds's 1980) measurements correlate very well qualitatively with Corke's (1990) observations. Moreover, in those early experiments the detuned subharmonic resonance (found later), was also probably present (because of its very large spectral width; see Section 4.3) but was not recognized.

In the theoretical field, many studies devoted to subharmonic resonances for more complicated boundary-layer flows took place during the past ten years. Reed (1984, 1985), El-Hady (1988), and Fischer & Dallmann (1987) studied 3D boundary layers; Nayfeh (1987b), El-Hady (1989), Reed et al (1990), Zurigat et al (1990), and Masad & Nayfeh (1990) investigated compressible boundary layers including the influence of pressure gradients and chemical reactions; Herbert & Bertolotti (1985) and Zelman & Maslennikova (1993) studied the effect of pressure gradients on the subharmonic resonance in incompressible boundary layers; El-Hady (1991) studied boundary layers with suction; Nayfeh & Ragab (1987) investigated boundary layers with a local separation zone induced by a roughness element; Zelman & Smorodsky (1991a,b) studied boundary layers with high shear layers. For reviews, see Nayfeh (1987b,c), Herbert (1988), and Reed & Saric (1989). All these studies used weak-nonlinear approaches or Floquet theory.

At the same time other theoretical approaches (including essentially nonlinear ones) were being developed. First of all, direct numerical simulation within the framework of the Navier-Stokes equations was carried out (for reviews see Fasel 1990 and Kleiser & Zang 1991). This approach gives a very good description of the N-regime of transition with resonant amplification of subharmonics and combination modes (see Figure 9). Smith & Stewart (1987) and Smith (1990) studied resonant-triad nonlinear interaction using asymptotic multi-structured analysis. Their analytical(!) solutions also displayed very good qualitative and even quantitative agreement with experimental data by Kachanov & Levchenko (1982).

Nonlinear effects observed at large amplitudes of subharmonic disturbances (usually at late stages of the N-breakdown), which result in a back influence on the fundamental wave and nonlinear coupling of all modes, were studied by Maslennikova & Zelman (1985), Herbert (1987), Zelman & Maslennikova (1990, 1993), Herbert & Crouch (1990), Crouch & Herbert (1993) (see also numerical simulations by Spalart & Yang 1987), and others. In most of these papers the experimental data quoted above were used for comparison of theories with observations. Two graphs, presented in Figure 11, demonstrate amplification curves calculated and measured for fundamental and subharmonic modes. In both cases rather a good agreement between theories and experiments is observed. In particular, Figure 11 (*left*) demonstrates possible rapid joint growth of both fundamental and subharmonic waves at late stages of the N-breakdown after crossing their amplification curves.

Herbert (1988), Rahmatullaev (1988), Crouch & Herbert (1993), and others studied threshold conditions for the beginning of the N-regime of breakdown. Figure 12 (from Crouch & Herbert 1993) presents a family of fundamental and (corresponding) subharmonic amplification curves

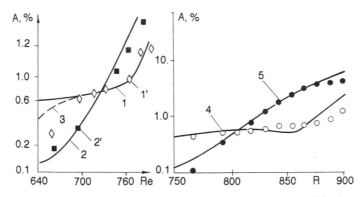

Figure 11 Comparison of measured (*points*) and calculated (*curves*) amplification of fundamental (1, 1′, 3, 4) and subharmonic (2, 2′, 5) amplitudes at more late stages of the N-breakdown (when $A_{1/2} > A_1$). [After Maslennikova & Zelman (1985, 1993) (*left*) and Crouch & Herbert 1993 (*right*).] Experiments: 1′, 2′—Saric et al (1984); 4, 5—Corke & Mangano (1989). 3—nonparallel theory.

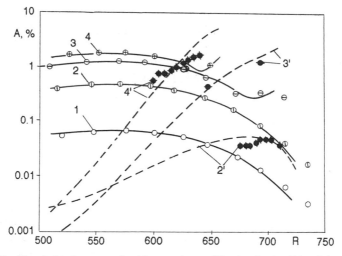

Figure 12 Threshold character of subharmonic amplification in the *N*-breakdown (after Crouch & Herbert 1993). Experimental points of Kachanov & Levchenko (1982). 1, 2, 3, 4—fundamental waves; 2′, 3′, 4′—subharmonics.

together with experimental points measured by Kachanov & Levchenko (1982) for various initial amplitudes of the fundamental wave. At its small amplitude the subharmonic resonance does not take place in theory (curve 1) and resonant amplification of the subharmonic is not observed in experiment. When the initial fundamental amplitude is higher (curves 2, 2′) resonance starts to amplify the subharmonic but it then, according to both theory and experiment, decays because of insufficient local amplitude of the fundamental wave. In this region resonant amplification cannot overcome the characteristic linear attenuation of the subharmonic mode. At higher initial fundamental amplitudes (curves 3, 3′, 4, 4′) the resonance leads to a strong subharmonic growth.

Thus, joint theoretical and experimental efforts of investigators from different countries resulted in clarifying the physical nature of the *N*-regime of boundary-layer breakdown, at least for the simplest case of flat-plate flow. The main mechanism of the *N*-breakdown was shown to be connected with rapid resonant amplification of 3D quasi-subharmonic modes through their interaction with a quasi-2D initial Tollmien-Schlichting wave amplified by primary linear instability of the flow.

5.   *K*-REGIME OF BREAKDOWN

Thus the nature of the *N*-regime of boundary-layer transition had been revealed soon after its first detection in 1976. At the same time, advances in understanding the causes of the K-transition were somewhat lacking. Theories were unable to describe the main features of the K-breakdown; experiments posed more questions than they provided answers. It is clear now that the difficulties in this field were connected with the much more complicated nature of the K-regime of boundary-layer transition compared with the N-transition. Moreover, understanding the K-regime had to await the discovery and explanation of the N-breakdown in order that new resonant ideas (developed for the description of this more simple case) could be applied to the subsequent investigation of the more complicated regime corresponding to the K-breakdown.

5.1   Early Studies and Ideas

The structure of K-breakdown was intensively studied experimentally beginning in the 1950s (see Section 3.1). This type of transition was clearly shown to be characterized by downstream growth of a spanwise modulation of both mean flow and disturbance amplitude, and the formation of peaks and valleys in the spanwise direction (see Klebanoff et al 1962 and Figures 2 and 3). This process was also shown to be accompanied by amplification of streamwise counter-rotating vortices which consisted of two components: one steady, the other oscillating (with mainly the fundamental frequency disturbed by higher harmonics). Klebanoff et al (1962) found that the magnitudes of both these components were similar. This attribute resulted in the pulsating character of the streamwise eddies: a phase of active vortex rotation (with magnitudes of spanwise velocity up to 6–7%) would almost completely disappear, with subsequent repetition of these phases with the fundamental period. Further downstream the powerful flashes (spikes) (see Figure 4) appeared suddenly in the peak spanwise positions where the disturbance amplitudes reached their local maxima. Spike magnitude increased up to 30–40%. The phenomenon of spike multiplication (doubling, as in Figure 4; tripling; etc) was also observed by Klebanoff et al (1962) further downstream within each period of the fundamental wave. This process was assumed to correspond to the beginning of the final laminar-flow breakdown. Rapid amplification of the spikes and their multiplication was attributed (by Klebanoff et al 1962 and subsequent investigators) to an "explosive" high-frequency secondary instability of the flow connected with the formation of inflection points and high-shear layers in the mean and instantaneous profiles of the longitudinal flow velocity. The 3D structure of all three components of the velocity field, including the instantaneous velocity profiles and high-shear layer, was carefully studied experimentally by Kovasznay et al (1962).

There are two main questions associated with the physical description of initial stages of the K-regime of boundary-layer breakdown, which were formulated on the basis of experimental data obtained in the fundamental studies mentioned above. Firstly, what is the cause of such rapid amplification of the pulsating streamwise vortices and of the corresponding strong spanwise modulation of the mean flow and disturbance amplitudes? Secondly, why do the flashes (spikes) appear in the peak positions and at a definite distance from the wall (in the external part of the boundary layer)?

One more issue was connected with a mysterious spanwise periodicity of the peak positions (often called the *preferred spanwise period*) observed as early as 1959 in experiments by Klebanoff & Tidstrom (1959). This observation motivated Klebanoff et al (1962) (and most of the subsequent experimentalists and theoreticians) to simulate a natural (i.e. uncontrolled) preexisting spanwise variation of the flow in a periodic way. However, much later it was shown experimentally (see Borodulin & Kachanov 1990) that this spanwise periodicity is not an inherent property of the K-breakdown, at least for the stage prior to spike formation (see also Kachanov 1987 and Section 7.2).

Early explanations of the main features of the K-breakdown mentioned above used mainly local, in time and space, notions. In particular, Klebanoff et al (1962) supposed that downstream amplification of spanwise modulation of the disturbances occurs because of varying local flow properties along the spanwise direction. The onset of spikes was also attributed to a local (both in time and space) high-frequency secondary instability of the flow (see Betchov 1960, Klebanoff et al 1962, and Section 6.2). Notions of local deformation and self-forcing of vortices associated with boundary-layer disturbances (i.e. instability eddies rather than instability waves) were often invoked to explain the phenomena observed. Some of these (and other) notions are still in use, others proved to be incorrect and have been rejected, but the most rapid advancement in the understanding of the physical nature of the K-regime of boundary-layer breakdown (as well as the N-regime) has been achieved in the 1970s and 1980s along quite another path. This path involved spectral notions, which are local in frequency-wavenumber space but essentially nonlocal in physical space.

One of the first attempts to describe the nonlinear wave interaction, and to explain the process of streamwise vortex formation, was undertaken by Benney & Lin (1960). They found that steady longitudinal eddies could be generated from a nonlinear superposition of a flat fundamental wave with a pair of oblique waves of the same frequency. This theory was criticized later by Stuart (1960) and others mainly because it did not take into account the real dispersion of instability waves. The problem of amplification of the steady streamwise vortices in initial stages of the K-breakdown was studied intensively in many subsequent theoretical studies (see

Antar & Collins 1975, Nelson & Craik 1977, Nayfeh 1987a, Asai & Nishioka 1989, and others). Itoh (1987) investigated this problem within the framework of weak-nonlinear resonance theory. He found that the existence of 2D waves with finite amplitude can induce 3D distortion with spanwise periodicity of the mean-flow field. Later Hall & Smith (1988) showed that an amplification of the streamwise vortices can be explained by their interaction with oblique instability waves within the framework of the Görtler vortex equations. These equations have singular solutions of large amplitudes which can exist in the absence of wall curvature. Hall & Smith (1988) found that a self-sustained growth of both longitudinal vortices and 3D instability waves can occur at high enough amplitudes of both modes (a strongly nonlinear interaction). These results were developed further by Hall & Smith (1989a,b) and very recently by Stewart & Smith (1992) and Walton & Smith (1992). The great interest in vortex amplification mechanisms is connected with their important role in the K-breakdown at subsequent stages of the transition development (see below and Sections 5.2 and 6).

A very important step in the theoretical description of the K-breakdown was made at an early stage in the study by Craik (1971) (see also Craik 1985). He developed the idea of a resonant interaction of a flat wave with two oblique subharmonics (Raetz 1959) to describe the process of rapid amplification of the spanwise modulation of the fundamental wave observed by Klebanoff et al (1962) (see Figure 3). He supposed that a flat second harmonic $(2\omega_1, 0)$ of the fundamental wave $(\omega_1, 0)$ could play the role of a main (forcing) wave in its interaction with its 3D subharmonics $(\omega_1, \pm\beta_r)$ which have the same frequencies as the primary fundamental mode. This idea was developed futher by Nayfeh & Bozatli (1979b) who found that the four-wave resonant interaction of waves $(\omega_1, 0)$, $(2\omega_1, 0)$, and $(\omega_1, \pm\beta_r)$ could occur and could result in a rapid amplification of the "subharmonics" (i.e. the 3D waves with the fundamental frequency).

As for the spikes, the main early idea (proposed for the first time by Betchov 1960 and Klebanoff et al 1962) was that local high-frequency (inflectional) secondary instabilities of the primary nonstationary flow was responsible for their generation. The essence of this mechanism consists of amplification of a packet of high-frequency fluctuations under the influence of an unstable inflectional instantaneous velocity profile with high-shear layer (Figure 13) which is formed (locally in time) in the peak position by a low-frequency primary wave (mainly by the disturbed fundamental mode). This approach was developed in a great number of subsequent studies (see Greenspan & Benney 1963, Gertsenshtein 1969, Landahl 1972, Zhigulyov et al 1976, Zelman & Smorodsky 1991a, and others, and also the review of Nayfeh 1987c). Despite the fact that this

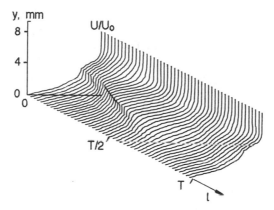

Figure 13 Low-frequency ($\omega \leqslant 3\omega_1$) instantaneous velocity profiles observed in peak position in the K-breakdown spike formation (Borodulin & Kachanov 1988).

was for twenty years the only explanation suggested for the spikes' onset, there were no direct experimental studies of its applicability to real boundary-layer transition (in the K-regime). The first such experiment was carried out in a plane channel flow (where almost the same K-type of transition is observed) by Nishioka et al (1980). These experiments confirmed the possibility of occurrence of a local secondary-instability mechanism (for more detail see Section 6.2).

Many problems in understanding the nature of the K-breakdown arose from the fact that the language of the analysis given by Klebanoff et al (1962), and in most subsequent experimental studies, concerned mainly local disturbance properties in time and space—a great difference from the wave-spectral notions developed in the theories in the 1960s through 1970s. This motivated a new set of experimental investigations of the K-breakdown in the 1980s. Simultaneously, new theoretical ideas and notions developed.

5.2 Recent Investigations

In 1980, in Novosibirsk, detailed experimental studies of the nature of the K-regime of boundary-layer transition were conducted under controlled conditions employing the technique of frequency and frequency-wavenumber complex Fourier analysis. The initial results, devoted to the deterministic stages of the K-breakdown, were presented by Kachanov et al (1984) (also published later in Kachanov et al 1989). Later results, devoted to flow randomization mechanisms, were published by Dryganets et al (1990). For the first time systematic information about the spectral (frequency and frequency-wavenumber) structure of the disturbances at nonlinear stages of their development was provided, thereby changing some traditional notions about the character of the K-breakdown.

In the experiments by Kachanov et al (1984, 1989) all typical manifestations of the K-regime found by Klebanoff et al (1962) and later investigators were reproduced, though some flow and disturbance parameters were somewhat changed (and varied within the experiments), and much new data were documented. One of the important conclusions of this work was that the spikes are not of stochastic (or irregular) character. Rather, they represent strictly deterministic, periodic structures (with conservative properties) generated by the flow as a result of gradual (not explosive) amplification of definite 3D harmonics of the fundamental wave and their phase synchronization.

Williams et al (1984) carried out an experimental study of early nonlinear stages of boundary-layer transition in a regime that probably corresponded to the K-breakdown. They investigated flat-plate flow in a water channel by means of a hot-film anemometer with a slant-film probe. This technique permitted them to carry out very detailed measurements of all three components of flow velocity and vorticity at the vortex loop formation stage (corresponding to the Λ-patterns mentioned above), which is observed in K-transition just before spike formation. Unfortunately the authors did not compare their results with previous hot-wire measurements, although their data complemented results of Klebanoff et al (1962), Kovasznay et al (1962), and others.

In the late 1970s through 1980s, Nishioka and colleagues obtained a set of interesting "microscopic" results for the K-breakdown in a plane channel flow. In particular, Nishioka et al (1980) studied experimentally the mechanism of local secondary instability and compared the results obtained with calculations within a linear approach. Asai & Nishioka (1989) investigated both numerically and experimentally the process of peak and valley formation. They found a threshold character for the K-transition in a plane channel flow (the same as in the boundary layer). It was shown that the critical value of the amplitude of a 2D fundamental wave necessary to initiate the K-breakdown, is approximately 1%.

On the basis of a detailed analysis of experimental data and results of weak-nonlinear theories a physical model of the disturbance development at deterministic stages of the K-regime of breakdown had been proposed by Kachanov (1987) (see also Kachanov 1990)—named the *wave-resonant concept* of K-breakdown. The causes of amplification of three-dimensionality and generation of spikes, and other characteristic features of the K-breakdown, were qualitatively or even quantitatively explained within the framework of this concept from the position of resonant interactions of definite 2D and 3D modes of frequency-wavenumber spectrum. This concept was substantiated by its indirect and direct (Borodulin & Kachanov 1989) comparison with experimental data (see Section 6.1).

The applicability of one more idea, namely the concept of local high-frequency secondary instability, to the explanation of the spike generation mechanism in the K-regime of boundary-layer breakdown was investigated in direct experiments by Borodulin & Kachanov (1988). These measurements corroborated, for the boundary-layer case, the conclusion of Nishioka et al (1980) about the occurence of the local instability mechanism in the K-breakdown at the stage of spike formation. But, at the same time, this concept was shown not to be responsible for the generation of the spikes themselves (see Section 6.2).

Dryganets et al (1990) studied experimentally later stages of the K-breakdown (after spike formation and multiplication) where flow randomization was observed. The amplification of random, quasi-stochastic disturbances in the K-regime was found to occur owing to a resonant interaction of deterministic waves with background disturbances. The process starts with the subharmonic resonance in the same way as in the N-regime of boundary-layer transition (see Section 6.5).

One more new approach for describing the K-regime (and maybe the developed turbulence) in the boundary layer is associated with discovery of coherent structures/solitons (CS-solitons). Soliton properties of spikes were found experimentally by Borodulin & Kachanov (1988) and later studied in more detail (Borodulin & Kachanov 1990, 1992) (see also Sections 6.3 and 6.4). CS-solitons are generated by the transitional flow in the K-regime in the peak position; they then move far downstream up to the region where nearly-developed turbulent flow is observed. Kachanov (1990, 1991) supposed that coherent structures, observed in the turbulent boundary layer, also represent the CS-solitons, or patterns close to them (see Section 6.6).

The development of all these (and other) ideas on the physical nature of the K-regime would have been impossible without rapid progress in theoretical studies and fruitful cooperation of theory and experiment. In the past decade, there were many theoretical studies devoted to this problem. Because our focus here is on experimental work, we only briefly review some of these theoretical studies and refer the reader to other recent reviews such as: Craik (1985), Zhigulyov & Tumin (1987), Nayfeh (1987a), Herbert (1988), Fasel (1990), Ryzhov (1990), Smith (1990), and Kleiser & Zang (1991) (see also Section 6).

Herbert (1985, 1988) developed a weak-nonlinear approach and Floquet theory for describing the K-breakdown of the boundary layer. He studied the *fundamental resonance* of modes $(\omega_1, 0)$, $(0, \pm \beta_r)$, and $(\omega_1, \pm \beta_r)$ as a main cause for initiating the K-breakdown. He found rather good (mainly qualitative) agreement with experiments by Klebanoff et al (1962) in amplification rates and y-profiles of rms disturbance amplitudes.

Numerical simulation of the experiments of Kachanov et al (1984, 1989) was carried out by Rist (1990) and Rist & Fasel (1991) on the basis of 3D Navier-Stokes equations. The results obtained showed excellent, qualitative and quantitative agreement with experimental data up to the stages of spike formation and multiplication. An example of this agreement is shown in Figure 14. Here two sets of amplitude and phase y-profiles for the fundamental wave and its harmonics (from $2\omega_1$ to $6\omega_1$) measured by Kachanov et al (1984, 1989) and calculated by Rist & Fasel (1991) are presented for the stage of K-breakdown where the spike starts to form. Except for a small difference in the y-scale connected with slightly bigger boundary-layer thickness observed in the experiment, agreement with calculations is almost complete. The same results are observed for the mean velocity profiles (Figure 14 *left*) and for numerous other distributions of various disturbance characteristics along all three spatial coordinates and time (for more detail see Rist 1990). Recently, Kloker & Bestek (1992) used direct numerical simulation to study the K-like regime of transition in a boundary with a strong adverse pressure gradient. The processes of disturbance development in the peak position were found to be rather similar to those observed in the case of zero pressure gradient. At the same time some essential differences were observed. In particular, the

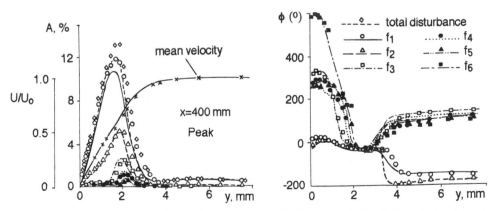

Figure 14 Comparison of y-profiles of mean velocity (*left*), amplitudes (*left*), and phases (*right*) of the fundamental wave; its harmonics and total disturbance intensity measured by Kachanov et al (1984, 1989) are shown as points; curves are from the calculations of Rist & Fasel (1991) for the K-transition. $x = 400$ mm, peak position.

disturbances developed rapidly in the former valleys (called the "co-peaks') with amplification rates higher than those found even in the peak positions. The formation of 3D high-shear layers was also observed there.

Rist & Fasel (1991) also conducted a comparative analysis of the role of the fundamental resonance and the Craik-Nayfeh-Bozatli resonance [between modes $(\omega_1, 0)$, $(2\omega_1, 0)$, and $(\omega_1, \pm\beta_r)$, used within the wave-resonant concept] in the process of amplification of the $(\omega_1, \pm\beta_r)$ mode. They found that both resonances played an important role in the beginning of the K-breakdown, and proposed to complement the wave-resonant concept (Kachanov 1987, 1990) by the fundamental resonance (see also Section 6.1). Good qualitative agreement with experimental observations was also obtained by other groups using direct numerical simulation (see Kleiser & Zang 1991).

Asymptotic theory (triple-deck, etc) represents one more theoretical approach which rapidly developed and provided a better description of nonlinear disturbance behavior in the K-breakdown (see, for example, Smith 1990). Collaboration of theoreticians and experimentalists in this field also gave very good results and led to the discovery of CS-solitons and description of some of their properties, at least in initial stages of their development (i.e. Kachanov & Ryzhov 1991, and Kachanov, Ryzhov & Smith 1993). It was found that the generation of coherent structures with very conservative properties (CS-solitons) is an inherent feature of the K-regime of boundary-layer transition (see Sections 6.3 and 6.4).

Critical threshold conditions for the beginning of the K-breakdown were studied by Rahmatullaev (1988) and Rahmatullaev & Shtern (1989) within the framework of the theory of bifurcations of self-oscillating regimes. Multi-mode regimes (including modes of fundamental and Craik-Nayfeh-Bozatli resonances) were studied. The threshold amplitudes of self-oscillations obtained were in quantitative agreement with experimental observations for both the K- and N-regimes of transition.

Thus, experimental and theoretical studies conducted during the past decade permitted us to extend our knowledge of the K-regime of transition and to explain some of the observed phenomena. A brief description of some important spectral features of the initial stages of K-breakdown is presented in the next section; the physical mechanisms of this type of transition is discussed in more detail in Sections 6.1–6.6.

5.3 Main Features of K-Breakdown from the Spectral Viewpoint

Kachanov et al (1984, 1989) have studied the K-breakdown from the spectral viewpoint. They found that that the disturbances remained deterministic and almost completely periodic until rather late stages of their development even in the peak positions. The disturbance spectra consisted solely of the fundamental wave and its harmonics up until the appearance of spikes and their multiplication. Thus all main spectral characteristics of these stages of the K-breakdown are attributed to amplitude and phase properties of the fundamental wave and its harmonics. As was shown by Klebanoff et al (1962), total disturbance intensity in the peak position has an explosive behavior when a probe moves downstream at constant distance from the wall. A rapid jump in the disturbance amplitude was observed just at the moment of spike onset (see the left panel of Figure 15). In practice, both the fundamental wave and its harmonics demonstrate the same behavior. However the explosive effect, as shown by Kachanov et al (1984, 1989), is connected with the evolution of wall-normal profiles of oscillations along the flow direction. The amplification curves of the harmonic amplitudes, determined at the maxima of their respective y-profiles (Figure 15 *right*), not only demonstrate no growth in the region of spike formation, but some of them actually decrease in this region. It was found that spike formation is not attributed to a jump in the disturbance amplitude, but to a synchronization of the harmonic phases in definite regions of space. Figure 16 (*top*) shows this synchronization in local wall-normal (*left*) and spanwise (*right*) distributions, observed in regions where

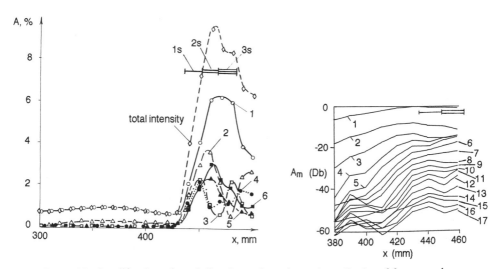

Figure 15 Amplification of total disturbance intensity and amplitudes of frequency harmonics ω_1, ω_2, ... ω_{17} (1, 2, ... 17) observed in the K-breakdown in peak position. (*Left*) along y = const. (Kachanov et al 1984). (*Right*) maximum amplitudes in y-profiles (Borodulin & Kachanov 1988). 1s, 2s, 3s: places of formation of 1st, 2nd, and 3rd spike.

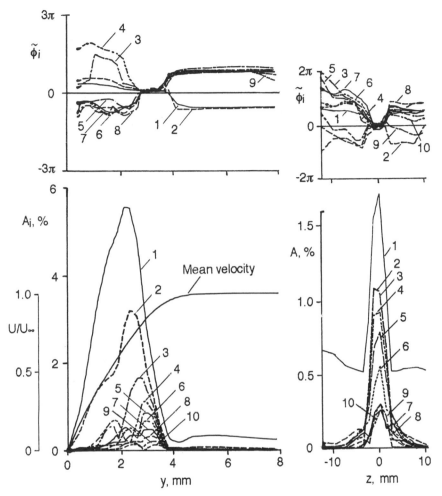

Figure 16 Profiles of harmonic amplitudes (*bottom*) and phases (*top*) measured along normal-to-wall (*left*) and spanwise (*right*) coordinates at the stage of developed spike (Borodulin & Kachanov 1992). 1, 2, ... 10: harmonics ω_1, ω_2, ... ω_{10}.

spikes were present in the oscilloscope traces. Corresponding profiles of the harmonic amplitudes (Figure 16 *bottom*) demonstrate local maxima in the spike position and strong spanwise localization of spikes in the peak neighborhood (*right*).

A frequency-wavenumber (ω, β) Fourier analysis conducted by Kachanov et al (1984, 1989) shows that, while the spikes are forming, the

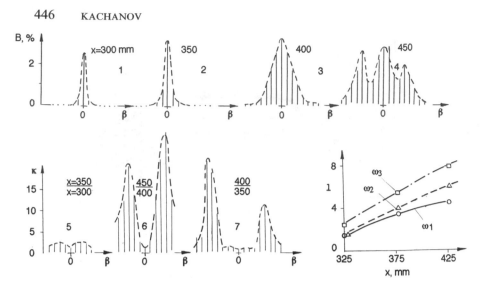

Figure 17 Downstream evolution of amplitudes *B* of frequency-wavenumber spectral modes (*top*) and their corresponding amplification factors (*bottom left*) for the fundamental frequency during spike formation. Numbers *l* of most amplified spectral modes ($n\omega_1$, $\pm l\beta_1$) as functions of ω and x (*bottom right*). 1, 2, 3, 4: $x = 300, 350, 400, 450$ mm; 5, 6, 7: $x = 325, 375, 425$ mm. $y = 1$ mm (Kachanov et al 1984).

amplitudes of three-dimensional modes of the frequency-wavenumber spectrum, having definite spanwise wavenumbers β, gradually amplify (Figure 17 *top*). Their amplification factors χ are 1 or 2 orders of magnitude greater than those for the plane waves (Figure 17 *bottom left*). Dependencies of the spanwise wavenumbers $\beta = l\beta_1$ (where β_1 is a fundamental wavenumber corresponding to a step of cellophane strips; see Figure 2) of the most amplified modes as functions of downstream coordinate x and frequency ω are also presented in Figure 17 (*bottom right*). Simultaneously, in the complex frequency-wavenumber spectra obtained at the y-coordinate of spike forming, a synchronization of the phases of amplified modes is observed (Figure 18).

These and other spectral features of the disturbance behavior in the K-breakdown needed explanation and interpretation.

6.   PHYSICAL NATURE OF THE K-BREAKDOWN

6.1   *Wave-Resonant Concept*

The *wave-resonant* (WR) concept has been developed by Kachanov (1987) on the basis of a detailed analysis of the experiments by Kachanov & Levchenko (1982) and Kachanov et al (1984, 1989) as well as of theoretical

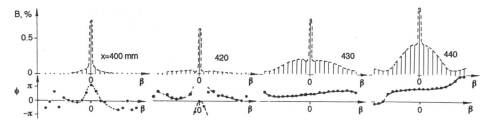

Figure 18   Evolution of frequency-wavenumber complex spectra at $y = y_s$. Synchronization of phases of frequency-wavenumber harmonics (for the fundamental frequency) during the process of spike formation (Kachanov et al 1984). 1, 2, 3, 4: x = 400, 420, 430, 440 mm.

results by Craik (1971), Zelman & Maslennikova (1982), Nayfeh & Bozatli (1979b), Herbert (1984), and other works.

The essence of the WR concept is explained in brief by Figure 19. The idea proposed by Craik (1971) and developed by Nayfeh & Bozatli (1979b)

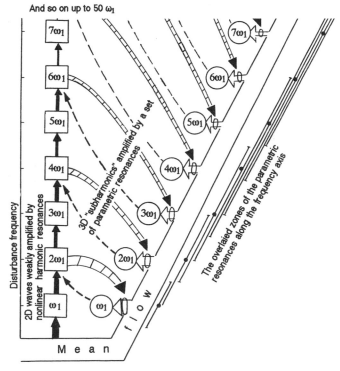

Figure 19   Sketch of the wave-resonant concept of *K*-breakdown (after Kachanov 1987, 1990).

(and in subsequent works) was taken as the basis of this concept. It consists of the following. The amplification of the 3D oscillations with the frequency of the fundamental wave ω_1 of type $(\omega_1, \pm\beta_r)$ could be initiated by a four-wave resonant interaction of waves $(\omega_1, 0)$, $(2\omega_1, 0)$, and $(\omega_1, \pm\beta_r)$. The WR concept proposes the generalization of this idea for the case of a great number of waves of type $(n\omega_1, 0)$ and $(m\omega_1, \pm l\beta_1)$, where $n = 2, 3, 4, 5, \ldots, m \approx n/2$ and $l\beta_1 \approx \beta_r$ are spanwise wavenumbers of the 3D priming disturbances which originated within the ranges of the resonant amplification; β_1 is the characteristic wavenumber of an initial spatial inhomogeneity.

It is supposed that these priming disturbances arise mainly as the result of a very weak (nonresonant) combination interaction of initial disturbances: primary wave $(\omega_1, 0)$ and small primary modulation of the mean flow $(0, \beta_1)$. A more detailed description of the WR concept is presented by Kachanov (1987).

As mentioned above (see Section 5.2), it is necessary to complement the WR concept by fundamental resonance, i.e. the resonant interaction between modes $(\omega_1, 0)$, $(\omega_1, \pm\beta_r)$, and $(0, \pm\beta_r)$ (see Rist & Fasel 1991). However, most likely, this resonance can strongly influence the amplification of fundamental frequency modes only, not the high-frequency harmonics, especially those with large spanwise wavenumbers. Moreover, the influence of steady vortices $(0, \pm\beta_r)$ is probably restricted by initial stages of the disturbance development. In particular, Rist & Fasel (1991; Figure 10)) show that at late stage ($x \approx 415$ mm) the Craik-Nayfeh-Bozatli ("subharmonic") resonance provides a better correlation with direct numerical simulation than either the fundamental resonance or a combined one.

It should be noted that the nonlinearity in subharmonic resonances (see Zelman & Maslennikova 1982, Spalart & Yang 1987, Herbert 1988) can result in an explosive amplification both of the fundamental wave and of the subharmonics; their phase synchronism is maintained (Kachanov & Levchenko 1982, Zelman & Maslennikova 1982) providing an additional mechanism for generating flat harmonics of the fundamental wave synchronized in phase. In Figure 19 these nonlinear connections are shown by dotted arrows (after Herbert 1988). As a result of the nonlinearities, when the amplitudes of the subharmonics $(m\omega_1, \pm l\beta_1)$ grow, a set of high-frequency waves can probably detach from thier base (the disturbances with the fundamental frequency ω_1) and exist separately.

Kachanov (1987) showed that if the cascade of resonances mentioned above occurs, this would inevitably result in the appearance of spikes, typical for the K-breakdown, located in the exact places of time and space (along y and z) observed experimentally. Most of the experimental results obtained by Klebanoff et al (1962), Kachanov et al (1984, 1989), and others could now be qualitatively or even quantitatively explained within the framework of the wave-resonant concept. However, direct evidence for the existence of the resonance cascade, postulated by the WR concept, was not obtained at that time (1986–87).

The evidence came later when Borodulin & Kachanov (1989) (see also Borodulin et al 1989) conducted their special experiments aimed at determining the correspondence between resonant-synchronism conditions and amplification rates for different modes of the frequency-wavenumber spectrum observed at the stage of spike formation. It was shown in these experiments (see also Kachanov 1990, 1991) that coincidence or closeness of the downstream phase velocities of the "subharmonics" ($n\omega_1$, $\pm l\beta_1$) with those of the corresponding "fundamental" (forcing) flat waves ($m\omega_1$, 0) (where $n \approx m/2$) were observed just in those regions of spanwise wavenumber ($\beta = l\beta_1$) where rapid growth of "subharmonics" took place; and vice versa: a rapid amplification of the 3D "subharmonics" ($n\omega_1$, $\pm l\beta_1$) (with increments of an order greater than linear) was observed just within those regions of the spanwise wavenumber [$\beta_1 \approx \pm(5-7)\beta_1$ for $n = 1, 2$] where the conditions of the resonant phase synchronism were satisfied exactly or approximately. It should be noted that the phase synchronism in the subharmonic triads is not only the condition for onset of the resonant amplification of the subharmonics, but also its effect, because of the phase locking phenomenon (observed by Kachanov & Levchenko 1982; see also Figure 7).

Thus, the results obtained by Borodulin & Kachanov (1989) showed that the system (cascade) of the parametric subharmonic resonances, postulated within the framework of the WR concept (see above and Figure 19), is actually observed in the K-regime of breakdown at the stage of spike formation. Very recent theoretical studies by Savenkov (1992, 1993) contain additional evidence supporting this conclusion. Savenkov used asymptotic theory to describe the development of linear and nonlinear disturbances in inlet pipe and channel flows. In particular, he found that when synchronism conditions for harmonic and subharmonic resonances are satisfied, both rapid (singular) amplification of disturbance magnitude and spike formation are observed in these flows at specific distances from the leading edge and only for certain instability wave modes.

6.2   Local Secondary Instability

For almost thirty years the concept of local high-frequency (inflectional) secondary (LHS) flow instability was the most widespread explanation for the cause of spike generation in the K-regime of breakdown (see for example Betchov 1960, Greenspan & Benney 1963, Gertsenshtein 1969, Landahl 1972, Zhigulyov et al 1976, Itoh 1981, Zelman & Smorodsky 1991a, and others, and for review Nayfeh 1987a). The essence of this concept was briefly described in Section 5.1. As mentioned above, the first direct experiment devoted to the corroboration of the LHS-instability mechanism was carried out in a flat channel flow by Nishioka et al (1980). The possibility of its occurence in the K-regime of breakdown had been confirmed. However, the question of whether this mechanism was responsible for the spikes' generation, had not been answered. [Later, Borodulin & Kachanov (1988) showed that it was not.]

Figure 20 (*left*) (taken from Borodulin & Kachanov 1988) shows a family of oscilloscope traces obtained at different distances from the wall in the region of a spike forming. The corresponding instantaneous low-frequency velocity profiles, superposed of the mean velocity profile and that of the first frequency harmonics (ω_1, $2\omega_1$, and $3\omega_1$) are shown in

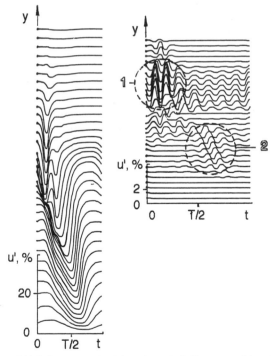

Figure 20 Sets of high-frequency (*right*) and total (*left*) traces with spikes measured at various distances from the wall (Borodulin & Kachanov 1988). 1: "upper" packets, 2: "lower" packets.

Figure 13. These inflectional profiles with high-shear layer are considered as causing the spikes, within the framework of the LHS-instability concept. The filtered high-frequency traces (obtained from the total signal shown in Figure 20 *left*), composed of harmonics higher than seventh, are shown in Figure 20 (*right*). Two groups of high-frequency secondary fluctuations are seen. The *upper packets*, observed far from the wall, correspond to the spikes (see Figure 20 *left*). Close to the wall other packets having smaller amplitudes (*lower packets*) were found (Figure 20 *right*).

It turned out (Figure 21 *left*) that the downstream group velocity of the lower packets (2) is practically equal to the velocity of the high-shear layer (3) in the low-frequency instantaneous profiles shown in Figure 13 (Landahl's 1972 criterion). Their phase was found to nearly coincide with that of the high-shear layer (Figure 21 *right*) and this coincidence was observed locally for all distances from the wall, i.e. the centers of the lower packets were constantly located within the region of high-shear and moved with that layer towards the wall with time. At the same time, all these and other characteristics of the upper packets (i.e. spikes) were seen to differ considerably from those of both lower packets and high-shear layer (Figure 21, curves 1).

Figure 15 (*right*) shows amplification curves of the fundamental wave and its harmonics, measured for the maxima in their y-profiles and corresponding to the upper packets, i.e. spikes (see Figure 16 *bottom left*). The curves demonstrate a wonderfully strict hierarchy of frequency harmonics. It was found that at the stages just before the first spike is formed, the amplitudes of frequency harmonics (within the range $3\omega_1$ to $17\omega_1$) decrease with frequency in exact accordance with a geometric progression with a factor from 0.4 to 0.8. None of the hillocks in the spectrum predicted by LHS-instability theories (for the unstable high-frequency secondary disturbances) were observed.

These and a number of other results obtained by Borodulin & Kachanov (1988) led them to the two main conclusions: (*a*) the mechanism of the LHS flow instability actually takes place in the K-regime of the boundary-layer transition (the same as found by Nishioka et al 1980 for the channel flow); however (*b*) the generation of the spikes is not directly connected with this mechanism. A very similar conclusion was later drawn by Zelman & Smorodsky (1991a) on the basis of a theoretical analysis of the influence of a shear layer (in the mean velocity profile) on the linear development of waves and wave packets. Of course these conclusions do not rule out the LHS instability's role in the breakdown process. It probably becomes important at later stages of the K-breakdown of a flat-plate boundary layer, or in some other boundary layers—for example over a concave wall (Swearingen & Blackwelder 1987) or a swept wing, where the local

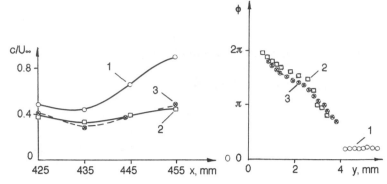

Figure 21 Comparison of downstream velocity (*left*) and characteristic phase (*right*) of upper (1) and lower (2) wave-packets with those of the high-shear layer (3) in the region of spike formation (Borodulin & Kachanov 1988).

inflectional profiles are generated by intensive steady vortices amplified by a primary instability.

6.3   *Identification of Flashes (Spikes) with Solitons*

Beginning with the pioneering work of Klebanoff et al (1962), generation of spikes in the K-regime of transition and their subsequent multiplication were usually attributed to the start of a stormy breakdown and randomization of the laminar flow. However, it was clearly shown by Kachanov et al (1984, 1989) and Borodulin & Kachanov (1988) that the multiplication of the spikes is of deterministic character and observed only at downstream displacement along the line y = constant. At the same time the spike by itself quickly moves away from the wall during its formation and then acquires a rather conservative form which remains essentially unchanged further downstream.

This property of the spikes is visually shown in Figure 22 obtained in experiments by Borodulin & Kachanov (1990) along a line $y = y_s$, where y_s is the position of the spike maximum. The traces are displayed for streamwise coordinate x from 400 to 595 mm. The last position corresponds to nearly-turbulent flow inside the boundary layer. Borodulin & Kachanov (1988) drew attention to the feature that "...despite the existence of the rather strong dispersion of the instability waves ... the spikes do not disperse but, on the contrary, gather in narrow flashes and propagate steadily downstream within the boundary layer almost without change of their form and amplitude...." On the basis of these and other observations they conclude that: "It is highly probable that the behaviour of

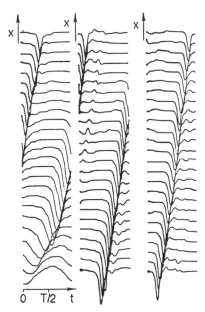

Figure 22   Formation and development of a spike-soliton in the *K*-breakdown measured by Borodulin & Kachanov (1990). Downstream coordinate grows (from bottom to top and from left to right) from $x = 400$ to 595 mm. $y = y_s$.

the spikes … can be described within the framework of a theory of solitons…," and the hypothesis was proposed that "….the spikes observed in the *K*-regime of transition, and described within the framework of the wave-resonant concept as weak-nonlinear wave packets, can be considered also as solitons…"

Independently, a mathematical basis for a possible explanation of the phenomenon discussed above was proposed. It was shown within the framework of asymptotic theory (Zhuk & Ryzhov 1982, Smith & Burggraf 1985) that the development of essentially nonlinear disturbances in the boundary layer may be described by the Benjamin-Ono equation which has solutions with soliton properties. Although these theoretical works laid the foundation for the following mathematical description of fully nonlinear fluctuations in the *K*-regime of boundary-layer transition, they did not contain a direct link to the nature of this regime and, in particular, the spikes. The first attempt to apply the soliton solutions of the Benjamin-Ono equation to the experimental investigation of large-amplitude disturbances was undertaken by Rothmayer & Smith (1987). The choice of a one-parameter family of solitons restricted the application to an analysis of alternative processes of laminar-turbulent transition, rather than to an investigation of spike properties in the K-regime. Another approach was used by Smith & Burggraf (1985), Conlisk et al (1987), and Ryzhov & Savenkov (1989), who calculated the development of 2D nonlinear wave packets within the framework of the theory of interacting boundary layers with a triple-deck structure. The results of the latter study correlated well, qualitatively, with the experimental data mentioned above, and tended to indicate soliton behavior of the central cycles of oscillation within the packet when their amplitude became sufficiently large. It was concluded that the eventual localization of high-frequency disturbances ultimately led to the soliton formation observed experimentally. Subsequent unification of the efforts of experimentalists and mathematicians permitted advances in the soliton approach and substantiation of the soliton nature of the spikes.

Why is the asymptotic approach so important for subsequent studies of the transition phenomenon? As mentioned above (see Sections 5.3, 6.1, and 6.2), the initial stages of K-breakdown can be described qualitatively, or even quantitatively, within a relatively simple weak-nonlinear approach (e. g. by the wave-resonant concept; Section 6.1) or, partially, even within linear notions (e. g. by local high-frequency secondary instability; Section 6.2). However, it is clear that the meaning of the weak-nonlinear notions tends to zero when the number of frequency harmonics tend to infinity and their amplitudes become close to each other. In place of this, a mathematical description of the essentially nonlinear behavior of the solitons (spikes) at subsequent stages of the transition can be achieved within the framework of a fully nonlinear theoretical model. The asymptotic theory mentioned above is just one of those approaches.

On the basis of theoretical studies listed above, Ryzhov (1990) analyzed a three-parameter soliton solution of the Benjamin-Ono equation and qualitatively compared its properties with those of experimentally-observed spikes. He found that this solution had a form very close to that of the spike and that its evolution with soliton amplitude tended to be the same. The spectrum of these Benjamin-Ono solutions was also very close to that observed experimentally: The amplitudes of harmonics decreased with frequency in accordance with a geometric progression (see Kachanov 1987) and the phases of wave valleys coincided with each other (i.e. were synchronized) (see Kachanov et al 1984, 1989).

The next step was taken by Kachanov & Ryzhov (1991) and Kachanov, Ryzhov & Smith (1993) when the properties of spikes were directly compared quantitatively with those of the Benjamin-Ono solitons. The form of the calculated solitons is shown in Figure 23 for a number of different values of an amplitude parameter Δ_i. Figure 24 compares four different

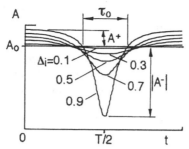

Figure 23 Form of a three-parameter soliton solution of the Benjamin-Ono equation for various values of amplitude parameter Δ_i considered by Ryzhov (1990) for describing spikes in the K-breakdown regime.

form parameters of spikes (see Figure 23) in theory and experiment as functions of the semi-swing of the oscillation A_m (as a measure of a soliton magnitude). Rather good qualitative correlation of all these (and other studied) form parameters is observed at the initial stage of the spike formation. The best agreement is seen for maximum positive (A^+) and negative ($|A^-|$) deviations of streamwise velocity disturbance within the spike-soliton (Figure 24 *left*) as well as for the soliton width τ_0 determined on the level of the mean deviation A_0 (Figure 24 *middle*). This result represents a unique case in which an *analytic* solution of essentially non-linear theory quantitatively describes the behavior of flow disturbances observed in a transition experiment! Form parameter γ_1 represents an asymmetry coefficient $(|A^-|-|A^+|)/(A^+ + A^-)$; q is the geometric pro-

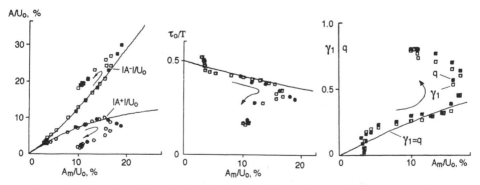

Figure 24 Experimental (*points*) and theoretical (*curves*) dependencies on magnitude $A_m = (A^+ - A^-)/2$ for different soliton form parameters (Kachanov & Ryzhov 1991; Kachanov, Ryzhov & Smith 1992). Arrows show downstream direction in the experiments.

gression factor mentioned above (Figure 24 *right*). The parameter γ_1 was shown to be exactly equal to q for the solitons of the Benjamin-Ono equation. In experiment they are also very close to each other and to the theoretical values (at a particular stage of the spike-soliton development). Arrows in Figure 24 show the downstream direction in the experiment. Rapid deviation of experimental points from the theoretical curves is, most probably, connected with 3D effects which become essential in the physical flow when the spike passes a certain stage of its development. The 2D theory, of course, can not describe this deviation. It is remarkable that the stage which is well described by theory extends in the experiment (Borodulin & Kachanov 1988) to a downstream position where the oscillation amplitude reaches values 25–30% which are of the same order as the maximal ones!

From the viewpoint of the WR concept (Section 6.1) the essentially three-dimensional character of spikes, observed in boundary-layer flow especially at late stages of their development, is closely connected with the specific dispersion characteristics of 2D and 3D instability waves in the flat-plate boundary layer. However, there exist other flows where unstable disturbances have no dispersion for 2D instability waves. One of these situations was studied by Savenkov (1992, 1993). He found that the formation and amplification of high-amplitude spikes consisted of two-dimensional, symmetric (for inlet channel flow) and axisymmetric (for inlet pipe flow) modes. Savenkov (1993) showed that these singular, rapidly growing solutions with spikes can transform into solitons when disturbances approach the second branch of the neutral stability curve. These soliton solutions were studied both numerically and asymptotically within long-wave and short-wave approaches. In the long-wave case the governing equation was reduced to the Korteweg-de Vries equation; in the short-wave case, to the Benjamin-Ono equation. In both asymptotics the soliton solutions—which had a spike-like form and represented concentrated vortices—were found and investigated.

Thus, combined experimental and theoretical studies, carried out during the past decade, demonstrate the soliton nature of coherent structures/spikes, at least for the initial (quasi-2D) stages of the spike development. The generation of *CS*-solitons is an inherent property of the *K*-transition of the boundary-layer (and, probably, flat-channel flow).

6.4   CS-*Solitons and Late Stages of* K-*Breakdown*

The most important feature of *CS*-solitons at late stages of their development is connected with the very essential role played by three-dimensional effects. Until recently there were no 3D theories (except for direct numerical simulation) that could describe this region of the transition.

Initial attempts at generalizing nonlinear asymptotic theory to 3D solitons propagating in the boundary layer were unsuccessful. However, recently a two-dimensional analog of the Benjamin-Ono equation, which is probably suitable for describing 3D CS-solitons, has been obtained by Shrira (1989) for a geophysical application (internal waves in a boundary layer near the ocean surface). The first qualitative comparison of soliton solutions of this equation with properties of 3D CS-solitons (Abramyan et al 1992) is rather hopeful. In this section we present some specific properties of the CS-solitons, observed at late stages of their development by Borodulin & Kachanov (1990, 1992).

A general 3D form of the spike-soliton is seen visually in Figure 25 (*left*) obtained at the stage where its formation is nearly complete ($x = 500$ mm). The downstream evolution of its typical normal-to-wall coordinate y_s/δ (determined at points y_s where the spike magnitude is maximal) is shown in Figure 26 (*left*). The spike moves quickly away from the wall and then propagates along the external edge of the boundary layer. Its downstream velocity c_s also increases almost to that of the free-stream (Figure 26 *right*). Both values saturate approximately near the point $x = 500$ mm. The downstream evolution of the soliton form is illustrated by Figures 27 and

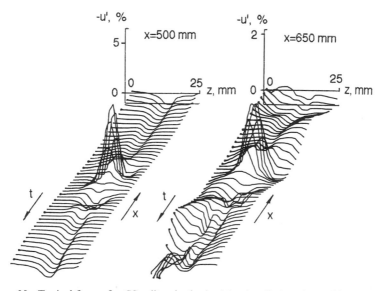

Figure 25 Typical form of a CS-soliton in the (t, z) [or (x, z)] plane (sets of instantaneous spanwise profiles of streamwise velocity disturbance) measured at $y = y_s$ at two downstream positions.

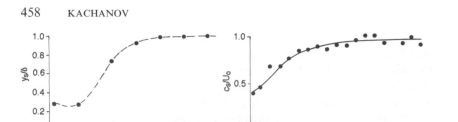

Figure 26   Downstream dependence of the y-position of soliton maximum y_s (*left*) and its downstream velocity c_s (*right*) measured by Borodulin & Kachanov (1990).

28. The spike magnitude $A_s = 2A_m$ (Figure 27 *left*) is stabilized past $x \approx 500$ mm and decreases slowly further downstream. The geometric progression factor q undergoes a rapid increase initially (Figure 27 *right*; see also Figure 24 *right*) but then also saturates (again at $x \approx 500$ mm) and then starts to decrease very slowly.   Figure 28 shows the downstream evolution of the spike-soliton width in time τ_1 and in space h_s (along the spanwise direction), determined at the level of a semi-swing of the oscillation. It is seen that τ_1 decreases rapidly during soliton formation but again becomes almost constant past $x \approx 480$ mm and has a very small value of about 0.08 of the fundamental period. Almost the same behavior is observed for the spanwise width of the spike h_s (curve 1). The formation of the *CS*-soliton is accompanied by its fast temporal and spanwise localization ("self-focusing"), which is connected with the rapid amplification of high-frequency 3D spectral models. The typical angle of the soliton spanwise dispersion at $x \geqslant 500$ mm is only about $0.2°$, i.e. 45 times (!) smaller than the corresponding angle for a linear wave packet (of small amplitude, produced by a point source) measured by Gilyov et al (1981, 1983) for the same free-stream speed and fundamental frequency (line 2 in Figure 28).

This very conservative behavior of spike properties is observed not only in physical space but also in Fourier space. Although Figure 16 shows

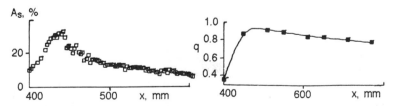

Figure 27   Downstream evolution of spike magnitude A_s (*left*) and geometric progression factor q (*right*) measured by Borodulin & Kachanov (1990). $y = y_s$.

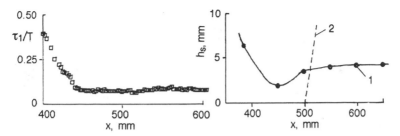

Figure 28 Downstream evolution of the temporal width of spike-soliton τ_1 (*left*) and its spanwise spatial width h_s (*right*, 1) (measured by Borodulin & Kachanov 1990); 2: angle of spanwise dispersion for corresponding linear wave-packet (Gilyov et al 1981, 1983).

distributions of amplitudes and phases of frequency harmonics observed at initial stages of spike formation, all the main properties of these disturbances remain practically the same further downstream up to very late stages in the breakdown process. In particular the harmonic amplitudes have very sharp peaks in spanwise distributions and exhibit local maxima in normal-to-wall profiles in the location where the spike is situated. The wave valleys of the harmonic phases are also synchronized in regions near the amplitude maxima.

Typical properties of a frequency-wavenumber spectrum (ω, β) (where β is a spanwise wavenumber) of the spike-soliton are illustrated by Figure 29 (*left*) where isolines of amplitudes (*bottom*) and phases (*top*) of spectral modes are shown. It is seen that the amplitudes of frequency-wavenumber modes decrease gradually with both frequency and absolute value of wavenumber. The phase part of the spectrum is presented by absolute values of the harmonic phases, equal to the phase detunings $|\Delta\phi|$ between the phase of the spike and the corresponding phases of the valleys of various frequency-wavenumber harmonics. Rather a deep "pit" observed in the center of the phase spectrum (Figure 29 *top left*) testifies to a strong phase coupling of almost every mode with a more or less noticeable amplitude (*bottom*).

Thus, at late stages of development the *CS*-soliton represents a coherent structure which is localized in time and space, and consists of a wide spectrum of strongly coupled frequency and frequency-wavenumber harmonics synchronized by their valleys.

It is interesting to note that all these (and other) main features of the *CS*-solitons remain practically unchanged further downstream, even during very late stages of the *K*-transition when the boundary layer becomes almost fully turbulent. This fact is illustrated by Figures 25 and 29: Figure 25 (*right*) demonstrates the 3D form of the spike-soliton

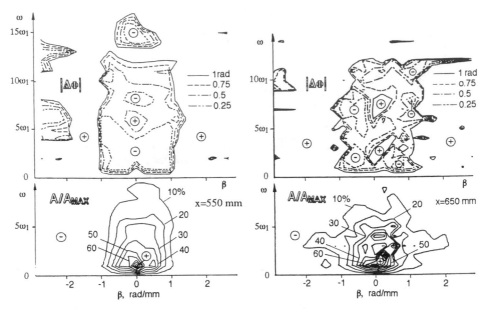

Figure 29 Isolines of the amplitudes of frequency-wavenumber harmonics A/A_{max} (*bottom*) and absolute values of phase detunings of their valleys with respect to spike phase $|\Delta\phi|$ (*top*) measured by Borodulin & Kachanov (1992) at $x = 550$ mm (*left*) and 650 mm (*right*). $y = y_s$.

observed at a very late stage of transition ($x = 650$ mm); Figure 29 (*right*) shows the corresponding form of the frequency-wavenumber spectrum. Comparison of these graphs with those obtained for the stage where the soliton has just formed ($x = 500$ mm in Figure 25 *left* and $x = 550$ mm in Figure 29 *left*) visually demonstrates the wonderfully strong conservatism of the *CS*-soliton properties mentioned above.

It is also interesting to look at the normal-to-wall structure of the disturbances observed at the peak position during late stages of the *K*-breakdown. A typical set of oscilloscope traces measured at various distances from the wall is shown in Figure 30. Two main features are seen: 1. Traces with single, double, etc spikes are observed in the external part of the boundary layer. 2. Closer to the wall the traces acquire a form of a "saw" with tooth-like jumps of the instantaneous streamwise velocity. The upper set of traces (with spikes) was shown to have very conservative properties. Traces with multiple spikes always appear together with a single spike and lie immediately under it, slightly closer to the wall. These spikes move downstream as an ensemble with practically the same speed

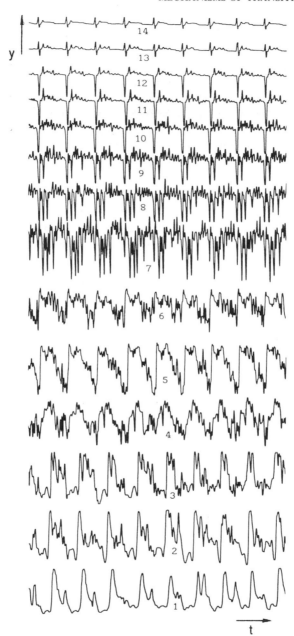

Figure 30 Oscilloscope traces of velocity disturbances at various distances from the wall obtained by Borodulin & Kachanov (1990) at $x = 550$ mm (developed 3D CS-soliton). 1, 2, ... 14 for $y/\delta = 0.026, 0.073, 0.119, 0.179, 0.3, 0.429, 0.776, 0.857, 0.951, 1.029, 1.12, 1.2, 1.31, 1.421.$ (Traces 1–6 have been reduced in vertical scale by a factor of 2.5.)

as the free-stream velocity. Most likely, all upper traces with spikes (from trace 7 to 14 in Figure 30) correspond to the same coherent structure CS-soliton, which thus has a fundamentally more complicated form than merely a single spike. (Solitons of such forms, with 2 or more humps are often observed in other flows; see for example Demyohin & Shkadov 1986.) However, it is not impossible for each spike to represent an independent (or almost independent) soliton of a relatively simple form (a single spike). In this case four (or more) solitons can be generated by a flow within each period of the fundamental wave in the K-breakdown. These simple spikes can interact and be coupled to each other. It is difficult to answer these questions without developing a theory that could describe these late (essentially 3D) stages of the formation and propagation of CS-solitons.

The bifurcation of disturbances into two types (lower and upper ones), mentioned in Section 6.2, is probably maintained at late stages. The upper disturbances are attributed to the spikes-solitons; the lower ones (observed in Figure 30 as the tooth-like jumps) are closely connected with a high-shear layer. This connection is visually illustrated by Figure 31 where the isolines of $\partial \hat{u}/\partial y$ (close to the instantaneous spanwise vorticity of the disturbances) are presented in the (y, t) [or (y, x)] plane for two consecutive downstream positions ($x = 500$ and 550 mm). Large points correspond to the positions of tooth-like structures (when the instantaneous streamwise velocity crosses the level of the mean velocity). It is seen that these structures coincide with a high-shear layer where the instantaneous spanwise vorticity has its maximum values. Small points mark the positions where first, second, third, and fourth spikes are observed (or can be distinguished) in the oscilloscope traces. Initially the spokes start to form from the region lying just under the high-shear layer (as seen from the third and fourth spike at $x = 500$ mm). But then, during their formation, they move quickly away from the wall and start to propagate independently of it with a speed that is almost twice as high as that of the high-shear layer (or tooth-like structures).

Figure 31 demonstrates one more interesting fact: Spikes are always observed in regions between pairs of counter-rotating vortices (the negative values of $\partial \hat{u}/\partial y$ are shown in Figure 31 by dotted lines). Most likely each spike has the form of a toroidal-like vortex with an axis of symmetry directed along the x-coordinate; the vortex pairs, then, represent sections of these toroidal vortices intersected by the x-y plane in the peak position. Strong rotation of these vortices result in large local negative values of the central streamwise flow velocity which appear as spikes in an oscilloscope trace when the vortices fly past a fixed hot-wire probe.

Thus, experimental studies (which correlate very well with numerical experiments by Rist 1990, 1992, and private communication) show that

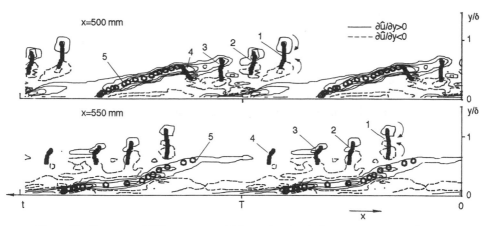

Figure 31 Contours of $\partial \hat{u}/\partial y$ = const. which are close to the instantaneous spanwise vorticity $\hat{\omega}_z$ (for disturbances only, without mean flow) measured by Borodulin & Kachanov (1992) in the (y, t) plane at $x = 500$ mm (*top*) and 550 mm (*bottom*). 1, 2, 3, 4: positions where 1st, 2nd, 3rd, and 4th spikes are observed in traces; 5: positions of tooth-like structures.

the CS-solitons at 3D stages of their development represent very localized (toroidal, most probably) vortices, observed in oscilloscope traces as a set of spikes. The spikes move downstream together within the external part of the boundary layer with the same (almost free-stream) velocity and remain essentially unchanged in form and spectral content for a distance of more then one hundred boundary-layer displacement thicknesses. Some of their properties are also in qualitative agreement with results of asymptotic theory [see, for example, Smith et al (1990), and Kachanov, Ryzhov & Smith (1993), and theory by Shrira (1989) and Abramyan et al (1992)].

6.5   *Flow Randomization in* K-*Breakdown*

Randomization in the K-regime of transition had originally been thought to begin with a "sudden, explosive" appearance of spikes. In light of the results presented above (see Sections 5.3, 6.1–6.4) this explanation can no longer be regarded as satisfactory. How does the randomization really occur? This problem was studied experimentally by Dryganets et al (1990). Some of the main results of this investigation are described below.

In Figure 32 families of traces of velocity fluctuations observed in the peak, under the spikes near the wall ($y = 1$ mm, $y/\delta \approx 0.2$), are shown for three subsequent downstream positions. (In particular the typical "saw" pattern of traces, mentioned in Section 6.4, is observed at $x = 510$ mm.) The periodic flow appearing at the spike-soliton formation stage ($x \approx 350$–500 mm) is seen to become gradually distorted by nonperiodic, random additions. These take the form of *stochastic antinodes* or *swellings* which have some unusual properties. For example, they are observed only in a statistical sense and are otherwise unobservable on any solitary trace. The stochastic antinodes do not represent turbulent regions typical for intermittence (which are not observed at all in either the K- or N-breakdowns of an initially periodic instability wave; see also Section 7.2). They do not contain fluctuations with frequencies higher than those in the initial periodic flow, and rather resemble the branching of a river delta.

What causes the amplification of these nonperiodic fluctuations? Kachanov (1985) proposed that they were caused by the same processes found in the N-regime of breakdown (see Section 4). Subsequent analysis confirms, in the main, this supposition. Typical behavior of quasi-subharmonic continuous-spectrum fluctuations, observed at the amplification stage of the random antinodes, are shown in Figure 33 (*top left*) together with the corresponding points of the phase trajectory in the plane of complex amplitude of this disturbance (*bottom left*). It is seen that at this stage the low-frequency disturbances represent oscillations with almost exactly subharmonic frequency, which have a random temporal amplitude modulation and almost constant phase, changing by π when the amplitude becomes zero. Such behavior is observed in the N-regime of breakdown (see Section 4.2 and Figure 7) and corresponds to amplification of quasi-subharmonic background disturbances under the action of a parametric resonant interaction with the fundamental wave. Dryganets et al (1990) carried out additional studies of the properties of the quasi-subharmonic fluctuations and fundamental wave, including their phase synchronism. They concluded that the amplification of these low-frequency continuous-spectrum disturbances is actually initiated with the same mechanism of subharmonic parametric resonance as in the N-regime of transition (for more details, see Kachanov 1991).

However, in contrast to the N-regime case, a great number of random

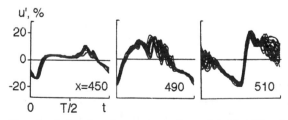

Figure 32 Families of velocity disturbance traces observed in the K-breakdown in peak position near the wall ($y/\delta \approx 0.2$). Formation of "random antinodes". (After Dryganets et al 1990.)

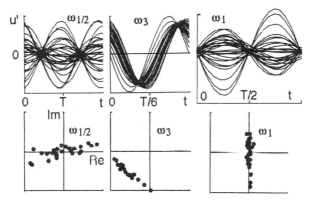

Figure 33 Typical behavior of continuous-spectrum quasi-random fluctuations observed at the stage of amplification of random antinodes near frequencies $\omega_1/2$ (*left*), $3\omega_1$ (*middle*), and ω_1 (*right*). Filtered traces (*top*) and corresponding points of phase trajectories (*bottom*) were measured by Dryganets et al (1990).

harmonics of the subharmonic wave ($m\omega_1/2$, with both odd and even m) are observed in the K-breakdown. Figure 33 illustrates this fact for fluctuations near frequencies ω_3 and ω_1. Figure 33 (*middle*) shows a family of traces (*top*) and points of phase trajectory (*bottom*) for total disturbances near the frequency ω_3. Figure 33 (*right*) corresponds only to random (continuous-spectrum) fluctuations near the frequency ω_1 without the presence of a deterministic fundamental wave (which has been subtracted from the total signal). Specific phase properties of these quasi-random disturbances (as well as the phase synchronism between those near ω_1 and deterministic wave ω_2) could most probably be explained in one of two ways: (*a*) either through the existence of a set of subharmonic-type resonances with the participation of random quasi-subharmonic disturbances of frequencies near $n\omega_1/2$ and deterministic forcing waves $n\omega_1$(Kachanov 1985), or by (*b*) a combination interaction of random quasi-subharmonics with harmonics of the fundamental wave (proposed for the N-breakdown by Kachanov et al 1977). This issue needs a more precise examination; perhaps both mechanisms are involved.

Note that the highest intensity of random, turbulent fluctuations amplified at late stages of the K-breakdown is observed inside the boundary layer in the region that is closer to the high-shear layer than to the spikes (Borodulin & Kachanov 1990). Although the relative role of each type of coherent structure (CS-solitons and near-wall structures) in the process of a background turbulence amplification is not quite clear now, the mechanism of local (inflectional) secondary instability of the high-shear layer (see Section 6.2), coupled with the wave-resonant notions discussed above, can probably account for the flow randomization at these stages of transition. An example of such an approach, conforming to a developed turbulent boundary layer, was reported by Landahl (1977).

6.6   Connection of K-Breakdown to Developed Turbulence

On the basis of the analysis of properties of the coherent disturbances observed in the K-regime and mechanisms of amplification (by them) of background continuous-spectrum fluctuations, Kachanov (1990, 1991) proposed a hypothesis on the structure of developed wall turbulence called the *resonant-soliton* concept. This formulation develops some existing notions and complements them with the new ideas and mechanisms that recently appeared during investigation of the K-breakdown.

The idea of generating turbulence in the boundary layer through the development of coherent structures was advanced long ago (Kline et al 1967). The notion of a continuous transition to turbulence taking place near the boundary of a viscous (laminar) sublayer was also well known (see, for example, Repik & Sosedko 1974). The resemblance of the coherent structures of turbulence observed near the wall—often called horseshoe-shaped vortices, with the Λ-vortices observed in transitional boundary layers was pointed out repeatedly; similarities between the accompanying inflectional instantaneous velocity profiles and high-shear layers were also noted (Blackwelder 1983). Landahl's (1967, 1977) proposed local mechanism for amplifying the background turbulence by these instantaneous profiles used mainly the same idea as the concept of local high-frequency secondary instability discussed above (see Section 6.2). However, many of these notions had been developed on a relatively superficial qualitative base and no theory of turbulence generation had resulted. The main problem was associated with lack of information on the physical mechanisms of coherent structure formation and the structures' interaction with the background turbulence. Although empirical knowledge in this field was continuously accumulated during the previous two decades, the very difficult experimental conditions for direct study of the developed turbulence restricted the quantity and quality of the data obtained. Many ideas promoted were of a local character and operated mainly with notions of vortex deformation and interaction but not wave development or interaction. [This was despite the fact that earlier Landahl (1967) had proposed a wave theory of turbulence which was later developed in a number of theoretical studies, including those where resonant wave interactions were investigated (see for example Zharov 1986)]. Recent development of both experimental and theoretical studies, devoted to transitional boundary layers, and operated with (in particular) spectral-wave notions, changed the situation.

Orszag & Patera (1983) made the first attempt to describe turbulent coherent structures as nonlinear wave packets consisting of resonantly interacting 2D and 3D waves. They compared their theoretical results, on a subharmonic resonant interaction of a 2D fundamental wave with a packet of 3D quasi-subharmonic disturbances, with experimental data of the incipient spots in channel flow (Carlson et al 1981) and also of coherent structures observed in the developed turbulent boundary layer (Brown & Thomas 1977). This model of nonlinear wave packets (in the spirit of the N-breakdown) is, of course, very simple. Nevertheless some qualitative correlation between it and experimental data was found. Sporadic generation of spikes by resonant wave interactions at random excitation of an inlet pipe flow was recently observed by Savenkov (1993). This result can also be considered as a very simple model for the appearance of the K-regime of breakdown in conditions close to developed turbulent flow.

During investigation of the K-transition (see above), the necessary complementary information about the mechanisms of coherent structure generation had been obtained. As a result, a rational foundation has appeared for developing not only a basis for the theory of coherent structures but also for a new theory of turbulence itself: the *resonant-soliton theory* (Kachanov 1990, 1991). The main ideas of this theory are described below.

Amplification of packets of relatively low-frequency fluctuations, leading to a phenomena analogous to the N- and K-regime of breakdown, takes place sporadically in the near-wall region of the turbulent boundary layer under the influence of the turbulent fluctuations initiated upstream. The K-regime of breakdown is of paramount importance. This regime appears when (and where) a localized flash (or a wave packet) of background disturbances reaches some critical threshold amplitude and has a sufficiently representative frequency and frequency-wavenumber spectrum (see Sections 5 and 6.1). The entrance to the K-regime initiates (near the wall) the cascade of resonant amplification of definite spectral modes described by the wave-resonant concept (see Section 6.1). The fact that the K-regime of breakdown occurs in both boundary-layer and flat channel flows testifies to its very weak sensitivity to the specific form of the mean velocity profile (see also Section 7.2). We also know that the resonant mechanisms of wave interactions have very large spectral width, i. e. they are very local in time and space (see Sections 4.3 and 6.1). Both these circumstances make possible the appearance of the K-regime mechanism in the laminar sublayer of the developed turbulent boundary layer. This mechanism leads to the formation, and simultaneous bifurcation, of two kinds of coherent structures observed in the K-transition: (*a*) *CS*-solitons and (*b*) near-wall tooth-like structures (connected with the high-shear layer; see Section 6.4 and Figure 31).

The *CS*-solitons behave in the same manner as in transitional flow (Sections 6.3 and 6.4). They move away from the wall, piercing the boundary layer during the formation process, and propagate downstream with nearly the free-stream speed, with virtually no change in their form. The most typical features of the *CS*-solitons are the spikes observed near the external edge of the boundary layer. These solitons correspond to typical eddies, i.e. the coherent structures observed in the outer part of the turbulent boundary layer. Simultaneously, in the near-wall region the formation of the mean and pulsating (traveling) streamwise vortices (see Section 5) leads to the development and downstream propagation of the high-shear layer. In oscilloscope traces, these appear as the tooth-like jumps of the streamwise instantaneous flow velocity and correspond to the near-wall tooth-like coherent structures or horseshoe-shaped vortices (see Section 6.4).

The interaction of the *CS*-solitons and near-wall structures with the turbulent noise leads (in the same way as in the *K*-transition) to the parametric resonant amplification of random background fluctuations, i.e. to the refueling of the turbulent energy from the mean flow directly to these continuous-spectrum disturbances owing to the *catalytic* influence of the coherent structures (see Sections 6.5 and 4.3). This catalytic mechanism promotes a long life for the coherent structures because of the absence (or weakness) of any back influence of the amplified background disturbances on them. However, the coherent structures are destroyed gradually, most likely by their nonlinear interaction with the turbulence. (This mechanism is, at present, relatively unstudied in transitional flow.) The lifetimes of *CS*-solitons are known to be longer (in transitional flow) than those of the tooth-like structures (or high-shear layer). The problem of how these structures interact has yet to be studied carefully in *K*-transitions.

Next we discuss some experimental evidence to support the resonant-soliton concept of developed turbulence.

Figures 34 (*a*, *c*, *e*) presents traces of the streamwise velocity component obtained experimentally by means of a conditional sampling technique by Fukunishi et al (1987), Makita et al (1987), and Thomas & Bull (1983), which demonstrate the properties of turbulent coherent structures. Corresponding traces for the coherent structures in the transitional boundary layer, observed in the *K*-regime by Borodulin & Kachanov (1988) and Kachanov et al (1984), are shown in Figure 34 (*b*, *d*, *f*). The graphs testify to the striking resemblance of these phenomena both near the wall and near the external edge of the boundary layer. In particular, despite an essential difference (about 1–2 orders) between the dimensional charac-

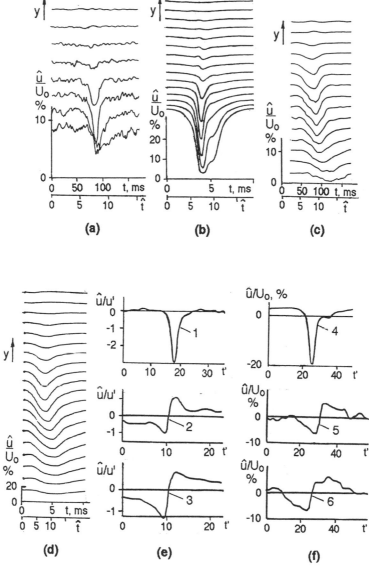

Figure 34 Properties of *CS*-solitons in the *K*-transition (*b*, *d*, *f*) and (ensemble averaged) coherent structures in the turbulent boundary layer (*a*, *c*, *e*). (*a*) and (*b*) are for the external part of boundary layer; (*c*) and (*d*) are at initial stages of the structure development; in (*e*) and (*f*), (1) and (4) are for developed external structures, wherease (2), (3), (5), and (6) are for near-wall structures. $\hat{t} = tU_o/\delta$; $t' = tU_o/\delta^*$. Sources: (*a*) Fukunishi et al 1987, (*b*) Borodulin & Kachanov 1988 (*c*) Makita et al 1987, (*d*) Borodulin & Kachanov 1988, (*e*) Thomas & Bull 1983, (*f*) Kachanov et al 1984. (After Kachanov 1989, 1990.)

teristic temporal and spatial scales of the structures in turbulent and transitional flows, the nondimensional ones are very close to each other. The amplitudes of the spikes are also very close—especially if one takes into account that the conditional sampling technique usually gives underestimated amplitudes (especially for the spikes) because of the influence of amplitude and phase noise produced by background turbulence. The downstream velocity of the tooth-like (near-wall) structures is typically equal to $0.67U_o$ (Thomas & Bull 1983) in the developed turbulent flow and about $(0.5-0.7)$ U_o (depending on y-coordinate) in the K-breakdown. A number of other characteristics of these structures are also very similar [for example, the form of a typical eddy found by Falco (1977), and that of a spike-soliton in the K-transition, both resemble toroidal vortices; see Section 6.4 and Figure 31).

7.   CONCLUDING REMARKS

7.1   *Types of Transition and Types of Theoretical Models*

This question was partially discussed in Section 4.3 and has been addressed in the review by Herbert (1988). Any discussion that relates to the types of transition (or theoretical approaches), such as the N-type, K-type, C-type, H-type, S-type and so on, is not only one of terminology but of fundamental principles.

We have to distinguish clearly the types of *transition* from the types of *resonances* or *theoretical models*. Unfortunately many investigators (mainly theoreticians) often do not make this distinction. As a result the *K-type of transition*, found by Klebanoff and others, is often called, for example, the *K-mode* or the *fundamental type* (meaning the fundamental resonance which can, probably, describe only one of the possible interaction mechanisms at the initial stage of the K-regime of transition). Craik's and Herbert's types of resonances (or types of theoretical approaches) are often called the C- and *H-types of transition*, while *both* of them well describe certain stages of the N-transition observed in experiment. The results of the weak-nonlinear approach by Zelman et al and Floquet theory by Herbert both correlate well with experimental data and simultaneously with each other despite their differing mathematical formulations. Moreover, the role of one or another type of nonlinear mechanism in each specific type of transition is often unclear and might be reevaluated as a result of subsequent studies. For example, for the case of K-transition there are two main resonant mechanisms for the explanation of the phenomena of amplification of 3D waves with fundamental frequency: the Craik-Nayfeh-Bozatli four-wave resonance and the fundamental five-wave resonance (see Section 5.2). As mentioned above (Sections 5.2 and 6.1), Rist & Fasel (1991) showed that none of these types of interaction can explain amplification rates of the $(\omega_1 \pm \beta_r)$ modes observed in direct numerical simulation [which is in good agreement with the experiment by Kachanov et al (1984, 1989)]. Only the combined influence of both these mechanisms yields the necessary growth rate of these modes.

It is believed that the types of transition represent very complicated multi-stage physical phenomena which can include different types of resonances and interactions, and can be described by different theoretical approaches. The transition as a whole, however, can be observed only in physical experiments (or in numerical ones incorporating a direct solution of the equations of fluid motion).

7.2 Intermittence and Turbulent Spots

It is interesting to note that neither intermittence nor turbulent spots are observed in either type of transition (K- or N-regimes) when the initial instability wave is strictly periodic in time! Meanwhile these phenomena are usually detected in "natural" (i.e. uncontrolled) conditions when the spectrum of background disturbances is more complicated and the instability wave has both amplitude and phase modulation in time. This modulation was simulated in many experiments either by excitation of localized wave packets (see for example Gaster & Grant 1975, Grek et al 1990, Cohen et al 1991) or by beating of two harmonic waves (Kachanov et al 1978b; Saric & Reynolds 1980; Corke 1990; Kachanov et al 1984, 1989; and others). The corresponding theoretical models were studied within a weak-nonlinear approach (see, for example, Zelman & Smorodsky 1991b) and direct numerical simulation (see, for example, Konzelmann 1990 and Fasel 1990). Savenkov (1993) studied the generation of spikes in random noise-like excitation disturbances. The experiments mentioned above showed that the temporal amplitude modulation of the initial instability wave can produce both the intermittence (i.e. the interchange in time of "laminar" and "turbulent" pieces of traces on the oscilloscope screen) and the turbulent spots (i.e. the spatially localized regions of turbulent behavior). This question was discussed in detail by Kachanov et al (1978b) within the framework of a very simple quasi-stationary, two-dimensional model. The onset of intermittence is probably connected, in most cases, with the temporal modulation of the instability wave as a local disturbance amplitude affects the local transitional process and local (in time) position of the spatial "transition point." This viewpoint correlates well with the very large spectral width of the subharmonic resonances (which play decisive roles in nonlinear stages) found in both experiment and theory. This large spectral width means that the resonant mechanisms of the boundary-layer breakdown are very local in time and in the flow direction (see Section 4.3). Only one or two periods of a wave are enough to provide the necessary conditions for rapid resonant amplification of this portion of the wave, which then occurs almost in the same way as if the wave were harmonic, i.e. unrestricted in time and x-coordinate. Very similar results were obtained theoretically by Savenkov (1993) who observed a sporadic formation of the single spikes in the case of random excitation of an inlet pipe flow.

Hence, nonlinear disturbance development inside each wave packet probably occurs mainly in the same way as it does in cases of strictly periodic initial instability waves, i. e. according to the N- and K-type of breakdown. At present there are a number of direct experimental observations [see, for example, Corke (1990) for the case of the N-regime and Kachanov et al (1984, 1989) for the K-regime] that corroborate this viewpoint.

However, intermittency does not imply the existence of turbulent spots. The latter exist not only in time but also in space and have (in the boundary layer) an essentially 3D character. Therefore some mechanism of nonlinear spanwise localization of the disturbance amplitudes must be invoked for their formation. Such a nonlinear mechanism is always observed in the K-breakdown and leads to the generation of CS-solitons and near-wall structures, and to the formation of local (in the spanwise direction) turbulent zones. This is probably the same mechanism that operates in most cases of "natural" boundary-layer transitions when turbulent spots are observed.

How the spots form in the N-regime of transition remains unclear, although a (not very strong) mechanism for the spanwise localization of the disturbances also exists in this case and is connected to the 3D character of the amplified quasi-subharmonic fluctuations. Possibly, a mechanism of spike formation similar to that observed in the K-regime can begin to work at a certain late stage of the N-breakdown. However, this point is completely unclear at present.

Borodulin & Kachanov (1990) explored the possibility of spatially localized (in the spanwise direction) regions of K-breakdown. They found no "preferred" spanwise wavelength of the disturbance modulation (Anders & Blackwelder 1980) (see also Kachanov 1987). The presence of simply a small hillock on the spanwise distribution of the initial, almost 2D, fundamental instability wave was sufficiant to develop a single (in space) peak (just downstream of this hillock) which had all of the typical attributes of the K-breakdown including CS-solitons and a high-shear layer. We note that the "preferred" spanwise wavelength λ_z observed in many experiments (and discussed for many years) differs in different studies (conducted at approximately the same flow and disturbance parameters but in different wind-tunnels), despite its weak sensitivity to initial conditions observed within each concrete experiment. For example, the measurements by Kozlov et al (1983) give $\lambda_z \approx 35$ mm (at $U_\infty = 8$ m/s) and a fundamental frequency parameter $F_1 \approx (60\text{--}140) \times 10^{-6}$. This value is 2–4 times greater than that found by Anders & Blackwelder (1980), who measured $\lambda_z \approx 8\text{--}16$ mm at $U_\infty \approx 3\text{--}15$ m/s and $F_1 \approx (80\text{--}240) \times 10^{-6}$. The belief that the "preferred" spanwise periodicity is not an inherent property of the K-breakdown was also discussed by Kachanov (1987) (in particular, from the viewpoint of the wave-resonant concept) and by Herbert (1988) (within the framework of an analysis of the energy balance).

The process of local formation (both in time and space) of a nonlinear wave packet and its development into a turbulent spot (corresponding, most probably, to the K-type of breakdown) was studied by Amini & Lespinard (1982) by means of a hot-wire. The results obtained were similar to those found by Klebanoff et al (1962). For example, the formation of structures closely resembling spikes was also detected. However, some distinctions from the usual K-breakdown (arising from the periodic excitation) was observed. Grek et al (1990) studied the same process by means of smoke-wire visualization. Theirs was (probably) the first experiment in which an isolated (in time and space) Λ-vortex was obtained. Successive stages of its downstream development were investigated.

Thus, we conclude that the only main condition that must be satisfied to generate turbulent spots is that the initial instability wave must have strong enough temporal modulation in its initial amplitude. It is not necessary for any strong spanwise modulation of the wave to exist—this can appear automatically as the result of the nonlinear development of the disturbances. Of course, the "microscopic mechanisms" of the onset of intermittence and turbulent spots have to be studied in more detail. In particular the problem of whether the K- and/or N-type of transition can be achieved inside localized wave packets has to be investigated in much more detail than in the previous work.

Despite all the results presented in this article concerned mainly with wall boundary layers, there is much evidence that very close physical mechanisms can operate in various other shear flows. The brightest example of this resemblance is the K-regime observed in flat-channel flow (Nishioka et al 1980). Savenkov (1992, 1993) found theoretically the formation of spikes (typical for the K-breakdown) connected with resonant wave interactions in inlet pipe and channel flows. Subharmonic-type resonances between the fundamental wave and low-frequency disturbances attributed to the N-regime of boundary-layer transition are also observed in various shear flows, such as in: the mixing layer (see, for example, Miksad 1972), flat-channel flow (Kozlov & Ramazanov 1984), separated flow (Boiko et al 1988), as well as in more exotic 3D boundary layers such as on a rotating disk (Balachandar et al 1990). The universality of the mechanism of subharmonic resonance for various shear flows was shown theoretically by Orszag & Patera (1983) and in other studies.

Consequently, many physical phenomena and mechanisms detected in the transitional boundary layer and discussed in the present article can also be recognized in other flows. Some of these mechanisms have been studied; others have not. But the deeper our research advances towards an understanding of developed turbulence, the clearer it becomes that the process of turbulence onset is essentially of a resonant nature.

ACKNOWLEDGMENTS

I am grateful to all my colleagues and coauthors, especially V. I. Borodulin, S. V. Dryganets, V. M. Gilyov, V. V.Kozlov, V. Y. Levchenko, A. Michalke, M. P. Ramazanov, O. S. Ryzhov, and F. T. Smith; each have made valuable contributions to our joint studies over many years. The results obtained have fundamentally influenced my work on this paper. I have benefited from numerous discussions with many of the authors of papers quoted here, in particular: H. Bestek, A. D. D. Craik, H. F. Fasel, A. A. Fernholz, A. V. Fyodorov, S. A. Gaponov, F. R. Hama, T. Herbert, L. Kleiser, B. G. B. Klingmann, M. T. Landahl, M. V. Morkovin, M. Nishioka, U. Rist, W. S. Saric, I. V. Savenkov, V. I. Shrira, Y. A. Stepanyants, V. N. Zhigulyov, M. B. Zelman, and many other scientists. I am also very grateful to the Editors† of the *Annual Review of Fluid Mechanics* who have done enormous work in preparing the manuscript of this paper and translating it from my Russian-English into English.

† The editors apologize for the delay in publishing this lengthy article. Owing to ongoing complications in communicating with scientists in the Former Soviet Union, it has proven difficult to commission articles from our FSU colleagues and to follow the normal editorial process.

Literature Cited

Abramyan, L. A., Stepanyants, Y. A., Shrira, V. I. 1992. Non-one-dimensional solitons in shear flows like a boundary layer. *Dokl. Akad. Nauk*³ 327(4–6) (In Russian)

Aizin, L. B., Polyakov, N. F. 1979. Acoustic generation of Tollmien-Schlichting waves over local unevenness of surfaces immersed in streams. *Preprint No. 17. USSR Acad. Sci., Sib. Div., Inst. Theoret. Appl. Mech., Novosibirsk* (In Russian)

Amini, J., Lespinard, G. 1982. Experimental study of an "incipient spot" in a transitional boundary layer. *Phys. Fluids* 25(10): 1743–50

Anders, J. B., Blackwelder, R. F. 1980. Longitudinal vortices in a transitioning boundary layer. See Eppler & Fasel 1980, pp. 110–19

Antar, B. N., Collins, F. G. 1975. Numerical calculation of finite amplitude effects in unstable laminar boundary layer. *Phys. Fluids* 18: 289–97

Arnal, D., Coustols, E., Juillen, J. C. 1984.

Etude expérimentale et théorique de la transition sur une aile en flèche infinie. *Rech. Aérosp.* 4: 275–90

Arnal, D., Michel, R., eds. 1990. *Laminar-Turbulent Transition.* Heidelberg: Springer-Verlag. 710 pp.

Asai, M., Nishioka, M. 1989. Origin of the peak-valley wave structure leading to wall turbulence. *J. Fluid Mech.* 208: 1–23

Balachandar, S., Street, C. L., Malik, R. 1990. Secondary instability in rotating disk flow. *AIAA Pap. No.* 90-1527

Barker, S. J., Jennings, C. 1977. The effect of wall heating upon transition in water boundary layers. *AGARD Symp. Laminar-Turbulent Transition. AGARD-CP-224, Pap. No.* 19

Benney, D. J., Lin, C. C. 1960. On the secondary motion induced by oscillations in a shear flow. *Phys. Fluids* 3: 656–57

Bertolotti, F. P., Herbert, T., Spalart, P. R. 1992. Linear and nonlinear stability of the Blasius boundary layer. *J. Fluid Mech.* 242: 441–74

Betchov, R. 1960. On the mechanism of turbulent transition. *Phys. Fluids* 3: 1026–27

Bippes, H. 1990. Instability features appearing on swept wing configurations. See Arnal & Michel 1990, pp. 419–30

Bippes, H. 1991. Experiments on transition in three-dimensional accelerated boundary layer flows. *Preprint reported in Conf. on Boundary Layer Transition & Control,* Cambridge, 1991. 17 pp.

Blackwelder, R. F. 1983. Analogies between transitional and turbulent boundary layers. *Phys. Fluids* 26(10): 2807–15

Boiko, A. V., Dovgal, A. V., Kozlov, V. V. 1988. Nonlinear interaction of disturbances at transition to turbulence in a zone of laminar-boundary-layer separation. *Izv. Sib. Otd. Akad. Nauk SSSR, Ser. Tekh. Nauk.*[1] 18(5): 44–50 (In Russian)

Boiko, A. V., Dovgal, A. V., Scherbakov, V. A., Simonov, O. A. 1991. Effects of laminar-turbulent transition in separation bubbles. In *Separated Flows and Jets,* ed. V. V. Kozlov, A. V. Dovgal, pp. 565–72. Berlin: Springer-Verlag

Borodulin, V. I., Dryganets, S. V., Kachanov, Y. S., Levchenko, V. Y., Ramazanov, M. P. 1989. Receptivity of transitional boundary layer to small background disturbances. *Int. Seminar: Problems of Simulation in Wind Tunnels.* 1: 107–16. Novosibirsk: Inst. Theor. & Appl. Mech. (In Russian)

Borodulin, V. I., Kachanov, Y. S. 1988. Role of the mechanism of local secondary instability in *K*-breakdown of boundary layer. *Izv. Sib. Otd. Akad. Nauk SSSR, Ser. Tekh. Nauk.*[1] 18: 65–77 (In Russian).

(Transl. *Sov. J. Appl. Phys.* 1989, 3(2): 70–81)

Borodulin, V. I., Kachanov, Y. S. 1989. Cascade of harmonic and parametric resonances in *K*-regime of boundary-layer breakdown. *Model Mekh. (Simulation in Mechanics)* 3(20), No. 2: 38–45 (In Russian)

Borodulin, V. I., Kachanov, Y. S. 1990. Experimental study of soliton-like coherent structures in boundary layer. *Proc. Scientific & Methodological Seminar on Ship Hydrodynamics, 19th Session.* 2: 99-1—99-10. Varna: Bulg. Ship Hydrodyn. Cent.

Borodulin, V. I., Kachanov, Y. S. 1992. Experimental study of soliton-like coherent structures. In *Eddy Structure Identification in Free Turbulent Shear Flows. IUTAM Symp.,* pp. XI.3.1–XI.3.9, Poitiers, 12–14 October, 1992

Brown, G. L., Thomas, A. S. W. 1977. Large structure in a turbulent boundary layer. *Phys. Fluids Suppl.* 20(10): S243–52

Bushnell, D. M., Malik, M. R. 1987. Application of stability theory to laminar flow control—progress and requirements. *Stability of Time Dependent & Spatially Varying Flows. Proc. Symp., Hampton,* 1985. New York

Carlson, D., Widnall, S. E., Peeters, M. F. 1981. A flow visualization study of transition in plane Poiseuille flow. *Flow Dyn. Res. Lab. (Dept. Aero. & Astro., MIT) Rep. No.* 81-3

Cohen, J., Breuer, K. S., Haritonidis, J. H. 1991. On the evolution of a wave-packet in a laminar boundary layer. *J. Fluid Mech.* 225: 575–606

Conlisk, A. T., Burggraf, O. R., Smith, F. T. 1987. Nonlinear neutral modes in the Blasius boundary layer. *Trans. ASME,* pp. 119–22, Cincinnati, Ohio

Corke, T. C. 1990. Effect of controlled resonant interactions and mode detuning on turbulent transition in boundary layers. See Arnal & Michel 1990, pp. 151–78

Corke, T. C., Mangano, R. A. 1989. Resonant growth of three-dimensional modes in transitioning Blasius boundary layers. *J. Fluid Mech.* 209: 93–150

Craik, A. D. D. 1971. Nonlinear resonant instability in boundary layers. *J. Fluid Mech.* 50: 393–413

Croswell, J. W. 1985. *On the energetics of primary and secondary instabilities in plane Poiseuille flow.* MS thesis. Verg. Polit. Inst. & State Univ., Blacksburg, W. Va.

Crouch, J. D., Herbert, T. 1993. Nonlinear evolution of secondary instabilities in boundary-layer transition. *Theoret. Comput. Fluid Dyn.* 4: 151–75

Craik, A. D. D. 1985. *Wave Interactions and*

Fluid Flows. Cambridge: Cambridge Univ. Press

Dallmann, U., Bieler, H. 1987. Analysis and simplified prediction of primary instability of three-dimensional boundary layer flows. *AIAA Pap. No.* 87-1338

Demyohin, E. A., Shkadov, V. Y. 1986. On the theory of solitons in systems with dissipation. *Izv. Akad. Nauk SSSR, Mekh. Zhidk. Gaza*[2] 3: 91–97 (In Russian)

Deyhle, H., Hohler, G., Bippes, H. 1993. Experimental investigation of instability wave propagation in a 3-D boundary-layer flow. Preprint. *AIAA J.* 4: 637–45

Dovgal, A. V., Kozlov, V. V. 1990. Hydrodynamic instability and receptivity of small separation regions. See Arnal & Michel 1990, pp. 523–31

Dovgal, A. V., Kozlov, V. V., Kosorygin, V. S., Ramazanov, M. P. 1981. Influence of disturbances on the structure of a flow near a separation zone. *Dokl. Akad. Nauk SSSR*[3] 258(1): 45–48 (In Russian)

Dovgal, A. V., Kozlov, V. V., Simonov, O. A. 1987. Experiments on hydrodynamic instability of boundary layers with separation. *Proc IUTAM Symp. on Boundary-Layer Separation,* ed. F. T. Smith, S. N. Brown, pp. 109–30. Springer-Verlag

Dovgal, A. V., Kozlov, V. V., Simonov, O. A. 1988. Stability of three-dimensional flow generated upon separation from a wall with an inflection. *Sov. J. Appl. Phys.* 2(4): 18–24

Dryganets, S. V., Kachanov, Y. S., Levchenko, V. Y., Ramazanov, M. P. 1990. Resonant flow randomization in K-regime of boundary layer transition. *Zh. Prikl. Mekh. Tekh. Fiz.*[4] 2: 83–94 (In Russian). (Transl. *J. Appl. Mech. Tech. Phys.* 1990, 31(2): 239–49)

El-Hady, N. M. 1988. Evolution of resonant wave triads in three-dimensional boundary layers *AIAA Pap. No.* 88-0405

El-Hady, N. M. 1989. Secondary instability of compressible boundary layer to subharmonic three-dimensional disturbances. *AIAA Pap. No.* 89-0035

El-Hady, N. M. 1991. Effect of suction on controlling the secondary instability of boundary layers. *Phys. Fluids A* 3(3): 393–402

Eppler, R., Fasel, H. eds. 1980. *Laminar-Turbulent Transition.* Berlin: Springer-Verlag. 432 pp.

Falco, R. E. 1977. Coherent motions in the outer region of turbulent boundary layer. *Phys. Fluids Suppl.* 20(10): S124–32

Fasel, H. 1990. Numerical simulation of instability and transition in boundary layer flows. See Arnal & Michel 1990, pp. 587–98

Fasel, H., Konzelmann, U. 1990. Non-parallel stability of a flat-plate boundary layer using the complete Navier-Stokes equations. *J. Fluid Mech.* 221: 311–47

Fasel, H. F., Rist, U., Konzelmann, U. 1987. Numerical investigation of the tree-dimensional development in boundary layer transition. *AIAA Pap. No.* 87-1203

Fischer, T. M., Dallmann, U. 1987. Theoretical investigation of secondary instability of three-dimensional boundary-layer flows. *AIAA Pap. No.* 87-1338

Fukunishi, Y., Sato, H., Inous, O. 1987. Study of developing process of coherent structures in turbulent boundary layer. *AIAA Pap. No.* 87-1253

Gaponov, S. A., Maslov, A. A. 1971. Stability of compressible boundary layer at subsonic speeds. *Izv. Sib. Otd. Akad. Nauk SSSR, Ser. Tekh. Nauk*[1] 3(1): 24–27 (In Russian)

Gaster, M., Grant, T. 1975. An experimental investigation of the formation and development of a wave packet in a laminar boundary layer. *Proc. R. Soc. London Ser. A* 347: 253–69

Gertsenshtein, S. Y. 1969. On stability of non-stationary unidirectional parallel flow of ideal fluid. *Izv. Akad. Nauk SSSR, Mekh. Zhidk. Gaza*[2] 2: 5–10 (In Russian)

Gilyov, V. M., Dovgal, A. V., Kachanov, Y. S., Kozlov, V. V. 1988. Development of spatial disturbances in a boundary layer with pressure gradient. *Izv. Akad. Nauk SSSR, Mekh. Zhidk. Gaza*[2] 3: 85–91 (In Russian)

Gilyov, V. M., Kachanov, Y. S., Kozlov, V. V. 1981. Development of a spatial wave packet in a boundary layer. *Preprint No.* 34. Inst. Theoret. Appl. Mech., Sib. Div. USSR Acad. Sci., Novosibirsk. 46 pp. (In Russian)

Gilyov, V. M., Kachanov, Y. S., Kozlov, V. V. 1983. Development of a spatial wave packet in a boundary layer. *Izv. Sib. Otd. Akad. Nauk SSSR, Ser. Tekh. Nauk*[1] 13(3): 27–37 (In Russian). (See also: Kachanov, Y. S. 1985. See Kozlov 1985, pp. 115–23)

Goldstein, M. E., Hultgren, L. S. 1989. Boundary-layer receptivity to long-wave free-stream disturbances. *Annu. Rev. Fluid Mech.* 21: 137–66

Grek, H. R., Kozlov, V. V., Ramazanov, M. P. 1990. Receptivity and stability of the boundary layer at a high turbulence level. See Arnal & Michel 1990, pp. 511–21

Greenspan, H. F., Benney, D. J. 1963. On shear-layer instability, breakdown and transition *J. Fluid Mech.* 15: 133–53

Hall, P., Smith, F. T. 1988. The nonlinear interaction of Tollmien-Schlichting waves and Taylor-Görtler vortices in curved channel flows. *Proc. R. Soc. London Ser. A* 417: 255

Hall, P., Smith, F. T. 1989a. Nonlinear Tollmien-Schlichting/ vortex interaction in boundary layers. *Eur. J. Mech. B/Fluids* 8: 179

Hall, P., Smith, F. T. 1989b. On strongly nonlinear vortex/wave interactions in boundary layer transition. *NASA Contr. Rep.* 181963, *ICASE Rep. No.* 89-82

Hama, F. R. 1959. Some transition patterns in axisymmetric boundary layers. *Phys. Fluids* 2(6): 664–67

Hama, F. R. 1963. Progressive deformation of a perturbed line vortex filament. *Phys. Fluids* 6: 526–34

Hama, F. R., Long, J. D., Hegarty, J. C. 1957. On transition from laminar to turbulent flow. *J. Appl. Phys.* 27: 388–94

Hama, F. R., Nutant, J. 1963. Detailed flow-field observations in the transition process in a thick boundary layer. *Proc.* 1963 *Heat Transfer & Fluid Mech. Inst.*, pp. 77–93. Palo Alto, Calif.: Stanford Univ. Press

Hefner, J. N., Bushnell, D. M. 1979. Application of stability theory to laminar flow control. *12th AIAA Fluid & Plasma Dyn. Conf.* Williamsburg, Virginia

Herbert, T. 1975. On finite amplitudes of periodic disturbances of the boundary layer along a flat plate. *Lect. Notes Phys.* 35: 212–17

Herbert, T. 1984. Analysis of the subharmonic route to transition in boundary layers. *AIAA Pap. No.* 84-0009

Herbert, T. 1985. Three-dimensional phenomena in the transitional flat-plate boundary layer. *AIAA Pap. No.* 85-0489

Herbert, T. 1987. On the mechanism of transition in boundary layers. *AIAA Pap. No.* 87-1201

Herbert, T. 1988. Secondary instability of boundary layers. *Annu. Rev. Fluid Mech.* 20: 487–526

Herbert, T., Bertolotti, F. P. 1985. Effects of pressure gradients on the growth of subharmonic disturbances in boundary layers. *Proc. Conf. on Low Reynolds Number Airfoil Aerodyn.*, p. 65, Univ. Notre Dame, Ind.

Herbert, T., Crouch, J. D. 1990. Threshold conditions for breakdown of laminar boundary layers. See Arnal & Michel 1990, pp. 93–101

Herbert, T., Morkovin, M. V. 1980. Dialogue on bridging some gaps in stability and transition research. See Eppler & Fasel 1980, pp. 47–72

Itoh, N. 1981. Secondary instability of laminar flows. *Proc. R. Soc. London Ser. A* 375(1763): 565–78

Itoh, N. 1987. Another route to the tree-dimensional development of Tollmien-Schlichting waves with finite amplitude. *J. Fluid Mech.* 181: 1–16

Itoh, N. 1990. Cross-flow instability of 3-D boundary layers on a flat plate. See Arnal & Michel 1990, pp. 359–68

Kachanov, Y. S. 1985. On resonant breakdown of laminar boundary layer. *Proc. Fifth Natl. Congr. on Theoret. & Appl. Mech., Actual and Topical Problems of Ship Hydro- and Aerodynamics*, 1985. 3: 71-1—71-11. Varna: Bulg. Ship Hydrodyn. Cent. (In Russian)

Kachanov, Y. S. 1987. On the resonant nature of the breakdown of a laminar boundary layer *J. Fluid Mech.* 184: 43–74

Kachanov, Y. S. 1990. Secondary and cascade resonant instabilities of boundary layers. Wave-resonant concept of a breakdown and its substantiation. See Arnal & Michel 1990, pp. 65–80

Kachanov, Y. S. 1991. Resonant-soliton nature of boundary layer transition. *Russ. J. Theor. Appl. Mech.* 1(2): 141–73

Kachanov, Y. S., Kozlov, V. V., Levchenko, V. Y. 1974a. Experimental investigation of the influence of cooling on the stability of laminar boundary layer. *Izv. Sib. Otd. Akad. Nauk SSSR, Ser. Tekh. Nauk*[1] 8(2): 75–79 (In Russian)

Kachanov, Y. S., Kozlov, V. V., Levchenko, V. Y. 1974b. Experimental study of laminar-boundary-layer stability on a wavy surface. *Izv. Sib. Otd. Akad. Nauk SSSR, Ser. Tekh. Nauk*[1] 13(3): 2–6 (In Russian)

Kachanov, Y. S., Kozlov, V. V., Levchenko, V. Y. 1975a. Generation and development of small disturbances in laminar boundary layer under the action of acoustic fields. *Izv. Sib. Otd. Akad. Nauk SSSR, Ser. Tekh. Nauk*[1] 13(3): 18–26 (In Russian)

Kachanov, Y. S., Kozlov, V. V., Levchenko, V. Y. 1975b. The development of small-amplitude oscillations in a laminar boundary layer. *Uch. Zap. TsAGI*[5] 6(5): 137–40 (In Russian) (Transl. *Fluid Mech. Sov. Res.* 1979, 8: 152–56)

Kachanov, Y. S., Kozlov, V. V., Levchenko, V. Y. 1977. Nonlinear development of a wave in a boundary layer. *Izv. Akad. Nauk SSSR, Mekh. Zhidk. Gaza*[2] 3: 49–53 (In Russian). (Transl. *Fluid Dyn.* 1978, 12: 383–90)

Kachanov, Y. S., Kozlov, V. V., Levchenko, V. Y. 1978a. Origin of Tollmien-Schlichting waves in boundary layer under the influence of external disturbances. *Izv. Akad. Nauk. SSSR, Mekh. Zhidk. Gaza*[2] 5: 85–94 (In Russian). (Transl. *Fluid Dyn.* 1979, 13: 704–11)

Kachanov, Y. S., Kozlov, V. V., Levchenko, V. Y. 1978b. Experiments on nonlinear interaction of waves in boundary layer.

478 KACHANOV

Preprint No. 16. Inst. Theoret. Appl. Mech., Siberian Div. USSR Acad. Sci., Novosibirsk, 35 pp. (In Russian). (See also Eppler & Fasel 1980, pp. 135–52)

Kachanov, Y. S., Kozlov, V. V., Levchenko, V. Y. 1982. Beginning of Turbulence in Boundary Layers. Novosibirsk: Nauka, Siberian Div. 152 pp. (In Russian)

Kachanov, Y. S., Kozlov, V. V., Levchenko, V. Y., Maksimov, V. P. 1979. Transformation of external disturbances into the boundary layer waves. Proc. Sixth Intl. Conf. on Numerical Methods in Fluid Dyn., pp. 299–307. Berlin: Springer-Verlag

Kachanov, Y. S., Kozlov, V. V., Levchenko, V. Y., Ramazanov, M. P. 1984. Experimental study of K-regime breakdown of laminar boundary layer. Preprint No. 9-84. Inst. Theoret. Appl. Mech., Siberian Div. USSR Acad. Sci., Novosibirsk (In Russian). (See also: Kachanov et al 1985 in Kozlov 1985, pp. 61–73)

Kachanov, Y. S., Kozlov, V. V., Levchenko, V. Y., Ramazanov, M. P. 1989. The nature of K-breakdown of laminar boundary layer. Izv. Sib. Otd. Akad. Nauk SSSR, Ser. Tekh. Nauk[1] 2: 124–58 (In Russian). (Transl. Sov. J. Appl. Phys. 1990, Vol. 4)

Kachanov, Y. S., Levchenko, V. Y. 1982. The resonant interaction of disturbances at laminar-turbulent transition in a boundary layer. Preprint No. 10-82. Inst. Theor. & Appl. Mech., Siberian Div. USSR Acad. Sci., Novosibirsk (In Russian). (See also J. Fluid Mech. 1984, 138: 209–47)

Kachanov, Y. S., Michalke, A. 1993. 3D instability of flat-plate boundary layer. Theory and experiment. Eur. J. Mech. B/Fluids. Submitted

Kachanov, Y. S., Ryzhov, O. S. 1991. Formation of solitons in transitional boundary layer, theory and experiment. Boundary Layer Transition & Control Conf., 8–12 April, 1991. Cambridge, UK. See also Sib. Fiz. Tekh. Z.[1] 1992, 1: 34–52 (In Russian)

Kachanov, Y. S., Ryzhov, O. S., Smith, F. T. 1993. Formation of solitons in transitional boundary layers: theory and experiments. J. Fluid Mech. 251: 273–97

Kachanov, Y. S., Tararykin, O. I. 1987. Experimental study of stability of a relaxing boundary layer. Izv. Sib. Otd. Akad. Nauk SSSR, Ser. Tekh. Nauk[1] 18(5): 9–19 (In Russian)

Kachanov, Y. S., Tararykin, O. I. 1990. The experimental investigation of stability and receptivity of a swept-wing flow. See Arnal & Michel 1990, pp. 499–509

Kachanov, Y. S., Tararykin, O. I., Fyodorov, A. V. 1989. Experimental simulation of swept-wing boundary layer in the region of secondary flow formation. Izv.

Sib. Otd. Akad. Nauk SSSR, Ser. Tekh. Nauk[1] 3: 44–53 (In Russian)

Kachanov, Y. S., Tararykin, O. I., Fyodorov, A. V. 1990. Investigation of stability to stationary boundary-layer disturbances in a model of a swept wing. Izv. Sib. Otd. Akad. Nauk SSSR, Ser. Tekh. Nauk[1] 5: 11–21 (In Russian)

Kelly, R. E. 1967. On the stability of an inviscid shear layer which is periodic in space and time. J. Fluid Mech. 27: 657–89

Kerschen, E. J. 1989. Boundary layer receptivity. AIAA Pap. No. 89-1109

Kerschen, E. J., Choudhari, M., Heinrich, R. A. 1990. Generation of boundary layer instability waves by acoustic and vortical freestream disturbances. See Arnal & Michel 1990, pp. 477–88

Klebanoff, P. S., Tidstrom, K. D. 1959. Evolution of amplified waves leading to transition in a boundary layer with zero pressure gradient. NACA Tech. Note D-195

Klebanoff, P. S., Tidstrom, K. D., Sargent, L. M. 1962. The three-dimensional nature of boundary-layer instability. J. Fluid Mech. 12: 1–34

Kleiser, L., Zang, T. A. 1991. Numerical simulation of transition in wall-bounded shear flows. Annu. Rev. Fluid Mech. 23: 495–537

Kline, S. J., Reynolds, W. C., Schraub, F. A., Runstadler, W. P. 1967. The structure of turbulent boundary layer. J. Fluid Mech. 30: 741–73

Klingmann, B. G. B., Boiko, A. V., Westin, K. J. A., Kozlov, V. V., Alfredsson, P. H. 1993. Experiments on the stability of Tollmien-Schlichting waves in the flat plate boundary layer. Eur. J. Mech. B/Fluids Submitted

Kloker, M., Bestek, H. 1992. Direct spatial simulation of laminar-turbulent transition in a boundary layer with strong adverse pressure gradient. Deutscher Luft- und Raumfahrtkongress/DGLR — Jahrestagung, 29 Sept.–2 Oct. 1992, Bremen

Knapp, C. F., Roache, P. J. 1968. A combined visual and hot-wire anemometer investigation of boundary-layer transition. AIAA J. 6: 29–36

Komoda, H. 1967. Non-linear development of disturbances in the laminar boundary layer. Phys. Fluids Suppl. 10: S87–94

Konzelmann, U. 1990. Numerische Untersuchungen zur raumlichen Entwicklung dreidimensionaler Wellen-paket in einer Plattengrenzschichtstromung. PhD thesis. Univ. Stuttgart

Kovasznay, L. S., Komoda, H., Vasudeva, B. R. 1962. Detailed flow field in transition. Proc. 1962 Heat Transfer & Fluid Mech. Inst., pp. 1–26. Palo Alto, Calif.: Stanford Univ. Press

Kozlov, L. F., Babenko, V. V. 1978. *Experimental Studies of the Boundary Layer.* Kiev: Naukova Dumka Publ. (In Russian)

Kozlov, V. V., ed. 1985. *Laminar-Turbulent Transition.* Berlin: Springer-Verlag. 759 pp.

Kozlov, V. V., Levchenko, V. Y., Saric, W. S. 1983. Formation of three-dimensional structures in a boundary layer at transition. *Preprint No. 10-83. Inst. Theoret. Appl. Mech., Siberian Div. USSR Acad. Sci., Novosibirsk* (In Russian)

Kozlov, V. V., Levchenko, V. Y., Scherbakov, V. A. 1978. Development of disturbances in boundary layer at suction through a slot. *Uch. Zap. TsAGI* 9(2): 99–105 (In Russian)

Kozlov, V. V., Ramazanov, M. P. 1984. Resonant interaction of disturbances in Poiseuille flow. *Dokl. Akad. Nauk SSSR* 275(6): 1346–49 (In Russian)

Kozlov, V. V., Ryzhov, O. S. 1990. Receptivity of boundary layers: asymptotic theory and experiment. *Proc. R. Soc. London Ser. A* 429: 341–73

Landahl, M. T. 1967. A wave-guide model for turbulent shear flow. *J. Fluid Mech.* 29: 441–59

Landahl, M. T. 1972. Wave mechanics of breakdown. *J. Fluid Mech.* 56(4): 755–802

Landahl, M. T. 1977. Dynamics of boundary layer turbulence and the mechanisms of drag redaction. *Phys. Fluids Suppl.* 20(10): S55–63

Levchenko, V. Y., Volodin, A. G., Gaponov, S. A. 1975. *Stability Characteristics of Boundary Layers.* Novosibirsk: Nauka. 313 pp. (In Russian)

Lin, C. C. 1955. *The Theory of Hydrodynamic Stability.* Cambridge: Cambridge Univ. Press

Loehrke, R. I., Morkovin, M. V., Fejer, A. A. 1975. Review—Transition in nonreversing oscillating boundary layers. *J. Fluid Eng., Trans. ASME* 97(4): 534–49

Lowell, K. L., Reshotko, E. 1974. Numerical study of the stability of a heated water boundary layer. *Case Western Reserve Univ., Rep. FTAS/TR-73-93.* Cleveland, Ohio

Mack, L. M. 1975. A numerical method for the prediction of high speed boundary-layer transition using linear theory. *NASA SP-347*, pp. 101–23

Makita, H., Sassa, K., Abe, M., Itabashi, A. 1987. Manipulation of an artificial large scale horse-shoe vortex by a thin plate placed in a turbulent boundary layer. *AIAA Pap. No.* 87-1232

Maksimov, V. P. 1979. Genesis of Tollmien-Schlichting wave in oscillating boundary layers. In *Development of Perturbations in Boundary Layers,* ed. V. Y. Levchenko,

pp. 68–75. Novosibirsk: Inst. Theoret. Appl. Mech., Sib. Div., USSR Acad. Sci. (In Russian)

Mangur, C. J. 1977. On the sensitivity of shear layers to sound. *AIAA Pap. No.* 77-1369

Masad, J. A., Nayfeh, A. H. 1990. On subharmonic instability of compressible boundary layers. See Arnal & Michel 1990, pp. 271–78

Maseev, L. M. 1968. Occurrence of three-dimensional perturbation in a boundary layer. *Fluid Dyn.* 3: 23–24

Maslennikova, I. I., Zelman, M. B. 1985. On subharmonic-type laminar-turbulent transition in boundary layer. See Kozlov 1985, pp. 21–28

Michalke, A. 1990. On the inviscid instability of wall-bounded velocity profiles close to separation. *Z. Flugwiss. Weltraumforch.* 14: 24–31

Michel, R., Arnal, D., Coustols, E., Juillen, J. C. 1985. Experimental and theoretical studies of boundary layer transition on a swept infinite wing. See Kozlov 1985, pp. 553–61

Miksad, R. W. 1972. Experiments on the non-linear stages of free-shear-layer transition. *J. Fluid Mech.* 59(1): 1–24

Morkovin, M. V. 1968. Critical evaluation of transition from laminar to turbulent shear layer with emphasis on hypersonically travelling bodies. *AFFDL Tech. Rep.* 68-149

Morkovin, M. V. 1969. On the many faces of transition. In *Viscous Drag Reduction,* ed. C. S. Wells, pp. 1–31. London: Plenum

Morkovin, M. V. 1984. Bypass transition to turbulence and research desiderata. In *Transition in Turbines,* pp. 161–204. NASA Conf. Publ. 2386

Morkovin, M. V., Reshotko, E. 1990. Dialogue on progress and issues in stability and transition research. See Arnal & Michel 1990, pp. 3–29

Murdock, J. W. 1980. The generation of Tollmien-Schlichting wave by a sound wave. *Proc. R. Soc. London Ser. A* 372: 1517

Nayfeh, A. H. 1987a. Nonlinear stability of boundary layers. *AIAA Pap. No.* 87-0044

Nayfeh, A. H. 1987b. On subharmonic instability of boundary layers. Proc. *ASME Forum on Unsteady Flow Separations, FED.* 52: 37–44

Nayfeh, A. H. 1987c. On secondary instabilities in boundary layers. In *Stability of Time Dependent and Spatially Varying Flows,* ed. D. L. Dwoyer, M. Y. Hussaini, pp. 18–42. Berlin: Springer-Verlag

Nayfeh, A. H., Bozatli, A. N. 1979a. Secondary instability in boundary-layer flows. *Phys. Fluids* 22: 805–13

Nayfeh, A. H., Bozatli, A. N. 1979b. Nonlinear wave interactions in boundary layers. *AIAA Pap. No.* 79-1456

Nayfeh, A. H., Ragab, S. A. 1987. Effect of a bulge on the secondary instability of boundary layers. *AIAA Pap. No.* 87-0045

Nelson, G. Craik, A. D. D. 1977. Growth of streamwise vorticity in unstable boundary layers. *Phys. Fluids* 20: 698–700

Nishioka, M., Asai, M., Iida, S. 1980. An experimental investigation of the secondary instability. See Eppler & Fasel 1980, pp. 37–46

Nishioka, M., Morkovin, M. V. 1986. Boundary-layer receptivity to unsteady pressure gradients: experiments and overview. *J. Fluid Mech.* 171: 219–61

Orszag, S. A., Patera, A. T. 1983. Secondary instability of wall-bounded shear flows. *J. Fluid Mech.* 128: 347–85

Raetz, G. S. 1959. A new theory of the cause of transition in fluid flows. *Norair Rep., NOR* 59-383 *(BLC*-121). Nawthorne, Calif.

Rahmatullaev, R. D. 1988. Bifurcation of three-dimensional flow birth in boundary layer. *Izv. Sib. Otd. Akad. Nauk SSSR, Ser. Tekh. Nauk*[1] 21(6): 126–31 (In Russian)

Rahmatullaev, R. D., Shtern, V. N. 1989. Threshold onset of three-dimensional structures at flow in flat channel. *Preprint No.* 196-89. *Inst. Thermophysics, Sib. Div. USSR Acad. Sci., Novosibirsk* (In Russian)

Reed, H. 1984. Wave-interactions in sweptwing flows. *AIAA Pap. No.* 84-1678

Reed, H. 1985. Disturbance-wave interactions in flows with crossflow. *AIAA Pap. No.* 85-0494

Reed, H. L., Saric, W. S. 1989. Stability of three-dimensional boundary layers. *Annu. Rev. Fluid Mech.* 21: 235–84

Reed, H. L., Stuckert, G., Balakumar, P. 1990. Stability of high-speed chemically reacting and three-dimensional boundary layers. See Arnal & Michel 1990, pp. 347–57

Repik, E. U., Sosedko, U. P. 1974. Studies of intermittent flow structure in near-wall region of turbulent boundary layer. In *Turbulent Flows*, ed. V. V. Struminsky. Moscow: Nauka. (In Russian)

Reshotko, E. 1976. Boundary-layer stability and transition. *Annu. Rev. Fluid Mech.* 8: 311–49

Rist, U. 1990. *Numerische Untersuchung der räumlichen, dreidimensionalen Störungsentwicklung beim Grenzschichtumschlag.* PhD thesis. Inst. A Mech. Univ. Stuttgart. 172 pp.

Rist, U., Fasel, H. 1991. Spatial three-dimensional numerical simulation of laminar-turbulent transition in a flat-plate boundary layer. *Boundary Layer Transition & Control Conf.*, pp. 25.1–25.9. Cambridge, UK: R. Aeronaut. Soc.

Rogler, H. L. 1977. The coupling between freestream disturbances driver oscillations, forced oscillations and stability waves in a spatial analysis of a boundary layer. *AGARD Symp. Laminar-Turbulent Transition. AGARD-CP*-244, *Pap. No.* 16. Copenhagen

Rogler, H. L., Reshotko, E. 1975. Disturbances in a boundary layer introduced by a low intensity array of vortices. *SIAM J. Appl. Mech.* 28(2): 431–62

Ross, J. A., Barnes, F. H., Burns, J. G., Ross, M. A. S. 1970. The flat plate boundary layer. Part 3. Comparison of theory with experiment. *J. Fluid Mech.* 43(4): 819–32

Rothmayer, A. P., Smith, F. T. 1987. Strongly nonlinear wavepackets in boundary layers. *Trans. ASME*, p. 67, Cincinnati, Ohio, June 1987

Ryzhov, O. S. 1990. On the formation of ordered vortex structures from unstable oscillation in a boundary layer. *Z. Vychisl. Mat. i Mat. Fiz. (J. Comput. Math. & Math. Phys.)* 29(12): 1804–14 (In Russian)

Ryzhov, O. S., Savenkov, I. V. 1989. Asymptotic approach to hydrodynamic stability theory. *Matemat. Model. (Mathematical Simulation)* 1(4): 61–68 (In Russian)

Santos, G. R., Herbert. T. 1986. Combination resonance in boundary layers. *Bull. Am. Phys. Soc.* 31: 1718

Saric, W. S., Carter, J. D., Reynolds, G. A. 1981. Computation and visualization of unstable-wave streaklines in a boundary layer. *Bull. Am. Phys. Soc.* 26: 1252

Saric, W. S., Kozlov, V. V., Levchenko, V. Y. 1984. Forced and unforced subharmonic resonance in boundary-layer transition. *AIAA Pap. No.* 84-0007

Saric, W. S., Nayfeh, A. H. 1977. Nonparallel stability of boundary layers with pressure gradients and suction. *AGARD Symp. Laminar-Turbulent Transition. AGARD-CP*-224, *Pap. No.* 6. Copenhagen

Saric, W. S., Reynolds, G. A. 1980. Experiments on the stability of nonlinear waves in a boundary layer. See Eppler & Fasel 1980, pp. 125–31

Saric, W. S., Thomas, A. S. W. 1984. Experiments on the subharmonic route to turbulence in boundary layers. In *Turbulence and Chaotic Phenomena in Fluids*, ed. T. Tatsumi, pp. 117–22. Amsterdam: North-Holland

Saric, W. S., Yeates, L. G. 1985. Generation of crossflow vortices in a three-dimensional flat-plate flow. See Kozlov 1985, pp. 429–37

Savenkov, I. V. 1992. Resonant amplification of two-dimensional disturbances in semi-infinite channel. *Z. Vychislit. Mat. i Mat. Fiz. (J. Comput. Math. and Math. Phys.)* 31(8): 1331–39 (In Russian)

Savenkov, I. V. 1993. Wave packets, resonant interactions and soliton formation in inlet pipe flow. *J. Fluid Mech.* 252: 1–30

Schlichting, H. 1979. *Boundary-Layer Theory.* New York: McGraw-Hill

Seifert, A. 1990. *On the interaction of low amplitude disturbances emanating from discrete points in a Blasius boundary layer.* PhD thesis. Tel-Aviv Univ. 118 pp.

Schneider, S. P. 1989. *Effects of controlled three-dimensional perturbations on boundary layer transition.* PhD thesis. Calif. Inst. Technol., Pasadena. 187 pp.

Schubauer, G. B. 1957. Mechanism of transition at subsonic speeds. In *Boundary layer Research Symposium*, ed. H. Görtler, p. 85. Berlin: Springer-Verlag

Schubauer, G. B., Klebanoff, P. S. 1956. Contribution on the mechanisms of boundary-layer transition. *NACA Rep.* 1289

Schubauer, G. B., Skramstad, H. K. 1947. Laminar boundary-layer oscillations and transition on a flat plate. *J. Res. Nat. Bur. Stand.* 38: 251–92

Shapiro, P. J. 1977. The influence of sound upon laminar boundary layer instability. *MIT Acoustic and Vibration Lab. Rep.* 83458-83560-1. Cambridge: MIT

Shrira, V. I. 1989. On "near-surface" waves of upper, quasi-uniform layer of ocean. *Dokl. Akad. Nauk SSSR*[3] 308(2): 732–36 (In Russian)

Sinder, B. A., Ferziger, J. H., Spalart, P. R., Reed, H. L. 1987. Local intermodal energy transfer of the secondary instability in plane channel. *AIAA Pap. No.* 97-1202

Smith, F. T. 1990. Nonlinear breakdown in boundary layer transition. See Arnal & Michel 1990, pp. 81–91

Smith, F. T., Burggraf, O. R. 1985. On the development of large-sized short-scaled disturbances in boundary layers. *Proc. R. Soc. London Ser. A* 399(1816): 25–55

Smith, F. T., Doorly, D. J., Rothmayer, A. P. 1990. On displacement-thickness, wall-layer and mid-flow scales in turbulent boundary layers, and slugs of vorticity in pipes and channels. *Proc. R. Soc. London Ser. A* 428: 255–81

Smith, F. T., Stewart, P. A. 1987. The resonant-triad nonlinear interaction in boundary-layer transition. *J. Fluid Mech.* 179: 227–52

Spalart, P. R., Yang, K. S. 1987. Numerical study of ribbon-induced transition in Blasius flow. *J. Fluid Mech.* 178: 345–65

Squire, H. B. 1933. On the stability for three-dimensional disturbances of viscous fluid flow between parallel walls. *Proc. R. Soc. London Ser. A* 142: 621–28

Stewart, P. A., Smith, F. T. 1987. Three-dimensional instabilities in steady and unsteady non-parallel boundary layers, including effects of Tollmien-Schlichting disturbances and cross flow. *Proc. R. Soc. London Ser. A* 409: 229–48

Stewart P. A., Smith, F. T. 1992. Three-dimensional nonlinear blow-up from a nearly planar initial disturbance, in boundary-layer transition: theory and experimental comparisons. *J. Fluid Mech.* 244: 649–76

Strazisar, A. J., Reshotko, E., Prahl, J. M. 1977. Experimental study of the stability of heated laminar boundary layer in water. *J. Fluid Mech.* 83(2): 225–47

Stuart, J. T. 1960. Nonlinear effects in hydrodynamic stability. *Proc. 10th Int. Congr. Appl. Mech.*, pp.63–67. Amsterdam: Elsevier

Swearingen, J. D., Blackwelder, R. F. 1987. The growth and breakdown of streamwise vortices in the presence of a wall. *J. Fluid Mech.* 182: 255–90

Tam, C. K. W. 1978. Excitation of instability waves in a two-dimensional shear layer by sound. *J. Fluid Mech.* 89(2): 357–71

Tani, J., Komoda, H. 1962. Boundary-layer transition in the presence of streamwise vortices. *J. Aerosp. Sci.* 29(4): 440

Thomas, A. S. W. 1987. Experiments on secondary instabilities in boundary layers. *Proc. 10th US Natl. Congr. Appl. Mech.* Austin, Tex. ASME

Thomas, A. S. W., Bull, M. K. 1983. On the role of wall-pressure fluctuations in deterministic motions in the turbulent boundary layer *J. Fluid Mech.* 128: 283–332

Thomas, A. S. W., Saric, W. S. 1981. Harmonic and subharmonic waves during boundary-layer transition. *Bull. Am. Phys. Soc.* 26: 1252

Tollmien, W. 1929. Über die Entstehung der Turbulenz. In *Nachr. Ges. Wiss.*, pp. 21–44. Göttingen: Math.-Phys. Klasse

Vasudeva, B. R. 1967. Boundary-layer instability experiment with localized disturbance. *J. Fluid Mech.* 29(4): 749–63

Volodin, A. G., Zelman, M. B. 1978. Three-wave resonant interaction of disturbances in a boundary layer. *Izv. Akad. Nauk SSSR, Mekh. Zhidk. Gaza*[2] 5: 78–84 (In Russian)

Walton, A. G., Smith, F. T. 1992. Properties of strongly nonlinear vortex/Tollmien-Schlichting-wave interactions. *J. Fluid Mech.* 244: 649–76

Williams, D. R., Fasel, H., Hama, F. R. 1984. Experimental determination of the three-dimensional vorticity field in the boundary-layer transition process. *J. Fluid Mech.* 149: 179–203

Wortmann, F. X. 1955. Untersuchung instabiler Grenzschichtschwingungen in einen Wasserkanal mit der Tellurmethode. In *50 Jahre Grenzschichtforschung*, ed. H. Görtler, W. Tollmien. Braunschweig: Vieweg

Wubben, F. J. M. 1990. Experimental investigation of Tollmien-Schlichting instability and transition in similar boundary layer flow in an adverse pressure gradient (Hartree $\beta = -0.14$). *Low Speed Lab., Dept. Aerosp. Eng., Delft Univ. Technol., LR-report* 604. The Netherlands

Wubben, F. J. M., Passchier, D. M., Van Ingen, J. L. 1990. Experimental investigation of Tollmien-Schlichting instability and transition in similar boundary layer flow in an adverse pressure gradient. See Arnal & Michel 1990, pp. 31–42

Zelman, M. B., Maslennikova, I. I. 1982. Resonant amplification of spatial disturbances in boundary layer. In *Instability of Sub- and Supersonic Flows*, ed. V. Y. Levchenko, pp. 5–15. Novosibirsk: Inst. Theor. Appl. Mech. (In Russian)

Zelman, M. B., Maslennikova, I. I. 1989. On the formation of spatial structure of the subharmonic regime of transition in Blasius flow. *Izv. Akad. Nauk SSSR, Mekh. Zhidk. Gaza*[2] 3: 77–81 (In Russian)

Zelman, M. B., Maslennikova, I. I. 1990. On the spatial structure of disturbances in boundary layer subharmonic transition. See Arnal & Michel 1990, pp. 137–42

Zelman, M. B., Maslennikova, I. I. 1993. Tollmien-Schlichting-wave resonant mechanism for subharmonic-type transition. *J. Fluid Mech.* 252: 499–78

Zelman, M. B., Smorodsky, B. V. 1991a. On linear evolution of disturbances in boundary layers with inflexional velocity profile. *Z. Prik. Mekh. Tekh. Fiz.*[4] 1: 50–55 (In Russian)

Zelman, M. B., Smorodsky, B. V. 1991b. On influence of inflexion in mean velocity profile on the resonant interaction of disturbances in boundary layer. *Z. Prik. Mekh. Tekh. Fiz.*[4] 2: 61–66 (In Russian)

Zharov, V. A. 1986. On wave theory of developed turbulent boundary layer. *Uch. Zap. TsAGI*[5] 17(5): 28–38 (In Russian)

Zhigulyov, V. N., Krikinskiy, A. N., Sidorenko, N. V., Tumin, A. M. 1976. On a mechanism of secondary instability and its role in a process of turbulent onset. In *Aeromekhanika*, pp. 118–40. Moscow: Nauka (In Russian)

Zhigulyov, V. N., Tumin, A. M. 1987. *Origin of Turbulence. Dynamic Theory of Generation and Development of Instabilities in Boundary Layers*. Novosibirsk: Nauka Publ., Sib. Div. 282 pp. (In Russian)

Zhuk, V. I., Ryzhov, O. S. 1982. On locally-inviscid disturbances in boundary layer with self-induced pressure. *Dokl. Akad. Nauk SSSR*[3] 263(1): 56–69 (In Russian)

Zurigat, Y. H., Nayfeh, A. H., Masad, J. A. 1990. Effect of pressure gradient on the stability of compressible boundary layers. *AIAA Pap. No.* 90-1451

NOTES

[1]This Russian journal: *Izvestiya Sibirskogo Otdeleniya Akademii Nauk SSSR* (Proceedings of Siberian Division of the USSR Academy of Sciences) is translated completely starting in 1987 as: *Soviet Journal of Applied Physics*. The Russian name of this journal was changed in 1990 to: *Sibirskiy Fiziko-Tekhnicheskiy Zurnal* (Siberian Physical-Technical J.).

[2]This Russian journal: *Izvestiya Akademii Nauk SSSR, Mekhanika Zhidkosti i Gaza* (Proceedings of the USSR Academy of Sciences, Fluid Mechanics) is translated completely into English as *Fluid Dynamics*.

[3]This Russian journal *Doklady Akademii Nauk SSSR* (Reports of the USSR Academy of Sciences) is translated completely into English.

[4]This Russian journal *Zhurnal Prikladnoy Mekhaniki i Tekhnicheskoy Fiziki* (Journal of Applied Mechanics and Technical Physics) is translated completely into English as: *Journal of Applied Mechanics and Technical Physics*.

[5]This Russian journal: *Uchyonye Zapiski TsAGI* (Scientific Notes of TsAGI) is translated completely into English as: *Fluid Mechanics—Soviet Research*.

Annu. Rev. Fluid Mech. 1994. 26 : 483–527

PARALLEL SIMULATION OF VISCOUS INCOMPRESSIBLE FLOWS

Paul F. Fischer

Division of Applied Mathematics, Brown University, Providence, Rhode Island 02912

Anthony T. Patera

Department of Mechanical Engineering, Massachusetts Institute of Technology, Cambridge, Massachusetts 02139

KEY WORDS: concurrent distributed computing, fluid flow simulation, numerical methods, parallel processing, scientific computation

1. INTRODUCTION

1.1 *Scope*

This article deals with the formulation, implementation, and application of parallel numerical algorithms for the simulation of incompressible viscous fluid flows. The article is intended to combine some aspects of a general introduction for the broader fluid dynamics community, some aspects of a review article appropriate for researchers active in at least numerical simulation, if not parallel processing, and lastly, some material particularly directed towards researchers already engaged in parallel computing.

Parallel processing is a "server" discipline to fluid mechanics, and can be discussed in a very broad context. However, although properly abstracted numerical methods are arguably universal, the relevance, efficiency, implementation, and disciplinary implications of a method depend strongly on physical context. We thus address the fluid dynamical "slice" of parallel computing, with emphasis on those aspects of parallel processing which, by necessity or merit, fall within the purview of the computational

483

0066–4189/94/0115–0483$05.00

fluid dynamicist. We attempt to include and attribute fluid-mechanics-relevant parallel research originating in the computer science, mathematics, physics, and broader continuum mechanics communities; reciprocally, researchers in these allied fields may find material in this review paper that bears on their own efforts. Many parallel concepts are, in fact, little more than common sense, and, as a result, have been coinvented or reinvented in many research domains; where possible and appropriate, we shall prefer the more fluid dynamical origin.

We have chosen to narrow our field of discussion to the flow of viscous, incompressible (Newtonian) fluids—rather than including, say, inviscid compressible flows—for several reasons. First, the quantity of material associated with incompressible flow simulations is already more than sufficient to constitute a review paper of digestible size. Second, the equations and quantities of interest in viscous incompressible fluid mechanics and inviscid compressible fluid mechanics are rather different, and the resulting numerical algorithms and parallel implementations reflect these differences. Third, the compressible fluid dynamics community is largely focused on a well-identified set of critical scientific and engineering problems; in contrast, the incompressible fluid dynamics community is spread over a much wider variety of physical and technological applications. As a result, first, methods for compressible flows are even more specialized with respect to methods for incompressible flows than the governing equations might indicate, and, second, many review articles of computational fluid dynamics originate in, and focus on, the compressible-flow research area. Among several recent parallel computational fluid dynamics reviews and compendia that consider both incompressible and compressible flows are Braaten (1993), Reinsch et al (1992), Simon (1992), and Karniadakis & Orszag (1993). In what follows, we will often use "fluid dynamics" to refer to *viscous incompressible* fluid dynamics; this incompressocentric terminology is adopted to simplify the prose, not to indicate relative importance.

1.2 *Motivation*

Fluid dynamics is, of course, not simply the solution of the Navier-Stokes equations for a particular configuration. Most broader problem statements, applied or fundamental, involve a vector of physical and technology-related parameters, **r**, which must be averaged over, eliminated by optimization, or varied. For instance, one might be interested in finding the permeability of a porous medium for different pore volumes and geometries, in minimizing the pumping power required to effect a prescribed heat transfer rate, in determining the class of geometries that admit convective rather than absolute instabilities, or in developing a turbulence model for a certain set of separated flows. In all these examples, the typically rather large parameter space precludes purely numerical solution; analytical, heuristic, and experimental data, as well as intuition, must be brought to bear if final goals are to be achieved.

Strictly numerical solutions to these larger questions are not viable because of the nonlinear, singular, and oft-temporal nature of fluid mechanical phenomena, and because of the decidedly pointwise nature of much computational inquiry. First, each subproblem datapoint (e.g. Navier-Stokes solution) for a particular parameter-vector value, $\mathbf{r}'$, is typically expensive—due to the many degrees-of-freedom required to adequately resolve the physics; second, the Navier-Stokes data points are often computationally decoupled—a random porous medium problem can not be effectively "continued" in concentration, nor can a direct simulation of turbulence be efficiently continued in Reynolds number; third, limited regularity information is available—rapid (e.g. bifurcation-induced) changes in, say, drag and heat transfer, are common. In summary, we can not afford to perform a large number of simulations, nor can we assume that a few solutions will be representative of the entire parameter space of interest.

We believe that parallel processing is uniquely positioned to move computational fluid dynamics beyond the single "can-do" calculation stage. To understand more clearly the role of parallel processing, we introduce a simple economic model of computation. For a fixed Navier-Stokes subproblem and associated algorithm, the two significant post-code-development computational costs are: the direct computational cost C_{dir}, which is directly (respectively inversely) proportional to the purchase cost c_P (respectively speed, s) of the machine; the indirect—or opportunity—cost C_{ind}, defined as wall-clock (or elapsed) computation time, which is inversely proportional to the speed of the machine. Consider two machines, the first capable of speed s_1 (measured in MFLOPS: millions of floating point operations per second), the second of speed $s_2 = s_1/\bar{P}$. We now introduce a third machine which is simply $\bar{P}$ identical processor-memory units of machine 2. Taking for the moment the most optimistic view that the numerical algorithm to be implemented is completely concurrent and noncommunicative, the time-to-compute C_{ind} on machine 3 will clearly be the same as for machine 1; more importantly, one unit of machine 2 will cost significantly less than $1/\bar{P}$ of the "technology-envelope" machine 1, and hence our computation on machine 3 will cost substantially less—in direct costs C_{dir}—than the same computation on machine 1.

This rather naive scenario has, in fact, been played out rather faithfully over the past decade. Vector supercomputers (machine 1) will soon be surpassed in performance (C_{ind}), and have already been bettered by an order of magnitude in performance-price considerations (C_{dir}), by a new class of scalable distributed-memory ensemble architectures (machine 3)—"multicomputers"—comprising tens to thousands of loosely connected commodity-chip units (machine 2). In the short term, these new architectures have precipitated a one-time "Schumpeterian" improvement in computational capability; this event is now past, and will be discussed no further in this paper. In the future, ensemble architectures will ensure that supercomputers enjoy the same growth curve as—and are able to directly exploit advances in—commodity-chip technology (Seitz 1992). In the past decade, commodity-chip technology has exhibited a 50-fold increase in speed s, and a 10-fold decrease in purchase cost c_P, resulting in rather remarkable 500-fold and 50-fold decreases in direct (C_{dir}) and indirect (C_{ind}) costs, respectively; there is every reason to believe that hardware cost and speed will continue to improve at a similar rate during the coming years. (We implicitly restrict our attention to scalable distributed-memory machines: Distributed-memory machines will be more likely to track the commodity-technology curve than shared-memory machines; methods that perform well on distributed-memory machines will also perform well on shared memory machines, and will, furthermore, be able to effectively exploit any hierarchical memory that might be present.)

It is not the case, however, that parallel computing will allow us to solve a given problem arbitrarily quickly. In particular, most respectable algorithms for fluid flow problems do not admit trivial concurrency; as the number of processors P increases, the computational grain size will decrease, and communication overhead and less-than-P-parallel operations will dominate. To achieve reasonable parallel efficiency we must consider "scaled speedup" (Gustafson et al 1988, Fox et al 1988), in which the number of processors requisitioned for a problem scales with the *size* of the problem; it follows that parallel computing will have the most significant proportional impact for large problems. Here large problem or many degrees-of-freedom must be interpreted in relation to the hardware ratio of the time-to-send a word to the time-to-calculate a result; if the past decade is any indication, this ratio will fluctuate, but may not change substantially over longer time scales. A fixed discrete problem optimally parallelized on $O(50)$ processors in 1994 may thus still be optimally parallelized on $O(50)$ processors in 2004.

We believe that future parallel computing investigations will lie between two polar scenarios. In Scenario I, we solve the same discrete problem in 1994 and 2004; the latter, by virtue of hardware advances, will probably incur only 1/500th the direct cost and (if as many processors are used in 2004 as in 1994) 1/50th the indirect cost of the former. Scenario I will permit greater, and more timely, variation in more parameters, thereby increasing both physical insight and engineering relevance. In Scenario II, we consider the treatment of problems of increasing complexity at fixed cost. It is plausible to assume that, even in the absence of algorithmic advances, parallel $\bar{P}$-processor 1994 and $10\bar{P}$-processor 2004 calculations with N and 500N degrees of freedom (or operations), respectively, will incur roughly the same direct *and* indirect costs (in fact, the increased grain size of our hypothetical 2004 calculation can be converted into further decreases in indirect costs). In contrast, a serial 2004 calculation with 500 times the degrees of freedom of a serial 1994 calculation would (presuming sufficient memory) probably exhibit roughly the same direct cost as the 1994 calculation, but a tenfold *increase* in the time to compute. Increasing numbers of processors applied to increasingly large problems will bring about improvements in reliability, relevance, and capability: Marginally resolved calculations will be repeated with a now-adequate number of degrees of freedom; highly simplified mathematical models will be replaced with more sophisticated models which reflect more of the physical ingredients embodied in the originating physical problem; models will be extended to include the new physical phenomena and additional scale complexity that inevitably accompanies the development of new technologies.

It should be noted that, for either scenario, for a given processor technology, parallel processing reduces only the indirect costs C_{ind}; in fact, direct costs C_{dir} will increase with P as $1/\eta(P)$, where $\eta(P)$ is the parallel efficiency. We can thus classify various applications of parallelism in terms of the relative importance attached to rapid turnaround. At one extreme, for throughput parallelism (facility utilization) the optimal choice—*memory permitting*—is relatively few processors. At the other extreme, industrial applications sensitive to time-to-market, interactive educational applications responsive to pedagogical concerns, visualization applications linked to creative modeling processes, and, ultimately, real-time control applications clearly place a premium on reduced indirect costs, even at the expense of increased direct costs.

It follows from Scenario II that, for the forseeable future, many individual Navier-Stokes calculations, and certainly many broader questions involving numerous fluid mechanics simulations, will remain expensive. Parallel processing will not alleviate the need for turbulence models, for further advances in numerical discretization and solution procedures, or for new approaches to incorporating large-scale computations into the engineering design (and perhaps even control) processes. The engineering applications are of particular, and renewed, interest: Parallel processing provides a unique mechanism by which to control the indirect (elapsed-time) costs of computing.

2. THE TECHNICAL CHALLENGE

2.1 *Algorithms and Architectures*

The major algorithmic challenge in parallel processing is the development of numerical methods that make effective use of parallel machines, that is, achieve reasonable parallel efficiency, and are respectable in terms of absolute operation count. In the final analysis, it is the product of these two factors that determines direct and indirect computational costs. Fortunately, for many problems, either the "best" serial algorithm can be effectively parallelized, or new equally respectable concurrent algorithms can be devised. Many researchers view the parallel algorithm development process in four stages. First, one selects a particular discretization and solution method, and represents the resulting computational tasks in some form of *computational graph*. Second, a *model architecture* is introduced which serves as a hardware-neutral algorithm-architecture interface. The model architecture may be specified in two fashions: explicitly, in terms of pseudo-hardware characteristics related to node (processor) performance, network topology (interconnection of nodes), and network performance; or implicitly, in terms of functional requirements. Third, a partition procedure P_1 is proposed which, based on heuristic analysis, complexity estimates, and perhaps optimization procedures, suggests a reasonable computational graph-to-model architecture mapping. Fourth, *implementation* on a real parallel machine is considered: This entails a second partition procedure P_2, which maps the model architecture to the actual architecture, thereby introducing *nonidealities* into the model-architecture parallel complexity estimates. Algorithms and model architectures, data and work partition schemes, parallel complexity estimates, and real-machine applications will be discussed in subsequent sections. We restrict ourselves here to an introductory characterization of model architectures and real machines.

A viable model architecture will balance algorithmic requirements and known machine characteristics: The former leads to simple complexity estimates and defines desirable or optimal performance; the latter ensures that model analyses will approximately reflect actual performance on some real machine. Most distributed-memory model architectures explicitly or implicitly assumed by many computational fluid dynamicists embody at least the following "baseline" ingredients:

1. the architecture comprises P (requisitioned) processor/memory nodes;
2. the P nodes are loosely interconnected by a topology described by a graph $G_0(V_0, E_0)$, where V_0, the vertex set, comprises the set of processing nodes, and E_0, the edge set, includes an edge between each pair of nodes connected by a physical communication channel;

3. the time to access local data and compute a result on a node is given by t_{calc};

4. the time to send m (32-bit) words between any two processors connected by an uncontested path in $G_0(V_0, E_0)$ is given by $mt_{comm}(m)$, where $t_{comm}(m)$, the time per word to send m (32-bit) words, is a decreasing function of m, and $t_{comm}(1) \gg t_{comm}(\infty) > t_{calc}$;

5. the time to perform and distribute the summation of P values distributed over P processors (a vector reduction, or gather/scatter operation) is dominated by $(const)t_{comm}(1) \log_2 P$.

We now discuss the extent to which actual machines satisfy these requirements. First, 1 is, by definition, satisfied, though in practice several model processors (often referred to as virtual parallel processors) may map to a single node of the actual machine. Second, 2 is satisfied so long as $G_0(V_0, E_0)$ can be embedded in, or "mapped to," the network of the actual machine, $G_A(V_A, E_A)$; G_0 is embedded in G_A, if, for every node in V_0 there exists a node in V_A, and, to every edge in E_0 we can assign an associated path in G_A such that two paths in G_A associated with two distinct edges in E_0 share no edges in E_A in common. For example, rings and meshes are easily embedded in hypercubes using binary-reflected Gray codes (e.g. Saad & Schultz 1985). Requirement 3 is, to some extent, definition; however, for actual machines, pipelining and vector considerations, as well as local memory access time—neglected in our model architecture—are critical to high performance. Requirement 4 is designed to reflect the relatively slow communication and message-startup latency typical of actual machines. It might appear unrealistic that the time to send m words is independent of the number of nodes ("hops") in the path linking the sending and receiving nodes; indeed, much early work on parallel processing focused on minimum-dilation mappings. However, in more recent "wormhole," or direct-connect, routers, messages are passed through intermediate nodes with effectively no interference, and distance is thus no longer a first-order *direct* concern. Lastly, requirement 5 can be satisfied by any number of "divide-and-conquer" tree-like embeddings in, for example, mesh, tree, or hypercube topologies $G_A(V_A, E_A)$.

The most common nonideality in the model architecture-to-actual machine mapping derives from the graph embedding issue. We emphasize that the critical concern is communication *contention*, in which parallel communication on the model architecture is thwarted on the actual machine by the presence of the same *physical* channel in the communication paths linking more than one pair of potentially simultaneously communicating nodes. Although more sophisticated routers can mitigate mapping-induced contention, routers do not necessarily eliminate the contention problem. Note that our term "nonideality" denotes only an imperfect match between the model and actual architectures; parallel efficiency is limited not only by nonidealities, but also by other overhead factors, such as communication, already reflected in the model architecture characterization.

We close this section with a very brief summary of a few existing machines; a more extensive catalog of this moving target can be found in Schreiber & Simon (1992). First, as regards number of processors, current machines comprise from tens to thousands of nodes, with from 8MB to 128MB per node; more critical than the actual number of nodes present in any particular system is intrinsic scalability of processors and memory, which most current distributed-memory machines enjoy. Second, as regards topology, parallel machines have, since incipience, explored a variety of network configurations, including the single bus, the ring, two-dimensional meshes, trees, and hypercubes; certain initially rejected configurations have been subsequently rehabilitated by advances in communication protocols. In general, the network interconnection scheme reflects performance, cost, and technology considerations (Dally 1987, Scott 1991). It is unlikely that final consensus has been reached: The Intel Scientific Supercomputers iPSC/860 (released 1990) and Paragon (released 1992) are based upon hypercube and two-dimensional mesh topologies, respectively; the Thinking Machines CM-2 (released 1987) and CM-5 (released 1991) are based upon hypercube and fat-tree topologies, respectively. Lastly, typical current hardware characteristics are: $t_{calc} \approx 0.1\mu s$, $t_{comm}(\infty)/t_{calc} \approx 10$, $t_{comm}(1)/t_{comm}(\infty) \approx 50\text{--}100$. Note that networked workstations comprise an alternative distributed-memory architecture which roughly honors the general requirements of our baseline model architecture; the hardware characteristics for networked workstations depend upon the platform and network components selected.

2.2   The Software Question

In this section we describe some *programming models* currently proposed; although the programming model and model architecture are, to some extent, distinct, limited program models can certainly restrict the available range of an architecture. We cursorily discuss three topics: program control, data control, and parallelization tools. *Program control* refers to the process by which already-identified concurrent tasks are assigned to available processors. The two most common models (Flynn 1972) are Single-Instruction-Multiple-Data (SIMD), in which a sequencer broadcasts a largely processor-uniform instruction to many processors acting in lockstep on different data, and Multiple-Instruction-Multiple-Data

(MIMD), in which typically self-synchronized individual processors run potentially completely different programs. An existing "data-parallel" SIMD machine is the CM-2 (the CM-5 supports both SIMD and MIMD constructs); current message-passing MIMD "multicomputers" include the iPSC/860 or Paragon, and numerous implementations of networked workstations. Algorithms are more readily programmed (and debugged) on SIMD machines; conversely, MIMD machines can more readily deal with computational inhomogeneities arising from, for example, adaptive mesh refinement or heterogeneous constitutive laws. In practice, many data-parallel computational fluid dynamics programs are constructed as Single-Program-Multiple-Data (SPMD), which combines some of the simplicity of SIMD with some of the flexibility of MIMD; SPMD programs are typically implemented on MIMD machines.

Program control concepts and implementations are relatively mature and sophisticated. There is, however, little consensus on the best approach to *data control*. Data control is the process by which data resident on one processor-memory unit is accessed by a second processor-memory unit. The spectrum of current approaches includes: explicit message passing extensions of standard languages, and libraries of common communication constructs (e.g. vector reduction), as exemplified by EXPRESS (ParaSoft 1992) and PVM (Beguilin et al 1992); tools for medium-level abstraction of data and data protocols, such as DIME (Williams & Glowinski 1989), PARTI (Sussman et al 1992), and PC (Scott & Bagheri 1990); and shared-memory programming models. There are clearly compromises to be made between run-time (performance) costs and pre-run-time (development) costs; in a shared-memory programming model for a distributed-memory machine, will messages be automatically, and optimally, bundled to reduce hardware communication latency?

The previous discussions of program and data control assume that the programmer dictates how tasks and data are to be distributed, with the software environment then facilitating implementation but not decision. In contrast, *parallelization tools* directly contribute to the decision process by which data and tasks are distributed. Parallelization tools may be: implicit in the programming model or programming language (e.g. FOR-TRAN 90), as is often the case for SIMD machines; semi-automatic, involving compiler directives; or fully automatic, as in parallelizing compilers. The ultimate goal is the autonomous synthesis, based only on architecture-free procedural statements, of appropriate data and work partitions. Automatic compilers are, unfortunately, currently unavailable. Furthermore, even if automatic compilers become available in the future, they will not be perfect, demanding further compromises between realized performance and development effort: Will an automatic parallelizing compiler distribute a high-order element over two processors, leading to unnecessary communication?

In summary, we remark that *no* tools or automatic compilers will be able to extract good parallel performance from a computational procedure that is fundamentally nonconcurrent. This implies that, quite independent of computer science advances, algorithmic issues—well within the purview of the computational fluid dynamicist—will remain important. Nevertheless, it is important to bear in mind that parallel processing will be advantageous only for those applications for which the benefit accrued from the solution more than balances both the run-time costs *and* the resources expended developing parallel implementations. Without significant progress in efficient, cross-platform standardized high-level tools, in particular shared-memory programming models and automatic parallelizing compilers, the fraction of problems that satisfy this criterion will remain small, and effectively serial computing will remain prevalent. (Third-party commercial software constitutes another solution to the parallel coding problem: a developer can amortize implementation costs over many users who, in turn, benefit from reduced run-time costs. At present, only a few commercial or in-house industrial (Braaten 1991) parallel incompressible fluid dynamics codes are available.) We hope parallel automation comes to pass, but focus here on algorithmic concerns which are, in any case, necessary, and both necessary and sufficient for sufficiently "important" problems.

3.   FINITE-ELEMENT DISCRETIZATIONS

3.1   *The Poisson Problem*

We begin by considering finite-element discretization of the Poisson problem, and then proceed, through the Stokes problem, to the full Navier-Stokes equations. The methods described are also applicable to finite-volume (Braaten 1991) and finite-difference (Gropp & Keyes 1992a, Cline & Schutt 1992) "nodal" representations. The Poisson model problem does not reflect many of the difficulties of even the Stokes problem, let alone the much more difficult Navier-Stokes equations. There are, however, good reasons to consider this problem: First, as regards computational constructs, iterative and direct solution procedures for the Poisson problem exercise many of the parallel kernels basic to solution of more general partial differential equations; second, the Poisson problem (for the pressure) is of direct interest in a number of fractional-step approaches to the full time-dependent *incompressible* Navier-Stokes equations; third, this problem serves as a convenient vehicle to illustrate many of the basic concepts of parallel processing.

3.1.1 ITERATIVE SOLUTION METHODS   *Structured problems*   We describe here simple unpreconditioned (or diagonal-preconditioned), conjugate-gradient iterative solution techniques (Golub & van Loan 1983). Although conjugate-gradient iteration is, in fact, computationally illustrative of a broad class of Richardson, steepest descent, and Krylov-space iterative techniques, a more rapid convergence rate requires more global operations and, hence, new parallel considerations (see Section 3.3).

We begin with a structured problem in a square; the regular domain may arise either because the original problem enjoys simple geometry, or because a global mapping procedure has been applied. We discretize the domain by N_{el} identical, square, Q_1 (first-order in each direction) Lagrangian finite elements, from which we derive roughly N_{el} associated global nodal degrees of freedom. Each iteration of the conjugate-gradient procedure comprises three basic computational tasks: (EV) evaluation of the action of the discrete operator on a known vector of finite-element nodal values; (UP) update of a vector of finite-element nodal values as the weighted sum of two known vectors; and (IN) calculation of the inner product between two vectors of finite-element nodal values.

For the *model architecture* we adopt the baseline system described in Section 2. Given the simple problem geometry, a natural choice for the model topology, G_0, is a two-dimensional $\sqrt{P} \times \sqrt{P}$ mesh, in which (almost) every processor is connected to four lattice neighbors. Note that, by virtue of pass-through routing, the mesh architecture accommodates $\log_2 P'$ vector reduction constructs over P'-processor *sub*networks; this capability can be important in the more complex algorithms described in subsequent sections. For the data and work distribution procedure P_1, one simply associates a subsquare of $\sqrt{N_{el}/P} \times \sqrt{N_{el}/P}$ contiguous elements, with associated finite-element nodes, to each of the processors, as shown in Figure 1. [As conjugate-gradient iteration is a "simultaneous displacement" approach, we need not consider coloring issues; the latter are, however, important in Gauss-Seidel-like methods (Harrar et al 1991).] Note that, by virtue of the element-based distribution, certain global finite-element nodes appear on more than one (and as many as four) processors. Given this homogeneous data partition, the work distribution then follows naturally as different instantiations of the same basic task (in "SPMD" style); this form of parallelism—data parallelism—is certainly the most prevalent mode of conceptualizing parallel algorithms.

The EV operation is performed in parallel by the now-standard global-matrix-free evaluation approach: (EV_1) the local elemental matrix-vector product is evaluated on each processor; (EV_2) an exchange/summation of finite-element nodal values on the edges of processor subsquares is effected, first to the east and west, then to the north and south. The four-node

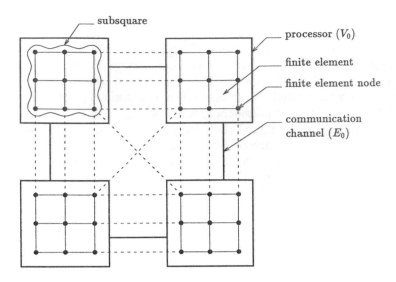

Figure 1 Mesh model architecture topology and element-to-processor distribution scheme for finite-element discretization of the Poisson problem in a square; here $P = 4$ and $N_{el} = 16$. Dashed lines indicate local nodes that represent the same global degree of freedom.

vertex summation is automatically and correctly effected by this two-stage edge summation procedure (Fox et al 1988). To perform the UP operation, given the redundant storage of global degrees of freedom, one simply updates the local copy of the finite-element nodes resident on each processor. Lastly, to perform the inner product IN: (IN_1) the local subsquare inner products are summed in parallel (dividing the contribution of each finite-element node by the number of processors on which a copy of the node resides); (IN_2) the $\log_2 P$ gather/scatter supported by the model architecture is evoked to perform the inter-processor sum. The parallel procedure can be synchronized, or can be loosely synchronized by the EV_2 and IN operations.

From the description of this algorithm and the model architecture definitions of Section 2 it follows that the solution time on P processors is given by

$$\tau_{CG}(P) = c_0 \sqrt{N_{el}} \{ \overbrace{[\delta_{1,P} + (1 - \delta_{1,P}) \lambda_1] c_1 t_{calc} N_{el}/P}^{EV_1, UP, IN_1}$$
$$+ (1 - \delta_{1,P}) \underbrace{[\lambda_2 c_2 t_{comm}(\sqrt{N_{el}/P}) \sqrt{N_{el}/P}}_{EV_2} + \underbrace{\lambda_3 c_3 t_{comm}(1) \ln P]}_{IN_2} \}, \quad (1)$$

where the c_i are constants independent of N_{el} and P, the prefactor $\sqrt{N_{el}}$ reflects the number of iterations required for convergence, and $\delta_{i,j}$ is the Kronecker delta symbol. The λ_i are, at present, unity, but may later reflect both additional parallel overhead in the model architecture and model-to-actual architecture nonidealities. Note that if the subsquare of finite elements is replaced with a single high-order (Lagrangian) spectral element (Fischer & Patera 1991, Watts et al 1992), the EV_2 communication term will remain *invariant*; it follows that high-order methods are more advantageous in the parallel than in the serial context, at least for smooth problems and algorithm-architectures which are communication-bound.

From (1) a simple expression for the "same-algorithm" parallel speedup, $S_{CG}(P) = \tau_{CG}(1)/\tau_{CG}(P)$, can be derived,

$$S_{CG}(P) = \frac{P}{\lambda_1 + \lambda_2 \dfrac{c_2 t_{comm}(\sqrt{N_{el}/P})}{c_1 t_{calc}} \dfrac{\sqrt{N_{el}/P}}{N_{el}/P} + \lambda_3 \dfrac{c_3 t_{comm}(1)}{c_1 t_{calc}} \dfrac{\ln P}{N_{el}/P}}. \qquad (2)$$

Estimates of this variety have been developed, and experimentally verified, by numerous authors (Fox et al 1988, Gustafson et al 1988, Gropp & Keyes 1988, Braaten 1991, Fischer & Patera 1991). First, we note that, in the limit in which t_{comm} is very small compared to t_{calc}, the speedup is simply P. This reflects first, the natural concurrency of the method, and second, the perfect load balance of the decomposition: All processors are equally occupied. If we ill-advisedly place $2N_{el}/P$ elements on one processor (and suitably decrement the number of elements on the other processors), we find $\lambda_1 = 2$, and a correspondingly reduced speedup of $P/2$. Second, we note that, even in the realistic case where t_{comm} is large compared to t_{calc}, the communication terms in the denominator of (2) can be controlled by proper choice of *granularity*. In particular, given that the algorithm is fundamentally local, and that the communication is effectively parallel, the surface-to-volume effect ensures that the second term in the denominator will be small so long as $\sqrt{N_{el}/P}$ is large compared to $t_{comm}(\sqrt{N_{el}/P})/t_{calc}$. Conversely, if we insist on fixing N_{el}, and consider the limit in which $P \to \infty$, we will have insufficient computation to cover the communication overhead, and the parallel efficiency, $\eta_{CG}(P) = S_{CG}(P)/P$, will tend to zero.

It is now well understood that the limit in which parallel processing is effective is not $P \to \infty$ for fixed N_{el}, but, rather, $P \to \infty$, $N_{el} \to \infty$, with the optimal granularity, N_{el}/P (qualitatively referred to as coarse, medium, or fine) determined by the algorithm and hardware "constants." To demonstrate this concept of scaled speedup (Fox et al 1988, Gustafson et al 1988), we now assume, for simplicity, that the granularity has been chosen such that the second term in the denominator of (2) is, indeed, negligible, and that the λ_i are unity. It then follows that the maximum speedup (minimum C_{ind}) is obtained for $P^{opt} = c_1 t_{calc} N_{el} / c_3 t_{comm}(1)$ processors, giving

$$\tau_{CG}^{opt} = \tau_{CG}(P^{opt})$$

$$= c_0 \sqrt{N_{el}} \left[c_3 t_{comm}(1) + c_3 t_{comm}(1) \ln \left(\frac{c_1}{c_3} \frac{t_{calc}}{t_{comm}(1)} N_{el} \right) \right]. \quad (3)$$

First, we note that the optimal number of processors increases with the size of the problem, N_{el}, and that the time per iteration on the optimal number of processors is, to within logarithmic corrections, independent of problem size. The solution time is not independent of problem size, however, since the number of conjugate-gradient iterations will increase as $\sqrt{N_{el}}$; improvements to this convergence rate require more implicit constructs which will, typically, decrease parallel efficiency. Second, we note that the optimal number of processors, and associated speedup, *increase* with decreasing computer speed, that is, increasing t_{calc}; this anomalous effect, as well as many others, arise when relative performance measures are incorrectly interpreted as absolute metrics (Bailey 1992). Although it is true that a slower computer appears to be a faster communicator, and hence can more effectively use more processors, the actual computation time, (3), will, of course, be larger the slower the processor.

We now turn to the last step of the implementation process: the consideration of nonidealities, primarily related to contention. As the simple mesh of Figure 1 is readily embedded in both meshes and hypercubes, most current machines would permit a contention-free mapping, P_2, from model to actual architecture. It is possible, however, that the particular hardware properties of the target machine will suggest alternative distribution schemes. For example, on a machine for which $t_{comm}(1) \gg t_{comm}(\infty)$, that is, on which latency completely dominates communication time, we would do better to partition our problem not into P boxes, but into P strips, each containing $\sqrt{N_{el}} \times \sqrt{N_{el}}/P$ elements: This strategy avoids (say) the first EV_2 exchange and, more generally, reduces the total number of messages initiated (Farhat & Lesoinne 1993).

It is important to note that nodal representations are not the only possible bases for finite-element spaces, and that other bases may lead to different forms of parallelism. In one such alternative approach (Griebel et al 1992, Griebel 1992), the finite-element space is represented by a hierarchical basis (related to wavelets), which then permits approximate construction of the discrete solution by a $\log_2 N_{el}$-subproblem combination technique similar to a systematic Richardson extrapolation. Each subproblem is an independent problem comprising (for the square model problem) either $\sqrt{N_{el}}$ or $\sqrt{N_{el}}/2$ total degrees of freedom. These sub-problems can thus be forked, in task-parallel, or control-parallel, style, to different processors. Although the procedure enjoys somewhat limited ($\log_2 N_{el}$) concurrency, communication requirements are low, and the technique is relatively simple to implement. This method has been implemented on a network of workstations (Griebel et al 1992), and extended to the Navier-Stokes equations in complex geometries through domain decomposition concepts (Griebel 1992).

Domain decomposition   As an intermediate step between structured problems and fully unstructured problems we consider domain decomposition procedures (Gropp & Keyes 1988, 1992a; Quarteroni 1991; Glowinski et al 1991). Domain decomposition typically connotes the association of groups of elements in roughly regular, few-sided, simply connected *subdomains*, the union of which reproduces the original discretization and geometry. Heterogeneous discretization and solution procedures that rely on the hierarchical domain decomposition framework are often denoted "domain decomposition" approaches. The domain decomposition notion allows us to extend the parallel framework for structured problems to globally unstructured problems for which a relatively regular subdomain definition can be prescribed. In some cases this subdomain definition takes the form of automatic partition of either a regular domain or a domain that admits a geometrically-induced partition; in other cases, the subdomain definition may derive from a semi-automatic mesh generation process.

An example of an irregular domain decomposition is shown in Figure 2, in which each circle represents a cylindrical obstacle, and the entire domain represents one realization of a random porous medium. For this problem the domain decomposition (dark lines in Figure 2) derives from cylinder-cum-generator Voronoi tesselation of the geometry (Sugihara & Iri 1989). A subdomain may include many low-order elements (as in Figure 2), or may comprise a single high-order element. Similarly, subdomains may be locally unstructured (as in Figure 2), providing greater flexibility in mesh and geometry generation, or locally structured, promoting efficient vectorization, pipelining, and potentially fast solution. Subdomains are typically selected prior to mesh generation, thereby permitting parallel mesh generation, albeit at the risk of less-than-perfect load balance.

The parallel implementation of conjugate-gradient iteration of domain decomposed finite-element discretizations follows closely the procedure for structured discretizations. The simplest approach is to replace the model architecture topology of Figure 1 with a new model architecture topology in which there is one processor for each subdomain, and one channel for each subdomain edge shared by two subdomains. The com-

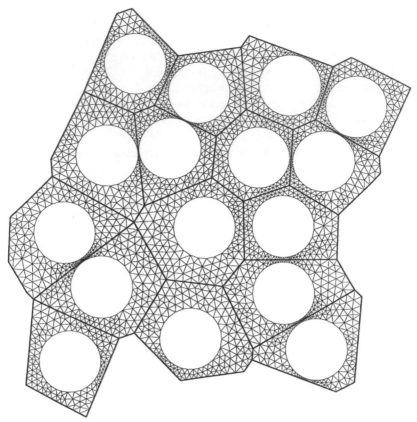

Figure 2 Unstructured triangular mesh and Voronoi domain decomposition for one real-ization of a fibrous porous medium; circles represent cylindrical obstacles (from Cruz 1993, with permission).

plexity estimate (1) then directly applies, save that: The first term, reflecting EV_1 (and IN_1) must be modified—i.e. λ_1 increased—to reflect the *maximum* number of elements per processor, with a corresponding decrease in speedup and efficiency; special vertex collection algorithms must now be effected, as the east-west north-south procedure will, in general, no longer be applicable. Proposed approaches to the general vertex summation problem include, in order of increasing efficiency (and complexity): distribution of all vertices to all processors, with global gather/scatter summation; division of vertices into regular vertices, which admit summation by an ordered sequence of edge summations, and special

vertices, which are subsequently treated by a reduced-order global gather/
scatter operation (Fischer & Patera 1991); and a routed gather/scatter
procedure which prunes the list of incomplete vertices as the requisite
summations are completed (Fox et al 1988). [It is also possible that certain
nonconforming approximations will obviate the need for vertex summation (Y. Maday, private communication).] The net effect of these vertex-
summation modifications is a larger $\ln P$ term in (1)—i.e. λ_3 is increased—
and a correspondingly reduced speedup. (For clarity and brevity we shall
present largely two-dimensional formulations; however, much of the work
reviewed considers fully three-dimensional problems. The extension of
the two-dimensional vertex-edge-area framework to the three-dimensional
vertex-edge-face-volume wireframe is perhaps the most challenging parallel task that must be undertaken.)

Lastly, as regards parallel implementation, the more irregular domain
decomposition model architecture will no longer obviously map to the
actual architecture of, say, a mesh-topology or hypercube-topology
machine: Contention nonidealities will appear, which will be reflected in
increased λ_2. Many researchers (e.g. Fischer & Patera 1991, Williams 1991),
assume, often on the basis of numerical experiments, that contention is
not a significant factor, perhaps at least partially due to the local and
homogeneous nature of finite-element constructs—this, as well as other
conclusions in this paper, may well change as various technology components improve at different rates. As regards model-to-actual architecture
distribution algorithms, P_2, it thus follows that: For single-subdomain to
single-processor mappings, a well designed domain decomposition will, by
construction, provide reasonable load balancing and relatively minimal
communication for effectively any subdomain-processor mapping; for
multiple subdomains per processor, the unstructured methods described
below may be applied at the subdomain level.

Unstructured problems   As a simple illustration of an unstructured problem we again refer to Figure 2, but now presume that only the mesh is
known, and no prior domain decomposition procedure is available. We
must therefore group the elements into "collections" (we reserve the term
subdomains for the regular structures described in the previous subsection)
which can then be mapped to model and finally actual processor configurations. Once these mappings are effected the parallel solution algorithm is largely similar to that described for domain decomposition discretizations—albeit with the potential for better load balance, but greater
contention (increased λ_2) and vertex (increased λ_3) contributions. We thus
focus on different techniques proposed for forming and mapping element
collections to actual architecture topologies (the model architecture top-

ology is less pivotal in unstructured problems since a general partition procedure must, in any event, be devised). Note that element-to-processor mapping techniques are effectively applied to the dual of the mesh graph, in which we associate a graph vertex to each finite element, and a graph edge to each pair of graph vertices associated with finite elements that share an elemental edge.

One class of techniques explicitly states the data and work distribution, or assignment, problem as an optimization problem, and then proceeds, for a particular discretization, to optimize the partition prior to performing the actual calculation. Current objective functions typically include (Fox et al 1988, Williams 1991, Farhat & Lesoinne 1993): a penalty on load imbalance, represented most simply as the maximum relative discrepancy in the number of elements assigned to each processor; a penalty on communication, represented, for example, as the total number of finite-element nodes on collection boundaries. The objective function might also penalize the number of collection edges in order to reflect latency and contention considerations. Earlier distribution objective functions also included mapping dilation effects.

The graph theoretic or combinatorial approach to the distribution problem reveals that the assignment problem is, at best, very difficult (Bokhari 1979). At the other extreme, reconfigurable hardware solutions to the problem appear—in retrospect—whimsical (Anagnostou et al 1989). At present, the most common approach to objective-function minimization is simulated annealing (Fox et al 1988, Williams 1991). Simulated annealing is a probabilistic descent method, in which the probability of accepting a randomly generated *cost-increasing* transition is proportional to a pseudo-temperature control variable. The pseudo-temperature is initially set high, so as to move the cost probability distribution rapidly to a stationary state, and then slowly reduced, in order to find the minimum-"energy" state. Simulated annealing is general and certainly simple to implement; however, the success and efficiency of the method depend strongly on the transition topology introduced and the rate of cooling selected. For parallel assignment problems the method appears reasonably successful, perhaps because the minimum is relatively broad, and the effect of less-than-global minima on parallel performance is therefore not great. Simulated annealing and other nondeterministic optimization procedures can also be gainfully employed to improve upon initial partitions generated by direct heuristic techniques (Vanderstraeten et al 1993).

A second class of techniques, perhaps the most widespread in current application, articulates the partition objective function only implicitly through the heuristic distribution procedure. Most of these techniques aim primarily to form aggregates which are of roughly the same size (for purposes of load balancing), and involve a minimal number of collection edges comprising a minimal number of finite-element (or, more generally, finite-volume or finite-difference) nodes. Once these aggregates are defined, the actual collection-to-processor mapping is typically less critical, assuming, once again, that routers will mitigate the effect of non-nearest-neighbor interactions and contention. Of these "implicit" heuristic approaches appropriate for iterative solvers we mention four: the greedy algorithm; recursive coordinate bisection; recursive graph bisection (or nested dissection); and recursive spectral bisection. Note that these algorithms themselves can be parallelized (Williams 1991, Ramamurti & Lohner 1993), which may be important for large problems, or for problems that require adaptive mesh refinement or continual mesh modification.

The greedy algorithm (Farhat & Lesoinne 1993) is the simplest and fastest: Starting with any finite element, neighboring elements are added until N_{el}/P have been included; one then proceeds to a new seed element, and the process is repeated. As the name implies, the algorithm is somewhat impulsive, though accumulation and seed heuristics can be developed that lead to reasonable aggregates. The recursive coordinate bisection method (Farhat & Lesoinne 1993) chooses a direction $\mathbf{q}$ (most plausibly one of the principal axes of the moment of inertia tensor), and sorts the elements into two groups according to projected location on the $\mathbf{q}$ axis; this process is then applied recursively to the resulting groups, until the requisite number of collections have been identified. Recursive graph bisection (Simon 1991) is similar to recursive coordinate bisection, except that distance is measured not with respect to geometric coordinate, but, rather, with respect to the shortest path in the mesh dual graph; nodes are sorted by, for example, the reverse Cuthill-McKee algorithm (George & Liu 1981). Lastly, we consider recursive spectral bisection (Simon 1991), which appears to yield somewhat better decompositions than the other heuristic methods, albeit at greater cost and complexity (possibly amortized over the ensuing parallel calculation). Recursive spectral bisection is based upon repeated application of a sorting procedure in which elements are ordered according to the magnitude of the components of the "Fiedler vector"—the eigenvector associated with the second smallest (in magnitude) eigenvalue of the mesh dual graph Laplacian matrix. For a regular lattice in a rectangular domain the graph Laplacian is simply the standard five-point finite-difference approximation with Neumann boundary conditions; spectral bisection will clearly divide the domain cleanly in half in the direction that minimizes the length of the cut.

The partition procedures described above largely ignore the possibility of subsequent solution-adaptive mesh refinement. Although one approach to refinement is to reappeal to a global partition procedure, the potentially large data transfers associated with this process suggest that at least certain aspects of adaptivity be anticipated in advance. At one extreme, one can randomly partition the elements into collections, i.e. geometrically scatter the finite elements so that subsequent local refinement will, perforce, be distributed over many processors (Fox & Otto 1986). While this distribution scheme can certainly lead to good load balancing, it also generates a large number of short messages, as well as the possibility of contention. [The concept of randomization in parallel processing appears often. Indeed, for general communication constructs it is proposed that random data distribution can reduce contention (Johan et al 1992). It is likely, however, that finite-element meshes contain more homogeneity and structure than the hypotheses of the random distribution arguments permit.] A more modest proposal for adaptivity, perhaps sufficient for incompressible flows, is to generate not P collections, but mP collections ($m \geq 1$), and to exchange collections (now subparcels) between maximally imbalanced processors as needed. This procedure has been applied to analogous distribution problems in Eulerian domain-parallelized rarefied gas Monte-Carlo direct simulations (McDonald 1991).

3.1.2 DIRECT SOLUTION METHODS   Although iterative solution methods are more naturally parallel than direct methods, efficient direct solution strategies have been effected. The main advantages of iterative methods are: ready parallelization due to locality, low memory requirements, and low operation counts for certain classes of problems (in particular in higher space dimensions). The disadvantage of iterative methods is that many of the most efficient iterative techniques are restricted to problems dominated by symmetric positive-definite contributions. In contrasts, the main advantage of direct methods (often in conjunction with nonlinear Newton iteration) is their robustness with respect to equation and discretization; direct methods apply to a wide range of often rather ill-conditioned problems. Not surprisingly, applications for the direct solvers described in this subsection center on very complex physics [e.g. non-Newtonian flows (Aggarwal & Keunings 1993)], for which iterative techniques remain relatively unreliable. The disadvantages of direct solvers are less transparent parallelization, and the rather rapid escalation of computational complexity and memory requirements with problem size; in the context of finite-element considerations, large three-dimensional calculations remain relatively rare due to the increased bandwidth associated with increasing space dimension. Iterative and direct strategies are not mutually exclusive: Many advanced solvers combine iterative and direct solvers, with various complete or incomplete forms of the latter serving as preconditioners for the former.

Direct methods take many forms, of which the most common are various *LU* embodiments of Gaussian elimination. We shall focus here on banded-matrix *LU* methods, which are equally appropriate for both structured and unstructured problems. [A number of more structured direct-solver techniques, such as ADI, are reviewed in Braaten (1993); parallel FFT considerations are presented in Section 5 of the present paper.] Even within the confines of banded Gaussian elimination there are several parallel options (with largely equivalent complexity). In the "scattered box" frontal approach (Fox et al 1988), an active elimination front of matrix entries is scattered across processors to ensure that row operations can proceed in parallel; in incomplete nested-dissection (Lucas et al 1987, Mu & Rice 1992, Keunings et al 1992, Mehrabi & Brown 1993), or multi-frontal, approaches, block elimination procedures are applied recursively. The latter appears more popular within the finite-element community, perhaps because the former necessitates a non-element-based distribution of data, thereby complicating the assembly process. The remainder of this sub-section focuses on incomplete nested dissection. We consider only the *LU* process since, although forward and back substitution admit less concurrency than factorization, the absolute operation count associated with substitution is sufficiently low that more-parallel factorization easily dominates less-parallel substitution [this is not the case for incomplete *LU* preconditioners, where the intrinsic dependencies of the substitution problem can be significant (Fatoohi 1993)].

Following the formulation of Lucas et al (1987), we reconsider the two-dimensional Poisson problem in a square and the associated model architecture mesh topology described in Section 3.1.1. We show in Figure 3*a* an incomplete nested-dissection ordering of the unknowns, in which the finite-element nodes in each of the "blocks" are numbered contiguously, and each of the blocks is numbered prior to any adjoining "separators." In Figure 3*b* we show the number of the processor on which the column associated with each degree of freedom of Figure 3*a* is (or will ultimately be) resident. Lastly, in Figure 4, we depict the elimination tree associated with the incomplete nested-dissection ordering. The elimination tree graphically illustrates the potential concurrency: All leaves at the same level of the tree afford simultaneous factorization. The parallel algorithm is thus: First, one proceeds with factorization of the blocks, in the process of which update information for the adjoining separators is also generated; next, one proceeds up the tree, eliminating all separators at each level in parallel. It is essential for optimal performance that, as the tree thins and the separator problem size grows, more processors be brought to bear on the individual separator problems. As the separator problem is effectively full, standard cooperative full-matrix *LU* methods can be exploited. For

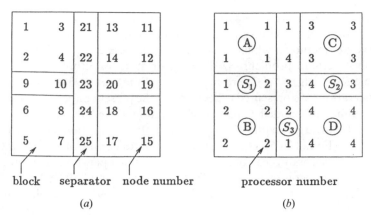

Figure 3 (*a*) Finite-element node numbering and (*b*) column-to-processor mapping for incomplete nested dissection solution of the Poisson problem in a square for $P = 4$ and $N_{el} = 16$ (after Lucas et al 1987).

example, suppose that the ith column of a N_{col}-column P'-processor separator matrix resides on processor $j = 1 + (i-1) \bmod P'$. One then: (*a*) proceeds to select each column, $i = 1, \ldots, N_{col}$, as the pivot column; (*b*) distributes the pivot column to all P' processors; and (*c*) updates, on each processor simultaneously, all resident columns $k > i$. Although earlier work assumed symmetric positive definite forms, more recent efforts have extended the incomplete nested-dissection algorithm to partial-pivoted nonsymmetric systems (Mu & Rice 1992, Keunings et al 1992, Mehrabi & Brown 1993). These methods have been implemented on the iPSC/2 and iPSC/860: The results to date appear very promising, and portend the extension of parallel fluid-dynamics calculations to problems of significant physical complexity.

We hazard here an approximate parallel complexity estimate that we believe roughly reflects the block (though not strip) incomplete nested-

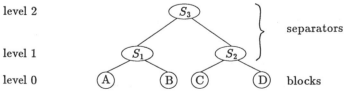

Figure 4 Elimination tree for the incomplete nested-dissection decomposition shown in Figure 3.

dissection algorithms described in the literature. In particular, the time to complete the LU decomposition for our $\sqrt{N_{el}} \times \sqrt{N_{el}}$ Poisson discretization on P processors ($P > 1$), $\tau_{ND}(P)$, is

$$\tau_{ND}(P) = e_1 t_{calc} N_{el}^2/P^2 + e_2 t_{calc} N_{el}^{3/2}/P + e_3 t_{comm}(\cdot) N_{el} \ln P, \tag{4}$$

where the e_i are constants independent of N_{el} and P, and $t_{comm}(\cdot)$ reflects a communication time intermediate between $t_{comm}(1)$ and $t_{comm}(\infty)$. The first term in (4) reflects the parallel block factorization (with minimum bandwidth ordering within blocks). The second term reflects the cooperative solution of the separator factorizations; here we are somewhat optimistic, and have neglected edge effects for the final columns of each separator problem. The third term reflects the separator column communications; here we are somewhat pessimistic, and have assumed a $\log_2 P$ broadcast even for two-processor separators. Note that a significant advantage of the incomplete nested-dissection block structure, compared to earlier procedures in which each finite-element node is treated as an individual leaf in the elimination tree, is the generation of fewer (longer) messages. From (4) it is seen that the optimal number of processors for minimum computation time will scale as $\sqrt{N_{el}}$, giving a minimum computation time which scales (to within logarithmic corrections) as N_{el}. The speedups with respect to serial factorization based on (complete) nested dissection (George & Liu 1981) and minimum-bandwidth orderings are thus $O(\sqrt{N_{el}})$ and $O(N_{el})$ respectively. If we compare these results to conjugate-gradient iteration (Equation 3), we see that, due to the higher level of concurrency, the operation count for conjugate gradients is somewhat better. However, (3) is specific to the particular model Poisson problem, while (4) is, of course, much more general, and applies with little qualification (e.g. to include vector-valued nodes) to most of the *two-dimensional* problems described in this paper.

The issues that arise if the square domain is replaced with a general finite-element discretization are closely related to the unstructured-induced issues discussed in the iterative context in Section 3.1.1. In particular, the finite element-to-block mapping objectives are very similar to the finite element-to-collection mapping objectives: The computational load must be balanced, the volume of data to be communicated should be minimized, and the number of distinct messages should be small. There are, however, certain additional conditions that do not arise in conjugate-gradient deliberations: Blocks that adjoin larger separators will incur more operations (Lucas et al 1987), and the aspect ratio of the blocks will affect the efficiency of the intra-block factorization (Farhat & Lesoinne 1993). The most natural heuristic approach to element-to-block mapping is, of course, recursive graph bisection. A number of different orderings and ordering procedures are discussed in Duff & Johnsson (1989); nested dissection performs consistently well, although elimination-tree height-minimizing algorithms are also advantageous.

In closing, we remark that nested-dissection methods are equivalent to a direct, recursive, application of the Schur complement construction which is central to iterative domain-decomposition solution approaches (Quarteroni 1991). Indeed, many domain-decomposition approaches can be viewed as iterative procedures for the solution of the skeletal Schur complement which arises at level 1 of incomplete nested dissection (see Figure 4); furthermore, to achieve subdomain-independent convergence, these iterative approaches often involve inter-subdomain operations (see Glowinski et al 1991 and Section 3.3) not dissimilar to the recursive aspects of nested dissection. Not surprisingly, the advantages of both direct nested dissection and iterative domain-decomposition methods derive from: from the parallel perspective, the relative infrequency with which the inter-domain residuals are updated, with most of the computational effort centered on intra-block operations; and from the absolute operation count perspective, the relative speed with which the small block subproblems can be solved, given the superlinearity of most inversion techniques with problem size (Gropp & Keyes 1992a).

3.2   The Stokes Problem

In this subsection we briefly comment on parallel solution of finite-element discretizations of the steady Stokes problem. Most (and, in the case of *direct* solvers, effectively all) of the parallel aspects of the Poisson problem extend directly to the steady Stokes problem, and we thus restrict our attention here to a description and example of one approach to parallel *iterative* Stokes solution (Fischer & Patera 1991)—the Uzawa procedure. An entirely different parallel approach to the steady Stokes problem, based on a boundary integral formulation, will be described in Section 4.

The essential concern in the finite-element discretization of the Stokes problem is the proper selection (Brezzi 1974)—or stabilization (Hughes & Franca 1987)—of compatible velocity and pressure spaces. The resulting discrete system takes the form of $\underline{A}\,\mathbf{u} - \mathbf{D}^{\mathrm{T}}\underline{p} = \underline{\mathbf{f}};\ -\mathbf{D}\cdot\mathbf{u} = 0$, where underline refers to finite-element nodal degrees of freedom, boldface refers to Cartesian components, $\mathbf{u}$ and p are the velocity and pressure, respectively, $\mathbf{f}$ is a prescribed force, $-\underline{A}$ is the discrete Laplacian operator, and $-\mathbf{D}^{\mathrm{T}}$ is the discrete gradient operator. In the Uzawa saddle-decoupling procedure (Cahouet & Chabard 1986, Maday et al 1993) one performs block Gaussian elimination to rewrite the discrete system as $-\underline{S}p = \mathbf{D}\cdot\underline{A}^{-1}\underline{\mathbf{f}};\ \underline{A}\,\mathbf{u} - \mathbf{D}^{\mathrm{T}}\underline{p} = \underline{\mathbf{f}}$, where $\underline{S} = \mathbf{D}\cdot\underline{A}^{-1}\mathbf{D}^{\mathrm{T}}$. The solution then proceeds by first applying conjugate iteration to the $\underline{S}$ system to find $\underline{p}$, and then applying conjugate iteration to the $\underline{A}$ system to recover $\mathbf{u}$. Note that each iteration of the conjugate gradient iteration of $\underline{S}$ requires an inner, hence nested, conjugate gradient solution of $\underline{A}$; the critical result is that, for finite elements that satisfy a uniform inf-sup condition, the condition number of $\underline{S}$ is independent of the mesh size, and, for not overly distorted domains, order unity.

As an example we consider the calculation of the permeability of a two-dimensional (fibrous) random porous medium comprising uniformly distributed circular cylindrical obstacles (Cruz 1993, Cruz & Patera 1993). For a particular volumetric concentration of cylinders c, the calculation proceeds in several stages: At the highest level, the size of the periodic supercell (containing many obstacles) is systematically extended until statistical quantities are invariant; at the middle level, a large number of supercell realizations (of which one is shown in Figure 2) are generated, and the permeability $\kappa(c)$ is computed as the sample mean of the individual normalized flowrates; at the lowest level, the individual-realization Stokes problem is solved. The lowest level evokes the parallel solution algorithm, which is based upon: Voronoi domain partition based on cylinder-center generators, automatic subdomain-based parallel mesh generation (Hecht & Saltel 1990), Taylor-Hood finite-element discretization by $\mathbf{P}_1$-$\mathbf{P}_2$ triangles, and Uzawa-based nested conjugate-gradient iterative solution of the discrete saddle problem. The algorithm is implemented on the Intel iPSC/860. A complete three-level calculation for a concentration of $c = 0.45$ requires order 50 hours of (typically 16-processor) computing; the particular 16-cylinder Stokes realization shown in Figure 2 comprises 11,100 degrees of freedom (roughly 45% hidden from view in the "nip" regions between close cylinders) and requires eight 16-processor minutes of computation at approximately 30 MFLOPS (32-bit). The current algorithm follows the least efficient vertex summation procedure, in which all vertices are stored on all processors, and accumulated through a global gather/scatter operation. The parallel efficiency remains high, however, because, at present, the within-processor unstructured finite-element evaluations make relatively poor use of the less-than-forgiving i860 processor— thereby increasing the effective t_{calc}. Although these algorithmic shortcomings are surmountable, they once again illustrate the potential pitfalls in reporting selected parallel metrics. [Note that *unity* parallel efficiency could be attained for this permeability problem by considering "Monte Carlo" parallelism, in which each Stokes realization is relegated to a different processor; unfortunately, there is often insufficient per-processor memory to pursue this throughput, or parametric, form of concurrency. Distributed-memory machines achieve adequate (and inexpensive) memory and paths to memory through aggregation.]

For a particle distribution in which essentially all geometrically possible configurations are equally likely, and a concentration of $c = 0.45$, Cruz finds for the permeability (nondimensionalized with respect to cylinder diameter) $\kappa = 0.00483$. This example is intended to illustrate several points. First, the broader problem of computing permeability is a multiple-partial-differential-equation problem (hundreds or thousands of individual Stokes flow problems are evoked for a single permeability data point, in the style of Scenario I of Section 1), placing new demands on parallel automation. Second, even for this simple Stokes flow, there would be significant computational economies in exploiting analytical bound solutions in "nip" regions in order to reduce the close-obstacle mesh requirements illustrated in Figure 2; most of the computational expense is attributable to the significant deterioration of the conditioning of the $\underline{S}$ operator due to very severe domain distortion. Third, for this two-dimensional Stokes problem, efficient semi-analytical methods can be developed to predict the permeability (Sangani & Yao 1988); the role of parallel processing in porous medium inquiries is clearly not in solving this particular problem more quickly, but, rather, in extending the model to include nonlinear effects and three-dimensionality (Scenario II of Section 1). Fourth, it is computationally expensive to predict the permeability for a particular concentration, and unlikely that the simulation procedure described will be exploited directly as a subroutine in a macroscale Darcy solution or a design exercise. The parallel code best serves not as an in situ procedure, but rather, as a method to generate or confirm higher-level models for $\kappa(c)$; this will be discussed further in Section 6.

3.3 *The Navier-Stokes Equations*

The incompressible Navier-Stokes problem remains challenging: The solutions at moderate and high Reynolds numbers exhibit a large range of spatial and temporal scales; the equations are nonlinear, nonsymmetric, and constrained. Two basic numerical strategies can be identified: a "coupled" approach, in which a full-implicit temporal treatment yields a nonlinear nonsymmetric elliptic system that must be solved at each time-step; and a "decoupled" approach, in which semi-implicit fractional time-stepping methods reduce the Navier-Stokes equations to a sequence of slightly more manageable subproblems. In essence, the coupled approach opts for infrequent (accuracy-limited) solution of a difficult problem; the decoupled approach prefers more frequent (stability-limited) inversion of somewhat simpler problems.

The critical ingredient in the coupled approach is the "inner" solution of the nonsymmetric sparse $N \times N$ linear system resulting from the application of an "outer" (typically Newton) iteration scheme to the nonlinear elliptic system engendered by full-implicit temporal discretization. *Direct* methods (see Section 3.1.2) are attractive in two space dimensions, however operation and memory requirements mount considerably in three dimensions. The most popular *iterative* approaches are Krylov-space techniques such as GMRES (Saad & Schultz 1986, Gropp & Keyes 1992a, Tezduyar et al 1992a). From the parallel perspective, GMRES differs from the simple conjugate-gradient procedure described in Section 3.1.1 in two essential aspects: the basis vectors for the Krylov subspace $K_j \in R^{N \times j}$ must be stored (here $j \in \{1, \ldots, j_{max}\}$ refers to iteration indexed with respect to the last GMRES restart); and a $j \times j$ reduced system must be inverted at each iteration to effect the requisite projection onto K_j. Since memory limitations constrain the maximal subspace dimension, j_{max}, the reduced system can be solved serially—on a host processor or on all processors—without undue degradation of parallel efficiency. For acceptable performance GMRES requires effective preconditioners; one class of preconditioners will be described below.

In the decoupled approach, operator splitting techniques are employed to break the Navier-Stokes equations into independent convection and unsteady Stokes subproblems. Convection can be treated explicitly (e.g. Moin & Kim 1982), implicitly (e.g. Williams & Glowinski 1989), or semi-implicitly by backwards differentiation along characteristics (Ewing & Russell 1981, Pironneau 1982). The method of characteristics can be recast as an Eulerian convection subintegration (Maday et al 1990) which avoids off-grid interpolation and affords simple parallel implementation. For the unsteady Stokes subproblem, two approaches are possible. First, a Uzawa-like technique similar to that described for the steady Stokes problem in Section 3.2 can be adopted (Cahouet & Chabard 1986, Williams & Glowinski 1989); unfortunately, for the *unsteady* Stokes problem, Poisson preconditioners are required to recover good conditioning properties. Second, a further splitting can be pursued to decouple the viscous and pressure contributions, thereby yielding: a set of Helmholtz problems for the velocity components [potentially coupled through complex boundary conditions (Ho & Patera 1991)]; and either a standard Poisson problem for the pressure (Chorin 1968, Temam 1984, Karniadakis et al 1991), or a consistent Poisson operator for the pressure in which boundary conditions are imposed directly on the velocity (Rønquist 1992). In summary, for fractional-step approaches, the elliptic problem associated with the incompressibility constraint dominates the work per timestep, while the convection operator (in conjunction with the singular viscous perturbation) governs the spatial resolution and acceptable timestep size.

For both the coupled and decoupled approaches the conditioning problems associated with elliptic contributions can be at least partially mitigated by new domain-decomposition-based two-level multigrid-like (McCormick 1992) schemes proposed and exercised by numerous research groups (e.g. Mandel 1990, Gropp & Keyes 1992a, Rønquist 1992, Tezduyar et al 1992b). Although these methods have, to date, been applied primarily to Poisson subproblems, Gropp & Keyes (1992a) apply the concept in the full GMRES coupled-strategy context to the calculation of flow over a backward facing step on the Intel iPSC/860. In general, the two-level methods balance an inter-subdomain coarse-grid operator to effect rapid transfer of long-range information with a largely intra-subdomain fine-grid preconditioner designed to accelerate convergence of the smaller scales. Unlike the reduced system in GMRES, the dimension of the coarse-grid operator is, although small compared to the total number of degrees of freedom in the discretization, at least as large as the number of processors P; it is therefore infeasible to solve the coarse-grid problem serially for large P, even if band structure can be exploited. Possible (direct-solution) parallel alternatives include: solving the coarse-grid problem on subsets of processors (Gropp 1992); or forming a row (or column) distribution of the coarse-grid inverse (once) and then calculating a parallel dense matrix-vector product (at each iteration) (Gropp & Keyes 1992a,b; Fischer 1993). If the LU decomposition dominates the forward and back substitution, as will be the case for problems in time-dependent geometry, cooperative LU techniques may also be considered (see Section 3.1.2).

For purposes of illustration we describe application of the two-level concept to the solution of the splitting-induced consistent Poisson problem, $\underline{E}\underline{p} = \underline{g}$, arising in m-subdomain finite-element or spectral-element discretization of the Navier-Stokes equations by a velocity-discontinuous pressure pair. Here $\underline{E} \in \mathsf{R}^{n \times n}$ is the consistent Poisson operator, $\underline{p} \in \mathsf{R}^n$ is the pressure, and $\underline{g}$ is the inhomogeneity reflecting contributions from other substeps of the splitting process. We follow the formulation of Rønquist (1992), in which the coarse-grid operator is folded into a global conjugate-gradient iteration through deflation (Nicolaides 1987, Mansfield 1988). The coarse (subscript c) and fine (subscript f) decomposition is effected through a subdomain-motivated prolongation operator $\underline{I} \in \mathsf{R}^{n \times m}$. The column space of the prolongation operator I is intended to approximate the span of the low eigenmodes of the $\underline{E}$ system; for this particular problem, $\underline{I}$ maps subdomain-piecewise-constant functions to the n nodes of the underlying finite-element representation. The pressure is then expressed as $\underline{p} = \underline{I}\underline{p}_c + \underline{p}_f$, leading to an algebraic reformulation of the original problem as solvable fine and coarse subproblems, $\underline{E}_f \underline{p}_f = \underline{g} - \underline{E}\underline{I}\underline{E}_c^{-1}\underline{I}^T\underline{g}$ and $\underline{E}_c\underline{p}_c = \underline{I}^T\underline{g} - \underline{I}^T\underline{E}\underline{p}_f$, respectively. Here $\underline{E}_f = \underline{E} - \underline{E}\underline{I}\underline{E}_c^{-1}\underline{I}^T\underline{E}$, and $\underline{E}_c = \underline{I}^T\underline{E}\underline{I}$. The fine system is then solved by conjugate-gradient iteration (with appropriate orthogonality conditions); note that each matrix-vector evaluation, $\underline{E}_f\underline{q}_f$, requires solution of a $(m \times m)$ coarse-grid system. Once $\underline{p}_f$ is established, the coarse-grid problem is solved for $\underline{p}_c$, and the procedure is complete. With appropriate application of a *local*, subdomain-based preconditioner to $\underline{E}_f$, the condition number of the fine system is significantly reduced relative to the originating $\underline{E}$ matrix.

We close this section with two Navier-Stokes calculations which embody many of the algorithms described above. The first example, due to Tezduyar et al (1992a), addresses three-dimensional wavy Taylor vortices in Taylor-Couette flow by finite-element GMRES-based coupled-strategy solution of the Navier-Stokes equations on the CM-200 and CM-5 machines. The calculations are reported for the case in which the inner cylinder, of radius R_i, is rotating with angular velocity ω, and the outer cylinder, of radius R_o, is stationary. The gap width is $R_o - R_i = 0.14R_i$, and the cylinder height is $3.0R_i$ (T. E. Tezduyar, private communication); zero normal velocity and zero shear stress boundary conditions are applied at the end-plates of the apparatus. The calculations reflect 38,400 hexahedral elements and 282,000 degrees of freedom. The gather/scatter operations required for inter-processor summation of nodal values (see Section 3.1.1) are effected by efficient vendor-supplied high-level communication constructs designed precisely for this purpose. The GMRES iteration is pursued with a maximal reduced-system order of $j_{max} = 30$; the reduced system is solved on the front-end processor in a time comprising less than 3% of the total solution time. The Tezduyar et al calculation achieves a sustained performance of over 2 GFLOPS (64-bit, input-output excluded) on 1024 "nodes" of the CM-200. Figure 5 shows the pressure contours in an unwrapped cylindrical cut, midway between the inner and outer cylin-

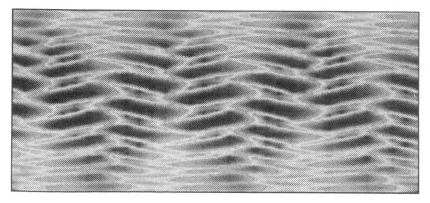

Figure 5   Pressure contours over an unwrapped cylindrical section in wavy Taylor-Couette flow at $Re = 1498$ (from Tezduyar et al 1992a, with permission).

ders at a Reynolds number of $Re = U(R_o - R_i)/v = 1498$, where U is the tangential velocity of the inner cylinder, $U = \omega R_i$, and v is the kinematic viscosity of the fluid. Other parallel computations of Taylor-Couette flow include those of Harrar et al (1991).

The second example, due to Fischer (1993), addresses the interaction of a flat-plate boundary layer with a hemispherical roughness element by spectral-element fractional-step solution of the Navier-Stokes equations on the i860-based Intel Delta machine. The hemisphere of radius R is centered at $(x, y, z) = (0, 0, 0)$; a Blasius profile with $\delta_{0.99} = 1.15R$ and freestream velocity $(U_\infty, 0, 0)$ is prescribed both as initial condition and as inflow profile at $(-8.4R, 0, 0)$. Symmetry boundary conditions are specified at $y = 0$, $y = -6.4R$, and $z = 6.5R$, and Neumann outflow boundary conditions are imposed at $x = 25.6R$. The spatial discretization comprises 512 ninth-order elements and approximately 375,000 gridpoints. The fractional-step method is based upon a characteristic method for convection, a Stokes splitting scheme, and the two-level deflated conjugate-gradient approach for the consistent-Poisson pressure problem. The calculation achieves a sustained performance of 5 GFLOPS (32-bit, input-output excluded) on 512 processors at a parallel efficiency of approximately 50%; fast vertex summation techniques (see Section 3.1) and distributed coarse-grid ($\underline{E}_c$) inverses are largely responsible for maintaining reasonable parallel efficiency. The computation requires 10 seconds per timestep, and roughly 25 minutes per shedding period. Figure 6 shows contours of spanwise vorticity (in the symmetry plane $y = 0$) and surfaces of streamwise vorticity (foreground) for a Reynolds number of $Re = U_\infty R/v = 500$. A horseshoe vortex forms upstream of the hemisphere, while hairpin vortices form periodically downstream of the hemisphere. The inset shows the time history of the velocity at two points in the wake of the hemisphere; the numerically predicted Strouhal number, fR/U_∞, of 0.16 is within the experimentally observed range of 0.13–0.23 (Acarlar & Smith 1987).

4. PARTICLE METHODS

In this section we discuss particle methods for Stokes flow, in which the particles are quite literally particles, and vortex methods, where the particles represent a discretization of the vorticity field. In both cases, the effective dimensionality of the problem is reduced with respect to grid-based discretizations. However, the nonlocality of the underlying spatial operators, which for grid-based techniques requires long-range coupling for effective solution, reappears in particle methods in the form of noncompact potentials. We do not consider potential-less particle techniques, such as cellular automata or rarefied-gas Monte-Carlo direct simulations

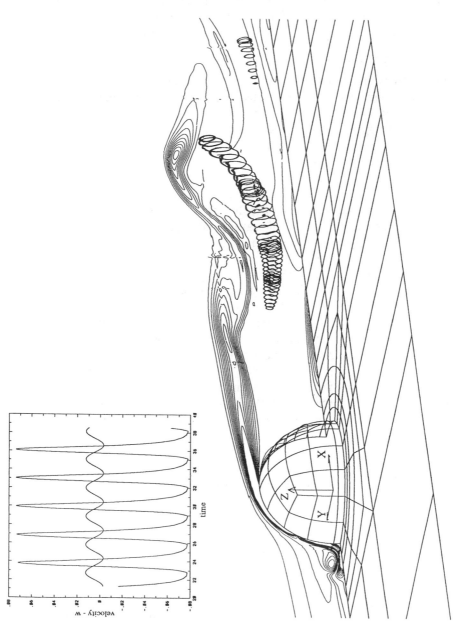

Figure 6 Spanwise (y) and streamwise (x) vorticity contours for flat-plate boundary-layer interaction with a hemispherical roughness element; time history of velocity in the wake (*inset*) reflects steady-periodic hairpin vortex formation (from Fischer 1993, with permission).

(McDonald 1991); from the parallel perspective, the former is analogous to a structured grid-based discretization, whereas the latter, at least for an Eulerian domain-decomposition particle distribution scheme, is similar to an unstructured and adaptive grid-based discretization.

4.1   *Stokes Flow*

The linearity and homogeneity of the Stokes equations admits an integral formulation in which the field problem is reduced to an integral equation defined over the surfaces of the domain. This fact is exploited in a number of different approaches to particulate problems, in which either the resistance (forces for prescribed motion) or mobility (motion for prescribed forces, as in sedimentation) of a configuration is required. One such approach, the range-completed double-layer technique (Fuentes & Kim 1992), has been applied in a parallel context. In particular, the double-layer representation results in a Fredholm integral equation of the second kind which, upon numerical quadrature and appropriate deflation, yields a dense matrix system of the form $\underline{\phi} = \underline{K}\underline{\phi} + \underline{b}$. As the eigenvalues of $\underline{K}$ are of magnitude less than unity, a Picard iteration, $\underline{\phi}^{n+1} = \underline{K}\underline{\phi}^n + \underline{b}$, can be pursued. The dense matrix-vector multiplication required at each iteration can be readily parallelized by row-to-processor partitions, in which each processor is responsible for a part of $\underline{\phi}^{n+1}$; however, the distribution of the $\underline{\phi}^n$ at each iteration requires all-to-all broadcasts. Fuentes & Kim reduce this communication overhead by recognizing that the off-diagonal entries of $\underline{K}$ decrease with distance from the diagonal: An asynchronous iterative procedure (Bertsekas 1983) is then developed, in which contributions far from the diagonal are updated only occasionally. Fuentes & Kim apply these techniques on the iPSC/860 to several multiple-sphere problems; the techniques can also be used in conjunction with the far-field fast summation methods described in the context of vortex methods below.

4.2   *Vortex Methods*

Vortex methods are effective discretizations for high Reynolds number unsteady flows (Chorin 1973). The primary advantages of vortex methods are a grid-free representation and automatic adaptivity, though, in practice, a grid may be evoked for solution of the requisite potential flow problem, and vortex redistribution may be occasionally required. The essential computational task in vortex methods is the classical N-body problem: the calculation of all pairwise interactions of N potentials, in which each potential is characterized by a prescribed functional form and an associated strength, α_i, $i = 1, \ldots, N$. Vortex methods are, in fact, so adaptive that two rather distinct subspecies thrive within the single state of California (and beyond). The two approaches differ, first, in the treatment of diffusion, and second, in the treatment of the N-body problem. We briefly describe each approach, in both instances for the more studied two-dimensional case.

In the first approach (Sethian et al 1992), diffusion is treated through a random walk, and the no-slip condition through the introduction of vortex sheets at the wall. At least partially as a result of the complexity associated with the latter, the N-body problem is treated not by fast summation techniques, but, rather, by a parallel technique—the replicated orrery—which can make effective use of $P > N$ processors. To begin, we describe a single digital orrery (Applegate et al 1985). The model architecture topology is the ring (or periodic one-dimensional mesh) shown in Figure 7a, with $P = N$ processors. For the data distribution procedure a Lagrangian vortex is assigned to each processor; more precisely, one begins with two copies of the vortex distribution, one static copy and one dynamic copy. The dynamic vortex set now effects a sequence of right shifts, as shown in Figure 7a: At each shift all processors, in parallel, update the resident partial sum to include the interaction between the (always resident) static vortex and the (currently resident) dynamic vortex; after N shifts all pairwise interactions are properly accumulated.

For the replicated orrery Sethian et al consider the 3-"cube" model architecture shown in Figure 7b, for which, in addition to the usual $\log_2 P$ gather/scatter requirement, $\log_2 P'$ summation over uncontended (e.g. co-axial) P'-processor subnetworks is also assumed. The dimensions of the cube are $(N/M) \times M \times M$, giving $P = NM$ processors in total. The static vortex distribution shown on the front face of the cube is replicated on all x–z planes; the initial dynamic vortex distribution is identical to the

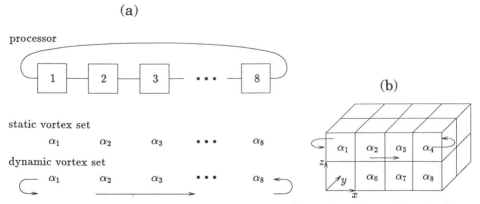

Figure 7 (*a*) Single and (*b*) replicated orrery structure for $N = 8$ vortices and $P = 8$ and $P = 16$ processors, respectively (after Sethian et al 1992).

static distribution, except that the entire pattern is shifted down one level in z for each level advanced in y. For this multiple ring construct all pairwise interactions are now completed after only N/M wraparound shifts in x (for all y and z) followed by a summation in y. The time to compute the N-body interactions on the model architecture is thus $\tau_{\text{orrery}}(P) = a_1 t_{\text{calc}} N^2/P + a_2 t_{\text{comm}}(P/N)\ln(P/N)$; it follows that optimal speedup (minimal time) is achieved for P scaling with N^2, consistent with the ideal (no-communication-penalty) replicated orrery concept. As regards model-to-actual architecture mapping, the algorithm described here, coupled with the necessary viscous and potential-flow corrections, has been implemented by Sethian et al on the CM-2, which readily supplies both the 3-cube embedding and the necessary $\log_2 P'$ axial summations. The adaptive nature of vortex methods poses little parallelization problem due to the Lagrangian vortex-to-processor distribution scheme; furthermore, new vortices are readily included by resizing the dimensions of the embedded 3-cube. The resulting simulation code has been applied to numerous moderate (order several thousand) Reynolds number flow problems, including complex unsteady flows in model engine-cylinder intake valves; typical simulations involve order tens of thousands of vortices and vortex sheets.

A second approach (Pepin & Leonard 1990, Clarke & Tutty 1992) to vortex methods exploits some variant of a viscous flux and redistribution scheme for the diffusive contributions, and fast summation techniques for the N-body problem. Very briefly, the fast N-body problem treatment is based upon: collecting physically proximate groups of vortices; expanding the potential contribution of each of the groups around an appropriate "centroid"; approximating the resulting sum for each group as a truncated (L-term) multipole expansion for distances large compared to the physical extent of the group; and calculating the effect of a group i on a group j far from group i either by 1. applying the group i truncated multipole expansion to each vortex in group j [serial complexity $O(N \log N)$] (Barnes & Hut 1986), or, 2. reexpanding the group i truncated multipole expansion at the centroid of group j as an L'-term Taylor series which then serves to update all group j vortices [serial complexity $O(N)$] (Greengard & Rohklin 1987).

Several possible parallel data distribution schemes are possible, all based on group-to-processor mapping: In the "Eulerian" approach, vortices are associated with fixed geometric subdomains (Greengard & Rohklin 1987); in the "Lagrangian-Eulerian" approach, the vortices are resorted by recursive coordinate bisection (in x–y or, perhaps preferably, based on moment-of-inertia considerations) every several timesteps in order to maintain geometrically coherent groups (Appel 1985, Pepin & Leonard 1990, Clarke & Tutty 1992). In some sense, the former distributes the necessary communication over many steps, whereas the latter opts for more substantial communication infrequently. Sorting for the Lagrangian-Eulerian procedure can either be done on the host, if sufficiently infrequent, or in parallel; parallelization of fast sorting techniques, such as quicksort, is simple if all degrees of freedom are initially on one processor—however truly distributed sorting procedures are less transparent (Fox et al 1988, Clarke & Tutty 1992). (Sorting routines are clearly good candidates for vendor libraries.) Although these partition approaches lead to good concurrency and minimum communication, the fast-summation vortex distribution is fundamentally more sensitive than typical analogous unstructured finite-element partition problems. First, equal numbers of vortices does not imply equal work; vortex groups that are large in area or centrally located will require more vortex-vortex updates than small or peripheral groups. Second, whereas for finite-element discretizations a poor element-to-processor mapping may require additional communication, for fast summation techniques an errant vortex-to-group assignment can compromise the multipole expansion. As for unstructured finite-element discretizations, once proper groups have been formed, the model-to-actual architecture mapping topological issues are largely secondary. As regards programming model, the algorithmic heterogeneity appears to prefer a MIMD implementation, though effective SIMD implementations for balanced nonadaptive discretizations have been developed (Zhao & Johnsson 1991).

The fast-summation techniques described have been implemented in Pepin & Leonard (1990) on the JPL MarkIII hypercube and in Clarke & Tutty (1992) on transputer systems. We present the results from Pepin (1990) for an 80,000-vortex calculation of the short-time startup flow past a cylinder at Reynolds number (based on freestream velocity and diameter D) of 3000. We show in Figure 8 the flow velocity in the x (freestream-flow) direction as a function of position downstream from the cylinder at different times T. Agreement with experiment and asymptotic theory is very good; the vortex calculation also reveals the $1/\sqrt{T}$ short-time singularity in drag. Quite independent of the particular import of this problem to understanding short-time, and more generally, unsteady vortex behavior, this calculation constitutes an example in which parallel processing supported, but did not subsume, a fluid dynamical inquiry.

5. FOURIER METHODS

Fourier methods have been the mainstay of turbulence calculations for two decades (Orszag 1972). In fact, the first example of parallel computation of viscous incompressible flow is the Fourier-based calculation undertaken

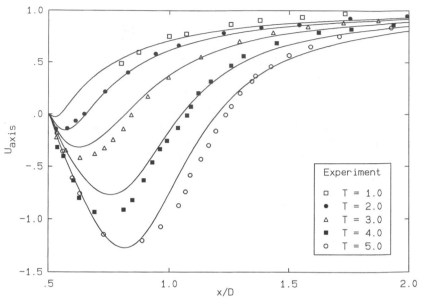

Figure 8 Comparison of the velocity on the symmetry axis for *Re* = 3000 obtained numerically by Pepin (1990) (*solid lines*) with experimental results of Bouard & Coutanceau (1980) (from Pepin 1990, with permission).

by Moin & Kim (1982) over a decade ago. Moin & Kim conducted a series of large-eddy turbulent channel flow simulations on the 64-processor ILLIAC IV, a precursor to current SIMD machines. The parallel algorithm developed can be viewed as an out-of-core variant of the transpose-based FFT techniques described below. Their maximum resolution ($63 \times 64 \times 128$) simulation remains a milestone calculation; we present in Figure 9 a result from this first parallel flow computation.

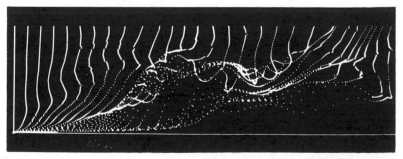

Figure 9 Wall-normal hydrogen-bubble flow visualization from ILLIAC IV large-eddy turbulent channel-flow simulation (from Moin & Kim 1982, with permission); inflectional instantaneous velocity profiles and strong shear layers corroborate experimental observations (Kim et al 1971).

In terms of computational complexity, the central issue in Fourier spectral methods is the multidimensional FFT. For homogeneous problems in R^d which admit Fourier decompositions in d spatial directions, all operations (in physical or wave space) are pointwise calculations and hence fully parallel. For problems having $d-k$ homogeneous directions (e.g. channel flow) the required operations in physical space are again pointwise, while the computations in wave space reduce to decoupled k-dimensional boundary value problems (Moin & Kim 1982, Mangiavacchi & Akhavan 1993, Karniadakis & Orszag 1993). The boundary value problems can be solved concurrently within each processor or, in memory-limited cases, within subsets of processors. In what follows, we consider the three-dimensional case in which all directions are homogeneous. Two approaches to the multidimensional FFT have been explored: the transpose approach, and the distributed FFT. As the recent review article of Karniadakis & Orszag (1993) focuses on parallel turbulence simulation, our remarks here shall be brief.

We begin by discussing the transpose approach (Jackson et al 1991, Wray & Rogallo 1992). In this approach, data are partitioned in (say) the z-direction as P two-dimensional slabs containing N^3/P words for each N^3-field in the computation. First, a two-dimensional FFT is carried out for each x–y plane in time $O(N^3 \log_2 N/P)$. Next, a transpose of the data is invoked such that the z-direction is now complete within each processor. Finally, the remaining (z-)FFT is computed. The only communication step is the transpose, or all-to-all (complete) exchange [more generally, efficient all-to-all broadcast procedures are often viable alternatives to reduction operations (e.g. Brunet & Johnsson 1992)]. We consider two limiting implementations, initially for hypercube architectures (Bokhari 1991). The first procedure is the "shuffle" algorithm, which comprises $\log_2 P$ exchange cycles during each of which each processor initiates a contention-free message of length $\frac{1}{2}N^3/P$. If latency is dominant (N^3/P appropriately small), the communication time is $O[t_{\mathrm{comm}}(1) \log_2 P]$. However, if each message transfer is bandwidth-limited (N^3/P appropriately large), the communication time for the shuffle algorithm will be $O[t_{\mathrm{comm}}(\infty)(N^3/P) \log_2 P]$. For the bandwidth-limited case, a second "dealing" complete-exchange strategy is preferred, in which each processor sends an N^3/P^2 parcel of data to each of the other processors over $P-1$ communication cycles. This procedure yields a transpose time in the bandwidth-limited case of $O[t_{\mathrm{comm}}(\infty)(N^3/P^2)(P-1)]$, which is clearly improved over the corresponding shuffle result. Finally, we point out that for mesh architectures, such as the Intel Delta, these algorithms will not be contention-free. For example, for the "dealing" strategy, it can be shown (Scott 1991, Wray & Rogallo 1992) that the optimal communication complexity on a mesh-based architecture is, for the bandwidth-limited case, $O[t_{comm}(\infty)$ $(N^3/P^2)P^{3/2}]$.

The transpose-based FFT algorithms require sufficient memory on each processor to hold an entire slab or pencil of data. However, large memory multiprocessors might not follow the optimal cost-performance scaling track (Seitz 1992). If we consider the case in which memory per processor, M, is fixed, simple scaling arguments lead to the conclusion that, since the *total* memory requirements scale as $N^3 \sim MP$, for sufficiently large P it will not be possible to accommodate $O(N^2)$, or even $O(N)$, data on each processor. In this instance it is of interest to consider the distributed FFT, as pursued by Pelz (1991a,b, 1992) in a series of NCUBE, iPSC/860, and CM-2 computations. This distributed FFT requires $(\log_2 P + 1)$ exchanges of length $\frac{1}{2}N^3/P$ (Swartztrauber 1987), yielding a communication complexity similar to that of the shuffle-based transpose. It appears that the relative advantage of the transpose and distributed FFT approaches will, in general, depend on communication, computation, and memory technology.

6. CONCLUSIONS

We briefly return to the application of parallel computations in engineering practice, in particular in those situations tending more towards Scenario II (see Section 1). Although Scenario I computations can be treated as modules directly integrable with other computational tools, such as mathematical programming libraries, design packages, and graphics systems, the direct and indirect costs associated with large (parallel) computations preclude such a context-independent formalism. This is well recognized: The benefits of direct simulations of turbulence derive not from the calculation data per se, but, rather, from the understanding that is subsequently engendered (e.g. Figure 9); this understanding may, in turn, lead to new optimization or control strategies, or perhaps new turbulence models. We close with an illustration of how the simulation-data reduction process might be systematized; other approaches to the synthesis problem include, for example, new (in fact, parallel) visualization systems (Sethian & Salem 1989).

As a simple example (Yesilyurt & Patera 1993b) we consider twodimensional flow in a plane channel of plate-separation $2h$ interrupted by an infinite regular periodic array of eddy-promoter cylinders of radius R placed a distance a from the bottom wall; the flow is assumed periodic on the scale of the cylinder x-separation, here $L = 6.666h$ (Karniadakis et al 1988). [Multiple-cylinder inflow-outflow parallel simulations and quiet experiments (Schatz et al 1991) indicate that the primary bifurcation is convective, not absolute; noisier experiments (Kozlu et al 1988), however, agree reasonably well with periodic calculations.] The flow is driven by a constant pressure gradient, $\overline{dp/dx}\hat{x}$, parallel to the channel walls; the Reynolds number is defined as $Re = -\overline{dp/dx}h^3/2\rho v^2$, where ρ and v are the fluid density and kinematic viscosity, respectively. An associated forced convection temperature problem is also introduced, in which the bottom and top plate support a temperature difference of ΔT (the eddy-promoters are assumed adiabatic). The Nusselt number for the flow, Nu, is defined as the time- and space-averaged heat flux through the bottom wall normalized by the conductive heat transfer in the absence of the eddy-promoter. This problem is a model for compact heat exchangers, in which, for example, pumping power and Nusselt number are to be minimized and maximized, respectively.

Yesilyurt & Patera perform 44 unsteady Navier-Stokes calculations for a fixed Reynolds number ($Re = 300$) and Prandtl number ($Pr = 1$), and for nondimensional cylinder placements, a/h, and cylinder radii, R/h, randomly and uniformly distributed over the triangular parameter—or *input*—domain of interest, $\Omega = \{0.1 \leq a/h \leq 1, 0.05 \leq R/h \leq a/h - 0.05\}$. Each (approximately 8000 degree-of-freedom) simulation, based on the parallel spectral-element fractional timestepping algorithm described in Section 3.3, costs roughly \$75 and requires 6 (16-node iPSC/860) hours to converge to the stationary temporal state needed for calculation of the Nusselt number. [Unsteady Navier-Stokes problems might benefit from parallelism in time; more-than-coarse-grain parallelism is not, however, evident, although several research groups are actively pursuing this topic (e.g. Horton 1991).] Next, 22 of the 44 $\{(a/h, R/h); Nu\}$ Navier-Stokes input-output pairs are used to construct a simulation *surrogate*—an approximate input-output model—for the Nusselt number, $\widetilde{Nu}(a/h, R/h)$, based on a scattered data interpolant (Renka 1988). Finally, the remaining 22 simulation results are evoked to *validate* this model over the input region of interest, $(a/h, R/h) \in \Omega$, that is, to ascertain how well $\widetilde{Nu}$ represents Nu. Additional information as to system behavior (e.g. prior work) is integrated into the surrogate prior to validation. [The surrogate here is static; various procedures for the construction of *dynamical* fluid mechanical simulation surrogates have also been studied extensively (Lumley 1967).] The final result is the approximate Nusselt number surface shown in Figure 10: for small a/h the flow is steady; for larger a/h both unsteady wall-mode Tollmien-Schlichting-like waves and steady wavy flows are observed. A simple validation procedure (Yesilyurt & Patera 1993a,b) guarantees that, with confidence level greater than 90%, the Nusselt number surrogate, $\widetilde{Nu}(a/h, R/h)$, will predict the output of the originating Navier-Stokes simulation, $Nu(a/h, R/h)$, with error less than 0.09 over more than 90% of the triangular parameter domain Ω.

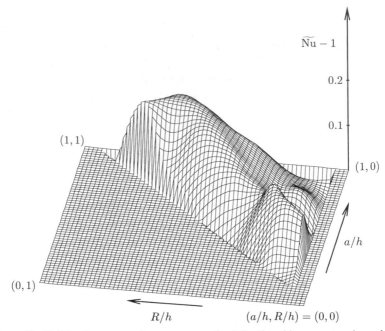

Figure 10   Validated surrogate (or response surface) for the eddy-promoter channel-flow Nusselt number as a function of nondimensional eddy-promoter placement (a/h) and radius (R/h) (from Yesilyurt & Patera 1993b, with permission).

The surrogate of Figure 10 can now be used in subsequent optimization studies—with quantifiable assurances as to fidelity—in lieu of the expensive originating parallel calculation. A slowly convergent optimization procedure will not exhaust resources proceeding through infeasible states, as the surrogate, unlike the Navier-Stokes simulation, can be evaluated effectively ad infinitum. Perhaps most importantly, inevitable mid-process modifications to design and optimization criteria and specifications can be accommodated by the surrogate *without* reappeal to the parent large-scale computation. The particular surrogate procedure presented here, and even surrogate procedures in general, may not be the best approach to exploiting large-scale parallel simulations in engineering practice. Nevertheless, it is clear that the effective incorporation of parallel simulations into synthesis—not just analysis—exercises, is a central problem for the coming decade; parallel processing can enable, and even stimulate, the requisite integration, but cannot, in itself, suggest appropriate computational paradigms.

ACKNOWLEDGMENTS

This review paper effort was supported by DARPA under Grant N00014-91-J-1889, by the ONR under Grants N00014-90-J-4124 and N00014-89-J-1610, by the NSF under Grant ASC-9107674, and by the AFOSR under Grant F49620-93-1-6090.

Literature Cited

Acarlar, M. S., Smith, C. R. 1987. A study of hairpin vortices in a laminar boundary layer. Part I. Hairpin vortices generated by a hemisphere protuberance. *J. Fluid Mech.* 175: 1–41

Aggarwal, R., Keunings, R. 1993. Finite element simulation of memory fluids on message-passing parallel computers. In *Parallel Computational Fluid Dynamics 92, Rutgers University*, ed. R. B. Pelz, A. Ecer, J. Hauser, pp. 1–8. Amsterdam: Elsevier

Anagnostou, G., Dewey, D., Patera, A. T. 1989. Geometry-defining processors for engineering design and analysis. *Visual Comput.* 5: 304–15

Appel, A. 1985. An efficient program for many-body simulation. *SIAM J. Sci. Stat. Comput.* 6: 85

Applegate, J. F., Douglas, M. R., Gursel, Y., Hunter, P., Scitz, C., Sussman, G. J. 1985. A digital orrery. *IEEE Trans. Comput.* 34(9): 822–31

Bailey, D. H. 1992. Misleading performance reporting in the supercomputing field. *NASA Rep. No. RNR-92-005*, NASA Ames Res. Cent.

Barnes, J., Hut, P. 1986. A hierarchical $O(N \log N)$ force-calculation algorithm. *Nature* 324: 446

Beguilin, A., Geist, A., Dongarra, J., Manchek, R. 1992. A user's guide to PVM parallel virtual machine. *ORNL/TM 11826*, Oak Ridge Natl. Lab., Oak Ridge, TN

Bertsekas, D. P. 1983. Distributed asynchronous computation of fixed points. *Math. Prog.* 27: 107

Bokhari, S. H. 1979. On the mapping problem. In *Proc. 1979 Int. Conf. Parallel Processing*, pp. 239–48

Bokhari, S. H. 1991. Complete exchange on the iPSC/860. *ICASE Rep. No. 91-4*, NASA Langley Res. Cent.

Bouard, R., Coutanceau, M. 1980. The early stage of development of the wake behind an impulsively started cylinder for $40 \leq Re \leq 10^4$. *J. Fluid Mech.* 101: 583

Braaten, M. E. 1991. Development of a par-allel computational fluid dynamics algorithm on a hypercube computer. *Int. J. Numer. Methods Fluids* 12: 947–63

Braaten, M. E. 1993. Applications of parallel computing in computational fluid dynamics: a review. In *Advances in Distributed Memory and Parallel Processing, Vol. 1: Applications*, ed. H. Tyrer. Ablex (Also GE Res. Dev. Cent. Tech. Inf. Ser. Rep. No. 89CRD121, July 1989.) In press

Brezzi, F. 1974. On the existence, uniqueness, and approximation of saddle-point problems arising from Lagrange multipliers. *RAIRO Anal. Numer.* 8: 129–51

Brunet, J.-P., Johnsson, S. L. 1992. All-to-all broadcast with applications on the Connection Machine. *Int. J. Supercomput. Appl.* 6(3): 241–56

Cahouet, J., Chabard, J. P. 1986. Multi-domains and multi-solvers finite element approach for the Stokes problem. In *Proc. Int. Symp. Innovative Numer. Methods in Eng., 4th*, ed. R. P. Shaw, pp. 317–22. Berlin: Springer-Verlag

Chorin, A. J. 1968. Numerical solution of the Navier-Stokes equations. *J. Math. Comput.* 22: 745

Chorin, A. J. 1973. Numerical study of slightly viscous flows. *J. Fluid Mech.* 57: 785–96

Clarke, N. R., Tutty, O. R. 1992. Two-dimensional Navier-Stokes flow simulation on MIMD processor arrays. In *Parallel Computing and Transputer Applications 92*, ed. M. Jane, J. L. Larriba, E. Oñate, B. Suárez, M. Valero, pp. 1323–32. IOS Press

Cline, D. D., Schutt, J. A. 1992. Large-scale simulation of the three-dimensional Navier-Stokes equations. In *Proc. Symp. High-Performance Computing for Flight Vehicles, Washington D.C.*

Cruz, M. E. 1993. *A parallel Monte-Carlo partial-differential-equation procedure for the analysis of multicomponent random media.* PhD thesis. Mass. Inst. Tech., Cambridge, MA

Cruz, M. E., Patera, A. T. 1993. A parallel Monte-Carlo partial-differential-equation procedure for the analysis of multi-component random media: application to thermal composites and porous media. *Int. J. Numer. Methods Eng.* Submitted

Dally, W. J. 1987. *A VLSI Architecture for Concurrent Data Structures.* Dordrecht: Kluwer

Duff, I. S., Johnsson, S. L. 1989. Node orderings and concurrency in structurally-symmetric sparse problems. In *Parallel Supercomputing: Methods, Algorithms and Applications,* ed. G. Carey, pp. 177–89. New York: Wiley

Ewing, R. E., Russell, T. F. 1981. Multistep Galerkin methods along characteristics for convection-diffusion problems. In *Advances in Computer Methods for Partial Differential Equations—IV, Rutgers University,* ed. R. Vichnevetsky, R. S. Stepleman, pp. 28–36. IMACS

Farhat, C., Lesoinne, M. 1993. Automatic partitioning of unstructured meshes for the parallel solution of problems in computational mechanics. *Int. J. Numer. Methods Eng.* 36(5): 745–64

Fatoohi, R. 1993. Adapting the INS3D-LU code to the CM2 and iPSC/860. *J. Supercomput.* Submitted

Fischer, P. F. 1993. Parallel domain decomposition for incompressible fluid dynamics. In *Proc. Int. Conf. on Domain Decomposition Methods in Science and Engineering, 6th,* ed. A. Quarteroni. Providence, RI: Am. Math. Soc. In press

Fischer, P. F., Patera, A. T. 1991. Parallel spectral element solution of the Stokes problem. *J. Comput. Phys.* 92(2): 380–421

Flynn, M. J. 1972. Some computer organizations and their effectiveness. *IEEE Trans. Comput.* C-21(9): 880–86

Fox, G. C., Johnson, M., Lyzenga, G., Otto, S. W., Salmon, J., Walker, D. 1988. *Solving Problems on Concurrent Processors: General Techniques and Regular Problems,* Vol. 1. Englewood Cliffs, NJ: Prentice-Hall

Fox, G. C., Otto, S. W. 1986. In *Hypercube Multiprocessors 1986,* ed. M. T. Heath, p. 244. Philadelphia: SIAM

Fuentes, Y. O., Kim, S. 1992. Parallel computational microhydrodynamics: communication scheduling strategies. *AIChE J.* 38(7): 1059–78

George, J. A., Liu, J. W. H. 1981. *Computer Solution of Large Sparse Positive Definite Systems.* Englewood Cliffs, NJ: Prentice-Hall

Glowinski, R., Kutznetsov, Yu., Meurant, G., Periaux, J., Widlund, O. B. ed. 1991. In *Proc. Int. Conf. Domain Decomposition Methods for Partial Differential Equations, 4th.* Philadelphia: SIAM

Golub, G. H., van Loan, C. F. 1983. *Matrix Computations.* Baltimore: Johns Hopkins Univ. Press

Greengard, L., Rokhlin, V. 1987. A fast algorithm for particle simulations. *J. Comput. Phys.* 73: 325

Griebel, M. 1993. Sparse grid multilevel methods, their parallelization, and their application to CFD. In *Parallel Computational Fluid Dynamics 92, Rutgers University,* ed. R. B. Pelz, A. Ecer, J. Hauser. Amsterdam: Elsevier

Griebel, M., Huber, W., Rüde, U., Störtkuhl, T. 1992. The combination technique for parallel sparse-grid-preconditioning or -solution of PDEs on workstation networks. In *Parallel Processing: CONPAR 92–VAPP V,* ed. L. Bougé, M. Cosnard, Y. Robert, D. Trystram, pp. 217–28. New York: Springer-Verlag

Gropp, W. D. 1992. Parallel computing and domain decomposition. In *Proc. Int. Conf. on Domain Decomposition Methods for Partial Differential Equations, 5th,* ed. D. E. Keyes, T. F. Chan, G. A. Meurant, J. S. Scroggs, R. G. Voigt, pp. 349–61. Philadelphia: SIAM

Gropp, W. D., Keyes, D. E. 1988. Complexity of parallel implementation of domain decomposition techniques for elliptic partial differential equations. *SIAM J. Sci. Stat. Comput.* 9: 312–26

Gropp, W. D., Keyes, D. E. 1992a. Domain decomposition methods in computational fluid dynamics. *Int. J. Numer. Methods Fluids* 14: 147–65

Gropp, W. D., Keyes, D. E. 1992b. Domain decomposition with local mesh refinement. *SIAM J. Sci. Stat. Comput.* 13: 967–93

Gustafson, J. L., Montry, G. R., Benner, R. E. 1988. Development of parallel methods for a 1024-processor hypercube. *SIAM J. Sci. Stat. Comput.* 9(4): 609–38

Harrar, D. L. II, Keller, H. B., Lin, D., Taylor, S. 1991. Parallel computation of Taylor-vortex flows. *CRPC Rep. No. 91-7,* Caltech, Pasadena, CA

Hecht, F., Saltel, E. 1990. Emc2: Editeur de maillages et de contours bidimensionnels. Manuel d'utilisation. *Rapp. Tech. No. 118.* Rocqueucourt, France: INRIA

Ho, L. W., Patera, A. T. 1991. Variational formulation of three-dimensional viscous free-surface flows: natural imposition of surface tension boundary conditions. *Int. J. Numer. Methods Fluids* 13: 691–98

Horton, G. 1991. *A time-parallel solution method for the Navier-Stokes equations.* PhD thesis. Univ. Erlangen-Nurnberg, Germany

Hughes, T. J. R., Franca, L. P. 1987. A new finite element formulation for computational fluid dynamics: VII. The Stokes problem with various well-posed boundary conditions: symmetric formulations that converge for all velocity/pressure spaces. *Comput. Methods Appl. Mech. Eng.* 65: 85–96

Jackson, E., She, Z., Orszag, S. A. 1991. A case study in parallel computing. I: Homogeneous turbulence on a hypercube. *J. Sci. Comput.* 6(1): 27–45

Johan, Z., Hughes, T. J. R., Mathur, K. K., Johnsson, S. L. 1992. A data-parallel finite element method for computational fluid dynamics on the Connection Machine system. *Comput. Methods Appl. Mech. Eng.* 99: 113–34

Karniadakis, G. E., Israeli, M., Orszag, S. A. 1991. High-order splitting methods for the incompressible Navier-Stokes equations. *J. Comput. Phys.* 97: 414–43

Karniadakis, G. E., Mikic, B. B., Patera, A. T. 1988. Minimum-dissipation transport enhancement by flow destabilization: Reynolds' analogy revisited. *J. Fluid Mech.* 192: 365–91

Karniadakis, G. E., Orszag, S. A. 1993. Nodes, modes, and flow codes. *Phys. Today* 46(3): 34–42

Kim, H. T., Kline, S. J., Reynolds, W. C. 1971. The production of turbulence near a smooth wall in a turbulent boundary layer. *J. Fluid Mech.* 50: 133

Keunings, R., Zone, O., Aggarwal, R. 1992. Parallel algorithms in computational rheology. In *Theoretical and Applied Rheology*, ed. P. Moldernaers, R. Keunings, pp. 274–76. Amsterdam: Elsevier

Kozlu, H., Mikic, B. B., Patera, A. T. 1988. Minimum-dissipation heat removal by scale-matched flow destabilization. *Int. J. Heat Mass Transfer* 31(10): 2023–32

Lucas, R. F., Blank, T., Tiemann, J. J. 1987. A parallel solution method for large sparse systems of equations. *IEEE Trans. Comput. Aided Des.* CAD-6(6): 981–91

Lumley, J. L. 1967. The structure of inhomogeneous turbulent flow. In *Atmospheric Turbulence and Radio Wave Propagation*, ed. A. M. Yaglom, V. I. Tatarski, pp. 166–78. Moscow: Nauka

Maday, Y., Meiron, D. I., Patera, A. T., Rønquist, E. M. 1993. Analysis of iterative methods for the steady and unsteady Stokes problem: application to spectral element discretizations. *SIAM J. Sci. Comput.* 14(2): 310–37

Maday, Y., Patera, A. T., Rønquist, E. M. 1990. An operator-integration-factor splitting method for time-dependent problems: application to incompressible fluid flow. *J. Sci. Comput.* 5(4): 263–92

Mandel, J. 1990. Hierarchical preconditioning and partial ortho-gonalization for the *p*-version of the finite element method. In *Proc. Int. Conf. on Domain Decomposition Methods in Science and Engineering, 3rd*, ed. T. F. Chan, R. Glowinski, J. Periaux, O. B. Widlund, pp. 141–56. Philadelphia: SIAM

Mangiavacchi, N., Akhavan, R. 1993. Direct numerical simulations of turbulent shear flows on distributed memory architectures. In *Proc. SIAM Conf. on Parallel Proc. for Sci. Comput.*, 6th, ed. R. F. Sincovec et al. Philadelphia: SIAM

Mansfield, L. 1988. On the use of deflation to improve the convergence of conjugate gradient iteration. *Comm. Appl. Numer. Methods.* 4: 151–56

McCormick, S. F. 1992. *Multilevel Projection Methods for Partial Differential Equations*. Philadelphia: SIAM

McDonald, J. D. 1991. Particle simulation in a multiprocessor environment. *Proc. AIAA 26th Thermophysics Conf., Honolulu*, AIAA 91-1366. Washington, DC: AIAA

Mehrabi, M. R., Brown, R. A. 1993. An incomplete nested dissection algorithm for parallel direct solution of finite element discretizations of partial differential equations. *J. Sci. Comput.* To appear

Moin, P., Kim, J. 1982. Numerical investigation of turbulent channel flow. *J. Fluid Mech.* 118: 341–77

Mu, M., Rice, J. R. 1992. A grid-based sub tree-subcube assignment strategy for solving partial differential equations on hypercubes. *SIAM J. Sci. Stat. Comput.* 13(3): 826–39

Nicolaides, R. A. 1987. Deflation of conjugate gradients with applications to boundary value problems. *SIAM J. Numer. Anal.* 24(2): 355–65

Orszag, S. A. 1972. Comparison of pseudo-spectral and spectral approximation. *Stud. Appl. Math.* 51: 253–59

ParaSoft Corp. 1992. Express Operating System. Pasadena, CA

Pelz, R. B. 1991a. The parallel Fourier pseudospectral method. *J. Comput. Phys.* 92(2): 296–312

Pelz, R. B. 1991b. Fourier spectral method on ensemble architectures. *Comput. Methods Appl. Mech. Eng.* 89: 529–42

Pelz, R. B. 1992. Hypercube FFT and Fourier pseudospectral method. In *Parallel Computational Fluid Dynamics: Implementations and Results Using Parallel Computers*, ed. H. D. Simon, pp. 189–214. Cambridge: MIT Press

Pepin, F. 1990. *Simulation of the flow past an impulsively started cylinder using a discrete vortex method.* PhD thesis. Caltech, Pasadena, CA

Pepin, F., Leonard, A. 1990. Concurrent

implementation of a fast vortex method. In *Proc. Distributed Memory Computing Conf., 5th, Charleston*, pp. 453–62. Los Alamitos, CA: IEEE Comput. Soc. Press

Pironneau, O. 1982. On the transport-diffusion algorithm and its applications to the Navier-Stokes equations. *Numer. Math.* 38: 309–32

Quarteroni, A. 1991. Domain decomposition and parallel processing for the numerical solution of partial differential equations. *Surv. Math. Ind.* 1: 75–118

Ramamurti, R., Löhner, R. 1993. A parallel implicit incompressible flow solver using unstructured meshes. *Int. J. Comp. Fluids.* Submitted

Reinsch, K. G., Schmidt, W., Ecer, A., Häuser, J., Periaux, J., eds. 1992. *Parallel Computational Fluid Dynamics 91.* Amsterdam: Elsevier

Renka, R. L. 1988. Algorithm 660: QSHEP2D: quadratic Shepard method for bivariate interpolation of scattered data. *Assoc. Comput. Mach. Trans. Math. Software* 14: 149–50

Rønquist, E. M. 1992. A domain decomposition method for elliptic boundary value problems: application to unsteady incompressible fluid flow. In *Proc. Int. Conf. on Domain Decomposition Methods for Partial Differential Equations, 5th*, ed. D. E. Keyes, T. F. Chan, G. A. Meurant, J. S. Scroggs, R. G. Voigt, pp. 545–57. Philadelphia: SIAM

Saad, Y., Schultz, M. H. 1985. Topological Properties of Hypercubes. *Res. Rep. No. YALEU/DCS/RR-389*, Yale Univ., New Haven, CT

Saad, Y., Schultz, M. H. 1986. GMRES: A generalized minimal residual algorithm for solving nonsymmetric linear systems. *SIAM J. Stat. Comput.* 7: 856–69

Sangani, A. S., Yao, C. 1988. Transport processes in random arrays of cylinders. II. Viscous flow. *Phys. Fluids* 31(9): 2435–44

Schatz, M. F., Tagg, R. P., Swinney, H. L., Fischer, P. F., Patera, A. T. 1991. Supercritical transition in plane channel flow with spatially periodic perturbations. *Phys. Rev. Lett.* 66(12): 1579–82

Schreiber, R., Simon, H. D. 1992. Towards the teraflops capability for CFD. In *Parallel Computational Fluid Dynamics—Implementations and Results Using Parallel Computers*, ed. H. D. Simon, pp. 331–60. Cambridge: MIT Press

Scott, D. S. 1991. Efficient all-to-all communication patterns in hypercube and mesh topologies. In *Proc. Distributed Memory Computing Conf., 6th*, ed. Q. Stout, pp. 398–403. Los Alamitos, CA: IEEE Comput. Soc. Press

Scott, R., Bagheri, B. 1990. Software environments for the parallel solution of partial differential equations. In *Computing Methods in Applied Sciencies and Engineering, IX*, ed. R. Glowinski, A. Lichnewsky, pp. 378–92. Philadelphia: SIAM

Seitz, C. L. 1992. Mosaic C: an experimental fine-grain multicomputer. In *Future Tendencies in Computer Science, Control, and Applied Mathematics, Proc. Int. Conf. Celebrating 25th Anniversary of INRIA*, ed. A. Bensoussan, J.-P. Verjus, LNCS 653. New York: Springer-Verlag

Sethian, J. A., Brunet, J.-P., Greenberg, A., Mesirov, J. P. 1992. Two-dimensional, viscous, incompressible flow in complex geometries on a massively parallel processor. *J. Comput. Phys.* 101(1): 185–206

Sethian, J. A., Salem, J. B. 1989. Animation of interactive fluid flow visualization tools on a data parallel machine. *Int. J. Supercomput. Appl.* 3(2): 10–39

Simon, H. D. 1991. Partitioning of unstructured problems for parallel processing. *Comput. Syst. Eng.* 2(2/3): 135–48

Simon, H. D., ed. 1992. *Parallel Computational Fluid Dynamics: Implementations and Results Using Parallel Computers.* Cambridge: MIT Press

Sugihara, K., Iri, M. 1989. VORONOI2 reference manual. *Res. Memo. RMI 89-04*, Dept. Math. Eng. and Inform. Phys. Faculty of Eng., Univ. Tokyo

Sussman, A., Saltz, J., Das, R., Gupta, S., Mavriplis, D., et al. 1992. PARTI primitives for unstructured and block structured problems. *NASA Contractor Rep. No. 189662, ICASE Interim Rep. No. 22*, NASA Langley Res. Cent.

Swartztrauber, P. N. 1987. Multiprocessor FFTs. *Parallel Comput.* 5: 197–210

Teman, R. 1984. *Navier-Stokes Equations. Theory and Numerical Analysis.* Amsterdam: North-Holland

Tezduyar, T. E., Aliabadi, S., Behr, M., Johnson, A., Mittal, S. 1992a. Massively parallel finite element computation of three dimensional flow problems. In *Proc. Japan Numerical Fluid Dynamics Symp., 6th, Chuo Univ., Tokyo, Japan* (Also *UMSI Rep. No. 92/224*, Univ. Minn., Minneapolis, MN)

Tezduyar, T. E., Behr, M., Mittal, S., Johnson, A. 1992b. Computation of unsteady incompressible flows with the stabilized finite element method—space-time formulations, iterative strategies and massively parallel implementation. In *New Methods in Transient Analysis*, ed. P. Smolinski, W. K. Liu, G. Hulbert, K. Tamma, AMD-Vol. 143, pp. 7–24. New York:

ASME (Also *UMSI Rep. No. 92/145*, Univ. of Minn., Minneapolis, MN)

Vanderstraeten, D., Zone, O., Keunings, R., Wolsey, L. 1993. Nondeterministic heuristics for automatic domain decomposition in direct parallel finite element calculations. In *Proc. SIAM Conf. Parallel Processing for Sci. Comput., 6th*, ed. R. F. Sincovec et al., pp. 929–32. Philadelphia: SIAM

Watts, R., Reeve, J. S., Tutty, O. R. 1992. A portable parallel implementation of a domain decomposed computational fluid dynamics algorithm. In *Parallel Computing and Transputer Applications 92*, ed. M. Jane, J. L. Larriba, E. Oñate, B. Suárez, M. Valero, pp. 1353–62. IOS Press

Williams, R. D. 1991. Performance of dynamic load balancing algorithms for unstructured mesh calculations. *Concurrency: Pract. Exp.* 3(5): 457–81

Williams, R. D., Glowinski, R. 1989. Distributed irregular finite elements. In *Numerical Methods in Laminar and Turbulent Flow*, ed. C. Taylor, P. Gresho, R. L. Sani, J. Hauser, pp. 3–14. Swansea, UK: Pineridge

Wray, A. A., Rogallo, R. S. 1992. Simulation of turbulence on the Intel Gamma and Delta. *NASA Tech. Memo.*, NASA Ames Res. Cent.

Yesilyurt, S., Patera, A. T. 1993a. Statistical modelling methods for deterministic computational systems. *J. Comput. Phys.* Submitted

Yesilyurt, S., Patera, A. T. 1993b. Surrogates for numerical simulations; optimization of eddy-promoter heat exchangers. *Comput. Methods Appl. Mech. Eng.* Submitted (also *ICASE Rep. No. 93-50*, NASA Langley Res. Cent.)

Zhao, F., Johnsson, S. L. 1991. The parallel multipole method on the Connection Machine. *SIAM J. Sci. Stat. Comput.* 12(6): 1420–37

Annu. Rev. Fluid Mech. 1994. 26:529–71

PULMONARY FLOW AND TRANSPORT PHENOMENA

J. B. Grotberg

Biomedical Engineering Department, Robert R. McCormick School of Engineering and Applied Science, Northwestern University, Evanston, Illinois 60208 and Department of Anesthesia, Northwestern University Medical School, Chicago, Illinois 60611

KEY WORDS: biofluid mechanics, pulmonary mechanics, respiratory flow, surface tension

INTRODUCTION

The lung is the respiratory organ whose development was the single most important event permitting the evolution of land animals (e.g. humans) from their ocean-dwelling cousins. Because of the presence and role of gases and liquids in the lung and immediate environment, its normal physiologic function, abnormal pathophysiology, assessment, and therapy involve a wide variety of fluid dynamical phenomena. The field was last reviewed in this forum by Pedley (1977) and a great deal of new research in pulmonary fluid dynamics has appeared since then. Here, we review six aspects of fluid mechanics and transport in the pulmonary system. These are functionally and anatomically related to the areas labeled in Figure 1: A) intra-airway gas flow and transport, B) surface tension phenomena, C) intrapleural fluid mechanics, D) flow limitation and wheezing, E) interstitial flows, and F) mucus transport.

The primary function of the lung is gas exchange, as it allows oxygen to move from the air into the blood and carbon dioxide to move out. The lung airways comprise a branching network of tubes which become narrower, shorter, and more numerous as they penetrate deeper into the lung. The trachea (radius 0.9 cm) divides into the right and left main bronchi which themselves bifurcate into the lobar, then segmental bronchi. The lung consists of a total of 23 such airway generations (the trachea

529

0066–4189/94/0115–0529$05.00

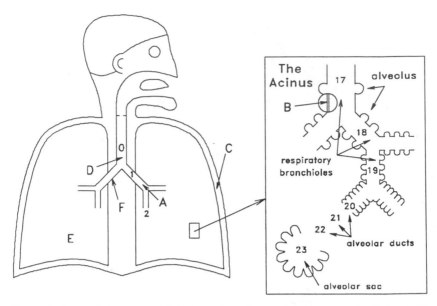

Figure 1 Anatomic locations of fluid dynamic and transport phenomena in the lung. A) Intra-airway gas flow and transport; B) surface tension phenomena; C) intrapleural fluid mechanics; D) flow limitation and wheezing; E) interstitial flows; F) mucus transport.

being generation zero), continuing down to the terminal air sacs, called alveoli, each with a radius of approximately 150 microns. The total surface area of the alveoli is about 100 m^2 in the adult, creating a very large interface for diffusional gas exchange. It must be pointed out that the walls of airways have viscoelastic properties so that the lengths and diameters of airways increase as the lung volume is increased. Models of the branching geometries from careful anatomical measurements include the symmetrical network of Weibel (1963) and the asymmetrical network of Horsefield et al (1971). The primary feature of airway geometry is the bifurcation of one tube into two smaller tubes (see Figure 2). From the data of Weibel (1963) and Haefeli-Bleuer & Weibel (1988), a typical bifurcation has the following characteristics: (*a*) $L/d_1 \approx 3.5$; (*b*) $d_2/d_1 \approx 0.79$ for generations 0–16; (*c*) a total area ratio of $2(d_2/d_1)^2 = 1.20$–1.25, for generations beyond the third; (*d*) a branching angle, ϕ, varying from $64°$ to $100°$; and (*e*) an axial radius of curvature R, between $5d_1$ and $10d_1$. For the first three generations, the total cross-sectional area decreases, with a minimum at generation 3. Modern views of the branching geometry now draw upon fractal theory and convenient equations relating airway dimensions to generation number may be found in the work of West et al (1986).

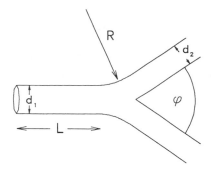

Figure 2   A model airway bifurcation. ϕ is the branching angle; R is the axial radius of curvature.

INTRA-AIRWAY FLOW AND TRANSPORT

Alternative Modes of Ventilation

The transport of respiratory gases received a great deal of attention in the previous decade due to the increasing awareness that animals could be mechanically ventilated with small tidal volumes at high frequency, essentially in a panting mode, with good results (Lunkenheimer et al 1972, Bohn et al 1980). The potential benefit of the small tidal volumes is to reduce the complications of conventional ventilators whose positive pressures can compromise cardiac function and damage the lung. High-frequency ventilation (HFV) utilizes a tidal volume (35–150 ml) much smaller than the anatomic dead space and a frequency (60–1800 breaths/min) much higher than the normal breathing rate. Two different modes of HFV include one in which inspiration is active with a passive expiratory phase, termed high-frequency jet ventilation (HFJV), and the other in which both phases are active, termed high-frequency oscillations (HFO). Also, the oscillations may be imposed at the airway opening or at the chest wall. Because the tidal volume is smaller than the dead space, CO_2 and O_2 cannot be transported directly to the gas-exchanging regions by pure convection. Therefore other transport processes are responsible for gas exchange. Experience with animals and humans indicates that gas exchange can be maintained under HFV, and that axial dispersion due to the interaction between radial diffusion and radially-nonuniform axial velocities is responsible for gas transport. In addition, experimental results of HFO (Mitzner et al 1983) suggest the existence of an optimal oscillation frequency, though whether or not tidal volume remained fixed in these preparations has been called into question. At the moment, HFV is used clinically in neonatal intensive care situations.

A less-studied alternative ventilatory mode is that of constant flow ventilation (CFV), in which a steady oxygen stream is directed into the lungs from a catheter placed near the distal end of the trachea. High flow rates of 2–3 liters/min/kg are required to keep normal blood gases in the absence of respiratory motions by the subject. Ingenito et al (1988) have proposed a conceptual model of this flow and transport system. In their model the transport regimes are divided into two parts: One is close enough to the jet to be dominated by turbulent convection and mixing; the other is more distal, dominated by diffusion, and enhanced by cardiac oscillations, which create cyclic air flow and hence improve dispersion by bouncing against the lungs. Gavriely et al (1992) showed that much lower flow rates (0.05–0.20 liters/min/kg) may be used in a dog model of marginal ventilatory movements—which could result from neuromuscular disease or spinal cord injury. This CFV mode decreases the work of breathing per unit of CO_2 outflow with dramatic reductions when coupled with simultaneous, imposed, external chest wall vibrations.

Oscillatory Flow and Transport in a Straight, Rigid Tube

The major aim of the various studies regarding HFV has been to characterize the transport by determining the effective diffusivity, D_{eff}, of the systems analyzed, either with measurements of mass flow rates and local baseline concentration gradients or with dispersing clouds of contaminant. The two approaches have their respective definitions of D_{eff} which converge as the cloud spreads axially (Chatwin 1975):

$$D_{eff} = \frac{\iint uc - D\frac{\partial c}{\partial x}}{(Area)\left(\frac{\partial c}{\partial x}\right)} \quad \text{or} \quad D_{eff} = \frac{1}{2}\frac{d\sigma^2}{dt}, \tag{1}$$

where u is the axial velocity, c is the solute concentration (e.g. CO_2), x is the axial coordinate, and σ^2 is the axial variance of the solute concentration.

The dispersion of solute in a straight tube with oscillatory flow has been considered by assuming an undisturbed linear concentration distribution along the axis. Harris & Goren (1967) first developed this theory, solving the Navier-Stokes equations and the convection-diffusion equation. They also performed experiments of oscillating liquid flow using HCl as a tracer contaminant. The linear axial profile took about 30 hours to establish before data could be taken in that study. Ultimately, their data and theory matched well. In particular, they studied how the transport rate varies with the Womersley parameter, $\alpha = a(\omega/v)^{1/2}$, and the amplitude parameter $A = d/a$, where a is the tube radius, $\omega = 2\pi f$, the angular frequency of oscillation, v the kinematic viscosity of the fluid, and d the stroke distance defined by $d = V_T/\pi a^2$, where V_T is the tidal volume of the tube. Typical parameter values for HFO given in Grotberg (1984) indicate that, for a tidal volume of 50 cc and a frequency of 10 Hz, $1 < A < 100$ (depending on the airway generation) and $1 < \alpha < 20$. Harris & Goren found that the transport is governed by the interaction of radial diffusion and axial convection (a form of Taylor dispersion) and increases monotonically with increasing α for fixed A and Schmidt number, $Sc = v/D$, where D is the molecular diffusivity. Though a complete functional dependence is given, simpler asymptotic forms of D_{eff} for their results are $D_{eff}/D \sim 1 + f(Sc)\alpha^4 A^2$ for $\alpha \ll 1$ and $D_{eff}/D \sim 1 + g(Sc)\alpha A^2$ for $\alpha \gg 1$, where f and g depend on Sc. The impetus for separating frequency and amplitude effects in this manner is somewhat clinical in that high frequency ventilators have an adjustable setting for both. Chatwin (1975) proposed a statistical approach for an initial cloud which, after long enough time, had locally linear axial profiles. His results are similar to those of Harris & Goren (1967). Watson (1983) proposed an identical theory to that of Harris & Goren (1967), with generalizations to more complicated cross-sectional shapes. In a companion paper to Watson's, Joshi et al (1983) measured transport of a methane gas tracer in oscillatory air flow with good correlation to the theory. The linear axial profile used in that study took about 20 hours to establish.

The dispersion of a localized contaminant distribution, such as an initial slug or a sudden discharge into the stream at a point, has been the focus of researchers interested in environmental applications such as pollutant discharge into rivers and tidal estuaries. Gill & Sankarasubramanian (1971) solved the convection-diffusion equation by expanding the local concentration in terms of the cross-sectionally averaged concentration and its axial derivatives. Although their theory is posed for general velocity fields, they only explored dispersion in steady flows. It is worth noting that the derivative-expansion method can be used to calculate the local concentration as well as the average concentration, and that it is applicable at smaller times than are required for the Taylor (1953) and Aris (1956) solutions, as shown by Yu (1981). Smith (1982) used his delay-diffusion equation approach to address the issue of small-time predictions for the dispersion coefficient, which could become negative over portions of the early cycles for an initial sudden discharge of contaminant. The contaminant slug is stretched during the first half of the cycle but may contract during the last half, so that the axial variance of the distribution has a negative time derivative. When time is large enough, $t > a^2/D$, however, these issues no longer arise—in the lung this criterion is usually met for gas transport. In aerosol transport, or in vascular applications, where D is small, Smith's approach would be more applicable.

Much faster experimental methods for determining D_{eff} have recently emerged. Gaver et al (1992) and Elad et al (1992) expanded on the unsteady bolus dispersion framework of Gill & Sankarasubramanian (1971). By injecting a bolus of argon contaminant gas and measuring the oscillatory decay of local concentration at the injection site, D_{eff} could be inferred with less than a minute of data, and this methodology has been used in vivo. The results agree with the constant gradient theories discussed above. Sharp et al (1991) also developed a rapid method for inferring D_{eff} in their curved tube experiments, discussed below, by following the progression of an oscillating tracer front and then curve-fitting their data to a one-dimensional, unsteady diffusion equation.

Branched Network Studies and Tapered Tubes

In efforts to examine the gas exchange during HFO, several investigators have performed bench-top experiments on models of multiple generations of branching tubes. Examples are Tarbell et al (1982) for $Sc \gg 1$ (dye transport in liquid systems) and Paloski et al (1987) for $Sc = O(1)$ (methane transport in gas). In each of these experiments the soluble tracer was introduced into the network system, cycled, and then D_{eff} inferred from the dispersion. The networks, however, were made of identical tubes so that the area ratio through the bifurcation was 2.0, rather than 1.25. Empirical relationships were sought between the measured transport coefficient and independent parameters, such as V_T, f, and Sc. Generally, axial dispersion is greater in these branching tube networks than in a straight tube. For example, if Sc is fixed near unity, Paloski et al (1987) found that $D_{eff}/D \sim A^{1.93}\alpha^{2.89}$ for $\alpha < 4$ and $D_{eff}/D \sim A^{1.93}\alpha 1.43$ for $\alpha > 4$. Paloski et al also present more complicated expressions for Sc ranging from 0.22 to 1.78, obtained by changing the background gas.

One explanation of the observed transport enhancement in networks, compared to oscillatory transport in straight tubes, is based on the work of Scherer & Haselton (1982) who focused on the asymmetry of axial velocity profiles on inspiration versus expiration resulting from the airway bifurcation geometry. They performed flow visualization experiments by injecting a slug of neutrally-buoyant particles into the branching region of a single bifurcation and then initiated their oscillations. Significant bidirectional steady streaming was observed along the axis, with a net zero bulk flow at any cross section of the system since the tidal volume is fixed. They measured dispersion of nondiffusible particles over a parameter range of frequency and tidal volumes. In an effort to explore this streaming effect, Tarbell et al (1982) investigated the dispersion coefficient for oscillating flow in a branching network similar to that of Scherer & Haselton (1982). They used a "nondiffusible" dye in water with $Sc = v/D \to \infty$, to highlight the purely convective phenomena relevant to the asymmetric flow field. Because D is so small, it is neglected in the dimensional analysis used to reduce the data. Their results indicate significant enhancement of transport over straight tube conditions. For transport of particles such as aerosols, these predictions are quite useful.

One must be careful not to apply large Sc results directly to CO_2 removal, however, where $Sc \approx 1$. The mass transport per unit cycle is related to the time average D_{eff}. We may decompose the velocity and concentration fields into their steady and periodic components, $u = u^s + u^p$ and $c = c^s + c^p$, respectively. The contribution of u^s to the steady component of D_{eff} must come from its product with c^s in Equation (1) above. The dependence of c^s on the cross-sectional coordinates is probably weak when $Sc \approx 1$ since the wall uptake of respiratory gases is small and cross-tube diffusion and mixing is fast. Since the integral of u^s across the cross section is zero, there would appear to be negligible contribution to D_{eff} from the steady streaming unless $Sc \gg 1$. As an example, Godleski & Grotberg (1988) examined oscillatory flow and mass transport in a tapered tube of small taper angle, ε. Their mass transport calculations demonstrated no contribution to D_{eff} from steady streaming. Transport could be better or worse compared to a straight tube depending on the parameter range, but this was due to the effect of the oscillatory velocity component having nonparallel streamlines interacting with the oscillatory concentration field. For example their asymptotic limits for $G_{ax}/D \sim B(l, \varepsilon)[1 + F_1(Sc, l)A^2\alpha^4]$ for $\alpha \ll 1$ and $G_{ax}/D \sim F_2(Sc, l, \varepsilon)A^2\alpha$ for $\alpha > 3$. Here we use G_{ax} as an overall effective conductance which characterizes the tube. D_{eff} itself is axially varying when the tube is tapered, so it does not reflect the entire effect. The dimensionless argument l represents geometric information and B, F_1, and F_2 are determined functions of the indicated arguments. One can see that the exponents of α and A are the same as for a straight tube, but the coefficients are such that transport is always better in a straight tube when $\alpha \gg 1$, when comparing tubes of the same initial radii. But when $\alpha \ll 1$, the tapered tube is better. Also, no local maxima of D_{eff} versus α were predicted.

Steady Streaming and Steady Pressure Gradients

An important aspect of the bidirectional steady streaming is that it coexists with an induced steady pressure gradient. Using a perturbation method for oscillating flow in a tapered channel with small taper angle, $\varepsilon \ll 1$, Grotberg (1984) showed that while the scale for axial velocities is ωd, the scale for lateral velocities is $\varepsilon \omega d$. The predicted steady pressure gradient has the higher pressure toward the wide end of the taper for all frequencies and this gradient increases with f. In addition, the steady streaming has a peculiar behavior when tidal volume is held fixed and f is increased. For small α the fluid near the center streams toward the wide end of the channel while fluid near the wall streams toward the narrow end. For $\alpha > 8$ or so, the centerline streaming reverses direction and goes toward the narrow end, just as as the wall layer does. An intermediate layer forms between them which flows to the wide end. The dimensionless Stokes layer thickness, $\delta_s/a = 1/\alpha$, is too thin to account for the intermediate drift layer, whose dimensionless thickness must be $1/(\varepsilon\alpha)$. Hence, this is a double boundary-layer problem for fixed tidal volume (see Figure 3). The predictions of pressure gradients are consistent with the observations by Simon et al (1984) of a steady pressure gradient in dog lungs during HFO, with larger pressures near the alveolar zone than at the bronchi and an increase of this gradient with larger f. This "hyperinflation" is a visible enlargement of the lungs to a new average size simply on the basis of increased oscillation frequency. Some physiologists attribute the hyperinflation to gas trapping due to limited flow out of the airways. The predicted details of the drift direction, including the reversal, are evident in the Scherer & Haselton (1982) data. The influence of this streaming on aerosol delivery is discussed in Briant et al (1992). Steady streaming and steady pressure gradients are also a consequence of airway flexibility (discussed below).

Curved Tube Flow and Transport

When a tube is curved, secondary flows are present in the cross section. For steady laminar flows, secondary motion makes the radial distribution

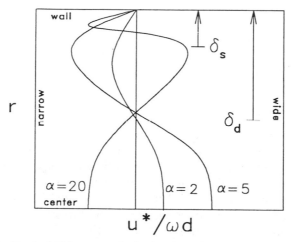

Figure 3   Steady drift in a tapered tube showing a double boundary layer for $\alpha \gg 1$.

of concentration more uniform, so that the flow is less effective in dispersing the solute axially. Axial dispersion is therefore expected to be decreased by secondary flows. This has been shown by a number of authors, e. g. Daskopoulos & Lenhoff (1988) by a modified method of moments and Johnson & Kamm (1986) by numerical simulations. Unlike the above authors who considered only the dynamical effects due to centrifugal force, Nunge et al (1972) employed the derivative-expansion method and also examined the geometrical effects of curvature. They found that axial dispersion in a curved tube during steady flow can be increased or decreased over the case of a straight tube, depending on the values of Reynolds number, Schmidt number, and δ.

For oscillatory flows in a curved tube, dispersion becomes more complicated. Eckmann & Grotberg (1988) examined mass transport in a curved tube with a steady, constant axial concentration gradient. For $\delta \ll 1$, their perturbation approach showed that tube curvature increases the rate of axial mass transport for all values of A and α. As α increases, the steady cross-sectional streamlines move from a pair of vortices to four vortices as has been found by other investigators (see Sudo et al 1992 and references therein). Because all curvature terms were kept in this analysis the centers of the vortices are considerably off-center. Also, when mass transport rates were plotted against α for certain values of A, they showed that a local maximum value of mass transport occurs. This maximum occurs because of the α-dependent phase relationship between the axial velocity and concentration fields, the lateral mixing being complicated by secondary flows. At certain frequencies there is a resonance between these two fields which optimizes transport. Experiments on dispersion in a curved tube were performed by Sharp et al (1991) using laser absorption for inferring methane tracer concentration which is averaged diametrically by the method. They demonstrate the enhancement of transport over the straight tube case and find a range where significant increases of D_{eff} with increasing α indicate the approach of a maximum. Pedley & Kamm (1988) described a model problem of secondary flow effects and found a convective resonance behavior responsible for a local maximum in axial transport. In their model, this resonance was between the main axial flow period and the secondary azimuthal convection period. Hence the appearance of an optimal frequency for transport during HFO, within a limited range of cycling frequencies, may have contributions from axial curvature effects due to the processes examined in these studies.

Jan et al (1989) studied oscillating flow in a model bifurcation with realistic geometry by flow visualization and particle tracking. A three-component classification scheme for flow phenomena is proposed using scaling arguments which balance, in pairs, the viscous, unsteady, and convective acceleration terms of either the axial or cross-stream momentum balance. The resulting velocity and length scales yield a measure of secondary to primary flow behavior, though how these scales may also balance the continuity equation is not discussed. An interesting experimental observation is that over part of the parameter range, secondary swirling flows follow the steady results of previous investigators such as Olson (1971) (reviewed in Pedley 1977). However when $\alpha > 10^{1/2}$ and $2 < A < 20$, the vortical flows are confined to the inner wall of the curved segment during inspiration and expiration leaving large portions of the cross section with only axial flow (see Figure 4).

Flexible Tube Flow and Transport

The effect of wall flexibility on gas exchange efficiency is particularly important, since, unlike curvature or tapering, airway flexibility changes with disease and also depends on the state of lung inflation. Patients with

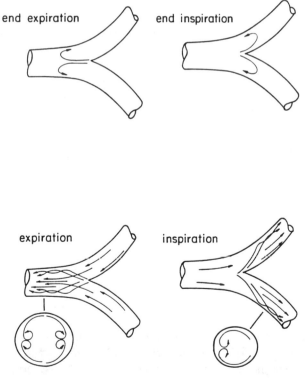

Figure 4   Wall-bounded secondary flows in a model bifurcation (from Jan et al 1989).

fibrosis and infant respiratory distress syndrome suffer from stiffer, less compliant, airways and supporting tissue. Aging and emphysema tend to increase the compliance. Also, the airways are more compliant as one moves deeper into the lung to the more distal generations. The first attempt at a global model of HFV was by Fredberg (1980) in which the steady, rigid tube dispersion of Taylor (1953) was incorporated into a network where local flows, and hence velocities, were dependent on airway compliance and resistance. An optimal frequency for transport was found in this approach due to mechanical resonance of the airways.

Dragon & Grotberg (1991) modeled volume-cycled oscillating flow in a flexible tube with transport. The important new parameter evolving from the interaction of the wave propagation and dispersion phenomena is $\kappa = Ea^2/(\rho_0 v^2)$, where E is the elastic modulus of the tube and ρ_0 is the tube wall density. One can think of $\kappa^{1/2}$ as a ratio of wall wave speed to fluid shear propagation speed. Typically κ ranges from 10^4 to 10^7. This model involves interactions of the oscillating flow, the corresponding traveling wave on the wall, and the unsteady concentration field. For $\kappa \to \infty$ the predictions coincide with those of Harris & Goren (1967). As κ is decreased, two important trends emerge. First, D_{eff} diminishes for all values of α and second, D_{eff} attains a local maximum at α_m in the HFO range. The value of α_m decreases with decreasing κ and coincides with a minimum phase angle between the wall position and the axial flow. Experimental observations appear to be in agreement with the trend of these results: A decrease in gas exchange efficiency due to radial airway motion was observed by Gavriely et al (1985) in cineradiobronchograms of dog lungs undergoing high frequency ventilation. Because air is shunted in the radial direction by the wall motion, axial transport of a diffusible species is reduced. According to the theory, both steady streaming and a steady pressure gradient, with higher pressures distally, are present—again consistent with the "hyperinflation" phenomena.

Hydon & Pedley (1993a) focus on flow driven only by lateral wall oscillations to assess its contribution to mass transport in a finite-length channel. The channel walls are kept parallel, a limit of long wave behavior, with prescribed motion. They are currently extending this work to include a simultaneous oscillatory axial pressure gradient to drive flows that better approximate HFV (Hydon & Pedley 1993b). The combined flows allow more or less axial transport depending on the phase angle between the vertical and axial pressure gradient oscillations. This phase angle is externally controllable in their model, unlike that found by Dragon & Grotberg where the wave propagation physics determines the phase angle.

Of the three effects discussed above (curvature, taper, flexibility) it appears that axial curvature may dominate the transport in many situations since D_{eff} always exceeds the straight tube case, as reflected in results of network transport experiments. Taper can help when $\alpha \ll 1$, but not when $Sc = O(1)$ because of steady streaming; flexibility always lowers transport compared to a rigid tube. Figure 5 shows a schematic comparison of transport effects due to either taper (Godleski & Grotberg 1988), axial curvature (Eckmann & Grotberg 1988), or flexibility (Dragon & Grotberg 1991).

Wall Uptake and Its Effects on Transport

In human airway transport, heat, water, and gases are inhaled and transported into and out of the airway walls and their liquid lining. It is very important to understand how these contaminants disperse axially and how heat and mass are absorbed radially. For example, when one breathes cool or dry air at high flow rates the removal of heat and water from the airway is known to provoke asthmatic attacks (exercise-induced asthma). Temperature measurements taken inside the airways during cold-air breathing showed that significant cooling could extend as deep as the 6th airway generation (McFadden et al 1985). Further studies have looked at the relative importance of heat versus water exchange in stimulating this bronchoconstriction (Gilbert et al 1987). Hanna & Scherer (1986) and Ingenito et al (1986) have developed one-dimensional models to account

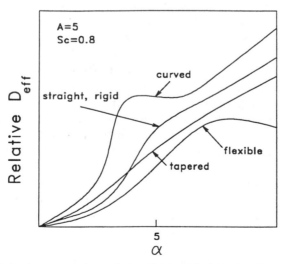

Figure 5   Schematic comparison of relative effective diffusivities in rigid, straight (Harris & Goren 1967), tapered (Godleski & Grotberg 1988), curved (Eckmann & Grotberg 1988), and flexible (Dragon & Grotberg 1991) tubes.

for the exchange of airway heat and water. They find that the ability of the airway tissue to replenish itself with heat and water from the vasculature is one of the most significant factors.

Another important example of wall transport is the uptake of gases and toxins. The lung is the major interface between the human body and the environment, and it processes about 10,000 liters of ambient atmospheric gas per day. Davidson & Schroter (1983) consider a 16-generation bronchial tree model by assuming steady, fully-developed flow in insulated liquid-lined tubes. These flow conditions were shown to be quasi-steady by Pedley et al (1970) when the breathing period is long enough to establish the steady boundary-layer thickness at the end of an airway, $\delta(L) = (\nu L/U)^{1/2}$. They show that this statement translates to a comparison of the steady boundary-layer thickness to the Stokes layer thickness such that quasi-steady behavior is expected when $\varepsilon_0 = [\delta(L)/\delta_s]^2 \ll 1$, which is true in most physiological instances, but not in HFO. Davidson & Schroter (1983) analyze the uptake of a soluble gas slug introduced into the trachea and show that the maximal absorption will peak in both the trachea and in a downstream segment, between generations 8 to 12. Grotberg et al (1990) developed a mathematical model for gas transport and absorption in liquid-lined tubes where chemical reactions with the toxin (ozone) occurred in the mucus, present in the first 13 generations. To determine the criterion for quasi-steady transport, they used a similar approach to show that this occurs when the ratio of steady to unsteady concentration boundary-layer thicknesses is small. This translates to $\varepsilon_0 Sc^{1/3} \ll 1$ and hence ignores the time variation of concentration with this restriction. By addressing developing, inspiratory flow and high and low Peclet number regimes, they simplified the analysis for small liquid-side radial transport compared to gas-side (appropriate for ozone) and found that the 17th and 18th generations receive the largest dose. Apparently the mucus reactants protect the upstream generations; the appearance of partially alveolated airway surfaces, with their very thin liquid films at these generations, leads to large uptake there. This result is consistent with test animal pathology and also with that of Miller et al (1985), who developed a one-dimensional model. However, their absolute dose was predicted to be much lower at these two generations.

In engineering contexts, Kurzweg (1985), using a multiscale expansion technique, found that heat transfer is enhanced during oscillatory flow in insulated tubes. Experiments by Peattie & Budwig (1989) noted this transport enhancement, as well. The problem for steady flow in pipes has been investigated by a few authors. For example, Sankarasubramanian & Gill (1973) studied the interphase mass transport due to an irreversible first-order reaction at the wall (i.e. the wall is conductive) and solved the convection-diffusion equation using the derivative-expansion method. Lungu & Moffatt (1982) considered the effect of wall conductance on heat diffusion and used Fourier transforms and a "mathematically preferred" average function to obtain a series solution. In both cases, the asymptotic solutions for large times showed that D_{eff} in the flow direction is a decreasing function of wall conductance. Mazumder & Samir (1992) recently studied the effect of wall conductance on axial dispersion in pulsatile tube flows. Using a numerical method they calculated D_{eff} for several values of wall conductance. The results show that D_{eff} decreases with wall conductance for all frequencies.

Turbulence in Oscillatory Pipe Flow

In addition to the laminar mixing phenomena, Fredberg (1980) has hypothesized that transport of gases through the airways could be substantially augmented by the chaotic velocity fluctuations associated with the generation of turbulence during high frequency ventilation. Previous investigators, including Hino et al (1976) and Ohmi et al (1982) have characterized transition to turbulence in oscillating tube flows using the parameter α and the Reynolds number Re, with length scale based on either a or $\sqrt{2\delta_s}$. If $U = \omega d$ is chosen as the velocity scale, then, in terms of the parameters we have defined above, $Re_\delta = \sqrt{2\alpha A}$ and $Re = 2\alpha^2 A$. For $\alpha \gg 1$ a Stokes layer develops in the tube and several studies of its stability have been made. The full time-dependent linear stability theory of von Kerczek & Davis (1974) predicts stability to infinitesimal disturbances in the range $0 < Re < 800$, $\alpha = 5.66$ for a flow confined between two infinite parallel plates. Yang & Yih (1977) consider theoretically the stability of a pressure-cycled oscillatory pipe flow subjected to small amplitude, axially-symmetric perturbations. Results of Floquet theory and numerical analysis show that the flows are stable for all modes and wavenumbers considered up to Reynolds numbers of 2000. Wu et al (1992) have presented a countering view by perturbing the Stokes flow with oblique waves; their analytical results suggest that explosive growth leading to turbulent bursts is possible. Akhavan et al (1991b) used a numerical approach for oscillatory channel flow and their findings suggest that transition to turbulence can be explained by a mechanism of secondary instability of two-dimensional, finite-amplitude waves to three-dimensional, infinitesimal disturbances. Three values of Reynolds numbers were tested ($Re_\delta \approx 200$, 500, and 1000); transition to turbulence at $Re_\delta \approx 500$ is suggested.

Several experiments concerning turbulence in oscillatory flows in pipes have been performed using a variety of techniques (Li 1954, Sergeev 1966, Merkli & Thomann 1975). In each case a critical value of $Re_\delta \approx 500$ is found. Turbulent bursts occurred during the decelerating phase of fluid motion, and these were always followed by relaminarization of the flow. Hino et al (1976) detected high-frequency velocity disturbances during deceleration at all radial locations investigated from the centerline to the wall for $Re_\delta = 550$, $\alpha < 5.5$. Eckmann & Grotberg (1991) measured transition to turbulence using LDV and hot-film methods and showed that for $500 < Re_\delta < 1310$ the core flow remains laminar while the Stokes layer becomes unstable during the deceleration phase of fluid motion. This turbulence is most intense near the solid boundary and is confined to this annular region. By comparing techniques, they showed that the appearance of turbulence at all transverse locations in previous studies using hot-films may be due to probe interference. Mello (1992) recently used numerical methods to analyze the stability of the oscillating pipe flow problem and found that the peak Reynolds stresses occurred in the boundary layer at the same phase of the oscillation as measured by Eckmann & Grotberg (1991).

Akhavan et al (1991a) also measured turbulence in the wall region using LDV instrumentation and found that there was good agreement with their theory. Another noninvasive method was developed by Kurzweg et al (1989) using streaming birefringence. Jan et al (1989) noted a turbulent regime in their flow studies of a model bifurcation, discussed above with regard to tube curvature. By flow visualization they found a turbulent boundary of $200 < Re_\delta < 500$ and observed the turbulence occurring during flow reversal (acceleration) rather than deceleration. The lower limit for this boundary may again implicate curvature effects as playing a dominant role in the airway bifurcation.

Normal Breathing

Normal breathing modalities did not receive the same level of scientific attention as HFV in the last fifteen years, but an important review of steady flows in a bifurcation, amongst other respiratory phenomena, is given in Pedley (1977). In addition, both steady and oscillatory flow dynamics are reviewed and new results reported by Chang & Menon (1985) in a branching model of the large airways using a realistic geometry. They show, for example, that oscillatory profiles, when taken at peak flow rate, compare favorably with steady profiles for $\alpha < 16$ (see Figure 6). Much farther down the tracheo-bronchial system are the alveolar ducts where gas exchange begins. Federspiel & Fredberg (1988) address the architecture of the alveolar ducts and the flow and transport that they provide. Their two-dimensional model consists of a central channel with axially periodic openings to contiguous alveoli whose shapes are circular in cross section. This geometry is shown in Figure 1 as the respiratory bronchioles in the acinus. Flow down the channel induces a vortical flow in each alveolus which modifies axial dispersion such that D_{eff} is less than D for low Peclet numbers (Pe) while for larger Pe, D_{eff} is larger than the equivalent Taylor-Aris result. Their approach incorporates a method-of-moments analysis developed by Brenner & Gaydos (1977) for dispersion in porous media.

Pulmonary Input Impedance

Independent of gas exchange goals, there are important research efforts to measure the pulmonary input impedance over a wide range of oscillatory frequencies, up to ~ 300 Hz (Jackson & Lutchen 1991). These acoustical oscillations, driven by loudspeakers, have tidal volumes of only a few milliliters. Both resonant and anti-resonant frequencies exist, but the first anti-resonant frequency appears to characterize airway geometric and material properties that may change with disease, allowing a passive test of pulmonary airway status. For uncooperative patients, such as infants, passive tests like this would be very useful. Bunk et al (1992) measured input impedance in a model airway bifurcation, demonstrating that the bifurcation geometry has little effect compared to the airway diameter. When the tidal volume increases, the bifurcation effects become more important as we have seen in HFV.

SURFACE TENSION PHENOMENA

A force that strongly influences pulmonary mechanical behavior is the surface tension arising at the interface of the liquid lining and the air within the lung. The lung regulates its surface tension by producing surface-tension-reducing substances, called surfactants, which lower the surface tension thus making the lung more compliant. Issues addressed here are the surface tension effects on closure and reopening of airways and also on transport of instilled liquid, as an aerosol or bolus, into the lung.

Airway Closure

The thin liquid lining that coats the airways may cause closing off of the small airways in the region of the small respiratory bronchioles, either by the formation of a liquid bridge or lens, or by provoking collapse of the flexible wall of such airways (Macklem et al 1970). Indeed, both phenomena may occur simultaneously (see Figure 7). This happens most frequently near the end of expiration when airway diameters are small. The closure and subsequent reopening of these airways contribute to the shape of the pressure-volume curve in cycled lungs, since the capacity to accommodate air volume is cyclically enhanced and diminished when the alveolar regions distal to the sites of closure are recruited and derecruited. Measuring the occurrence of airway closure is one component of a standard

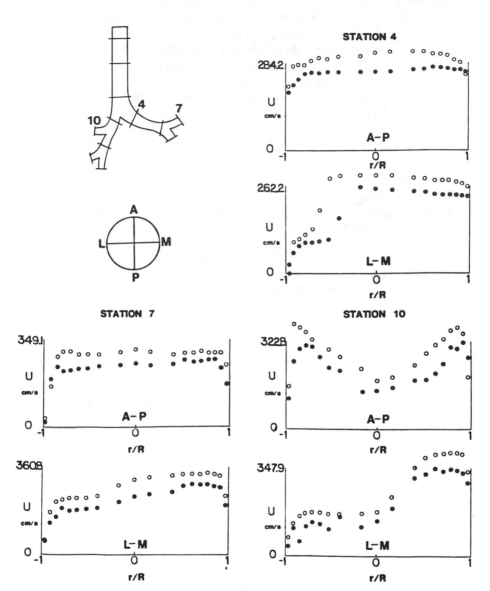

Figure 6 Comparison of oscillatory velocity profiles at inspiratory peak vs steady inspiratory flow at similar *Re*. ● $Re_{\text{peak, oscillatory}} = 8332$, $\alpha = 16$. ○, $Re_{\text{steady}} = 8846$. (From Chang & Menon 1985.)

pulmonary function test: the nitrogen washout test. When this test shows early airway closure, it is often inferred that there is inhomogeneous ventilation distribution within the lung.

The stability of a liquid film coating the inside of a tube was first analyzed by Goren (1962) who assumed an inviscid core fluid, conditions similar to the pulmonary airways. If we call the radius of the tube a, and the radius of the interface b, then $\varepsilon = (a-b)/a$ is the dimensionless film thickness which is of primary importance in determining stability. Goren's linear stability theory showed that for $R \to 0$, the critical wavelengths had a lower limit defined by $2\pi b/\lambda_c < 0.672$ if $\varepsilon = 0.33$ and $2\pi b/\lambda_c < 0.707$ as $\varepsilon \to 0$. These values are similar to those found by Rayleigh (1879) for the dynamic instability of an unconfined inviscid liquid thread. Thus the presence of a rigid boundary does not have much influence on the critical wavelength criterion, provided the wavelength is long enough. The length of the tube can limit the length of allowable waves, however, and such a consideration is important in human airways where the airway length may determine the unstable mode. The boundary has important effects on the growth rate of instabilities and, because of the boundary, one needs to determine a critical value for ε, above which the instability will lead to formation of a liquid bridge or lens. Goldsmith & Mason (1963) measured initial growth rates of capillary instabilities and closing times for liquid bridge formation in small capillary tubes and found the time to closure, $t_c = 10.1$ for $\varepsilon = 0.19$ and $t_c = 7.2$ for $\varepsilon = 0.25$ (assuming that their Figure 7 caption contains inadvertently transposed descriptions of these two growth curves) where $t_c^* = \varepsilon^{-3} a t_c / \sigma$ is a time scale proposed much later by Hammond (1983).

Everett & Haynes (1972) were the first to address the notion of a critical film thickness, by using thermodynamic principles of effective surface area minimization. Their static stability theory indicates that the liquid volume for lens formation is $V \geqslant V_c = 5.47 a^3$. If $V < V_c$, then the fluid will assume

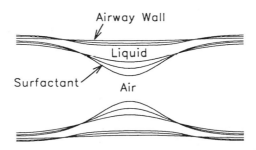

Figure 7 Capillary-elastic instability of a liquid-lined flexible tube as a model of airway closure.

unduloid-shaped collars lining the wall but the core fluid will be continuous. Everett & Haynes verified their theory by condensing water from the gas phase onto the inside of a carefully cooled capillary tube and monitoring the progression of first unduloid shapes and then the formation of a lens. V_c was estimated from the axial length of the observed lens immediately after formation and found to agree with their theoretical prediction to a very high degree of accuracy. Hammond (1983) applied lubrication theory to the stability of viscous films coating rigid tubes and solved the nonlinear evolution equation for the interface position in the limit $\varepsilon \to 0$, which included the linearized Young-Laplace equation. He tracked the evolution of unduloids as proposed by Everett & Haynes (1972) but could not predict lenses because of his linearized form of the normal-stress boundary condition (Young-Laplace equation). This form does not give an accurate approximation for the curvature for thick films—the ones that are likely to form lenses. Gauglitz & Radke (1988) used a modification of Hammond's analysis and developed a more accurate form of the Young-Laplace equation for thicker films so that ε appears explicitly in evolution equations. They computed $\varepsilon_c = 0.12$, while their experiments indicate an average value of $\varepsilon_c \approx 0.09$ as determined from the number of lenses formed from a long bubble.

Kamm & Schroter (1989) performed experiments on the stability of a viscous film coating a vertical cylindrical tube by injecting small quantities of oil from the top and found that $V_c = 5.6a^3$, which yielded $\varepsilon_c \approx 0.16$ for $2\pi a/L = 1.05$, where L is an airway length and $L/a = 6$. Their experimental value of V_c agrees well with the predictions and experiments of Everett & Haynes (1972). Using anatomical data of airway diameters and lengths, and estimates of liquid lining dimensions, Kamm & Schroter make predictions of lung volumes that would lead to airway closure, i.e. the "closing volume" which is a clinically measured quantity, and of the airway generation in which this closure would occur. By assuming the initial film thickness in each airway to be 10 microns at total lung capacity (TLC) (full expansion), they show that during expiration the first airways to close will be just proximal to the terminal bronchioles ($\sim$ generation 16–17) and that the closure will occur at 23% TLC. For a 5 micron initial thickness, the closing volume was at 12% TLC in the same airway generation. It is common to find closing volumes in this range. Johnson et al (1991) investigated the closure flow by using a quasi-one-dimensional model that includes inertia and also a full Navier-Stokes numerical solution. Choosing a lining thickness that was above the critical value, $\varepsilon = 0.2$, they found that $t_c \approx 38$ when they set $2\pi a/L = 1.05$.

In an independent approach, Halpern & Grotberg (1992) used the lubrication theory approach of Hammond (1983) and the modified Young-

Laplace equation of Gauglitz & Radke (1988) to analyze the stability of a liquid-lined flexible tube in an effort to understand the combined capillary-elastic instability. They derived nonlinear coupled evolution equations for the air-liquid and wall-liquid interfaces. A new parameter arises in this analysis: the flexibility parameter $\Gamma = (1-\gamma^2)\sigma/Eh_0$, which is the ratio of the surface tension (σ) forces to the tube elastic forces (E = elastic modulus, γ = Poisson ratio, h_0 = wall thickness). Halpern & Grotberg showed that ε_c decreases with increasing Γ, as the tube softens. For example, ε_c decreases from 0.12 for $\Gamma = 0$ (rigid tube) to 0.06 for $\Gamma = 0.5$, and the corresponding closure times are also reduced. The tube wall is drawn inward by the growing capillary instability, accentuating the destabilizing transverse curvature of the air-liquid interface and reducing the distance to close. Also, the fluid feeding the growing liquid bulge experiences less shear resistance since thinning near the borders of the bulge is not as severe when the wall is allowed to move. They also found that closure time increases as λ/a decreases below ~ 15 (for the chosen parameter values) and also as ε decreases. When parameters corresponding to those of Goldsmith & Mason (1963) are used, there is very good agreement with the t_c found in their experiments.

Both the latter two theories, Halpern & Grotberg (1992) and Johnsen et al (1991), have been further developed to include the effects of surfactants (Halpern & Grotberg 1993, Otis et al 1993). Both theories show that the dynamics of the lens formation is slowed by the presence of surfactants since their concentration is increased in the region of largest transverse curvature, due both to surface area reduction there and convection from the thinning regions. Closure time can be increased by a factor of four, for example using human pulmonary parameter estimates (from roughly 0.060 sec to 0.240 sec). Halpern & Grotberg (1993) also show that ε_c is significantly increased in the presence of surfactants, particularly for the rigid tube limit ($\Gamma \to 0$), a result not obtainable from the V_c predictions of Everett & Haynes (1972) which assumed the absence of surfactants. This is consistent with the observation of Liu et al (1991) who demonstrated that surfactants can be responsible for opening small tubes plugged with a liquid. The implication is that surfactants, if added exogenously to individuals with closed airways, could open those airways and improve lung function and ventilation. This, in fact, is one of the important clinical effects of exogenous surfactant replacement therapy discussed below. Figure 8 shows this relationship for three choices of surfactant activity: surfactant free ($\delta = 0$), normal surfactant conditions ($\delta = 8$), and a surfactant deficiency ($\delta = 0.4$). As Γ increases the surfactant stabilization effects are still present, until $\Gamma \approx 0.7$ where wall collapse begins to dominate and the curves coalesce.

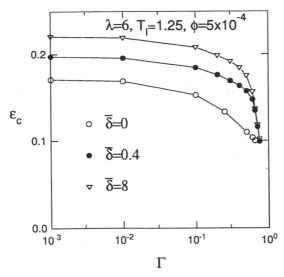

Figure 8 Critical liquidfilm thickness, ε_c, vs Γ, the relative measure of interfacial to elastic forces, for three levels of surfactant activity: surfactant free ($\delta = 0$), normal surfactant conditions ($\delta = 8$), and a surfactant deficiency ($\delta = 0.4$)

Airway Reopening and Crackles

Once the airway is closed, we need to consider how it reopens, presumably during inspiration when the airways are pulled outward radially and air pressure can build on the proximal side of the lens. This involves the novel work of Gaver et al (1990) who have approached the problem as the penetration of a finger of air into a flexible channel which is filled with a viscous fluid. This is the classical two-phase displacement in a Hele-Shaw cell, but with flexible walls. In their experiments, a soft polyethylene tube was filled with an oil and then flattened. Air was introduced at one end which progressed as a moving front into the tube which is open behind the front and flattened ahead of it. The opened walls remain coated with the liquid. The opening front travels at a speed U which depends on the system parameters (Figure 9). When the capillary number $Ca = \mu U/\sigma$ is small, they found that the opening pressure is $\sim 8\sigma/R$, where R is the tube radius. In this regime there is an apparent yield pressure which must be exceeded before the front will propagate; decreasing wall tension makes the yield pressure even higher. When $Ca > 0.5$, viscous forces become increasingly important in determining the relationship between pressure and Ca. Gaver et al conclude that airway closure can continue, even with rather vigorous inspiration, if the time scale for opening, based on U and a local length scale, is long enough. Higher values of μ and σ increase the time scale and would contribute to keeping airways closed. Here again we see that larger surface tensions not only contribute to closure, but also to maintaining closure.

The opening of a finite length lens is likely to be abrupt since the liquid bridge ruptures and the airway walls snap open. This process has long been implicated as the source of crackling sounds in the lung which are heard with a stethoscope. Such sounds occur in various lung conditions, many of which are characterized by abnormally high liquid volumes in the airways or abnormally high surface tension, and can be diagnosed. Loudon & Murphy (1991) provide a good review of the clinical and physiological state of the art for lung sounds.

Dynamic Surfactant Spreading

Hyaline membrane disease is the manifestation of surfactant deficiency in prematurely born infants (Avery & Mead 1959). These patients have stiff lungs which are difficult to ventilate and have a propensity for closure of small airways and alveoli, a condition known as atelectasis. An effective technique for the treatment of hyaline membrane disease is to deliver surfactant externally, either directly through an endotracheal tube, or through inhalation of surfactant in aerosol form (Lewis et al 1991). Since the barrier between pulmonary capillaries and the liquid layer in alveoli is so thin, aerosol inhalation of medications is also used as a method of rapid drug delivery in a variety of clinical settings. While most work in pulmonary aerosol mechanics focuses on the deposition pattern within the

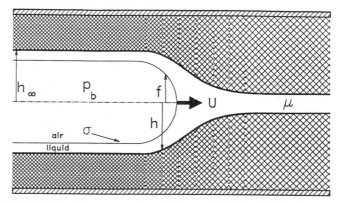

Figure 9 Propagation of an air finger into a collapsed, liquid-filled tube (after Gaver et al 1990).

bronchial network, the details of the business end of droplet–liquid lining interaction is a relatively new consideration.

When a drop or bolus of surfactant comes into contact with a clean liquid substrate, the large surface tension difference at its edge, expressed by the positive value of the spreading coefficient, $S = \sigma_0 - \sigma_m$, causes it to spread. Here σ_m represents the combined surface tensions of the drop/substrate and drop/gas boundaries and σ_0 the higher surface tension of the substrate/gas boundary. Spreading is achieved through the development, ahead of the bulk of the drop, of a thin monomolecular layer of surfactant on the surface of the substrate, along which the jump in surface tension is accommodated, as described by DiPietro et al (1978) for oil spreading over a deep fluid (the ocean). For thin layers, as found in the lungs, a Marangoni flow is induced. Its dynamics were studied in Borgas & Grotberg (1988) and Gaver & Grotberg (1990) using a lubrication theory for an initially long surfactant monolayer, length L_0, compared to the undisturbed substrate film thickness, height H_0. The quantity $\varepsilon = H_0/L_0 \ll 1$, and if diffusion of the surfactant on the surface of the film, D_s, is sufficiently slow, as is the case with biological surfactants, the flow induces large deformations in the film height. The leading edge of the insoluble monolayer behaves like an advancing rigid plate, and the abrupt transition to undisturbed conditions just ahead of the monolayer produces a shock discontinuity (in the limit $D_s \to 0$) in film height (a new type of fluid dynamical shock), with the film elevated beneath the leading edge of the monolayer to almost $2H_0$. To accommodate this elevation the film thins in regions closer to the center of the drop.

Figure 10 shows a typical sequence of film shapes, due to an advancing surfactant monolayer, including the development of film rupture. Of particular importance in the context of surfactant replacement therapy are estimates of spreading rates, either of a planar front of surfactant advancing along the mucus lining of an airway, or of a droplet of surfactant spreading over the film covering an alveolar wall. In a systematic mathematical analysis, the work in Jensen & Grotberg (1992) showed that the length of the film increases with time proportional to: $t^{1/4}$ for a drop, which is consistent with the theory of Gaver & Grotberg (1990) and experiments of Gaver & Grotberg (1992); $t^{1/2}$ for a front which is consistent with the theory of Borgas & Grotberg (1988) and the experiments and theory of Ahmad & Hansen (1972); and $t^{1/3}$ for a strip (2-D drop) which is consistent with the results of Espinosa (1991). Here the dimensionless time variable is $t = \varepsilon^2 St^*/\mu H_0$.

A typical wave shape as it progresses in time is seen in Figure 10. The fluid wells up under the wave as the surface tension gradient drags fluid forward from the rear faster than can be accommodated by the forward

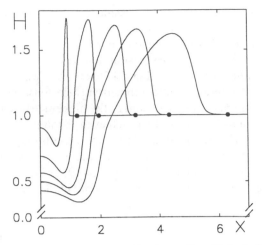

Figure 10 Spreading surfactant front on a thin viscous film. Circles indicate leading edge of surfactant.

velocity of the shock. The thinnest part of the film is approaching $x = 2$. Although capillarity is a weak force, it is also important in stabilizing the film against instabilities due to van der Waals forces. These forces, too, are very weak, unless the film becomes extremely thin (less than 1000 Å), in which case they encourage depressions in the film to grow, and may ultimately cause the film to rupture, here near $x = 2$ (Ruckenstein & Jain 1974, Williams & Davis 1982). Film rupture has a dramatic effect since it halts the spreading process as observed in the experiments of Weh & Linde (1973), Keshgi & Scriven (1991), and Gaver & Grotberg (1992) and predicted in Jensen & Grotberg (1992). Hence it may be important to establish the conditions under which rupture occurs; for example, if the surfactant is too strong the surface tension gradients may become very large and force H to zero rather rapidly. When considering design of surfactant agents, one would like to make a strong surfactant to improve spreading, but the mixture cannot be too strong or it may cut itself off from spreading, altogether. The presence of endogenous surfactant, already on the interface, can ameliorate this potential problem. Also, the cellular boundary can supply fluid to the film in response to the transmural pressure gradient that a nearly-rupturing locus would cause. Also possible are fingering instabilities at the edge of the bulk droplet; these have been observed experimentally by Troian et al (1989) and modeled by Troian et al (1990).

INTRAPLEURAL FLUID FLOW

The liquid of the pleural space provides lubrication between the visceral pleura covering the lung and the parietal pleura covering the chest wall. This liquid space, whose normal thickness is on the order of 10–30 microns, is subject to large volume changes during pleural effusion and may be a significant route for the clearance of pulmonary edema (Broaddus et al 1990). Most investigators agree that the pressure gradient during apnea (breath-holding), from the top of the pleural space to the bottom of the pleural space, is somewhat less than hydrostatic, and that the intrapleural liquid flows downward, following the intrapleural pressure gradients. Indeed, this has been appreciated from experimental data for over thirty years. There are currently two differing, but not necessarily exclusive, conceptual models of pleural liquid mechanics. In one mechanism, the observed downward flow through the pleural space is thought to be due to the generation of liquid at the top of the pleural space, and the extraction of liquid at the bottom of the pleural space and the draining action of gravity acting directly on the liquid (Lai-Fook & Rodarte 1991). This direct effect of gravity appears doubtful since the pleural liquid has a constant density. Miserocchi et al (1988) share many of Lai-Fook's views, but argue that pleural liquid pressure is lower than pleural surface pressure due to areas of contact between the visceral and parietal pleurae, though these have not been observed as yet.

A Proposed Model and Problem Formulation

Although certain aspects of existing theories hold some merit, the theories are not detailed fluid mechanical models of the pleural space nor do they account for all the forces involved, e. g. lung buoyancy. The theory of Grotberg & Glucksberg (1993) is a new approach toward understanding the transient and steady state behavior of intrapleural flows. The lung, containing tissue, blood, and air, has a density of approximately 0.2 g/cc compared to the pleural liquid density of ~ 1 g/cc. The key idea is that the lung tends to rise within the pleural liquid, forming a squeeze film which forces fluid downward, hence accounting for the observed flow in the direction of gravity. To model this phenomenon, consider a pair of initially concentric horizontal cylinders separated by a small liquid-filled gap which corresponds to an animal experiment with the animal placed horizontally. The inner cylinder represents the lung while the outer cylinder represents the chest wall. Choosing reasonable values of the parameters for a rabbit we find that lubrication theory applies, simplifying the analysis. The balance of buoyancy versus squeeze-film pressure forces on the inner cylinder leads to a solution for the offset displacement of the cylinder centers, $Y(T)$

(Figure 11*a*), for several initial positions. Figure 11*b* shows a selected pressure difference, $\Delta P(T) = p(\phi = -\pi/4, T) - p(\phi = +\pi/4, T)$ scaled on the hydrostatic pressure between these points. The ratio of lung density to intrapleural liquid density is γ, which typically has a value of 0.2 in air-filled lungs. Essentially all known data on intrapleural pressure gradient measurements for the past 30 years fall within the values shown in this figure.

Results

The Grotberg & Glucksberg model predicts that the lung floats upwards so that the film is thicker at the bottom, consistent with observations by Lai-Fook & Kaplowitz (1985). Furthermore, flows will diminish and the pressure gradient will approach hydrostatic as an apneic (no ventilation) state continues, consistent with reports by Miserocchi et al (1988). Miserocchi et al also determined that the half-life of the pressure gradient transient induced by a prone-to-supine change in posture is on the order of 5 minutes, consistent with the time scale $T_0 = R_0/U = 300$ sec in our example. The velocity scale is $U = \rho_l(1-\gamma)gd_0^2/12\mu$, where the concentric gap width is d_0.

An increase in pleural liquid volume has been shown to cause the pressure gradients to approach hydrostatic conditions more rapidly (Miserocchi et al 1983). For our model, this is accomplished by increasing d_0 so that U is larger and all transients are faster. Recent measurements (Miserocchi et al 1993) have shown that the vertical pressure gradient of apneic rabbits is significantly closer to hydrostatic after the lungs were partially filled with perflubron, a perfluorocarbon of specific gravity 2,

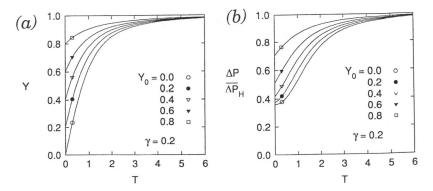

Figure 11   (*a*) Center offset $Y(T)$ and (*b*) pressure difference $\Delta P(T)/\Delta P_H$ for several initial positions Y_0.

which reduced the buoyancy by increasing γ. This is also consistent with our model, which stimulated that experiment.

Discussion of the Model

How is this less-than-hydrostaic ΔP maintained at steady state conditions? The above model is always transient, but it is not difficult to imagine how Y and ΔP could appear to be time-independent during respiration or converge to a steady value during apnea. During breathing one possibility is that the motion of breathing resets the position of Y to some intermediate value allowing the lung to float upward in between breaths. This is a high frequency disturbance to the system since the breathing period, ~ 3 seconds, is very short compared to T_0. There can be steady streaming flows and pressures when the cylinder radii are oscillating and the cylinders are off-center. It is also possible that the known leakage flows crossing the cylinders, due to transmural pressure gradients, influence the film pressure field and its motion to keep the lung in an intermediate position. A coupling of these effects may also come into play. For example, leakage into the film at the top, which is known to occur from the lung side, would be augmented transiently during inspiration when intrapleural pressure becomes quite negative. Then, during expiration this new fluid can be squeezed downward around the lung, both by wall motion and by buoyancy, where there is known leakage out of the film through the chest wall side.

FLOW LIMITATION AND WHEEZING

During a forceful expiration the flow rate is known to reach a limiting value, even if the effort is increased (Fry & Hyatt 1960). This flow limitation is apparent in normal people, but individuals with obstructive lung disease such as asthma and emphysema may experience this phenomenon earlier during expiration and their limit is well below normal. Indeed, in severe situations these patients may be flow-limited during most of expiration. The measurement of flow and lung volume during maximal effort, a common clinical test of pulmonary function, yields the maximal expiratory flow-volume curve.

One-Dimensional Steady Flow in Collapsible Tubes

The typical bench-top model experiment used to investigate flow limitation consists of the fluid conveyed along a rigid pipe with a flexible section (Figure 12a). Both the driving pressure, $P_u - P_d$, and the transmural pressure, $P_u - P_e$, are controlled by using either a pressure chamber around the flexible tube or a system of flow resistors upstream and downstream of the tube. The pressure conditions on the extrapulmonary (but intra-thoracic) bronchi are equivalent to keeping $P_u - P_e$ fixed while increasing $P_u - P_d$. That is because the intrapleural pressure is involved in setting both P_u and P_e (see Figure 12c). The narrowest portion of the tube near the downstream end has a rather flattened elliptical shape in cross section. One concern of experimentalists and theorists alike is the geometry of such a situation in real airways: These do not have the sudden change in elastic properties that the tethering to a rigid tube imposes.

Analyses of expiratory flow limitation have generally rested on one-dimensional models of steady flow through a flexible tube (Lambert & Wilson 1972, 1973). When the transmural pressure difference is critical, these tubes partially collapse into an elliptical cross section. The flow becomes limited at the narrowest portion of the tube (Figure 12a), the "choke point"—a term borrowed from compressible gas dynamics and not a reference to emergency respiratory situations. It was recognized by Dawson & Elliott (1977) and Shapiro (1977) that flow limitation was

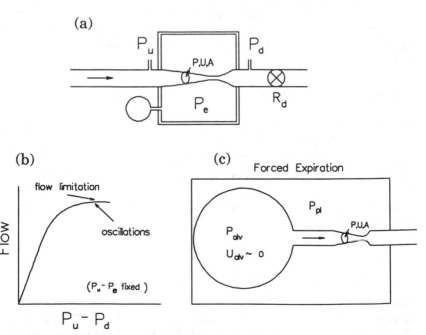

Figure 12 (*a*) A typical experimental system for flow limitation studies; (*b*) results of flow vs driving pressure indicating flow limitation and oscillations; (*c*) relationship of pressures to forced expiration.

related to the wave speed of fluid-elastic waves on the tube wall. In such theories, the airway elastic behavior is lumped into a tube-law, relating cross-sectional area of the tube to its transmural pressure, $A(P)$. The one-dimensional momentum balance relates P to an average velocity U, which may be expressed in terms of A and the flow rate Q, which is fixed, hence satisfying continuity. From the analysis a wave speed C emerges, such that $C^2 = (A/\rho)dP/dA$, and flow limitation occurs where $S = U/C = 1$. Here, ρ is the fluid density. This criterion is analogous to either the Mach number in compressible gas flow or the Froude number in open channel flow reaching unity. A more compliant tube yields a smaller C, and hence smaller U and Q. This is a common problem in emphysema, for example. At the location where $S = 1$, a fluid-elastic shock develops and subcritical ($S < 1$) and supercritical ($S > 1$) flow regimes may be identified (Shapiro 1977). Of course, C depends on the lung's state of inflation as well as airway generation, i.e. axial position. Also, the characteristics of the parenchymal lung tissue (alveolar walls) which surround and tether many of the airways are very important. Consequently, overall expiratory flow is limited by those airways where $S = 1$ at the lowest flow. In some instances this may be the distal trachea or within the first few generations. Elad et al (1988) have developed an elaborate model of forced expiration for the entire lung by including many details of geometry and material properties in the one-dimensional approach.

Elliott & Dawson (1977) used a liquid flow system similar to Figure 12a to test the wave-speed theory of flow limitation and found good correlation of the data with the predictions. They observed oscillations at flow limitation, and this emphasizes that liquid flows in thin-walled tubes are not an optimal system for modeling air flow in pulmonary airways. The ratio of wall mass to fluid mass is $O(1)$ with a liquid and $O(10^2-10^3)$ with a gas, so that oscillations may be dominated by wall inertia in an airway whereas a liquid flow model would be unable to test that effect on flow limitation.

Self-Excited Oscillations and Flow Limitation

Figure 12b shows the flow limitation phenomena and its relationship to self-excited oscillations of the coupled tube-fluid system. When the fluid used is a liquid, there tend to be relaxation oscillations of the choke point as it moves upstream, opens, and then reforms downstream. The process repeats cyclically and has been referred to as "milking" (Bertram 1982). This mode is somewhat different from the higher frequency oscillations that are referred to as flutter. In that case the point of collapse does not appear to move upstream as long as flow conditions are fixed. In Gavriely et al (1989) a simultaneous study of flow limitation and flow-induced tube oscillations was made using air flow and liquid flow. In those experiments, oscillations only occurred in the presence of flow limitation. Depending on the driving and transmural pressures, a wide range of oscillation frequencies were observed, 260–750 Hz, all from the same tube conveying air. Those measurements show the power spectrum of the oscillations to contain a very sharp peak at the fundamental frequency, similar to human measurements of wheezing breath sounds (Gavriely et al 1984).

A clinical observation is that individuals with obstructive lung disease usually wheeze due to their limited flow rates (Kraman 1983, Gavriely et al 1987; see Figure 12b). Such wheezes have been measured in human subjects with a high degree of resolution both for diseased lungs and normal ones, and even occur in a constant-flow, constant-volume lung preparation (Gavriely & Grotberg 1988). They appear as a sharp spike in the power spectrum. Apparently wheezing only occurs in the presence of flow limitation, although there is evidence of flow limitation occurring without wheezing. This implies that, in such instances, nonsteady flow, involving dynamics of the tube wall, is the appropriate model for investigating the coupled phenomena of flow limitation and oscillations, first proposed in Grotberg & Davis (1980). The results of the critical flow speed in the air-flow experiments of Gavriely et al (1989) are intriguing in that they occurred for $S \approx 0.3$, not $S = 1$, again raising the issue of whether or not the steady wave speed theory of flow limitation is appropriate in unsteady systems with large wall inertia. Since the definition of C excludes wall density, it likely overestimates to a significant degree the pertinent wall wave speed.

There have been a number of theoretical studies of flow through flexible tubes that examine flow-induced oscillations. One set of theories may be categorized as *lumped parameter models*, in which the distributed properties of the fluid and wall are reduced to an analogy of a self-excited oscillator (Conrad 1969, Katz et al 1969, Bertram & Pedley 1982). The other set of theories uses *continuum modeling* which either address aerodynamic flutter phenomena or flow separation at the choke point as in Cancelli & Pedley (1985). One aim of both sets of models is to predict the critical fluid speed at which oscillations begin and the oscillation frequency. The continuum models are also capable of determining the wavelength of the oscillations.

There was quite a controversy in the tube-flutter (continuum) modeling literature during this time, since the earlier theories predicted static divergence (steady buckling) of the instability whenever the wall had damping (Weaver & Paidoussis 1977, Matsuzaki & Fung 1979). However, experiments (for example, those in Weaver & Paidoussis 1977) clearly showed oscillations. These models used inviscid flow theory, but Grotberg & Reiss (1984), modeling a two-dimensional, flexible channel-flow system, showed that the controversy may be resolved by including fluid viscous effects.

Their model of aeroelastic flutter used a fluid friction factor—a hydraulic approximation which leads to traveling wave flutter (TWF) instabilities of frequency ω_c, wavelength λ_c occurring at a critical flow speed, U_c. The fluid friction allows the wall position and pressure to be out of phase, thus leading to a time-periodic instability in the form of flutter waves. Both the critical flutter frequencies and critical fluid speeds found in this theory are consistent with the experimental results of Gavriely et al (1989), with $S \approx 0.3$ (for a more thorough discussion, see Grotberg & Gavriely 1989). One interpretation of this discrepancy with the wave-speed theory of flow limitation ($S = 1$) is that the wall inertia strongly influences the pressure-flow relationship by making the system less stable. It is likely that con-siderable energy loss occurs due to the oscillation, both in the fluid and in the visco-elastic solid, a mechanism that would reduce the mean flow response to driving pressure. For fluid speeds above this value, the fre-quency and oscillation amplitude increase, i.e. it is a supercritical bifur-cation. Later, Grotberg & Shee (1985) showed that this hydraulic approxi-mation is equivalent to an Orr-Sommerfeld system in the limit of vanishing fluid and wall viscosities as long as their ratio remains $O(1)$.

Walsh et al (1991) have focused on the tracheal geometry in a partially collapsed state. Their model involves a detailed description of undamped, curved-shell flutter theory with one-dimensional, inviscid fluid mechanical equations. The axial wall displacements are assumed to be much larger than the lateral wall displacements, in the absence of longitudinal tethering constraints. Therefore axial wall inertia plays a key role in the long-wave flutter instabilities they find, which can occur subcritically. A cautionary note is that frictionless systems such as this and the model of Webster et al (1985) can produce linear instabilities which turn out to be mathematically singular when examined in the nonlinear regime, as discussed in Grotberg & Shee (1985).

In the separated-flow model of Cancelli & Pedley (1985), it is the effect of fluid viscosity that leads to oscillations They examine separated flow just downstream of the choke point and find that energy loss and pressure recovery in this region are required for oscillations. The flow separation point oscillates longitudinally, and their massless wall allows the choke point to follow this pressure disturbance in axial motion. It moves upstream when a high-pressure wave advances from the outlet region and it moves downstream when a low-pressure wave follows. The oscillations are sustained only when the flow is supercritical; this is because upstream wave propagation through the constriction is not permitted when $S > 1$. Hence the pressure peak is caught by the pressure trough. The period of oscillation is related to the speed of traveling waves, back and forth, between the choke point and outlet.

It is interesting to speculate that coupled modes of oscillation may exist in which flutter, which can occur for $S < 1$, and flow separation oscillations might reinforce one another and bring into play both the sudden area change at the choke point and the aeroelastic behavior of the oncoming flow through the narrowing, nearly two-dimensional channel. Another possible source of interaction is vortex shedding, which, like flow separation, relies on the sudden expansion of the cross-sectional area that exists in traditional tube experiments. Again, this sudden expansion may or may not reflect real airway geometry during flow limitation and wheezing.

Effects of Viscous Profiles

More detailed fluid modeling for flow in flexible channels would involve the full Orr-Sommerfeld system. This has been studied in several contexts for single flexible walls in which TWF, the basis of the aeroelastic flutter theories, and modifications of the Tollmien-Schlichting instability (TSI) are examined. In the early work of Benjamin (1963) and Landahl (1962), their calculations for flow over a single flexible boundary show stabilization of the TSI, so that as the wall flexibility is increased the neutral stability curve shifts to higher critical Reynolds numbers. The subject of TSI, TWF, and their combinations for single flexible walls has been carefully explored by Carpenter & Garrad (1985, 1986) and compared to earlier classification schemes of Benjamin and Landahl (waves of class A, B, and C). A review of flow over single compliant walls is given by Riley et al (1988).

A similar stabilization of the TSI was seen by Hains & Price (1962) for Poiseuille flow through a flexible channel. Very recently, there have been some nonlinear studies of this flow (Rotenberry & Saffman 1990, Rotenberry 1992). These studies imply that wall flexibility increases the critical Reynolds number beyond $Re_c = 5772$, the value found for fully-developed, rigid channel flow. For entrance flow into a rigid channel, Re_c increases from 5772 to much higher values as the boundary-layer thickness decreases (Chen & Sparrow 1967). For example, if $x/Re = 0.01$, then $Re_c = 12,000$. Here x is scaled on the channel half-width. These Reynolds number values are may not be attained in forced expiration, so TSI is an unlikely candidate for self-induced oscillations related to wheezing. Of course, the geometry of the choked area may lower this critical Reynolds number to a physiologic range.

INTERSTITIAL FLUID FLOW

Under physiologic conditions there is a net filtration of fluid out of the small blood vessels in the lung and into the pulmonary interstitium. The interstitium is an acellular, gelatinous matrix that forms part of the scaffolding that gives structure to the lung tissue. From here, the fluid may exit the lung via lymphatic drainage or possibly through the intrapleural liquid system. It also may accumulate in the gel and ultimately in the alveolar air sacs if there is too much fluid. This latter condition is known as pulmonary edema, a common complication in heart attacks and other cardiac conditions. Transport of microvascular fluid through the pulmonary interstitium to the lymphatic system, as well as storage of this volume load in the interstitial space, serve as hedges against the sequence of events leading to pulmonary edema and alveolar flooding (Taylor & Parker 1985).

Flow of liquid through the lung interstitium, from the sites of filtration to the routes of clearance, depends upon the resistance and compliance of the interstitial matrix (Lai-Fook & Toporoff 1980). The interstitium is composed of collagen, hyaluronic acid, and proteoglycans that compose the extracellular matrix. Its macromolecular structure is such that it traps water and also forms a porous network through which water is transported. The macromolecular matrix is the structural source of the interstitium's elastic properties. It has been shown, e.g. Glucksberg & Bhattacharya (1989) amongst others, that the compliance (local volume increase per local pressure increase) of the pulmonary interstitium varies with its state of hydration. There appears to be a level of hydration below which interstitium has low compliance and above which, compliance rises dramatically. Hydration of the interstitial gel may cause widening of the distances between matrix constituents which may exceed the limit of effectiveness of intramolecular interactions, disentangle molecular constituents, or otherwise disrupt the structural integrity of the matrix and thus diminish the ability of the interstitium to generate elastic restoring forces.

The pulmonary interstitium, like other fibrous connective tissue, may be considered to be a poroelastic material in which the interstitial resistance is coupled to the elastic properties of the interstitial matrix. The relation of interstitial resistance and stiffness to interstitial fluid flow has largely been limited to studies involving global measurements in whole lung preparations. Lung interstitial resistance and compliance have been estimated by a variety of techniques (Taylor & Parker 1985). Although liquid flows through the matrix, there is resistance to this flow which is a function of pore size and liquid viscosity. It has been suggested by several authors, e.g. Lai-Fook (1988), that nonedematous interstitium is "tighter" with a small average pore size and therefore high resistance. Hydrated interstitium is "looser," with larger average pore size and lower resistance. Simple geometric considerations predict that a large decrease in resistance occurs in hydration due to an increase in pore diameter.

Ford et al (1991) developed a new technique to study local transport phenomena in pulmonary interstitium. Two micropipettes are positioned such that their tips are placed adjacent to each other in the subpleural perialveolar interstitium. One micropipette is used to inject a small bolus of fluid, while the other simultaneously records interstitial pressure. By modeling the interstitium as a homogeneous, linearly poroelastic material, they were able to infer both local compliance and local flow resistance from the decaying pressure signal.

MUCUS TRANSPORT

The airways are coated with liquid and in the first 15 or so generations it is a bi-layer consisting of a serous subphase covered by a mucus blanket. The serous fluid is regarded as Newtonian while the mucus is clearly non-Newtonian and possesses a yield stress while exhibiting characteristics of shear thinning, visco-elasticity, and thixotropy (Hwang et al 1969, Davis 1973, Powell et al 1974). The combined thickness of the bi-layer is on the order of 5–10 microns in the larger airways under normal circumstances. The liquid serves to protect the underlying cells from drying and to trap inhaled particles and microbes that present a danger to the lung. The clearance of the mucus layer and its contents constitutes a primary pulmonary defense mechanism. The clearance is achieved by gravity, shear stress from the air flow, and ciliary propulsion (see Figure 13).

Cough

Many of us take for granted the ability to generate a strong cough and expel mucus from our lungs. However several diseases, such as those that affect the neuro-muscular system, severely limit the necessary effort leaving

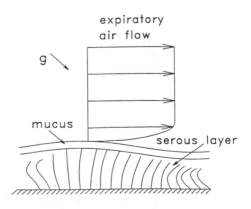

Figure 13   Mucus transport due to gravity, air-flow induced shear stress, and ciliary motion.

the individual prone to infection. It is routine to find weak coughs in patients with severe chest wall or abdominal pain as follows surgery to those areas, and clearance of airway liquid to avoid pneumonia is a central theme of their post-operative care. Respiratory air flow induces a shear stress on the liquid lining and Blake (1975) predicts that even during normal breathing the average stress is mouthward, due to the asymmetry of velocity profiles, and large enough to compete in importance with ciliary propulsion and gravity-driven flow. The most extreme shear stress condition of course occurs during cough where intrapleural pressures exceed 100 mm Hg and the average air speeds, U, of the turbulent flow are estimated to be larger than 200 m/sec (Ross et al 1955). Cough is especially important when airways have accumulated abnormal amounts of liquids whose rheological properties may be altered (Leith 1977). Accumulations and rheological changes may occur in asthma, chronic bronchitis, and cystic fibrosis (Shimura et al 1988, Yeates et al 1976).

Scherer & Burtz (1978) simulated the annular two-phase flow dynamics of cough in bench-top experiments where sudden blasts of air from a pressurized tank were introduced into a flexible, but not collapsible, tube lined with a Newtonian liquid layer. They achieved peak Reynolds numbers from 10^4 to 10^5 which correlate to values in the collapsed trachea during cough. The fraction of original liquid lining, m, expelled from the tube decreases monotonically with increasing values of the dimensionless group $\mu/\rho U^2\tau$, where τ is the time constant of pressure decay in the air tank during the cough. During the process, high-speed motion pictures revealed large-amplitude waves on the air-liquid interface and droplets first breaking off, and then exiting with the air flow. King et al (1985) used a similar approach but employed non-Newtonian fluids relevant to human mucus. Clearance was lowered for fluids with higher elasticity but comparable dynamic viscosities. In a following paper, Soland et al (1987) repeated these experiments in a flexible-walled system and found that clearance was enhanced compared to their rigid apparatus when the cough flow was created by negative pressure. The negative pressure causes partial wall collapse, simulating tracheal narrowing due to increased transmural pressure differences. The reduction of lumenal cross-sectional area increases the gas speeds and shear—one cause of the enhancement. A second cause is the downstream traveling wall-wave created by the flow which adds to the axial transport of the adherent mucus layer and accentuates the disturbances leading to droplet formation.

Basser et al (1989) have also examined non-Newtonian effects in a rigid-walled channel and focused primarily on the role of yield stress by using off-the-shelf mayonnaise in their experiments. Mayonnaise has many rheological similarities to mucus; its yield stress, $\tau_y = 610$ dynes/cm^2, compares particularly well to that of tracheal mucus as measured by Davis (1973). Basser et al explain mucus clearance as an instability, similar to an avalanche, where the local fluid shear exceeds the yield stress, in this case because of the air flow over the liquid. Just prior to motion, the fluid shear is highest at the underlying wall. Their high-speed motion pictures indicate that once the yield stress is exceeded in a small region, the mucus there flows and piles up forming a finite-width wave front which propagates at a relatively constant speed of 2.5 m/s, while spreading laterally across the rectangular test section. The wave crest rises until it touches the upper wall, almost blocking the air flow entirely. Then the pressure builds up dramatically until there is a large clearance event which moves a significant volume of mayonnaise. Rather than droplet formation and entrainment as a major transport mechanism for Newtonian fluids, their data suggest that this bulk transport is the chief mechanism during a cough. From the analysis a key parameter evolves: the yielding number $= \Delta P H/2L\tau_y$, where H is the channel height and $\Delta P/L$ is the pressure gradient along the channel. Its critical value for initiating motion is $\Delta P_{crit}H/2L\tau_y = 1/(1+h/H)$, where h is the liquid height, though no transport was detected for $h/H < 0.1$. This formula fits the data in a general way, despite the scatter. With modest assumptions, they estimate a human tracheal wall shear $\tau_w = 2$ dynes/cm^2 for normal breathing, well below τ_y, while during a cough $\tau_w = 1692$ dynes/cm^2 which is $\sim 3\tau_y$. Hence, the avalanche mechanism is available for coughs, but not during quiet breathing.

Some investigators have connected the general interest in high-frequency ventilation with potential methods of enhancing mucus transport. If the expiratory time is made reasonably smaller than the inspiratory time for such ventilators, then the mouthward shear stress dominates the liquid flow, aiding clearance (Chong et al 1987). External chest wall compressions were also shown to be useful for promoting mucus transport (King et al 1983), and mucus shear thinning due to the imposed fluid oscillations appear to be partly responsible for the enhancement.

Ciliary Transport

Cilia line the airways of the proximal bronchial tree and are responsible for propelling the mucus layer mouthward. Each cilium beats, causing its tip to undergo an elliptical orbit. There is coordination of the beating yielding a metachronal wave disturbance whose geometry propagates away from the mouth (see Figure 13). Generally, the beat frequency is about 15 Hz and the wavelength is 30 microns producing an average mucus velocity of 0.2 mm/s. The power stroke towards the mouth occurs when the cilium is straightest and its tip contacts the mucus layer. Analytical models of ciliary propulsion have sometimes treated the cilia as a continuous wall undergoing a traveling-wave motion (Blake 1971, Ross & Corrsin 1974).

This assumption is justified in part by the inter-ciliary distance of 0.3 microns and ciliary diameter of 0.3 microns, both comparable to the wavelength. For example, in Ross & Corrsin (1974) each wall point undergoes an elliptic trajectory in Lagrangian space and the resulting Eulerian description leads to a nonsinusoidal wave on an extensible wall. These boundary effects are sufficient to produce a net mouthward displacement of mucus and are likely to be the important mechanism when the periciliary serous layer is so thick that the cilia tips cannot touch the mucus. Their two-layer model predicts mucus motion similar to an elastic slab and the displacements per cycle are relatively insensitive to mucus viscosity changes but decrease significantly with serous layer viscosity increases. Models of the ciliary interaction with the mucus layer, such as that of Silberberg (1990), rely again on its elastic slab behavior. Given the mucus relaxation time of approximately 1 sec compared to the perturbation period of 0.067 sec, this is reasonable. Just how ciliary motion delivers momentum to the fluid bi-layer has not been analyzed.

The liquid lining of the acinus region is single-layered and considered Newtonian. However, there are no cilia at this level, so its outflow must rely on some other mechanism. One proposal is that surface tension gradients pull fluid from the alveolar region mouthward. These gradients arise because the alveoli produce surfactants, lowering the interfacial tension there, while the alveolar ducts do not. Davis et al (1974) solved the problem of a liquid-lined, finite length tube with an insoluble surfactant at the air-liquid interface. Their steady-state solutions predict an axially-varying film height, thickest at the low-surface-tension (σ_1) end while thinning gradually along the tube to the high-surface-tension (σ_2) end. Surface tension influences the transport by its axial gradient, but also by its absolute value which sets the normal stress jump across the interface. In the thick-layer region, the transverse surface curvature is larger than that found at the other end, so the liquid pressure experiences a contribution whose gradient is opposite to the flow, tending to slow it down. The thin film approximation, $\varepsilon \to 0$, is handled as a singular perturbation with a boundary layer at the low-surface-tension end, whose length is $O(\Sigma \varepsilon^2)$, where $\Sigma = \sigma_1/(\sigma_2 - \sigma_1)$. For the lung application this is about 1% of the tube length. The outer solution is exactly the planar solution where surface tension only enters the problem through its gradients. Of course, we know that if the layer becomes thick enough, closure may occur and these steady solutions are not stable.

PERSPECTIVE AND FUTURE DIRECTIONS

An obvious distinction between this review and Pedley's (1977) is that, during the intervening years, the field of respiratory biofluid mechanics has extended beyond aerodynamic studies of intra-airway gas flow. It now encompasses many more phenomena which fall into several other subdisciplines of fluid mechanics. Among these are thin-film flows, surface-tension driven flows, interfacial instabilities, elasto-hydrodynamic lubrication, flows in poro-elastic media, flows of non-Newtonian fluids, aero-elastic instabilities, and low Reynolds number flows. Because the lung contains both gases and liquids, undergoing steady and unsteady flows, in geometries whose length scales vary from centimeter to submicron levels, the governing dimensionless parameters span a very wide range. The result is a rich environment for fluid dynamical research where a multitude of flow regimes and physical interactions exist, often simultaneously. External (e.g. mechanical ventilation) and self-directed (e.g. altered breathing patterns) manipulations of the respiratory system add to this variety in frequently occurring physiological venues, much more so than in the cardiovascular system, for example.

Clear indications of progress in pulmonary biofluid mechanics come also from its wider acceptance in the spectrum of disciplines it embraces: applied mathematics, fluid mechanics, physiology, and clinical medicine. This is evident from the many different journals and reviews (see the references) where important fluid dynamical observations, experiments, and analyses are made, each with its unique viewpoint and relevance to the lung. With this "ease" of cross-disciplinary communication comes the investigator's responsibility of outlining the biological application and the fundamental scientific questions for all to appreciate, while acknowledging the original literature, regardless of its relationship to the intended audience.

The future for research in this field will likely entail more in-depth studies of the broadened agenda mentioned above, as well as newer pursuits such as the interaction of flow and transport with pulmonary cells and gas exchange by partial liquid (perfluorocarbon) ventilation, a method that holds some promise of opening closed airways and alveoli with imposed liquid flows while delivering oxygen and removing carbon dioxide.

ACKNOWLEDGMENTS

Support for this work comes from NIH grants HL41126 and HL01818 and NSF grant CTS 9013083.

Literature Cited

Ahmad, J., Hansen, R. S. 1972. A simple quantitative treatment of the spreading of monolayers on thin liquid films. *J. Colloid Interface Sci.* 38: 601–4

Akhavan, R., Kamm, R. D., Shapiro, A. H. 1991a. An investigation of transition to turbulence in bounded oscillatory Stokes flow. Part 1, Experiments. *J. Fluid Mech.* 225: 395–422

Akhavan, R., Kamm, R. D., Shapiro, A. H.

1991b. An investigation of transition to turbulence in bounded oscillatory Stokes flow. Part 2, Numerical simulations. *J. Fluid Mech.* 225: 423–44

Aris, R. 1956. On the dispersion of a solute in a fluid flowing through a tube. *Proc. R. Soc. London Ser. A* 235: 67–77

Avery, M. E., Mead, J. 1959. Surface properties in relation to atelectasis and hyaline membrane disease. *Am. J. Dis. Child.* 97: 517–23

Basser, P. J., McMahon, T. A., Griffith, P. 1989. The mechanism of mucus clearance in cough. *J. Biomech. Eng.* 111: 288–97

Benjamin, T. B. 1963. The threefold classification for unstable disturbances in flexible surfaces bounding inviscid flows. *J. Fluid Mech.* 16: 436–50

Bertram, C. D. 1982. Two modes of instability in a thick-walled collapsible tube conveying a flow. *J. Biomech.* 15: 223–24

Bertram, C. D., Pedley, T. J. 1982. A mathematical model of unsteady collapsible tube behavior. *J. Biomech.* 15: 39–50

Blake, J. 1971. A spherical envelope approach to ciliary propulsion. *J. Fluid Mech.* 46: 199–208

Blake, J. 1975. On the movement of mucus in the lung. *J. Biomech.* 8: 179–90

Bohn, D. J., Miyasaka, K., Marchak, E. B., Thompson, W. K., Froese, A. B., Bryan, A. C. 1980. Ventilation by high-frequency oscillation. *J. Appl. Physiol.* 48: 710–16

Borgas, M. S., Grotberg, J. B. 1988. Monolayer flow on a thin film. *J. Fluid Mech.* 193: 151–70

Brenner, H., Gaydos, L. J. 1977. The constrained Brownian movement of spherical particles in cylindrical pores of comparable radius: models of the diffusive and convective transport of solute molecules in membranes and porous media. *J. Colloid Interface Sci.* 58: 312–56

Broaddus, V. C., Wiener-Kronish, J. P., Staub, N. C. 1990. Clearance of lung edema into the pleural space of volume-loaded anesthetized sheep. *J. Appl. Physiol.* 68: 2623–30

Briant, J. K., Frank, D. D., James, A. C., Eyler, L. L. 1992. Numerical simulation of aerosol particle transport by oscillating flow in respiratory airways. *Ann. Biomed. Eng.* 20: 573–81

Bunk, D. A., Federspiel, W. J., Jackson, A. C. 1992. Influence of bifurcations on forced oscillations in an airway model. *J. Biomech. Eng.* 114: 216–21

Cancelli, C., Pedley, T. J. 1985. A separated flow model for collapsible tube oscillations. *J. Fluid Mech.* 157: 375–404

Carpenter, P. W., Garrad, A. D. 1985. The hydrodynamic stability of flow over Kramer-type compliant surfaces. Part 1.

Tollmien-Schlichting instabilities. *J. Fluid Mech.* 155: 465–510

Carpenter, P. W., Garrad, A. D. 1986. The hydrodynamic stability of flow over Kramer-type compliant surfaces. Part 2. Flow induced surface instabilities. *J. Fluid Mech.* 170: 199–232

Chang, H. K., Menon, A. S. 1985. Air flow dynamics in the human airways. In *Aerosols in Medicine. Principles, Diagnosis and Therapy*, ed. F. Moren, M. T. Newhouse, M. B. Dolovich, chap. 4. New York: Elsevier

Chatwin, P. C. 1975. On the longitudinal dispersion of passive contaminant in oscillatory flows in tubes. *J. Fluid Mech.* 71: 513–27

Chen, T. S., Sparrow, E. M. 1967. Stability of the developing laminar flow in a parallel plate channel. *J. Fluid Mech.* 30: 209–24

Chong, S. K., Iglesias, A. J., Sackner, M. A. 1987. Mucus clearance by two-phase gas-liquid flow mechanism: asymmetric periodic flow model. *J. Appl. Physiol.* 62: 959–71

Conrad, W. A. 1969. Pressure-flow relationships in collapsible tubes. *IEEE Trans. Biomed. Eng.* 16: 284–95

Daskopoulos, P. H., Lenhoff, A. M. 1988. Dispersion coefficient for laminar flow in curved tubes. *AIChE J.* 34: 2052–58

Davidson, M. R. , Schroter, R. C. 1983. A theoretical model of absorption of gases by the bronchial wall. *J. Fluid Mech.* 129: 313–35

Davis, S. S. 1973. Rheological examination of sputum and saliva and the effect of drugs. In *Rheology of Biological Systems*, ed. H. L. Gabelnick, M. Litt, pp. 157–94. Springfield, Ill: Thomas

Davis, S. H., Liu, A.-K., Sealy, G. R. 1974. Motion driven by surface-tension gradients in a tube lining. *J. Fluid Mech.* 62: 737–52

Dawson, S. V., Elliott, E. A. 1977. Wave-speed limitation on expiratory flow: a unifying concept. *J. Appl. Physiol.* 43: 498–515

DiPietro, N. D., Huh, C., Cox, R. G. 1978. The hydrodynamics of the spreading on one liquid on the surface of another. *J. Fluid Mech.* 84: 529–49

Dragon, C. A., Grotberg, J. B. 1991. Oscillatory flow and mass transport in a flexible tube. *J. Fluid Mech.* 231: 135–55

Eckmann, D. M., Grotberg, J. B. 1988. Oscillatory flow and mass transport in a curved tube. *J. Fluid Mech.* 188: 509–27

Eckmann, D. M., Grotberg, J. B.. 1991. Experiments on transition to turbulence in oscillatory pipe flow. *J. Fluid Mech.* 222: 329–50

Elad, D., Halpern, D., Grotberg, J. B. 1992.

Gas dispersion in volume-cycled tube flow I. Theory. *J. Appl. Physiol.* 72: 312–20

Elad, D. E., Kamm, R. D., Shapiro, A. H. 1988. Mathematical simulation of forced expiration. *J. Appl. Physiol.* 65: 14–25

Elliott, E. A., Dawson, S. V. 1977. Test of wave speed theory of flow limitation in elastic tubes. *J. Appl. Physiol.* 43: 516–22

Espinosa, F. F. 1991. *Spreading of surfactant in a small pulmonary airway.* MS thesis. MIT, Cambridge, Mass.

Everett, D. H., Haynes, J. M. 1972. Model studies of capillary condensation 1. Cylindrical pore model with zero contact angle. *J. Colloid Interface Sci.* 38: 125–37

Federspiel, W. J. and J. J. Fredberg. 1988. Axial dispersion in respiratory bronchioles and alveolar ducts. *J. Appl. Physiol.* 64: 2614–21

Ford, T. R., Sachs, J. R., Grotberg, J. B., Glucksberg, M. R. 1991. Perialveolar interstitial resistance and compliance in isolated rat lung. *J. Appl. Physiol.* 70: 2750–56

Fredberg, J. J. 1980. Augmented diffusion in the airways can support pulmonary gas exchange. *J. Appl. Physiol.* 49: 232–38

Fry, D. L., Hyatt, R. E. 1960. Pulmonary mechanics. A unified analysis of the relationship between pressure, volume and gas flow in the lungs of normal and diseased human subjects. *Am. J. Med.* 29: 672–89

Gauglitz, P. A., Radke, C. J. 1988. An extended evolution equation for liquid film breakup in cylindrical capillaries. *Chem. Eng. Sci.* 43: 1457–65

Gaver, D. P., Grotberg, J. B. 1990. The dynamics of a localized surfactant on a thin film. *J. Fluid Mech.* 213: 127–48

Gaver, D. P., Grotberg, J. B. 1992. Droplet spreading on a thin viscous film. *J. Fluid Mech.* 235: 399–414

Gaver, D. P., Samsel, W., Solway, J. 1990. Effects of surface tension and viscosity on airway reopening. *J. Appl. Physiol.* 69: 74–85

Gaver, D. P. III, Solway, J., Punjabi, N., Elad, D., Grotberg, J. B., Gavriely, N. 1992. Gas dispersion in volume-cycled tube flow II. Tracer-bolus experiments. *J. Appl. Physiol.* 72: 321–31

Gavriely, N., Eckmann, D., Grotberg, J. B. 1992. Gas exchange by intratracheal insufflation in a ventilatory dog model. *J. Clin. Invest.* 90: 2376–83

Gavriely, N., Grotberg, J. B. 1988. Flow limitation and wheezes in a constant flow and volume lung preparation. *J. Appl. Physiol.* 64: 17–20

Gavriely, N., Kelly, K. B., Grotberg, J. B., Loring, S. H. 1987. Forced expiratory wheezes are a manifestation of airway flow limitation. *J. Appl. Physiol.* 62: 2398–403

Gavriely N., Palti, Y., Alroy, G., Grotberg, J. B. 1984. Measurements and theory of wheezing breath sounds. *J. Appl. Physiol.* 57: 481–92

Gavriely, N., Shee, T. R., Cugell, D. W., Grotberg, J. B. 1989. Flutter in flow limited collapsible tubes: a mechanism for generation of wheezes. *J. Appl. Physiol.* 66: 2251–61

Gavriely, N., Solway, J., Drazen, J. M., Slutsky, A. S., Brown, R., et al. 1985. Radiographic visualization of airway wall movement during oscillatory flow in dogs. *J. Appl. Physiol.* 58: 645–52

Gilbert, I. A., Fouke, J. M., McFadden, E. R. Jr. 1987. Heat and water flux in the intrathoracic airways and exercise-induced asthma. *J. Appl. Physiol.* 63: 1681–91

Gill, W. N., Sankarasubramanian, R. 1971. Dispersion of a non-uniform slug in time-dependent flow. *Proc. R. Soc. London Ser. A* 322: 101–17

Glucksberg, M. R., Bhattacharya, J. 1989. Effect of dehydration on interstitial pressures in the isolated dog lung. *J. Appl. Physiol.* 67: 839–45

Godleski, D. A., Grotberg, J. B. 1988. Convection-diffusion interaction for oscillatory flow in a tapered tube. *J. Biomech. Eng.* 110: 283–91

Goldsmith, H. L., Mason, S. G. 1963. The flow of suspensions through tubes II. Single large bubbles. *J. Colloid Sci.* 18: 237–61

Goren, S. L. 1962. The instability of an annular thread of fluid. *J. Fluid Mech.* 12: 309–19

Grotberg, J. B. 1984. Volume-cycled oscillatory in a tapered channel. *J. Fluid Mech.* 141: 249–64

Grotberg, J. B., Davis, S. H. 1980. Fluid-dynamic flapping of a collapsible channel: sound generation and flow limitation. *J. Biomech.* 13: 219–30

Grotberg, J. B., Gavriely, N. 1989. Flutter in collapsible tubes: a theoretical model of wheezes. *J. Appl. Physiol.* 66: 2262–73

Grotberg, J. B., Glucksberg, M. R. 1993. A buoyancy-driven squeeze-film model of intrapleural fluid dynamics: basic concepts. *J. Appl. Physiol.* In press

Grotberg, J. B., Reiss, E. L. 1984. Subsonic flapping flutter. *J. Sound Vib.* 92: 349–61

Grotberg, J. B., Shee, T. R. 1985. Compressible-flow channel flutter. *J. Fluid Mech.* 159: 175–93

Grotberg, J. B., Sheth, B. V., Mockros, L. F. 1990. An analysis of pollutant gas transport and absorption in pulmonary airways. *J. Biomech. Eng.* 112: 168–76

Haefeli-Bleuer, B., Weibel, E. R. 1988. Morphometry of the human pulmonary acinus. *Anat. Rec.* 220: 401–14

Hains, F. D., Price, J. F. 1962. Effect of a flexible wall on the stability of Poiseuille flow. *Phys. Fluids* 5: 365

Halpern, D., Grotberg, J. B. 1992. Fluid-elastic instabilities of liquid-lined flexible tubes. *J. Fluid Mech.* 244: 615–32

Halpern, D., Grotberg, J. B. 1993. Surfactant effects on fluid-elastic instabilities of liquid-lined flexible tubes: a model of airway closure. *J. Biomech. Eng.* 115: 271–77

Hammond, P. S. 1983. Nonlinear adjustment of a thin annular film of viscous fluid surrounding a thread of another within a circular pipe. *J. Fluid Mech.* 137: 363–84

Hanna, L. M., Scherer, P. W. 1986. A theoretical model of localized heat and water vapor transport in the human respiratory tract. *J. Biomech. Eng.* 108: 19–27

Harris, H. G., Goren, S. L. 1967. Axial diffusion in a cylinder with pulsed flow. *Chem. Eng. Sci.* 22: 1571–76

Hino, M., Sawamoto, M., Takasu, S. 1976. Experiments on transition to turbulence in an oscillatory pipe flow. *J. Fluid Mech.* 75: 193–207

Horsfield, K., Dart, D., Olson, D. E., Filley, G. F., Cumming, G. 1971. Models of the human bronchial tree. *J. Appl. Physiol.* 31: 207–17

Hwang, S. H., Litt, M., Forsman, W. C. 1969. Rheological properties of mucus. *Rheol. Acta* 8: 438–48

Hydon, P. E., Pedley, T. J. 1993a. Axial dispersion in a channel with oscillating walls. *J. Fluid Mech.* 249: 535–55

Hydon, P. E., Pedley, T. J. 1993b. Dispersion in oscillatory channel flow with coupled transverse wall motion. *Eur. J. Fluids B* In press

Ingenito, E., Kamm, R. D., Watson, J. W., Slutsky, A. S. 1988. Model of constant flow ventilation in a dog lung. *J. Appl. Physiol.* 64: 2150–59

Ingenito, E. P., Solway, J., McFadden, E. R. Jr., Pichurko, B. M., Cravalho, E. G. , Drazen, J. M. 1986. Finite difference analysis of respiratory heat transfer. *J. Appl. Physiol.* 61: 2252–59

Jackson, A. C., Lutchen, K. R. 1991. Physiological basis for resonant frequencies in respiratory system impedances in dogs. *J. Appl. Physiol.* 70: 1051–8

Jan, D. L., Shapiro, A. H., Kamm, R. D. 1989. Some features of oscillatory flow in model bifurcations. *J. Appl. Physiol.* 67: 147–59

Jensen, O., Grotberg, J. B. 1992. Insoluble surfactant spreading on a thin viscous film: shock evolution and film rupture. *J. Fluid Mech.* 240: 259–88

Johnson, M., Kamm, R. D. 1986. Numerical studies of steady flow dispersion at low Dean number in a gently curving tube. *J. Fluid Mech.* 172: 329–45

Johnson, M., Kamm, R. D., Ho, L. W., Shapiro, A. H., Pedley, T. J. 1991. The nonlinear growth of surface-tension driven instabilities of a thin annular film. *J. Fluid Mech.* 233: 141–56

Joshi, C.H., Kamm, R. D., Drazen, J. M., Slutsky, A. S. 1983. An experimental study of gas exchange in laminar oscillatory flow. *J. Fluid Mech.* 133: 245–54

Kamm, R. D., Schroter, R. C. 1989. Is airway closure caused by a thin liquid instability? *Respir. Physiol.* 75: 141–56

Katz, A.I., Chen, Y., Moreno, A. H. 1969. Flow through a collapsible tube, experimental analysis and mathematical model. *Biophys. J.* 9: 1261–79

Keshgi, H. S., Scriven, L. E. 1991. Dewetting: nucleation and growth of dry regions. *Chem. Eng. Sci.* 46: 519–26

King, M., Brock, G., Lundell, C. 1985. Clearance of mucus by simulated cough. *J. Appl. Physiol.* 58: 1776–82

King, M., Phillips, D. M., Gross, D., Vartian, V., Chang, H. K., Zidulka, A. 1983. Enhanced tracheal mucus clearance with high frequency chest wall compression. *Am. Rev. Respir. Dis.* 128: 511–15

Kraman, S. S. 1983. The forced expiratory wheeze: its site of origin and possible association with lung compliance. *Respiration* 44: 189–96

Kurzweg, U. H. 1985. Enhanced heat conduction in oscillating viscous flows within parallel-plate channels. *J. Fluid Mech.* 156: 291–300

Kurzweg, U. H., Lindgren, E. R., Lothrop, B. 1989. Onset of turbulence in oscillating flow at low Womersley number. *Phys. Fluids A* 1: 1972–75

Lai-Fook, S. J. 1988. Pressure-flow behavior of pulmonary interstitium. *J. Appl. Physiol.* 64: 2372–80

Lai-Fook, S. J., Kaplowitz, M. R. 1985. Pleural space thickness in situ by light microscopy in five mammalian species. *J. Appl. Physiol.* 59: 603–10

Lai-Fook, S. J., Rodarte, J. R. 1991. Pleural pressure distribution and its relationship to lung volume and interstitial pressure. *J. Appl. Physiol.* 70: 967–78

Lai-Fook, S. J., Toporoff, B. 1980. Pressure-volume behavior of perivascular interstitium measured in isolated dog lung. *J. Appl. Physiol.* 48: 939–46

Lambert, R. K., Wilson, T. A. 1972. Flow limitation in a collapsible tube. *J. Appl. Physiol.* 33: 150–53

Lambert, R. K., Wilson, T. A. 1973. A model for the elastic properties of the lung and their effect on expiratory flow. *J. Appl. Physiol.* 34: 34–48

Landahl, M. T. 1962. On the stability of a laminar incompressible boundary layer over a flexible surface. *J. Fluid Mech.* 13: 607–32

Leith, D. E. 1977. Cough. In *Lung Biology in Health and Disease, Respiratory Defense Mechanisms*, ed. J. E. Brain, D. F. Proctor, L. M. Reid, Vol. 5, Part II, Chap. 15, pp. 545–92. New York: Dekker

Lewis, J., Ikegami, M., Higuchi, R., Jobe, A., Absolom, D. 1991. Nebulized vs. instilled exogenous surfactant in an adult lung injury model. *J. Appl. Physiol.* 71: 1270–76

Li, H. 1954. Stability of oscillatory laminar flow along a wall. Beach Erosion Bd., *US Army Corps Eng. Tech. Memo.* 47

Liu, M., Wang, L., Li, E., Enhorning, G. 1991. Pulmonary surfactant will secure free airflow through a narrow tube. *J. Appl. Physiol.* 71: 742–48

Loudon, R. G., Murphy, L. H. 1991. Lung sounds. In *The Lung: Scientific Foundations*, ed. R. G. Crystal, J. B. West, pp. 1011–19. New York: Raven

Lungu, E.M., Moffatt, H. K. 1982. The effect of wall conductance on heat diffusion in duct flow. *J. Eng. Math.* 16: 121–36

Lunkenheimer, P.P., Rafflenbeul, W., Kellar, H., Frank, I., Dickhut, H. H. Fuhrmann, C. 1972. Application of tracheal pressure oscillations as a modification of "Diffusional Respiration." *Br. J. Anaesth.* 44: 627

Macklem, P. T., Proctor, D. F., Hogg, J. C. 1970. The stability of peripheral airways. *Respir. Physiol.* 8: 191–203

Matsuzaki, Y., Fung, Y. C. 1979. Non-linear stability analysis of a two-dimensional model of an elastic tube conveying a compressible flow. *Trans. ASME E: J. Appl. Mech.* 46: 31–36

Mazumder, B. S., Samir, K. D. 1992. Effect of boundary reaction on solute dispersion in pulsatile flow through a tube. *J. Fluid Mech.* 239: 523–49

McFadden, E. R. Jr., Pichurko, B. M., Bowman, H. F., Ingenito, E., Burns, S., et al. 1985. Thermal mapping of the airways in humans. *J. Appl. Physiol.* 58: 564–70

Mello, D. 1992. *The stability of oscillatory flow in a circular pipe.* PhD thesis. Brown Univ., Providence, RI

Merkli, P., Thomann, H. 1975. Transition to turbulence in oscillating pipe flow. *J. Fluid Mech.* 68: 567–75

Miller, F.J., Overton, J. H., Jaskot, R. H., Menzel, D. B. 1985. A model of regional uptake of gaseous pollutants in the lung. *Tox. Appl. Pharm.* 79: 11–27

Miserocchi, G., Del Fabbro, M., Venturoli, D., Glucksberg, M. R., Grotberg, J. B. 1993. *Gravity dependent distribution of pleural liquid pressure (PFB) filled lungs.* Presented at Vth Int. Soc. for Blood Substitutes Annu. Meet., San Diego, CA

Miserocchi, G., Negrini, D., Mariani, E., Passafaro, M. 1983. Reabsorption of a saline- and plasma-induced hydrothorax. *J. Appl. Physiol.* 54: 1574–88

Miserocchi, G., Negrini, D., Pistolesi, M., Bellina, C. R., Gilardi, M. C., et al 1988. Intrapleural liquid flow down a gravity-dependent hydraulic pressure gradient. *J. Appl. Physiol.* 64: 577–84

Mitzner, W., Permutt, S., Wienmann, G. 1983. Gas transport during high-frequency ventilation: theoretical model and experimental validation. *Ann. Biomed. Eng.* 12: 407–19

Nunge, R. L., Lin, T. S., Gill, W. N. 1972. Laminar dispersion in curved tubes and channels. *J. Fluid Mech.* 51: 363–83

Ohmi, M., Iguchi, M., Kakehashi, K., Masuda, T. 1982. Transition to turbulence and velocity distribution in an oscillating pipe flow. *Bull. JSME* 25: 365–71

Olson, D. E. 1971. *Fluid mechanics relevant to respiration: flow within curved or elliptical tubes and bifurcating systems.* PhD thesis. Imperial College, London

Otis, D. R., Johnson, M., Pedley, T. J., Kamm, R. D. 1993. The role of pulmonary surfactant in airway closure. *J. Appl. Physiol.* In press

Paloski, W. H., Slosberg, R. B., Kamm, R. D. 1987. Effects of gas properties and waveform asymmetry on gas transport in a branching tube network. *J. Appl. Physiol.* 62: 892–901

Peattie, R. A., Budwig, R. 1989. Heat transfer in laminar, oscillatory flow in cylindrical and conical tubes. *Int. J. Heat Mass Transfer* 32: 923–34

Pedley, T. J. 1977. Pulmonary fluid dynamics. *Annu. Rev. Fluid Mech.* 9: 229–74

Pedley, T. J., Kamm, R. D. 1988. The effect of secondary motion on axial transport in oscillatory tube flow. *J. Fluid Mech.* 193: 347–67

Pedley, T. J., Schroter, R. C., Sudlow, M. F. 1970. Energy losses and pressure drops in models of human airways. *Respir. Physiol.* 9: 371–86

Powell, R. L., Aharonson, E. F., Schwarz, W. H., Proctor, D. F., Adams, G. K., Reasor, M. 1974. Rheological behavior of normal tracheobronchial mucus of canines. *J. Appl. Physiol.* 37: 447–51

Rayleigh, Lord. 1879. On the instability of cylindrical fluid surfaces. In *Scientific Papers*, 3: 594–96. Cambridge: Cambridge Univ. Press

Riley, J. J., Gad-el-Hak, M., Metalfe, R.

W. 1988. Compliant coatings. *Annu. Rev. Fluid Mech.* 20: 393–420

Ross, B. B., Gramiak, R., Rahn, H. 1955. Physical dynamics of the cough mechanism. *J. Appl. Physiol.* 8: 264–69

Ross, S. M., Corrsin, S. 1974. Results of an analytical model of mucociliary pumping. *J. Appl. Physiol.* 37: 333–40

Rotenberry, J. M. 1992. Finite-amplitude shear waves in a channel with compliant boundaries. *Phys. Fluids A* 4: 270–6

Rotenberry, J. M., Saffman, P. G. 1990. Effect of compliant boundaries on weakly nonlinear shear waves in channel flow. *SIAM J. Appl. Math.* 50: 361–94

Ruckenstein, E., Jain, R. K. 1974. Spontaneous rupture of thin liquid films. *Chem. Soc. Faraday Trans. II* 70: 132–47

Sankarasubramanian, R., Gill, W. M. 1973. Unsteady convective diffusion with interphase mass transfer. *Proc. R. Soc. London Ser. A* 333: 115–32

Scherer, P. W., Burtz, L. 1978. Fluid mechanical considerations relevant to coughing. *J. Biomech.* 11: 183–87

Scherer, P. W., Haselton, F. R. 1982. Convection exchange in oscillatory flow through bronchial-tree models. *J. Appl. Physiol.* 53: 1023–33

Sergeev, S. I. 1966. Fluid oscillations in pipes at moderate Reynolds numbers. *Trans. Sov. Fluid Dyn.* 1: 121–22

Shapiro A. H. 1977. Steady flow in collapsible tubes. *J. Biomech. Eng.* 99: 126–47

Sharp, M. K., Kamm, R. D., Shapiro, A. H., Kimmel, E., Karniadakis, G. E. 1991. Dispersion in a curved tube during oscillatory flow. *J. Fluid Mech.* 223: 537–63

Shimura, S., Sasaki, T., Sasaki, H., Takishima, T., Umeya, K. 1988. Viscoclastic properties of bronchorrhea sputum in bronchial asthmatics. *Biorheology* 25: 173–79

Silberberg, A. 1990. On mucociliary transport. *Biorheology* 27: 295–307

Simon, B. A., Weinmann, G. G., Mitzner, W. 1984. Mean airway pressure and alveolar pressure during high-frequency ventilation. *J. Appl. Physiol.* 57: 1069–78

Smith, R. 1982. Contaminant dispersion in oscillatory flow. *J. Fluid Mech.* 114: 379–98

Soland, V., Brock, G., King, M. 1987. Effect of airway wall flexibility on clearance by simulated cough. *J. Appl. Physiol.* 63: 707–12

Sudo, K., Sumida, M., Yamane, R. 1992. Secondary motion of fully developed oscillatory flow in a curved pipe. *J. Fluid Mech.* 237: 189–208

Tarbell, J. M., Ultman, J. S., Durlofsky, L. 1982. Oscillatory dispersion in a branching tube network. *J. Biomech. Eng.* 104: 338–42

Taylor, G. I. 1953. Dispersion of solute matter in solvent flowing through a tube. *Proc. R. Soc. London Ser. A* 291: 186–203

Taylor, A. E., Parker, J. C. 1985. Pulmonary interstitial spaces and lymphatics. In *Handbook of Physiology*, Sect. 3, The Respiratory System, Vol. 1, Circulation and Nonrespiratory Functions, ed. A. P. Fishman, A. B. Fisher, pp. 167–230, Bethesda, MD: Am. Physiol. Soc.

Troian, S. M., Herbolzheimer, E., Safran, S. A. 1990. Model for the fingering instability of spreading surfactant drops. *Phys. Rev. Lett.* 65: 333–36

Troian, S. M., Wu, X. L., Safran, S. A. 1989. Fingering instability in thin wetting films. *Phys. Rev. Lett.* 62: 1496–99

von Kerczek, C., Davis, S. H. 1974. Linear stability theory of oscillatory Stokes layers. *J. Fluid Mech.* 62: 753–73

Walsh, C., Sullivan, P. A., Hanson, J. S. 1991. Subcritical flutter in collapsible tube flow: a model of expiratory flow in the trachea. *J. Biomech. Engng.* 113: 21–26

Watson, E. J. 1983. Diffusion in oscillatory pipe flow. *J. Fluid Mech.* 133: 233–44

Weaver D. S., Paidoussis, M. P. 1977. On collapse and flutter phenomena in thin tubes conveying fluid. *J. Sound Vib.* 50: 117–32

Webster, P. M., Sawatzky, R. P., Hoffstein, V., Leblanc, R., Hinchey, M. J., Sullivan, P.A. 1985. Wall motion in expiratory flow limitation: choke and flutter. *J. Appl. Physiol.* 59: 1304 12

Weh, L., Linde, H. 1973. Kraterfömige Oberflächenstöraungen in Anstrichfilmen, verursacht durch Silikonölzusätze. *Plaste Kautsch.* 20: 849–60 (In German)

Weibel, E.R. 1963. *Morphometry of the Human Lung.* New York: Academic

West, B. J., Bhargava, V., Goldberger, A. L. 1986. Beyond the principle of similitude: renormalization in the bronchial tree. *J. Appl. Physiol.* 60: 1089–97

Williams, M. B., Davis, S. H. 1982. Nonlinear theory of film rupture. *J. Colloid Interface Sci.* 90: 220–28

Wu, X., Lee, S. S., Cowley, S. J. 1992. On the nonlinear three-dimensional instability of Stokes layers and other shear layers to pairs of oblique waves. *NASA Tech. Memo.* 105918. ICOMP-92-20

Yang, W. H., Yih, C.-S. 1977. Stability of time-periodic flows in a circular pipe. *J. Fluid Mech.* 82: 497–505

Yeates, D. B., Sturgess, J. M., Kahn, S. R., Levison, H., Aspin, N. 1976. Mucociliary transport in the trachea of patients with cystic fibrosis. *Arch. Dis. Child.* 51: 28–33

Yu, J. S. 1981. Dispersion in laminar flow through tubes by simultaneous diffusion and convection. *J. Appl. Mech.* 48: 217–23

Annu. Rev. Fluid Mech. 1994. 26 : 573–616

VORTEX INTERACTIONS WITH WALLS[1]

T. L. Doligalski

U. S. Army Research Office, Research Triangle Park, North Carolina

C. R. Smith and J. D. A. Walker

Department of Mechanical Engineering and Mechanics,
Lehigh University, Bethlehem, Pennsylvania

KEY WORDS: unsteady separation, dynamic stall, turbulence

1. INTRODUCTION

Situations where an effectively irrotational freestream contains regions of concentrated vorticity are common in external aerodynamics, where vortices are known to play an important role in the flight of modern helicopters (Carr 1988) and aircraft (Cunningham 1989, Mabey 1989). Vortices may arise as a consequence of shedding from some upstream surface or near certain three-dimensional boundary geometries that act to promote vortex formation. Examples of the former type of vortex generation include: 1. vortices that trail from the tips of airfoils (Harvey & Perry 1971) and control surfaces on submarines (Lugt 1983), 2. transverse vortices shed from maneuvering airfoil surfaces and helicopter blades in a process known as dynamic stall (McCroskey 1982, Francis & Keesee 1985, Carr 1988), and 3. shedding from stationary obstacles. Geometry-induced creation can occur in any situation where a flow along a wall approaches a surface-mounted obstacle; examples include: 1. airframe features such as wing/body junctions, 2. conning towers on submarines, and 3. computer chips mounted on electrical circuit boards. Similar geometries are en-

573

countered in a variety of internal flows, such as branching pipes, and in turbine and compressor passages (Verdon 1989, 1992). The cited examples involve macroscopic large-scale motions, but important vortex motions are also prevalent at much smaller scale. Concentrated small vortices of the hairpin type are known to be an important influence in a boundary layer undergoing transition and appear to be the dominant feature in the production of turbulence near solid surfaces (Falco 1991, Grass et al 1991, Robinson 1991, Smith et al 1991b).

Vortices have been described by Kuchemann (1965) as "the sinews and muscles of fluid motion," and there is a rich, and seemingly endless, variety of complex phenomena associated with them. For this reason there has been intense interest for many decades in developing a basic understanding of the motion of vortices and their mutual interactions. Theoretical results, modern computational schemes, and inviscid vortex interactions have been described in a number of review articles (Saffman & Baker 1979, Leonard 1985, Shariff & Leonard 1992). This article is concerned with the physical processes that occur when a vortex is in proximity to a solid wall. It should be noted that related and sometimes similar phenomena also occur when vortices approach a free surface, especially when the interface has surface contamination (Bernal & Kwon 1989, Sarpkaya & Suthon 1991, Tryggvason et al 1991, Lugt & Ohring 1992, Sarpkaya 1992); however, attention will be restricted here to situations where the fluid must satisfy the no-slip condition at the boundary.

When a vortex is convected near a solid surface, a viscous response is generally provoked in the near-wall flow. A vortex Reynolds number Re_v may be defined in terms of the vortex circulation Γ and the kinematic viscosity v according to $Re_v = \Gamma/(2\pi v)$. For low Re_v the viscous response tends to be relatively muted (Peace & Riley 1983, Walker et al 1987). However for sufficiently high Re_v, a sequence of events initiates in the boundary layer, which culminates in an abrupt eruption of surface fluid and usually leads to the formation of a new vortex structure. The behavior is generic in that: (a) it is induced for a wide variety of vortex configurations such as vortex rings (Walker et al 1987, Ersoy & Walker 1987, Chu & Falco 1988, Falco 1991), vortex pairs (Ersoy & Walker 1985, 1986), and convected rectilinear filaments (Doligalski & Walker 1984); and (b) it occurs over an extensive range of Reynolds numbers (Walker et al 1987, Chuang & Conlisk 1989, Peridier et al 1991b). In general, such eruptive events develop because any vortex in proximity to a wall induces a persistent adverse pressure gradient on the boundary layer, in a frame of reference which moves with the vortex. Recent theoretical results (Elliott et al 1983, Cowley et al 1991) imply that steady motion at high Re_v is not possible under such circumstances and that an abrupt eruption of the boundary layer is to be anticipated. The process is initiated by a local concentration of the vorticity field within the boundary layer, which stimulates a rapidly-rising thin spire of fluid that ultimately interacts strongly with the external flow. These spires typically contain concentrated vorticity and often roll up into vortex structures in a subsequent strong viscous-inviscid interaction. The phenomena are regenerative in the sense that "parcels" of concentrated vorticity are ejected from near the surface to form new vortex structures. A number of realizations of this phenomenon will be described in this article.

In Section 2 the nature of this type of interaction in two-dimensional flows is described; more complex three-dimensional realizations are discussed in Section 3.

2. TWO-DIMENSIONAL VORTEX INTERACTIONS

2.1 *Introduction*

Many of the general features of the flow induced by a moving vortex can be appreciated through consideration of a simple model problem consisting of a two-dimensional vortex convected in a uniform flow above a wall. To illustrate that the behavior is generic, two limiting cases will be discussed. The first is the rectilinear vortex in which the vorticity is tightly concentrated in a small core having dimensions $O(Re_v^{-1/2})$, as discussed by Ting & Tung (1965). Suppose that a rectilinear vortex of negative rotation $-\kappa$ is located a distance a above an infinite plane wall, as shown schematically in Figure 1 (the circulation is related to vortex strength κ by

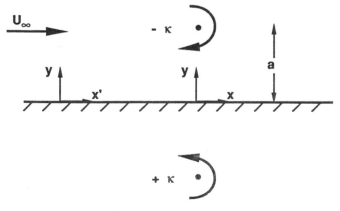

Figure 1 Schematic diagram of a rectilinear vortex in a uniform flow above a wall; the coordinate x is in a system moving with the vortex.

$\Gamma = 2\pi\kappa$). Inviscid theory (Milne-Thomson 1962) predicts that the vortex will be convected to the left with self-induced speed $V_s = \kappa/2a$ by the field of its image below the plate. If a uniform flow of speed U_o is superimposed to the right, the net result is a negative-rotation vortex convected to the right at constant height above the wall with speed

$$V_c = U_o - \frac{\kappa}{2a}. \tag{1}$$

A fractional convection rate α may be defined according to

$$\alpha = \frac{V_c}{U_o} = 1 - \frac{\kappa}{2aU_o}, \tag{2}$$

with the principal range of interest being $0 \le \alpha < 1$. The relative vortex speed α is influenced by the strength κ and distance from the wall a, with smaller values of α (either increased κ and/or decreased a) generally indicative of a more dramatic viscous response in the surface flow (Doligalski & Walker 1984). The limiting case $\alpha = 0$ corresponds to a vortex that is sufficiently strong and/or close enough to the wall so that it remains stationary in the uniform flow; for $\alpha < 0$, the vortex propagates upstream. On the other hand, the limit $\alpha \to 1$ describes a very weak vortex and/or one that is far from the surface. For values of $\alpha > 1$, the vortex has positive rotation and advances to the right with a speed in excess of U_o.

In the range of interest, $\alpha = 0.75$ is a critical value insofar as the viscous response at the surface is concerned, and this can be understood from two perspectives. First, consider the laboratory frame of reference. Let (x', y) denote Cartesian coordinates measuring distance along the plate and normal to it, respectively, from an origin fixed on the plate, with corresponding velocity components (u, v); these quantities are made dimensionless with respect to a and U_o, respectively. A dimensionless streamfunction (with respect to $U_o a$) defined by $u = \partial\tilde{\psi}/\partial y$, $v = -\partial\tilde{\psi}/\partial x'$ is given by

$$\tilde{\psi} = y + (1-\alpha)\log\left[\frac{x^2+(y-1)^2}{x^2+(y+1)^2}\right], \tag{3}$$

(Doligalski & Walker 1984); here $x = x' - \alpha t$, where x denotes streamwise distance in a frame of reference moving uniformly with the vortex and t is dimensionless time (with respect to a/U_o). The instantaneous streamline patterns, as seen by an observer in the laboratory frame, are shown in Figure 2 for various values of the fractional convection rate α. For $\alpha > 0.75$, there is a stagnation point aloft and below the vortex center, with the stagnation point approaching the surface as $\alpha \to 0.75$. For $\alpha < 0.75$, there are two apparent stagnation points near the surface which bracket a region of back flow. As α decreases, the stagnation points move progressively outboard of the vortex center, and eventually approach limiting values of $x = \pm\sqrt{3}$ as $\alpha \to 0$. The closed curve defining $\psi = 0$ is known as the Kelvin oval; half of the Kelvin oval appears in Figure 2d for $\alpha = 0$. Near the wall, the streamwise velocity distribution $\tilde{U}_\infty$ imposed by the moving vortex on the boundary layer is obtained by differentiation of Equation (3) with respect to y and subsequently taking the limit $y \to 0$, i.e.

$$\tilde{U}_\infty = 1 - \frac{4(1-\alpha)}{x^2+1}. \tag{4}$$

Consequently, as x decreases from upstream infinity $(x \to \infty)$, $\tilde{U}_\infty$ decreases monotonically from 1 to a minimum of $4\alpha - 3$ beneath the vortex center at $x = 0$; as x then decreases toward downstream infinity $(x \to -\infty)$, $\tilde{U}_\infty$ increases monotonically to 1. It follows that a region of adverse pressure gradient occurs for $x < 0$. For $\alpha < 0.75$, two stagnation points at $x = \pm\sqrt{3-4\alpha}$ bound a zone of negative $\tilde{U}_\infty$, and the vortex is strong enough and/or close enough to the wall so that an observer in the laboratory frame sees reverse flow near the wall below the vortex center. The streamwise velocity distribution induced near the surface for the case $\alpha \to 0$ is shown in Figure 3, where it may be noted that the absolute minimum velocity at $x = 0$ is -3.

In a frame of reference that convects uniformly with the vortex, the inviscid motion is steady and the wall appears to move uniformly to the left with constant speed. The moving reference frame provides a natural description of the phenomena, especially for evaluation of the boundary-layer response. A streamfunction in the moving frame may be obtained from Equation (3) by superimposing a uniform speed $-\alpha$ to the left. An observer in this frame sees a velocity at infinity of $(1-\alpha)U_0$, which is convenient to adopt as a representative speed in place of U_0. Using this nondimensionalization, the streamfunction in the convected frame is

$$\psi = \frac{\tilde{\psi} - \alpha y}{1-\alpha} = y + \log\left[\frac{x^2+(y-1)^2}{x^2+(y+1)^2}\right]. \tag{5}$$

Consequently, the streamline pattern for all values of α in the convected frame is the same as shown in Figure 2d for $\alpha = 0$, and the dimensionless mainstream velocity induced by the vortex on the boundary layer in the convected frame is

$$U_\infty(x) = 1 - \frac{4}{x^2+1}, \tag{6}$$

for all fractional convection rates α. This distribution is shown in Figure

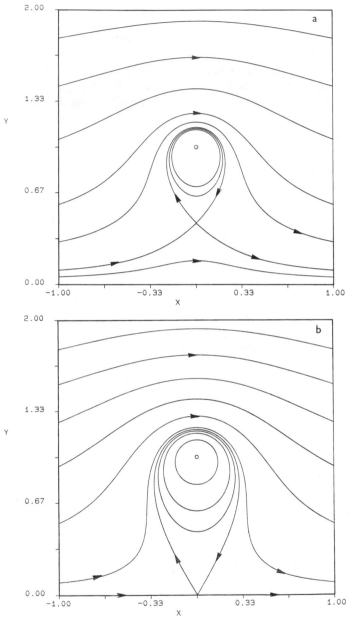

Figure 2 Instantaneous streamline patterns induced by a convected vortex as seen by an observer in the laboratory frame of reference. (*a*) $\alpha = 0.80$, (*b*) $\alpha = 0.75$, (*c*) $\alpha = 0.70$, (*d*) $\alpha = 0$. Also, (*d*) represents the streamline patterns for all α in a frame of reference convecting with the vortex.

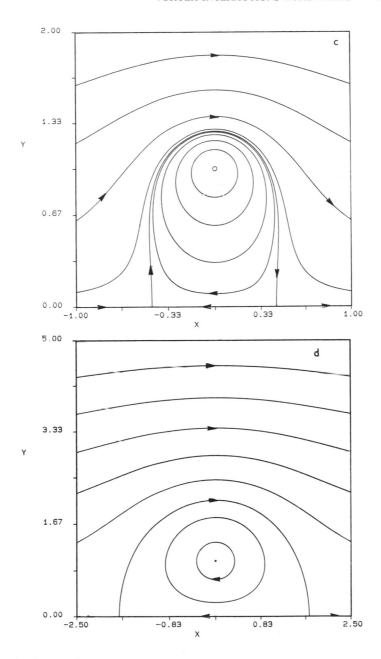

Figure 2—(continued)

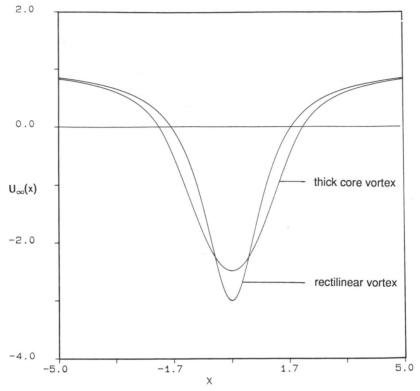

Figure 3 Streamwise velocity distribution induced near the wall by a rectilinear and a thick core vortex in a frame of reference convecting with the vortex.

3 where it may be noted that for $x < 0$ there is a region of deceleration (and hence adverse pressure gradient), which is most severe immediately to the left of the vortex center.

The fact that $\alpha = 0.75$ represents a critical convection rate can also be appreciated from the perspective of the moving frame, where the wall appears to move uniformly to the left with velocity $- V_c$. The dimensionless wall velocity [with respect to $(1 - \alpha)U_o$] is

$$u_w = -\beta; \qquad \beta = \frac{\alpha}{1 - \alpha}. \tag{7}$$

For $\alpha = 0.75$, $u_w = -3$ and thus the wall speed exactly balances the maximum velocity of -3 induced by the vortex at the boundary-layer edge at $x = 0$ (cf Equation 6). For all values of $\alpha < 0.75$, the magnitude of the wall speed is less than 3, and it emerges that the boundary-layer response is dominated by the influence of the adverse pressure gradient imposed by the external flow; this response eventually develops as a sharply focused eruption of the surface flow (Doligalski & Walker 1984). On the other hand, when $\alpha > 0.75$, the magnitude of the wall speed is greater than 3 (in dimensionless variables), and the influence of the rapidly moving wall appears to dominate the boundary-layer development. In such cases, the action of the vortex also provokes significant boundary-layer growth in the region to the left of the vortex center. However, the growth now occurs over a range of $x = O(1)$ and is much less dramatic than the explosive "spiky" response characteristic of the range $\alpha < 0.75$.

The rectilinear vortex represents a limiting case of vortex motion in two dimensions where the vorticity is tightly concentrated in a small core. Batchelor (1967, p. 534) has discussed another limit in which the vorticity is proportional to the streamfunction and is distributed over a finite circular area, outside of which the motion is irrotational (see also Doligalski & Walker 1984). It can be shown that the streamline pattern associated with such motion is similar to that due to the rectilinear vortex in Figure 2d, with the zero streamline being a semi-circle as compared to a Kelvin oval, and with stagnation points located at ± 2.081 as opposed to $\pm\sqrt{3}$ respectively. The velocity induced near the surface is very similar to the velocity induced by the rectilinear vortex as shown in Figure 3. The minimum velocity is achieved at $x = 0$ and is $u_{min} = 1/J_0(\lambda_1) = -2.483$, as opposed to -3 for the rectilinear vortex; here J_0 denotes a Bessel function. A critical convection rate occurs when $u_{min} = -\beta$, and it can be shown that this is at $\alpha_c = 0.7108$ (as compared to $\alpha_c = 0.75$ for a rectilinear vortex). It is evident that both the rectilinear and thick core vortices induce similar motion near the surface. Since the nature of the viscous response at the surface depends only on the mainstream velocity distribution outside the boundary layer and the relative motion of the wall, the general features described by Doligalski & Walker (1984) can be expected to apply for a wide range of vortex configurations. The characteristics of this viscous response are described next.

2.2 The Viscous Response

At high Reynolds numbers, laminar boundary layers exhibit a strong proclivity to abruptly develop a sharply-focused eruption in regions of adverse pressure gradient. The modern terminology for such events is *separation* and implies a process of boundary-layer detachment from the surface. These eruptions are characterized by a rapidly-rising thin plume of fluid across which intense gradients in the vorticity field develop in a direction tangential to the surface. Modern theoretical and computational studies have identified the critical triggering events and flow structure associated with the initiation of separation; but because separation develops abruptly and at very small spatial scales, the birth of the process is extremely difficult to pinpoint in an experiment. However, a variety of experimental studies have clearly documented the latter stages of boundary-layer separation in which the eruptive plume is often observed to roll up into a new vortex structure. These separation events, which culminate in such an unsteady viscous-inviscid interaction, represent a fundamental process at high Reynolds number wherein the boundary-layer vorticity is first concentrated into a very narrow band (in the streamwise direction) and then ejected an $O(1)$ distance from the surface. Since this sequence of events commonly occurs in vortex-induced separation, the general features of the process will now be described.

The basic eruptive phenomenon was first discovered by Van Dommelen (1981) and Van Dommelen & Shen (1980, 1982) in their study of the boundary-layer development on a circular cylinder impulsively started from rest. This classical problem had previously been considered by many authors, who all agreed on the nature of the early boundary-layer development (Riley 1975). Soon after initiation of the motion, a phenomenon (classically referred to as separation) occurs wherein a point of zero shear moves up the back of the cylinder, with a region of reversed flow occurring between it and the rear stagnation point. For high Reynolds number flows, the question of how the boundary layer would ultimately interact with the external flow had been controversial (Riley 1975). Sears & Telionis (1975) suggested that a singularity would form in the boundary-layer solution at finite time in an event signaling the onset of interaction; they argued that the term "separation" should be reserved for events where the boundary layer actually starts to break away from the surface. Their proposals were controversial, although at the time it was clear that a variety of authors were experiencing great difficulty in extending numerical solutions of the boundary-layer equations beyond a certain stage. The subsequent work of Van Dommelen & Shen (1980, 1982) definitively showed that a singularity formed at finite time, and for the first time revealed the basic fluid dynamics associated with the event. In the process, fluid particles near the point of zero shear, but above the wall (Van Dommelen & Shen 1980), experience strong compression in the tangential direction and consequently become extremely elongated in a direction normal to the wall (on the boundary-layer scale); this has the effect of focusing the local boundary-layer vorticity into a narrow "spike" that moves abruptly and rapidly away from the surface. On the boundary-layer scale, the phenomenon appears as an explosively-growing spike in displacement thickness having a lateral thickness which approaches zero as $Re \to \infty$; on the scale of the external flow, the event appears as a sharply focused concentration of vorticity near the surface that is about to knife into the mainstream. The modern definition of unsteady separation (Cowley et al 1991) is related to the concepts proposed by Sears & Telionis (1975) and implies an event in which a hitherto thin viscous boundary layer starts to interact strongly with the external flow and thereby detach from the surface.

A subsequent study by Elliott et al (1983) showed that the development of a spike is a generic response for two-dimensional boundary layers exposed to a persistent adverse pressure gradient. An analytic solution describing the terminal state of the boundary layer (just prior to inter-action) was obtained, whose structure is shown schematically in Figure 4. In the classical hierarchal approach (Smith 1982), the pressure distribution near the surface is evaluated from a solution of the Euler equations and then prescribed for the boundary-layer calculation; for impulsively-started problems, this procedure is valid until the boundary layer starts to develop strong localized outflows and a tendency toward interaction. An unsteady separation singularity occurs in the boundary-layer solution as a conse-quence of attempting to impose a prescribed pressure gradient for an indefinite period of time. The separation singularity should not be regarded as inconsequential, however, and its structure is of particular interest in understanding the new physics that must come into play as the interaction starts to evolve. Let t_s denote the finite time at which a singularity forms in the boundary-layer solution for any given unsteady boundary-layer flow. As indicated in Figure 4, the boundary layer upstream and down-stream of the separation (region I) bifurcates with a layer II remaining close to the wall and a layer IV growing explosively from the surface at a

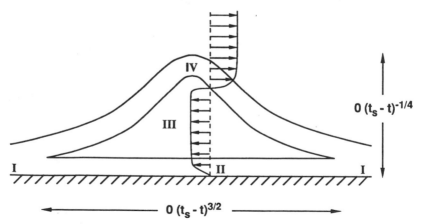

Figure 4 Schematic diagram of the structure of a boundary layer approaching separation (not to scale).

rate proportional to $(t_s - t)^{-1/4}$; note that the shear-layer structure in Figure 4 is substantially magnified in the tangential direction. The governing equations in the central zone III are inviscid and nonlinear and control the dynamics of the separation; they describe a motion with almost zero vorticity which is thinning in the tangential direction as $(t_s - t)^{3/2}$. In physical terms, a fluid particle is being sharply compressed in the streamwise direction, and because volume must be conserved in an incompressible fluid, the particle must become very elongated in a direction normal to the surface. Across the upper shear layer IV, significant changes in velocity and vorticity occur, and as the process focuses into a spike, the shear layer is driven rapidly outward toward the inviscid region.

The revelations of the work of Van Dommelen & Shen (1980, 1982) and Elliott et al (1983) that a "spike-like" response was not only a proper boundary-layer behavior, but also the expected response in high Reynolds number flows, have wide ranging implications for a number of flow applications (particularly boundary-layer turbulence and transition), some of which will be discussed in this article. For vortex-induced separation, the initiation of displacement thickness spikes had been reported by Doligalski & Walker (1978, 1984), Ersoy & Walker (1985, 1986), and Walker et al (1987) in computational studies of the boundary layers induced by a convected rectilinear vortex, counter-rotating vortex pairs, and vortex rings, respectively, using an Eulerian formulation. However, as discussed by Van Dommelen & Shen (1980, 1982), it proves to be essentially impossible to accurately resolve the phenomena involved using conventional methods based on the Eulerian description of the fluid motion; this is because the eruption takes place in a progressively narrowing zone as the interaction develops, and ultimately cannot be resolved using a fixed spatial mesh. In principle, some type of time-dependent, adaptive mesh could be considered; however, because separation develops abruptly and at locations that are not predictable a priori, the construction of an appropriate Eulerian algorithm is very challenging. Van Dommelen & Shen (1980, 1982) adopted an entirely new approach and reformulated the cylinder problem in Lagrangian coordinates, wherein the trajectories of many fluid particles are evaluated as functions of their initial position and time. This formulation is especially suitable for separating boundary layers, since a multitude of fluid particles are convected into the eruptive zone, which is therefore well resolved. At present, Lagrangian methods are the only schemes that have successfully computed unsteady boundary-layer development all the way to the singular terminal boundary-layer state (Cowley et al 1991, Peridier et al 1991a). Figure 5 (Peridier et al 1991a) shows computed results for the temporal development of displacement thickness induced by a rectilinear vortex above an infinite plane wall in an otherwise stagnant fluid; here a sharply focused eruption was found to develop abruptly, along with a spike having an internal shear-layer structure similar to that shown in Figure 4.

The terminal singular state that develops in the boundary-layer solution is an indication that new physics must come into play just prior to the realization of the singularity. Elliott et al (1983) address this issue and formulate a problem which will be referred to as the *first interactive stage* (see also Peridier et al 1991a). It was demonstrated that when $t_s - t = O(Re^{-2/11})$, the external flow starts to respond to the developing eruption, giving rise to pressure changes that depend on a Cauchy integral involving the developing boundary-layer displacement thickness. Interacting boundary-layer formulations (IBL) have historically provided accurate representations of steady flows at high Reynolds numbers (Smith 1982), but usually are based on the premise that changes in the external flow field produced by the thickening boundary layer are relatively small. In unsteady separation, and especially in vortex-induced separation, experiments show that changes induced in the external motion are substantial with thin, intense spires that penetrate an $O(1)$ distance from the wall before rolling up into new vortex structures (Walker et al 1987, Smith et al 1991b). Consequently, conventional IBL methods can be expected to describe the evolution of the eruption for only a limited period of time. Indeed, Smith (1988) has shown that a new interactive singularity occurs

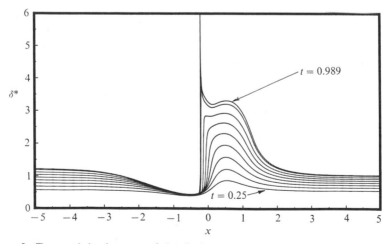

Figure 5 Temporal development of the displacement thickness for vortex induced separation (Peridier et al 1991a); plotted curves are at $t = 0.25$ to 0.95 at 0.10 intervals and $t_s = 0.989$; the vortex is of positive rotation above the boundary layer with center at $x = 0$.

in the IBL formulations. Recent computational results for vortex-induced separation at finite, but large, values of Re closely confirm the singular structure of Smith (1988) (Peridier et al 1991b). It may be noted, however, that the effects of interaction do not alter the focusing characteristics of the eruption, and a three-region structure similar to that shown in Figure 4 still develops (Smith 1988). During this interactive phase, new features develop; both the wall shear stress and the pressure field exhibit sharp streamwise variations in the vicinity of the eruption (Peridier et al 1991b). As discussed by Hoyle et al (1991), normal pressure gradients start to influence the process next, and a subsequent stage involving vortex formation is suggested.

It is evident that the eruptive process involves a number of complex stages and flow physics, and that the mathematical description at large Reynolds numbers is complicated. However, continuing work should eventually result in the development of procedures that can compute the evolution of strong unsteady separations culminating in the ejection of vorticity an $O(1)$ distance from the wall. At the same time, the eruptive process has been observed in a number of carefully constructed experiments of vortex-induced separation (Walker et al 1987, Smith et al 1991b, Haidari & Smith 1990, Greco 1991). Because vortices generally are in motion, and because the induced eruptions tend to be violent and abrupt, considerable care must be exercised in constructing the experiment and in performing the flow visualization at the right time and place.

2.3   Vortex-Induced Separation

A canonical problem that illustrates many of the features of vortex-induced separation is a rectilinear vortex above an infinite plane wall in an otherwise stagnant fluid. Inviscid theory predicts that such a vortex will move with constant speed and at a constant height from the wall under the influence of its image vortex below the plate (Walker 1978). However, a secondary eddy soon develops in the boundary layer near the wall as a consequence of the adverse pressure gradient induced by the vortex. Instantaneous streamlines in the boundary layer at a dimensionless time of 0.75 after the initiation of the motion are shown in Figure 6 in a frame of reference that moves uniformly with the vortex (Peridier et al 1991a); note that the center of the primary vortex above the boundary layer is at $x = 0$ and its rotation is counterclockwise. It is evident that the secondary eddy contains fluid recirculating in the clockwise direction and is not attached to the surface; however, when viewed from a laboratory frame of reference, the onset of recirculation appears as an eddy attached to the surface which subsequently appears to move along the wall (Harvey & Perry 1971, Walker 1978). It may be noted that the events of interest develop and take place

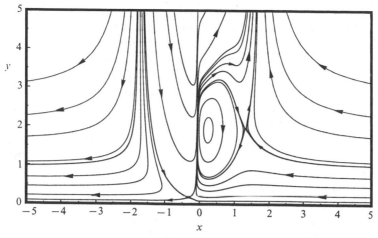

Figure 6 Instantaneous streamlines in the boundary layer at $t = 0.75$ showing a secondary eddy (Peridier et al 1991a).

in a frame of reference moving with the vortex, and are most easily understood from this perspective. As described in Section 2.2, the evolution of recirculation in the surface flow heralds the onset of a subsequent focused eruption and separation of the boundary layer as shown by Peridier et al (1991a,b). A hint of this may be seen in Figure 6 where the streamlines are beginning to congregate on the left side of the secondary eddy; the focusing continues with time leading to the formation of the displacement thickness "spike" shown in Figure 5. This problem has direct relevance to the behavior of aircraft trailing vortices near a ground plane and has been studied experimentally by Harvey & Perry (1971); these authors were among the first to observe that the boundary layer on the ground responds through a focused eruption that culminates in the ejection of a secondary vortex far from the surface. Similar phenomena are observed when a vortex ring approaches a solid wall in an otherwise stagnant fluid (Cerra & Smith 1983, Didden & Ho 1985, Ersoy & Walker 1987, Walker et al 1987, Chu & Falco 1988, Lim 1989); again an unsteady separation of the surface layer leads to the ejection of secondary vortex rings, as well as subsequent tertiary rings in some cases. These ejected vortex structures are observed to: 1. have a strength comparable to the primary vortex and 2. interact in an inviscid manner with the primary vortex so as to significantly alter its motion. The entire process documented in these studies shows clearly how one vortex can provoke the creation of another, through an unsteady viscous-inviscid interaction with the near-wall flow; it is highly suggestive of how regeneration of new vorticity occurs in turbulent and transitional flows near walls.

The inviscid flow associated with a second canonical problem—namely a rectilinear vortex convecting at a fractional rate in a uniform flow above a wall—has been described in Section 2.1. Numerical solutions for the boundary-layer flow induced by such vortex motion have been obtained by Doligalski & Walker (1984) who show that a recirculation zone develops in the region $x < 0$ (to the left of the vortex center) and that a spike in displacement thickness starts to form for situations where $\alpha < 0.75$. The widest recirculation zone is realized for $\alpha \to 0$, and as $\alpha \to 0.75$, the tangential dimensions of the recirculation zone diminish. Recently, A. T. Degani (1993, private communication) has recalculated the boundary-layer development for this problem in Lagrangian coordinates. He finds that as α increases toward 0.75, the location of the separation x_s moves progressively toward the stagnation point $x = -3^{1/2}$ and the strength of the eruption, as measured by the velocity jump across the upper shear layer shown in Figure 4, decreases. In a frame of reference moving with the vortex, the wall moves to the left with increasing speed β (cf Equation 7) as α increases and, finally, for a critical value β_c, with α slightly less than 0.75, the influence of the moving wall appears to dominate the boundary-layer development and a "spiky" response does not occur. An essentially similar suppression of separation by a moving wall has also been found by Degani & Walker (1992) in their study of the boundary layer on an impulsively started translating and rotating cylinder; here the dimensionless mainstream velocity is given by $U_\infty = -2 \sin x$, where x measures distance from the rear stagnation point. If the dimensionless speed of the wall is again denoted by β, Degani & Walker (1992) find that a sharp boundary-layer eruption occurs for $\beta \geqslant 0$, but is eventually suppressed as β approaches 2 from below. Both sets of results indicate that once the wall speed is sufficiently high, the evolution of recirculating flow is suppressed, and consequently the sharp eruptive response of the boundary layer is abated. For the rectilinear vortex, when $\alpha > 0.75$, the boundary-layer response in the region of adverse pressure gradient now consists of a much more gradual, but significant growth. Eventually, a rollover of the thickening boundary layer into another vortex structure is also expected in this situation, and this has been confirmed in the experiments reported by Doligalski et al (1980) for a parent vortex having a fractional convection rate of $\alpha \approx 0.95$.

Vortex-induced eruptions have been observed in a variety of other circumstances. It is known that an important aspect of noise generation and unsteady loading due to flow-induced vibrations is the interaction of vortices with solid surfaces; examples include wake-body interactions caused by shedding from upstream tubes in heat exchangers, the rotating blades in turbomachinery that pass through the wakes of stationary blade rows, and vortices shed from various features of aircraft that subsequently move over the airframe surface. Ziada & Rockwell (1982) and Kaykayoglu & Rockwell (1986) have studied vortices that originate in a mixing layer and subsequently convect over the surface of a wedge; a similar configuration was considered by Lucas & Rockwell (1984), where the incident vortices develop from a self-excited jet flow. In all situations, the development of surface-layer eruptions and the formation of secondary vortices was observed, even though the vortices in some configurations are relatively weak, with the vortex Reynolds numbers being low to moderate. Interactions of vortices with cavities (Tang & Rockwell 1983), with elliptic leading edges (Gursul & Rockwell 1990), and with various obstacles (Homa et al 1988) all show similar processes of secondary vortex generation. Indeed as Homa et al (1988) discuss, the process of development of secondary vortices is relatively independent of body shape and generally produces secondary vortices of comparable circulation to the incident primary vortices.

In recent years there have been a number of efforts to calculate the vortex-induced interactions through numerical solutions of the full Navier-Stokes equations (Peace & Riley 1983, Orlandi 1990, Orlandi & Jimenez 1991). These simulations have been carried out in the low to moderate Reynolds number range and become progressively difficult to perform with good accuracy as the Reynolds number increases; they do, however, show eruptions of the surface layer and formation of secondary vortices in a manner similar to that seen in experiments. At the same time, it should be remarked that there can be little expectation that standard Eulerian simulations, based on the full Navier-Stokes equations, can be extended to the high Reynolds number regime commonly encountered in practical flow applications. Theoretical considerations (as outlined in Section 2.2) indicate that new innovative algorithms (possibly of the Lagrangian type) will be required to describe, in a sensible way, the sharply-focused eruptive phenomena characteristic of the high Reynolds number regime.

2.4   *Interactions in Dynamic Stall*

Dynamic stall is a term describing a sequence of events that develop in the flow field near an airfoil undergoing a pitching motion, thereby increasing the effective angle of attack of the oncoming stream. In this process, a vortex forms and resides above the airfoil for a brief period. This primary stall vortex appears to be "connected" in some sense with the moving airfoil, and during this portion of the cycle dramatic increases in lift are observed. This has led some authors (e.g. Francis & Keesee 1985) to suggest that innovative use of unsteady effects could lead to more maneuverable aircraft. The penalty associated with the primary stall vortex is that it soon detaches and convects into the wake, leading to full stall; this event is characterized by an abrupt decrease in lift and a strong pitching moment— both of which can have serious consequences.

Dynamic stall has been extensively investigated because of its historical importance in helicopter applications (see, for example, Carr 1988). In forward flight, the rotorcraft blades that are translating in the direction of flight (the advancing side) typically experience a relatively high-speed external flow at a small angle of attack. By contrast the blades on the retreating side (those moving opposite to the direction of flight) see a relatively lower-speed external mainstream. Consequently, to achieve lift comparable to the advancing blades, the retreating blades are usually pitched to a higher angle of attack. This strategy is generally effective, but can be disastrous if dynamic stall occurs. A large body of literature on dynamic stall exists (see the reviews by McCroskey 1982 and Carr 1988), particularly in relation to rotorcraft applications; the phenomenon also occurs on compressor blades in turbomachinery and on wind turbines used to generate electricity. Unfortunately, despite a wealth of detailed experimentation, many aspects of the basic physics are far from understood. As Crighton (1985) remarks, "There are many more reports of experiments on oscillating airfoils . . . no consistent pattern emerges; indeed it rapidly becomes apparent that experiments done in this area with whatever facilities happen to be available and without an underlying theoretical framework are bound to lead to the present totally unsatisfactory situation."

Relatively recent advances in theoretical understanding of unsteady viscous flows has led to a clearer appreciation of the dynamics of such stall events. A complete review of dynamic stall will not be undertaken here; rather the focus will be on two key aspects: 1. the process that leads to the formation of the primary dynamic stall vortex and 2. the mechanism whereby the vortex finally breaks contact with the airfoil. Both issues are important. In some situations, such as helicopters, it is desirable to suppress the initial formation of the vortex altogether. On the other hand, for a modern agile aircraft undergoing a maneuver, the presence of the dynamic stall vortex is potentially beneficial since relatively high levels of lift are possible (Francis & Keesee 1985, Helin 1990). The challenge here is to determine a means to maintain contact between the vortex and the airfoil during the entire maneuver. Rational control methods for influencing the unsteady processes on the upper airfoil surface can only be developed once the sequence of events leading to dynamic stall is well understood.

The Reynolds number associated with a pitching airfoil in a stream of speed U_∞ is usually defined as $Re_c = U_\infty c/v$, where c is the chord length. Early flow visualization experiments at low Reynolds numbers revealed that as the angle of attack is increased from some initial value (usually zero), a thin region of reversed flow develops near the trailing edge and subsequently spreads toward the leading edge of the airfoil. Although this expanding zone of reverse flow was initially believed to be significant, subsequent experiments and numerical simulations showed that the critical events at higher Reynolds numbers take place in the vicinity of the leading edge. To a certain extent this is to be expected, since the boundary layer on the downstream side of the leading edge is exposed to an increasingly severe adverse pressure gradient as the airfoil pitches up. It might well be anticipated that as the angle of attack increases, the boundary-layer flow will eventually be unable to negotiate a path along the contour of the airfoil surface and must interact strongly with the external flow.

In recent years, a number of numerical simulations of the flow produced by a pitching airfoil have been carried out for moderate Reynolds numbers ranging from $Re_c = 5000$ (Shih et al 1992) to $Re_c = 45,000$ (Ghia et al 1992a 1992b). For the most part, these simulations exhibit phenomena similar to those observed in recent experimental studies of dynamic stall (Metwally 1990, Shih et al 1992, Gendrich et al 1993), wherein a primary stall vortex is shed from the leading edge region and subsequently convects over the airfoil surface. The simulations become increasingly difficult to perform with good accuracy as the Reynolds number Re_c increases; the most thoroughly documented case is for $Re_c \simeq 10^4$ (Visbal & Shang 1989, Osswald et al 1991, Ghia et al 1992a, Gendrich et al 1993, Knight & Ghosh Choudhuri 1993), which is much lower than typical Reynolds numbers encountered in actual applications of dynamic stall. Nevertheless, the flow development is of interest, and in Figure 7, some computed results of Knight & Ghosh Choudhuri (1993) are shown near the leading edge of a NACA 0012 airfoil, which is in the process of being smoothly pitched up from an angle of $\alpha = 0°$. At an angle of attack of $\alpha = 19.5°$ (Figure 7a), notice that a region of recirculation has developed on the upper surface in the form of a closed recirculating eddy which is not attached to the wall. The streamline patterns are plotted in a frame of reference that is moving with the airfoil and exhibit a behavior similar to vortex-induced separation (Walker 1978). In flows with moving walls, the development of such topologies is common (Doligalski & Walker 1984, Ece et al 1984), and this generally heralds the eventual development of a sharp eruption and boundary-layer separation if the Reynolds number is sufficiently high. However, for the moderate Reynolds number of Figure 7, a sharp response does not occur, and the recirculating eddy gradually grows in a direction normal to the wall as the airfoil continues to pitch. It should be noted that the observed streamline patterns depend upon the frame of reference of the observer. For example, in the case of vortex-induced separation, the onset of recirculation appears first as an eddy detached from the wall, when viewed from a frame of reference moving with the vortex (Walker 1978); however, when viewed from the laboratory frame, reversed flow appears to develop as a separation bubble that is attached to the wall. An analogous behavior applies to Figure 7, and when viewed from the laboratory frame, the onset of recirculation appears to be an eddy attached to the surface (D. R. Knight, private communication). Such bubbles are frequently mentioned in descriptions of dynamic stall (see, for example, Currier & Fung 1992, Chandrasekhara et al 1993).

For moderate Reynolds number, the first appearance of a recirculating eddy seems to represent the birth of a dynamic stall vortex; this growing primary vortex (shown in Figure 7a) gives rise to a smooth interaction in which the external flow is gradually displaced. At a later stage, shown in Figure 7b, the primary vortex induces a second and more significant interaction; this occurs in the flow near the airfoil surface through the development of a secondary eddy. It may be noted that by this stage, the principal influence on the viscous flow between the primary vortex and the airfoil surface is the vortex-induced adverse pressure gradient impressed to the left of the stall-vortex center; at this point of development, the fluid near the surface is effectively shielded by the primary vortex from other unsteady influences associated with the pitching airfoil. The evolution of this secondary eddy is essentially similar to that induced by a vortex above a plane wall (Walker 1978, Peridier et al 1991a). The next phase in the dynamic stall is a compression of the vorticity field near the right side of the secondary eddy and a violent narrow-band eruption of the surface flow. This event is a strong, unsteady viscous-inviscid interaction in which the secondary eddy is torn from the surface and ejected into the mainstream. As the erupting plume is deflected downstream, it wraps over the top of the primary vortex which then "breaks contact" or detaches from the airfoil surface (Ghia et al 1992a). This event is an unsteady separation phenomenon in the sense described in Section 2.2 and is important because detachment of the primary stall vortex initiates airfoil stall and degradation of lift.

In actual applications involving dynamic stall, the Reynolds numbers are usually significantly higher than 10^4 and, consequently, a somewhat different evolution near the leading edge is expected from that indicated in Figure 7. As the airfoil pitches up at high Re_c, a detached eddy, of the type shown in Figure 7a in a frame of reference moving with the airfoil, will develop initially near the leading edge, and with this event, a line of zero vorticity is present in the flow field. At this stage two necessary

$\alpha = 19.5°$

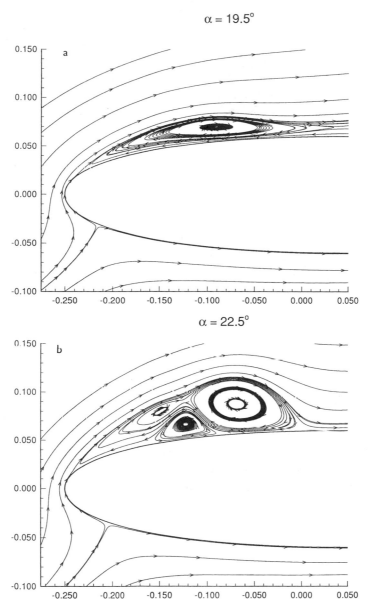

$\alpha = 22.5°$

Figure 7   Instantaneous streamlines near the leading edge of an airfoil pitching at constant rate (Knight & Ghosh Choudhuri 1993).

conditions for a narrow-band eruption of the surface layer are satisfied, namely: 1. a zero vorticity line and 2. a local zone of adverse pressure gradient. Consequently, once Re_c is sufficiently large, a "spike-like" boundary-layer response can be expected on the upstream side of the recirculation zone in much the same manner as described by Peridier et al (1991a,b). Figure 8 shows a schematic representation of the various stages of such a process. As discussed in Section 2.2, this type of response is generic in high Reynolds number flows and gives rise to a thin plume containing significant levels of vorticity, which rapidly leaves the surface as shown in Figure 8c. This process may be conceptualized as a window of narrow streamwise extent opening at the top of the boundary layer and allowing concentrated vorticity to spill out into the external flow. As the plume moves farther away from the surface, a deflection in the downstream direction is expected with an ultimate roll-up into a primary dynamic stall vortex, as shown in Figure 8d. The birth of this process is very difficult to discern in experiments at high Reynolds number since the boundary layer is very thin and the triggering events for the process shown in Figure 8 (particularly Figures 8b and c) occur at small spatial scales deep within the boundary layer. However, various key aspects of the developing interaction can be observed. Acharya & Metwally (1992) have inferred instantaneous distributions of surface vorticity flux from detailed measurements of the surface pressure distribution on a pitching airfoil for $Re_c = O(10^5)$. The results show a "spike" in vorticity flux near the leading edge which apparently is solely responsible for feeding the formation of the dynamic stall vortex (cf Figures 8c and d). Eventually a second spike in surface flux is found as the primary stall vortex provokes separation of the surface layer and the interaction shown in Figure 8e, which Acharya & Metwally (1992) argue acts to "cut off the dynamic-stall vortex from its source of vorticity." It may be noted that sharp variations in the pressure gradient are known to be characteristic of a separating boundary layer (Smith 1988, Conlisk 1989, Peridier et al 1991b). Further evidence for the physical process sketched in Figure 8 may be found in recent visualization experiments reported by Metwally (1990), Karim (1992), and Acharya et al (1993) in the range Re = 30,000–120,000; these clearly show a sharp plume that originates near the leading edge and rolls up into the primary stall vortex. Some representative smoke visualization results for $Re_c = 30,000$, with the airfoil pitching at a constant rate are shown in Figure 9. Here the visualization has been carried out using smoke wires near the surface as well as in advance of the airfoil; the instantaneous angle of attack is shown for each figure. In Figures 9a and b, a thin spire of fluid that originates near the leading edge is visualized. The subsequent roll-up of this spire into a primary stall vortex may be seen in Figure 9c. Consequently, at

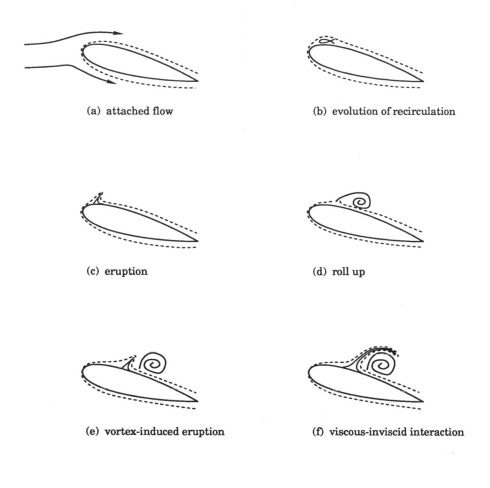

(a) attached flow

(b) evolution of recirculation

(c) eruption

(d) roll up

(e) vortex-induced eruption

(f) viscous-inviscid interaction

(g) stall

Figure 8   Schematic of stages of dynamic stall at high Reynolds number.

sufficiently high Re_c, the birth of the primary stall vortex is rather different from that shown in Figure 7 for moderate Reynolds numbers; at high Re_c, initiation of the process occurs through the roll-up of a shear layer that originates from a spike-like eruption (or boundary-layer separation) that has its roots in the onset of recirculation near the leading edge. The conjectured sequence of events bears a similarity to experimental descriptions of "bursting" of a leading edge separation bubble (Currier & Fung 1992, Chandrasekhara et al 1993). Once the primary stall vortex has formed, the process of detachment (Figures 8e,f) is the same as at moderate Reynolds numbers. The primary vortex provokes separation of the surface layer which leads to a strong viscous-inviscid interaction wherein an erupting plume from the surface wraps around the primary stall vortex, which then quickly detaches from the airfoil surface as shown in Figure 8g (see also Figure 9d).

3.   THREE-DIMENSIONAL INTERACTIONS

3.1   *Introduction*

Three-dimensional vortex/surface interactions are common in a variety of situations, but are less well understood than in two-dimensions, principally because the interactions involve complex, evolving flow topologies in three dimensions (Ersoy & Walker 1987, Pedrizzetti 1992). However, a common underlying thread is that three-dimensional vortices in proximity to a solid wall expose the near-wall flow to persistent zones of adverse pressure gradient. Again, the central interest here is in the high Reynolds number regime, and it should be expected that eruptive responses will again be provoked in the surface flow. A recent theoretical study by Van Dommelen & Cowley (1990) (see also Cowley et al 1991) suggests that a generic surface response occurs in three dimensions, wherein boundary-layer separation takes place along a crescent-shaped ridge, and gives rise to a sharply focused tongue of vorticity leaving the surface (as opposed to a "knife-edge" in two dimensions). However, the structure of the erupting ridge is similar to the two-dimensional case in any plane normal to both the wall and the contour of the base of the ridge (Van Dommelen & Cowley 1990). Recent experimental studies show the evolution of these eruptive tongues in a variety of circumstances, and that a tendency for roll-up of these tongues into new vortex structures is common. In the following sections, recent work on vortex interactions in three-dimensional flows will be described.

3.2   *End-Wall Vortices*

Whenever a boundary-layer flow along a surface encounters an obstruction on the surface (such as a cylinder or wing), a complex, three-dimensional

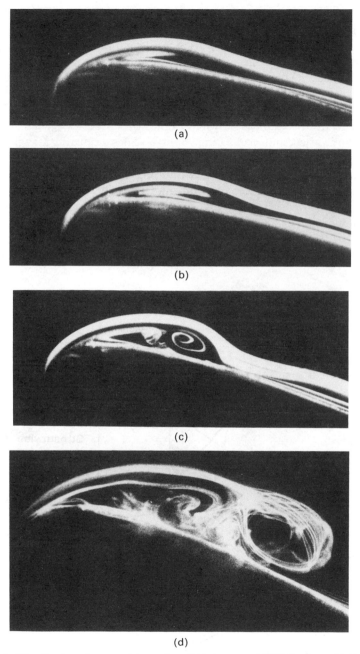

Figure 9   Visualization of stages of dynamic stall (Acharya et al 1993) for $Re_c = 30,000$ with the airfoil pitching at constant rate. Angle of attack (*a*) 23°, (*b*) 24°, (*c*) 26°, (*d*) 30°.

flow field develops in the vicinity of the obstacle. The approach flow encounters a strong adverse streamwise pressure gradient, which coupled with the cross-stream pressure gradients generated by curvature of the external flow around the obstruction, results in a concentration of the boundary-layer vorticity into a system of discrete vortices near the end-wall juncture; extensions of these vortices engirdle the obstruction to form "necklace" or "horseshoe" vortices, with the "legs" of the vortices extending downstream, as shown in Figure 10. Examples of such flows are many and varied, and span a range of flow scales, occurring at: wing/body junctions on aircraft; conning-tower and control-surface junctions on submarines and ships; base flows near buildings, pilings, and support poles; and the mounting surfaces of electronic components, such as computer chips.

Juncture flows have been extensively examined for both laminar and turbulent flows (Baker 1979, Thomas 1987, Greco 1991), and necklace vortex behavior can be categorized into several regimes of steady and

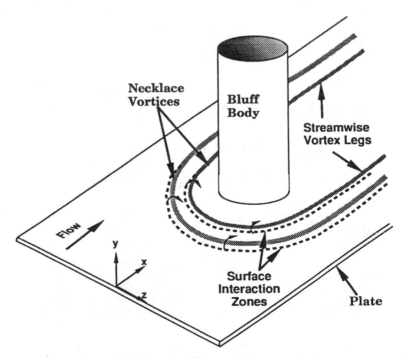

Figure 10   Schematic of flow structure near a juncture.

unsteady behavior that are often described in terms of Reynolds numbers based on a length characteristic of the obstacle and/or the approach boundary layer (Greco 1991). However, the details of the surface interaction induced by necklace vortices have only been examined recently, and these studies (Greco 1991, Smith et al 1991a) show that strong surface interactions develop both upstream of the obstacle and in proximity to the streamwise legs. It may be noted that although the obstacle shown in Figure 10 is a circular cylinder, essentially the same type of interactions are observed for a wide variety of shapes such as airfoils, turbine blades, and even rectangular block obstacles (Seal et al 1993). Recent experimental and computational studies (Greco 1991; Visbal 1991a,b; Seal et al 1993) of vortex interactions have shown the generation and concentration of vorticity, of opposite sign to the original (or primary) necklace vortices, in the end-wall surface flow beneath a necklace vortex. The generated vorticity is in the form of a secondary vortex similar to that observed in vortex-induced separation (Peridier et al 1991a,b) and dynamic stall (see Section 2.4). For low Reynolds numbers, the secondary vortices are laminar and generally observed to be stable; in such cases, a secondary vortex develops for each primary necklace vortex and a steady vortex pattern of two, four, or six alternating sign vortices occurs, with the number of vortices growing with increasing plate Reynolds number (Baker 1979; Visbal 1991a,b).

As the plate Reynolds number increases beyond a critical value, the necklace vortex pattern becomes inherently unsteady. The primary necklace vortices are clearly observed to form in the region of adverse pressure gradient upstream of the obstacle, and then apparently provoke a focused eruption of the surface flow. By this process, the primary vortex is "released" and convected downstream toward the obstruction, in a manner reminiscent of the breakaway of the dynamic stall vortex from an airfoil described in Section 2.4. Each necklace vortex interacts with the surface flow as it moves toward the obstruction, generating opposite sign vorticity, which quickly focuses into a narrow region of vorticity that subsequently ejects from the wall, wrapping around and engirdling the primary vortex. Figure 11 shows instantaneous vorticity contours from a recent particle image velocimetry (PIV) study (Seal et al 1993), illustrating the ejected vorticity (solid contours) due to interactions of periodic necklace vortices (circular, dashed contours) in front of a rectangular body on the symmetry plane. The ejected vorticity cross-cancels with the vorticity of the primary vortex, apparently reducing the strength of the primary vortex (see also Visbal 1991b). This vorticity-generation/ejection process appears to be perpetuated as the primary vortex approaches the obstruction, until the primary vortex reaches a position where the induced image velocity apparently balances the local mean velocity, and the weakened primary vortex is effectively brought to rest.

The necklace vortices engirdle the obstruction, with the downstream trailing extensions of the vortices appearing as streamwise "legs," which move periodically inward toward the symmetry plane, as shown in Figure 10. As fluid particles near the wall are convected downstream, they experience an adverse cross-stream pressure gradient, and at successive downstream stations, the near-wall flow can be expected to undergo a spatial development similar to that observed by Harvey & Perry (1971) and similar to the temporal sequence described by Peridier et al (1991a) and Doligalski & Walker (1984). Indeed as Smith et al (1991a,b) show, the necklace vortex legs provoke a strong interaction, wherein sharply eruptive spires of surface fluid develop in proximity to the streamwise legs. Figure 12*a* shows a schematic of this vortex-induced eruptive behavior as viewed from behind the cylinder and looking upstream. In Figure 12*b* experimental end-view visualization of the interaction is shown (Smith et al 1991a); here the vortex cores have been marked by a hydrogen bubble wire aloft upstream, while the spires are marked by a bubble wire essentially on the surface. Note that the eruptive spires develop in a frame of reference moving with the vortex and are observed to move in conjunction with the necklace vortex as it washes inward toward the symmetry plane. Note also that a spire is just about to form and project well away from the surface near the less

FLOW ➡️

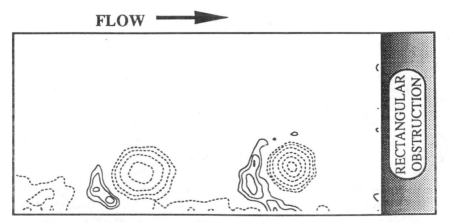

Figure 11 Instantaneous vorticity contours from PIV study of transient necklace vortices on symmetry plane of bluff body junction region; note ejection of positive vorticity (*solid contours*) from surface by negative vorticity (*dashed contours*) necklace vortices. (Seal et al 1993.)

mature vortex to the left in Figure 12b. When observed in plan view, these fluid spires appear as extended narrow streamwise regions of low-speed fluid, reminiscent of low-speed streaks observed in the wall region of turbulent boundary layers (Smith et al 1991b). As the eruptive ridge of fluid penetrates the outer higher-velocity flow, it is observed to rollover, forming a series of hairpin vortices, which interact with the streamwise vortex leg and contribute to a spreading and dispersion of the coherency of the leg. A second hydrogen bubble visualization, shown in Figure 12c is somewhat downstream of that in Figure 12b and shows the subsequent rollover of the eruptive spire into a hairpin vortex.

As the Reynolds number is progressively increased, a destabilization of the necklace vortices sets in, wherein azimuthal waviness or "kinks" occur along the vortex in a manner strikingly similar to the azimuthal undulations noted for destabilized vortex rings (Widnall & Sullivan 1973, Walker et al 1987). As a consequence of the shear near the surface, these kinks are accentuated by Biot-Savart effects (in a process similar to that described by Smith et al 1991b); this causes portions of a distorted necklace vortex to make a close approach to the surface, thereby hastening a surface interaction. The end result is a localized, three-dimensional interaction which culminates in the rapid focusing of surface vorticity and the ejection of thin tongues of surface fluid in a response analogous to that predicted by Van Dommelen & Cowley (1990). These thin tongues of fluid rapidly rollover into new hairpin vortices, which appear to amalgamate with the parent necklace vortices to produce a complex transitional-type flow. It should be noted that the end-wall boundary layer passes into transition at a stage when the upstream boundary layer is still laminar. Consequently, the mechanisms for transition near the obstacle appear entirely associated with the necklace vortices and are significantly different from the usual transitional routes observed in flat-plate transition.

With further increases in the Reynolds number, the end-wall and the approach boundary-layer flows become fully turbulent, which modifies the character of the vortex formation significantly, resulting in the development of a dominant single horseshoe vortex, which is rather larger and more chaotic than the laminar necklace vortices. Essentially all previous measurements of the characteristics of such turbulent vortices have been either time-mean or single point measurements and suggest the existence of a large steady horseshoe vortex wrapped around the obstacle. In recent times, temporal measurements have indicated that the horseshoe vortex behaves in a quasi-periodic unsteady manner (Devenport & Simpson 1990, Fleming et al 1992). Recent visualization and particle image velocimetry results (Seal et al 1993) in a low Reynolds number flow suggest that the physics of the turbulent vortex behavior is intimately connected with the

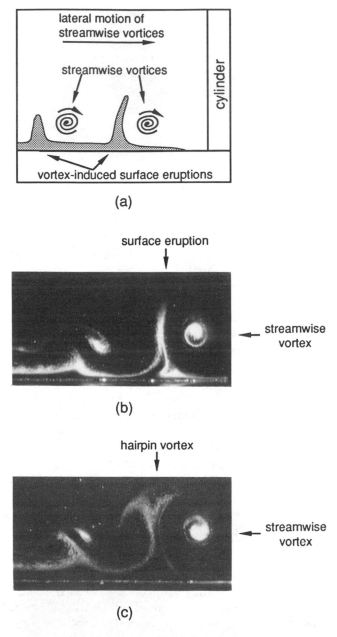

(a)

(b)

(c)

Figure 12 End-view of necklace vortex legs and induced surface eruptions. (*a*) Schematic of interaction. (*b*) Visualization of initial eruptions. (*c*) Visualization of breakdown of eruptive spire to a hairpin vortex. (Smith et al 1991a.)

surface interaction phenomena discussed in this paper. In general, the turbulent horseshoe vortex is observed to induce the generation of secondary vortices, which grow and are subsequently ejected from the surface in a strong, narrow-band eruption of surface fluid, essentially consistent with the process of vortex-induced separation discussed in Section 2.3. An example of this process is illustrated in Figure 13, which consists of a series of four particle image symmetry-plane visualizations of the interaction of the horseshoe vortex in the junction region of a rectangular body. The visualizations were done in water using 12 μm particles and a rapid-scanning laser beam, scanning from below the approach surface and along the symmetry plane. Note that the flow is left-to-right with the block just

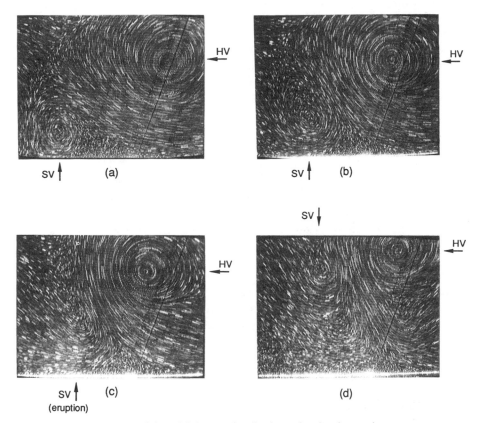

Figure 13   Four sequential particle image visualizations of surface interaction on symmetry plane of a turbulent horseshoe vortex in the junction region of a rectangular body. Flow is left-to-right with body just to the right of the field-of-view. $Re_0 = 700$; each picture taken 0.25 seconds apart. (Seal et al 1993.)

HV $\Rightarrow$ horseshoe vortex; SV $\Rightarrow$ secondary vortex.

out of the image to the right; the lower wall appears as the bright line at the bottom of the images. The horseshoe vortex appears at the upper right of each frame, moving temporally within the image due to interaction with the secondary vortices, similar to the movement observed in the ring vortex study of Walker et al (1987). In Figure 13a, note the development of a secondary vortex of opposite sign vorticity in the lower left (upstream) corner of the image, adjacent to the plate. With time, this vortex begins to be compressed (Figure 13b), followed by a sudden rapid compression and ejection from the surface (Figure 13c), with a reappearance of the vestiges of the secondary vortex aloft (as shown in Figure 13d in the upper part of the image just to left of center). Although it is difficult to appreciate from the static images, the video sequence (from which the photographs in Figure 13 were taken) clearly illustrates the focusing and eruption of a narrow region of vorticity-bearing fluid (Figures 13c,d), consistent with the unsteady separation processes discussed in Sections 2.2 and 2.3. Subsequent interaction of the ejected flow, with both the horseshoe vortex and the freestream, gives rise to complicated, three-dimensional behavior which is speculated to be part of the reason for the observed quasi-periodic motion detected by Devenport & Simpson (1990). The dynamics of such juncture flows are not revealed by mean measurements, and it is clear that much more work will be required to appreciate the turbulent dynamics and structure of these important flows. However, it is evident at this stage that the mechanisms of turbulence production in the end-wall region are significantly different than on a flat surface (Smith et al 1991b), and this brings into question the viability of trying to tailor conventional turbulence models for such regions.

3.3   *Interactions Associated with Rotorcraft*

As discussed by Mabey (1989) and Cunningham (1989), unsteady three-dimensional vortical flows are known to be an important feature of the flow fields associated with modern aircraft; unfortunately, the accurate numerical simulation of viscous flows at high Reynolds number is not possible at present. Some progress has been made in the moderate Reynolds number range (see, for example, Stanek & Visbal 1991); these calculations reveal complex three-dimensional vortex/surface interactions producing secondary vortices and ejection of vorticity from the surface. By and large such research is still in its infancy, and much remains to be learned about the various topologies and possible separation processes.

Another important occurrence of vortex/surface interactions is for rotorcraft, where all portions of a helicopter experience the influence of vortex motion during various phases of flight (Sheridan & Smith 1980). It may be noted that two types of vortices are shed by the rotors, namely: 1.

transverse vortices which can subsequently interact with the following rotor blades, potentially producing almost two-dimensional interactions of the type described in Section 2, and 2. tip vortices that trail from the end of the rotors. Because the rotor produces a down wash, the tip vortices descend and impinge on the fuselage. The interactions induced by these vortices are important, and have received considerable attention in recent years because they appear to be associated with high levels of noise, as well as with complex and sometimes unexpected unsteady forces on the helicopter. Brand et al (1989) and Liou et al (1990) report observations consistent with the evolution of secondary vortices near the airframe when the primary tip vortices approach the fuselage. Because the radius of curvature of the tip vortices is relatively large, the tip vortices that approach the fuselage are almost two-dimensional, and Affes et al (1992) and Affes & Conlisk (1993) model the start of the interaction with the fuselage using an initially straight vortex convected downward toward a circular cylinder; once near the cylinder, the vortex develops a three-dimensional distortion. Affes et al (1992) find good agreement with experimental measurements for the surface pressure distribution, and definitively show that large local pressure perturbations are directly associated with the approach of the tip vortex. Affes & Conlisk (1993) have considered the boundary-layer response; they show that a secondary eddy forms, which is concentrated near the top of the airframe as the tip vortex approaches and, further, that a strong separation process is initiated. These results have been confirmed in a recent experimental study by Affes et al (1993) in which a simplified two-blade rotor above a cylindrical fuselage was used. The tip vortices were marked with smoke, visualized using a laser sheet, and recorded on video. A frame from this video sequence is shown in Figure 14, which shows the tip vortex impacting on the cylinder at a location under the retreating rotor blades and where the tip vortex makes its closest approach to the cylinder. The clear spire shown near the surface is indicative of a process of surface-layer separation and the evolution of a secondary vortex. Affes et al (1993) also recorded instantaneous pressure measurements on the cylinder surface, and these are shown in Figure 15 at exactly the instant depicted in Figure 14. The pressure minimum to the right is a characteristic signal associated with the approach of the tip vortex to the airframe, while the sharp pressure variation to the left was closely identified with the separation process suggested in Figure 14. The development of sharp spatial variations or spikes in the surface pressure distribution has also been noted recently by Bi et al (1993) in their studies of rotor tip vortex interactions with a fuselage in low-speed forward flight. These experimental measurements (see also Acharya & Metwally 1992) seem to confirm the theoretical predictions, which show that significant intense

primary
vortex

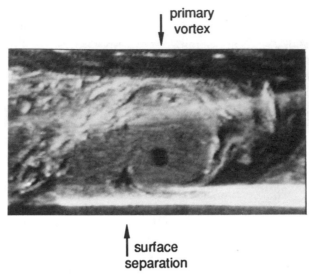

surface
separation

Figure 14   Visualization of rotor-tip vortex/fuselage interaction showing surface-layer separation (Affes et al 1993).

variations develop in the surface pressure as the boundary layer starts to separate and evolve toward an eruption (Conlisk 1989, Peridier et al 1991b). The observed processes appear to be very similar to the phenomena described in Sections 2 and 3.2. Furthermore, it is possible that the eruption on the top of the fuselage and the subsequent interaction between the primary tip vortex and the spire leads to the "destruction" of the primary vortex core conjectured by Brand et al (1989); during the interaction process the primary tip vortex appears to lose definition, such that it is difficult to observe identifiable vortices below the airframe.

3.4   Transition and Turbulence (*Hairpin Vortices*)

Hairpin-type vortices develop from the deformation of local vorticity layers in a bounded shear flow and commonly occur in a variety of situations, such as near small surface protrusions, through local fluid injection into a boundary layer, and in transitional and fully-turbulent boundary layers. Hairpin vortices are widely believed to be a basic flow structure of turbulent boundary layers (Falco 1991, Grass et al 1991, Robinson 1991, Smith et al 1991b). The structure originally described by Theodorsen (1952) is a symmetric vortex having two legs in the shape of a hairpin. However, the majority of naturally-occurring vortices are expected to be asymmetric (or one-legged); these have been described

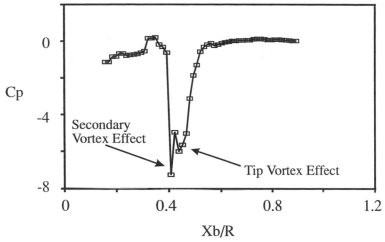

Figure 15   Surface pressure distribution induced by interaction shown in Figure 14 (Affes et al 1993).

by Robinson as "quasi-streamwise vortices," the terminology implying a vortex that has a significant component aligned in the streamwise direction. Regardless of the configuration, hairpin vortices interact with the viscous flow near the surface to generate eruptions of surface fluid and new hairpin vortices, thus providing regeneration and the introduction of new vorticity into the boundary layer.

In turbulent flows, hairpin vortices appear to form in a number of ways, resulting in multiple hairpin vortices which coalesce into complex tangles of vorticity which are difficult to behaviorally interrogate. Several recent experiments have been carried out to determine the basic dynamics of hairpin vortices in isolation from other background effects; these experiments examined both shedding from hemispherical protuberances (Acarlar & Smith 1987a), and controlled injection through the bounding surface (Acarlar & Smith 1987b, Haidari & Smith 1990, Smith et al 1991b) to selectively generate and examine the development and surface interaction of both single and periodically-shed streets of hairpin vortices. In the most detailed study to date, Haidari & Smith (1990) employed discrete surface injection to create a single hairpin vortex within an otherwise laminar boundary layer; using split-screen video viewing and analysis of detailed hydrogen bubble visualization, the development and interaction of these hairpin vortices were then closely observed and assessed as the vortices advected downstream. Although the interaction processes are complicated and take place rather abruptly, careful introduction of the visualization medium and positioning of the viewing perspective established that a hairpin vortex moving in proximity to a surface stimulates three-dimensional, eruptive events which give rise to the generation of secondary hairpin vortices. This generation process is illustrated schematically in Figure 16 (modified from Smith et al 1991b), which shows the basic sequence for the interaction induced by a symmetric hairpin vortex in motion above a wall. In the region between the vortex legs and behind the vortex head, the local pressure gradient is adverse in both the streamwise and lateral directions, and hence conditions are ripe for an eventual eruption of the surface layer in a high Reynolds number flow (Van Dommelen & Cowley 1990). For the symmetric hairpin vortex shown in Figure 16, separation of the surface layer is expected in three locations: behind the vortex head, where the adverse streamwise pressure gradient is augmented by the spanwise effects of the adjacent vortex legs, and where each trailing leg makes its closest approach to the surface. Only the separation of the surface layer near the lower leg is shown in Figure 16a. One noteworthy aspect associated with hairpin vortices is that the trailing portion of the hairpin leg moves continuously toward the wall as the vortex is convected in the shear flow (Hon & Walker 1991, Smith et al 1991b); this is important because the basic eruptive effect is enhanced the closer the vortex is to the surface (see Sections 2.1 and 2.3).

Although the shape and timing of the eruptions near the head and legs vary, the general process is the same, leading to the creation of another hairpin vortex in a process summarized briefly as follows. In a frame of reference moving with a section of the vortex, the hairpin-induced pressure gradient initiates separation of the surface layer along U-shaped ridges containing elevated vorticity levels, with the top of the U facing downstream (as indicated by the dotted lines in Figure 16a). The surface layer separates in a thin tongue or ridge, indicated by the shaded regions in Figure 16, and penetrates into regions of higher velocity; a roll-over into a subsequent vortex then starts as shown in Figure 16b, with the roll-over starting at the tip of the erupting tongue and spreading outward along the separating ridge. The roll-over rapidly culminates in the formation of the new secondary hairpin vortices shown in Figure 16c. This general pattern of surface interaction has been shown clearly by Haidari & Smith (1990), and the onset of this process is consistent with the general theory of Van Dommelen & Cowley (1990).

Upon generation of secondary hairpin vortices, the flow becomes much more complex due to the mutual interactions and amalgamation of the primary and secondary hairpin vortices. However, the same general processes of vortex/surface fluid interaction continue, providing a process of growth and expansion of the original three-dimensional vortex structure.

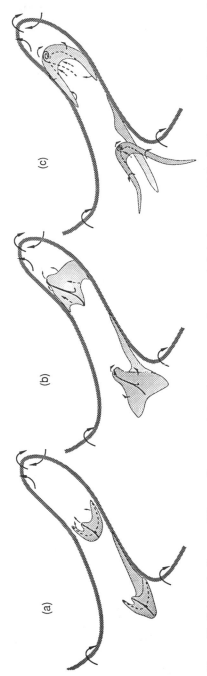

Figure 16 Schematic diagram of the eruptive process induced by a convected symmetric hairpin vortex. (*a*) Surface layer separation. (*b*) Onset of roll over. (*c*) Secondary hairpin creation.

Indeed, as shown by Haidari & Smith (1990), once the plate Reynolds number is sufficiently high, a turbulent spot eventually develops from a single hairpin vortex through the regenerative process described here. A similar process of vortex-induced hairpin or "horseshoe" formation also appears to have been observed within a turbulent boundary layer by Li & Lian (1992), even though their experiment was not done within a controlled flow environment; their work emphasizes that similar processes of hairpin-type vortex generation occur without the presence of vortex symmetry. The present process of generation applies generically for asymmetric hairpin vortices, with the exception that the bulk of the eruptive activity is expected near the trailing vortex legs. As discussed by Smith et al (1991b) in a detailed evaluation of direct simulations of turbulent channel flow, this process provides a dynamical explanation for what Robinson (1991) described as the evolution of "new vortical arches" adjacent to existing "quasi-streamwise vortices." It should be noted that while the secondary vortices are expected to be almost symmetric at birth, as shown in Figure 16c, the background environment in a fully turbulent flow is expected to rapidly distort such vortices into asymmetric vortices (as observed by Robinson 1991).

Essentially similar eruptive and regenerative phenomena are seen in bypass and end-stage transition; the former terminology implies transition provoked directly by mainstream disturbances (i.e. vortices), while the latter term implies the later stages of natural transition, once amplification of Tollmein-Schlichting waves has given rise to a process of hairpin vortex formation. Figure 17a shows a hydrogen bubble visualization in a water channel for a transitional boundary layer at $Re_x \approx 62,500$, tripped with a circular rod. In this visualization, a transverse hydrogen bubble wire was placed essentially on the surface, and the generated bubble sheet illuminated with a light sheet located just downstream and transverse to the flow direction; the visualization is seen in end view, looking upstream. In this highly magnified visualization, a thin eruptive spire is isolated by the light sheet, which is indicative of a rapid penetration of wall-layer fluid well into the boundary layer. Note that these types of events are the most commonly observed behavior in this transitional-type flow. Figure 17b is a smoke visualization from a video by Wallace et al (1990), taken in a low-speed wind tunnel of a tripped, low Reynolds number ($Re_\theta = 892$) turbulent flow. Here smoke was introduced through two upstream, transverse slots and illuminated with a transverse laser light sheet almost immediately downstream of the second slot. Similar to Figure 17a, Figure 17b reveals a series of thin, eruptive spires (the thin, bright, vertical projections) developing near the wall and projecting well into the boundary layer. The surface-layer eruptions shown in Figures 17a and 17b are believed to be slices through the eruptive tongues shown schematically in Figure 16 (Smith et al 1991b). It may be noted that these types of surface eruptions are the most dominant near-wall event observed in end-view visualizations of transitional-type flows, as shown here, as well as in fully turbulent boundary layers (Smith et al 1991b).

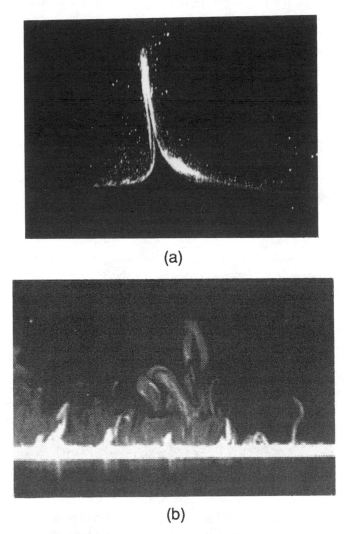

(a)

(b)

Figure 17   End-view visualizations of surface eruption in both a transitional and turbulent boundary layer. (*a*) Thin, eruptive spire adjacent to surface in a tripped, transitional boundary layer (hydrogen bubble visualization). (*b*) Series of eruptive surface spires in a turbulent boundary layer (surface smoke visualization) from Wallace et al (1990).

4. SUMMARY

The response of a viscous boundary layer to vortex motion has been described in this article, with an emphasis on flows at high Reynolds number. In general, a boundary-layer eruption occurs in such circumstances in the form of a thin spire containing significant levels of vorticity, which usually rolls up into one or more secondary vortices. This regenerative process of vortex-induced separation is evidently a common phenomenon occurring in a wide variety of applications, some of which have been described here. Because these eruptions develop and take place in moving reference frames and involve sharply focused eruptions that occur abruptly (at times and locations that are not easily predicted a priori), experimental and computational research on such phenomena is extremely challenging. It is evident that conventional experimental techniques, wherein measurements are taken at fixed points in space, are virtually useless in attempting to detect such phenomena. Progress can be made using combinations of innovative flow visualization and particle image velocimetry; however, a moving reference frame is often crucial to proper interpretation of vortex-induced events. It is also worth noting that a substantial increase in resolution may be required in current PIV methods in order to capture the abrupt narrow-band eruptions.

From a computational point of view, it seems likely that new and innovative numerical schemes (possibly of the Lagrangian type) will be required in order to effectively penetrate the unsteady, high Reynolds number regime, where sharply focused, abrupt eruptions of the boundary layer are prevalent. It is highly unlikely that current methods based on solutions of the full Navier-Stokes equations, which employ either fixed spatial meshes or spectral schemes, can be extended in any meaningful way to the high Reynolds number regime, for either the phenomena discussed in this article or the simulation of transition and turbulence (see, for example, Zang 1991). What ultimately is required is a scheme that can sense the onset of eruptive events and respond quickly enough to concentrate computational resources into the erupting region; this numerical process must continue as the erupting vorticity leaves the surface and eventually rolls up into a new vortex. One potentially attractive possibility is a hybrid interactive scheme in which an unsteady boundary layer computation is interactively coupled to numerical solutions of the unsteady Euler equations; the central difficulty that must be overcome is how to structure an accurate algorithm that permits communication between both flow regions when strong interactions are required. This high Reynolds number problem may well be regarded as one of the most important unsolved problems in fluid mechanics.

ACKNOWLEDGMENTS

We would like to thank AFOSR for its long-standing support of vortex/ surface interactions and support under Grants No. 93-1-0217 and 91-0069. We would also like to thank Mr. J. P. Fitzgerald, Mr. C. V. Seal and Mrs. JoAnn Casciano for their assistance in preparation of the figures and text.

Literature Cited

Acarlar, M. S., Smith, C. R. 1987a. A study of hairpin vortices in a laminar boundary layer, Part I. Hairpin vortices generated by a hemisphere protuberance. *J. Fluid Mech.* 175: 1–41

Acarlar, M. S., Smith, C. R. 1987b. A study of hairpin vortices in a laminar boundary layer, Part II. Hairpin vortices generated by fluid injection. *J. Fluid Mech.* 175: 43–83

Acharya, M., Karim, M. A., Metwally, M. H. 1993. Development of the dynamic stall vortex over a pitching airfoil. Submitted

Acharya, M., Metwally, M. H. 1992. Unsteady pressure field and vorticity production over a pitching airfoil. *AIAA J.* 30: 403–11

Affes, H., Conlisk, A. T. 1993. The three-dimensional boundary-layer flow due to a vortex filament outside a circular cylinder. *AIAA Pap.* 93-0212

Affes, H., Conlisk, A. T., Kim, J. M., Komerath, N. M. 1992. An experimental and analytical study of interaction of a vortex with an airframe. *AIAA Pap.* 92-0319

Affes, H., Xiao, Z., Conlisk, A. T., Kim, J. M., Komerath, N. M. 1993. The three-dimensional boundary-layer flow due to a rotor-tip vortex. *AIAA Pap.* 93-3081

Baker, C. J. 1979. The laminar horseshoe vortex. *J. Fluid Mech.* 95: 347–67

Batchelor, G. K. 1967. *An Introduction to Fluid Dynamics.* Cambridge: Cambridge Univ. Press

Bi, N.-P., Leishman, J. G., Crouse, G. L. 1993. Investigation of rotor tip vortex interactions with a body. *J. Aircraft* 30: To appear

Bernal, L. P., Kwon, J. T. 1989. Vortex ring dynamics at a free surface. *Phys. Fluids A* 1: 449–51

Brand, A., Komerath, N. M., McMahon, H. M. 1989. Results from laser sheet visualization of a periodic rotor wake. *J. Aircraft* 26: 438–43

Carr, L. 1988. Progress in analysis and prediction of dynamic stall. *J. Aircraft* 25: 6–17

Cerra, A. W., Smith, C. R. 1983. Experimental observations of vortex ring interaction with the fluid adjacent to a surface. *Rep. FM-4*, Dept. Mech. Eng. and Mechanics, Lehigh Univ., Bethlehem, Pa.

Chandrasekhara, M. S., Carr, L. W., Wilder, M. C. 1993. Interferometric investigations of compressible dynamic stall over a transiently pitching airfoil. *AIAA Pap.* 93-0211

Chu, C. C., Falco, R. E. 1988. Vortex ring/ viscous wall layer interaction model of the turbulence production process near walls. *Exp. Fluids* 6: 305–15

Chuang, F. S., Conlisk, A. T. 1989. The effect of interaction on the boundary layer induced by a convected rectilinear vortex. *J. Fluid Mech.* 200: 337–65

Conlisk, A. T. 1989. The pressure field in intense vortex-boundary layer interactions. *AIAA Pap.* 89-0293

Cowley, S. J., Van Dommelen, L. L., Lam, S. T. 1991. On the use of Lagrangian variables in descriptions of unsteady boundary-layer separation. *Phil. Trans. R. Soc. London Ser. A* 333: 343–78

Crighton, D. G. 1985. The Kutta condition in unsteady flow. *Annu. Rev. Fluid Mech.* 17: 411–45

Cunningham, A. M. 1989. Practical problems: airplanes. *Prog. Astronaut. Aeronaut.* 120: 75–132

Currier, J., Fung, F.-Y. 1992. Analysis of the onset of dynamic stall. *AIAA J.* 30: 2469–77

Degani, A. T., Walker, J. D. A. 1992. Calculation of unsteady separation from stationary and moving walls. In *Proc. IUTAM Symp. on Bluff-Body Wakes, Dynamics and Instabilities, Göttingen, Germany, Sept. 7–11,* ed. H. Eckelmann, J. M. R. Graham, P. Huerre, P. A. Monkewitz, pp. 23–26. Berlin: Springer-Verlag

Devenport, W. J., Simpson, R. L. 1990. Time-dependent and time-averaged turbulence structure near the nose of a wingbody junction. *J. Fluid Mech.* 210: 23–55

Didden, N., Ho, C.-M. 1985. Unsteady separation in a boundary layer produced by an impinging jet. *J. Fluid Mech.* 160: 235–56

Doligalski, T. L., Smith, C. R., Walker, J. D. A. 1980. A production mechanism for turbulent boundary layer flows. *Prog. Astronaut. Aeronaut.* 72: 47–72

Doligalski, T. L., Walker, J. D. A. 1978. Shear layer breakdown due to vortex motion. *AFOSR Workshop on Coherent Structures in Turbulent Boundary Layers*, ed. C. R. Smith, D. E. Abbott, pp. 288–339. Bethlehem, PA: Lehigh Univ.

Doligalski, T. L., Walker, J. D. A. 1984. Boundary layer induced by a convected two-dimensional vortex. *J. Fluid Mech.* 139: 1–28

Ece, M. C., Doligalski, T. L., Walker, J. D. A. 1984. The boundary layer on an impulsively started rotating and translating cylinder. *Phys. Fluids.* 27: 1077–89

Elliott, J. W., Cowley, S. J., Smith, F. T. 1983. Breakdown of boundary layers: (i) on moving surfaces; (ii) in self-similar unsteady flows: (iii) in fully unsteady flow. *Geophys. Astrophys. Fluid Dyn.* 25: 77–138

Ersoy, S., Walker, J. D. A. 1985. Viscous flow induced by counter-rotating vortices. *Phys. Fluids* 28: 2687–98

Ersoy, S., Walker, J. D. A. 1986. Flow induced by a vortex pair. *AIAA J.* 24: 1597–605

Ersoy, S., Walker, J. D. A. 1987. The boundary layer due to a three-dimensional vortex loop. *J. Fluid Mech.* 185: 569–98

Falco, R. E. 1991. A coherent structure model of the turbulent boundary layer and its ability to predict Reynolds number dependence. *Phil. Trans. R. Soc. London Ser. A* 336: 103–29

Fleming, J. F., Simpson, R. L., Devenport, W. J. 1992. An experimental study of a turbulent wing-body junction and wake flow. *AIAA Pap.* 92-0434

Francis, M. S., Keesee, J. E. 1985. Airfoil dynamic stall performance with large-amplitude motions. *AIAA J.* 23: 1653–59

Gendrich, C. P. , Koochesfahani, M. M., Visbal, M. R. 1993. Initial acceleration effects on the flowfield development around rapidly pitching airfoils. *AIAA Pap.* 93-0438

Ghia, K. N., Yang, J., Osswald, G. A., Ghia, U. 1992a. Study of the role of unsteady separation in the formation of dynamic stall vortex. *AIAA Pap.* 92-0196

Ghia, K. N., Yang, J., Osswald, G. A., Ghia, U. 1992b. Physics of forced unsteady flow for a NACA 0015 airfoil undergoing constant-rate pitch-up motion. *Fluid Dyn. Res.* 10: 351–69

Grass, A. J., Stuart, R. J., Mansour-Tehrani, M. 1991. Vortical structures and coherent motion in turbulent flow over smooth and rough boundaries. *Phil. Trans. R. Soc. London Ser. A* 336: 35–66

Greco, J. J. 1991. *Turbulent flow structure behavior in the vicinity of a cylinder-flat plate juncture.* MS thesis. Dept. Mech. Eng. and Mechanics, Lehigh Univ. 158 pp.

Gursul, I., Rockwell, D. 1990. Vortex street impinging upon an elliptical leading edge. *J. Fluid Mech.* 211: 211–42

Haidari, A. H., Smith, C. R. 1990. Generation and growth of single hairpin vortices. *Rep. FM*-16, Dept. Mech. Eng. and Mechanics, Lehigh Univ., Bethlehem, Pa. 228 pp.

Harvey, J. K., Perry, F. J. 1971. Flowfield produced by trailing vortices in the vicinity of the ground. *AIAA J.* 9: 1659–60

Helin, H. E. 1990. The relevance of unsteady aerodynamics for highly maneuverable and agile aircraft. In *Numerical and Physical Aspects of Aerodynamic Flows IV*, ed. T. Cebeci, pp. 229–37. Berlin: Springer-Verlag

Homa, J., Lucas, M., Rockwell, D. 1988. Interaction of impulsively generated vortex pairs with bodies. *J. Fluid Mech.* 197: 571–94

Hon, T. L., Walker, J. D. A. 1991. Evolution of hairpin vortices in a shear flow. *Comput. Fluids* 20: 343–58

Hoyle, J. M., Smith, F. T., Walker, J. D. A. 1991. On sublayer eruption and vortex formation. *Comput. Phys. Commun.* 65: 151–57

Karim, M. A. 1992. *Experimental investigation of the formation and control of the dynamic stall vortex over a pitching airfoil.* MS thesis. Ill. Inst. Tech.

Kaykaykoglu, R., Rockwell, D. 1986. Unstable jet-edge interaction. Part 1. Instantaneous pressure fields at a single frequency. *J. Fluid Mech.* 169: 125–49

Knight, D. R., Ghosh Choudhuri, P. 1993. 2-D unsteady leading edge separation on a pitching airfoil. *AIAA Pap.* 93-2977

Kuchemann, D. 1965. Report on IUTAM symposium on concentrated vortex motion in fluids. *J. Fluid Mech.* 21: 1–20

Leonard, A. 1985. Computing three-dimensional incompressible flows with vortex elements. *Annu. Rev. Fluid Mech.* 17: 523–59

Li, K. W., Lian, Q. X. 1992. The ways of formation of the horseshoe vortex in turbulent boundary layers. *Acta Mech. Sin.* 24: 145–52

Lim, T. T. 1989. An experimental study of a vortex ring interaction with an inclined wall. *Exp. Fluids* 7: 453–63

Liou, S. G., Komerath, N. M., McMahon, H. M. 1990. Velocity field of a cylinder in the wake of a rotor in forward flight. *J. Aircraft* 27: 804–9

Lucas, M., Rockwell, D. 1984. Self-excited

jet: upstream modulation and multiple frequencies. *J. Fluid Mech.* 147: 333–52

Lugt, H. J. 1983. *Vortex Flow in Nature and Technology.* New York: Wiley

Lugt, H., Öhring, S. 1992. The oblique ascent of a viscous vortex pair toward a free surface. *J. Fluid Mech.* 236: 461–76

Mabey, D. G. 1989. Physical phenomena associated with unsteady transonic flows. *Prog. Astronaut. Aeronaut.* 120: 1–55

McCroskey, W. J. 1982. Unsteady airfoils. *Annu. Rev. Fluid Mech.* 14: 285–311

Metwally, M. H. 1990. *Investigation and control of the unsteady flow field over a pitching airfoil.* PhD dissertation. Ill. Inst. Tech.

Milne-Thomson, L. M. 1962. *Theoretical Hydrodynamics.* London: MacMillan

Orlandi, P. 1990. Vortex dipole rebound from a wall. *Phys. Fluids A* 2: 1429–36

Orlandi, P., Jiménez, J. 1991. A model for bursting of near wall vortical structures in boundary layers. *Eighth Symposium on Turbulent Shear Flows,* Tech. Univ. Munich, Sept. 9–11, 28-1-1–6

Osswald, G. A., Ghia, K. N., Ghia, U. 1991. Simulation of dynamic stall phenomenon using unsteady Navier-Stokes equations. *Comput. Phys. Commun.* 65: 209–18

Peace, A. J., Riley, N. 1983. A viscous vortex pair in ground effect. *J. Fluid Mech.* 129: 409–26

Pedrizzetti, G. 1992. Close interaction between a vortex filament and a rigid sphere. *J. Fluid Mech.* 245: 701–22

Peridier, V. J., Smith, F. T., Walker, J. D. A. 1991a. Vortex-induced boundary-layer separation. Part I. The limit problem Re → ∞. *J. Fluid Mech.* 232: 99–131

Peridier, V. J., Smith, F. T., Walker, J. D. A. 1991b. Vortex-induced boundary-layer separation. Part II. Unsteady interacting boundary-layer theory. *J. Fluid Mech.* 232: 133–65

Riley, N. 1975. Unsteady laminar boundary layers. *SIAM Rev.* 17: 274–97

Robinson, S. K. 1991. Coherent motions in the turbulent boundary layer. *Annu. Rev. Fluid. Mech.* 23: 601–39

Saffman, P. G., Baker, G. R. 1979. Vortex inteactions. *Annu. Rev. Fluid Mech.* 11: 95–122

Sarpkaya, T. 1992. Brief review of some time-dependent flows. *J. Fluids Eng.* 114: 283–98

Sarpkaya, T., Suthon, P. 1991. Interaction of a vortex couple with a free surface. *Exp. Fluids.* 11: 205–17

Seal, C. V., Smith, C. R., Akin, O., Rockwell, D. 1993. Instantaneous velocity and vorticity characteristics in the juncture region of a rectangular block-flat plate. *J. Fluid Mech.* Submitted

Sears, W. R., Telionis, D. P. 1975. Boundary-layer separation in unsteady flow. *SIAM J. Appl. Math.* 28: 215–35

Shariff K., Leonard, A. 1992. Vortex rings. *Annu. Rev. Fluid Mech.* 24: 235–79

Sheridan, P. F., Smith, R. F. 1980. Interactional aerodynamics—a new challenge to helicopter technology. *J. Am. Helicopter Soc.* 25: 3–21

Shih, C., Lourenco, L., Van Dommelen, L., Krothapalli, A. 1992. Unsteady flow past an airfoil pitching at constant rate. *AIAA J.* 30: 1153–61

Smith, C. R., Fitzgerald, J. P., Greco, J. J. 1991a. Cylinder end-wall vortex dynamics. *Phys. Fluids A* 3: 2031

Smith, C. R., Walker, J. D. A., Haidari, A. H., Sobrun, U. 1991b. On the dynamics of near-wall turbulence. *Phil. Trans. R. Soc. London Ser. A* 336: 131–75

Smith, F. T. 1982. On the high Reynolds number theory of laminar flows. *IMA J. Appl. Math.* 28: 207–81

Smith, F. T. 1988. Finite-time breakup can occur in any unsteady interacting boundary layer. *Mathematica* 35: 256–73

Stanek, M. J., Visbal, M. R. 1991. Investigation of vortex development on a pitching slender body of revolution. *AIAA Pap.* 91-3273. To appear in *J. Aircraft* 30(5), 1993

Tang, Y.-P., Rockwell, D. 1983. Instantaneous pressure fields at a corner associated with vortex impingement. *J. Fluid Mech.* 126: 187–204

Ting, L., Tung, C. 1965. Motion and decay of a vortex in a non-uniform stream. *Phys. Fluids* 8: 1039–51

Theodorsen, T. 1952. Mechanism of turbulence. *Proc. 2nd Midwestern Conf. on Fluid Mech., Bull. No.* 149, Ohio State Univ., Columbus, Ohio

Thomas, A. S. W. 1987. The unsteady characteristics of laminar juncture flow. *Phys. Fluids* 30: 283–85

Tryggvason, G., Unverdi, S. O., Song, M., Abdollahi-Alibeik, J. 1991. Interaction of vortices with a free surface and density interfaces. *Lect. Appl. Math.* 28: 679–99

Van Dommelen, L. L. 1981. *Unsteady boundary-layer separation.* PhD dissertation. Cornell Univ.

Van Dommelen, L. L., Cowley, J. J. 1990. On the Lagrangian description of unsteady boundary layer separation. Part I. General theory. *J. Fluid Mech.* 210: 593–626

Van Dommelen, L. L., Shen, S. F. 1980. The spontaneous generation of the singularity in a separating boundary layer. *J. Comput. Phys.* 38: 125–40

Van Dommelen, L. L., Shen, S. F. 1982 The genesis of separation. *Proc. Symp. on Numerical and Physical Aspects of Aero-*

dynamic Flow, Long Beach, Calif., ed. T. Cebeci, pp. 293–311

Verdon, J. M. 1989. Unsteady aerodynamics for turbomachinery applications. *Prog. Astronaut. Aeronaut.* 120: 287–347

Verdon, J. M. 1992. Unsteady aerodynamic methods for turbomachinery aeroelastic and aeroacoustic applications. *AIAA Pap.* 92-0011

Visbal, M. R. 1991a. The laminar horseshoe vortex system formed at a cylinder/plate juncture. *AIAA Pap.* 91-1826

Visbal, M. R. 1991b. Structure of laminar juncture flows. *AIAA J.* 29: 1273–82

Visbal, M. R., Shang, J. S. 1989. Investigation of the flow structure around a rapidly pitching airfoil. *AIAA J.* 27: 1044–51

Walker, J. D. A. 1978. The boundary layer due to rectilinear vortex. *Proc. R. Soc. London Ser. A.* 359: 167–88

Walker, J. D. A., Smith, C. R., Doligalski, T. L., Cerra, A. W. 1987. Impact of a vortex ring on a wall. *J. Fluid Mech.* 181: 99–140

Wallace, J. M., Balint, J. L., Ayrault, M. 1990. *Flow visualization study of the effects of trip type on the structure of the turbulent boundary layer.* Video Tape Pres., Turbulence Lab., Dept. Mech. Eng., Univ. Md.

Widnall, S. E., Sullivan, J. P. 1973. On the stability of vortex rings. *Proc. R. Soc. London Ser. A.* 332: 335–53

Zang, T. A. 1991. Numerical simulation of the dynamics of turbulent boundary layers: perspectives of a transition simulator. *Phil. Trans. R. Soc. London Ser. A* 336: 95–102

Ziada, S., Rockwell, D. 1982. Vortex-leading-edge interaction. *J. Fluid Mech.* 118: 79–107

Annu. Rev. Fluid Mech. 1994. 26 : 617–59

DYNAMICS OF COUPLED OCEAN-ATMOSPHERE MODELS:
The Tropical Problem

J. David Neelin

Department of Atmospheric Sciences, University of California,
Los Angeles, California 90024

Mojib Latif

Max-Planck-Institut für Meteorologie, D-20146 Hamburg 13,
Federal Republic of Germany

Fei-Fei Jin

Department of Meteorology, University of Hawaii at Manoa, Honolulu,
Hawaii 96822

KEY WORDS: Ocean-atmosphere interaction, climate variability, El Niño/ Southern Oscillation

INTRODUCTION

Large-scale ocean-atmosphere interaction plays a crucial role in natural climate variability on a broad range of time scales and in anthropogenic climate change. The development of coupled ocean-atmosphere models is thus widely regarded as essential for simulating, understanding, and predicting the global climate system. Although these efforts typically benefit from years of previous work with atmospheric and oceanic models, coupling the two components represents a major step because of the new interactions introduced into the system. These can produce new phenomena, not found in either medium alone, the mechanisms for which present exciting theoretical problems. The removal of artificial negative feedbacks

617

0066–4189/94/0115–0617$05.00

produced by fixed boundary conditions in the uncoupled case also provides a stringent test of physical processes represented in both component models.

Pioneering work on coupling oceanic and atmospheric general circulation models (GCMs) began during the late 1960s and the 1970s (Manabe & Bryan 1969, Bryan et al 1975, Manabe et al 1975, Manabe et al 1979, Washington et al 1980). The difficulties encountered in obtaining accurate climate simulations with these models were sufficient that use of such coupled GCMs (CGCMs) did not gain momentum until the late 1980s and early 1990s. While the anthropogenic warming problem drove the development of global models (e.g. Gates et al 1985, Schlesinger et al 1985, Sperber et al 1987, Bryan et al 1988, Manabe & Stouffer 1988, Washington & Meehl 1989, Stouffer et al 1989, Manabe et al 1990, Cubasch et al 1992, Manabe et al 1992), evidence that ocean-atmosphere interaction is responsible for the El Niño/Southern Oscillation (ENSO) phenomenon provided a driving force in the development of models aimed at the tropical regions, both CGCMs and less complex models.

In this article, we consider the dynamics of coupled models relating to internal variability of the climate system that arises through ocean-atmosphere interaction. We focus on the tropical problem because it has been more thoroughly studied than the extratropical problem, and the crucial role of coupling has been clearly demonstrated. The field has developed to a stage that can be well summarized, and where short-range climate prediction is becoming a reality. A briefer section provides an indication of developments for the problem of coupled extratropical variability, which is in its infancy.

Despite the importance of coupled models to the study of anthropogenic global warming, we do not address this question beyond providing an indication of some of the difficulties these models face. It is the subject of many articles (e.g. Mitchell 1989, Houghton et al 1990, Gates et al 1992 and references therein) and merits a separate review. For other general references on coupled models, we note a review of global CGCMs (Meehl 1990a), a textbook on the tropical problem (Philander 1990), edited volumes on climate modeling (Trenberth 1993, Schlesinger 1990), and selected conference proceedings (Nihoul 1985, 1990; Charnock & Philander 1989).

COUPLED OCEAN-ATMOSPHERE MODELS

A hierarchy of complexity exists in climate models, the most complex being the atmospheric, oceanic, and coupled general circulation models (AGCMs, OGCMs, and CGCMs; for these and other acronyms, see Table 1). GCMs are generally based on the primitive equations (a filtered version of the Navier-Stokes equations; e.g. Washington & Parkinson 1986), with detailed parameterizations of sub-grid-scale processes (e.g. turbulent mixing, and for AGCMs radiative transfer and moist convection). These attempt to simulate an approximation to both the climatology and natural variability. A variety of models based on further approximations are used for particular applications; often these are formulated as anomaly models about a specified climatology. Coupling considerations tend to be similar—we outline the procedures as applied to GCMs. The class of models often used in global warming studies in which the ocean acts only as a heat capacitor—and has no active dynamics—is not discussed.

For climate time scales, a division of the coupled system at the ocean-atmosphere interface is not easy to defend. Incoming solar (shortwave) radiation is primarily absorbed at the ocean surface and energy is lost through evaporation, infrared (longwave) radiation, and sensible heat fluxes to the atmosphere, which in turn re-emits longwave radiation to space. The one-dimensional equilibrium of these processes (and the strength of the negative feedback to perturbations from this equilibrium) provides a first approximation to the climate, modified of course by three-dimensional transports and feedbacks in both media. Interrupting this exchange at the ocean surface is questionable on time scales longer than a few months (shorter for some phenomena). Historically, however, this division permitted atmospheric and oceanic modelers to concentrate purely on problems in their respective media, as necessitated by the complexity of these subsystems. Since the parameterization of sub-grid-scale processes is one of the most crucial aspects of climate modeling, this separate development may be partially justified by arguing that the difference in density and effective heat capacity is sufficient that individual parameterizations of fast sub-grid processes may be developed initially in uncoupled models. The limitations of this approach will no doubt be re-examined when coupled models reach a more mature stage. Surface heat flux boundary conditions for uncoupled ocean models are particularly problematic (e.g. Bretherton 1982, Seager et al 1988) since the negative feedback on sea surface temperature (SST) involves the atmospheric response.

Table 1   Acronyms used in the text

ENSO	El Niño/Southern Oscillation
GCM	General Circulation Model
AGCM/CGCM/OGCM	Atmospheric/Coupled/Ocean GCM
HCM	Hybrid Coupled Model
ICM	Intermediate Coupled Model
SSO regime	Standing-SST Oscillatory regime
SST	Sea Surface Temperature

In a typical coupling scheme for an ocean-atmosphere model, the ocean model passes SST to the atmosphere, while the atmosphere passes back heat flux components, freshwater flux, and horizontal momentum fluxes (*surface stress*—oceanographic usage refers only to stress tensor components associated with vertical fluxes of horizontal momentum). Land temperature is necessarily computed interactively, with parameterizations ranging from the zero heat-capacity approximation to more complex land-surface models (e.g. Dickinson 1983). The numerical coupling interval (over which interfacial variables are averaged before being passed) is chosen for computational convenience or to satisfy assumptions of physical parameterizations. Although heat fluxes are calculated using the atmospheric boundary-layer parameterizations based on SST from the previous interval, the important dependence of heat flux on SST is retained as long as the heat flux coupling interval is sufficiently small.

The atmospheric response to SST is rapidly redistributed vertically, especially in convective regions, and is nonlocal horizontally on time scales longer than dynamical adjustment times—on the order of a few days to a month. For most purposes, the atmosphere can be assumed to be in statistical equilibrium with given SST (and land/ice/snow) boundary conditions on time scales longer than a season. The ocean responds on a wide range of time scales, from days (for some features of the mixed layer) to millenia (for the deep-ocean thermal adjustment). It is thus common to characterize the ocean as having the *memory* of the system. For global coupled models where the deep ocean is integrated to equilibrium, asynchronous coupling techniques are sometimes used (e.g. Manabe et al 1979).

Climate drift—i.e. departure of the model climatology from the observed (and from the climate simulated by the component models in uncoupled tests)—is a common problem in coupled models. It often appears as a slow adjustment away from initial conditions towards an internal equilibrium, hence the term "drift;" it may also refer to cases of faster adjustment and to the error at equilibrium. Although numerics contribute, climate drift arises primarily from the cumulative effects of errors in the sub-grid parameterizations; as such the process of correcting it based on careful physical arguments can be slow and painstaking. In cases where the sources of drift are well separated from mechanisms governing the geophysical phenomena of interest, it has been argued (e.g. Manabe & Stouffer 1988, Sausen et al 1988) that correcting the drift by a *flux correction* may permit progress even with an imperfect model. Roughly speaking, the model's equilibrium climatology of all or some of the interfacial variables is subtracted and replaced with observed values that are passed between subsystems; effectively the model is only used to compute anomalies from climatology. The success of flux-correction techniques depends on the problem.

Coupled models designed for the tropical problem do not treat the deep-ocean thermohaline circulation which maintains cold waters at depth. Typically the ocean basin is simply interrupted at some latitude, using a sponge layer (with temperature and salinity strongly constrained toward climatological values) to avoid effects of the artificial boundaries propagating into the region of interest via wall-trapped Kelvin waves. Observed climatological SST is specified in the ocean regions which are not actively modeled. Other models simulate the upper ocean only, with motionless deep waters (e.g. Gent & Cane 1989). Additional design specifications for the tropical problem include the use of sufficiently high-resolution ocean components to resolve equatorial wave dynamics with characteristic meridional scales of order 2° latitude. For the global problem, the ocean models typically are used with coarser resolution because of the necessity of very long integrations for equilibration.

THE TROPICAL PROBLEM

Background

Ocean-atmosphere interaction is particularly amenable to study in the tropics because at large scales each medium is strongly controlled by the boundary conditions imposed by the other. The upper ocean circulation is largely determined by the past history of the wind stress with little internal variability; likewise the major features of the tropical atmospheric circulation are determined by the SST, with internal variability largely confined to time scales less than 1–2 months. This contrasts to the midlatitude situation where internal variability of both atmosphere and ocean is large.

THE BJERKNES HYPOTHESIS   Because the ENSO phenomenon is the largest signal in interannual climate variability, it has dominated the literature; here we bring in other aspects of the tropical problem where possible. The reigning paradigm for ENSO dynamics is that it arises through ocean-atmosphere interaction in the tropical Pacific (although its influence extends globally and interactions with other parts of the climate system are by no means excluded), as first hypothesized by Bjerknes (1969). The essence of Bjerknes' postulate still stands as the basis of present day work—that ENSO arises as a self-sustained cycle in which anomalies of SST in the Pacific cause the trade winds to strengthen or slacken, and that this in turn drives the changes in ocean circulation that produce anomalous SST. Within this paradigm, one may still distinguish a variety of mechanisms that potentially contribute to the maintainance and time scale of the cycle; these have provided challenges for both theory and simulation.

MODEL HIERARCHY   Beginning at about the same time as Bjerknes' hypothesis was formulated, the foundations for modeling the tropical coupled system were laid through the study of the individual physical components. The dynamics of the equatorial ocean response to wind stress were examined in shallow-water models representing the upper ocean (e.g. Moore 1968; Cane & Sarachik 1977, 1981; McCreary 1976), modified shallow-water models (e.g. Cane 1979, Schopf & Cane 1983), and ocean general circulation models (OGCMs; e.g. Philander & Pacanowski 1980, Philander 1981). And in the atmosphere, it was demonstrated semi-empirically that simple atmospheric models with steady, damped shallow-water dynamics could provide a reasonable approximation to the low-level tropical atmospheric response to SST anomalies (e.g. Matsuno 1966, Gill 1980, Gill & Rasmusson 1983). There is still disagreement as to the best formulation of these simple atmospheric models (Zebiak 1986, Lindzen & Nigam 1987, Neelin & Held 1987, Neelin 1989a, Allen & Davey 1993) but their simulation of anomalous wind-stress feedbacks to the ocean from given SST is given credence by AGCM simulations (e.g. Lau 1985, Palmer & Mansfield 1986, Mechoso et al 1987, Shukla & Fennessey 1988, and references therein).

As a result of the development of complementary models of varying degrees of complexity, the tropical coupled problem has benefited from a full hierarchy of models. The basis for a more quantitative understanding of coupled ocean-atmosphere interaction was initially provided by coupled models constructed from variations on modified shallow-water ocean and simple atmospheric models: both in simple linear versions (Lau 1981a; Philander et al 1984; Gill 1985; Hirst 1986, 1988; Wakata & Sarachik 1991; Neelin 1991) and in nonlinear versions (e.g. Cane & Zebiak 1985, Anderson & McCreary 1985, Zebiak & Cane 1987, Battisti 1988, Battisti & Hirst 1989, Schopf & Suarez 1988, Yamagata & Masumoto 1989, Graham & White 1990). The simplest linear shallow-water models, together with some useful models that condense the dynamics even further, are loosely referred to as *simple models*, while the more complex and carefully parameterized of the modified shallow-water models are often referred to as *intermediate coupled models* (ICMs). The next step up the model hierarchy, in order of increasing complexity, is the *hybrid coupled models* or *HCMs*. These consist of an ocean GCM coupled to a simpler atmospheric model (e.g. Neelin 1989b, 1990, Latif & Villwock 1990, Barnett et al 1993), the justification being that the ocean contains both the memory and limiting nonlinearity of the system—the atmosphere is thus treated as the fast component of a stiff system. The most complex models are the coupled GCMs in which both components include relatively complete sub-grid parameterization packages (e.g. Philander et al 1989, 1992; Lau et al 1992; Sperber & Hameed 1991; Gordon 1989; Meehl 1990b; Nagai et al 1992; Mechoso et al 1993; Neelin et al 1992). It should be noted that the divisions in the hierarchy are not sharp and some of the lowest-resolution CGCMs may not be much more complex than the best ICMs. Many of these models produce interannual variability through coupled interactions which have significant parallels to ENSO dynamics.

Our approach here is to summarize basic phenomenological features from a modeler's point of view (i.e. we do not attempt a complete review of the large observational literature) in the *Observations* section, and then to present a cross-section of model results in the *Simulation* section, which includes selected intermediate models as well as CGCMs and HCMs. The *Theory* section makes use of intermediate and simple models to outline basic mechanisms of interaction, describes the manner in which different mechanisms combine and contribute to the sensitivity of the coupled system, and details the current understanding of the bifurcation structure. Many of the theoretical considerations prove useful in understanding results of the more complex models.

Observations

Aspects of El Niño and the Southern Oscillation were known individually long before any connection was made. The term "El Niño," which originated with Peruvian fishermen, now refers to strong warmings of surface waters through the eastern and central equatorial Pacific that last about a year (e.g. Rasmusson & Carpenter 1982, Deser & Wallace 1990). Although it is common to refer to these as "events," they exhibit a distinct oscillatory behavior now understood to be part of a low frequency cycle. The Southern Oscillation was discovered by Walker (1923), and its global scale was inferred early on (Belarge 1957) from correlation maps of sea-level pressure anomalies which exhibit anomalies of opposite sign in the eastern and western hemisphere. The larger scale of this pattern relative to the SST anomaly is typical of the atmosphere's nonlocal response to boundary conditions. The strong relationship between interannual variability of SST and sea-level pressure may be seen in Figure 1.

As a background to understanding ENSO-related interannual variability, a brief description of the time-mean circulation is required. Differential forcing of the atmosphere by the SST boundary condition thermodynamically drives direct circulation cells: convection tends to organize roughly over the warmest SST, producing regions of strong surface convergence (known as intertropical convergence zones). The zonally-symmetric (i.e. averaged around latitude circles) component of this circulation

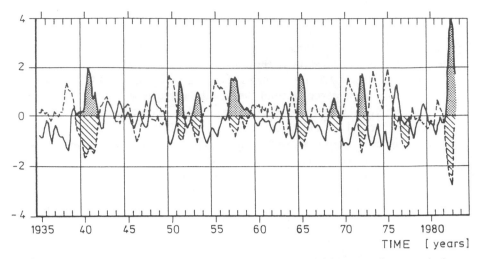

Figure 1 Time series of the Southern Oscillation Index, which measures the atmospheric sea-level pressure gradient across the tropical Pacific basin (*dashed curve*), and sea surface temperature (SST) anomalies at Puerto Chicama, Peru (*solid curve*). Both series are normalized by their standard deviation; shading indicates major ENSO warm phases (high SST, low Southern Oscillation Index). After Rasmusson (1984).

is referred to as the *Hadley circulation*, the zonally-asymmetric component as the *Walker circulation*. The Hadley circulation contributes an easterly (i.e. westward) component to tropical surface winds. This is strongly reinforced over the tropical Pacific by the Walker circulation driven by the strong SST gradient across the basin between the warm waters in the west and the cooler eastern waters.

The westward wind stress has a strong impact on the ocean circulation. The input of momentum is balanced, in a vertical average, largely by pressure gradients in the upper ocean. A sea-level gradient of about 40 cm across the Pacific is compensated by a slope in the *thermocline* (the interface that separates the well-mixed, warm surface waters from the cold waters at deeper levels) which slopes upward to the east. Within the upper ocean, the differential deposition of stress by vertical viscosity drives westward surface currents along the equator, and Ekman drift due to the Coriolis force to either side of the equator drives a narrow band of upwelling along the equator, especially under the regions of strong easterlies in the eastern/central Pacific. The combination of upwelling and shallow thermocline produces the *equatorial cold tongue* in the east, while the deep thermocline in the west is associated with warm SST—the western Pacific *warm pool*.

The important dependence of SST in the equatorial cold tongue region on wind-driven ocean dynamics (rather than just on air-sea heat exchange) and the Walker circulation response to anomalies in the SST pattern form the key elements of the Bjerknes hypothesis. Consider an initial positive SST anomaly in the eastern equatorial Pacific. This anomaly reduces the zonal SST gradient and hence the strength of the Walker circulation, resulting in weaker trade winds at the equator. This leads to a deeper thermocline and reduced currents and produces higher surface temperatures in the cold tongue region, further reducing the SST gradient in a positive feedback which can lead to instability of the climatological state via ocean-atmosphere interaction. The cyclic nature of the unstable mode depends on the time scales of response within the ocean. The details of what produces the cycle are subtle, as elaborated in the *Theory* section, but a concise observational picture motivated by theoretical considerations is provided by Latif et al (1993b).

Figure 2 shows characteristic anomaly patterns of three crucial quantities: zonal wind stress, SST, and the depth of the thermocline or upper ocean heat content, as measured by depth of the $20°C$ isotherm. The patterns represent an estimate of the dominant coupled ENSO mode as obtained by principal oscillation pattern analysis (Hasselmann 1988)—specifically, the leading eigenvector of the system matrix obtained by fitting a first-order Markov process to the data, where oscillations are represented by the cycle of patterns in temporal quadrature. The right panels show conditions during the warm phase of the ENSO cycle, i.e. during El Niño (the cold phase simply has reversed signs under this technique). Most of the tropical Pacific is covered by anomalously warm surface waters (Figure 2*d*), with maximum anomalies in the eastern equatorial Pacific. These SST anomalies are highly consistent with the patterns obtained by other techniques, including the well-known Rasmusson & Carpenter (1982) composites. The positive SST anomaly is accompanied by a westerly (eastward) zonal wind stress anomaly (Figure 2*b*) which reduces the mean Walker Circulation. Consistent with this feature, the tilt in the thermocline is reduced as indicated by the negative anomalies in the upper ocean heat content which are centered off the equator (Figure 2*f*).

The phase differences necessary to maintain the oscillation exist between sea surface temperature and wind on the one hand and upper ocean heat content on the other. As described in the *Theory* section, the ocean is not in equilibrium with the atmosphere and carries information associated with past winds that permits continuous oscillations. This feature is clearly seen during the transition phase in upper ocean heat content (Figure 2*e*) which shows a pronounced equatorially-trapped signal in the western Pacific. This signal appears not to be related to the contemporaneous winds (Figure 2*a*), but rather was generated by anomalous eastward winds of the preceding cold phase (Figure 2*b*, but with reversed signs). Equatorial wave dynamics dictate that the heat content anomalies at latitudes larger than a few degrees propagate westward and reflect at the western maritime boundary into the equatorial wave guide. The transition phase SST (Figure 2*c*) does not show a clear signal; variations in SST can therefore be described to first order as a standing oscillation. Thus, it is the subsurface memory of the ocean that is crucial to ENSO (see e.g. Latif & Graham 1992 and Graham & White 1990 for additional observations).

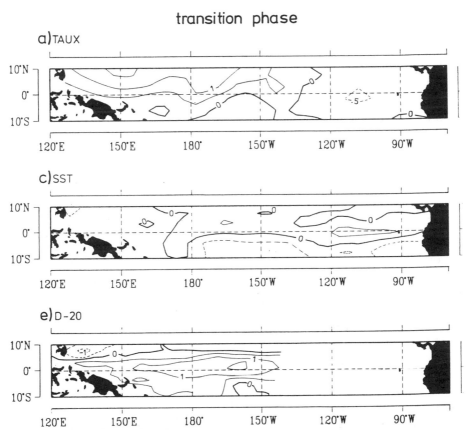

Figure 2   Spatial patterns of the dominant mode of ENSO variability as represented by the leading principal oscillation pattern (see text). The oscillation is represented by two time phases in quadrature during the cycle: transition phase (panels *a, c, e*) and extreme phase (panels *b, d, f*). (*a*), (*b*) wind stress anomaly, (*c*), (*d*) sea surface temperature anomaly, (*e*), (*f*) heat content anomaly as measured by the depth of the 20°C isotherm (blank areas in the eastern Pacific are due to lack of subsurface data). After Latif et al (1993b).

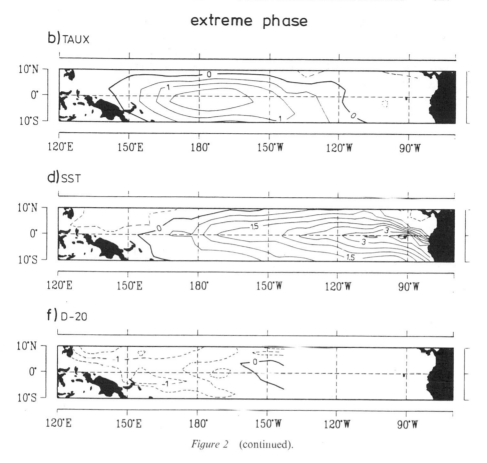

Figure 2 (continued).

The transition phase zonal wind stress (Figure 2a) shows a pronounced westerly anomaly centered over the northwestern Pacific so that the evolution in zonal wind stress is also characterized by a slowly eastward-propagating feature. The role of this propagation in maintaining the ENSO cycle, however, is still a controversial issue. Several authors have argued that this feature indicates a link to circulation systems over India, in particular the Monsoon (e.g. Barnett 1983).

A complementary view of the oceanic side of this feedback is provided by time-longitude plots of SST and a measure of thermocline depth anomalies along the equator (Figure 3). The time series is limited by the length of the records of ocean subsurface temperature. Even without statistical techniques, it is easy to pick out the dominant standing oscillation pattern in SST (although some hints of propagation may be noted—see e.g. Gill & Rasmusson 1983, Barnett et al 1991), and the characteristic signature of subsurface memory—the lead of the heat content anomalies in the western part of the basin relative to the eastern part. Several coupled ocean-atmosphere models simulate variability patterns to those described above.

There is evidence that the spectral peak associated with ENSO may have a quasi-biennial component in addition to the dominant low-frequency

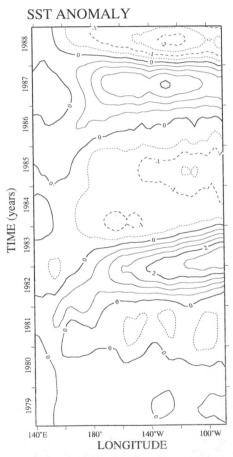

SST ANOMALY

Figure 3   Time-longitude plot of observed anomalies along the equator. (*Left*) SST (contour interval 0.5°C). (*Right*) heat content integrated above 275 m (contour interval 100°C m). The data have been low-pass filtered to remove variability on time scales smaller than 17 months. Data sets are described in Reynolds (1988) and Barnett et al (1993), respectively.

HEAT CONTENT ANOMALY

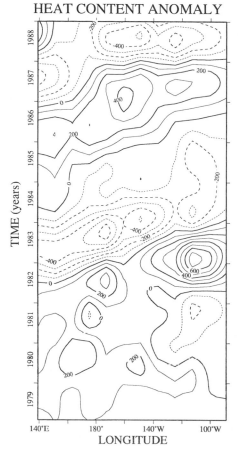

Figure 3 (continued).

(3–6 year) component, e.g. Rasmusson et al (1990), Latif et al (1993b). Spatial structures and interactions between assumed spectral bands have been examined, e.g. by Barnett 1991, Ropelewski et al 1992. For further discussion of ENSO observations see, for instance, Cane (1986), Rasmusson & Wallace (1983), Philander (1990), and references therein. Discussion of the seasonal cycle and interannual variability in the tropical Atlantic may be found in Lamb et al (1986), Lough (1986), Wolter (1989), Servain & Legler (1986), Philander & Chao (1991), Houghton & Tourre (1992), and Servain (1991), while Zebiak (1993) gives evidence that the latter may in part share similar dynamics to ENSO.

Models and Simulation

AN INTERMEDIATE MODEL   The intermediate coupled model of Cane & Zebiak (1985, with Zebiak & Cane 1987; collectively CZ hereafter) has proven influential in ENSO studies and has provided the first successful ENSO forecasts with a coupled model (see *Prediction* section). A version of the ocean component is described in the *Theory* section. The atmosphere (Zebiak 1986) is one of several simple atmosphere models which attempt to improve on that of Gill (1980); drawbacks include lack of a moisture budget and formulation with discontinuous derivatives, but similar results are obtained with different atmospheric models (Jin & Neelin 1993a, N. Graham, personal communication). Figure 4 shows the SST and thermocline depth anomalies over one period of the simulated ENSO cycle from the linearized version of the CZ model used by Battisti & Hirst (1989) to examine the essential dynamics. The typical stationary oscillation in SST may be seen, with the lead of the western-basin thermocline-depth anomaly relative to the eastern basin characterizing the subsurface memory. The details of the transition between west and east differ from those observed because the simulated winds are shifted relative to observed winds, but the cycle is not strongly sensitive to this. Simulated ENSO events tend to resemble each other strongly in this model, and Battisti (1988) and CZ disagree over the degree of irregularity that can be generated by internal model dynamics, but there is reasonable consensus that basic elements of ENSO dynamics are captured.

INTERCOMPARISON OF GCM SIMULATIONS   A recent comparison (Neelin et al 1992) of the tropical simulations of seventeen coupled ocean-atmosphere models, contributed by a dozen institutions worldwide, represents a snapshot at a relatively early stage of a rapidly developing field. We review some of the results, with the caveat that in the brief time since their collection, several of the models have made great progress in the accuracy of simulation and new models have been developed which are not yet published. The comparison was intended to give a feel for the sensitivity of the system modeled (possible in part because the models were not yet optimized), to point out common problems, and to provide a forum for discussing the broad range of coupled-model behavior.

The models were selected on the basis of having at least one component of sufficient complexity to be called a GCM, (i.e. CGCMs and HCMs), with two representatives of the ICMs—those of Cane & Zebiak (1985) and Schopf & Suarez (1988), the latter differing from a GCM principally by lack of a moisture budget. Some of the models are global, designed for global warming studies; others have a dynamically-active ocean only in the tropical Pacific, and were designed for the tropical problem. SST was

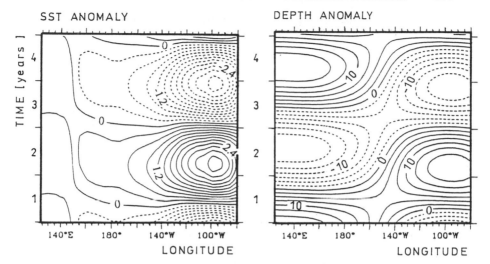

Figure 4   Time-longitude plot of anomalies along the equator from the Battisti & Hirst (1989) linearized version of the Cane & Zebiak (1985) intermediate coupled model. (*Left*) SST (contour interval 0.3°C). (*Right*) thermocline depth (contour interval 2.5 m). After Battisti & Hirst (1989).

chosen as the principle variable of comparison because of its crucial role in mediating the interactions.

Table 2 provides a summary of the results, augmented with more recent results where published, roughly classified according to the type of inter-annual variability and the simulation of climate in the equatorial Pacific. Models are listed as in Neelin et al (1992); the most closely related independent references available are Endoh et al (1991), Gent & Tribbia (1993), Gates et al (1985), Gordon (1989), Latif et al (1988), Latif et al (1993a), Lau et al (1992), Meehl (1990b), Neelin (1990), Philander et al (1992), Schopf & Suarez (1988), Sperber & Hameed (1991), Zebiak & Cane (1987). Many of the models exhibit climate drift. Some of the models, especially those with simplified atmospheres, sidestep this problem by flux correction. The category "Modest drift" as used here means only that the degree of drift in SST was relatively small by current (subjective) standards and comparable to that of uncoupled components. Interannual variability is weaker than observed in many of the models—the category "Weak interannual variability" means too weak to be classified.

Climate drift occurs in a variety of forms. A general cooling of large parts of the ocean basin is the most common form of slow drift. Fast climate drift is characteristically a coupled-dynamical effect leading to an

Table 2 Summary of models grouped according to common behavior for both tropical climatology and interannual variability as reflected in the sea surface temperature field (after Neelin et al 1992)

Variability	Climate		
	Modest drift	Flux corrected	Other
Weak interannual variability		Latif et al-1 Neelin-2[b, c] Oberhuber et al-1[d]	Gordon & Ineson-2[a] Gates et al[d, e] Cubasch et al[d, e] Oberhuber et al-2[d, e]
Interannual variability with zonal propagation of SST anomalies	Lau et al[d]	Neelin-1[b]	Meehl & Washington[a, d, f] Gates & Sperber[a, d, f] Tokioka et al[a, f]
Interannual variability with standing SST anomalies	Philander et al Gent & Tribbia Nagai et al (1992)[g] Mechoso et al (1993)[g]	Zebiak & Cane[b] Allaart et al[b, c]	Schopf & Suarez[a, b] Latif & Sterl[a]

[a] Slow cooling of warm regions.
[b] Model with simplified atmospheric component.
[c] Multiple climate states known or suspected.
[d] Model with global-domain ocean component.
[e] Weak zonal gradient; weak cold tongue.
[f] Cold tongue extended or cold tongue/warm pool boundaries displaced.
[g] Dates given for recently added references; otherwise see Neelin et al (1992).

overly-weak or overly-strong equatorial cold tongue. Three-dimensional feedbacks between SST, convection zones, wind stress, and ocean circulation qualitatively similar to those responsible for El Niño are seen to play a role in creating such drift or in exacerbating weaknesses in parameterizations controlling one-dimensional, vertical-column processes such as cloud-radiative interaction or vertical mixing. We note many situations where the position of the cold tongue migrates or extends within the basin, with a warm pool developing in the eastern part of the basin in some instances. The observed convection zone in the eastern Pacific stays north of the equator in all seasons; in some models it migrates across the equator with season. The similarities between the fast mode of climate drift to interannual phenomena of comparable time scale implies that, unlike numerical weather prediction—in which correction of climate drift was only addressed as the models matured—interannual climate forecasting with coupled GCMs must address the accurate simulation of certain aspects of the climatology at a relatively early stage.

We find that there is little relation between the presence of climate drift and the existence of significant interannual variability, so long as the cold tongue is present somewhere in the basin. Interannual variability tends to come in two varieties: cases in which anomalies in SST, wind, etc propagate in the longitudinal direction along the equator and cases in which anomalies develop as a standing oscillation in the cold tongue region. In the latter case, fine ocean model resolution is required near the equator and subsurface memory due to oceanic adjustment processes is believed to determine the time scales; in the former case, coarse ocean model resolution does not preclude interannual oscillations and the time scales of ocean wave dynamics are not essential to the period.

Figure 5 provides an example of interannual variability from one of the first coupled GCMs with a high-resolution tropical ocean component (Philander et al 1989, 1992; Philander et al in Table 2). While the spectrum of interannual time scales may not exactly match that observed (possibly due to the removal of the seasonal cycle in this model for hypothesis-testing purposes), the spatial form, again with dominant standing oscillations in SST and with subsurface phase lags, is reasonably close to the observed form; Chao & Philander (1993) also compare these results to the uncoupled ocean component forced with obsrved winds to provide a longer surrogate time series for the subsurface anomalies. A number of other CGCMs have variations on this spatial form, some having clearer propagation characteristics in SST, combined with significant subsurface phase lags (e.g. Nagai et al 1992, Latif et al 1993a).

The rich variety of coupled phenomena found in these models serves as an indication of the sensitivity of the coupled system and lends support to qualitative arguments that coupled feedbacks are crucial in establishing tropical climate features. Even the most important features, such as the extent and position of the equatorial cold tongue and western Pacific warm pool, are not guaranteed to be reproduced in coupled GCMs. The lack of robustness in these features does not necessarily imply major faults in the models since coupled feedbacks can turn a small deficiency in one of the components into a significant departure in the coupled climatology. For example, a tendency of the atmospheric model to give slightly weak easterlies can result in a weaker cold tongue which in turn further weakens the Trades. In some models this can lead to a permanent warm state, although in others, weak AGCM stresses do not adversely affect either climatology or interannual oscillations.

Because the behavior of the coupled system can be qualitatively different (and difficult to anticipate) from that of the individual components, coupling should be regarded as a crucial part of the testing and development procedure for AGCMs and OGCMs being used for climate studies. In particular, the simulation of the warm-pool/cold-tongue configuration in the equatorial Pacific can represent a stringent test of the combined effects

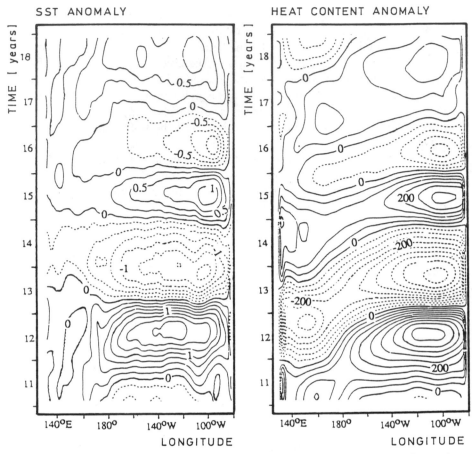

Figure 5 Time-longitude plot of anomalies along the equator from the Philander et al (1992) coupled GCM. (*Left*) SST (contour interval 0.25°C). (*Right*) heat content integrated above 300 m (contour interval 50 C° m). After Chao & Philander (1993). The data have been low-pass filtered to remove time scales less than 24 months.

of vertical-mixing parameterizations, interactive cloud-radiative schemes, and surface-flux parameterizations with the three-dimensional dynamics. The rate of improvement of recent model versions (both those in the table and currently unpublished models) is particularly encouraging in this respect.

Theory

CONTEXT AND HISTORY   The considerable differences in the nature of the coupled variability produced by the different models above is related to the sensitivity and the rich variety of flow regimes found in ICMs and simple models, which exhibit multiple mechanisms of coupled interaction. The character of the interannual variability in nonlinear models is largely determined by the first bifurcation from the climate state (Neelin 1990, Münnich et al 1991)—in other words by the leading unstable mode of the system linearized about the climatological state. Many of the most pressing questions about the range of coupled variability found in coupled models can thus be addressed by understanding the relation between flow regimes in the linear problem. To keep this multiparameter bifurcation problem tractable, the key is to choose a few crucial parameters that capture the range of behavior of interest, and to map out the connections among regimes close to that of the real system and those that provide useful simplifications.

In the literature, the search for simple prototype systems to provide conceptual analogs for the modes of coupled variability has led in a number of apparently contradictory directions, and it is desirable to bring these together. We approach this by presenting first a version of the CZ ICM scaled to highlight parameters used to show these connections succinctly. We derive three important simple models from this and discuss the differing idealizations. We then return to the ICM to show how the simple models relate to the connections between eigenmodes in the coupled parameter space. This completes the discussion of the primary bifurcation, i.e. how the period and spatial form of the ENSO cycle are determined and its maintenance through instability of the climatological state. We conclude with a discussion of higher bifurcations and describe what is known about the sources of irregularity in the ENSO cycle.

In ordering the presentation to emphasize a unified view, the historical aspects are necessarily simplified, so we preface with a brief overview of the literature (see also McCreary & Anderson 1991, Ghil et al 1991). Early theoretical work includes low-order models by McWilliams & Gent (1978) and some nonrotating coupled cases (Lau 1981a). Models by McCreary (1983) and McCreary & Anderson (1984) have often been omitted from recent citation because of the use of a discontinuous switch in their atmosphere, but elements of their discussion of basin adjustment processes have been incorporated in later work. Philander et al (1984) presented the first linear instability study in a coupled modified-shallow-water system, and refinements and additional mechanisms were elaborated numerically in Gill (1985), Yamagata (1985), Hirst (1986, 1988), Battisti & Hirst (1989), Wakata & Sarachik (1991), and analytically in Neelin (1991). Nonlinear solutions in ICMs were introduced in Cane & Zebiak (1985) and Zebiak & Cane (1987), in a regime now felt to approximate that of the observed, and by Anderson & McCreary (1985), Yamagata & Masumoto (1989) in a different regime; hints at regime connections may be found in Xie et al (1989) and Wakata & Sarachik (1991).

Much of the terminology used in these papers is based on the Rossby and Kelvin modes of the uncoupled ocean in an infinite or periodic basin, presumably because these are most familiar to oceanographers. A significant step toward thinking in terms of the fully coupled problem was advanced by Schopf & Suarez (1988) and Suarez & Schopf (1988) using a simple model with a single spatial variable to explain the oscillation in their ICM; Battisti & Hirst (1989) showed that a version of this model could be fitted to a number of important aspects of the oscillation in the CZ model, and that the Hopf-bifurcation regime was the physically relevant one. Referred to hereafter as the *SSBH delayed-oscillator model*, it consists of a differential-delay equation representing the time evolution of SST averaged over a small eastern equatorial box, with a net growth tendency representing local positive feedback mechanisms due to coupling and a delayed negative feedback representing the equatorial-wave adjustment process; whether the latter can be interpreted literally in terms of off-equatorial Rossby wave packets reflecting from the western boundary has been the subject of debate (Graham & White 1988, Battisti 1989, Chao & Philander 1993). The model is designed to represent the regime in which SST variability occurs as a standing oscillation in the strongly-coupled eastern basin, and in which time scales of ocean wave dynamics provide the memory of the system essential to the oscillation.

On the other hand, there exists a large class of coupled regimes in which ocean wave dynamics is not essential to interannual oscillation. In an idealized limit (the *fast-wave limit*), coupled modes are associated with the time derivative of the SST equation, and hence referred to as *SST modes*. These do involve subsurface ocean dynamics, but the time-dependence of this component is secondary. A distorted-physics method for testing this (involving articicial multipliers on selected OGCM time derivatives) was employed in Neelin (1991) to show the relevance of this limit to oscillations in one flow regime of an HCM. Hirst (1986, 1988) and Neelin (1991) showed, by numerical and analytical methods respectively, that a number of physical processes cooperate in the destabilization of SST modes whereas they compete in terms of the direction of propagation. Propagation is essential to the period in these modes and they provide a good prototype for slowly-propagating modes in a number of intermediate models and GCMs (e.g. Anderson & McCreary 1985, Yamagata & Masumoto 1989, Meehl 1990b, Lau et al 1992).

Because the SSBH delayed-oscillator model is based on the SST equation, it was natural to hypothesize that nonpropagating SST modes away from the fast-wave limit might be perturbed by wave time scales to produce standing oscillations. Such a connection is inherent in the analysis of Wakata & Sarachik (1991) in which the relation between a propagating regime of Hirst (1988) and a standing oscillation regime is demonstrated. In an apparent contradiction, two models aimed at producing more rigorous derivations of the SSBH delayed oscillator (Cane et al 1990 plus Münnich et al 1991, MCZ hereafter; and Schopf & Suarez 1990) emphasize a rather different limit. These models also assume that the coupling occurs at a single point rather than across all or most of the basin. SST-mode solutions in the fast-wave limit allow an analytical approach to the spatial structure of the coupled modes, inclusion of several growth mechanisms, and a determination of their relation to propagating regimes, but at the cost of eliminating subsurface memory. Jin & Neelin (1993a,b) and Neelin & Jin (1993; collectively JN hereafter) outlined the complementarity between these approaches and the usefulness of analytical prototypes which include solutions for the spatial structure of the coupled modes in various limits.

INTERMEDIATE COUPLED MODEL   We present here the JN "stripped-down" version of the CZ ICM, as a basis for deriving simpler models and discussing flow regimes. We nondimensionalize to bring out a few *primary parameters* from among the many lurking in the coupled system. These are:

μ: the relative coupling coefficient—strength of the wind-stress feedback from the atmosphere per unit SST anomaly, scaled to be order unity for the strongest realistic coupling; for $\mu = 0$ the model is uncoupled.

δ: the relative adjustment time coefficient—measures the ratio of the time scale of oceanic adjustment by wave dynamics to the time scale of adjustment of SST by coupled feedback and damping processes. It is scaled to be order unity at standard values of dimensional coefficients.

δ_s: surface-layer coefficient. This parameter governs the strength of feedbacks due to vertical-shear currents and upwelling, (u_s, v_s, w_s), created by viscous transfer between the surface layer and the rest of the thermocline. As $\delta_s \to 0$ the effects of these feedbacks become negligible.

A modified shallow-water model with an embedded, fixed-depth mixed layer (Cane 1979, Schopf & Cane 1983) provides the ocean-dynamics component:

$$(\delta\partial_t + \varepsilon_m)u'_m 2 - yv'_m + \partial_x h' = \tau'$$

$$yu'_m + \partial_y h' = 0$$

$$(\delta\partial_t + \varepsilon_m)h' + \partial_x u'_m + \partial_y v'_m = 0 \qquad (1)$$

$$\varepsilon_s u_s' - y v_s' = \delta_s \tau'$$

$$\varepsilon_s v_s + y u_s' = 0, \qquad (2)$$

where latitude, y, appears due to the nondimensionalized Coriolis force and the equations are applied here to departures (primed quantities), from a specified climatology (denoted by an overbar). Anomalous vertical mean currents above the thermocline, (u_m', v_m'), and thermocline depth, h', are governed by the shallow-water component in the long-wave approximation (1), with suitable boundary conditions at basin boundaries (Gill & Clarke 1974); vertical shear currents, (u_s', v_s'), are governed by local viscous equations (2). Both are driven by the zonal wind stress anomaly, τ'. The damping rates, ε_m and ε_s are not treated as primary parameters because the former is small and the latter can be largely absorbed into δ_s. For a more formal scaling see JN; for justification of several approximations, see Cane (1979) and CZ. Vertical velocities are given by the divergence of the horizontal velocities and the values of surface currents and upwelling into the surface layer by the sum of anomalous mean and shear contributions plus the climatology: $u = \bar{u} + u_m' + u_s'$, $w = \bar{w} + w_m' + w_s'$.

Because SST serves as a key interfacial variable, careful parameterization of processes that affect SST are largely responsible for the success of the CZ model. The direct effects of temperature variations in the surface layer on pressure gradients are neglected in (1), but a prognostic equation for SST is carried separately which contains all the essential nonlinearity of the CZ model:

$$\partial_t T + u \partial_x T + \mathsf{H}(w) w (T - T_{sub}) + v \partial_y T + \varepsilon_T (T - T_0) = 0 \qquad (3)$$

in nondimensional form. Here T is total SST and H is an analytic version of the Heaviside function due to upstream differencing into the surface layer. The Newtonian cooling represents all physical processes that bring SST towards a radiative-convective-mixing equilibrium value, T_0. The subsurface temperature field, T_{sub}, characterizes values upwelled from the underlying shallow-water layer and is parameterized nonlinearly on the thermocline depth—deeper thermocline results in warmer T_{sub}. Motivated by the fact that the strongest SST response to upwelling, advection, and thermocline depth change are confined to a fairly narrow band along the equator for the phenonena of interest, Neelin (1991) applied this equation to the SST in an equatorial band, where each of the variables in (3) need only be evaluated at the equator, and where the $v \partial_y T$ term is replaced by a suitable upstream differencing. In the JN ICM, this captures all the essential behavior of the CZ model while permitting a number of analytical results to be generated in special cases.

The simple atmospheric models that provide a zeroth-order approximation to the wind-stress response to SST anomalies can be written

$$\tau' = \mu A(T'; x, y), \tag{4}$$

where μ is the coupling coefficient and $A(T'; x, y)$ is a linear but nonlocal function of T' over the entire basin. For a Gill (1980) model with a specified meridional profile of the forcing appropriate to the assumed SST y-dependence, A is a simple integral operator.

Coupling is carried out by a version of flux correction: running the ocean model with observed climatological wind stress to define the ocean climatological state ($\bar{u}$, $\bar{w}$, $\bar{T}$, . . .), then defining SST anomalies, T', with respect to this. A known climatological solution to the coupled system is thus constructed. For sufficiently small coupling, this state is unique and stable; interannual variability must arise by bifurcations from this state as μ increases.

USEFUL LIMITS   We introduce terminology for limits that are useful for understanding how coupled modes relate to simpler cases and for comparing various theoretical models. The *weak-coupling limit* is reached at small μ, i.e. little wind-stress feedback per unit SST anomaly; these modes are found not to be good prototypes for fully coupled modes. At large μ, one obtains the *strong-coupling limit*. When the time scale of dynamical adjustment of the ocean is small compared to the time scale of SST change by coupled processes (i. e. small δ), one has the *fast-wave limit*; which is very useful for generating analytical results that provide understanding of spatial structure and growth characteristics. The *fast-SST limit* is reached at large δ; this is the converse to the fast-wave limit, i.e. sea surface temperature adjusts quickly compared to ocean dynamical processes.

In the uncoupled case or in the fast-wave limit, the modes of the ICM, linearized about its climatology, separate into a set associated purely with the time derivatives of the shallow-water equations, referred to as *ocean-dynamics modes*, and a set associated purely with the time derivative of the SST equation, referred to as *SST modes*. In an uncoupled, zonally-bounded basin, the ocean-dynamics modes consist of a set of ocean-basin modes (Cane & Moore 1981) and a scattering spectrum (JN). At low frequencies and basin scales, the ocean-dynamics modes are very different from the Rossby and Kelvin modes of the infinite-basin case. In the coupled system, the distinction between corresponding coupled modes is maintained in the idealized fast-wave and fast-SST limits, but in most of the parameter space the coupled modes will have a mixed nature, for which we use the descriptive term *mixed SST/ocean-dynamics modes* when it is necessary to be specific; otherwise the term *coupled modes* is taken to imply this.

WIND-DRIVEN OCEAN RESPONSE   Extensive theory exists for the adjustment of the uncoupled shallow-water ocean to time-varying winds (see Moore & Philander 1978, Cane & Sarachik 1983, McCreary 1985 for reviews). Much of it is phrased in terms of adjustment to abrupt changes in wind; a much better prototype for understanding interannual coupled oscillations is the case of forcing by low-frequency, time-periodic winds (Cane & Sarachik 1981, Philander & Pacanowski 1981). Figure 6 shows a time-longitude plot of thermocline perturbations along the equator for such a case. The western Pacific leads the eastern Pacific by between 90 and 180

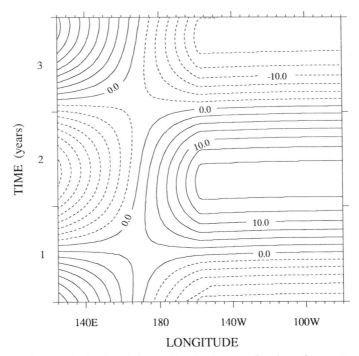

Figure 6   Time-longitude plot of thermocline depth anomalies along the equator from a shallow-water ocean model forced by specified wind stress: constant in longitude through the western half of the basin, zero in the eastern half, Gaussian in latitude (scale 5°), and periodic in time with period 3 years and amplitude 0.2 dyne/cm². Following Cane & Sarachik (1981) with specified modifications, and with frictional damping of time scale 0.5 yr in the ocean. Contour interval 2.5 m.

degrees in temporal phase. The fast-wave limit case, in which the ocean approaches equilibrium with the wind, would correspond to 180° phase difference between these, which would remove the apparent slow eastward "propagation." It should be emphasized that this is not a wave propagation in the sense of any individual free wave of the system, but rather the sum total of the ocean response which is not quite in equilibrium with the wind forcing. The slight departures from equilibrium, as measured by the difference from a 180° phase lag, characterize the oceanic memory which is so important to interannual variability.

SIMPLE MODELS: POINT COUPLING   One special case where solutions of (1) can be carried forward is if the surface-layer feedbacks are dropped ($\delta_s = 0$) and if it is assumed that *coupling to the atmosphere occurs at a single point*, e.g. at the eastern boundary. The wind-stress magnitude is taken to be proportional to SST at that point and the spatial form of wind stress is fixed, for instance to be a patch of very small longitudinal extent (here placed at mid-basin for ease of presentation), with Gaussian y-dependence of curvature α. We give here an extended version of the MCZ model (or Schopf & Suarez 1990) using this approximation; the SST equation (3) and ocean shallow-water dynamics (1) can be reduced to

$$\delta^{-1}\partial_t T' + [(T' - T'_{\text{sub}}(h'))] = 0 \tag{5a}$$

$$h'(t) - \sum_{j=1}^{\infty} a_j(\varepsilon_m)h'(t-4j) + \mu \sum_{j=0}^{\infty} b_j(\alpha, \varepsilon_m)T'(t-1/2-2j), \tag{5b}$$

where T' and h' are SST and thermocline depth anomalies at the eastern point, respectively, and the coefficients a_j, b_j summarize information about the ocean dynamics, boundary conditions, and parameters. The reduction of ocean dynamics to sums over discrete transit times results, of course, from the point-wise coupling assumptions. In contrast to (1), here time has been normalized by the time scale characterizing ocean dynamics (the Kelvin-wave basin-crossing time), so δ appears in the SST equation and the integer lag dependences on the past history of h' and T' are due to wave transit times across the basin with reflection at basin boundaries. This rescaling is because the MCZ model has been used primarily in the fast-SST limit ($\delta \to \infty$), which results in dropping the time derivative in (5a). In this case, (5) becomes an iterated map of high order for modes related to the time derivatives of the shallow-water equations, which yielded the lags.

A simpler delay equation which has proven influential in the field can be derived from the above by a series of simplifications which cannot be rigorously justified but which retain essential features of the dynamics:

(*i*) Set all $a_j = 0$ in (5b) while retaining the b_j, which amounts to removing eastern-boundary wave reflections while keeping those at the western boundary. This does irreparable damage to the uncoupled ocean dynamics, so the usefulness of this simplification depends on the coupling dominating the spatial structure of the mode—we show below how this comes about in an ICM. (*ii*) Move the wind stress to the point of coupling to SST (the position of the eastern boundary is now immaterial); this is reasonable when the coupled frequency is much less than the Kelvin transit time across the separation. (*iii*) Truncate the sum over b_j to only two terms— a single westward Rossby wave plus an eastward Kelvin wave. Because the series converges slowly, this can only be justified qualitatively. We present the model linearized about the climatology, since this determines the period and the location of the bifurcation:

$$\partial_t T' = (\mu b_0 - 1)T'(t) - \mu b_1 T'(t - 4\delta) \tag{6}$$

This is the *SSBH delayed-oscillator model*. Note that we have restored the time nondimensionalization used in (1–3). Nonlinear versions are straightforward to derive from (5) and differ significantly from that given by SSBH for more realistic $T_{sub}(h')$; $dT_{sub}/dh'|_0 = 1$ is used without loss of generality for this simple case. When the model is uncoupled ($\mu = 0$), the solution is a purely decaying mode whose eigenvalue is determined by the SST equation, i.e. an SST mode, in contrast to the MCZ model which also has uncoupled ocean modes. In the fast-wave limit ($\delta \to 0$), the model has stationary (i.e. nonoscillatory) SST modes which become unstable for coupling above $\mu = (b_0 - b_1)^{-1}$. For realistic values of δ, this unstable mode becomes oscillatory due to the adjustment time scales of subsurface ocean dynamics, here represented by a single delay. The SSBH model may thus be summarized as an SST-mode whose growth can be understood from the fast-wave limit, perturbed to give oscillation by aspects of ocean dynamics which are *not* characteristic of the uncoupled case. We will show below that this interpretation can be carried over to an ICM.

On the other hand, consider the modes of the extended MCZ model (Equation 5), linearized about climatology, with time dependence $\exp(\sigma t)$. The infinite series in (5b) can be summed exactly (Cane & Sarachik 1981) under certain conditions (note the contrast to the severe truncation of the SSBH model which has sometimes been interpreted too strongly in terms of individual waves). Equation (5b) becomes, in the simplest case:

$$[\sinh(2\sigma)/\sinh(\sigma)]h' = \mu T' \tag{5b'}$$

In the fast-SST limit, there is a singularity at $\mu = 2$, with two equal eigenvalues, demarcating the boundary between oscillatory eigenvalues below and stationary eigenvalues above, one of which is strongly growing.

In this case, the singularity leads to a codimension 2 (double-zero) bifurcation. When any other destabilizing process is added, the bifurcation is oscillatory (Hopf); with damping and no other destabilizing process it is stationary (transcritical)—but note that stationary bifurcations must be treated with caution since they are not robust to relaxation of assumptions used to construct the climatological state.) The oscillatory case is the one that applies to ENSO, but it is worth asking why the ocean dynamics "break" from oscillatory behavior, as would be expected of wave-related modes in a bounded basin, into a growing stationary mode. Consider the case where coupled feedback processes are very strong; then local interactions dominate nonlocal wave-propagation processes yielding pure growth— the mode grows too fast to be affected by weak return signals from the western boundary. The transition has to occur at moderate coupling. The remarkable feature which will be shown in an ICM below is that the stationary mode, even in the fast-SST limit, shares more characteristics with the SST mode in the fast-wave limit than with the uncoupled ocean modes. In fact, at strong coupling, the stationary mode eigensurface is continuously connected across the whole range of δ.

SIMPLE MODELS: FAST-WAVE LIMIT   Although the time scale of subsurface dynamics is the dominant factor in setting the period of ENSO, the strong simplifications that occur in the fast-wave limit permit insight into spatial structure. Setting $\delta = 0$ in (1), i.e. assuming that oceanic adjustment occurs fast compared to other time scales, and considering that the damping ε_m is very weak, reduces the shallow-water equations to Sverdrup balance along the equator:

$$\partial_x h' = \tau' \tag{7}$$

with negligible vertical mean currents. The off-equatorial ocean solution plays a significant role which can be summarized in boundary conditions to (7) suitably derived as the limit to wave adjustment processes, as discussed in JN and in Hao et al (1993), both of which provide further analysis of this fast-wave limit case. The multiple coupled feedback mechanisms can be seen from a linearized version of the SST equation (3), with h' given by combining Equations (7) and (4) (see Neelin 1991).

A number of physical mechanisms contribute to destabilization of SST modes. However, these mechanisms tend to compete in terms of whether the mode will be purely growing or will propagate slowly along the equator. For instance, a warm SST anomaly will lead to westerly wind anomalies above and to the west of the warm SST, which will lead to eastward current anomalies and reduced upwelling and thus to a warming of SST which will both enhance the original anomaly and cause it to shift slightly westward. In the ICM, these feedbacks are controlled by δ_s. On the other hand, the thermocline slope will tend to be reduced below and slightly to the east of the SST anomaly. The subsurface waters being carried to the surface will be warmer than normal, thus tending to enhance the initial anomaly. Since both the thermocline response and wind-stress response are nonlocal, the shape of the anomalies will evolve to satisfy basin boundary conditions, leading for a broad range of parameters to a stationary growing mode. Analytical results for both propagating and stationary cases can be obtained in the fast-wave limit (JN); Hao et al 1993 give nonlinear solutions. The mode with the largest spatial scale in the basin is always the most unstable, with SST and wind structure similar to observations. The larger SST anomalies in the east and central basin are produced partly by the shape of the climatological upwelling, and partly by purely dynamical effects, with east-west asymmetry introduced by the latitudinal derivative of the Coriolis force. The analytical results also indicate the role of the eastern boundary in keeping the mode from propagating; the point-coupling models emphasize the role of the western boundary on the ocean—but for spatial structure, eastern boundary effects enter mainly through the atmosphere.

The feedback loop described above sounds very similar to that described in the Bjerknes hypothesis. It gives the mechanism maintaining interannual variability and, suitably extended by the analytical results, the spatial form. However, it only gives the interannual period in regimes with coherent zonal propagation along the equator. This is a good prototype for the slowly-propagating modes in a number of coupled models (e.g. Meehl 1990b, Lau et al 1992), but to understand how this mode relates to the observed system, it is essential to see how time scales of subsurface ocean dynamics perturb it in the vicinity of the stationary regime to produce oscillations with standing-oscillatory SST anomalies.

PARAMETER DEPENDENCE OF LEADING MODES IN AN ICM   A global picture of the connection of coupled modes in the ICM (1)–(3) can be delineated by tracing the behavior of the few leading (fastest-growing or slowest-decaying) eigenmodes as a function of parameters μ and δ, beginning with $\delta_s = 0$ for simplicity. In the fast-wave limit ($\delta = 0$), a stationary (i.e. purely growing) SST mode becomes unstable. Its spatial structure is suggestive: It looks like the warm phase of Figure 4, except that the thermocline component has eastern and western parts of the basin exactly out of phase, so there is no oscillation. As one moves from the fast-wave limit to realistic relative-time-scale ratios (larger δ) one finds that this stationary eigenmode is scarcely changed. In fact, for coupling values stronger than a certain threshold (where coupled processes dominate those associated with oceanic wave propagation, as discussed above), the eigensurface extends without substantial change from the fast-wave limit all the way to the fast-SST limit. This is pivotal in understanding the coupled system because 1. it allows the spatial form and growth mechanisms of important coupled regimes to be understood from the fast-wave limit; and 2. it implies that modes associated with ocean dynamics must connect somehow to this strongly-growing mode.

To illustrate how this happens, Figure 7 shows a typical slice through parameter space as a function of coupling, for a realistic value of δ. The eigenvalues of the five leading modes are plotted as dots on the complex plane (growth-rate, frequency), for evenly spaced coupling values in the range $\mu = 0$ to 0.8. Left-right symmetry occurs because oscillatory modes always exist as conjugate pairs. The strongly growing stationary (i.e. non-oscillatory) branch in the strong-coupling range is indicated as "stationary regime" on the figure, since it is the only unstable mode in this parameter range. This is the mode that is so closely related to the SST mode in the fast-wave limit. At a slightly lower coupling value a singularity occurs where this is converted into an oscillatory mode—this singularity (corresponding to a codimension-2 double-zero bifurcation) extends as a curve in the μ-δ parameter plane connecting the eigensurface associated with the gravest SST mode to surfaces that are associated with ocean dynamics modes at low coupling. To the lower-coupling side of this singularity one finds the regime with oscillations that have a standing SST component (denoted in the figure as "standing-SST oscillatory regime'; SSO regime), corresponding to that shown in Figure 4 for the CZ ICM. The spatial form is similar to the stationary SST mode, and the mode is destabilized by the same coupling mechanisms, but subsurface oceanic dynamics provide the memory for the oscillation, characterized by temporal phase lag of the thermocline across the basin as in observations (Figure 3). This regime extends across a large range of δ, from $\delta = O(1)$ to the fast-SST limit.

In contrast, the connection of this standing-SST oscillation regime to the uncoupled case is complicated. In Figure 7, the SSO regime eigensurface eventually connects to one of the modes from the discretized scattering spectrum, but as it does so the mode rapidly changes in spatial form. When one includes variations in δ, one finds that the standing-oscillation regime connects, not to a single mode from the uncoupled oceanic dynamics spectrum, but to a series of them: The low coupling end of the branch attaches first to the lowest-frequency scattering mode (as in Figure 7), then to sequentially higher-frequency scattering modes, and finally at large δ, near the fast-SST limit, it connects to the gravest ocean basin mode, much as in the MCZ point-coupling model. These successive connections are

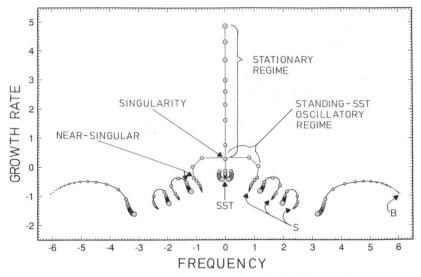

Figure 7 Eigenvalues of the five leading modes of the Jin & Neelin (1993a) intermediate coupled model as a function of coupling coefficient, μ, for a realistic value of the relative time scale coefficient, $\delta = 1.5$. Dots give frequency and growth rate of each mode on the complex plane, with dot size representing coupling for constant increments from $\mu = 0$ to $\mu = 0.8$. Eigenvalues trace out continuous paths as a function of coupling (indicated by interpolated lines for clarity). Uncoupled modes (ocean basin mode B, discretized scattering modes S, and an SST mode) are indicated at $\mu = 0$ (smallest dots). The modes have mixed character for larger μ: The purely growing mode which produces the *stationary regime*, indicated over the range of large μ, is closely related to the stationary SST mode; this is connected at a singularity to the important *standing-SST oscillatory regime* which extends over a range of moderate coupling values.

accomplished by a sequence of additional singularities; a "near-singular" point is shown, where the SSO regime connects to the next scattering mode at larger δ. However as δ varies, the characteristics of the SSO regime are almost completely insensitive to which uncoupled ocean mode it is attached to; its properties are fundamentally deterined by the coupling and it is thus best approached conceptually from the strong-coupling side.

It is thus much simpler to view the standing-oscillation regime as an extension of the strongly-growing stationary mode towards lower coupling, where ocean dynamics begin to regain some aspects of wavelike behavior. In this interpretation, one begins by understanding the spatial form and instability mechanisms of the mode in the fast-wave limit at fairly strong coupling. As one follows the stationary mode out to realistic values of the relative time-scale parameter and down to moderate coupling, it retains its form but acquires a frequency associated with "picking up" a degree of freedom from among the low-frequency part of the scattering spectrum on the low-coupling side. This view is consistent, in terms of physical content, with the original interpretation of the SSBH delayed oscillator model (Equation 6) (as long as it is understood that the subsurface memory is not associated with individual waves). Furthermore, the smooth connection from the fast-wave limit to the fast-SST limit implies that the seemingly contradictory approaches to the problem represented by (6) and (5b′) are just alternative approximations to the same eigensurface.

Finally, to make the connection to propagating regimes such as occur in some of the models, which may be relevant to the differences in evolution of certain ENSO events, consider reintroducing a third parameter, such as δ_s. As this changes, it is easy to move smoothly and gradually from the standing-oscillation regime to a regime of the mixed SST/ocean-dynamics modes where propagation occurs during parts of the cycle and contributes to the period (JN, Kleeman 1993). The standing-oscillation regime provides the clearest case emphasizing the role of subsurface dynamics in determining periodicity; the fast-wave-limit propagating cases provide alternate simple cases in which periodicity is provided by zonal phase lags. Between these continuously connected regimes, both characteristics can coexist within the same coupled mode. There is thus no contradiction between evidence for importance of subsurface dynamics in the ENSO cycle and indications of other contributing mechanisms.

TRANSITIONS TO IRREGULARITY   The modeling consensus is thus that ENSO dynamics are fundamentally oscillatory. In particular, for models whose uncoupled components have no internal variability, interannual variability arises as a forward Hopf bifurcation of the coupled system, yielding a limit cycle. The obvious question is then the source of irregularity in the observed cycle: (a) transition to chaotic behavior by higher bifurcations associated with the coupled dynamics, and/or (b) stochastic forcing by atmospheric "noise" from shorter-lived phenomena which do not depend on coupling?

With regard to internal dynamics, CZ pointed out early on that their model achieved a degree of irregularity through deterministic coupled dynamics alone. Disagreement by Battisti & Hirst (1989) over whether this was due to the CZ numerical implementation seems to have been settled in the larger picture in favor of the original finding; for instance, the smoothly posed, simpler version of JN also possesses irregular regimes. Explicit discussion of the bifurcation structure of the coupled system and secondary bifurcations to regimes of complex behavior was given in an HCM in Neelin (1990), but the first clear demonstration of a bifurcation sequence into chaotic behavior was given by MCZ in the point-coupling model (5). An unfortunate footnote must be added for Vallis (1986), who attempted to raise these questions in an ad hoc model (now thought to lack essential physics) and instead illustrated the dangers of spurious chaos due to highly-truncated numerics (Vallis 1988). As to the scenario for the transition to chaos, MCZ cite the Ruelle-Takens-Newhouse scenario (e.g. Eckmann 1981) as a possibility, based upon their observing irregular behavior subsequent to one period doubling. However, it is clear that the presence of parametric forcing by the annual cycle plays an important role in the prevalence of chaotic regimes in parameter space (Zebiak & Cane 1987, MCZ) and in the widespread frequency-locked regimes (CZ, Battisti & Hirst 1989, Schopf & Suarez 1990, Barnett et al 1993). A "Devil's staircase" scenario (e.g. Jensen et al 1984) is among current postulates (F.-F. Jin et al, personal communication; E. Tziperman et al, personal communication).

On the stochastic forcing side, early discussions of ENSO were often phrased in terms of random wind events initiating an El Niño warm phase. Among modelers this has given way to the view that stochastic forcing more likely disrupts the cycle (or possibly excites a weakly-decaying oscillatory mode, if below the Hopf bifurcation). Zebiak (1989) indicates that such effects have only a minor impact in the CZ ICM; Latif & Villwock (1990) and T. P. Barnett et al (personal communication) indicate that the effects of randomized atmospheric forcing can be considerable on an uncoupled tropical ocean model. Problems in quantifying the importance of stochastic effects involve estimation of spatial coherence, which is extremely important to ocean response and to separation of the stochastic component from the atmospheric deterministic response to SST.

Prediction and Predictability

The underlying periodic aspects of ENSO and the above theoretical considerations imply a good deal of ENSO predictability. A hierarchy of ENSO prediction schemes has been developed which includes statistical and physical models (Inoue & O'Brien 1984, Cane et al 1986, Graham et al 1987a,b, Xu & von Storch 1990, Goswami & Shukla 1991, Keppene & Ghil 1992, Barnston & Ropelewski 1992, Latif et al 1993b, Penland & Magorian 1993). A more complete list of references can be found in the review papers by Barnett et al (1988) and Latif et al (1993c). The most succesful schemes, the coupled ocean-atmosphere models, show significant skill in predicting ENSO even at lead times beyond one year. Figure 8 shows the anomaly correlation of the observed with the predicted SST anomalies averaged over the region of greatest variability for the CZ ICM—the first coupled model used for ENSO forecasts. Comparable

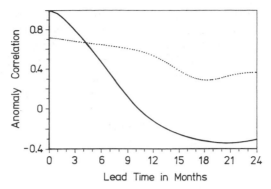

Figure 8   Skill scores of prediction ensembles as a function of lead time for forecasts by the Cane & Zebiak (1985) intermediate coupled model (*dotted curve*), compared with skill obtained by assuming persistence of anomalies (*solid curve*). The measure is correlation of predicted and observed SST averaged over the region of largest ENSO anomalies ($\pm 5°$ latitude, 150°W to 90°W longitude), during the period 1972 to 1991. Note that 0-month lead can differ from observed because SST data are not used in the initialization. Data from S. Zebiak (personal communication); for methodology see Cane et al (1986).

results have recently been obtained with CGCMs (Latif et al 1993b) and HCMs (Barnett et al 1993).

At lead times of a few months, the coupled models do not beat the persistence forecast which assumes that the SST anomalies remain constant throughout the forecast period. This is due to the fact that up to present no ocean observations are used in the initialization of the coupled models. Instead, the observed wind stresses are used to spin up to the ocean component, but errors in the forcing and the model formulation manifest themselves as considerable errors in the initial SST anomaly fields. Thus significant improvement of the forecasts at small lead times can be achieved by assimilating in situ ocean obervations (e.g. Leetmaa & Ji 1989) which are becoming increasingly available (e.g. Hayes et al 1991), and/or obser-vations from space (e.g. Tai et al 1989). In the case of coupled GCMs, a further reduction of climate drift will greatly aid this process. Upper limits on predictability are an area of current investigation (Blumenthal 1991, Goswami & Shukla 1991, Keppenne & Ghil 1992).

THE EXTRATROPICAL PROBLEM

Already in the late 1950s and early 1960s possible large-scale air-sea interactions in midlatitudes over both the Pacific and Atlantic Oceans were described by several authors (e.g. Namias 1959, Bjerknes 1964). Theoretical work by Hasselmann (1976) showed that the ocean can convert the white noise forcing by the atmosphere into a red noise SST spectrum through its large heat capacity. Such low-frequency SST anomalies can potentially feed back onto the atmospheric circulation. Recent modeling results suggest that the midlatitudinal atmosphere may indeed be significantly influenced by midlatitudinal SST anomalies, especially on interdecadal time scales (e.g. Hense et al 1990). The response characteristics, however, appear to be much more complex than in the tropics. Since the understanding of the extratropical problem is still at a rather low level, we restrict ourselves to pointing out four of the most important differences from the tropical problem.

First, the midlatitudinal circulation is influenced not only by midlatitudinal but also by tropical SST anomalies, as shown in many observational and modeling studies (e.g. Shukla & Wallace 1983, Lau 1985). A characteristic response pattern, the Pacific/North-America pattern, describing the response of the atmospheric winter circulation to tropical Pacific SST anomalies associated with the extremes in the ENSO cycle, has been identified (Wallace & Gutzler 1981) and exploited for short-range climate predictions for the North Pacific/North American region (Barnett & Preisendorfer 1987).

Second, both atmosphere and ocean have a much higher level of uncoupled internal variability. Uncoupled ocean models can produce decadal- or centennial-scale variations (e.g. Weaver et al 1991, Mikolajewicz & Maier-Reimer 1990). The effect of slowly-varying ocean boundary conditions on the atmosphere can be overwhelmed by the atmospheric noise level; for instance, Lau (1985), comparing AGCM runs with observed and climatological SST, found that observed SST variations produced a significant increase in atmospheric variability only in the tropics, while the midlatitudinal atmosphere exhibits a realistic level of interannual variability even in the case with climatological SST (Lau 1981b). Both effects make it difficult to assess the role of coupling on observed interannual to interdecadal variability (e.g. Gordon et al 1992). For instance, Delworth et al (1993) provide an analysis in a coupled GCM integration of Atlantic interdecadal variability involving changes in the overturning thermohaline circulation and advection-induced changes in density. While these phenomena have signatures in SST and air temperature, they are hypothesized to be uncoupled oceanic phenomena, driven by stochastic forcing from the atmosphere.

Third, the response of the general circulation in midlatitudes to SST anomalies (tropical and extratropical) is highly nonlinear (Kushnir & Lau 1992), while the response of the tropical atmosphere can be approximated by linear dynamics (e.g. Gill 1980). Experiments with general circulation models provide an opportunity to further investigate the relationship between extratropical SST anomalies and atmospheric flow regimes. Palmer & Sun (1985) investigated the reponse of the atmosphere to SST anomalies in the northwestern Atlantic. They showed that the model response was consistent with data and concluded that a positive feedback between ocean and atmosphere is possible during certain times of the year which might contribute to persistent climate anomalies. Some evidence of impacts of extratropical SST anomalies on the general circulation is also provided by Lau & Nath (1990) and Kushnir & Lau (1992) but the relationships between midlatitudinal SSTs and atmospheric indices appear to be far more complicated than in the tropics, in part due to the importance of transient disturbances to the time-averaged response.

Finally, direct effects of local air-sea heat exchange on the ocean play a more active role at midlatitudes than in the tropics where SST anomalies result primarily from variations in the surface wind stress. Persistent large-scale midlatitudinal SST fluctuations can be identified in Atlantic, Pacific, and global time series (Wallace & Jiang 1987, Wallace et al 1990, Folland et al 1991, Ghil & Vautard 1991). These anomalies may be driven by anomalies in the surface heat flux (e.g. Alexander 1992a, Cayan 1992), at least on monthly-to-interannual time scales. Kushnir (1993) argues that ocean circulation is important on longer time scales. Part of the interannual variability of SST in the North Pacific is linked to the ENSO phenomenon (Weare et al 1976, Luksch et al 1990, Alexander 1992a) and results from changes in the atmospheric circulation over the North Pacific in response to tropical SST anomalies. During an El Niño (warm) phase, for instance, an anomalous low-pressure system develops over the North Pacific, thereby strengthening the Aleutian Low. The changes in surface wind stress and more importantly those in surface heat flux force negative SST anomalies in the central North Pacific; the reverse occurs during an ENSO cold phase. The temperature near the American west coast tends to vary in phase with the tropical anomaly and is probably related in part to coastal Kelvin waves, which are generated by the reflection of equatorial Kelvin wave packets. Anomalous warm air advection in response to the strengthening of the Aleutian Low also plays a significant role in the generation of these anomalies. Pitcher et al (1988) show that these North Pacific SST anomalies can contribute a considerable atmospheric response; on the other hand, Alexander (1992b) finds that the local ocean-atmosphere feedback tends to act as a damping on the North Pacific SST response to ENSO.

SUMMARY AND DISCUSSION

The past decade has seen our knowledge of ocean-atmosphere interaction for the tropical problem go from the level of hypothesis to that of a field with rapidly developing theoretical and numerical modeling components. Theory for the El Niño/Southern Oscillation phenomenon has reached a relatively mature level for understanding the mechanisms contributing to the maintenance and period of the cycle, as they relate to the primary bifurcation from the climate state in models of different levels of complexity. The relationship between several regimes of interannual variability found in models has been largely understood, as has the complementary relationship between simple prototypes for the modes of coupled variability. These illustrate both the importance of subsurface ocean dynamics in providing the memory of the system, and the fundamental impact of coupling in determining the spatial character of these modes. The exact mechanism of the two apparent time scales in the ENSO signal and the dominant sources of irregularity in the cycle are not yet understood, although hypotheses have been posed in terms of the higher bifurcations of the coupled system or stochastic forcing due to uncoupled variability.

Models that capture the primary bifurcation in a realistic regime have been used to skillfully predict ENSO-related tropical Pacific SST anomalies at lead times out to a year. The potential for predictability beyond this is not yet known; a major area of current endeavor is ascertaining to what degree such tropical predictability can translate into useful midlatitude climate predictions on seasonal-to-interannual time scales.

Simulation of tropical climate and ENSO-related variability with coupled GCMs is improving at a rapid rate. The climate drift and variety of regimes of variability in earlier versions of these models are characteristic of the sensitivity of the coupled system and provide an apt demonstration that a coupled model is more complex than the sum of its uncoupled components. Because three-dimensional feedbacks tended to exacerbate relatively small errors in physical parameterizations in some of the early versions, small improvements in these parameterizations have in several cases provided highly encouraging improvements in simulation. This rapid learning curve for the tropical problem is partly the result of not needing to explicitly simulate the global thermohaline circulation which maintains the deep-ocean temperature and salinity through high-latitude convective sinking. Coupled GCMs for phenomena involving this circulation may have a longer development time to achieve accurate simulation without flux correction.

Exciting new areas within the tropical problem include: ocean-atmosphere interactions within the Atlantic and Indian basins, multi-basin interactions, and possible interactions with neighboring land processes (e.g. Southeast Asian and Indian Monsoon circulations, Tibetan plateau snow cover, Sahel rainfall, and South American convergence zones). Monsoon-ENSO interactions have already received considerable speculation (e.g. Yasunari & Seki 1992, Webster & Yang 1992); given the complexity of coupled processes in the tropical Pacific basin alone, unraveling next-order linkages to other subsystems will be a true challenge to models at all levels in the hierarchy. Circumstantial evidence from the coupled GCMs suggests the importance of coupled interactions in maintaining major features of the tropical climate and seasonal cycle, for instance, the warm-pool/cold-tongue configuration in the Pacific, and that the mechanisms involved may be qualitatively similar to those active in interannual variability. Developing a theoretical understanding of how these apply to the climatology would be a valuable asset both from a conceptual point of view, and for distinguishing the plausible from the speculative in tropical aspects of global-change scenarios.

While the midlatitude coupled problem is complicated by large internal variability of both atmosphere and ocean, there is reason to hope that the enthusiasm and experience that have accumulated for coupled interactions in the tropics will be carried to higher latitudes. There is growing attention to internal climate variability at decadal and longer time scales both in the tropics and extratropics, due to its importance in the problem of detection of anthropogenic warming and as a new frontier in simulation and theory. This will no doubt lead to a plethora of hypothesized mechanisms which may take decades to refute or verify due to the lack of long observational time series of dynamically-important quantities. Nonetheless, we can look forward to the need for a review of coupled ocean-atmosphere dynamics for the extratropical problem and new aspects of the tropical problem within a relatively few years.

ACKNOWLEDGMENTS

Support for this study has been provided by National Science Foundation ATM-9215090, Presidential Young Investigator award ATM-9158294, National Oceanographic and Atmospheric Administration NA16RC0178/26GP00114-01, CEC Environmental Programme EV5V-CT-0121, the Max-Planck-Institut für Meteorologie, and the Max-Planck Society (partial support during a sabbatical leave by JDN at MPIM). We thank M. Grunert and W. Weibel for assistance with graphics, and many colleagues for constructive comments.

Literature Cited

Alexander, M. A. 1992a. Midlatitude atmosphere-ocean interaction during El Niño. Part I: the North Pacific. *J. Climate* 5: 944–58

Alexander, M. A. 1992b. Midlatitude atmosphere-ocean interaction during El Niño. Part II: the Northern hemisphere atmosphere. *J. Climate* 5: 959–72

Allen, M. R., Davey, M. K. 1993. Empirical parameterization of tropical ocean-atmosphere coupling: the "inverse Gill problem." *J. Climate* 6: 509–30

Anderson, D. L. T., McCreary, J. P. 1985. Slowly propagating disturbances in a coupled ocean-atmosphere model. *J. Atmos. Sci.* 42: 615–29

Barnett, T. P. 1983. Interaction of the monsoon and Pacific Trade Wind system at interannual time scales. Part I: the equatorial zone. *Mon. Weather Rev.* 111: 756–73

Barnett, T. P. 1991. The interaction of multiple time scales in the tropical climate system. *J. Climate* 4: 269–85

Barnett, T. P., Graham, N., Cane, M., Zebiak, S., Dolan, S., O'Brien, J., Legler, D. 1988. On the prediction of El Niño of 1986–1987. *Science* 241: 192–96

Barnett, T. P., Latif, M., Graham, N., Flügel, M., Pazan, S., White, W. 1993. ENSO and ENSO-related predictability. Part I: prediction of equatorial Pacific sea surface temperature with a hybrid coupled ocean-atmosphere model. *J. Climate* 6: 1545–66

Barnett, T. P., Latif, M., Kirk, E., Roeckner, E. 1991. On ENSO physics. *J. Climate* 4: 487–515

Barnett, T. P., Preisendorfer, R. 1987. Origins and levels of monthly forecast skill for United States surface air temperatures determined by canonical correlation analysis. *Mon. Weather Rev.* 115: 1825–50

Barnston, A. G., Ropelewski, C. F. 1992. Prediction of ENSO episodes using canonical correlation analysis. *J. Climate* 5: 1316–45

Battisti, D. S. 1988. The dynamics and thermodynamics of a warming event in a coupled tropical atmosphere/ocean model. *J. Atmos. Sci.* 45: 2889–919

Battisti, D. S. 1989. On the role of off-equatorial oceanic Rossby waves during ENSO. *J. Phys. Oceanogr.* 19: 551–59

Battisti, D. S., Hirst, A. C. 1989. Interannual variability in the tropical atmosphere/ocean system: influence of the basic state, ocean geometry and nonlinearity. *J. Atmos. Sci.* 46: 1687–712

Berlage, H. P. 1957. Fluctuations in the general atmospheric circulation of more than one year, their nature and prognostic value. *K. Ned. Meteor. Inst. Meded. Verh.* 69: 152 pp.

Bjerknes, J. 1964. Atlantic air-sea interaction. *Adv. Geophys.* 10: 1–82

Bjerknes, J. 1969. Atmospheric teleconnections from the equatorial Pacific. *Mon. Weather Rev.* 97: 163–72

Blumenthal, M. B. 1991. Predictability of a coupled ocean-atmosphere model, *J. Climate* 4: 766–84

Bretherton, F. P. 1982. Ocean climate modeling. *Prog. Oceanogr.* 11: 93–129

Bryan, K., Manabe, S., Pacanowski, R. C. 1975. A global ocean-atmosphere climate model. Part II: the ocean circulation. *J. Phys. Oceanogr.* 5: 30–46

Bryan, K., Manabe, S., Spelman, M. J. 1988. Interhemispheric asymmetry in the transient response of a coupled ocean-atmosphere model to a CO_2 forcing. *J. Phys. Oceanogr.* 18: 851–67

Cane, M. A. 1979. The response of an equatorial ocean to simple wind stress patterns: I. Model formulation and analytic results. *J. Mar. Res.* 37: 233–52

Cane, M. A. 1986. El Niño. *Annu. Rev. Earth Planet. Sci.* 14: 43–70

Cane, M. A., Moore, D. W. 1981. A note on low-frequency equatorial basin modes. *J. Phys. Oceanogr.* 11: 1578–84

Cane, M. A., Münnich, M., Zebiak, S. E. 1990. A study of self-excited oscillations of the tropical ocean-atmosphere system. Part I: linear analysis. *J. Atmos. Sci.* 47: 1562–77

Cane, M. A., Sarachik, E. S. 1977. Forced baroclinic ocean motions: II. The linear equatorial bounded case. *J. Mar. Res.* 35: 395–432

Cane, M. A., Sarachik, E. S. 1981. The response of a linear baroclinic equatorial ocean to periodic forcing. *J. Mar. Res.* 39: 651–93

Cane, M. A., Sarachik, E. S. 1983. Equatorial oceanography. *Rev. Geophys. Space Phys.* 21: 1137–48

Cane, M. A., Zebiak, S. E. 1985. A theory for El Niño and the Southern Oscillation. *Science* 228: 1084–87

Cane, M. A., Zebiak, S. E., Dolan, S. C. 1986. Experimental forecasts of El Niño. *Nature* 321: 827–32

Cayan, D. R. 1992. Latent and sensible heat flux anomalies over the northern oceans: driving the sea surface temperature. *J. Phys. Oceanogr.* 22: 859–81

Chao, Y., Philander, S. G. H. 1993. On the structure of the Southern Oscillation. *J. Climate* 6: 450–69

Charnock, H., Philander, S. G. H., eds. 1989. The dynamics of the coupled atmosphere and ocean. *Proc. of a Royal Society Discussion Meeting.* London: R. Soc. 315 pp.

Cubasch, U., Hasselmann, K., Höck, H., Maier-Reimer, E., Mikolajewicz, U., et al. 1992. Time-dependent greenhouse warming computations with a coupled ocean-atmosphere model. *Climate Dyn.* 8: 55–69

Delworth, T. L., Manabe, S., Stouffer, R. J. 1993. Interdecadal variations of the thermohaline circulation in a coupled ocean-atmosphere model. *J. Climate* In press

Deser C., Wallace J. M. 1990. Large-scale atmospheric circulation features of warm and cold episodes in the tropical Pacific. *J. Climate* 3: 1254–81

Dickinson, R. E. 1983. Land surface processes and climate-surface albedos and energy balance. *Adv. Geophys.* 25: 305–53

Eckmann, J. P. 1981. Roads to turbulence in dissipative dynamical systems. *Rev. Mod. Phys.* 53: 643–53

Endoh, M., Tokioka, T., Nagai, T. 1991. Tropical Pacific sea surface temperature variations in a coupled atmosphere-ocean general circulation model. *J. Mar. Syst.* 1: 293–98

Folland, C., Owen, J., Ward, M. N., Colman, A. 1991. Prediction of seasonal rainfall in the Sahel region using empirical and dynamical methods. *J. Forecasting* 10: 21–56

Gates, W. L., Han, Y. J., Schlesinger, M. E. 1985. The global climate simulated by a coupled atmosphere-ocean general circulation model: preliminary results. See Nihoul 1985, 40: 131–51

Gates, W. L., Mitchell, J. F. B., Boer, G. J., Cubasch, U., Meleshko, V. P. 1992. Climate modeling, climate prediction and model validation. In *Climate Change 1992: The supplementary report to the Intergovernmental Panel on Climate Change Scientific Assessment*, ed. J. T. Houghton, B.A. Callander, S.K. Varney, pp. 97–133. Cambridge: Univ. Cambridge Press

Gent, P. R., Cane, M. A. 1989. A reduced gravity, primitive equation model of the upper equatorial ocean. *J. Comput. Phys.* 81: 444–80

Gent, P. R., Tribbia, J. J. 1993. Simulation and predictability in a coupled TOGA model. *J. Climate* In press

Ghil, M., Kimoto, M., Neelin, J. D. 1991. Nonlinear dynamics and predictability in the atmospheric sciences. *Rev. Geophys.*, Suppl., pp. 46–55, U.S. Natl. Rep. to the Int. Union of Geodesy and Geophys. 1987–1990

Ghil, M., Vautard, R. 1991. Interdecadal oscillations and the warming trend in global temperature time series. *Nature* 350: 324–27

Gill, A. E., 1980. Some simple solutions for heat induced tropical circulation. *Q. J. R. Meteorol. Soc.* 106: 447–62

Gill, A. E., 1985. Elements of coupled ocean-atmosphere models for the tropics. See Nihoul 1985, 40: 303–28

Gill, A. E., Clarke, A. J. 1974. Wind-induced upwelling, coastal current, and sea-level changes. *Deep-Sea Res.* 21: 325–45

Gill, A. E., Rasmusson, E. M. 1983. The 1982–1983 climate anomaly in the equatorial Pacific. *Nature* 306: 229–34

Gordon, C. 1989. Tropical ocean-atmosphere interactions in a coupled model. *Phil. Trans. R. Soc. London Ser. A* 329: 207–23

Gordon, A. L., Zebiak, S. E., Bryan, K. 1992. Climate variability and the Atlantic Ocean. *EOS, Trans. Am. Geophys. Union* 73(15): 161

Goswami, B. N., Shukla, J. 1991. Predictability of a coupled ocean-atmosphere model. *J. Climate* 4: 3–22

Graham, N. E., Michaelsen, J., Barnett, T. P. 1987a. An investigation of the El Niño-Southern Oscillation cycle with statistical models. 1. Predictor field characteristics. *J. Geophys. Res.* 92: 14,251–70

Graham, N. E., Michaelsen, J., Barnett, T. P. 1987b. An investigation of the El Niño-Southern Oscillation cycle with statistical models. 2. Model results. *J. Geophys. Res.* 92: 14,271–89

Graham, N. E., White, W. B. 1988. The El Niño cycle: Pacific ocean-atmosphere system. *Science* 240: 1293–302

Graham, N. E., White, W. B. 1990. The role of the western boundary in the ENSO cycle: experiments with coupled models. *J. Phys. Oceanogr.* 20: 1935–48

Hao, Z., Neelin, J. D., Jin, F.-F. 1993. Nonlinear tropical air-sea interaction in the fast-wave limit. *J. Climate* 6: 1523–44

Hasselmann, K. 1976. Stochastic climate models. Part I: theory. *Tellus* 28: 289–305

Hasselmann, K. 1988. PIPs and POPs: the reduction of complex dynamical systems using Principal Interaction and Principal Oscillation Patterns. *J. Geophys. Res.* 93D: 11,015–21

Hayes, S. P., Mangum, L. J., Picaut, L. J., Sumi, A., Takeuchi, K. 1991. TOGA-TAO: a moored array for real-time measurements in the tropical Pacific ocean. *Bull. Am. Meteorol. Sci.* 72: 339

Hense, A., Glowienka-Hense, R., von Storch, H., Stähler, U. 1990. Northern Hemisphere atmospheric response to changes of Atlantic Ocean SST on decadal time scales: a GCM experiment. *Climate Dyn.* 4: 157–74

Hirst, A. C. 1986. Unstable and damped equatorial modes in simple coupled ocean-atmosphere models. *J. Atmos. Sci.* 43: 606–30

Hirst, A. C. 1988. Slow instabilities in tropical ocean basin-global atmosphere models. *J. Atmos. Sci.* 45: 830–52

Houghton, J. T., Jenkins, G. J., Ephraums, J. J., eds. 1990. *Climate Change: The Intergovernmental Panel on Climate Change Scientific Assessment.* Cambridge: Univ. Cambridge Press. 364 pp.

Houghton, R. W., Tourre, Y. 1992. Characteristics of low frequency sea surface temperature fluctuations in the tropical Atlantic. *J. Climate* 5: 765–71

Inoue, M., O'Brien, J. J. 1984. A forecasting model for the onset of a major El Niño. *Mon. Weather Rev.* 112: 2326–37

Jensen, M. H., Bak, P., Bohr, T. 1984. Transition to chaos by interaction of resonances in dissipative systems. Part I. Circle maps. *Phys. Rev. A* 30: 1960–69

Jin, F.-F., Neelin, J. D. 1993a. Modes of interannual tropical ocean-atmosphere interaction—a unified view. Part I: numerical results. *J. Atmos. Sci.* 50: 3477–503

Jin, F.-F., Neelin, J. D. 1993b. Modes of interannual tropical ocean-atmosphere interaction—a unified view. Part III: analytical results in fully-coupled cases. *J. Atmos. Sci.* 50: 3523–40

Keppenne, C. L., Ghil, M. 1992. Adaptive filtering and prediction of the Southern Oscillation index. *J. Geophys. Res.* 97: 20,449–54

Kleeman, R. 1993. On the dependence of hindcast skill on ocean thermodynamics in a coupled ocean-atmosphere model. *J. Climate* In press

Kushnir, Y. 1993. Interdecadal variations in North Atlantic sea surface temperature and associated atmospheric conditions. *J. Climate* 6: In press

Kushnir, Y., Lau, N. C. 1992. The general circulation model response to a North Pacific SST anomaly: dependence on time scale and pattern polarity. *J. Climate* 4: 271–83

Lamb, P. J., Peppler, R. A., Hastenrath, S. 1986. Interannual variability in the tropical Atlantic. *Nature* 322: 238–40

Latif, M., Barnett, T. P., Cane, M. A., Flügel, M., Graham, N. E., et al. 1993c. A review of ENSO prediction studies. *Climate Dyn.* In press

Latif, M., Biercamp, J., von Storch, H. 1988. The response of a coupled ocean-atmosphere general circulation model to wind bursts. *J. Atmos. Sci.* 45: 964–79

Latif, M., Graham, N. E. 1992. How much predictive skill is contained in the thermal structure of an OGCM? *J. Phys. Oceanogr.* 22: 951–62

Latif, M., Sterl, A., Maier-Reimer, E., Junge, M. M. 1993a. Climate variability in a coupled GCM. Part I: the tropical Pacific. *J. Climate* 6: 5–21

Latif, M., Sterl, A., Maier-Reimer, E., Junge, M. M. 1993b. Structure and predictability of the El Niño/Southern Oscillation phenomenon in a coupled ocean-atmosphere general circulation model. *J. Climate* 6: 700–8

Latif, M., Villwock, A. 1990. Interannual variability as simulated in coupled ocean-atmosphere models. *J. Mar. Syst.* 1: 51–60

Lau, K. M. 1981a. Oscillations in a simple equatorial climate system. *J. Atmos. Sci.* 38: 248–61

Lau, N. C. 1981b. A diagnostic study of recurrent meteorological anomalies appearing in a 15-year simulation with a GFDL general circulation model. *Mon. Weather Rev.* 109: 2287–311

Lau, N. C. 1985. Modelling the seasonal dependence of the atmospheric response to observed El Niños in 1962–76. *Mon. Weather Rev.* 113: 1970–96

Lau, N. C., Nath, M. J. 1990. A general circulation model study of the atmospheric response to extratropical SST anomalies observed during 1950–79. *J. Climate* 3: 965–89

Lau, N. C., Philander, S. G. H., Nath, M. J. 1992. Simulation of El Niño/Southern Oscillation phenomena with a low-resolution coupled general circulation model of the global ocean and atmosphere. *J. Climate* 5: 284–307

Leetmaa, A., Ji, M. 1989. Operational hindcasting of the tropical Pacific. *Dyn. Atmos. Oceans* 13: 465–90

Lindzen, R. S., Nigam, S. 1987. On the role of sea surface temperature gradients in forcing low level winds and convergence in the tropics. *J. Atmos. Sci.* 44: 2418–36

Lough, J. M. 1986. Tropical Atlantic sea surface temperature and rainfall variations in Subsaharan Africa. *Mon. Weather Rev.* 114: 561–70

Luksch, U., von Storch, H., Maier-Reimer, E. 1990. Modeling North Pacific SST anomalies as a response to anomalous atmospheric forcing. *J. Mar. Systems* 1: 155–68

Manabe, S., Bryan, K. 1969. Climate calculations with a combined ocean-atmosphere model. *J. Atmos. Sci.* 26: 786–89

Manabe, S., Bryan, K., Spelman, M. J. 1975. A global ocean-atmosphere climate model. Part I: the atmospheric circulation. *J. Phys. Oceanogr.* 5: 3–29

Manabe, S., Bryan, K., Spelman, M. J. 1979. A global ocean-atmosphere climate model with seasonal variation for future studies of climate sensitivity. *Dyn. Atmos. Oceans* 3: 393–426

Manabe, S., Bryan, K., Spelman, M. J. 1990. Transient response of a global ocean-atmosphere model to a doubling of atmospheric carbon dioxide. *J. Phys. Oceanogr.* 20: 722–49

Manabe, S., Spelman, M. J., Stouffer, R. J. 1992. Transient responses of a coupled ocean-atmosphere model to gradual changes of atmospheric CO_2. Part II: seasonal response. *J. Climate* 5: 105–26

Manabe, S., Stouffer, R. J. 1988. Two stable equilibria of a coupled ocean-atmosphere model. *J. Climate* 1: 841–66

Matsuno, T. 1966. Quasi-geostrophic motions in the equatorial area. *J. Meteorol. Soc. Jpn.* Ser. II 44: 25–43

McCreary, J. P. 1976. Eastern tropical ocean response to changing wind systems with application to El Niño. *J. Phys. Oceanogr.* 6: 632–45

McCreary, J. P. 1983. A model of tropical ocean-atmosphere interaction. *Mon. Weather Rev.* 111: 370–87

McCreary, J. P. 1985. Modeling equatorial ocean circulation. *Annu. Rev. Fluid Mech.* 17: 359–409

McCreary, J. P., Anderson, D. L. T. 1984. A simple model of El Niño and the Southern Oscillation. *Mon. Weather Rev.* 112: 934–46

McCreary J. P., Anderson D. L. T. 1991. An overview of coupled ocean-atmosphere models of El Niño and the Southern Oscillation. *J. Geophys. Res.* 96: 3125–50

McWilliams, J. C., Gent, P. R. 1978. A coupled air and sea model for the tropical Pacific. *J. Atmos. Sci.* 35: 962–89

Mechoso, C. R., Kitoh, A., Moorthi, S., Arakawa, A. 1987. Numerical simulations of the atmospheric response to a sea surface temperature anomaly over the equatorial eastern Pacific ocean. *Mon. Weather Rev.* 115: 2936–56

Mechoso, C. R., Ma, C.-C., Farrara, J. D., Spahr, J., Moore, R. W. 1993. Parallelization and distribution of a coupled atmosphere-ocean general circulation model. *Mon. Weather Rev.* In press

Meehl, G. A. 1990a. Development of global coupled ocean-atmosphere general circulation models. *Climate Dyn.* 5: 19–33

Meehl, G. A. 1990b. Seasonal cycle forcing of El Niño-Southern Oscillation in a global coupled ocean-atmosphere GCM. *J. Climate* 3: 72–98

Mikolajewicz, U., Maier-Reimer, E. 1990. Internal secular variability in an ocean general circulation model. *Climate Dyn.* 4: 145–56

Mitchell, J. F. B. 1989. The "greenhouse effect" and climate change. *Rev. Geophys.* 27: 115–39

Moore, D. W. 1968. *Planetary-gravity waves in an equatorial ocean.* PhD thesis. Harvard Univ. 207 pp.

Moore, D. W., Philander, S. G. H. 1978. Modeling of the tropical oceanic circulation. In *The Sea*, ed. E. D. Goldberg et al, 6: 319–61. New York: Wiley-Intersci.

Münnich, M., Cane, M. A., Zebiak, S. E. 1991. A study of self-excited oscillations in a tropical ocean-atmosphere system. Part II: nonlinear cases. *J. Atmos. Sci.* 48: 1238–48

Nagai, T., Tokioka, T., Endoh, M., Kitamura, Y. 1992. El Niño–Southern oscillation simulated in an MRI atmosphere-ocean coupled general circulation model. *J. Climate* 5: 1202–33

Namias, J. 1959. Recent seasonal interactions between North Pacific waters and the overlying atmospheric circulation. *J. Geophys. Res.* 64: 631–46

Neelin, J. D. 1989a. A note on the interpretation of the Gill model. *J. Atmos. Sci.* 46: 2466–68

Neelin, J. D. 1989b. Interannual oscillations in an ocean general circulation model coupled to a simple atmosphere model. *Phil. Trans. R. Soc. London Ser. A* 329: 189–205

Neelin, J. D. 1990. A hybrid coupled general circulation model for El Niño studies. *J. Atmos. Sci.* 47: 674–93

Neelin, J. D. 1991. The slow sea surface temperature mode and the fast-wave limit: analytic theory for tropical interannual oscillations and experiments in a hybrid coupled model. *J. Atmos. Sci.* 48: 584–606

Neelin, J. D., Held, I. M. 1987. Modelling tropical convergence based on the moist static energy budget. *Mon. Weather Rev.* 115: 3–12

Neelin, J. D., Jin, F.-F. 1993. Modes of interannual tropical ocean-atmosphere interaction—a unified view. Part II: analytical results in the weak-coupling limit. *J. Atmos. Sci.* 50: 3504–22

Neelin, J. D., Latif, M., Allaart, M. A. F., Cane, M. A., Cubasch, U., et al. 1992. Tropical air-sea interaction in general circulation models. *Climate Dyn.* 7: 73–104

Nihoul, J. C. J., ed. 1985. *Coupled Ocean-Atmosphere Models.* Amsterdam: Elsevier Oceanogr. Ser.

Nihoul, J. C. J., ed. 1990. *Coupled Ocean-Atmosphere Modeling.* Amsterdam: Elsevier. 313 pp.

Palmer, T. N., Mansfield, D. A. 1986. A study of wintertime circulation anomalies during past El Niño events using a high-resolution general circulation model. Part I: influence of model climatology. *Q. J. R. Meteorol. Soc.* 112: 613–38

Palmer, T. N., Sun, Z.-B. 1985. A modeling and observational study of the relationship of sea surface temperature in the northwestern Atlantic and the atmospheric general circulation. *Q. J. R. Meteorol. Soc.* 111: 947–75

Penland, C., Magorian, T. 1993. Prediction of Niño-3 sea surface temperatures using linear inverse modeling. *J. Climate* 6: 1067–76

Philander, S. G. H. 1981. The response of the equatorial oceans to a relaxation of the trade winds. *J. Phys. Oceanogr.* 11: 176–89

Philander, S. G. H. 1990. *El Niño, La Niña, and the Southern Oscillation.* San Diego: Academic. 293 pp.

Philander, S. G. H., Chao, Y. 1991. On the contrast between the seasonal cycles of the equatorial Atlantic and Pacific oceans. *J. Phys. Oceanogr.* 21: 1399–406

Philander, S. G. H., Lau, N. C., Pacanowski, R. C., Nath, M. J. 1989. Two different simulations of Southern Oscillation and El Niño with coupled ocean-atmosphere general circulation models. *Phil. Trans. R. Soc. London Ser. A* 329: 167–78

Philander, S. G. H., Pacanowski, R. C. 1980. The generation of equatorial currents. *J. Geophys. Res.* 85: 1123–36

Philander, S. G. H., Pacanowski, R. C. 1981. Response of the equatorial ocean to periodic forcing. *J. Geophys. Res.* 86: 1903–16

Philander, S. G. H., Pacanowski, R. C., Lau, N. C., Nath, M. J. 1992. Simulation of ENSO with a global atmospheric GCM coupled to a high-resolution, tropical Pacific ocean GCM. *J. Climate* 5: 308–29

Philander, S. G. H., Yamagata, T., Pacanowski, R. C. 1984. Unstable air-sea interactions in the tropics. *J. Atmos. Sci.* 41: 604–13

Pitcher, E. J., Blackmon, M. L., Bates, G. T., Muñoz, S. 1988. The effect of North Pacific sea surface temperature anomalies on the January climate of a general circulation model. *J. Atmos. Sci.* 45: 173–88

Rasmusson, E. M. 1984. El Niño: the ocean/atmosphere connection. *Oceanus* 27: 5–13

Rasmusson, E. M., Carpenter, T. H. 1982. Variations in tropical sea surface temperature and surface wind fields associated with the Southern Oscillation/El Niño. *Mon. Weather Rev.* 110: 354–84

Rasmusson, E. M., Wallace, J. M 1983. Meteorological aspects of El Niño/Southern Oscillation. *Science* 222: 1195–202

Rasmusson, E. M., Wang, X., Ropelewski, C. F. 1990. The biennial component of ENSO variability. *J. Mar. Syst.* 1: 71–96

Ropelewski, C. F., Halpert, M. S., Wang, X. 1992. Observed tropospheric biennial variability and its relation to the Southern Oscillation. *J. Climate* 5: 594–614

Sausen, R., Barthels, K., Hasselmann, K. 1988. Coupled ocean-atmosphere models with flux correction. *Climate Dyn.* 2: 154–63

Schlesinger, M. E., ed. 1990. *Climate-Ocean Interaction.* Dordrecht: Kluwer. 385 pp.

Schlesinger, M. E., Gates, W. L., Han, Y. J. 1985. The role of the ocean in CO_2-induced climate change: preliminary results from the OSU coupled atmosphere-ocean general circulation model. See Nihoul 1985, 40: 447–78

Schopf, P. S., Cane, M. A. 1983. On equatorial dynamics, mixed layer physics and sea surface temperature. *J. Phys. Oceanogr.* 13: 917–35

Schopf, P. S., Suarez M. J. 1988. Vacillations in a coupled ocean-atmosphere model. *J. Atmos. Sci.* 45: 549–66

Schopf, P. S., Suarez, M. J. 1990. Ocean wave dynamics and the time scale of ENSO. *J. Phys. Oceanogr.* 20: 629–45

Seager, R., Zebiak, S. E., Cane, M. A. 1988. A model of the tropical Pacific sea surface temperature climatology. *J. Geophys. Res.* 93: 1265–80

Servain, J. 1991. Simple climatic indices for the tropical Atlantic Ocean and some applications. *J. Geophys. Res.* 96: 15137–46

Servain, J., Legler, D. M. 1986. Empirical orthogonal function analyses of tropical Atlantic sea surface temperature and wind stress: 1964–1979. *J. Geophys. Res.* 91: 14181–91

Shukla, J., Fennessey, M. J. 1988. Prediction of time-mean atmospheric circulation and rainfall: influence of Pacific sea surface temperature anomaly. *J. Atmos. Sci.* 45: 9–28

Shukla, J., Wallace, J. M. 1983. Numerical simulation of the atmospheric response to equatorial Pacific sea surface temperature anomalies. *J. Atmos. Sci.* 40: 1613–30

Sperber, K. R., Hameed, S. 1991. Southern Oscillation in the OSU coupled upper ocean-atmosphere GCM. *Climate Dyn.* 6: 83–97

Sperber, K. R., Hameed, S., Gates, W. L., Potter, G. L. 1987. Southern Oscillation simulated in a global climate model. *Nature* 329: 140–42

Stouffer, R. J., Manabe, S., Bryan, K. 1989. Interhemispheric asymmetry in climate response to a gradual increase of atmospheric CO_2. *Nature* 342: 660–62

Suarez, M. J., Schopf, P. S. 1988. A delayed action oscillator for ENSO. *J. Atmos. Sci.* 45: 3283–87

Tai, C. K., White, W. B., Pazan S. E. 1989. Geosat crossover analysis in the tropical Pacific. 2. Verification analysis of altimetric sea level maps with expendable bathythermograph and island sea level data. *J. Geophys. Res.* 94C: 897–908

Trenberth, K. E., ed. 1993. *Climate System Modeling.* New York: Cambridge Univ. Press. 817 pp.

Vallis, G. K., 1986: El Niño: a chaotic dynamical system? *Science* 232: 243–45

Vallis, G. K. 1988. Conceptual models of El Niño and the Southern Oscillation. *J. Geophys. Res.* 93C: 13,979–91

Wakata, Y., Sarachik, E. S. 1991. Unstable coupled atmosphere-ocean basin modes in the presence of a spatially varying basic state. *J. Atmos. Sci.* 48: 2060–77

Walker, G. T. 1923. Correlation in seasonal variations of weather, VIII: a preliminary study of world weather. *Mem. Indian Meteorol. Dep.* 24: 75–131

Wallace, J. M., Gutzler, D. S. 1981. Teleconnections in the potential height field during the Northern Hemisphere winter. *Mon. Weather Rev.* 109: 784–812

Wallace, J. M., Jiang, Q. 1987. On the observed structure of the interannual variability of the atmosphere/ocean climate system. In *Atmospheric and Oceanic Variability*, ed. H. Cattle, pp. 17–43. London: Roy. Meteorol. Soc.

Wallace, J. M., Smith, C., Jiang, Q. 1990. Spatial patterns of atmosphere-ocean interaction in the northern winter. *J. Climate* 3: 990–98

Washington, W. M., Meehl, G. A. 1989. Climate sensitivity due to increased CO_2: experiments with a coupled atmosphere and ocean general circulation model. *Climate Dyn.* 4: 1–38

Washington, W. M., Parkinson, C. L. 1986. *An Introduction to Three-Dimensional Climate Modeling*. Mill Valley, Calif: Univ. Sci. Books. 422 pp.

Washington, W. M., Semtner, A. J., Meehl, G. A., Knight, D. J., Mayer, T. A. 1980. A general circulation experiment with a coupled atmosphere, ocean and sea ice model. *J. Phys. Oceanogr.* 10: 1887–908

Weare, B., Navato, A., Newell, R. F. 1976. Empirical Orthogonal analysis of Pacific Ocean sea surface temperatures. *J. Phys. Oceanogr.* 6: 671–78

Weaver, A. J., Sarachik, E. S., Marotzke, J. 1991. Freshwater flux forcing of decadal and interdecadal oceanic variability. *Nature* 353: 836–38

Webster, P. J., Yang, S. 1992. Monsoon and ENSO: selectively interactive systems. *Q. J. R. Meteorol. Soc.* 118: 877–926

Wolter, K. 1989. Modes of tropical circulation, Southern Oscillation, and Sahel rainfall anomalies. *J. Climate* 8: 149–72

Xie, S.-P., Kubokawa, A., Hanawa, K. 1989. Oscillations with two feedback processes in a coupled ocean-atmosphere model. *J. Climate* 2: 946–64

Xu, J.-S., von Storch, H. 1990. Predicting the state of the Southern Oscillation using principal-oscillation-pattern analysis. *J. Climate* 3: 1316–29

Yamagata, T. 1985. Stability of a simple air-sea coupled model in the tropics. See Nihoul 1985, pp. 637–57

Yamagata, T., Masumoto, Y. 1989. A simple ocean-atmosphere coupled model for the origin of warm El Niño Southern Oscillation event. *Phil. Trans. R. Soc. London Ser. A* 329: 225–36

Yasunari, T., Seki, Y. 1992. Role of Asian monsoon on the interannual variability of the global climate system. *J. Meteorol. Soc. Jpn.* 70: 177–89

Zebiak, S. E. 1986. Atmospheric convergence feedback in a simple model for El Niño. *Mon. Weather Rev.* 114: 1263–71

Zebiak, S. E. 1989. On the 30–60 day oscillation and prediction of El Niño. *J. Climate* 2: 1381–87

Zebiak, S. E. 1993. Air-sea interaction in the equatorial Atlantic region. *J. Climate* 6: 1567–86

Zebiak, S. E., Cane, M. A. 1987. A model El Niño Southern Oscillation. *Mon. Weather Rev.* 115: 2262–78

SUBJECT INDEX

661

CUMULATIVE INDEXES

CONTRIBUTING AUTHORS, VOLUMES 1–26

683

CHAPTER TITLES, VOLUMES 1–26

687

COMPRESSIBLE FLUIDS